Achieve Your Potential

The authors have developed specific content in MyMathLab® to ensure that you have many resources to help you achieve success in mathematics - and beyond! The MyMathLab features described here will help you:

- **Brush up** on math skills and concepts you may have forgotten
- **Maintain** new concepts as you move through your math course
- **Develop** skills that will help with your transition from college to your profession

Adaptive Study Plan

The Study Plan will help you study more efficiently and effectively.

Your performance and activity are assessed continually in real time, providing a personalized experience based on your individual needs.

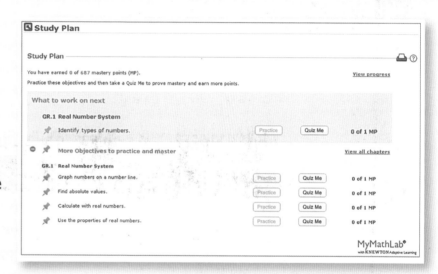

Skills for Success

The Skills for Success Module supports your continued success in college.

This module provides tutorials and guidance on a variety of topics, including transitioning to college, online learning, time management, and more.

Additional content is provided to help with the development of professional skills, such as resume writing or creation and interview preparation.

Getting Ready

Are you frustrated when you know you learned a math concept in the past, but you can't quite remember the skill when it's time to use it? Don't worry!

The authors have included Getting Ready material so you can brush up on forgotten material efficiently by taking a quick skill review quiz to pinpoint the areas where you need help.

Then a personalized homework assignment provides additional practice on those forgotten concepts, right when you need it.

Ongoing Review

Cumulative Review Tests (in the MyMathLab Ready to Go course) and Cumulative Review Assignments in MyMathLab help students synthesize and maintain concepts.

Scan the code or visit **www.mymathlab.com/success-stories** to learn how students across the country have used MyMathLab to achieve their potential and be successful in their math courses.

Essentials of College Algebra

ELEVENTH EDITION

Essentials of College Algebra

ELEVENTH EDITION

Margaret L. Lial
American River College

John Hornsby
University of New Orleans

David I. Schneider
University of Maryland

Callie J. Daniels
St. Charles Community College

PEARSON

Boston   Columbus   Indianapolis   New York   San Francisco   Upper Saddle River
Amsterdam   Cape Town   Dubai   London   Madrid   Milan   Munich   Paris   Montréal   Toronto
Delhi   Mexico City   São Paulo   Sydney   Hong Kong   Seoul   Singapore   Taipei   Tokyo

Editor-in-Chief: Anne Kelly

Executive Content Editor: Christine O'Brien

Editorial Assistant: Judith Garber

Senior Managing Editor: Karen Wernholm

Senior Production Project Manager: Beth Houston

Digital Assets Manager: Marianne Groth

Media Producer: Christina Maestri

Software Development: Kristina Evans, MathXL; John Flanagan, TestGen

Senior Marketing Manager: Peggy Sue Lucas

Marketing Assistant: Justine Goulart

Senior Author Support/Technology Specialist: Joe Vetere

Rights and Permissions Advisor: Aptara, Inc.

Image Manager: Rachel Youdelman

Procurement Specialist: Debbie Rossi

Associate Director of Design: Andrea Nix

Senior Designer: Beth Paquin

Text Design, Production Coordination, Composition: Cenveo® Publisher Services

Illustrations: Cenveo® Publisher Services, Laserwords, Network Graphics

Cover Design: Jenny Willingham

Cover Image: Rafal Olechowski/Fotolia

For permission to use copyrighted material, grateful acknowledgment is made to the copyright holders on page I-12, which is hereby made part of this copyright page.

Many of the designations used by manufacturers and sellers to distinguish their products are claimed as trademarks. Where those designations appear in this book, and Pearson Education was aware of a trademark claim, the designations have been printed in initial caps or all caps.

Library of Congress Cataloging-in-Publication Data

Lial, Margaret L.
        Essentials of college algebra/Margaret L. Lial, John Hornsby, David I. Schneider, Callie J. Daniels.
        --11th edition.
        pages cm.
        Includes indexes.
        ISBN 0-321-91225-X (student edition)
1. Algebra--Textbooks. I. Hornsby, John, 1949-II. Schneider, David I. III. Daniels, Callie J. IV. Title.
        QA154.3.L546 2015
        512.9--dc23   2013024179

Copyright © 2015, 2011, 2008 Pearson Education, Inc.

All rights reserved. No part of this publication may be reproduced, stored in a retrieval system, or transmitted, in any form or by any means, electronic, mechanical, photocopying, recording, or otherwise, without the prior written permission of the publisher. Printed in the United States of America. For information on obtaining permission for use of material in this work, please submit a written request to Pearson Education, Inc., Rights and Contracts Department, 501 Boylston Street, Suite 900, Boston, MA 02116, fax your request to 617-671-3447, or e-mail at http://www.pearsoned.com/legal/permissions.htm.

6 16

www.pearsonhighered.com

ISBN-10: 0-321-91225-X
ISBN-13: 978-0-321-91225-1

To Faith Elizabeth Varnado—you are the light of our lives
E.J.H.

To Kurtis, Clayton, and Grady
C.J.D.

On March 16, 2012, the mathematics education community lost one of its most influential members with the passing of our beloved mentor, colleague, and friend Marge Lial. On that day, Marge lost her long battle with ALS. Throughout her illness, Marge showed the remarkable strength and courage that characterized her entire life.

We would like to share a few comments from among the many messages we received from friends, colleagues, and others whose lives were touched by our beloved Marge:

"What a lady"   "A great friend"
"A remarkable person"   "Sorely missed but so fondly remembered"
"Gracious to everyone"   "Even though our crossed path was narrow, she made an impact
"One of a kind"   and I will never forget her."
"Truly someone special"   "There is talent and there is Greatness. Marge was truly Great."
"A loss in the mathematical world" "Her true impact is almost more than we can imagine."

In the world of college mathematics publishing, Marge Lial was a rock star. People flocked to her, and she had a way of making everyone feel like they truly mattered. And to Marge, they did. She and Chuck Miller began writing for Scott Foresman in 1970. Three years before her passing she told us that she could no longer continue because "just getting from point A to point B" had become too challenging. That's our Marge—she even gave a geometric interpretation to her illness.

It has truly been an honor and a privilege to work with Marge Lial these past twenty years. While we no longer have her wit, charm, and loving presence to guide us, so much of who we are as mathematics educators has been shaped by her influence. We will continue doing our part to make sure that the Lial name represents excellence in mathematics education. And we remember daily so many of the little ways she impacted us, including her special expressions, "Margisms" as we like to call them. She often ended emails with one of them— the single word "Onward."

We conclude with a poem by Callie Daniels describing Marge's influence on her and many others in the mathematics community.

Your courage inspires me
Your strength...impressive
Your wit humors me
Your vision...progressive

Your determination motivates me
Your accomplishments pave my way
Your vision sketches images for me
Your influence will forever stay.

Thank you, dearest Marge.
Knowing you and working with you has been a divine gift.

Onward.

John Hornsby
David Schneider
Callie Daniels

Contents

Preface xiii
Resources for Success xvi

R Review of Basic Concepts 1

R.1 Sets 2
Basic Definitions ■ Operations on Sets

R.2 Real Numbers and Their Properties 8
Sets of Numbers and the Number Line ■ Exponents ■ Order of Operations ■ Properties of Real Numbers ■ Order on the Number Line ■ Absolute Value

R.3 Polynomials 21
Rules for Exponents ■ Polynomials ■ Addition and Subtraction ■ Multiplication ■ Division

R.4 Factoring Polynomials 32
Factoring Out the Greatest Common Factor ■ Factoring by Grouping ■ Factoring Trinomials ■ Factoring Binomials ■ Factoring by Substitution

R.5 Rational Expressions 41
Rational Expressions ■ Lowest Terms of a Rational Expression ■ Multiplication and Division ■ Addition and Subtraction ■ Complex Fractions

R.6 Rational Exponents 49
Negative Exponents and the Quotient Rule ■ Rational Exponents ■ Complex Fractions Revisited

R.7 Radical Expressions 59
Radical Notation ■ Simplified Radicals ■ Operations with Radicals ■ Rationalizing Denominators

Test Prep 70 ■ Review Exercises 73 ■ Test 77

1 Equations and Inequalities 79

1.1 Linear Equations 80
Basic Terminology of Equations ■ Solving Linear Equations ■ Identities, Conditional Equations, and Contradictions ■ Solving for a Specified Variable (Literal Equations)

1.2 Applications and Modeling with Linear Equations 86
Solving Applied Problems ■ Geometry Problems ■ Motion Problems ■ Mixture Problems ■ Modeling with Linear Equations

1.3 Complex Numbers 97
Basic Concepts of Complex Numbers ■ Operations on Complex Numbers

1.4 Quadratic Equations 104

Solving a Quadratic Equation ▪ Completing the Square ▪
The Quadratic Formula ▪ Solving for a Specified Variable ▪
The Discriminant

Chapter 1 Quiz (Sections 1.1–1.4) **113**

1.5 Applications and Modeling with Quadratic Equations 114

Geometry Problems ▪ Using the Pythagorean Theorem ▪
Height of a Projected Object ▪ Modeling with Quadratic Equations

1.6 Other Types of Equations and Applications 124

Rational Equations ▪ Work Rate Problems ▪ Equations with Radicals ▪
Equations with Rational Exponents ▪ Equations Quadratic in Form

Summary Exercises on Solving Equations **136**

1.7 Inequalities 137

Linear Inequalities ▪ Three-Part Inequalities ▪ Quadratic Inequalities ▪
Rational Inequalities

1.8 Absolute Value Equations and Inequalities 149

Basic Concepts ▪ Absolute Value Equations ▪ Absolute Value Inequalities ▪
Special Cases ▪ Absolute Value Models for Distance and Tolerance

Test Prep **156** ▪ Review Exercises **160** ▪ Test **166**

2 | Graphs and Functions 169

2.1 Rectangular Coordinates and Graphs 170

Ordered Pairs ▪ The Rectangular Coordinate System ▪ The Distance
Formula ▪ The Midpoint Formula ▪ Graphing Equations in Two Variables

2.2 Circles 180

Center-Radius Form ▪ General Form ▪ An Application

2.3 Functions 188

Relations and Functions ▪ Domain and Range ▪
Determining Whether Relations Are Functions ▪ Function Notation ▪
Increasing, Decreasing, and Constant Functions

2.4 Linear Functions 203

Graphing Linear Functions ▪ Standard Form $Ax + By = C$ ▪ Slope ▪
Average Rate of Change ▪ Linear Models

Chapter 2 Quiz (Sections 2.1–2.4) **216**

2.5 Equations of Lines and Linear Models 217

Point-Slope Form ▪ Slope-Intercept Form ▪ Vertical and Horizontal Lines ▪
Parallel and Perpendicular Lines ▪ Modeling Data ▪
Solving Linear Equations in One Variable by Graphing

Summary Exercises on Graphs, Circles, Functions, and Equations **230**

2.6   Graphs of Basic Functions 231
Continuity ▪ The Identity, Squaring, and Cubing Functions ▪
The Square Root and Cube Root Functions ▪ The Absolute Value Function ▪
Piecewise-Defined Functions ▪ The Relation $x = y^2$

2.7   Graphing Techniques 242
Stretching and Shrinking ▪ Reflecting ▪ Symmetry ▪
Even and Odd Functions ▪ Translations

Chapter 2 Quiz (Sections 2.5–2.7) **257**

2.8   Function Operations and Composition 258
Arithmetic Operations on Functions ▪ The Difference Quotient ▪
Composition of Functions and Domain

Test Prep **271** ▪ Review Exercises **275** ▪ Test **280**

3 | Polynomial and Rational Functions 283

3.1   Quadratic Functions and Models 284
Quadratic Functions ▪ Graphing Techniques ▪ Completing the Square ▪
The Vertex Formula ▪ Quadratic Models

3.2   Synthetic Division 300
Synthetic Division ▪ Evaluating Polynomial Functions Using the
Remainder Theorem ▪ Testing Potential Zeros

3.3   Zeros of Polynomial Functions 306
Factor Theorem ▪ Rational Zeros Theorem ▪ Number of Zeros ▪
Conjugate Zeros Theorem ▪ Finding Zeros of a Polynomial Function ▪
Descartes' Rule of Signs

3.4   Polynomial Functions: Graphs, Applications, and Models 318
Graphs of $f(x) = ax^n$ ▪ Graphs of General Polynomial Functions ▪
Behavior at Zeros ▪ Turning Points and End Behavior ▪
Graphing Techniques ▪ Intermediate Value and Boundedness Theorems ▪
Approximating Real Zeros ▪ Polynomial Models

Summary Exercises on Polynomial Functions, Zeros, and Graphs **337**

3.5   Rational Functions: Graphs, Applications, and Models 339
The Reciprocal Function $f(x) = \frac{1}{x}$ ▪ The Function $f(x) = \frac{1}{x^2}$ ▪
Asymptotes ▪ Steps for Graphing Rational Functions ▪
Rational Function Models

Chapter 3 Quiz (Sections 3.1–3.5) **359**

Summary Exercises on Solving Equations and Inequalities **359**

3.6   Variation 361
Direct Variation ▪ Inverse Variation ▪ Combined and Joint Variation

Test Prep **370** ▪ Review Exercises **375** ▪ Test **380**

4 | **Inverse, Exponential, and Logarithmic Functions** 383

4.1 Inverse Functions **384**
One-to-One Functions ▪ Inverse Functions ▪ Equations of Inverses ▪ An Application of Inverse Functions to Cryptography

4.2 Exponential Functions **397**
Exponents and Properties ▪ Exponential Functions ▪ Exponential Equations ▪ Compound Interest ▪ The Number *e* and Continuous Compounding ▪ Exponential Models

4.3 Logarithmic Functions **413**
Logarithms ▪ Logarithmic Equations ▪ Logarithmic Functions ▪ Properties of Logarithms

Summary Exercises on Inverse, Exponential, and Logarithmic Functions **426**

4.4 Evaluating Logarithms and the Change-of-Base Theorem **427**
Common Logarithms ▪ Applications and Models with Common Logarithms ▪ Natural Logarithms ▪ Applications and Models with Natural Logarithms ▪ Logarithms with Other Bases

Chapter 4 Quiz (Sections 4.1–4.4) **439**

4.5 Exponential and Logarithmic Equations **439**
Exponential Equations ▪ Logarithmic Equations ▪ Applications and Models

4.6 Applications and Models of Exponential Growth and Decay **451**
The Exponential Growth or Decay Function ▪ Growth Function Models ▪ Decay Function Models

Summary Exercises on Functions: Domains and Defining Equations **462**

Test Prep **465** ▪ Review Exercises **468** ▪ Test **472**

5 | **Systems and Matrices** 475

5.1 Systems of Linear Equations **476**
Linear Systems ▪ Substitution Method ▪ Elimination Method ▪ Special Systems ▪ Applying Systems of Equations ▪ Solving Linear Systems with Three Unknowns (Variables) ▪ Using Systems of Equations to Model Data

5.2 Matrix Solution of Linear Systems **493**
The Gauss-Jordan Method ▪ Special Systems

5.3 Determinant Solution of Linear Systems **504**
Determinants ▪ Cofactors ▪ Evaluating $n \times n$ Determinants ▪ Determinant Theorems ▪ Cramer's Rule

5.4 **Partial Fractions** 516
Decomposition of Rational Expressions ▪ Distinct Linear Factors ▪
Repeated Linear Factors ▪ Distinct Linear and Quadratic Factors ▪
Repeated Quadratic Factors

Chapter 5 Quiz (Sections 5.1–5.4) **522**

5.5 **Nonlinear Systems of Equations** 523
Solving Nonlinear Systems with Real Solutions ▪
Solving Nonlinear Systems with Nonreal Complex Solutions ▪
Applying Nonlinear Systems

Summary Exercises on Systems of Equations **533**

5.6 **Systems of Inequalities and Linear Programming** 534
Solving Linear Inequalities ▪ Solving Systems of Inequalities ▪
Linear Programming

5.7 **Properties of Matrices** 545
Basic Definitions ▪ Adding Matrices ▪ Special Matrices ▪
Subtracting Matrices ▪ Multiplying Matrices ▪ Applying Matrix Algebra

5.8 **Matrix Inverses** 558
Identity Matrices ▪ Multiplicative Inverses ▪ Solving Systems
Using Inverse Matrices

Test Prep **569** ▪ Review Exercises **575** ▪ Test **580**

Glossary **583**
Solutions to Selected Exercises **S-1**
Answers to Selected Exercises **A-1**
Index of Applications **I-1**
Index **I-5**
Photo Credits **I-12**

Preface

As authors, we have called upon our classroom experiences, use of MyMathLab, suggestions from users and reviewers, and many years of writing to provide tools that will support learning and teaching. This new edition of *Essentials of College Algebra* continues our effort to provide a sound pedagogical approach through logical development of the subject matter. This approach forms the basis for all of the Lial team's instructional materials available from Pearson Education, in both print and technology forms.

Our goal is to produce a textbook that will be an integral component of the student's experience in learning college algebra. With this in mind, we have provided a textbook that students can read more easily, which is often a difficult task, given the nature of mathematical language. We have also improved page layouts for better flow, provided additional side comments, and updated many figures.

We realize that today's classroom experience is evolving and that technology-based teaching and learning aids have become essential to address the ever-changing needs of instructors and students. As a result, we've worked to provide support for all classroom types—traditional, hybrid, and online. In the 11th edition, text and online materials are more tightly integrated than ever before. This enhances flexibility and ease of use for instructors and increases success for students. See pages xvi–xvii for descriptions of these materials.

NEW TO THE 11TH EDITION

- Enhancing the already well-respected exercises, approximately 750 are new or modified, and hundreds present updated real-life data. In addition, the MyMathLab course has expanded coverage of all exercise types appearing in the exercise sets, as well as the mid-chapter Quizzes and Summary Exercises.

- In **Chapter 1** we have rewritten the introduction to the set of complex numbers and have included a new diagram illustrating the relationships among its subsets. We have also expanded the discussion of solving absolute value equations and inequalities.

- In **Chapter 2** we have prepared new art for shrinking and stretching graphs, added new exercises on translations, and expanded the exercises involving the difference quotient. We now use open intervals when discussing increasing, decreasing, and constant behavior of functions, and we now define intercepts as ordered pairs.

- **Chapter 3** has undergone a particularly extensive revision. We have updated the opening discussion by providing a table of examples of polynomial functions (constant, linear, quadratic, cubic, quartic) and have identified the degree and leading coefficient. There is an increased emphasis on displaying all possibilities for positive, negative, and nonreal complex zeros in a table format, with graphical representations to illustrate them. We now have a new summary exercise set on solving both equations and inequalities (linear, quadratic, polynomial, rational, and miscellaneous).

- In **Chapter 4** we have added more work with translations of exponential and logarithmic functions, and we have included additional examples and exercises on solving equations that involve these functions.

- In **Chapter 5** we now present the determinant theorems within the exposition, and we have written new examples and exercises showing how they are used.

- For visual learners, numbered **Figure** and **Example** references within the text are set using the same typeface as the figure and bold print for the example. This makes it easier for the students to identify and connect them. We also have increased our use of a "drop down" style, when appropriate, to distinguish between simplifying expressions and solving equations, and we have added many more explanatory side comments. Interactive figures with accompanying exercises and explorations are now available and assignable in MyMathLab.

FEATURES OF THIS TEXT

SUPPORT FOR LEARNING CONCEPTS

We provide a variety of features to support students' learning of the essential topics of college algebra. Explanations that are written in understandable terms, figures and graphs that illustrate examples and concepts, graphing technology that supports and enhances algebraic manipulations, and real-life applications that enrich the topics with meaning all provide opportunities for students to deepen their understanding of mathematics. These features help students make mathematical connections and expand their own knowledge base.

- **Examples** Numbered examples that illustrate the techniques for working exercises are found in every section. We use traditional explanations, side comments, and pointers to describe the steps taken—and to warn students about common pitfalls. Some examples provide additional graphing calculator solutions, although these can be omitted if desired.

- **Now Try Exercises** Following each numbered example, the student is directed to try a corresponding odd-numbered exercise (or exercises). This feature allows for quick feedback to determine whether the student has understood the principles illustrated in the example.

- **Real-Life Applications** We have included hundreds of real-life applications, many with data updated from the previous edition. They come from fields such as business, entertainment, sports, biology, astronomy, geology, and environmental studies.

- **Function Boxes** Beginning in Chapter 2, functions provide a unifying theme throughout the text. Special function boxes (for example, see page 232) offer a comprehensive, visual introduction to each type of function and also serve as an excellent resource for reference and review. Each function box includes a table of values, traditional and calculator-generated graphs, the domain, the range, and other special information about the function. These boxes are now assignable in MyMathLab.

- **Figures and Photos** Today's students are more visually oriented than ever before, and we have updated the figures in this edition to a greater extent than in our previous few editions. Interactive figures with accompanying exercises and explorations are now available and assignable in MyMathLab.

- **Use of Graphing Technology** We have integrated the use of graphing calculators where appropriate, although *this technology is completely optional and can be omitted without loss of continuity.* We continue to stress that graphing calculators support understanding but that students must first master the underlying mathematical concepts. Exercises that require their use are marked with an icon.

■ **Cautions and Notes**   Text that is marked **CAUTION** warns students of common errors, and NOTE comments point out explanations that should receive particular attention.

■ **Looking Ahead to Calculus**   These margin notes offer glimpses of how the topics currently being studied are used in calculus.

SUPPORT FOR PRACTICING CONCEPTS

This text offers a wide variety of exercises to help students master college algebra. The extensive exercise sets provide ample opportunity for practice, and the exercise problems increase in difficulty so that students at every level of understanding are challenged. The variety of exercise types promotes understanding of the concepts and reduces the need for rote memorization.

■ **Exercise Sets**   We have revised many drill and application exercises for better pairing of corresponding even and odd exercises, and answers to the odd exercises are provided in the Student Edition. In addition to these, we include writing exercises ▤, optional graphing calculator problems ▱, and multiple-choice, matching, true/false, and completion exercises. Those marked *Concept Check* focus on conceptual thinking. *Connecting Graphs with Equations* exercises challenge students to write equations that correspond to given graphs. Finally, MyMathLab offers Pencast solutions for selected Connecting Graphs with Equations problems.

■ **Relating Concepts Exercises**   Appearing in selected exercise sets, these groups of exercises are designed so that students who work them in numerical order will follow a line of reasoning that leads to an understanding of how various topics and concepts are related. All answers to these exercises appear in the student answer section, and these exercises are now assignable in MyMathLab.

■ **Complete Solutions to Selected Exercises**   Exercise numbers marked ▨ indicate that a full worked-out solution appears at the back of the text. These are often exercises that extend the skills and concepts presented in the numbered examples.

SUPPORT FOR REVIEW AND TEST PREP

Ample opportunities for review are found within the chapters and at the ends of chapters. Quizzes that are interspersed within chapters provide a quick assessment of students' understanding of the material presented up to that point in the chapter. Chapter "Test Preps" provide comprehensive study aids to help students prepare for tests.

■ **Quizzes**   Students can periodically check their progress with in-chapter quizzes that appear in all chapters, beginning with Chapter 1. All answers, with corresponding section references, appear in the student answer section. These quizzes and additional cumulative chapter review exercises are now assignable in MyMathLab.

■ **Summary Exercises**   These sets of in-chapter exercises give students the all-important opportunity to work *mixed* review exercises, requiring them to synthesize concepts and select appropriate solution methods. The summary exercises are now assignable in MyMathLab.

■ **End-of-Chapter Test Prep**   Following the final numbered section in each chapter, the Test Prep provides a list of **Key Terms**, a list of **New Symbols** (if applicable), and a two-column **Quick Review** that includes a section-by-section summary of concepts and examples. This feature concludes with a comprehensive set of **Review Exercises** and a **Chapter Test.** The Test Prep, Review Exercises, and Chapter Test are assignable in MyMathLab.

■ **Glossary**   A comprehensive glossary of important terms drawn from the entire book follows Chapter 5.

Resources for Success

MyMathLab® **Online Course** (access code required)

MyMathLab delivers **proven results** in helping individual students succeed. It provides **engaging experiences** that personalize, stimulate, and measure learning for each student. And it comes from an **experienced partner** with educational expertise and an eye on the future. MyMathLab helps prepare students and gets them thinking more conceptually and visually through the following features:

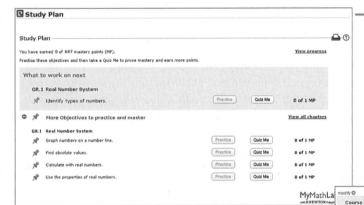

─○ Adaptive Study Plan

The Study Plan makes studying more efficient and effective for every student. Performance and activity are assessed continually in real time. The data and analytics are used to provide personalized, content–reinforcing concepts that target each student's strengths and weaknesses.

Getting Ready ○─

Students refresh prerequisite topics through assignable skill review quizzes and personalized homework integrated in MyMathLab.

─○ Video Assessment

Video assessment is tied to key Example/Solution videos to check students' conceptual understanding of important math concepts.

Enhanced Graphing ○─ Functionality

New functionality within the graphing utility allows graphing of 3-point quadratic functions, 4-point cubic graphs, and transformations in exercises.

Skills for Success Modules are integrated within the MyMathLab course to help students succeed in collegiate courses and prepare for future professions.

Cumulative Review Tests (in the MyMathLab Ready to Go course) and **Cumulative Review Assignments** in MyMathLab help students synthesize and maintain concepts.

Instructor Resources

Additional resources can be downloaded from **www.pearsonhighered.com** or hardcopy resources can be ordered from your sales representative.

Ready to Go MyMathLab® Course

Now it is even easier to get started with MyMathLab. The Ready to Go MyMathLab course option includes author-chosen preassigned homework, integrated review, and more.

Annotated Instructor's Edition

This edition provides answers in the margins to most text exercises, as well as helpful Teaching Tips and Classroom Examples.

TestGen®

TestGen® (www.pearsoned.com/testgen) enables instructors to build, edit, print, and administer tests using a computerized bank of questions developed to cover all the objectives of the text.

PowerPoint® Lecture Slides

These fully editable slides correlate to the textbook.

Instructor's Solutions Manual

This manual includes fully worked solutions to all textbook exercises.

Instructor's Testing Manual

This manual includes diagnostic pretests, chapter tests, final exams, and additional test items, with answers provided.

Student Resources

The following are additional resources to help improve student success.

Student's Solutions Manual

This manual provides detailed solutions to all odd-numbered text exercises.

Video Lectures

Quick Review videos cover key definitions and procedures from each section. Example/Solution videos walk students through the detailed solution process for every example in the textbook.

Additional Skill and Drill Manual

This manual provides additional practice and test preparation for students.

MyNotes

This note-taking guide is available in MyMathLab or packaged with text and access code. It offers structure for student reading and understanding of the textbook. It includes textbook examples along with ample space for students to write solutions and notes.

MyClassroomExamples

This alternate version of a note-taking guide is available in MyMathLab and offers structure for classroom lecture. It includes Classroom Examples along with ample space for students to write solutions and notes.

ACKNOWLEDGMENTS

We wish to thank the following individuals who provided valuable input into this edition of the text.

Elaine Fitt – Bucks County Community College

Eric Matsuoka – Leeward Community College

Kristina Sampson – Lonestar College – Cypress

David Schweitzer – University of Central Florida

Jinhua Tao – Central Missouri University

Our sincere thanks to those individuals at Pearson Education who have supported us throughout this revision: Greg Tobin, Anne Kelly, Beth Houston, and Christine O'Brien. Terry McGinnis continues to provide behind-the-scenes guidance for both content and production. We have come to rely on her expertise during all phases of the revision process. Marilyn Dwyer of Cenveo® Publisher Services, with the assistance of Carol Merrigan, provided excellent production work. Special thanks go out to Abby Tanenbaum for updating data in applications, and to Chris Heeren and Paul Lorczak for their excellent accuracy-checking. We thank Lucie Haskins, who provided an accurate index, and Becky Troutman, who revised our Index of Applications. We appreciate the valuable suggestions for Chapter 5 that Mary Hill of *College of Dupage* made during our meeting with her.

As an author team, we are committed to providing the best possible college algebra course to help instructors teach and students succeed. As we continue to work toward this goal, we welcome any comments or suggestions you might send, via e-mail, to math@pearson.com.

John Hornsby
David I. Schneider
Callie J. Daniels

R Review of Basic Concepts

Positive and negative numbers, used to represent gains and losses on a board such as this one, are examples of *real numbers* encountered in applications of mathematics.

R.1 Sets

R.2 Real Numbers and Their Properties

R.3 Polynomials

R.4 Factoring Polynomials

R.5 Rational Expressions

R.6 Rational Exponents

R.7 Radical Expressions

R.1 Sets

- Basic Definitions
- Operations on Sets

Basic Definitions   A **set** is a collection of objects. The objects that belong to a set are called the **elements,** or **members,** of the set. In algebra, the elements of a set are usually numbers. Sets are commonly written using **set braces,** { }. For example, the set containing the elements 1, 2, 3, and 4 is written as follows.

$$\{1, 2, 3, 4\}$$

Since the order in which the elements are listed is not important, this same set can also be written as $\{4, 3, 2, 1\}$ or with any other arrangement of the four numbers.

To show that 4 is an element of the set $\{1, 2, 3, 4\}$, we use the symbol $\in$.

$$4 \in \{1, 2, 3, 4\}$$

Since 5 is *not* an element of this set, we place a slash through the symbol $\in$.

$$5 \notin \{1, 2, 3, 4\}$$

It is customary to name sets with capital letters. If S is used to name the set above, then we write it as follows.

$$S = \{1, 2, 3, 4\}$$

Set S was written by listing its elements. Set S might also be described as

"the set containing the first four counting numbers."

In this example, the notation $\{1, 2, 3, 4\}$, with the elements listed between set braces, is briefer than the verbal description.

The set F, consisting of all fractions between 0 and 1, is an example of an **infinite set,** one that has an unending list of distinct elements. A **finite set** is one that has a limited number of elements. The process of counting its elements comes to an end. Some infinite sets can be described by listing. For example, the set of numbers N used for counting, called the **natural numbers,** or the **counting numbers,** can be written as follows.

$$N = \{1, 2, 3, 4, \ldots\} \quad \text{Natural (counting) numbers}$$

The three dots (*ellipsis points*) show that the list of elements of the set continues according to the established pattern.

Sets are often written using a variable to represent an arbitrary element of the set. For example,

$$\{x \mid x \text{ is a natural number between 2 and 7}\} \quad \text{Set-builder notation}$$

(which is read "the set of all elements x such that x is a natural number between 2 and 7") uses **set-builder notation** to represent the set $\{3, 4, 5, 6\}$. The numbers 2 and 7 are *not* between 2 and 7.

EXAMPLE 1   **Using Set Notation and Terminology**

Identify each set as *finite* or *infinite*. Then determine whether 10 is an element of the set.

(a) $\{7, 8, 9, \ldots, 14\}$ **(b)** $\left\{1, \frac{1}{4}, \frac{1}{16}, \frac{1}{64}, \ldots\right\}$

(c) $\{x \mid x \text{ is a fraction between 1 and 2}\}$

(d) $\{x \mid x \text{ is a natural number between 9 and 11}\}$

SOLUTION

(a) The set is finite, because the process of counting its elements 7, 8, 9, 10, 11, 12, 13, and 14 comes to an end. The number 10 does belong to the set, and this is written as follows.

$$10 \in \{7, 8, 9, \ldots, 14\}$$

(b) The set is infinite, because the ellipsis points indicate that the pattern continues forever. In this case,

$$10 \notin \left\{1, \tfrac{1}{4}, \tfrac{1}{16}, \tfrac{1}{64}, \ldots\right\}.$$

(c) Between any two distinct natural numbers there are infinitely many fractions, so this set is infinite. The number 10 is not an element.

(d) There is only one natural number between 9 and 11, namely 10. So the set is finite, and 10 is an element.

✔ *Now Try Exercises 1, 3, 5, and 7.*

EXAMPLE 2 **Listing the Elements of a Set**

Use set notation, and write the elements belonging to each set.

(a) $\{x \mid x \text{ is a natural number less than } 5\}$

(b) $\{x \mid x \text{ is a natural number greater than 7 and less than } 14\}$

SOLUTION

(a) The natural numbers less than 5 form the set $\{1, 2, 3, 4\}$.

(b) This is the set $\{8, 9, 10, 11, 12, 13\}$.

✔ *Now Try Exercise 15.*

When we are discussing a particular situation or problem, the **universal set** (whether expressed or implied) contains all the elements included in the discussion. The letter U is used to represent the universal set. The **null set,** or **empty set,** is the set containing no elements. We write the null set by either using the special symbol $\emptyset$, or else writing set braces enclosing no elements, $\{\ \}$.

CAUTION Do not combine these symbols. $\{\emptyset\}$ *is not the null set.*

Every element of the set $S = \{1, 2, 3, 4\}$ is a natural number. S is an example of a *subset* of the set N of natural numbers, and this is written

$$S \subseteq N.$$

By definition, set A is a **subset** of set B if every element of set A is also an element of set B. For example, if $A = \{2, 5, 9\}$ and $B = \{2, 3, 5, 6, 9, 10\}$, then $A \subseteq B$. However, there are some elements of B that are not in A, so B is not a subset of A, which is written

$$B \nsubseteq A.$$

By the definition, every set is a subset of itself. Also, by definition, $\emptyset$ is a subset of every set.

If A is any set, then $\emptyset \subseteq A$.

Figure 1 shows a set A that is a subset of set B. The rectangle in the drawing represents the universal set U. Such diagrams are called **Venn diagrams.**

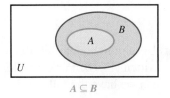

$A \subseteq B$

Figure 1

Two sets A and B are equal whenever $A \subseteq B$ and $B \subseteq A$. Equivalently, $A = B$ if the two sets contain exactly the same elements. For example,

$$\{1, 2, 3\} = \{3, 1, 2\}$$

is true, since both sets contain exactly the same elements. However,

$$\{1, 2, 3\} \neq \{0, 1, 2, 3\},$$

since the set $\{0, 1, 2, 3\}$ contains the element 0, which is not an element of $\{1, 2, 3\}$.

EXAMPLE 3 Examining Subset Relationships

Let $U = \{1, 3, 5, 7, 9, 11, 13\}$, $A = \{1, 3, 5, 7, 9, 11\}$, $B = \{1, 3, 7, 9\}$, $C = \{3, 9, 11\}$, and $D = \{1, 9\}$. Determine whether each statement is *true* or *false*.

(a) $D \subseteq B$　　**(b)** $B \subseteq D$　　**(c)** $C \nsubseteq A$　　**(d)** $U = A$

SOLUTION

(a) All elements of D, namely 1 and 9, are also elements of B, so D is a subset of B, and $D \subseteq B$ is true.

(b) There is at least one element of B (for example, 3) that is not an element of D, so B is *not* a subset of D. Thus, $B \subseteq D$ is false.

(c) C is a subset of A, because every element of C is also an element of A. Thus, $C \subseteq A$ is true, and as a result, $C \nsubseteq A$ is false.

(d) U contains the element 13, but A does not. Therefore, $U = A$ is false.

✔ *Now Try Exercises 43, 45, 53, and 55.*

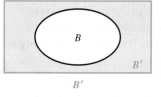

B'

Figure 2

Operations on Sets　Given a set A and a universal set U, the set of all elements of U that do not belong to set A is called the **complement** of set A. For example, if set A is the set of all students in your class 30 years old or older, and set U is the set of all students in the class, then the complement of A would be the set of all the students in the class younger than age 30. The complement of set A is written A' (read **"A-prime"**). The Venn diagram in **Figure 2** shows a set B. Its complement, B', is in color.

EXAMPLE 4 Finding the Complement of a Set

Let $U = \{1, 2, 3, 4, 5, 6, 7\}$, $A = \{1, 3, 5, 7\}$, and $B = \{3, 4, 6\}$. Find each set.

(a) A'　　**(b)** B'　　**(c)** $\emptyset'$　　**(d)** U'

SOLUTION

(a) Set A' contains the elements of U that are not in A. Thus, $A' = \{2, 4, 6\}$.

(b) $B' = \{1, 2, 5, 7\}$　　**(c)** $\emptyset' = U$　　**(d)** $U' = \emptyset$

✔ *Now Try Exercise 79.*

Given two sets A and B, the set of all elements belonging both to set A **and** to set B is called the **intersection** of the two sets, written $A \cap B$. For example, if $A = \{1, 2, 4, 5, 7\}$ and $B = \{2, 4, 5, 7, 9, 11\}$, then we have the following.

$$A \cap B = \{1, 2, 4, 5, 7\} \cap \{2, 4, 5, 7, 9, 11\} = \{2, 4, 5, 7\}$$

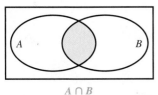

$A \cap B$

Figure 3

The Venn diagram in **Figure 3** shows two sets A and B. Their intersection, $A \cap B$, is in color.

Two sets that have no elements in common are called **disjoint sets.** If A and B are any two disjoint sets, then $A \cap B = \emptyset$. For example, there are no elements common to both $\{50, 51, 54\}$ and $\{52, 53, 55, 56\}$, so these two sets are disjoint.

$$\{50, 51, 54\} \cap \{52, 53, 55, 56\} = \emptyset$$

> **EXAMPLE 5** **Finding the Intersection of Two Sets**
>
> Find each of the following.
>
> **(a)** $\{9, 15, 25, 36\} \cap \{15, 20, 25, 30, 35\}$
>
> **(b)** $\{2, 3, 4, 5, 6\} \cap \{1, 2, 3, 4\}$
>
> **(c)** $\{1, 3, 5\} \cap \{2, 4, 6\}$
>
> **SOLUTION**
>
> **(a)** $\{9, 15, 25, 36\} \cap \{15, 20, 25, 30, 35\} = \{15, 25\}$
> The elements 15 and 25 are the only ones belonging to both sets.
>
> **(b)** $\{2, 3, 4, 5, 6\} \cap \{1, 2, 3, 4\} = \{2, 3, 4\}$
>
> **(c)** $\{1, 3, 5\} \cap \{2, 4, 6\} = \emptyset$ Disjoint sets

☑ *Now Try Exercises 59, 65, and 75.*

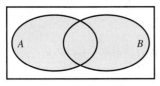

$A \cup B$

Figure 4

The set of all elements belonging to set A ***or*** to set B (or to both) is called the **union** of the two sets, written $A \cup B$. For example, if $A = \{1, 3, 5\}$ and $B = \{3, 5, 7, 9\}$ then we have the following.

$$A \cup B = \{1, 3, 5\} \cup \{3, 5, 7, 9\} = \{1, 3, 5, 7, 9\}$$

The Venn diagram in **Figure 4** shows two sets A and B. Their union, $A \cup B$, is in color.

> **EXAMPLE 6** **Finding the Union of Two Sets**
>
> Find each of the following.
>
> **(a)** $\{1, 2, 5, 9, 14\} \cup \{1, 3, 4, 8\}$
>
> **(b)** $\{1, 3, 5, 7\} \cup \{2, 4, 6\}$
>
> **(c)** $\{1, 3, 5, 7, \ldots\} \cup \{2, 4, 6, \ldots\}$
>
> **SOLUTION**
>
> **(a)** Begin by listing the elements of the first set, $\{1, 2, 5, 9, 14\}$. Then include any elements from the second set that are not already listed.
>
> $$\{1, 2, 5, 9, 14\} \cup \{1, 3, 4, 8\} = \{1, 2, 3, 4, 5, 8, 9, 14\}$$
>
> **(b)** $\{1, 3, 5, 7\} \cup \{2, 4, 6\} = \{1, 2, 3, 4, 5, 6, 7\}$
>
> **(c)** $\{1, 3, 5, 7, \ldots\} \cup \{2, 4, 6, \ldots\} = N$ Natural numbers

☑ *Now Try Exercises 73 and 77.*

The **set operations** are summarized below.

Set Operations

Let A and B be sets, with universal set U.

The **complement** of set A is the set A' of all elements in the universal set that do *not* belong to set A.

$$A' = \{x \mid x \in U, \quad x \notin A\}$$

The **intersection** of sets A and B, written $A \cap B$, is made up of all the elements belonging to both set A *and* set B.

$$A \cap B = \{x \mid x \in A \text{ and } x \in B\}$$

The **union** of sets A and B, written $A \cup B$, is made up of all the elements belonging to set A *or* to set B.

$$A \cup B = \{x \mid x \in A \text{ or } x \in B\}$$

R.1 Exercises

Identify each set as finite *or* infinite. *Then determine whether* 10 *is an element of the set.* **See Example 1.**

1. $\{4, 5, 6, \ldots, 15\}$

2. $\{1, 2, 3, 4, 5, \ldots, 75\}$

3. $\{1, \frac{1}{2}, \frac{1}{4}, \frac{1}{8}, \ldots\}$

4. $\{4, 5, 6, \ldots\}$

5. $\{x \mid x \text{ is a natural number greater than } 11\}$

6. $\{x \mid x \text{ is a natural number greater than or equal to } 10\}$

7. $\{x \mid x \text{ is a fraction between 1 and } 2\}$

8. $\{x \mid x \text{ is an even natural number}\}$

Use set notation, and list all the elements of each set. **See Example 2.**

9. $\{12, 13, 14, \ldots, 20\}$

10. $\{8, 9, 10, \ldots, 17\}$

11. $\{1, \frac{1}{2}, \frac{1}{4}, \ldots, \frac{1}{32}\}$

12. $\{3, 9, 27, \ldots, 729\}$

13. $\{17, 22, 27, \ldots, 47\}$

14. $\{74, 68, 62, \ldots, 38\}$

15. $\{x \mid x \text{ is a natural number greater than 7 and less than } 15\}$

16. $\{x \mid x \text{ is a natural number not greater than } 4\}$

Insert $\in$ *or* $\notin$ *in each blank to make the resulting statement true.* **See Examples 1 and 2.**

17. $6 \underline{\hspace{0.6cm}} \{3, 4, 5, 6\}$

18. $9 \underline{\hspace{0.6cm}} \{3, 2, 5, 9, 8\}$

19. $-4 \underline{\hspace{0.6cm}} \{4, 6, 8, 10\}$

20. $-12 \underline{\hspace{0.6cm}} \{3, 5, 12, 14\}$

21. $0 \underline{\hspace{0.6cm}} \{2, 0, 3, 4\}$

22. $0 \underline{\hspace{0.6cm}} \{0, 5, 6, 7, 8, 10\}$

23. $\{3\} \underline{\hspace{0.6cm}} \{2, 3, 4, 5\}$

24. $\{5\} \underline{\hspace{0.6cm}} \{3, 4, 5, 6, 7\}$

25. $\{0\} \underline{\hspace{0.6cm}} \{0, 1, 2, 5\}$

26. $\{2\} \underline{\hspace{0.6cm}} \{2, 4, 6, 8\}$

27. $0 \underline{\hspace{0.6cm}} \emptyset$

28. $\emptyset \underline{\hspace{0.6cm}} \emptyset$

Determine whether each statement is true *or* false. ***See Examples 1–3.***

29. $3 \in \{2, 5, 6, 8\}$ **30.** $6 \in \{-2, 5, 8, 9\}$

31. $1 \in \{3, 4, 5, 11, 1\}$ **32.** $12 \in \{18, 17, 15, 13, 12\}$

33. $9 \notin \{2, 1, 5, 8\}$ **34.** $3 \notin \{7, 6, 5, 4\}$

35. $\{2, 5, 8, 9\} = \{2, 5, 9, 8\}$ **36.** $\{3, 0, 9, 6, 2\} = \{2, 9, 0, 3, 6\}$

37. $\{5, 8, 9\} = \{5, 8, 9, 0\}$ **38.** $\{3, 7, 12, 14\} = \{3, 7, 12, 14, 0\}$

39. $\{x \,|\, x \text{ is a natural number less than } 3\} = \{1, 2\}$

40. $\{x \,|\, x \text{ is a natural number greater than } 10\} = \{11, 12, 13, \dots\}$

Let $A = \{2, 4, 6, 8, 10, 12\}, \;\; B = \{2, 4, 8, 10\}, \;\; C = \{4, 10, 12\}, \;\; D = \{2, 10\}, \;\; and$
$U = \{2, 4, 6, 8, 10, 12, 14\}.$

Determine whether each statement is true *or* false. ***See Example 3.***

41. $A \subseteq U$ **42.** $C \subseteq U$ **43.** $D \subseteq B$ **44.** $D \subseteq A$

45. $A \subseteq B$ **46.** $B \subseteq C$ **47.** $\emptyset \subseteq A$ **48.** $\emptyset \subseteq \emptyset$

49. $\{4, 8, 10\} \subseteq B$ **50.** $\{0, 2\} \subseteq D$ **51.** $B \subseteq D$ **52.** $A \subseteq C$

Insert $\subseteq$ *or* $\nsubseteq$ *in each blank to make the resulting statement true.* ***See Example 3.***

53. $\{2, 4, 6\}$ _____ $\{3, 2, 5, 4, 6\}$ **54.** $\{1, 5\}$ _____ $\{0, -1, 2, 3, 1, 5\}$

55. $\{0, 1, 2\}$ _____ $\{1, 2, 3, 4, 5\}$ **56.** $\{5, 6, 7, 8\}$ _____ $\{1, 2, 3, 4, 5, 6, 7\}$

57. $\emptyset$ _____ $\{1, 4, 6, 8\}$ **58.** $\emptyset$ _____ $\emptyset$

Determine whether each statement is true *or* false. ***See Examples 4–6.***

59. $\{5, 7, 9, 19\} \cap \{7, 9, 11, 15\} = \{7, 9\}$

60. $\{8, 11, 15\} \cap \{8, 11, 19, 20\} = \{8, 11\}$

61. $\{2, 1, 7\} \cup \{1, 5, 9\} = \{1\}$

62. $\{6, 12, 14, 16\} \cup \{6, 14, 19\} = \{6, 14\}$

63. $\{3, 2, 5, 9\} \cap \{2, 7, 8, 10\} = \{2\}$ **64.** $\{8, 9, 6\} \cup \{9, 8, 6\} = \{8, 9\}$

65. $\{3, 5, 9, 10\} \cap \emptyset = \{3, 5, 9, 10\}$ **66.** $\{3, 5, 9, 10\} \cup \emptyset = \{3, 5, 9, 10\}$

67. $\{1, 2, 4\} \cup \{1, 2, 4\} = \{1, 2, 4\}$ **68.** $\{1, 2, 4\} \cap \{1, 2, 4\} = \emptyset$

69. $\emptyset \cup \emptyset = \emptyset$ **70.** $\emptyset \cap \emptyset = \emptyset$

Let $U = \{0, 1, 2, 3, 4, 5, 6, 7, 8, 9, 10, 11, 12, 13\}, \;\; M = \{0, 2, 4, 6, 8\},$
$N = \{1, 3, 5, 7, 9, 11, 13\}, \;\; Q = \{0, 2, 4, 6, 8, 10, 12\}, \;\; and \;\; R = \{0, 1, 2, 3, 4\}.$

Use these sets to find each of the following. Identify any disjoint sets. ***See Examples 4–6.***

71. $M \cap R$ **72.** $M \cap U$ **73.** $M \cup N$

74. $M \cup R$ **75.** $M \cap N$ **76.** $U \cap N$

77. $N \cup R$ **78.** $M \cup Q$ **79.** N'

80. Q' **81.** $M' \cap Q$ **82.** $Q \cap R'$

83. $\emptyset \cap R$ **84.** $\emptyset \cap Q$ **85.** $N \cup \emptyset$

86. $R \cup \emptyset$ **87.** $(M \cap N) \cup R$ **88.** $(N \cup R) \cap M$

89. $(Q \cap M) \cup R$ **90.** $(R \cup N) \cap M'$ **91.** $(M' \cup Q) \cap R$

92. $Q \cap (M \cup N)$ **93.** $Q' \cap (N' \cap U)$ **94.** $(U \cap \emptyset') \cup R$

R.2 Real Numbers and Their Properties

- Sets of Numbers and the Number Line
- Exponents
- Order of Operations
- Properties of Real Numbers
- Order on the Number Line
- Absolute Value

Sets of Numbers and the Number Line   As mentioned in the previous section, the set of **natural numbers** is written in set notation as follows.

$$\{1, 2, 3, 4, \ldots\}$$ Natural numbers (Section R.1)

Including 0 with the set of natural numbers gives the set of **whole numbers.**

$$\{0, 1, 2, 3, 4, \ldots\}$$ Whole numbers

Including the negatives of the natural numbers with the set of whole numbers gives the set of **integers.**

$$\{\ldots, -3, -2, -1, 0, 1, 2, 3, \ldots\}$$ Integers

Integers can be **graphed** on a **number line.** See **Figure 5.** Every number corresponds to one and only one point on the number line, and each point corresponds to one and only one number. The number associated with a given point is called the **coordinate** of the point. This correspondence forms a **coordinate system.**

Origin

Graph of the Set $\{-3, -1, 0, 1, 3, 5\}$

Figure 5

The result of dividing two integers (with a nonzero divisor) is called a *rational number,* or *fraction.* A **rational number** is an element of the set defined as follows.

$$\left\{\frac{p}{q}\middle|\ p \text{ and } q \text{ are integers and } q \neq 0\right\}$$ Rational numbers

The set of rational numbers includes the natural numbers, the whole numbers, and the integers. For example, the integer -3 is a rational number because it can be written as $\frac{-3}{1}$. Numbers that can be written as repeating or terminating decimals are also rational numbers. For example, $0.\overline{6} = 0.66666\ldots$ represents a rational number that can be expressed as the fraction $\frac{2}{3}$.

Graph of the Set of Real Numbers

Figure 6

The set of all numbers that correspond to points on a number line is the **real numbers,** shown in **Figure 6.** Real numbers can be represented by decimals. Since every fraction has a decimal form—for example, $\frac{1}{4} = 0.25$—real numbers include rational numbers.

Some real numbers cannot be represented by quotients of integers. These numbers are **irrational numbers.** The set of irrational numbers includes $\sqrt{3}$ and $\sqrt{5}$. Another irrational number is π, which is *approximately* equal to 3.14159. The numbers in the set $\left\{-\frac{2}{3}, 0, \sqrt{2}, \sqrt{5}, \pi, 4\right\}$ can be located on a number line, as shown in **Figure 7.**

$\sqrt{2}$, $\sqrt{5}$, and π are irrational. Since $\sqrt{2}$ is approximately equal to 1.41, it is located between 1 and 2, slightly closer to 1.

Figure 7

The sets of numbers discussed so far are summarized as follows.

Sets of Numbers

Set	Description
Natural numbers	$\{1, 2, 3, 4, \ldots\}$
Whole numbers	$\{0, 1, 2, 3, 4, \ldots\}$
Integers	$\{\ldots, -3, -2, -1, 0, 1, 2, 3, \ldots\}$
Rational numbers	$\left\{\frac{p}{q}\middle| p \text{ and } q \text{ are integers and } q \neq 0\right\}$
Irrational numbers	$\{x\,|\,x \text{ is real but not rational}\}$
Real numbers	$\{x\,|\,x \text{ corresponds to a point on a number line}\}$

EXAMPLE 1 **Identifying Sets of Numbers**

Let $A = \left\{-8, -6, -\frac{12}{4}, -\frac{3}{4}, 0, \frac{3}{8}, \frac{1}{2}, 1, \sqrt{2}, \sqrt{5}, 6\right\}$. List the elements from A that belong to each set.

(a) Natural numbers   **(b)** Whole numbers   **(c)** Integers

(d) Rational numbers   **(e)** Irrational numbers   **(f)** Real numbers

SOLUTION

(a) Natural numbers: 1 and 6   **(b)** Whole numbers: 0, 1, and 6

(c) Integers: -8, -6, $-\frac{12}{4}$ (or -3), 0, 1, and 6

(d) Rational numbers: -8, -6, $-\frac{12}{4}$ (or -3), $-\frac{3}{4}$, 0, $\frac{3}{8}$, $\frac{1}{2}$, 1, and 6

(e) Irrational numbers: $\sqrt{2}$ and $\sqrt{5}$

(f) All elements of A are real numbers.

> ✔ *Now Try Exercises 1, 11, and 13.*

Figure 8 shows the relationships among the subsets of the real numbers.

Real Numbers

| Rational numbers $\frac{4}{9}, -\frac{5}{8}, \frac{11}{7}$ | Irrational numbers $\sqrt{2}$ |

Integers $-11, -6, -3, -2, -1$

Whole numbers 0

Natural numbers 1, 2, 3, 4, 5, 37, 40

$\sqrt{15}$ $-\sqrt{8}$ π $\frac{\pi}{4}$

Figure 8

Exponents The product $2 \cdot 2 \cdot 2$ can be written as 2^3, where the 3 shows that three factors of 2 appear in the product.

Exponential Notation

If n is any positive integer and a is any real number, then the nth power of a is written using exponential notation as follows.

$$a^n = \underbrace{a \cdot a \cdot a \cdot \ldots \cdot a}_{n \text{ factors of } a}$$

That is, a^n means the product of n factors of a. The integer n is the **exponent,** a is the **base,** and a^n is a **power** or an **exponential expression** (or simply an **exponential**). Read a^n as **"a to the nth power,"** or just **"a to the nth."**

EXAMPLE 2 Evaluating Exponential Expressions

Evaluate each exponential expression, and identify the base and the exponent.

(a) 4^3 **(b)** $(-6)^2$ **(c)** -6^2 **(d)** $4 \cdot 3^2$ **(e)** $(4 \cdot 3)^2$

SOLUTION

(a) $4^3 = \underbrace{4 \cdot 4 \cdot 4}_{3 \text{ factors of } 4} = 64$ The base is 4 and the exponent is 3.

(b) $(-6)^2 = (-6)(-6) = 36$ The base is -6 and the exponent is 2.

(c) $-6^2 = -(6 \cdot 6) = -36$ — Notice that parts (b) and (c) are different.
The base is 6 and the exponent is 2.

(d) $4 \cdot 3^2 = 4 \cdot 3 \cdot 3 = 36$ The base is 3 and the exponent is 2.
$3^2 = 3 \cdot 3$, NOT $3 \cdot 2$

(e) $(4 \cdot 3)^2 = 12^2 = 144$ — $(4 \cdot 3)^2 \neq 4 \cdot 3^2$
The base is $4 \cdot 3$, or 12, and the exponent is 2.

☑ *Now Try Exercises 17, 19, 21, and 23.*

Order of Operations When a problem involves more than one operation symbol, we use the following order of operations.

Order of Operations

If grouping symbols such as parentheses, square brackets, absolute value bars, or fraction bars are present, begin as follows.

Step 1 Work separately above and below each **fraction bar**.

Step 2 Use the rules below within each set of **parentheses** or **square brackets**. Start with the innermost set and work outward.

If no grouping symbols are present, follow these steps.

Step 1 Simplify all **powers** and **roots**. *Work from left to right.*

Step 2 Do any **multiplications** or **divisions** in order. *Work from left to right.*

Step 3 Do any **negations**, **additions**, or **subtractions** in order. *Work from left to right.*

EXAMPLE 3 Using Order of Operations

Evaluate each expression.

(a) $6 \div 3 + 2^3 \cdot 5$ **(b)** $(8 + 6) \div 7 \cdot 3 - 6$

(c) $\dfrac{4 + 3^2}{6 - 5 \cdot 3}$ **(d)** $\dfrac{-(-3)^3 + (-5)}{2(-8) - 5(3)}$

SOLUTION

(a) $6 \div 3 + 2^3 \cdot 5 = 6 \div 3 + 8 \cdot 5$ Evaluate the exponential.

$= 2 + 8 \cdot 5$ Divide. Multiply or divide *in order from left to right.*

$= 2 + 40$ Multiply.

$= 42$ Add.

(b) $(8 + 6) \div 7 \cdot 3 - 6 = 14 \div 7 \cdot 3 - 6$ Work inside parentheses.

> Be careful to divide *before* multiplying here.

$= 2 \cdot 3 - 6$ Divide.

$= 6 - 6$ Multiply.

$= 0$ Subtract.

(c) $\dfrac{4 + 3^2}{6 - 5 \cdot 3} = \dfrac{4 + 9}{6 - 15}$ Evaluate the exponential and multiply.

$= \dfrac{13}{-9}, \quad \text{or} \quad -\dfrac{13}{9}$ Add and subtract; $\frac{a}{-b} = -\frac{a}{b}$.

(d) $\dfrac{-(-3)^3 + (-5)}{2(-8) - 5(3)} = \dfrac{-(-27) + (-5)}{2(-8) - 5(3)}$ Evaluate the exponential.

$= \dfrac{27 + (-5)}{-16 - 15}$ Multiply.

$= \dfrac{22}{-31}, \quad \text{or} \quad -\dfrac{22}{31}$ Add and subtract; $\frac{a}{-b} = -\frac{a}{b}$.

✔ *Now Try Exercises 25, 27, and 33.*

EXAMPLE 4 **Using Order of Operations**

Evaluate each expression for $x = -2$, $y = 5$, and $z = -3$.

(a) $-4x^2 - 7y + 4z$ **(b)** $\dfrac{2(x - 5)^2 + 4y}{z + 4}$ **(c)** $\dfrac{\dfrac{x}{2} - \dfrac{y}{5}}{\dfrac{3z}{9} + \dfrac{8y}{5}}$

SOLUTION

> Use parentheses around substituted values to avoid errors.

(a) $-4x^2 - 7y + 4z = -4(-2)^2 - 7(5) + 4(-3)$ Substitute: $x = -2$, $y = 5$, and $z = -3$.

$= -4(4) - 7(5) + 4(-3)$ Evaluate the exponential.

$= -16 - 35 - 12$ Multiply.

$= -63$ Subtract.

(b) $\dfrac{2(x - 5)^2 + 4y}{z + 4} = \dfrac{2(-2 - 5)^2 + 4(5)}{-3 + 4}$ Substitute: $x = -2$, $y = 5$, and $z = -3$.

$= \dfrac{2(-7)^2 + 20}{1}$ Work inside parentheses. Then multiply and add.

$= 2(49) + 20$ Evaluate the exponential.

$= 98 + 20$ Multiply.

$= 118$ Add.

(c) $\dfrac{\dfrac{x}{2} - \dfrac{y}{5}}{\dfrac{3z}{9} + \dfrac{8y}{5}} = \dfrac{\dfrac{-2}{2} - \dfrac{5}{5}}{\dfrac{3(-3)}{9} + \dfrac{8(5)}{5}}$ Substitute: $x = -2$, $y = 5$, and $z = -3$.

$= \dfrac{-1 - 1}{-1 + 8}, \quad \text{or} \quad -\dfrac{2}{7}$ Simplify the fractions.

✔ *Now Try Exercises 35, 43, and 45.*

Properties of Real Numbers The following basic properties can be generalized to apply to expressions with variables.

Properties of Real Numbers

Let a, b, and c represent real numbers.

Property	Description
Closure Properties $a + b$ is a real number. ab is a real number.	The sum or product of two real numbers is a real number.
Commutative Properties $a + b = b + a$ $ab = ba$	The sum or product of two real numbers is the same regardless of their order.
Associative Properties $(a + b) + c = a + (b + c)$ $(ab)c = a(bc)$	The sum or product of three real numbers is the same no matter which two are added or multiplied first.
Identity Properties There exists a unique real number 0 such that $a + 0 = a$ and $0 + a = a.$ There exists a unique real number 1 such that $a \cdot 1 = a$ and $1 \cdot a = a.$	The sum of a real number and 0 is that real number, and the product of a real number and 1 is that real number.
Inverse Properties There exists a unique real number $-a$ such that $a + (-a) = 0$ and $-a + a = 0.$ If $a \neq 0$, there exists a unique real number $\frac{1}{a}$ such that $a \cdot \dfrac{1}{a} = 1$ and $\dfrac{1}{a} \cdot a = 1.$	The sum of any real number and its negative is 0, and the product of any nonzero real number and its reciprocal is 1.
Distributive Properties $a(b + c) = ab + ac$ $a(b - c) = ab - ac$	The product of a real number and the sum (or difference) of two real numbers equals the sum (or difference) of the products of the first number and each of the other numbers.

The **multiplication property of zero** says that $0 \cdot a = a \cdot 0 = 0$ for all real numbers a.

CAUTION With the commutative properties, the *order* changes, but with the associative properties, the *grouping* changes.

Commutative Properties	Associative Properties
$(x + 4) + 9 = (4 + x) + 9$	$(x + 4) + 9 = x + (4 + 9)$
$7 \cdot (5 \cdot 2) = (5 \cdot 2) \cdot 7$	$7 \cdot (5 \cdot 2) = (7 \cdot 5) \cdot 2$

EXAMPLE 5 **Simplifying Expressions**

Use the commutative and associative properties to simplify each expression.

(a) $6 + (9 + x)$ **(b)** $\dfrac{5}{8}(16y)$ **(c)** $-10p\left(\dfrac{6}{5}\right)$

SOLUTION

(a) $6 + (9 + x) = (6 + 9) + x$ Associative property

$ = 15 + x$ Add.

(b) $\dfrac{5}{8}(16y) = \left(\dfrac{5}{8} \cdot 16\right)y$ Associative property

$\phantom{\dfrac{5}{8}(16y)} = 10y$ Multiply.

(c) $-10p\left(\dfrac{6}{5}\right) = \dfrac{6}{5}(-10p)$ Commutative property

$\phantom{-10p\left(\dfrac{6}{5}\right)} = \left[\dfrac{6}{5}(-10)\right]p$ Associative property

$\phantom{-10p\left(\dfrac{6}{5}\right)} = -12p$ Multiply.

✔️ *Now Try Exercises 63 and 65.*

Figure 9 helps to explain the distributive property. The area of the entire region shown can be found in two ways, as follows.

$$4(5 + 3) = 4(8) = 32$$

or $$4(5) + 4(3) = 20 + 12 = 32$$

The result is the same. This means that

$$4(5 + 3) = 4(5) + 4(3).$$

Geometric Model of the Distributive Property

Figure 9

EXAMPLE 6 **Using the Distributive Property**

Rewrite each expression using the distributive property and simplify, if possible.

(a) $3(x + y)$ **(b)** $-(m - 4n)$ **(c)** $\dfrac{1}{3}\left(\dfrac{4}{5}m - \dfrac{3}{2}n - 27\right)$ **(d)** $7p + 21$

SOLUTION

(a) $3(x + y) = 3x + 3y$

(b) $-(m - 4n) = -1(m - 4n)$ *Be careful with the negative signs.*

$ = -1(m) + (-1)(-4n)$

$ = -m + 4n$

(c) $\dfrac{1}{3}\left(\dfrac{4}{5}m - \dfrac{3}{2}n - 27\right) = \dfrac{1}{3}\left(\dfrac{4}{5}m\right) + \dfrac{1}{3}\left(-\dfrac{3}{2}n\right) + \dfrac{1}{3}(-27)$

$\phantom{\dfrac{1}{3}\left(\dfrac{4}{5}m - \dfrac{3}{2}n - 27\right)} = \dfrac{4}{15}m - \dfrac{1}{2}n - 9$

(d) $7p + 21 = 7p + 7 \cdot 3$

$ = 7(p + 3)$ Distributive property in reverse

✔️ *Now Try Exercises 67, 69, and 71.*

$-\sqrt{5}$ is to the left of $-\frac{11}{7}$ on the number line, so $-\sqrt{5} < -\frac{11}{7}$, and $\sqrt{20}$ is to the right of π, indicating that $\sqrt{20} > \pi$.

Figure 10

Figure 11

Order on the Number Line   If the real number a is to the left of the real number b on a number line, then

> **a is less than b,**   written $a < b$.

If a is to the right of b, then

> **a is greater than b,**   written $a > b$.

> The inequality symbol must point toward the lesser number.

Figure 10 illustrates this with several pairs of numbers. Statements involving these symbols, as well as the symbols less than or equal to, $\leq$, and greater than or equal to, $\geq$, are called **inequalities**. The inequality $a < b < c$ says that b is **between** a and c since $a < b$ and $b < c$.

Absolute Value   The distance on the number line from a number to 0 is called the **absolute value** of that number. The absolute value of the number a is written $|a|$. For example, the distance on the number line from 5 to 0 is 5, as is the distance from -5 to 0. See **Figure 11**. Therefore, both of the following are true.

$$|5| = 5 \quad \text{and} \quad |-5| = 5$$

NOTE   *Since distance cannot be negative, the absolute value of a number is always positive or* **0.**

The algebraic definition of absolute value follows.

Absolute Value

Let a represent a real number.

$$|a| = \begin{cases} a & \text{if } a \geq 0 \\ -a & \text{if } a < 0 \end{cases}$$

That is, the absolute value of a positive number or **0** *equals that number, while the absolute value of a negative number equals its negative (or opposite).*

EXAMPLE 7   **Evaluating Absolute Values**

Evaluate each expression.

(a) $\left|-\dfrac{5}{8}\right|$ **(b)** $-|8|$ **(c)** $-|-2|$ **(d)** $|2x|$, for $x = \pi$

SOLUTION

(a) $\left|-\dfrac{5}{8}\right| = \dfrac{5}{8}$ **(b)** $-|8| = -(8) = -8$

(c) $-|-2| = -(2) = -2$ **(d)** $|2\pi| = 2\pi$

> ✔ *Now Try Exercises 83 and 87.*

Absolute value is useful in applications where only the *size* (or magnitude), not the *sign,* of the difference between two numbers is important.

EXAMPLE 8 **Measuring Blood Pressure Difference**

Systolic blood pressure is the maximum pressure produced by each heartbeat. Both low blood pressure and high blood pressure may be cause for medical concern. Therefore, health care professionals are interested in a patient's "pressure difference from normal," or P_d.

If 120 is considered a normal systolic pressure, then

$$P_d = |P - 120|, \quad \text{where } P \text{ is the patient's recorded systolic pressure.}$$

Find P_d for a patient with a systolic pressure, P, of 113.

SOLUTION
$$P_d = |P - 120|$$
$$= |113 - 120| \quad \text{Let } P = 113.$$
$$= |-7| \quad \quad \text{Subtract.}$$
$$= 7 \quad \quad \quad \text{Definition of absolute value}$$

☑ *Now Try Exercise 89.*

Properties of Absolute Value

Let a and b represent real numbers.

Property	Description
1. $\|a\| \geq 0$	The absolute value of a real number is positive or 0.
2. $\|-a\| = \|a\|$	The absolute values of a real number and its opposite are equal.
3. $\|a\| \cdot \|b\| = \|ab\|$	The product of the absolute values of two real numbers equals the absolute value of their product.
4. $\dfrac{\|a\|}{\|b\|} = \left\|\dfrac{a}{b}\right\| \quad (b \neq 0)$	The quotient of the absolute values of two real numbers equals the absolute value of their quotient.
5. $\|a + b\| \leq \|a\| + \|b\|$ (the triangle inequality)	The absolute value of the sum of two real numbers is less than or equal to the sum of their absolute values.

LOOKING AHEAD TO CALCULUS

One of the most important definitions in calculus, that of the **limit,** uses absolute value. (The symbols ϵ (epsilon) and δ (delta) are often used to represent small quantities in mathematics.)

Suppose that a function f is defined at every number in an open interval I containing a, except perhaps at a itself. Then the limit of $f(x)$ as x approaches a is L, written

$$\lim_{x \to a} f(x) = L,$$

if for every $\epsilon > 0$ there exists a $\delta > 0$ such that $|f(x) - L| < \epsilon$ whenever $0 < |x - a| < \delta$.

To illustrate these properties, see the following.

$$|-15| = 15 \text{ and } 15 \geq 0 \quad \text{Property 1}$$
$$|-10| = 10 \text{ and } |10| = 10, \text{ so } |-10| = |10|. \quad \text{Property 2}$$
$$|5x| = |5| \cdot |x| = 5|x| \text{ since 5 is positive.} \quad \text{Property 3}$$
$$\left|\frac{2}{y}\right| = \frac{|2|}{|y|} = \frac{2}{|y|}, \quad y \neq 0 \quad \text{Property 4}$$

To illustrate the triangle inequality, we let $a = 3$ and $b = -7$.

$$|a + b| = |3 + (-7)| = |-4| = 4$$
$$|a| + |b| = |3| + |-7| = 3 + 7 = 10$$

Thus, $\quad |a + b| \leq |a| + |b|.$ \quad Property 5

> **NOTE** As seen in **Example 9(b)**, absolute value bars can also act as symbols of inclusion. Remember this when applying the rules for order of operations.

EXAMPLE 9 **Evaluating Absolute Value Expressions**

Let $x = -6$ and $y = 10$. Evaluate each expression.

(a) $|2x - 3y|$

(b) $\dfrac{2|x| - |3y|}{|xy|}$

SOLUTION

(a)
$$
\begin{aligned}
|2x - 3y| &= |2(-6) - 3(10)| &&\text{Substitute.}\\
&= |-12 - 30| &&\text{Work inside absolute value bars. Multiply.}\\
&= |-42| &&\text{Subtract.}\\
&= 42 &&\text{Definition of absolute value}
\end{aligned}
$$

(b)
$$
\begin{aligned}
\dfrac{2|x| - |3y|}{|xy|} &= \dfrac{2|-6| - |3(10)|}{|-6(10)|} &&\text{Substitute.}\\[2mm]
&= \dfrac{2 \cdot 6 - |30|}{|-60|} &&|-6| = 6;\ \text{multiply.}\\[2mm]
&= \dfrac{12 - 30}{60} &&\text{Multiply. } |30| = 30, |-60| = 60\\[2mm]
&= \dfrac{-18}{60} &&\text{Subtract.}\\[2mm]
&= -\dfrac{3}{10} &&\text{Write in lowest terms; } \tfrac{-a}{b} = -\tfrac{a}{b}.
\end{aligned}
$$

✔ *Now Try Exercises 93 and 95.*

Distance between Points on a Number Line

If P and Q are points on a number line with coordinates a and b, respectively, then the distance $d(P, Q)$ between them is given by the following.

$$d(P, Q) = |b - a| \quad \text{or} \quad d(P, Q) = |a - b|$$

Figure 12

That is, the distance between two points on a number line is the absolute value of the difference between their coordinates in either order. See **Figure 12.**

EXAMPLE 10 **Finding the Distance between Two Points**

Find the distance between -5 and 8.

SOLUTION Use the first formula above, with $a = -5$ and $b = 8$.

$$|b - a| = |8 - (-5)| = |8 + 5| = |13| = 13$$

Alternatively, for $a = 8$ and $b = -5$, we obtain the same result.

$$|b - a| = |(-5) - 8| = |-13| = 13$$

✔ *Now Try Exercise 105.*

R.2 Exercises

1. *Concept Check* Match each number from Column I with the letter or letters of the sets of numbers from Column II to which the number belongs. There may be more than one choice, so give all choices.

I		II	
(a) 0	**(b)** 34	**A.** Natural numbers	**B.** Whole numbers
(c) $-\frac{9}{4}$	**(d)** $\sqrt{36}$	**C.** Integers	**D.** Rational numbers
(e) $\sqrt{13}$	**(f)** 2.16	**E.** Irrational numbers	**F.** Real numbers

 2. Explain why no answer in **Exercise 1** can contain both D and E as choices.

Concept Check *Decide whether each statement is* true *or* false. *If it is false, tell why.*

3. Every integer is a whole number.

4. Every natural number is an integer.

5. Every irrational number is an integer.

6. Every integer is a rational number.

7. Every natural number is a whole number.

8. Some rational numbers are irrational.

9. Some rational numbers are whole numbers.

10. Some real numbers are integers.

Let set $A = \left\{ -6, -\frac{12}{4}, -\frac{5}{8}, -\sqrt{3}, 0, \frac{1}{4}, 1, 2\pi, 3, \sqrt{12} \right\}$. *List all the elements of A that belong to each set.* **See Example 1.**

11. Natural numbers

12. Whole numbers

13. Integers

14. Rational numbers

15. Irrational numbers

16. Real numbers

Evaluate each expression. **See Example 2.**

17. -2^4

18. -3^5

19. $(-2)^4$

20. $(-2)^6$

21. $(-3)^5$

22. $(-2)^5$

23. $-2 \cdot 3^4$

24. $-4 \cdot 5^3$

Evaluate each expression. **See Example 3.**

25. $-2 \cdot 5 + 12 \div 3$

26. $9 \cdot 3 - 16 \div 4$

27. $-4(9 - 8) + (-7)(2)^3$

28. $6(-5) - (-3)(2)^4$

29. $\left(4 - 2^3\right)\left(-2 + \sqrt{25}\right)$

30. $\left(5 - 3^2\right)\left(\sqrt{16} - 2^3\right)$

31. $\left(-\frac{2}{9} - \frac{1}{4}\right) - \left[-\frac{5}{18} - \left(-\frac{1}{2}\right)\right]$

32. $\left[-\frac{5}{8} - \left(-\frac{2}{5}\right)\right] - \left(\frac{3}{2} - \frac{11}{10}\right)$

33. $\dfrac{-8 + (-4)(-6) \div 12}{4 - (-3)}$

34. $\dfrac{15 \div 5 \cdot 4 \div 6 - 8}{-6 - (-5) - 8 \div 2}$

Evaluate each expression for $p = -4$, $q = 8$, *and* $r = -10$. **See Example 4.**

35. $-p^2 - 7q + r^2$

36. $-p^2 - 2q + r$

37. $\dfrac{q + r}{q + p}$

38. $\dfrac{p + r}{p + q}$

39. $\dfrac{3q}{r} - \dfrac{5}{p}$

40. $\dfrac{3r}{q} - \dfrac{2}{r}$

41. $\dfrac{5r}{2p - 3r}$

42. $\dfrac{3q}{3p - 2r}$

43. $\dfrac{\dfrac{q}{2} - \dfrac{r}{3}}{\dfrac{3p}{4} + \dfrac{q}{8}}$

44. $\dfrac{\dfrac{q}{4} - \dfrac{r}{5}}{\dfrac{p}{2} + \dfrac{q}{2}}$

45. $\dfrac{-(p + 2)^2 - 3r}{2 - q}$

46. $\dfrac{-(q - 6)^2 - 2p}{4 - p}$

47. $\dfrac{3p + 3(4 + p)^3}{r + 8}$

48. $\dfrac{5q + 2(1 + p)^3}{r + 3}$

Identify the property illustrated in each statement. Assume all variables represent real numbers. **See Examples 5 and 6.**

49. $6 \cdot 12 + 6 \cdot 15 = 6(12 + 15)$

50. $8(m + 4) = 8m + 32$

51. $(t - 6) \cdot \left(\dfrac{1}{t - 6}\right) = 1$, if $t - 6 \neq 0$

52. $\dfrac{2 + m}{2 - m} \cdot \dfrac{2 - m}{2 + m} = 1$, if $m \neq 2$ or -2

53. $(7.5 - y) + 0 = 7.5 - y$

54. $1 \cdot (3x - 7) = 3x - 7$

55. $5(t + 3) = (t + 3) \cdot 5$

56. $-7 + (x + 3) = (x + 3) + (-7)$

57. $(5x)\left(\dfrac{1}{x}\right) = 5\left(x \cdot \dfrac{1}{x}\right)$

58. $(38 + 99) + 1 = 38 + (99 + 1)$

59. $5 + \sqrt{3}$ is a real number.

60. 5π is a real number.

61. Is there a commutative property for subtraction? That is, in general, is $a - b$ equal to $b - a$? Support your answer with examples.

62. Is there an associative property for subtraction? That is, does $(a - b) - c$ equal $a - (b - c)$ in general? Support your answer with examples.

Simplify each expression. **See Examples 5 and 6.**

63. $\dfrac{10}{11}(22z)$

64. $\left(\dfrac{3}{4}r\right)(-12)$

65. $(m + 5) + 6$

66. $8 + (a + 7)$

67. $\dfrac{3}{8}\left(\dfrac{16}{9}y + \dfrac{32}{27}z - \dfrac{40}{9}\right)$

68. $-\dfrac{1}{4}(20m + 8y - 32z)$

Use the distributive property to rewrite sums as products and products as sums. **See Example 6.**

69. $8p - 14p$ **70.** $15x - 10x$ **71.** $-4(z - y)$ **72.** $-3(m + n)$

Concept Check Use the distributive property to calculate each value mentally.

73. $72 \cdot 17 + 28 \cdot 17$

74. $32 \cdot 80 + 32 \cdot 20$

75. $123\dfrac{5}{8} \cdot 1\dfrac{1}{2} - 23\dfrac{5}{8} \cdot 1\dfrac{1}{2}$

76. $17\dfrac{2}{5} \cdot 14\dfrac{3}{4} - 17\dfrac{2}{5} \cdot 4\dfrac{3}{4}$

Concept Check Decide whether each statement is true *or* false. *If false, correct the statement so it is true.*

77. $|6 - 8| = |6| - |8|$

78. $|(-3)^3| = -|3^3|$

79. $|-5| \cdot |6| = |-5 \cdot 6|$

80. $\dfrac{|-14|}{|2|} = \left|\dfrac{-14}{2}\right|$

81. $|a - b| = |a| - |b|$, if $b > a > 0$

82. If a is negative, then $|a| = -a$.

Evaluate each expression. **See Example 7.**

83. $|-10|$ **84.** $|-15|$ **85.** $-\left|\dfrac{4}{7}\right|$

86. $-\left|\dfrac{7}{2}\right|$ **87.** $-|-8|$ **88.** $-|-12|$

Solve each problem. **See Example 8.**

89. *Blood Pressure Difference* Calculate the P_d value for a woman whose actual systolic pressure is 116 and whose normal value should be 125.

90. *Systolic Blood Pressure* If a patient's P_d value is 17 and the normal pressure for his gender and age should be 130, what are the two possible values for his systolic blood pressure?

Let x = −4 and y = 2. Evaluate each expression. **See Example 9.**

91. $|3x - 2y|$ **92.** $|2x - 5y|$ **93.** $|-3x + 4y|$ **94.** $|-5y + x|$

95. $\dfrac{2|y| - 3|x|}{|xy|}$ **96.** $\dfrac{4|x| + 4|y|}{|x|}$ **97.** $\dfrac{|-8y + x|}{-|x|}$ **98.** $\dfrac{|x| + 2|y|}{-|x|}$

Justify each statement by giving the correct property of absolute value from this section. Assume all variables represent real numbers.

99. $|m| = |-m|$ **100.** $|-k| \geq 0$

101. $|9| \cdot |-6| = |-54|$ **102.** $|k - m| \leq |k| + |-m|$

103. $|12 + 11r| \geq 0$ **104.** $\left|\dfrac{-12}{5}\right| = \dfrac{|-12|}{|5|}$

Find the given distances between points P, Q, R, and S on a number line, with coordinates −4, −1, 8, and 12, respectively. **See Example 10.**

105. $d(P, Q)$ **106.** $d(P, R)$ **107.** $d(Q, R)$ **108.** $d(Q, S)$

Concept Check Determine what signs on values of x and y would make each statement true. Assume that x and y are not 0. (You should be able to work these mentally.)

109. $xy > 0$ **110.** $x^2y > 0$ **111.** $\dfrac{x}{y} < 0$

112. $\dfrac{y^2}{x} < 0$ **113.** $\dfrac{x^3}{y} > 0$ **114.** $-\dfrac{x}{y} > 0$

Solve each problem.

115. *Golf Scores* Phil Mickelson won the 2010 Masters Golf Tournament with a total score that was 16 under par, and Zach Johnson won the 2007 tournament with a total score that was 1 above par. Using −16 to represent 16 below par and +1 to represent 1 over par, find the difference between these scores (in either order) and take the absolute value of this difference. What does this final number represent? (*Source:* www.masters.org)

116. *Total Football Yardage* During his 16 years in the NFL, Marcus Allen gained 12,243 yd rushing, 5411 yd receiving, and −6 yd returning fumbles. Find his total yardage (called *all-purpose yards*). Is this the same as the sum of the absolute values of the three categories? Explain. (*Source: The Sports Illustrated Sports Almanac.*)

Blood Alcohol Concentration The blood alcohol concentration (BAC) of a person who has been drinking is approximated by the following expression.

   number of oz × % alcohol × 0.075 ÷ body weight in lb − hr of drinking × 0.015

(*Source: Lawlor, J., Auto Math Handbook: Mathematical Calculations, Theory, and Formulas for Automotive Enthusiasts, HP Books.*)

117. Suppose a policeman stops a 190-lb man who, in 2 hr, has ingested four 12-oz beers (48 oz), each having a 3.2% alcohol content. Calculate the man's BAC to the nearest thousandth. Follow the order of operations.

118. Find the BAC to the nearest thousandth for a 135-lb woman who, in 3 hr, has drunk three 12-oz beers (36 oz), each having a 4.0% alcohol content.

119. Calculate the BACs in **Exercises 117 and 118** if each person weighs 25 lb more and the rest of the variables stay the same. How does increased weight affect a person's BAC?

120. Predict how decreased weight would affect the BAC of each person in **Exercises 117 and 118.** Calculate the BACs if each person weighs 25 lb less and the rest of the variables stay the same.

Passer Rating for NFL Quarterbacks *The current system of rating passers in the National Football League, adopted in 1973, is based on four performance components: completions, touchdowns, yards gained, and interceptions, as percentages of the number of passes attempted. It uses the following formula.*

$$\text{Rating} = \frac{\left(250 \cdot \dfrac{C}{A}\right) + \left(1000 \cdot \dfrac{T}{A}\right) + \left(12.5 \cdot \dfrac{Y}{A}\right) + 6.25 - \left(1250 \cdot \dfrac{I}{A}\right)}{3},$$

where A = attempted passes, C = completed passes, T = touchdown passes, Y = yards gained passing, and I = interceptions.

In addition to the weighting factors appearing in the formula, the four category ratios are limited to nonnegative values with the following maximums.

$$0.775 \text{ for } \frac{C}{A}, \quad 0.11875 \text{ for } \frac{T}{A},$$

$$12.5 \text{ for } \frac{Y}{A}, \quad 0.095 \text{ for } \frac{I}{A}$$

Archie Manning, father of NFL quarterbacks Peyton and Eli, signed this photo for author Hornsby's son, Jack.

Exercises 121–132 give the 2010 regular season statistics for the top twelve quarterbacks. Use the formula to determine the rating for each. Round each answer to the nearest tenth.

Quarterback, Team	A Att	C Comp	T TD	Y Yds	I Int
121. Tom Brady, NE	492	324	36	3900	4
122. Philip Rivers, SD	541	357	30	4710	13
123. Aaron Rodgers, GB	475	312	28	3922	11
124. Michael Vick, PHI	372	233	21	3018	6
125. Ben Roethlisberger, PIT	389	240	17	3200	5
126. Josh Freeman, TB	474	291	25	3451	6
127. Joe Flacco, BAL	489	306	25	3622	10
128. Matt Cassel, KC	450	262	27	3116	7
129. Matt Schaub, HOU	574	365	24	4370	12
130. Peyton Manning, IND	679	450	33	4700	17
131. Matt Ryan, ATL	571	357	28	3705	9
132. Drew Brees, NO	658	448	33	4620	22

Source: www.nfl.com

133. Steve Young, of the San Francisco 49ers, set a full season rating record of 112.8 in 1994 and held that record until Peyton Manning surpassed it in 2004. (As of 2011, Manning's all-time record holds.) If Manning had 336 completions, 49 touchdowns, 10 interceptions, and 4557 yards, for 497 attempts, what was his rating in 2004?

134. Refer to the passer rating formula and determine the highest rating possible (considered a "perfect" passer rating).

R.3 Polynomials

- Rules for Exponents
- Polynomials
- Addition and Subtraction
- Multiplication
- Division

Rules for Exponents   From **Section R.2,** the notation a^m (where m is a positive integer and a is a real number) means that a appears as a factor m times. In the same way, a^n (where n is a positive integer) means that a appears as a factor n times. In the product $a^m \cdot a^n$, the base a would appear $m + n$ times, so the **product rule** states the following.

$$a^m \cdot a^n = a^{m+n} \quad \text{Product rule}$$

Also consider the expression $(2^5)^3$, which can be written as follows.

$$(2^5)^3 = 2^5 \cdot 2^5 \cdot 2^5 \qquad \text{Definition of exponent}$$
$$= 2^{5+5+5}, \quad \text{or} \quad 2^{15} \qquad \text{Generalization of the product rule}$$

The exponent 15 could have been obtained by multiplying 5 and 3. This example suggests the first of the **power rules** below. The others are found in a similar way.

$$\textbf{1. } (a^m)^n = a^{mn} \qquad \textbf{2. } (ab)^m = a^m b^m \qquad \textbf{3. } \left(\frac{a}{b}\right)^m = \frac{a^m}{b^m} \quad (b \neq 0)$$

Power rules for positive integers m and n and real numbers a and b

Rules for Exponents

For all positive integers m and n and all real numbers a and b, the following rules hold.

Rule	Description
Product Rule $a^m \cdot a^n = a^{m+n}$	When multiplying powers of like bases, keep the base and add the exponents.
Power Rule 1 $(a^m)^n = a^{mn}$	To raise a power to a power, multiply the exponents.
Power Rule 2 $(ab)^m = a^m b^m$	To raise a product to a power, raise each factor to that power.
Power Rule 3 $\left(\dfrac{a}{b}\right)^m = \dfrac{a^m}{b^m} \quad (b \neq 0)$	To raise a quotient to a power, raise the numerator and the denominator to that power.

EXAMPLE 1   Using the Product Rule

Find each product.

(a) $y^4 \cdot y^7$

(b) $(6z^5)(9z^3)(2z^2)$

SOLUTION

(a) $y^4 \cdot y^7 = y^{4+7} = y^{11}$ Product rule: Keep the base and add the exponents.

(b) $(6z^5)(9z^3)(2z^2) = (6 \cdot 9 \cdot 2) \cdot (z^5 z^3 z^2)$ Commutative and associative properties **(Section R.2)**

$$= 108z^{5+3+2} \qquad \text{Multiply. Apply the product rule.}$$
$$= 108z^{10} \qquad \text{Add.}$$

✔ *Now Try Exercises 3, 5, and 7.*

EXAMPLE 2 **Using the Power Rules**

Simplify. Assume all variables represent nonzero real numbers.

(a) $(5^3)^2$ **(b)** $(3^4x^2)^3$ **(c)** $\left(\dfrac{2^5}{b^4}\right)^3$ **(d)** $\left(\dfrac{-2m^6}{t^2z}\right)^5$

SOLUTION

(a) $(5^3)^2 = 5^{3(2)} = 5^6$ Power rule 1

(b) $(3^4x^2)^3 = (3^4)^3(x^2)^3$ Power rule 2

             $= 3^{4(3)}x^{2(3)}$ Power rule 1

             $= 3^{12}x^6$

(c) $\left(\dfrac{2^5}{b^4}\right)^3 = \dfrac{(2^5)^3}{(b^4)^3}$ Power rule 3

           $= \dfrac{2^{15}}{b^{12}}$ Power rule 1

(d) $\left(\dfrac{-2m^6}{t^2z}\right)^5 = \dfrac{(-2m^6)^5}{(t^2z)^5}$ Power rule 3

           $= \dfrac{(-2)^5(m^6)^5}{(t^2)^5z^5}$ Power rule 2

           $= \dfrac{-32m^{30}}{t^{10}z^5}$,   or $-\dfrac{32m^{30}}{t^{10}z^5}$ Evaluate $(-2)^5$. Then use Power rule 1.

✔ *Now Try Exercises 11, 17, and 19.*

CAUTION The expressions mn^2 and $(mn)^2$ are **not** equivalent. The second power rule can be used only with the second expression:

$$(mn)^2 = m^2n^2.$$

A zero exponent is defined as follows.

Zero Exponent

For any nonzero real number a, $a^0 = 1$.

That is, any nonzero number with a zero exponent equals 1.

To illustrate why a^0 is defined to equal 1, consider the product

$$a^n \cdot a^0, \quad \text{for} \quad a \neq 0.$$

We want the definition of a^0 to be consistent so that the product rule applies. Now apply this rule.

$$a^n \cdot a^0 = a^{n+0} = a^n$$

The product of a^n and a^0 must be a^n, and thus a^0 is acting like the identity element 1. So for consistency, we *define* a^0 to equal 1. (**0^0 is undefined.**)

EXAMPLE 3 **Using the Definition of a^0**

Evaluate each power.

(a) 4^0 **(b)** $(-4)^0$ **(c)** -4^0 **(d)** $-(-4)^0$ **(e)** $(7r)^0$

SOLUTION

(a) $4^0 = 1$ Base is 4.   **(b)** $(-4)^0 = 1$ Base is -4.

(c) $-4^0 = -(4^0) = -1$ Base is 4.   **(d)** $-(-4)^0 = -(1) = -1$ Base is -4.

(e) $(7r)^0 = 1,\ r \neq 0$ Base is $7r$.

✔ *Now Try Exercise 23.*

Polynomials   Any collection of numbers or variables joined by the basic operations of addition, subtraction, multiplication, or division (except by 0), or the operations of raising to powers or taking roots, formed according to the rules of algebra, is an **algebraic expression.**

$$-2x^2 + 3x, \quad \frac{15y}{2y - 3}, \quad \sqrt{m^3 - 64}, \quad (3a + b)^4 \quad \text{Algebraic expressions}$$

The product of a real number and one or more variables raised to powers is a **term.** The real number is the **numerical coefficient,** or just the **coefficient,** of the variables. The coefficient of the variable in $-3m^4$ is -3, while the coefficient in $-p^2$ is -1. **Like terms** are terms with the same variables each raised to the same powers.

$$-13x^3, \quad 4x^3, \quad -x^3 \quad \text{Like terms} \qquad 6y, \quad 6y^2, \quad 4y^3 \quad \text{Unlike terms}$$

A **polynomial** is defined as a term or a finite sum of terms, with only positive or zero integer exponents permitted on the variables. If the terms of a polynomial contain only the variable x, then the polynomial is a **polynomial in x.**

$$5x^3 - 8x^2 + 7x - 4, \quad 9p^5 - 3, \quad 8r^2, \quad 6 \quad \text{Polynomials}$$

The terms of a polynomial cannot have variables in a denominator.

$$9x^2 - 4x + \frac{6}{x} \quad \text{Not a polynomial}$$

The **degree of a term** with one variable is the exponent on the variable. For example, the degree of $2x^3$ is 3, and the degree of $17x$ (that is, $17x^1$) is 1. The greatest degree of any term in a polynomial is the **degree of the polynomial.** For example,

$$4x^3 - 2x^2 - 3x + 7$$

has degree 3, because the greatest degree of any term is 3. A nonzero constant such as -6, equivalent to $-6x^0$, has degree 0. (The polynomial 0 has no degree.)

A polynomial can have more than one variable. A term containing more than one variable has degree equal to the sum of all the exponents appearing on the variables in the term. For example, $-3x^4y^3z^5$ has degree $4 + 3 + 5 = 12$. The degree of a polynomial in more than one variable is equal to the greatest degree of any term appearing in the polynomial. By this definition, the polynomial

$$2x^4y^3 - 3x^5y + x^6y^2$$

has degree 8, because the x^6y^2 term has the greatest degree, 8.

A polynomial containing exactly three terms is a **trinomial.** A two-term polynomial is a **binomial.** A single-term polynomial is a **monomial.**

EXAMPLE 4 **Classifying Polynomials**

The table classifies several polynomials.

Polynomial	Degree	Type
$9p^7 - 4p^3 + 8p^2$	7	Trinomial
$29x^{11} + 8x^{15}$	15	Binomial
$-10r^6s^8$	14	Monomial
$5a^3b^7 - 3a^5b^5 + 4a^2b^9 - a^{10}$	11	None of these

✔ *Now Try Exercises 27, 29, 33, and 35.*

Addition and Subtraction Since the variables used in polynomials represent real numbers, a polynomial represents a real number. This means that all the properties of the real numbers mentioned in **Section R.2** hold for polynomials. In particular, the distributive property holds.

$$3m^5 - 7m^5 = (3 - 7)m^5 = -4m^5 \quad \text{Distributive property}$$

Thus, polynomials are added by adding coefficients of like terms, and they are subtracted by subtracting coefficients of like terms.

EXAMPLE 5 **Adding and Subtracting Polynomials**

Add or subtract, as indicated.

(a) $(2y^4 - 3y^2 + y) + (4y^4 + 7y^2 + 6y)$

(b) $(-3m^3 - 8m^2 + 4) - (m^3 + 7m^2 - 3)$

(c) $(8m^4p^5 - 9m^3p^5) + (11m^4p^5 + 15m^3p^5)$

(d) $4(x^2 - 3x + 7) - 5(2x^2 - 8x - 4)$

SOLUTION

(a) $(2y^4 - 3y^2 + y) + (4y^4 + 7y^2 + 6y)$ $\boxed{y = 1y}$

$= (2 + 4)y^4 + (-3 + 7)y^2 + (1 + 6)y$ Add coefficients of like terms.

$= 6y^4 + 4y^2 + 7y$ Work inside parentheses.

(b) $(-3m^3 - 8m^2 + 4) - (m^3 + 7m^2 - 3)$

$= (-3 - 1)m^3 + (-8 - 7)m^2 + [4 - (-3)]$ Subtract coefficients of like terms.

$= -4m^3 - 15m^2 + 7$ Simplify.

(c) $(8m^4p^5 - 9m^3p^5) + (11m^4p^5 + 15m^3p^5) = 19m^4p^5 + 6m^3p^5$

(d) $4(x^2 - 3x + 7) - 5(2x^2 - 8x - 4)$

$= 4x^2 - 4(3x) + 4(7) - 5(2x^2) - 5(-8x) - 5(-4)$

         Distributive property **(Section R.2)**

$= 4x^2 - 12x + 28 - 10x^2 + 40x + 20$ Multiply.

$= -6x^2 + 28x + 48$ Add like terms.

✔ *Now Try Exercises 39 and 41.*

As shown in **Examples 5(a), (b), and (d)**, polynomials in one variable are often written with their terms in **descending order** (or descending degree). Thus, the term of greatest degree is first, the one with the next greatest degree is next, and so on.

Multiplication   One way to find the product of two polynomials, such as $3x - 4$ and $2x^2 - 3x + 5$, is to treat $3x - 4$ as a single expression and use the distributive property.

$$(3x - 4)(2x^2 - 3x + 5)$$
$$= (3x - 4)(2x^2) - (3x - 4)(3x) + (3x - 4)(5)$$
$$= 3x(2x^2) - 4(2x^2) - 3x(3x) - (-4)(3x) + 3x(5) - 4(5)$$
$$= 6x^3 - 8x^2 - 9x^2 + 12x + 15x - 20$$
$$= 6x^3 - 17x^2 + 27x - 20$$

Another method is to write such a product vertically, similar to the method used in arithmetic for multiplying whole numbers.

$$2x^2 - 3x + 5$$

Place like terms in the same column.

$$3x - 4$$
$$\overline{-8x^2 + 12x - 20} \longleftarrow -4(2x^2 - 3x + 5)$$
$$\underline{6x^3 - 9x^2 + 15x} \longleftarrow 3x(2x^2 - 3x + 5)$$
$$6x^3 - 17x^2 + 27x - 20 \quad \text{Add in columns.}$$

EXAMPLE 6   **Multiplying Polynomials**

Multiply $(3p^2 - 4p + 1)(p^3 + 2p - 8)$.

SOLUTION

This process is an ordered method of applying the distributive property.

$$3p^2 - 4p + 1$$
$$p^3 + 2p - 8$$
$$\overline{-24p^2 + 32p - 8} \longleftarrow -8(3p^2 - 4p + 1)$$

Write like terms in columns.

$$6p^3 - 8p^2 + 2p \quad \longleftarrow 2p(3p^2 - 4p + 1)$$
$$\underline{3p^5 - 4p^4 + p^3} \quad \longleftarrow p^3(3p^2 - 4p + 1)$$
$$3p^5 - 4p^4 + 7p^3 - 32p^2 + 34p - 8 \quad \text{Add in columns.}$$

✔ *Now Try Exercise 53.*

The **FOIL method** is a convenient way to find the product of two binomials. The memory aid **FOIL** (for **F**irst, **O**utside, **I**nside, **L**ast) gives the pairs of terms to be multiplied to find the product, as shown in the next example.

EXAMPLE 7   **Using the FOIL Method to Multiply Two Binomials**

Find each product.

(a) $(6m + 1)(4m - 3)$ **(b)** $(2x + 7)(2x - 7)$ **(c)** $r^2(3r + 2)(3r - 2)$

SOLUTION

$$\overset{\text{F}}{} \quad \overset{\text{O}}{} \quad \overset{\text{I}}{} \quad \overset{\text{L}}{}$$

(a) $(6m + 1)(4m - 3) = 6m(4m) + 6m(-3) + 1(4m) + 1(-3)$
$$= 24m^2 - 14m - 3 \qquad {-18m + 4m = -14m}$$

(b) $(2x + 7)(2x - 7) = 4x^2 - 14x + 14x - 49 \qquad \text{FOIL}$
$$= 4x^2 - 49 \qquad \text{Combine like terms.}$$

(c) $r^2(3r + 2)(3r - 2) = r^2(9r^2 - 6r + 6r - 4) \qquad \text{FOIL}$
$$= r^2(9r^2 - 4) \qquad \text{Combine like terms.}$$
$$= 9r^4 - 4r^2 \qquad \text{Distributive property}$$

✔ *Now Try Exercises 45 and 47.*

In **Example 7(a)**, the product of two binomials is a trinomial, while in **Examples 7(b) and (c)**, the product of two binomials is a binomial. *The product of two binomials of the forms $x + y$ and $x - y$ is always a binomial.* The squares of binomials, $(x + y)^2$ and $(x - y)^2$, are also special products.

Special Products

Product of the Sum and Difference of Two Terms	$(x + y)(x - y) = x^2 - y^2$
Square of a Binomial	$(x + y)^2 = x^2 + 2xy + y^2$
	$(x - y)^2 = x^2 - 2xy + y^2$

EXAMPLE 8 Using the Special Products

Find each product.

(a) $(3p + 11)(3p - 11)$ **(b)** $(5m^3 - 3)(5m^3 + 3)$

(c) $(9k - 11r^3)(9k + 11r^3)$ **(d)** $(2m + 5)^2$

(e) $(3x - 7y^4)^2$

SOLUTION

$\boxed{(3p)^2 = 3^2 p^2, \text{ not } 3p^2}$

(a) $(3p + 11)(3p - 11) = (3p)^2 - 11^2$ $(x + y)(x - y) = x^2 - y^2$

$= 9p^2 - 121$ Power rule 2

(b) $(5m^3 - 3)(5m^3 + 3) = (5m^3)^2 - 3^2$ $(x - y)(x + y) = x^2 - y^2$

$= 25m^6 - 9$ Power rules 2 and 1

(c) $(9k - 11r^3)(9k + 11r^3) = (9k)^2 - (11r^3)^2$

$= 81k^2 - 121r^6$

(d) $(2m + 5)^2 = (2m)^2 + 2(2m)(5) + 5^2$ $(x + y)^2 = x^2 + 2xy + y^2$

$= 4m^2 + 20m + 25$ Power rules 2 and 1; Multiply.

$\boxed{\text{Remember the middle term, } 2(2m)(5).}$

(e) $(3x - 7y^4)^2 = (3x)^2 - 2(3x)(7y^4) + (7y^4)^2$ $(x - y)^2 = x^2 - 2xy + y^2$

$= 9x^2 - 42xy^4 + 49y^8$ Power rule 2; Multiply.

✔ *Now Try Exercises 59, 61, 63, and 65.*

CAUTION See **Examples 8(d) and (e)**. *The square of a binomial has three terms.* Do **not** give $x^2 + y^2$ as the result of expanding $(x + y)^2$, or $x^2 - y^2$ as the result of expanding $(x - y)^2$.

$$(x + y)^2 = x^2 + 2xy + y^2$$

$$(x - y)^2 = x^2 - 2xy + y^2$$

Remember to include the middle term.

EXAMPLE 9 **Multiplying More Complicated Binomials**

Find each product.

(a) $[(3p - 2) + 5q][(3p - 2) - 5q]$ **(b)** $(x + y)^3$ **(c)** $(2a + b)^4$

SOLUTION

(a) $[(3p - 2) + 5q][(3p - 2) - 5q]$

$= (3p - 2)^2 - (5q)^2$ Product of the sum and difference of two terms

$= 9p^2 - 12p + 4 - 25q^2$ Square both quantities.

(b) $(x + y)^3 = (x + y)^2(x + y)$

> This does *not* equal $x^3 + y^3$.

$= (x^2 + 2xy + y^2)(x + y)$ Square $x + y$.

$= x^3 + 2x^2y + xy^2 + x^2y + 2xy^2 + y^3$ Multiply.

$= x^3 + 3x^2y + 3xy^2 + y^3$ Combine like terms.

(c) $(2a + b)^4 = (2a + b)^2(2a + b)^2$

$= (4a^2 + 4ab + b^2)(4a^2 + 4ab + b^2)$ Square each $2a + b$.

$= 16a^4 + 16a^3b + 4a^2b^2 + 16a^3b + 16a^2b^2$ Distributive property

$\quad + 4ab^3 + 4a^2b^2 + 4ab^3 + b^4$

$= 16a^4 + 32a^3b + 24a^2b^2 + 8ab^3 + b^4$ Combine like terms.

✔ *Now Try Exercises 69, 73, and 75.*

Division   The quotient of two polynomials can be found with an algorithm (that is, a step-by-step procedure) for long division similar to that used for dividing whole numbers. ***Both polynomials must be written in descending order to use this algorithm.***

EXAMPLE 10 **Dividing Polynomials**

Divide $4m^3 - 8m^2 + 5m + 6$ by $2m - 1$.

SOLUTION

$4m^3$ divided by $2m$ is $2m^2$.
$-6m^2$ divided by $2m$ is $-3m$.
$2m$ divided by $2m$ is 1.

$$
\begin{array}{r}
2m^2 - 3m + 1 \\
2m - 1 \overline{) 4m^3 - 8m^2 + 5m + 6} \\
\end{array}
$$

$\underline{4m^3 - 2m^2}$ ⟵ $2m^2(2m - 1) = 4m^3 - 2m^2$

> To subtract, add the opposite.

$-6m^2 + 5m$ ⟵ Subtract. Bring down the next term.

$\underline{-6m^2 + 3m}$ ⟵ $-3m(2m - 1) = -6m^2 + 3m$

$2m + 6$ ⟵ Subtract. Bring down the next term.

$\underline{2m - 1}$ ⟵ $1(2m - 1) = 2m - 1$

7 ⟵ Subtract. The remainder is 7.

Thus, $\dfrac{4m^3 - 8m^2 + 5m + 6}{2m - 1} = 2m^2 - 3m + 1 + \dfrac{7}{2m - 1}$.

> Remember to add $\frac{\text{remainder}}{\text{divisor}}$.

✔ *Now Try Exercise 91.*

When a polynomial has a missing term, we allow for that term by inserting a term with a 0 coefficient for it.

EXAMPLE 11 **Dividing Polynomials with Missing Terms**

Divide $3x^3 - 2x^2 - 150$ by $x^2 - 4$.

SOLUTION Both polynomials have missing first-degree terms. Insert each missing term with a 0 coefficient.

$$
\begin{array}{r}
3x - 2 \qquad\qquad \text{— Missing term} \\
x^2 + 0x - 4{\overline{\smash{\big)}\,3x^3 - 2x^2 + 0x - 150}} \\
\underline{3x^3 + 0x^2 - 12x} \\
-2x^2 + 12x - 150 \\
\underline{-2x^2 + 0x + 8} \\
12x - 158 \leftarrow \text{Remainder}
\end{array}
$$

Missing term ———

Insert placeholders for missing terms.

The division process ends when the remainder is 0 or the degree of the remainder is less than that of the divisor. Since $12x - 158$ has lesser degree than the divisor $x^2 - 4$, it is the remainder. Thus, the entire quotient is written as follows.

$$
\frac{3x^3 - 2x^2 - 150}{x^2 - 4} = 3x - 2 + \frac{12x - 158}{x^2 - 4}
$$

✔ *Now Try Exercise 93.*

R.3 Exercises

Simplify each expression. See Example 1.

1. $(-4x^5)(4x^2)$ **2.** $(3y^4)(-6y^3)$ **3.** $n^6 \cdot n^4 \cdot n$

4. $a^8 \cdot a^5 \cdot a$ **5.** $9^3 \cdot 9^5$ **6.** $4^2 \cdot 4^8$

7. $(-3m^4)(6m^2)(-4m^5)$ **8.** $(-8t^3)(2t^6)(-5t^4)$ **9.** $(5x^2y)(-3x^3y^4)$

10. *Concept Check* Decide whether each expression has been simplified correctly. If not, correct it. Assume all variables represent nonzero real numbers.

  (a) $(mn)^2 = mn^2$ **(b)** $y^2 \cdot y^5 = y^7$ **(c)** $\left(\dfrac{k}{5}\right)^3 = \dfrac{k^3}{5}$ **(d)** $3^0 y = 0$

  (e) $4^5 \cdot 4^2 = 16^7$ **(f)** $(a^2)^3 = a^5$ **(g)** $cd^0 = 1$ **(h)** $(2b)^4 = 8b^4$

Simplify each expression. Assume variables represent nonzero real numbers. See Examples 1–3.

11. $(2^2)^5$ **12.** $(6^4)^3$ **13.** $(-6x^2)^3$

14. $(-2x^5)^5$ **15.** $-(4m^3n^0)^2$ **16.** $-(2x^0y^4)^3$

17. $\left(\dfrac{r^8}{s^2}\right)^3$ **18.** $\left(\dfrac{p^4}{q}\right)^2$ **19.** $\left(\dfrac{-4m^2}{tp^2}\right)^4$

20. $\left(\dfrac{-5n^4}{r^2}\right)^3$ **21.** $-\left(\dfrac{x^3y^5}{z}\right)^0$ **22.** $-\left(\dfrac{p^2q^3}{r^3}\right)^0$

*Match each expression in Column I with its equivalent in Column II. **See Example 3.***

	I	II		I	II
23.	(a) 6^0	**A.** 0	**24.**	(a) $3p^0$	**A.** 0
	(b) -6^0	**B.** 1		(b) $-3p^0$	**B.** 1
	(c) $(-6)^0$	**C.** -1		(c) $(3p)^0$	**C.** -1
	(d) $-(-6)^0$	**D.** 6		(d) $(-3p)^0$	**D.** 3
		E. -6			**E.** -3

25. Explain why $x^2 + x^2$ is not equivalent to x^4.

26. Explain why $(x + y)^2$ is not equivalent to $x^2 + y^2$.

*Identify each expression as a polynomial or not a polynomial. For each polynomial, give the degree and identify it as a monomial, binomial, trinomial, or none of these. **See Example 4.***

27. $-5x^{11}$ **28.** $-4y^5$ **29.** $6x + 3x^4$

30. $-9y + 5y^3$ **31.** $-7z^5 - 2z^3 + 1$ **32.** $-9t^4 + 8t^3 - 7$

33. $15a^2b^3 + 12a^3b^8 - 13b^5 + 12b^6$ **34.** $-16x^5y^7 + 12x^3y^8 - 4xy^9 + 18x^{10}$

35. $\dfrac{3}{8}x^5 - \dfrac{1}{x^2} + 9$ **36.** $\dfrac{2}{3}t^6 + \dfrac{3}{t^5} + 1$

37. 5   **38.** 9

*Find each sum or difference. **See Example 5.***

39. $(5x^2 - 4x + 7) + (-4x^2 + 3x - 5)$

40. $(3m^3 - 3m^2 + 4) + (-2m^3 - m^2 + 6)$

41. $2(12y^2 - 8y + 6) - 4(3y^2 - 4y + 2)$

42. $3(8p^2 - 5p) - 5(3p^2 - 2p + 4)$

43. $(6m^4 - 3m^2 + m) - (2m^3 + 5m^2 + 4m) + (m^2 - m)$

44. $-(8x^3 + x - 3) + (2x^3 + x^2) - (4x^2 + 3x - 1)$

*Find each product. **See Examples 6–8.***

45. $(4r - 1)(7r + 2)$ **46.** $(5m - 6)(3m + 4)$

47. $x^2\left(3x - \dfrac{2}{3}\right)\left(5x + \dfrac{1}{3}\right)$ **48.** $m^3\left(2m - \dfrac{1}{4}\right)\left(3m + \dfrac{1}{2}\right)$

49. $4x^2(3x^3 + 2x^2 - 5x + 1)$ **50.** $2b^3(b^2 - 4b + 3)$

51. $(2z - 1)(-z^2 + 3z - 4)$ **52.** $(3w + 2)(-w^2 + 4w - 3)$

53. $(m - n + k)(m + 2n - 3k)$ **54.** $(r - 3s + t)(2r - s + t)$

55. $(2x + 3)(2x - 3)(4x^2 - 9)$ **56.** $(3y - 5)(3y + 5)(9y^2 - 25)$

57. $(x + 1)(x + 1)(x - 1)(x - 1)$ **58.** $(t + 4)(t + 4)(t - 4)(t - 4)$

*Find each product. **See Examples 8 and 9.***

59. $(2m + 3)(2m - 3)$ **60.** $(8s - 3t)(8s + 3t)$ **61.** $(4x^2 - 5y)(4x^2 + 5y)$

62. $(2m^3 + n)(2m^3 - n)$ **63.** $(4m + 2n)^2$ **64.** $(a - 6b)^2$

65. $(5r - 3t^2)^2$ **66.** $(2z^4 - 3y)^2$

67. $[(2p - 3) + q]^2$ **68.** $[(4y - 1) + z]^2$

69. $[(3q + 5) - p][(3q + 5) + p]$ **70.** $[(9r - s) + 2][(9r - s) - 2]$

71. $[(3a + b) - 1]^2$ **72.** $[(2m + 7) - n]^2$ **73.** $(y + 2)^3$

74. $(z - 3)^3$ **75.** $(q - 2)^4$ **76.** $(r + 3)^4$

Perform the indicated operations. See Examples 5–9.

77. $(p^3 - 4p^2 + p) - (3p^2 + 2p + 7)$ **78.** $(x^4 - 3x^2 + 2) - (-2x^4 + x^2 - 3)$

79. $(7m + 2n)(7m - 2n)$ **80.** $(3p + 5)^2$

81. $-3(4q^2 - 3q + 2) + 2(-q^2 + q - 4)$ **82.** $2(3r^2 + 4r + 2) - 3(-r^2 + 4r - 5)$

83. $p(4p - 6) + 2(3p - 8)$ **84.** $m(5m - 2) + 9(5 - m)$

85. $-y(y^2 - 4) + 6y^2(2y - 3)$ **86.** $-z^3(9 - z) + 4z(2 + 3z)$

Perform each division. See Examples 10 and 11.

87. $\dfrac{-4x^7 - 14x^6 + 10x^4 - 14x^2}{-2x^2}$ **88.** $\dfrac{-8r^3s - 12r^2s^2 + 20rs^3}{-4rs}$

89. $\dfrac{4x^3 - 3x^2 + 1}{x - 2}$ **90.** $\dfrac{3x^3 - 2x + 5}{x - 3}$

91. $\dfrac{6m^3 + 7m^2 - 4m + 2}{3m + 2}$ **92.** $\dfrac{10x^3 + 11x^2 - 2x + 3}{5x + 3}$

93. $\dfrac{x^4 + 5x^2 + 5x + 27}{x^2 + 3}$ **94.** $\dfrac{k^4 - 4k^2 + 2k + 5}{k^2 + 1}$

Relating Concepts

For individual or collaborative investigation *(Exercises 95–98)*

The special products can be used to perform selected multiplications. On the left, we use $(x + y)(x - y) = x^2 - y^2$. *On the right,* $(x - y)^2 = x^2 - 2xy + y^2$.

$51 \times 49 = (50 + 1)(50 - 1)$	$47^2 = (50 - 3)^2$
$= 50^2 - 1^2$	$= 50^2 - 2(50)(3) + 3^2$
$= 2500 - 1$	$= 2500 - 300 + 9$
$= 2499$	$= 2209$

Use special products to evaluate each expression.

95. 99×101 **96.** 63×57 **97.** 102^2 **98.** 71^2

Solve each problem.

99. *Geometric Modeling* Consider the figure, which is a square divided into two squares and two rectangles.

(a) The length of each side of the largest square is $x + y$. Use the formula for the area of a square to write the area of the largest square as a power.

(b) Use the formulas for the area of a square and the area of a rectangle to write the area of the largest square as a trinomial that represents the sum of the areas of the four figures that comprise it.

(c) Explain why the expressions in parts (a) and (b) must be equivalent.

(d) What special product formula from this section does this exercise reinforce geometrically?

100. *Geometric Modeling* Use the reasoning process of **Exercise 99** and the accompanying figure to geometrically support the distributive property. Write a short paragraph explaining this process.

101. *Volume of the Great Pyramid* An amazing formula from ancient mathematics was used by the Egyptians to find the volume of the frustum of a square pyramid, as shown in the figure. Its volume is given by

$$V = \frac{1}{3}h(a^2 + ab + b^2),$$

where b is the length of the base, a is the length of the top, and h is the height. (*Source:* Freebury, H. A., *A History of Mathematics,* Macmillan Company, New York.)

(a) When the Great Pyramid in Egypt was partially completed to a height h of 200 ft, b was 756 ft, and a was 314 ft. Calculate its volume at this stage of construction.

(b) Try to visualize the figure if $a = b$. What is the resulting shape? Find its volume.

(c) Let $a = b$ in the Egyptian formula and simplify. Are the results the same?

102. *Volume of the Great Pyramid* Refer to the formula and the discussion in **Exercise 101.**

(a) Use $V = \frac{1}{3}h(a^2 + ab + b^2)$ to determine a formula for the volume of a pyramid with square base of length b and height h by letting $a = 0$.

(b) The Great Pyramid in Egypt had a square base of length 756 ft and a height of 481 ft. Find the volume of the Great Pyramid. Compare it with the 273-ft-tall Superdome in New Orleans, which has an approximate volume of 100 million ft^3. (*Source: Guinness Book of World Records.*)

(c) The Superdome covers an area of 13 acres. How many acres does the Great Pyramid cover? (*Hint:* 1 acre = 43,560 ft^2)

(Modeling) Number of Farms in the United States The graph shows the number of farms in the United States since 1940 for selected years. The polynomial

$$0.001147x^2 - 4.5905x + 4595$$

provides a good approximation of the number of farms for these years by substituting the year for x and evaluating the polynomial. For example, if x = 1960, the value of the polynomial is approximately 3.9, which differs from the data in the bar graph by only 0.1.

Evaluate the polynomial for each year and then give the difference from the value in the graph.

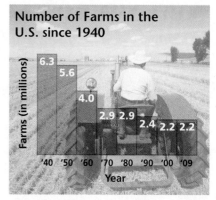

Source: U.S. Department of Agriculture.

103. 1940

104. 1970

105. 1990

106. 2009

Concept Check Perform each operation mentally.

107. $(0.25^3)(400^3)$ **108.** $(24^2)(0.5^2)$ **109.** $\dfrac{4.2^5}{2.1^5}$ **110.** $\dfrac{15^4}{5^4}$

R.4 Factoring Polynomials

- Factoring Out the Greatest Common Factor
- Factoring by Grouping
- Factoring Trinomials
- Factoring Binomials
- Factoring by Substitution

The process of finding polynomials whose product equals a given polynomial is called **factoring.** Unless otherwise specified, we consider only integer coefficients when factoring polynomials. For example, since

$$4x + 12 = 4(x + 3)$$

both 4 and $x + 3$ are **factors** of $4x + 12$ and $4(x + 3)$ is a **factored form** of $4x + 12$.

A polynomial with variable terms that cannot be written as a product of two polynomials of lower degree is a **prime polynomial.** A polynomial is **factored completely** when it is written as a product of prime polynomials.

Factoring Out the Greatest Common Factor To factor $6x^2y^3 + 9xy^4 + 18y^5$, we look for a monomial that is the greatest common factor (GCF) of the three terms.

$$6x^2y^3 + 9xy^4 + 18y^5 = 3y^3(2x^2) + 3y^3(3xy) + 3y^3(6y^2) \quad \text{GCF} = 3y^3$$

$$= 3y^3(2x^2 + 3xy + 6y^2) \quad \text{Distributive property (Section R.2)}$$

EXAMPLE 1 **Factoring Out the Greatest Common Factor**

Factor out the greatest common factor from each polynomial.

(a) $9y^5 + y^2$
(b) $6x^2t + 8xt + 12t$
(c) $14(m + 1)^3 - 28(m + 1)^2 - 7(m + 1)$

SOLUTION

(a) $9y^5 + y^2 = y^2(9y^3) + y^2(1) \quad \text{GCF} = y^2$

$$= y^2(9y^3 + 1) \quad \text{Distributive property}$$

Remember to include the 1. ⟋   Original polynomial

CHECK Multiply out the factored form: $y^2(9y^3 + 1) = \overbrace{9y^5 + y^2}$. ✓

(b) $6x^2t + 8xt + 12t = 2t(3x^2 + 4x + 6) \quad \text{GCF} = 2t$

CHECK $2t(3x^2 + 4x + 6) = 6x^2t + 8xt + 12t$ ✓

(c) $14(m + 1)^3 - 28(m + 1)^2 - 7(m + 1)$

$$= 7(m + 1)[2(m + 1)^2 - 4(m + 1) - 1] \quad \text{GCF} = 7(m + 1)$$

$$= 7(m + 1)[2(m^2 + 2m + 1) - 4m - 4 - 1] \quad \begin{array}{l}\text{Square } m + 1 \\ \text{(Section R.3);} \\ \text{distributive property}\end{array}$$

Remember the middle term. ⟋

$$= 7(m + 1)(2m^2 + 4m + 2 - 4m - 4 - 1) \quad \text{Distributive property}$$

$$= 7(m + 1)(2m^2 - 3) \quad \text{Combine like terms.}$$

☑ *Now Try Exercises 3, 9, and 15.*

CAUTION   In **Example 1(a),** the 1 is essential in the answer, since

$$y^2(9y^3) \neq 9y^5 + y^2.$$

Factoring can always be checked by multiplying.

Factoring by Grouping   When a polynomial has more than three terms, it can sometimes be factored using **factoring by grouping.** Consider this example.

$$
\overbrace{ax + ay}^{\substack{\text{Terms with}\\\text{common}\\\text{factor }a}} + \overbrace{6x + 6y}^{\substack{\text{Terms with}\\\text{common}\\\text{factor } 6}} = (ax + ay) + (6x + 6y) \quad \text{Group the terms so that each group has a common factor.}
$$

$$
= a(x + y) + 6(x + y) \qquad \text{Factor each group.}
$$

$$
= (x + y)(a + 6) \qquad \text{Factor out } x + y.
$$

It is not always obvious which terms should be grouped. In cases like the one above, group in pairs. Experience and repeated trials are the most reliable tools.

EXAMPLE 2   **Factoring by Grouping**

Factor each polynomial by grouping.

(a) $mp^2 + 7m + 3p^2 + 21$ **(b)** $2y^2 + az - 2z - ay^2$

(c) $4x^3 + 2x^2 - 2x - 1$

SOLUTION

(a) $mp^2 + 7m + 3p^2 + 21 = (mp^2 + 7m) + (3p^2 + 21)$ Group the terms.

$$
= m(p^2 + 7) + 3(p^2 + 7) \qquad \text{Factor each group.}
$$

$$
= (p^2 + 7)(m + 3) \qquad \begin{array}{l}p^2 + 7 \text{ is a}\\\text{common factor.}\end{array}
$$

CHECK $(p^2 + 7)(m + 3) = mp^2 + 3p^2 + 7m + 21$ FOIL **(Section R.3)**

$$
= mp^2 + 7m + 3p^2 + 21 \;\checkmark \quad \begin{array}{l}\text{Commutative}\\\text{property (Section R.2)}\end{array}
$$

(b) $2y^2 + az - 2z - ay^2 = 2y^2 - 2z - ay^2 + az$ Rearrange the terms.

$$
= (2y^2 - 2z) + (-ay^2 + az) \quad \text{Group the terms.}
$$

$$
\boxed{\begin{array}{c}\text{Be careful with}\\\text{signs here.}\end{array}} \;\; = 2(y^2 - z) - a(y^2 - z) \quad \begin{array}{l}\text{Factor out 2 and } -a\\\text{so that } y^2 - z \text{ is a}\\\text{common factor.}\end{array}
$$

$$
= (y^2 - z)(2 - a) \qquad \text{Factor out } y^2 - z.
$$

(c) $4x^3 + 2x^2 - 2x - 1 = (4x^3 + 2x^2) + (-2x - 1)$ Group the terms.

$$
= 2x^2(2x + 1) - 1(2x + 1) \qquad \text{Factor each group.}
$$

$$
= (2x + 1)(2x^2 - 1) \qquad \text{Factor out } 2x + 1.
$$

☑ *Now Try Exercises 19 and 21.*

Factoring Trinomials   As shown here, factoring is the opposite of multiplication.

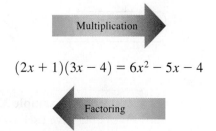

$$
(2x + 1)(3x - 4) = 6x^2 - 5x - 4
$$

One strategy in factoring trinomials requires using the FOIL method in reverse.

EXAMPLE 3 **Factoring Trinomials**

Factor each trinomial, if possible.

(a) $4y^2 - 11y + 6$ **(b)** $6p^2 - 7p - 5$

(c) $2x^2 + 13x - 18$ **(d)** $16y^3 + 24y^2 - 16y$

SOLUTION

(a) To factor this polynomial, we must find integers a, b, c, and d such that

$$4y^2 - 11y + 6 = (ay + b)(cy + d). \quad \text{FOIL}$$

Using FOIL, we see that $ac = 4$ and $bd = 6$. The positive factors of 4 are 4 and 1 or 2 and 2. Since the middle term has a negative coefficient, we consider only negative factors of 6. The possibilities are -2 and -3 or -1 and -6.

Now we try various arrangements of these factors until we find one that gives the correct coefficient of y.

$$(2y - 1)(2y - 6) = 4y^2 - 14y + 6 \quad \text{Incorrect}$$
$$(2y - 2)(2y - 3) = 4y^2 - 10y + 6 \quad \text{Incorrect}$$
$$(y - 2)(4y - 3) = 4y^2 - 11y + 6 \quad \text{Correct}$$

Therefore, $4y^2 - 11y + 6$ factors as $(y - 2)(4y - 3)$.

CHECK $(y - 2)(4y - 3) = 4y^2 - 3y - 8y + 6 \quad \text{FOIL}$

$$= 4y^2 - 11y + 6 \checkmark \quad \text{Original polynomial}$$

(b) Again, we try various possibilities to factor $6p^2 - 7p - 5$. The positive factors of 6 could be 2 and 3 or 1 and 6. As factors of -5 we have only -1 and 5 or -5 and 1.

$$(2p - 5)(3p + 1) = 6p^2 - 13p - 5 \quad \text{Incorrect}$$
$$(3p - 5)(2p + 1) = 6p^2 - 7p - 5 \quad \text{Correct}$$

Thus, $6p^2 - 7p - 5$ factors as $(3p - 5)(2p + 1)$.

(c) If we try to factor $2x^2 + 13x - 18$ as above, we find that none of the pairs of factors gives the correct coefficient of x.

$$(2x + 9)(x - 2) = 2x^2 + 5x - 18 \quad \text{Incorrect}$$
$$(2x - 3)(x + 6) = 2x^2 + 9x - 18 \quad \text{Incorrect}$$
$$(2x - 1)(x + 18) = 2x^2 + 35x - 18 \quad \text{Incorrect}$$

Additional trials are also unsuccessful. Thus, this trinomial cannot be factored with integer coefficients and is prime.

(d) $16y^3 + 24y^2 - 16y = 8y(2y^2 + 3y - 2) \quad$ Factor out the GCF, $8y$.

$$= 8y(2y - 1)(y + 2) \quad \text{Factor the trinomial.}$$

> Remember to include the common factor in the final form.

✔ *Now Try Exercises 25, 27, 29, and 31.*

NOTE In **Example 3**, we chose positive factors of the positive first term. We could have used two negative factors, but the work is easier if positive factors are used.

Each of the special patterns for multiplication given in **Section R.3** can be used in reverse to get a pattern for factoring. Perfect square trinomials can be factored as follows.

Factoring Perfect Square Trinomials

$$x^2 + 2xy + y^2 = (x + y)^2$$
$$x^2 - 2xy + y^2 = (x - y)^2$$

EXAMPLE 4   **Factoring Perfect Square Trinomials**

Factor each trinomial.

(a) $16p^2 - 40pq + 25q^2$ **(b)** $36x^2y^2 + 84xy + 49$

SOLUTION

(a) Since $16p^2 = (4p)^2$ and $25q^2 = (5q)^2$, we use the second pattern shown in the box with $4p$ replacing x and $5q$ replacing y.

$$16p^2 - 40pq + 25q^2 = (4p)^2 - 2(4p)(5q) + (5q)^2$$
$$= (4p - 5q)^2$$

Make sure that the middle term of the trinomial being factored, $-40pq$ here, is twice the product of the two terms in the binomial $4p - 5q$.

$$-40pq = 2(4p)(-5q)$$

Thus, $16p^2 - 40pq + 25q^2$ factors as $(4p - 5q)^2$.

CHECK $(4p - 5q)^2 = 16p^2 - 40pq + 25q^2$ ✓   Multiply.

(b) $36x^2y^2 + 84xy + 49$ factors as $(6xy + 7)^2$ ⟵ $\boxed{2(6xy)(7) = 84xy}$

CHECK Square $6xy + 7$: $(6xy + 7)^2 = 36x^2y^2 + 84xy + 49.$ ✓

☑ *Now Try Exercises 41 and 45.*

Factoring Binomials   Check first to see whether the terms of a binomial have a common factor. If so, factor it out. The binomial may also fit one of the following patterns.

Factoring Binomials

Difference of Squares	$x^2 - y^2 = (x + y)(x - y)$
Difference of Cubes	$x^3 - y^3 = (x - y)(x^2 + xy + y^2)$
Sum of Cubes	$x^3 + y^3 = (x + y)(x^2 - xy + y^2)$

CAUTION   *There is no factoring pattern for a sum of squares in the real number system.* In particular, $x^2 + y^2$ **does not factor as** $(x + y)^2$, for real numbers x and y.

> **EXAMPLE 5** **Factoring Differences of Squares**

Factor each polynomial.

(a) $4m^2 - 9$ (b) $256k^4 - 625m^4$ (c) $(a + 2b)^2 - 4c^2$

(d) $x^2 - 6x + 9 - y^4$ (e) $y^2 - x^2 + 6x - 9$

SOLUTION

(a) $4m^2 - 9 = (2m)^2 - 3^2$ Write as a difference of squares.

             $= (2m + 3)(2m - 3)$ Factor.

Check by multiplying.

(b) $256k^4 - 625m^4 = (16k^2)^2 - (25m^2)^2$ Write as a difference of squares.

(Don't stop here.)$= (16k^2 + 25m^2)(16k^2 - 25m^2)$ Factor.

                $= (16k^2 + 25m^2)(4k + 5m)(4k - 5m)$ Factor $16k^2 - 25m^2$.

    CHECK $(16k^2 + 25m^2)(4k + 5m)(4k - 5m)$

                $= (16k^2 + 25m^2)(16k^2 - 25m^2)$ Multiply the last two factors.

                $= 256k^4 - 625m^4$ ✓   Original polynomial

(c) $(a + 2b)^2 - 4c^2 = (a + 2b)^2 - (2c)^2$ Write as a difference of squares.

                $= [(a + 2b) + 2c][(a + 2b) - 2c]$ Factor.

                $= (a + 2b + 2c)(a + 2b - 2c)$

Check by multiplying.

(d) $x^2 - 6x + 9 - y^4 = (x^2 - 6x + 9) - y^4$ Group terms.

                $= (x - 3)^2 - y^4$ Factor the trinomial.

                $= (x - 3)^2 - (y^2)^2$ Write as a difference of squares.

                $= [(x - 3) + y^2][(x - 3) - y^2]$ Factor.

                $= (x - 3 + y^2)(x - 3 - y^2)$

Check by multiplying.

(e) $y^2 - x^2 + 6x - 9 = y^2 - (x^2 - 6x + 9)$ Factor out the negative sign and group the last three terms.

(Be careful with signs. This is a perfect square trinomial.)

                $= y^2 - (x - 3)^2$ Write as a difference of squares.

                $= [y - (x - 3)][y + (x - 3)]$ Factor.

                $= (y - x + 3)(y + x - 3)$ Distributive property

Check by multiplying.

                                 ✔ *Now Try Exercises 51, 55, 57, and 59.*

CAUTION When factoring as in **Example 5(e),** be careful with signs. Inserting an open parenthesis following the minus sign requires changing the signs of all of the following terms.

EXAMPLE 6 **Factoring Sums or Differences of Cubes**

Factor each polynomial.

(a) $x^3 + 27$ **(b)** $m^3 - 64n^3$ **(c)** $8q^6 + 125p^9$

SOLUTION

(a) $x^3 + 27 = x^3 + 3^3$ Write as a sum of cubes.

$\qquad = (x + 3)(x^2 - 3x + 3^2)$ Factor.

$\qquad = (x + 3)(x^2 - 3x + 9)$ Apply the exponent.

(b) $m^3 - 64n^3 = m^3 - (4n)^3$ Write as a difference of cubes.

$\qquad = (m - 4n)\left[m^2 + m(4n) + (4n)^2\right]$ Factor.

$\qquad = (m - 4n)(m^2 + 4mn + 16n^2)$ Simplify.

(c) $8q^6 + 125p^9 = (2q^2)^3 + (5p^3)^3$ Write as a sum of cubes.

$\qquad = (2q^2 + 5p^3)\left[(2q^2)^2 - 2q^2(5p^3) + (5p^3)^2\right]$ Factor.

$\qquad = (2q^2 + 5p^3)(4q^4 - 10q^2p^3 + 25p^6)$ Simplify.

☑ *Now Try Exercises 61, 63, and 65.*

Factoring by Substitution   We introduce a new technique for factoring.

EXAMPLE 7 **Factoring by Substitution**

Factor each polynomial.

(a) $10(2a - 1)^2 - 19(2a - 1) - 15$ **(b)** $(2a - 1)^3 + 8$

(c) $6z^4 - 13z^2 - 5$

SOLUTION

(a) $10(2a - 1)^2 - 19(2a - 1) - 15$

$\qquad = 10u^2 - 19u - 15$ Replace $2a - 1$ with u so that $(2a - 1)^2$ becomes u^2.

$\qquad = (5u + 3)(2u - 5)$ Factor.

> Don't stop here. Replace u with $2a - 1$.

$\qquad = \left[5(2a - 1) + 3\right]\left[2(2a - 1) - 5\right]$ Replace u with $2a - 1$.

$\qquad = (10a - 5 + 3)(4a - 2 - 5)$ Distributive property

$\qquad = (10a - 2)(4a - 7)$ Simplify.

$\qquad = 2(5a - 1)(4a - 7)$ Factor out the common factor.

(b) $(2a - 1)^3 + 8 = u^3 + 8$ Replace $2a - 1$ with u.

$\qquad = u^3 + 2^3$ Write as a sum of cubes.

$\qquad = (u + 2)(u^2 - 2u + 4)$ Factor.

$\qquad = \left[(2a - 1) + 2\right]\left[(2a - 1)^2 - 2(2a - 1) + 4\right]$
     Replace u with $2a - 1$.

$\qquad = (2a + 1)(4a^2 - 4a + 1 - 4a + 2 + 4)$
     Add, and then multiply.

$\qquad = (2a + 1)(4a^2 - 8a + 7)$ Combine like terms.

(c) $6z^4 - 13z^2 - 5 = 6u^2 - 13u - 5$ Replace z^2 with u.

> Remember to make the final substitution.

$\qquad = (2u - 5)(3u + 1)$ Use FOIL to factor.

$\qquad = (2z^2 - 5)(3z^2 + 1)$ Replace u with z^2.

☑ *Now Try Exercises 79 and 83.*

R.4 Exercises

Factor out the greatest common factor from each polynomial. **See Examples 1 and 2.**

1. $12m + 60$

2. $15r - 27$

3. $8k^3 + 24k$

4. $9z^4 + 81z$

5. $xy - 5xy^2$

6. $5h^2j + hj$

7. $-4p^3q^4 - 2p^2q^5$

8. $-3z^5w^2 - 18z^3w^4$

9. $4k^2m^3 + 8k^4m^3 - 12k^2m^4$

10. $28r^4s^2 + 7r^3s - 35r^4s^3$

11. $2(a + b) + 4m(a + b)$

12. $6x(a + b) - 4y(a + b)$

13. $(5r - 6)(r + 3) - (2r - 1)(r + 3)$

14. $(4z - 5)(3z - 2) - (3z - 9)(3z - 2)$

15. $2(m - 1) - 3(m - 1)^2 + 2(m - 1)^3$

16. $5(a + 3)^3 - 2(a + 3) + (a + 3)^2$

17. *Concept Check* When directed to completely factor the polynomial $4x^2y^5 - 8xy^3$, a student wrote $2xy^3(2xy^2 - 4)$. When the teacher did not give him full credit, he complained because when his answer is multiplied out, the result is the original polynomial. Give the correct answer.

18. *Concept Check* Kurt factored $16a^2 - 40a - 6a + 15$ by grouping and obtained $(8a - 3)(2a - 5)$. Callie factored the same polynomial and gave an answer of $(3 - 8a)(5 - 2a)$. Which answer is correct?

Factor each polynomial by grouping. **See Example 2.**

19. $6st + 9t - 10s - 15$

20. $10ab - 6b + 35a - 21$

21. $2m^4 + 6 - am^4 - 3a$

22. $15 - 5m^2 - 3r^2 + m^2r^2$

23. $p^2q^2 - 10 - 2q^2 + 5p^2$

24. $20z^2 - 8x + 5pz^2 - 2px$

Factor each trinomial, if possible. **See Examples 3 and 4.**

25. $6a^2 - 11a + 4$

26. $8h^2 - 2h - 21$

27. $3m^2 + 14m + 8$

28. $9y^2 - 18y + 8$

29. $15p^2 + 24p + 8$

30. $9x^2 + 4x - 2$

31. $12a^3 + 10a^2 - 42a$

32. $36x^3 + 18x^2 - 4x$

33. $6k^2 + 5kp - 6p^2$

34. $14m^2 + 11mr - 15r^2$

35. $5a^2 - 7ab - 6b^2$

36. $12s^2 + 11st - 5t^2$

37. $12x^2 - xy - y^2$

38. $30a^2 + am - m^2$

39. $24a^4 + 10a^3b - 4a^2b^2$

40. $18x^5 + 15x^4z - 75x^3z^2$ **41.** $9m^2 - 12m + 4$

42. $16p^2 - 40p + 25$

43. $32a^2 + 48ab + 18b^2$

44. $20p^2 - 100pq + 125q^2$

45. $4x^2y^2 + 28xy + 49$

46. $9m^2n^2 + 12mn + 4$

47. $(a - 3b)^2 - 6(a - 3b) + 9$

48. $(2p + q)^2 - 10(2p + q) + 25$

49. *Concept Check* Match each polynomial in Column I with its factored form in Column II.

I	II
(a) $x^2 + 10xy + 25y^2$	**A.** $(x + 5y)(x - 5y)$
(b) $x^2 - 10xy + 25y^2$	**B.** $(x + 5y)^2$
(c) $x^2 - 25y^2$	**C.** $(x - 5y)^2$
(d) $25y^2 - x^2$	**D.** $(5y + x)(5y - x)$

50. *Concept Check* Match each polynomial in Column I with its factored form in Column II.

I	II
(a) $8x^3 - 27$	**A.** $(3 - 2x)(9 + 6x + 4x^2)$
(b) $8x^3 + 27$	**B.** $(2x - 3)(4x^2 + 6x + 9)$
(c) $27 - 8x^3$	**C.** $(2x + 3)(4x^2 - 6x + 9)$

Factor each polynomial. ***See Examples 5 and 6.***

51. $9a^2 - 16$ **52.** $16q^2 - 25$ **53.** $x^4 - 16$

54. $y^4 - 81$ **55.** $25s^4 - 9t^2$ **56.** $36z^2 - 81y^4$

57. $(a + b)^2 - 16$ **58.** $(p - 2q)^2 - 100$

59. $p^4 - 625$ **60.** $m^4 - 1296$

61. $8 - a^3$ **62.** $27 - r^3$

63. $125x^3 - 27$ **64.** $8m^3 - 27n^3$

65. $27y^9 + 125z^6$ **66.** $27z^9 + 64y^{12}$

67. $(r + 6)^3 - 216$ **68.** $(b + 3)^3 - 27$

69. $27 - (m + 2n)^3$ **70.** $125 - (4a - b)^3$

71. *Concept Check* Which of the following is the correct complete factorization of $x^4 - 1$?

 A. $(x^2 - 1)(x^2 + 1)$ **B.** $(x^2 + 1)(x + 1)(x - 1)$

 C. $(x^2 - 1)^2$ **D.** $(x - 1)^2(x + 1)^2$

72. *Concept Check* Which of the following is the correct factorization of $x^3 + 8$?

 A. $(x + 2)^3$ **B.** $(x + 2)(x^2 + 2x + 4)$

 C. $(x + 2)(x^2 - 2x + 4)$ **D.** $(x + 2)(x^2 - 4x + 4)$

Relating Concepts

For individual or collaborative investigation (Exercises 73–78)

The polynomial $x^6 - 1$ can be considered either a difference of squares or a difference of cubes. **Work Exercises 73–78 in order,** *to connect the results obtained when two different methods of factoring are used.*

73. Factor $x^6 - 1$ by first factoring as a difference of squares, and then factor further by using the patterns for a sum of cubes and a difference of cubes.

74. Factor $x^6 - 1$ by first factoring as a difference of cubes, and then factor further by using the pattern for a difference of squares.

75. Compare your answers in **Exercises 73 and 74.** Based on these results, what is the factorization of $x^4 + x^2 + 1$?

76. The polynomial $x^4 + x^2 + 1$ cannot be factored using the methods described in this section. However, there is a technique that enables us to factor it, as shown here. Supply the reason why each step is valid.

$$
\begin{aligned}
x^4 + x^2 + 1 &= x^4 + 2x^2 + 1 - x^2 &&\underline{\hspace{4cm}} \\
&= (x^4 + 2x^2 + 1) - x^2 &&\underline{\hspace{4cm}} \\
&= (x^2 + 1)^2 - x^2 &&\underline{\hspace{4cm}} \\
&= (x^2 + 1 - x)(x^2 + 1 + x) &&\underline{\hspace{4cm}} \\
&= (x^2 - x + 1)(x^2 + x + 1) &&\underline{\hspace{4cm}}
\end{aligned}
$$

77. Compare your answer in **Exercise 75** with the final line in **Exercise 76.** What do you notice?

78. Factor $x^8 + x^4 + 1$ using the technique outlined in **Exercise 76.**

Factor each polynomial by substitution. **See Example 7.**

79. $7(3k - 1)^2 + 26(3k - 1) - 8$ **80.** $6(4z - 3)^2 + 7(4z - 3) - 3$

81. $9(a - 4)^2 + 30(a - 4) + 25$ **82.** $4(5x + 7)^2 + 12(5x + 7) + 9$

83. $m^4 - 3m^2 - 10$ **84.** $a^4 - 2a^2 - 48$

Factor by any method. **See Examples 1–7.**

85. $4b^2 + 4bc + c^2 - 16$ **86.** $(2y - 1)^2 - 4(2y - 1) + 4$

87. $x^2 + xy - 5x - 5y$ **88.** $8r^2 - 3rs + 10s^2$

89. $p^4(m - 2n) + q(m - 2n)$ **90.** $36a^2 + 60a + 25$

91. $4z^2 + 28z + 49$ **92.** $6p^4 + 7p^2 - 3$

93. $1000x^3 + 343y^3$ **94.** $b^2 + 8b + 16 - a^2$

95. $125m^6 - 216$ **96.** $q^2 + 6q + 9 - p^2$

97. $64 + (3x + 2)^3$ **98.** $216p^3 + 125q^3$

99. $(x + y)^3 - (x - y)^3$ **100.** $100r^2 - 169s^2$

101. $144z^2 + 121$ **102.** $(3a + 5)^2 - 18(3a + 5) + 81$

103. $(x + y)^2 - (x - y)^2$ **104.** $4z^4 - 7z^2 - 15$

105. Are there any conditions under which a sum of squares can be factored? If so, give an example.

106. *Geometric Modeling* Explain how the figures give geometric interpretation to the formula $x^2 + 2xy + y^2 = (x + y)^2$.

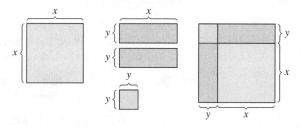

Factor each polynomial over the set of rational number coefficients.

107. $49x^2 - \dfrac{1}{25}$ **108.** $81y^2 - \dfrac{1}{49}$

109. $\dfrac{25}{9}x^4 - 9y^2$ **110.** $\dfrac{121}{25}y^4 - 49x^2$

Concept Check *Find all values of b or c that will make the polynomial a perfect square trinomial.*

111. $4z^2 + bz + 81$ **112.** $9p^2 + bp + 25$

113. $100r^2 - 60r + c$ **114.** $49x^2 + 70x + c$

R.5 Rational Expressions

■ Rational Expressions

■ Lowest Terms of a Rational Expression

■ Multiplication and Division

■ Addition and Subtraction

■ Complex Fractions

Rational Expressions The quotient of two polynomials P and Q, with $Q \neq 0$, is a **rational expression.**

$$\frac{x + 6}{x + 2}, \quad \frac{(x + 6)(x + 4)}{(x + 2)(x + 4)}, \quad \frac{2p^2 + 7p - 4}{5p^2 + 20p} \quad \text{Rational expressions}$$

The **domain** of a rational expression is the set of real numbers for which the expression is defined. Because the denominator of a fraction cannot be 0, the domain consists of all real numbers except those that make the denominator 0. We find these numbers by setting the denominator equal to 0 and solving the resulting equation. For example, in the rational expression

$$\frac{x + 6}{x + 2},$$

the solution to the equation $x + 2 = 0$ is excluded from the domain. Since this solution is -2, the domain is the set of all real numbers x not equal to -2, or

$$\{x \mid x \neq -2\}. \quad \text{Set-builder notation (Section R.1)}$$

If the denominator of a rational expression contains a product, we determine the domain with the **zero-factor property** (covered in more detail in **Section 1.4**), which states that $ab = 0$ if and only if $a = 0$ or $b = 0$.

EXAMPLE 1 **Finding the Domain**

Find the domain of the rational expression.

$$\frac{(x + 6)(x + 4)}{(x + 2)(x + 4)}$$

SOLUTION

$$(x + 2)(x + 4) = 0 \quad \text{Set the denominator equal to zero.}$$

$$x + 2 = 0 \quad \text{or} \quad x + 4 = 0 \quad \text{Zero-factor property}$$

$$x = -2 \quad \text{or} \quad x = -4 \quad \text{Solve each equation.}$$

The domain is the set of real numbers *not equal to* -2 or -4, written

$$\{x \mid x \neq -2, -4\}.$$

☑ *Now Try Exercises 1, 3, and 7.*

Lowest Terms of a Rational Expression A rational expression is written in **lowest terms** when the greatest common factor of its numerator and its denominator is 1. We use the following **fundamental principle of fractions.**

Fundamental Principle of Fractions

$$\frac{ac}{bc} = \frac{a}{b} \quad (b \neq 0, c \neq 0)$$

EXAMPLE 2 Writing Rational Expressions in Lowest Terms

Write each rational expression in lowest terms.

(a) $\dfrac{2x^2 + 7x - 4}{5x^2 + 20x}$ 　　　　　　　　　**(b)** $\dfrac{6 - 3x}{x^2 - 4}$

SOLUTION

(a) $\dfrac{2x^2 + 7x - 4}{5x^2 + 20x} = \dfrac{(2x - 1)(x + 4)}{5x(x + 4)}$ 　　Factor. **(Section R.4)**

$\qquad\qquad\qquad = \dfrac{2x - 1}{5x}$ 　　Fundamental principle and lowest terms

To determine the domain, we find values of x that make the *original* denominator $5x^2 + 20x$ equal to 0, and exclude them.

$$5x^2 + 20x = 0 \qquad \text{Set the denominator equal to 0.}$$

$$5x(x + 4) = 0 \qquad \text{Factor.}$$

$$5x = 0 \quad \text{or} \quad x + 4 = 0 \qquad \text{Zero-factor property}$$

$$x = 0 \quad \text{or} \qquad x = -4 \qquad \text{Solve each equation.}$$

The domain is $\{x \mid x \neq 0, -4\}$. *From now on, we will assume such restrictions when writing rational expressions in lowest terms.*

(b) $\dfrac{6 - 3x}{x^2 - 4} = \dfrac{3(2 - x)}{(x + 2)(x - 2)}$ 　　Factor.

$\qquad\qquad\quad = \dfrac{3(2 - x)(-1)}{(x + 2)(x - 2)(-1)}$ 　　$2 - x$ and $x - 2$ are opposites. Multiply numerator and denominator by -1.

$\qquad\qquad\quad = \dfrac{3(2 - x)(-1)}{(x + 2)(2 - x)}$ 　　$(x - 2)(-1) = -x + 2 = 2 - x$

Be careful with signs.

$\qquad\qquad\quad = \dfrac{-3}{x + 2}$ 　　Fundamental principle and lowest terms

Working in an alternative way would lead to the equivalent result $\dfrac{3}{-x - 2}$.

☑ *Now Try Exercises 13 and 17.*

LOOKING AHEAD TO CALCULUS

A standard problem in calculus is investigating what value an expression such as $\frac{x^2 - 1}{x - 1}$ approaches as x approaches 1. We cannot do this by simply substituting 1 for x in the expression since the result is the indeterminate form $\frac{0}{0}$. When we factor the numerator and write the expression in lowest terms, it becomes $x + 1$. Then by substituting 1 for x, we get $1 + 1 = 2$, which is called the **limit** of $\frac{x^2 - 1}{x - 1}$ as x approaches 1.

CAUTION The fundamental principle requires a pair of common *factors*, one in the numerator and one in the denominator. *Only after a rational expression has been factored can any common factors be divided out.*

For example, 　$\dfrac{2x + 4}{6} = \dfrac{2(x + 2)}{2 \cdot 3} = \dfrac{x + 2}{3}$. 　Factor first, and then divide.

Multiplication and Division We now multiply and divide fractions.

Multiplication and Division

For fractions $\frac{a}{b}$ and $\frac{c}{d}$ ($b \neq 0, d \neq 0$), the following hold.

$$\frac{a}{b} \cdot \frac{c}{d} = \frac{ac}{bd} \quad \text{and} \quad \frac{a}{b} \div \frac{c}{d} = \frac{a}{b} \cdot \frac{d}{c} \quad (c \neq 0)$$

*That is, to find the product of two fractions, multiply their numerators to find the numerator of the product. Then multiply their denominators to find the denominator of the product. To divide two fractions, multiply the **dividend** (the first fraction) by the reciprocal of the **divisor** (the second fraction).*

EXAMPLE 3 **Multiplying or Dividing Rational Expressions**

Multiply or divide, as indicated.

(a) $\dfrac{2y^2}{9} \cdot \dfrac{27}{8y^5}$

(b) $\dfrac{3m^2 - 2m - 8}{3m^2 + 14m + 8} \cdot \dfrac{3m + 2}{3m + 4}$

(c) $\dfrac{3p^2 + 11p - 4}{24p^3 - 8p^2} \div \dfrac{9p + 36}{24p^4 - 36p^3}$

(d) $\dfrac{x^3 - y^3}{x^2 - y^2} \cdot \dfrac{2x + 2y + xz + yz}{2x^2 + 2y^2 + zx^2 + zy^2}$

SOLUTION

(a) $\dfrac{2y^2}{9} \cdot \dfrac{27}{8y^5} = \dfrac{2y^2 \cdot 27}{9 \cdot 8y^5}$ Multiply fractions.

$= \dfrac{2 \cdot 9 \cdot 3 \cdot y^2}{9 \cdot 2 \cdot 4 \cdot y^2 \cdot y^3}$ Factor.

$= \dfrac{3}{4y^3}$ Fundamental principle

Although we usually factor first and then multiply the fractions (see parts (b)–(d)), we did the opposite here. Either order is acceptable.

(b) $\dfrac{3m^2 - 2m - 8}{3m^2 + 14m + 8} \cdot \dfrac{3m + 2}{3m + 4}$

$= \dfrac{(m - 2)(3m + 4)}{(m + 4)(3m + 2)} \cdot \dfrac{3m + 2}{3m + 4}$ Factor.

$= \dfrac{(m - 2)(3m + 4)(3m + 2)}{(m + 4)(3m + 2)(3m + 4)}$ Multiply fractions.

$= \dfrac{m - 2}{m + 4}$ Fundamental principle

(c) $\dfrac{3p^2 + 11p - 4}{24p^3 - 8p^2} \div \dfrac{9p + 36}{24p^4 - 36p^3}$

$= \dfrac{(p + 4)(3p - 1)}{8p^2(3p - 1)} \div \dfrac{9(p + 4)}{12p^3(2p - 3)}$ Factor.

$= \dfrac{(p + 4)(3p - 1)}{8p^2(3p - 1)} \cdot \dfrac{12p^3(2p - 3)}{9(p + 4)}$ Multiply by the reciprocal of the divisor.

$= \dfrac{12p^3(2p - 3)}{9 \cdot 8p^2}$ Divide out common factors. Multiply fractions.

$= \dfrac{3 \cdot 4 \cdot p^2 \cdot p(2p - 3)}{3 \cdot 3 \cdot 4 \cdot 2 \cdot p^2}$ Factor.

$= \dfrac{p(2p - 3)}{6}$ Fundamental principle

(d) $\dfrac{x^3 - y^3}{x^2 - y^2} \cdot \dfrac{2x + 2y + xz + yz}{2x^2 + 2y^2 + zx^2 + zy^2}$

$= \dfrac{(x - y)(x^2 + xy + y^2)}{(x + y)(x - y)} \cdot \dfrac{2(x + y) + z(x + y)}{2(x^2 + y^2) + z(x^2 + y^2)}$ — Factor. Group terms and factor.

$= \dfrac{(x - y)(x^2 + xy + y^2)}{(x + y)(x - y)} \cdot \dfrac{(2 + z)(x + y)}{(2 + z)(x^2 + y^2)}$ — Factor by grouping. **(Section R.4)**

$= \dfrac{x^2 + xy + y^2}{x^2 + y^2}$ — Multiply fractions; fundamental principle

> ✔ *Now Try Exercises 23, 33, and 37.*

Addition and Subtraction   We now add and subtract fractions.

Addition and Subtraction

For fractions $\dfrac{a}{b}$ and $\dfrac{c}{d}$ ($b \neq 0, d \neq 0$), the following hold.

$$\dfrac{a}{b} + \dfrac{c}{d} = \dfrac{ad + bc}{bd} \quad \text{and} \quad \dfrac{a}{b} - \dfrac{c}{d} = \dfrac{ad - bc}{bd}$$

That is, to add (or subtract) two fractions in practice, find their least common denominator (LCD) and change each fraction to one with the LCD as denominator. The sum (or difference) of their numerators is the numerator of their sum (or difference), and the LCD is the denominator of their sum (or difference).

Finding the Least Common Denominator (LCD)

Step 1 Write each denominator as a product of prime factors.

Step 2 Form a product of all the different prime factors. Each factor should have as exponent the *greatest* exponent that appears on that factor.

EXAMPLE 4   **Adding or Subtracting Rational Expressions**

Add or subtract, as indicated.

(a) $\dfrac{5}{9x^2} + \dfrac{1}{6x}$

(b) $\dfrac{y}{y - 2} + \dfrac{8}{2 - y}$

(c) $\dfrac{3}{(x - 1)(x + 2)} - \dfrac{1}{(x + 3)(x - 4)}$

SOLUTION

(a) $\dfrac{5}{9x^2} + \dfrac{1}{6x}$

Step 1 Write each denominator as a product of prime factors.

$$9x^2 = 3^2 \cdot x^2$$

$$6x = 2^1 \cdot 3^1 \cdot x^1$$

Step 2 For the LCD, form the product of all the prime factors, with each factor having the greatest exponent that appears on it.

Greatest exponent on 3 is 2. Greatest exponent on x is 2.

$$\text{LCD} = 2^1 \cdot 3^2 \cdot x^2$$

$$= 18x^2$$

Write the given expressions with this denominator, and then add.

$$\frac{5}{9x^2} + \frac{1}{6x} = \frac{5 \cdot 2}{9x^2 \cdot 2} + \frac{1 \cdot 3x}{6x \cdot 3x} \qquad \text{LCD} = 18x^2$$

$$= \frac{10}{18x^2} + \frac{3x}{18x^2} \qquad \text{Multiply.}$$

$$= \frac{10 + 3x}{18x^2} \qquad \text{Add the numerators.}$$

Always check to see that the answer is in lowest terms.

(b) $\dfrac{y}{y-2} + \dfrac{8}{2-y}$ We arbitrarily choose $y - 2$ as the LCD.

$$= \frac{y}{y-2} + \frac{8(-1)}{(2-y)(-1)} \qquad \begin{array}{l}\text{Multiply the second expression by } -1 \text{ in}\\ \text{both the numerator and the denominator.}\end{array}$$

$$= \frac{y}{y-2} + \frac{-8}{y-2} \qquad \text{Simplify.}$$

$$= \frac{y-8}{y-2} \qquad \text{Add the numerators.}$$

We could use $2 - y$ as the common denominator instead of $y - 2$.

$$\frac{y(-1)}{(y-2)(-1)} + \frac{8}{2-y} \qquad \begin{array}{l}\text{Multiply the first expression by } -1 \text{ in}\\ \text{both the numerator and the denominator.}\end{array}$$

$$= \frac{-y}{2-y} + \frac{8}{2-y}, \quad \text{or} \quad \frac{8-y}{2-y} \qquad \boxed{\begin{array}{l}\text{This equivalent}\\ \text{expression results.}\end{array}}$$

(c) $\dfrac{3}{(x-1)(x+2)} - \dfrac{1}{(x+3)(x-4)}$ The LCD is $(x-1)(x+2)(x+3)(x-4)$.

$$= \frac{3(x+3)(x-4)}{(x-1)(x+2)(x+3)(x-4)} - \frac{(x-1)(x+2)}{(x+3)(x-4)(x-1)(x+2)}$$

$$= \frac{3(x^2 - x - 12) - (x^2 + x - 2)}{(x-1)(x+2)(x+3)(x-4)} \qquad \begin{array}{l}\text{Multiply in the numerators, and}\\ \text{then subtract them.}\end{array}$$

                                                  $\boxed{\text{Be careful with signs.}}$

$$= \frac{3x^2 - 3x - 36 - x^2 - x + 2}{(x-1)(x+2)(x+3)(x-4)} \qquad \text{Distributive property } \textbf{(Section R.2)}$$

$$= \frac{2x^2 - 4x - 34}{(x-1)(x+2)(x+3)(x-4)} \qquad \text{Combine like terms in the numerator.}$$

             ✔ ***Now Try Exercises 47, 51, and 59.***

CAUTION *When subtracting fractions where the second fraction has more than one term in the numerator, as in Example 4(c), be sure to distribute the negative sign to each term.* Use parentheses as in the second step to avoid an error.

Complex Fractions   The quotient of two rational expressions is a **complex fraction.** There are two methods for simplifying a complex fraction.

EXAMPLE 5   **Simplifying Complex Fractions**

Simplify each complex fraction. In part (b), use two methods.

(a) $\dfrac{6 - \dfrac{5}{k}}{1 + \dfrac{5}{k}}$

(b) $\dfrac{\dfrac{a}{a+1} + \dfrac{1}{a}}{\dfrac{1}{a} + \dfrac{1}{a+1}}$

SOLUTION

(a) Method 1 for simplifying uses the identity property for multiplication. We multiply both numerator and denominator by the LCD of all the fractions, k.

$$\frac{6 - \dfrac{5}{k}}{1 + \dfrac{5}{k}} = \frac{k\left(6 - \dfrac{5}{k}\right)}{k\left(1 + \dfrac{5}{k}\right)} = \frac{6k - k\left(\dfrac{5}{k}\right)}{k + k\left(\dfrac{5}{k}\right)} = \frac{6k - 5}{k + 5}$$

> Distribute k to *all* terms within the parentheses.

(b) $\dfrac{\dfrac{a}{a+1} + \dfrac{1}{a}}{\dfrac{1}{a} + \dfrac{1}{a+1}} = \dfrac{\left(\dfrac{a}{a+1} + \dfrac{1}{a}\right)a(a+1)}{\left(\dfrac{1}{a} + \dfrac{1}{a+1}\right)a(a+1)}$

For Method 1, multiply both numerator and denominator by the LCD of all the fractions, $a(a+1)$.

(Method 1)

$$= \frac{\dfrac{a}{a+1}(a)(a+1) + \dfrac{1}{a}(a)(a+1)}{\dfrac{1}{a}(a)(a+1) + \dfrac{1}{a+1}(a)(a+1)}$$

Distributive property

$$= \frac{a^2 + (a+1)}{(a+1) + a}$$

Multiply.

$$= \frac{a^2 + a + 1}{2a + 1}$$

Combine like terms.

$\dfrac{\dfrac{a}{a+1} + \dfrac{1}{a}}{\dfrac{1}{a} + \dfrac{1}{a+1}} = \dfrac{\dfrac{a^2 + 1(a+1)}{a(a+1)}}{\dfrac{1(a+1) + 1(a)}{a(a+1)}}$

For Method 2, find the LCD, and add terms in the numerator and denominator of the complex fraction.

(Method 2)

$$= \frac{\dfrac{a^2 + a + 1}{a(a+1)}}{\dfrac{2a + 1}{a(a+1)}}$$

Combine terms in the numerator and denominator.

$$= \frac{a^2 + a + 1}{a(a+1)} \cdot \frac{a(a+1)}{2a + 1}$$

Definition of division

> The result is the same as in Method 1.

$$= \frac{a^2 + a + 1}{2a + 1}$$

Multiply fractions, and write in lowest terms.

> ✔️ *Now Try Exercises 61 and 73.*

R.5 Exercises

*Find the domain of each rational expression. **See Example 1.***

1. $\dfrac{x+3}{x-6}$

2. $\dfrac{2x-4}{x+7}$

3. $\dfrac{3x+7}{(4x+2)(x-1)}$

4. $\dfrac{9x+12}{(2x+3)(x-5)}$

5. $\dfrac{12}{x^2+5x+6}$

6. $\dfrac{3}{x^2-5x-6}$

7. $\dfrac{x^2-1}{x+1}$

8. $\dfrac{x^2-25}{x-5}$

9. $\dfrac{x^3-1}{x-1}$

10. *Concept Check* Use specific values for x and y to show that in general, $\frac{1}{x}+\frac{1}{y}$ is not equivalent to $\frac{1}{x+y}$.

*Write each rational expression in lowest terms. **See Example 2.***

11. $\dfrac{8x^2+16x}{4x^2}$

12. $\dfrac{36y^2+72y}{9y^2}$

13. $\dfrac{3(3-t)}{(t+5)(t-3)}$

14. $\dfrac{-8(4-y)}{(y+2)(y-4)}$

15. $\dfrac{8k+16}{9k+18}$

16. $\dfrac{20r+10}{30r+15}$

17. $\dfrac{m^2-4m+4}{m^2+m-6}$

18. $\dfrac{r^2-r-6}{r^2+r-12}$

19. $\dfrac{8m^2+6m-9}{16m^2-9}$

20. $\dfrac{6y^2+11y+4}{3y^2+7y+4}$

21. $\dfrac{x^3+64}{x+4}$

22. $\dfrac{y^3-27}{y-3}$

*Find each product or quotient. **See Example 3.***

23. $\dfrac{15p^3}{9p^2}\div\dfrac{6p}{10p^2}$

24. $\dfrac{8r^3}{6r}\div\dfrac{5r^2}{9r^3}$

25. $\dfrac{2k+8}{6}\div\dfrac{3k+12}{2}$

26. $\dfrac{5m+25}{10}\div\dfrac{6m+30}{12}$

27. $\dfrac{x^2+x}{5}\cdot\dfrac{25}{xy+y}$

28. $\dfrac{y^3+y^2}{7}\cdot\dfrac{49}{y^4+y^3}$

29. $\dfrac{4a+12}{2a-10}\div\dfrac{a^2-9}{a^2-a-20}$

30. $\dfrac{6r-18}{9r^2+6r-24}\div\dfrac{4r-12}{12r-16}$

31. $\dfrac{p^2-p-12}{p^2-2p-15}\cdot\dfrac{p^2-9p+20}{p^2-8p+16}$

32. $\dfrac{x^2+2x-15}{x^2+11x+30}\cdot\dfrac{x^2+2x-24}{x^2-8x+15}$

33. $\dfrac{m^2+3m+2}{m^2+5m+4}\div\dfrac{m^2+5m+6}{m^2+10m+24}$

34. $\dfrac{y^2+y-2}{y^2+3y-4}\div\dfrac{y^2+3y+2}{y^2+4y+3}$

35. $\dfrac{x^3+y^3}{x^3-y^3}\cdot\dfrac{x^2-y^2}{x^2+2xy+y^2}$

36. $\dfrac{x^2-y^2}{(x-y)^2}\cdot\dfrac{x^2-xy+y^2}{x^2-2xy+y^2}\div\dfrac{x^3+y^3}{(x-y)^4}$

37. $\dfrac{xz-xw+2yz-2yw}{z^2-w^2}\cdot\dfrac{4z+4w+xz+wx}{16-x^2}$

38. $\dfrac{ac+ad+bc+bd}{a^2-b^2}\cdot\dfrac{a^3-b^3}{2a^2+2ab+2b^2}$

39. *Concept Check* Which of the following rational expressions is equivalent to -1? In choices A, B, and D, $x\neq-4$, and in choice C, $x\neq4$. (*Hint:* There may be more than one answer.)

A. $\dfrac{x-4}{x+4}$ **B.** $\dfrac{-x-4}{x+4}$ **C.** $\dfrac{x-4}{4-x}$ **D.** $\dfrac{x-4}{-x-4}$

40. Explain how to find the least common denominator of several fractions.

Perform each addition or subtraction. See Example 4.

41. $\dfrac{3}{2k} + \dfrac{5}{3k}$

42. $\dfrac{8}{5p} + \dfrac{3}{4p}$

43. $\dfrac{1}{6m} + \dfrac{2}{5m} + \dfrac{4}{m}$

44. $\dfrac{8}{3p} + \dfrac{5}{4p} + \dfrac{9}{2p}$

45. $\dfrac{1}{a} - \dfrac{b}{a^2}$

46. $\dfrac{3}{z} + \dfrac{x}{z^2}$

47. $\dfrac{5}{12x^2y} - \dfrac{11}{6xy}$

48. $\dfrac{7}{18a^3b^2} - \dfrac{2}{9ab}$

49. $\dfrac{17y + 3}{9y + 7} - \dfrac{-10y - 18}{9y + 7}$

50. $\dfrac{7x + 8}{3x + 2} - \dfrac{x + 4}{3x + 2}$

51. $\dfrac{1}{x + z} + \dfrac{1}{x - z}$

52. $\dfrac{m + 1}{m - 1} + \dfrac{m - 1}{m + 1}$

53. $\dfrac{3}{a - 2} - \dfrac{1}{2 - a}$

54. $\dfrac{4}{p - q} - \dfrac{2}{q - p}$

55. $\dfrac{x + y}{2x - y} - \dfrac{2x}{y - 2x}$

56. $\dfrac{m - 4}{3m - 4} - \dfrac{5m}{4 - 3m}$

57. $\dfrac{4}{x + 1} + \dfrac{1}{x^2 - x + 1} - \dfrac{12}{x^3 + 1}$

58. $\dfrac{5}{x + 2} + \dfrac{2}{x^2 - 2x + 4} - \dfrac{60}{x^3 + 8}$

59. $\dfrac{3x}{x^2 + x - 12} - \dfrac{x}{x^2 - 16}$

60. $\dfrac{p}{2p^2 - 9p - 5} - \dfrac{2p}{6p^2 - p - 2}$

Simplify each expression. See Example 5.

61. $\dfrac{1 + \dfrac{1}{x}}{1 - \dfrac{1}{x}}$

62. $\dfrac{2 - \dfrac{2}{y}}{2 + \dfrac{2}{y}}$

63. $\dfrac{\dfrac{1}{x + 1} - \dfrac{1}{x}}{\dfrac{1}{x}}$

64. $\dfrac{\dfrac{1}{y + 3} - \dfrac{1}{y}}{\dfrac{1}{y}}$

65. $\dfrac{1 + \dfrac{1}{1 - b}}{1 - \dfrac{1}{1 + b}}$

66. $\dfrac{2 + \dfrac{2}{1 + x}}{2 - \dfrac{2}{1 - x}}$

67. $\dfrac{\dfrac{1}{a^3 + b^3}}{\dfrac{1}{a^2 + 2ab + b^2}}$

68. $\dfrac{\dfrac{1}{x^3 - y^3}}{\dfrac{1}{x^2 - y^2}}$

69. $\dfrac{m - \dfrac{1}{m^2 - 4}}{\dfrac{1}{m + 2}}$

70. $\dfrac{y + \dfrac{1}{y^2 - 9}}{\dfrac{1}{y + 3}}$

71. $\dfrac{\dfrac{3}{p^2 - 16} + p}{\dfrac{1}{p - 4}}$

72. $\dfrac{\dfrac{6}{x^2 - 25} + x}{\dfrac{1}{x - 5}}$

73. $\dfrac{\dfrac{y + 3}{y} - \dfrac{4}{y - 1}}{\dfrac{y}{y - 1} + \dfrac{1}{y}}$

74. $\dfrac{\dfrac{x + 4}{x} - \dfrac{3}{x - 2}}{\dfrac{x}{x - 2} + \dfrac{1}{x}}$

75. $\dfrac{\dfrac{1}{x + h} - \dfrac{1}{x}}{h}$

76. $\dfrac{\dfrac{-2}{x + h} - \dfrac{-2}{x}}{h}$

77. $\dfrac{\dfrac{1}{(x + h)^2 + 9} - \dfrac{1}{x^2 + 9}}{h}$

78. $\dfrac{\dfrac{2}{(x + h)^2 + 16} - \dfrac{2}{x^2 + 16}}{h}$

(Modeling) Distance from the Origin of the Nile River The Nile River in Africa is about 4000 mi *long. The Nile begins as an outlet of Lake Victoria at an altitude of* 7000 ft *above sea level and empties into the Mediterranean Sea at sea level* (0 ft). *The distance from its origin in thousands of miles is related to its height above sea level in thousands of feet* (x) *by the following rational expression.*

$$\frac{7 - x}{0.639x + 1.75}$$

For example, when the river is at an altitude of 600 ft, x = 0.6 *(thousand), and the distance from the origin is*

$$\frac{7 - 0.6}{0.639(0.6) + 1.75} \approx 3, \quad which\ represents\ 3000\ mi.$$

(Source: World Almanac and Book of Facts.)

79. What is the distance from the origin of the Nile when the river has an altitude of 7000 ft?

80. What is the distance from the origin of the Nile when the river has an altitude of 1200 ft?

(Modeling) Cost-Benefit Model for a Pollutant In situations involving environmental pollution, a **cost-benefit model** expresses cost in terms of the percentage of pollutant removed from the environment. Suppose a cost-benefit model is expressed as

$$y = \frac{6.7x}{100 - x},$$

where y is the cost in thousands of dollars of removing x percent of a certain pollutant. Find the value of y for each given value of x.

81. x = 75 (75%)   **82.** x = 95 (95%)

R.6 Rational Exponents

- Negative Exponents and the Quotient Rule
- Rational Exponents
- Complex Fractions Revisited

Negative Exponents and the Quotient Rule In **Section R.2,** we justified the definition $a^0 = 1$ for $a \neq 0$ using the product rule for exponents. Suppose that n is a positive integer, and we wish to define a^{-n} to be consistent with the application of the product rule. Consider the product $a^n \cdot a^{-n}$, and apply the rule.

$$a^n \cdot a^{-n} = a^{n+(-n)} \quad \text{Product rule}$$

$$= a^0 \quad n \text{ and } -n \text{ are additive inverses.}$$

$$= 1 \quad \text{Definition of } a^0$$

The expression a^{-n} acts as the *reciprocal* of a^n, which is written $\frac{1}{a^n}$. Thus, these two expressions must be equivalent.

Negative Exponent

Suppose that a is a nonzero real number and n is any integer.

$$a^{-n} = \frac{1}{a^n}$$

EXAMPLE 1 **Using the Definition of a Negative Exponent**

Evaluate each expression. In parts (d) and (e), write the expression without negative exponents. Assume all variables represent nonzero real numbers.

(a) 4^{-2} **(b)** -4^{-2} **(c)** $\left(\dfrac{2}{5}\right)^{-3}$ **(d)** $(xy)^{-3}$ **(e)** xy^{-3}

SOLUTION

(a) $4^{-2} = \dfrac{1}{4^2} = \dfrac{1}{16}$ **(b)** $-4^{-2} = -\dfrac{1}{4^2} = -\dfrac{1}{16}$

(c) $\left(\dfrac{2}{5}\right)^{-3} = \dfrac{1}{\left(\frac{2}{5}\right)^3} = \dfrac{1}{\frac{8}{125}} = 1 \div \dfrac{8}{125} = 1 \cdot \dfrac{125}{8} = \dfrac{125}{8}$

Multiply by the reciprocal of the divisor.

(d) $(xy)^{-3} = \dfrac{1}{(xy)^3}$, or $\dfrac{1}{x^3y^3}$ **(e)** $xy^{-3} = x \cdot \dfrac{1}{y^3} = \dfrac{x}{y^3}$

Base is xy. Base is y.

✔ **Now Try Exercises 3, 5, 7, 9, and 11.**

CAUTION *A negative exponent indicates a reciprocal, not a sign change of the expression.*

Example 1(c) showed the following.

$$\left(\frac{2}{5}\right)^{-3} = \frac{125}{8} = \left(\frac{5}{2}\right)^3$$

We can generalize this result. If $a \neq 0$ and $b \neq 0$, then for any integer n, the following is true.

$$\left(\frac{a}{b}\right)^{-n} = \left(\frac{b}{a}\right)^n$$

The **quotient rule** for exponents follows from the definition of exponents.

Quotient Rule

Suppose that m and n are integers and a is a nonzero real number.

$$\frac{a^m}{a^n} = a^{m-n}$$

That is, when dividing powers of like bases, keep the same base and subtract the exponent of the denominator from the exponent of the numerator.

CAUTION When applying the quotient rule, be sure to subtract the exponents in the correct order. Be careful especially when the exponent in the denominator is negative, and avoid sign errors.

> **EXAMPLE 2** Using the Quotient Rule

Simplify each expression. Assume all variables represent nonzero real numbers.

(a) $\dfrac{12^5}{12^2}$ **(b)** $\dfrac{a^5}{a^{-8}}$ **(c)** $\dfrac{16m^{-9}}{12m^{11}}$ **(d)** $\dfrac{25r^7z^5}{10r^9z}$

SOLUTION

Use parentheses to avoid errors.

(a) $\dfrac{12^5}{12^2} = 12^{5-2} = 12^3$

(b) $\dfrac{a^5}{a^{-8}} = a^{5-(-8)} = a^{13}$

(c) $\dfrac{16m^{-9}}{12m^{11}} = \dfrac{16}{12} \cdot m^{-9-11}$

$= \dfrac{4}{3}m^{-20}$

$= \dfrac{4}{3} \cdot \dfrac{1}{m^{20}}$

$= \dfrac{4}{3m^{20}}$

(d) $\dfrac{25r^7z^5}{10r^9z} = \dfrac{25}{10} \cdot \dfrac{r^7}{r^9} \cdot \dfrac{z^5}{z^1}$

$= \dfrac{5}{2}r^{-2}z^4$

$= \dfrac{5z^4}{2r^2}$

✔ *Now Try Exercises 15, 21, 23, and 25.*

The rules for exponents in **Section R.3** were stated for positive integer exponents and for zero as an exponent. Those rules continue to apply in expressions involving negative exponents, as seen in the next example.

> **EXAMPLE 3** Using the Rules for Exponents

Simplify each expression. Write answers without negative exponents. Assume all variables represent nonzero real numbers.

(a) $3x^{-2}(4^{-1}x^{-5})^2$ **(b)** $\dfrac{12p^3q^{-1}}{8p^{-2}q}$ **(c)** $\dfrac{(3x^2)^{-1}(3x^5)^{-2}}{(3^{-1}x^{-2})^2}$

SOLUTION

(a) $3x^{-2}(4^{-1}x^{-5})^2 = 3x^{-2}(4^{-2}x^{-10})$ Power rules **(Section R.3)**

$= 3 \cdot 4^{-2} \cdot x^{-2+(-10)}$ Rearrange factors; product rule **(Section R.3)**

$= 3 \cdot 4^{-2} \cdot x^{-12}$ Simplify the exponent on x.

$= \dfrac{3}{16x^{12}}$ Write with positive exponents.

(b) $\dfrac{12p^3q^{-1}}{8p^{-2}q} = \dfrac{12}{8} \cdot \dfrac{p^3}{p^{-2}} \cdot \dfrac{q^{-1}}{q^1}$

$= \dfrac{3}{2} \cdot p^{3-(-2)}q^{-1-1}$ Quotient rule

$= \dfrac{3}{2}p^5q^{-2}$ Simplify the exponents.

$= \dfrac{3p^5}{2q^2}$ Write with positive exponents.

(c) $\dfrac{(3x^2)^{-1}(3x^5)^{-2}}{(3^{-1}x^{-2})^2} = \dfrac{3^{-1}x^{-2}3^{-2}x^{-10}}{3^{-2}x^{-4}}$ Power rules

$$= \dfrac{3^{-1+(-2)}x^{-2+(-10)}}{3^{-2}x^{-4}} = \dfrac{3^{-3}x^{-12}}{3^{-2}x^{-4}}$$ Product rule

$$= 3^{-3-(-2)}x^{-12-(-4)} = 3^{-1}x^{-8}$$ Quotient rule

Be careful with signs.

$$= \dfrac{1}{3x^8}$$ Write with positive exponents.

✔ *Now Try Exercises 29, 33, 35, and 37.*

CAUTION Notice the use of the power rule $(ab)^n = a^n b^n$ in **Example 3(c):**

$$(3x^2)^{-1} = 3^{-1}(x^2)^{-1} = 3^{-1}x^{-2}.$$

*Remember to apply the exponent to the numerical coefficient **3**.*

Rational Exponents   The definition of a^n can be extended to rational values of n by defining $a^{1/n}$ to be the nth root of a. By one of the power rules of exponents (extended to a rational exponent),

$$(a^{1/n})^n = a^{(1/n)n} = a^1 = a,$$

which suggests that $a^{1/n}$ is a number whose nth power is a.

The Expression $a^{1/n}$

$a^{1/n}$, n **Even**   If n is an *even* positive integer, and if $a > 0$, then $a^{1/n}$ is the positive real number whose nth power is a. That is, $(a^{1/n})^n = a$. (In this case, $a^{1/n}$ is the principal nth root of a. See **Section R.7.**)

$a^{1/n}$, n **Odd**   If n is an *odd* positive integer, and a is *any nonzero real number*, then $a^{1/n}$ is the positive or negative real number whose nth power is a. That is, $(a^{1/n})^n = a$.

For all positive integers n, $0^{1/n} = 0$.

EXAMPLE 4   **Using the Definition of $a^{1/n}$**

Evaluate each expression.

(a) $36^{1/2}$ **(b)** $-100^{1/2}$ **(c)** $-(225)^{1/2}$ **(d)** $625^{1/4}$

(e) $(-1296)^{1/4}$ **(f)** $-1296^{1/4}$ **(g)** $(-27)^{1/3}$ **(h)** $-32^{1/5}$

SOLUTION

(a) $36^{1/2} = 6$ because $6^2 = 36$.   **(b)** $-100^{1/2} = -10$

(c) $-(225)^{1/2} = -15$ **(d)** $625^{1/4} = 5$

(e) $(-1296)^{1/4}$ is not a real number.   **(f)** $-1296^{1/4} = -6$

(g) $(-27)^{1/3} = -3$ **(h)** $-32^{1/5} = -2$

✔ *Now Try Exercises 39, 41, and 45.*

The notation $a^{m/n}$ must be defined in such a way that all the previous rules for exponents still hold. For the power rule to hold, $(a^{1/n})^m$ must equal $a^{m/n}$. Therefore, $a^{m/n}$ is defined as follows.

The Expression $a^{m/n}$

Let m be any integer, n be any positive integer, and a be any real number for which $a^{1/n}$ is a real number.

$$a^{m/n} = (a^{1/n})^m$$

EXAMPLE 5 **Using the Definition of $a^{m/n}$**

Evaluate each expression.

(a) $125^{2/3}$ **(b)** $32^{7/5}$ **(c)** $-81^{3/2}$ **(d)** $(-27)^{2/3}$ **(e)** $16^{-3/4}$ **(f)** $(-4)^{5/2}$

SOLUTION

(a) $125^{2/3} = (125^{1/3})^2$

$\qquad = 5^2,\quad$ or $\quad 25$

(b) $32^{7/5} = (32^{1/5})^7$

$\qquad = 2^7,\quad$ or $\quad 128$

(c) $-81^{3/2} = -(81^{1/2})^3$

$\qquad = -9^3,\quad$ or $\quad -729$

(d) $(-27)^{2/3} = [(-27)^{1/3}]^2$

$\qquad = (-3)^2,\quad$ or $\quad 9$

(e) $16^{-3/4} = \dfrac{1}{16^{3/4}}$

$\qquad = \dfrac{1}{(16^{1/4})^3}$

$\qquad = \dfrac{1}{2^3},\quad$ or $\quad \dfrac{1}{8}$

(f) $(-4)^{5/2}$ is not a real number. This is because $(-4)^{1/2}$ is not a real number.

✔ *Now Try Exercises 49, 53, and 55.*

NOTE For all real numbers a, integers m, and positive integers n for which $a^{1/n}$ is a real number, $a^{m/n}$ can be interpreted as follows.

$$a^{m/n} = (a^{1/n})^m \quad \text{or} \quad a^{m/n} = (a^m)^{1/n}$$

So $a^{m/n}$ can be evaluated either as $(a^{1/n})^m$ or as $(a^m)^{1/n}$.

$\qquad 27^{4/3} = (27^{1/3})^4 = 3^4 = 81$

$\qquad\qquad\qquad\qquad\qquad$ The result is the same.

or $\qquad 27^{4/3} = (27^4)^{1/3} = 531{,}441^{1/3} = 81$

The earlier results for integer exponents also apply to rational exponents.

Definitions and Rules for Exponents

Suppose that r and s represent rational numbers. The results here are valid for all positive numbers a and b.

Product rule	$a^r \cdot a^s = a^{r+s}$	**Power rules** $\quad (a^r)^s = a^{rs}$
Quotient rule	$\dfrac{a^r}{a^s} = a^{r-s}$	$(ab)^r = a^r b^r$
Negative exponent	$a^{-r} = \dfrac{1}{a^r}$	$\left(\dfrac{a}{b}\right)^r = \dfrac{a^r}{b^r}$

EXAMPLE 6 **Using the Rules for Exponents**

Simplify each expression. Assume all variables represent positive real numbers.

(a) $\dfrac{27^{1/3} \cdot 27^{5/3}}{27^3}$ **(b)** $81^{5/4} \cdot 4^{-3/2}$ **(c)** $6y^{2/3} \cdot 2y^{1/2}$

(d) $\left(\dfrac{3m^{5/6}}{y^{3/4}}\right)^2 \left(\dfrac{8y^3}{m^6}\right)^{2/3}$ **(e)** $m^{2/3}\left(m^{7/3} + 2m^{1/3}\right)$

SOLUTION

(a) $\dfrac{27^{1/3} \cdot 27^{5/3}}{27^3} = \dfrac{27^{1/3+5/3}}{27^3}$ Product rule

$= \dfrac{27^2}{27^3}$ Simplify.

$= 27^{2-3}$ Quotient rule

$= 27^{-1}, \quad \text{or} \quad \dfrac{1}{27}$ Negative exponent

(b) $81^{5/4} \cdot 4^{-3/2} = \left(81^{1/4}\right)^5 \left(4^{1/2}\right)^{-3}$ **(c)** $6y^{2/3} \cdot 2y^{1/2} = 12y^{2/3+1/2}$

$= 3^5 \cdot 2^{-3}$ $= 12y^{7/6}$

$= \dfrac{3^5}{2^3}, \quad \text{or} \quad \dfrac{243}{8}$

(d) $\left(\dfrac{3m^{5/6}}{y^{3/4}}\right)^2 \left(\dfrac{8y^3}{m^6}\right)^{2/3} = \dfrac{9m^{5/3}}{y^{3/2}} \cdot \dfrac{4y^2}{m^4}$ Power rules

$= 36m^{5/3-4}y^{2-3/2}$ Quotient rule

$= 36m^{-7/3}y^{1/2}$ Simplify the exponents.

$= \dfrac{36y^{1/2}}{m^{7/3}}$ Simplify.

(e) $m^{2/3}\left(m^{7/3} + 2m^{1/3}\right) = m^{2/3} \cdot m^{7/3} + m^{2/3} \cdot 2m^{1/3}$ Distributive property (Section R.2)

> Do *not* multiply the exponents.

$= m^{2/3+7/3} + 2m^{2/3+1/3}$ Product rule

$= m^3 + 2m$ Simplify.

✔ *Now Try Exercises 61, 63, 67, 69, and 75.*

EXAMPLE 7 **Factoring Expressions with Negative or Rational Exponents**

Factor out the least power of the variable or variable expression. Assume all variables represent positive real numbers.

(a) $12x^{-2} - 8x^{-3}$ **(b)** $4m^{1/2} + 3m^{3/2}$ **(c)** $(y-2)^{-1/3} + (y-2)^{2/3}$

SOLUTION

(a) The least exponent on $12x^{-2} - 8x^{-3}$ is -3. Since 4 is a common numerical factor, factor out $4x^{-3}$.

$12x^{-2} - 8x^{-3} = 4x^{-3}\left(3x^{-2-(-3)} - 2x^{-3-(-3)}\right)$ Factor.

$= 4x^{-3}(3x - 2)$ Simplify the exponents.

Check by multiplying on the right.

LOOKING AHEAD TO CALCULUS
The technique of **Example 7(c)** is used often in calculus.

(b) $4m^{1/2} + 3m^{3/2} = m^{1/2}(4 + 3m)$ Factor out $m^{1/2}$.

To *check*, multiply $m^{1/2}$ by $4 + 3m$.

(c) $(y - 2)^{-1/3} + (y - 2)^{2/3} = (y - 2)^{-1/3}[1 + (y - 2)]$
$$= (y - 2)^{-1/3}(y - 1)$$

> ✔ *Now Try Exercises 83, 89, and 93.*

Complex Fractions Revisited   Negative exponents are sometimes used to write complex fractions. Recall that complex fractions are simplified either by first multiplying the numerator and denominator by the LCD of all the denominators, or by performing any indicated operations in the numerator and the denominator and then using the definition of division for fractions.

EXAMPLE 8   **Simplifying a Fraction with Negative Exponents**

Simplify $\dfrac{(x + y)^{-1}}{x^{-1} + y^{-1}}$. Write the result with only positive exponents.

SOLUTION $\dfrac{(x + y)^{-1}}{x^{-1} + y^{-1}} = \dfrac{\dfrac{1}{x + y}}{\dfrac{1}{x} + \dfrac{1}{y}}$ Definition of negative exponent

$$= \dfrac{\dfrac{1}{x + y}}{\dfrac{y + x}{xy}}$$ Add fractions in the denominator. (Section R.5)

$$= \dfrac{1}{x + y} \cdot \dfrac{xy}{x + y}$$ Multiply by the reciprocal of the denominator of the complex fraction.

$$= \dfrac{xy}{(x + y)^2}$$ Multiply fractions.

> ✔ *Now Try Exercise 99.*

CAUTION   Remember that if $r \neq 1$, then $(x + y)^r \neq x^r + y^r$. In particular, this means that $(x + y)^{-1} \neq x^{-1} + y^{-1}$.

R.6   Exercises

Concept Check In Exercises 1 and 2, match each expression in Column I with its equivalent expression in Column II. Choices may be used once, more than once, or not at all.

I	II	I	II
1. (a) 4^{-2}	**A.** 16	**2. (a)** 5^{-3}	**A.** 125
(b) -4^{-2}	**B.** $\dfrac{1}{16}$	**(b)** -5^{-3}	**B.** -125
(c) $(-4)^{-2}$	**C.** -16	**(c)** $(-5)^{-3}$	**C.** $\dfrac{1}{125}$
(d) $-(-4)^{-2}$	**D.** $-\dfrac{1}{16}$	**(d)** $-(-5)^{-3}$	**D.** $-\dfrac{1}{125}$

Write each expression with only positive exponents and evaluate if possible. Assume all variables represent nonzero real numbers. See Example 1.

3. $(-4)^{-3}$

4. $(-5)^{-2}$

5. -5^{-4}

6. -7^{-2}

7. $\left(\dfrac{1}{3}\right)^{-2}$

8. $\left(\dfrac{4}{3}\right)^{-3}$

9. $(4x)^{-2}$

10. $(5t)^{-3}$

11. $4x^{-2}$

12. $5t^{-3}$

13. $-a^{-3}$

14. $-b^{-4}$

Perform the indicated operations. Write each answer using only positive exponents. Assume all variables represent nonzero real numbers. See Examples 2 and 3.

15. $\dfrac{4^8}{4^6}$

16. $\dfrac{5^9}{5^7}$

17. $\dfrac{x^{12}}{x^8}$

18. $\dfrac{y^{14}}{y^{10}}$

19. $\dfrac{r^7}{r^{10}}$

20. $\dfrac{y^8}{y^{12}}$

21. $\dfrac{6^4}{6^{-2}}$

22. $\dfrac{7^5}{7^{-3}}$

23. $\dfrac{4r^{-3}}{6r^{-6}}$

24. $\dfrac{15s^{-4}}{5s^{-8}}$

25. $\dfrac{16m^{-5}n^4}{12m^2n^{-3}}$

26. $\dfrac{15a^{-5}b^{-1}}{25a^{-2}b^4}$

27. $-4r^{-2}(r^4)^2$

28. $-2m^{-1}(m^3)^2$

29. $(5a^{-1})^4(a^2)^{-3}$

30. $(3p^{-4})^2(p^3)^{-1}$

31. $\dfrac{(p^{-2})^0}{5p^{-4}}$

32. $\dfrac{(m^4)^0}{9m^{-3}}$

33. $\dfrac{(3pq)q^2}{6p^2q^4}$

34. $\dfrac{(-8xy)y^3}{4x^5y^4}$

35. $\dfrac{4a^5(a^{-1})^3}{(a^{-2})^{-2}}$

36. $\dfrac{12k^{-2}(k^{-3})^{-4}}{6k^5}$

37. $\dfrac{(5x)^{-2}(5x^3)^{-3}}{(5^{-2}x^{-3})^3}$

38. $\dfrac{(8y^2)^{-4}(8y^5)^{-2}}{(8^{-3}y^{-4})^2}$

Evaluate each expression. See Example 4.

39. $169^{1/2}$

40. $121^{1/2}$

41. $16^{1/4}$

42. $625^{1/4}$

43. $\left(-\dfrac{64}{27}\right)^{1/3}$

44. $\left(-\dfrac{8}{27}\right)^{1/3}$

45. $(-4)^{1/2}$

46. $(-64)^{1/4}$

Concept Check In Exercises 47 and 48, match each expression from Column I with its equivalent expression from Column II. Choices may be used once, more than once, or not at all.

<table>
<tr><td colspan="2" align="center">**I**</td><td colspan="2" align="center">**II**</td></tr>
<tr><td>**47. (a)** $\left(\dfrac{4}{9}\right)^{3/2}$</td><td>**48. (a)** $\left(\dfrac{8}{27}\right)^{2/3}$</td><td>**A.** $\dfrac{9}{4}$</td><td>**B.** $-\dfrac{9}{4}$</td></tr>
<tr><td>**(b)** $\left(\dfrac{4}{9}\right)^{-3/2}$</td><td>**(b)** $\left(\dfrac{8}{27}\right)^{-2/3}$</td><td>**C.** $-\dfrac{4}{9}$</td><td>**D.** $\dfrac{4}{9}$</td></tr>
<tr><td>**(c)** $-\left(\dfrac{9}{4}\right)^{3/2}$</td><td>**(c)** $-\left(\dfrac{27}{8}\right)^{2/3}$</td><td>**E.** $\dfrac{8}{27}$</td><td>**F.** $-\dfrac{27}{8}$</td></tr>
<tr><td>**(d)** $-\left(\dfrac{4}{9}\right)^{-3/2}$</td><td>**(d)** $-\left(\dfrac{27}{8}\right)^{-2/3}$</td><td>**G.** $\dfrac{27}{8}$</td><td>**H.** $-\dfrac{8}{27}$</td></tr>
</table>

Perform the indicated operations. Write each answer using only positive exponents. Assume all variables represent positive real numbers. See Examples 5 and 6.

49. $8^{2/3}$

50. $27^{4/3}$

51. $100^{3/2}$

52. $64^{3/2}$

53. $-81^{3/4}$

54. $(-32)^{-4/5}$

55. $\left(\dfrac{27}{64}\right)^{-4/3}$

56. $\left(\dfrac{121}{100}\right)^{-3/2}$

57. $3^{1/2} \cdot 3^{3/2}$

58. $6^{4/3} \cdot 6^{2/3}$

59. $\dfrac{64^{5/3}}{64^{4/3}}$

60. $\dfrac{125^{7/3}}{125^{5/3}}$

61. $y^{7/3} \cdot y^{-4/3}$

62. $r^{-8/9} \cdot r^{17/9}$

63. $\dfrac{k^{1/3}}{k^{2/3} \cdot k^{-1}}$

64. $\dfrac{z^{3/4}}{z^{5/4} \cdot z^{-2}}$

65. $\dfrac{(x^{1/4}y^{2/5})^{20}}{x^2}$

66. $\dfrac{(r^{1/5}s^{2/3})^{15}}{r^2}$

67. $\dfrac{(x^{2/3})^2}{(x^2)^{7/3}}$

68. $\dfrac{(p^3)^{1/4}}{(p^{5/4})^2}$

69. $\left(\dfrac{16m^3}{n}\right)^{1/4}\left(\dfrac{9n^{-1}}{m^2}\right)^{1/2}$

70. $\left(\dfrac{25^4a^3}{b^2}\right)^{1/8}\left(\dfrac{4^2b^{-5}}{a^2}\right)^{1/4}$

71. $\dfrac{p^{1/5}p^{7/10}p^{1/2}}{(p^3)^{-1/5}}$

72. $\dfrac{z^{1/3}z^{-2/3}z^{1/6}}{(z^{-1/6})^3}$

Solve each applied problem.

73. *(Modeling) Holding Time of Athletes* A group of ten athletes were tested for isometric endurance by measuring the length of time they could resist a load pulling on their legs while seated. The approximate amount of time (called the **holding time**) that they could resist the load was given by the formula

$$t = 31{,}293w^{-1.5},$$

where w is the weight of the load in pounds and the holding time t is measured in seconds. (*Source:* Townend, M. Stewart, *Mathematics in Sport,* Chichester, Ellis Horwood Limited.)

(a) Determine the holding time for a load of 25 lb.

(b) When the weight of the load is doubled, by what factor is the holding time changed?

74. *Duration of a Storm* Suppose that meteorologists approximate the duration of a particular storm by using the formula

$$T = 0.07D^{3/2},$$

where T is the time in hours that a storm of diameter D (in miles) lasts.

(a) The National Weather Service reports that a storm 4 mi in diameter is headed toward New Haven. How long can the residents expect the storm to last?

(b) After weeks of dry weather, a thunderstorm is predicted for the farming community of Apple Valley. The crops need at least 1.5 hr of rain. Local radar shows that the storm is 7 mi in diameter. Will it rain long enough to meet the farmers' need?

Find each product. Assume all variables represent positive real numbers. **See Example 6(e) in this section and Example 7 in Section R.3.**

75. $y^{5/8}(y^{3/8} - 10y^{11/8})$

76. $p^{11/5}(3p^{4/5} + 9p^{19/5})$

77. $-4k(k^{7/3} - 6k^{1/3})$

78. $-5y(3y^{9/10} + 4y^{3/10})$

79. $(x + x^{1/2})(x - x^{1/2})$

80. $(2z^{1/2} + z)(z^{1/2} - z)$

81. $(r^{1/2} - r^{-1/2})^2$

82. $(p^{1/2} - p^{-1/2})(p^{1/2} + p^{-1/2})$

Factor, using the given common factor. Assume all variables represent positive real numbers. See Example 7.

83. $4k^{-1} + k^{-2}$; k^{-2} **84.** $y^{-5} - 3y^{-3}$; y^{-5} **85.** $4t^{-2} + 8t^{-4}$; $4t^{-4}$

86. $5r^{-6} - 10r^{-8}$; $5r^{-8}$ **87.** $9z^{-1/2} + 2z^{1/2}$; $z^{-1/2}$ **88.** $3m^{2/3} - 4m^{-1/3}$; $m^{-1/3}$

89. $p^{-3/4} - 2p^{-7/4}$; $p^{-7/4}$ **90.** $6r^{-2/3} - 5r^{-5/3}$; $r^{-5/3}$

91. $-4a^{-2/5} + 16a^{-7/5}$; $4a^{-7/5}$ **92.** $-3p^{-3/4} - 30p^{-7/4}$; $-3p^{-7/4}$

93. $(p+4)^{-3/2} + (p+4)^{-1/2} + (p+4)^{1/2}$; $(p+4)^{-3/2}$

94. $(3r+1)^{-2/3} + (3r+1)^{1/3} + (3r+1)^{4/3}$; $(3r+1)^{-2/3}$

95. $2(3x+1)^{-3/2} + 4(3x+1)^{-1/2} + 6(3x+1)^{1/2}$; $2(3x+1)^{-3/2}$

96. $7(5t+3)^{-5/3} + 14(5t+3)^{-2/3} - 21(5t+3)^{1/3}$; $7(5t+3)^{-5/3}$

97. $4x(2x+3)^{-5/9} + 6x^2(2x+3)^{4/9} - 8x^3(2x+3)^{13/9}$; $2x(2x+3)^{-5/9}$

98. $6y^3(4y-1)^{-3/7} - 8y^2(4y-1)^{4/7} + 16y(4y-1)^{11/7}$; $2y(4y-1)^{-3/7}$

Perform all indicated operations and write each answer with positive integer exponents. See Example 8.

99. $\dfrac{a^{-1} + b^{-1}}{(ab)^{-1}}$ **100.** $\dfrac{p^{-1} - q^{-1}}{(pq)^{-1}}$

101. $\dfrac{r^{-1} + q^{-1}}{r^{-1} - q^{-1}} \cdot \dfrac{r - q}{r + q}$ **102.** $\dfrac{x^{-2} + y^{-2}}{x^{-2} - y^{-2}} \cdot \dfrac{x + y}{x - y}$

103. $\dfrac{x - 9y^{-1}}{(x - 3y^{-1})(x + 3y^{-1})}$ **104.** $\dfrac{a - 16b^{-1}}{(a + 4b^{-1})(a - 4b^{-1})}$

Simplify each rational expression. Use factoring, and refer to Section R.5 as needed. Assume all variable expressions represent positive real numbers.

105. $\dfrac{(x^2 + 1)^4(2x) - x^2(4)(x^2 + 1)^3(2x)}{(x^2 + 1)^8}$ **106.** $\dfrac{(y^2 + 2)^5(3y) - y^3(6)(y^2 + 2)^4(3y)}{(y^2 + 2)^7}$

107. $\dfrac{4(x^2 - 1)^3 + 8x(x^2 - 1)^4}{16(x^2 - 1)^3}$ **108.** $\dfrac{10(4x^2 - 9)^2 - 25x(4x^2 - 9)^3}{15(4x^2 - 9)^6}$

109. $\dfrac{2(2x - 3)^{1/3} - (x - 1)(2x - 3)^{-2/3}}{(2x - 3)^{2/3}}$ **110.** $\dfrac{7(3t + 1)^{1/4} - (t - 1)(3t + 1)^{-3/4}}{(3t + 1)^{3/4}}$

Concept Check   *Answer each question.*

111. If the lengths of the sides of a cube are tripled, by what factor will the volume change?

112. If the radius of a circle is doubled, by what factor will the area change?

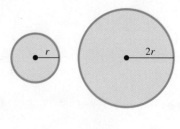

Concept Check   *Calculate each value mentally.*

113. $0.2^{2/3} \cdot 40^{2/3}$ **114.** $0.1^{3/2} \cdot 90^{3/2}$ **115.** $\dfrac{2^{2/3}}{2000^{2/3}}$ **116.** $\dfrac{20^{3/2}}{5^{3/2}}$

R.7 Radical Expressions

- Radical Notation
- Simplified Radicals
- Operations with Radicals
- Rationalizing Denominators

Radical Notation In **Section R.6** we used rational exponents to express roots. An alternative notation for roots is **radical notation.**

Radical Notation for $a^{1/n}$

Suppose that a is a real number, n is a positive integer, and $a^{1/n}$ is a real number.

$$\sqrt[n]{a} = a^{1/n}$$

Radical Notation for $a^{m/n}$

Suppose that a is a real number, m is an integer, n is a positive integer, and $\sqrt[n]{a}$ is a real number.

$$a^{m/n} = \left(\sqrt[n]{a}\right)^m = \sqrt[n]{a^m}$$

In the radical $\sqrt[n]{a}$, the symbol $\sqrt[n]{}$ is a **radical symbol,** the number a is the **radicand,** and n is the **index.** We use the familiar notation $\sqrt{a}$ instead of $\sqrt[2]{a}$ for the square root.

For even values of n (square roots, fourth roots, and so on), when a is positive, there are two nth roots, one positive and one negative. In such cases, the notation $\sqrt[n]{a}$ represents the positive root, the **principal nth root.** We write the **negative root** as $-\sqrt[n]{a}$.

EXAMPLE 1 **Evaluating Roots**

Write each root using exponents and evaluate.

(a) $\sqrt[4]{16}$ **(b)** $-\sqrt[4]{16}$ **(c)** $\sqrt[5]{-32}$

(d) $\sqrt[3]{1000}$ **(e)** $\sqrt[6]{\dfrac{64}{729}}$ **(f)** $\sqrt[4]{-16}$

SOLUTION

(a) $\sqrt[4]{16} = 16^{1/4}$

$\phantom{\sqrt[4]{16}} = 2$

(b) $-\sqrt[4]{16} = -16^{1/4}$

$\phantom{-\sqrt[4]{16}} = -2$

(c) $\sqrt[5]{-32} = (-32)^{1/5}$

$\phantom{\sqrt[5]{-32}} = -2$

(d) $\sqrt[3]{1000} = 1000^{1/3}$

$\phantom{\sqrt[3]{1000}} = 10$

(e) $\sqrt[6]{\dfrac{64}{729}} = \left(\dfrac{64}{729}\right)^{1/6}$

$\phantom{\sqrt[6]{\dfrac{64}{729}}} = \dfrac{2}{3}$

(f) $\sqrt[4]{-16}$ is not a real number.

✔ Now Try Exercises 1, 3, 7, and 11.

EXAMPLE 2 Converting from Rational Exponents to Radicals

Write in radical form and simplify. Assume all variable expressions represent positive real numbers.

(a) $8^{2/3}$ **(b)** $(-32)^{4/5}$ **(c)** $-16^{3/4}$ **(d)** $x^{5/6}$

(e) $3x^{2/3}$ **(f)** $2p^{1/2}$ **(g)** $(3a + b)^{1/4}$

SOLUTION

(a) $8^{2/3} = \left(\sqrt[3]{8}\right)^2$ **(b)** $(-32)^{4/5} = \left(\sqrt[5]{-32}\right)^4$ **(c)** $-16^{3/4} = -\left(\sqrt[4]{16}\right)^3$

$\qquad = 2^2$ $\qquad\qquad\qquad = (-2)^4$ $\qquad\qquad\qquad = -(2)^3$

$\qquad = 4$ $\qquad\qquad\qquad\quad = 16$ $\qquad\qquad\qquad\quad = -8$

(d) $x^{5/6} = \sqrt[6]{x^5}$ $\qquad\qquad\qquad\qquad\qquad$ **(e)** $3x^{2/3} = 3\sqrt[3]{x^2}$

(f) $2p^{1/2} = 2\sqrt{p}$ $\qquad\qquad\qquad\qquad\qquad$ **(g)** $(3a + b)^{1/4} = \sqrt[4]{3a + b}$

✔ *Now Try Exercises 13, 15, and 17.*

LOOKING AHEAD TO CALCULUS
In calculus, the "power rule" for derivatives requires converting radicals to rational exponents.

CAUTION It is not possible to "distribute" exponents over a sum, so in **Example 2(g)**, $(3a + b)^{1/4}$ *cannot be written as* $(3a)^{1/4} + b^{1/4}$.

$$\sqrt[n]{x^n + y^n} \text{ is not equivalent to } x + y.$$

(For example, let $n = 2$, $x = 3$, and $y = 4$ to see this.)

EXAMPLE 3 Converting from Radicals to Rational Exponents

Write in exponential form. Assume all variable expressions represent positive real numbers.

(a) $\sqrt[4]{x^5}$ **(b)** $\sqrt{3y}$ **(c)** $10\left(\sqrt[5]{z}\right)^2$

(d) $5\sqrt[3]{(2x^4)^7}$ **(e)** $\sqrt{p^2 + q}$

SOLUTION

(a) $\sqrt[4]{x^5} = x^{5/4}$ **(b)** $\sqrt{3y} = (3y)^{1/2}$ **(c)** $10\left(\sqrt[5]{z}\right)^2 = 10z^{2/5}$

(d) $5\sqrt[3]{(2x^4)^7} = 5(2x^4)^{7/3}$ $\qquad\qquad\qquad$ **(e)** $\sqrt{p^2 + q} = (p^2 + q)^{1/2}$

$\qquad\qquad\qquad = 5 \cdot 2^{7/3}x^{28/3}$

✔ *Now Try Exercises 19 and 21.*

We *cannot* simply write $\sqrt{x^2} = x$ for all real numbers x. For example, if $x = -5$, then

$$\sqrt{x^2} = \sqrt{(-5)^2} = \sqrt{25} = 5 \ne x.$$

To take care of the fact that a negative value of x can produce a positive result, we use absolute value. For any real number a, the following holds.

$$\sqrt{a^2} = |a|$$

For example, $\sqrt{(-9)^2} = |-9| = 9$ and $\sqrt{13^2} = |13| = 13$.

We can generalize this result to any *even* nth root.

Evaluating $\sqrt[n]{a^n}$

If n is an *even* positive integer, then $\sqrt[n]{a^n} = |a|$.

If n is an *odd* positive integer, then $\sqrt[n]{a^n} = a$.

EXAMPLE 4 **Using Absolute Value to Simplify Roots**

Simplify each expression.

(a) $\sqrt{p^4}$　　　　　**(b)** $\sqrt[4]{p^4}$　　　　　**(c)** $\sqrt{16m^8r^6}$

(d) $\sqrt[6]{(-2)^6}$　　　　**(e)** $\sqrt[5]{m^5}$　　　　**(f)** $\sqrt{(2k + 3)^2}$

(g) $\sqrt{x^2 - 4x + 4}$

SOLUTION

(a) $\sqrt{p^4} = \sqrt{(p^2)^2}$　　　**(b)** $\sqrt[4]{p^4} = |p|$　　　**(c)** $\sqrt{16m^8r^6} = |4m^4r^3|$

$\qquad\quad = |p^2|$　　　　　　　　　　　　　　　　　　　$= 4m^4|r^3|$

$\qquad\quad = p^2$

(d) $\sqrt[6]{(-2)^6} = |-2|$　　　**(e)** $\sqrt[5]{m^5} = m$　　　**(f)** $\sqrt{(2k + 3)^2} = |2k + 3|$

$\qquad\qquad\quad = 2$

(g) $\sqrt{x^2 - 4x + 4} = \sqrt{(x - 2)^2}$

$\qquad\qquad\qquad\quad = |x - 2|$

☑ *Now Try Exercises 27, 29, and 31.*

NOTE　When working with variable radicands, we will *usually* assume that all variables in radicands represent only nonnegative real numbers.

The following rules for working with radicals are simply the power rules for exponents written in radical notation.

Rules for Radicals

Suppose that a and b represent real numbers, and m and n represent positive integers for which the indicated roots are real numbers.

Rule	Description
Product rule $\sqrt[n]{a} \cdot \sqrt[n]{b} = \sqrt[n]{ab}$	The product of two roots is the root of the product.
Quotient rule $\sqrt[n]{\dfrac{a}{b}} = \dfrac{\sqrt[n]{a}}{\sqrt[n]{b}}$ $(b \neq 0)$	The root of a quotient is the quotient of the roots.
Power rule $\sqrt[m]{\sqrt[n]{a}} = \sqrt[mn]{a}$	The index of the root of a root is the product of their indexes.

EXAMPLE 5 **Simplifying Radical Expressions**

Simplify. Assume all variable expressions represent positive real numbers.

(a) $\sqrt{6} \cdot \sqrt{54}$ **(b)** $\sqrt[3]{m} \cdot \sqrt[3]{m^2}$ **(c)** $\sqrt{\dfrac{7}{64}}$

(d) $\sqrt[4]{\dfrac{a}{b^4}}$ **(e)** $\sqrt[7]{\sqrt[3]{2}}$ **(f)** $\sqrt[4]{\sqrt{3}}$

SOLUTION

(a) $\sqrt{6} \cdot \sqrt{54} = \sqrt{6 \cdot 54}$ Product rule   **(b)** $\sqrt[3]{m} \cdot \sqrt[3]{m^2} = \sqrt[3]{m^3}$

$\qquad\qquad\qquad = \sqrt{324}, \quad \text{or} \quad 18$ $= m$

(c) $\sqrt{\dfrac{7}{64}} = \dfrac{\sqrt{7}}{\sqrt{64}}$ Quotient rule   **(d)** $\sqrt[4]{\dfrac{a}{b^4}} = \dfrac{\sqrt[4]{a}}{\sqrt[4]{b^4}}$

$\qquad\quad = \dfrac{\sqrt{7}}{8}$ $\qquad\quad = \dfrac{\sqrt[4]{a}}{b}$

(e) $\sqrt[7]{\sqrt[3]{2}} = \sqrt[21]{2}$ Power rule   **(f)** $\sqrt[4]{\sqrt{3}} = \sqrt[4 \cdot 2]{3} = \sqrt[8]{3}$

✔️ *Now Try Exercises 37, 41, 45, and 65.*

> **NOTE** Converting to rational exponents shows why these rules work.
>
> $$\sqrt[7]{\sqrt[3]{2}} = (2^{1/3})^{1/7} = 2^{(1/3)(1/7)} = 2^{1/21} = \sqrt[21]{2} \quad \text{Example 5(e)}$$

Simplified Radicals   In working with numbers, we prefer to write a number in its simplest form. For example, $\frac{10}{2}$ is written as 5, and $-\frac{9}{6}$ is written as $-\frac{3}{2}$. Similarly, expressions with radicals can be written in their simplest forms.

Simplified Radicals

An expression with radicals is simplified when all of the following conditions are satisfied.

1. The radicand has no factor raised to a power greater than or equal to the index.

2. The radicand has no fractions.

3. No denominator contains a radical.

4. Exponents in the radicand and the index of the radical have greatest common factor 1.

5. All indicated operations have been performed (if possible).

EXAMPLE 6 **Simplifying Radicals**

Simplify each radical.

(a) $\sqrt{175}$ **(b)** $-3\sqrt[5]{32}$ **(c)** $\sqrt[3]{81x^5y^7z^6}$

SOLUTION

(a) $\sqrt{175} = \sqrt{25 \cdot 7}$ **(b)** $-3\sqrt[5]{32} = -3\sqrt[5]{2^5}$

$\qquad\quad = \sqrt{25} \cdot \sqrt{7}$ $\qquad\qquad = -3 \cdot 2$

$\qquad\quad = 5\sqrt{7}$ $\qquad\qquad = -6$

(c) $\sqrt[3]{81x^5y^7z^6} = \sqrt[3]{27 \cdot 3 \cdot x^3 \cdot x^2 \cdot y^6 \cdot y \cdot z^6}$ Factor. **(Section R.4)**

$= \sqrt[3]{(27x^3y^6z^6)(3x^2y)}$ Group all perfect cubes.

$= 3xy^2z^2\sqrt[3]{3x^2y}$ Remove all perfect cubes from the radical.

☑ *Now Try Exercises 33 and 51.*

Operations with Radicals   Radicals with the same radicand and the same index, such as $3\sqrt[4]{11pq}$ and $-7\sqrt[4]{11pq}$, are **like radicals.** On the other hand, examples of **unlike radicals** are as follows.

$$2\sqrt{5} \quad \text{and} \quad 2\sqrt{3} \quad \text{Radicands are different.}$$

$$2\sqrt{3} \quad \text{and} \quad 2\sqrt[3]{3} \quad \text{Indexes are different.}$$

We add or subtract like radicals by using the distributive property. ***Only like radicals can be combined.*** Sometimes we need to simplify radicals before adding or subtracting.

EXAMPLE 7   **Adding and Subtracting Radicals**

Add or subtract, as indicated. Assume all variables represent positive real numbers.

(a) $3\sqrt[4]{11pq} + \left(-7\sqrt[4]{11pq}\right)$ **(b)** $\sqrt{98x^3y} + 3x\sqrt{32xy}$

(c) $\sqrt[3]{64m^4n^5} - \sqrt[3]{-27m^{10}n^{14}}$

SOLUTION

(a) $3\sqrt[4]{11pq} + \left(-7\sqrt[4]{11pq}\right) = -4\sqrt[4]{11pq}$

(b) $\sqrt{98x^3y} + 3x\sqrt{32xy} = \sqrt{49 \cdot 2 \cdot x^2 \cdot x \cdot y} + 3x\sqrt{16 \cdot 2 \cdot x \cdot y}$ Factor.

$= 7x\sqrt{2xy} + 3x(4)\sqrt{2xy}$ Remove all perfect squares from the radicals.

$= 7x\sqrt{2xy} + 12x\sqrt{2xy}$ Multiply.

$= (7x + 12x)\sqrt{2xy}$ Distributive property **(Section R.2)**

$= 19x\sqrt{2xy}$ Add.

(c) $\sqrt[3]{64m^4n^5} - \sqrt[3]{-27m^{10}n^{14}} = \sqrt[3]{(64m^3n^3)(mn^2)} - \sqrt[3]{(-27m^9n^{12})(mn^2)}$

$= 4mn\sqrt[3]{mn^2} - (-3)m^3n^4\sqrt[3]{mn^2}$

$= 4mn\sqrt[3]{mn^2} + 3m^3n^4\sqrt[3]{mn^2}$

$= (4mn + 3m^3n^4)\sqrt[3]{mn^2}$ ⟵ This *cannot* be simplified further.

☑ *Now Try Exercises 69 and 73.*

If the index of the radical and an exponent in the radicand have a common factor, we can simplify the radical by first writing it in exponential form. We simplify the rational exponent, and then write the result as a radical again, as shown in **Example 8** on the next page.

EXAMPLE 8 **Simplifying Radicals**

Simplify each radical. Assume all variables represent positive real numbers.

(a) $\sqrt[6]{3^2}$ **(b)** $\sqrt[6]{x^{12}y^3}$ **(c)** $\sqrt[9]{\sqrt{6^3}}$

SOLUTION

(a) $\sqrt[6]{3^2} = 3^{2/6}$

$= 3^{1/3}$

$= \sqrt[3]{3}$

(b) $\sqrt[6]{x^{12}y^3} = (x^{12}y^3)^{1/6}$

$= x^2 y^{3/6}$

$= x^2 y^{1/2}$

$= x^2 \sqrt{y}$

(c) $\sqrt[9]{\sqrt{6^3}} = \sqrt[9]{6^{3/2}}$

$= (6^{3/2})^{1/9}$

$= 6^{1/6}$

$= \sqrt[6]{6}$

✔ *Now Try Exercises 63 and 67.*

In **Example 8(a),** we simplified $\sqrt[6]{3^2}$ as $\sqrt[3]{3}$. However, to simplify $(\sqrt[6]{x})^2$, the variable x must represent a nonnegative number. For example, consider the statement

$$(-8)^{2/6} = [(-8)^{1/6}]^2.$$

This result is not a real number, since $(-8)^{1/6}$ is not a real number. On the other hand,

$$(-8)^{1/3} = -2.$$

Here, even though $\frac{2}{6} = \frac{1}{3}$,

$$(\sqrt[6]{x})^2 \neq \sqrt[3]{x}.$$

If a is nonnegative, then it is always true that $a^{m/n} = a^{mp/(np)}$. Simplifying rational exponents on negative bases should be considered case by case.

EXAMPLE 9 **Multiplying Radical Expressions**

Find each product.

(a) $\left(\sqrt{7} - \sqrt{10}\right)\left(\sqrt{7} + \sqrt{10}\right)$ **(b)** $\left(\sqrt{2} + 3\right)\left(\sqrt{8} - 5\right)$

SOLUTION

(a) $\left(\sqrt{7} - \sqrt{10}\right)\left(\sqrt{7} + \sqrt{10}\right) = \left(\sqrt{7}\right)^2 - \left(\sqrt{10}\right)^2$ Product of the sum and difference of two terms **(Section R.3)**

$= 7 - 10$ $\left(\sqrt{a}\right)^2 = a$

$= -3$ Subtract.

(b) $\left(\sqrt{2} + 3\right)\left(\sqrt{8} - 5\right) = \sqrt{2}\left(\sqrt{8}\right) - \sqrt{2}(5) + 3\sqrt{8} - 3(5)$

FOIL **(Section R.3)**

$= \sqrt{16} - 5\sqrt{2} + 3\left(2\sqrt{2}\right) - 15$ Multiply; $\sqrt{8} = 2\sqrt{2}$.

$= 4 - 5\sqrt{2} + 6\sqrt{2} - 15$ Simplify.

$= -11 + \sqrt{2}$ Combine like terms.

✔ *Now Try Exercises 77 and 83.*

Rationalizing Denominators The third condition for a simplified radical requires that no denominator contain a radical. We achieve this by **rationalizing the denominator**—that is, multiplying by a form of 1.

EXAMPLE 10 **Rationalizing Denominators**

Rationalize each denominator.

(a) $\dfrac{4}{\sqrt{3}}$

(b) $\sqrt[4]{\dfrac{3}{5}}$

SOLUTION

(a) $\dfrac{4}{\sqrt{3}} = \dfrac{4}{\sqrt{3}} \cdot \dfrac{\sqrt{3}}{\sqrt{3}}$ Multiply by $\frac{\sqrt{3}}{\sqrt{3}}$ (which equals 1).

$\qquad = \dfrac{4\sqrt{3}}{3}$ $\sqrt{a} \cdot \sqrt{a} = a$

(b) $\sqrt[4]{\dfrac{3}{5}} = \dfrac{\sqrt[4]{3}}{\sqrt[4]{5}}$ Quotient rule

The denominator will be a rational number if it equals $\sqrt[4]{5^4}$. That is, four factors of 5 are needed under the radical. Since $\sqrt[4]{5}$ has just one factor of 5, three additional factors are needed, so multiply by $\dfrac{\sqrt[4]{5^3}}{\sqrt[4]{5^3}}$.

$$\dfrac{\sqrt[4]{3}}{\sqrt[4]{5}} = \dfrac{\sqrt[4]{3} \cdot \sqrt[4]{5^3}}{\sqrt[4]{5} \cdot \sqrt[4]{5^3}} \qquad \text{Multiply by } \tfrac{\sqrt[4]{5^3}}{\sqrt[4]{5^3}}.$$

$$= \dfrac{\sqrt[4]{3 \cdot 5^3}}{\sqrt[4]{5^4}} \qquad \text{Product rule}$$

$$= \dfrac{\sqrt[4]{375}}{5} \qquad \text{Simplify.}$$

☑ *Now Try Exercises 55 and 59.*

EXAMPLE 11 **Simplifying Radical Expressions with Fractions**

Simplify each expression. Assume all variables represent positive real numbers.

(a) $\dfrac{\sqrt[4]{xy^3}}{\sqrt[4]{x^3y^2}}$

(b) $\sqrt[3]{\dfrac{5}{x^6}} - \sqrt[3]{\dfrac{4}{x^9}}$

SOLUTION

(a) $\dfrac{\sqrt[4]{xy^3}}{\sqrt[4]{x^3y^2}} = \sqrt[4]{\dfrac{xy^3}{x^3y^2}}$ Quotient rule

$\qquad = \sqrt[4]{\dfrac{y}{x^2}}$ Simplify the radicand.

$\qquad = \dfrac{\sqrt[4]{y}}{\sqrt[4]{x^2}}$ Quotient rule

$\qquad = \dfrac{\sqrt[4]{y}}{\sqrt[4]{x^2}} \cdot \dfrac{\sqrt[4]{x^2}}{\sqrt[4]{x^2}}$ Rationalize the denominator.

$\qquad = \dfrac{\sqrt[4]{x^2y}}{x}$ $\sqrt[4]{x^2} \cdot \sqrt[4]{x^2} = \sqrt[4]{x^4} = x$

LOOKING AHEAD TO CALCULUS

Another standard problem in calculus is investigating the value that an expression such as $\dfrac{\sqrt{x^2 + 9} - 3}{x^2}$ approaches as x approaches 0. This cannot be done by simply substituting 0 for x, since the result is $\frac{0}{0}$. However, by **rationalizing the numerator,** we can show that for $x \neq 0$ the expression is equivalent to $\dfrac{1}{\sqrt{x^2 + 9} + 3}$. Then, by substituting 0 for x, we find that the original expression approaches $\frac{1}{6}$ as x approaches 0.

(b) $\sqrt[3]{\dfrac{5}{x^6}} - \sqrt[3]{\dfrac{4}{x^9}} = \dfrac{\sqrt[3]{5}}{\sqrt[3]{x^6}} - \dfrac{\sqrt[3]{4}}{\sqrt[3]{x^9}}$ Quotient rule

$\qquad\qquad = \dfrac{\sqrt[3]{5}}{x^2} - \dfrac{\sqrt[3]{4}}{x^3}$ Simplify the denominators.

$\qquad\qquad = \dfrac{x\sqrt[3]{5}}{x^3} - \dfrac{\sqrt[3]{4}}{x^3}$ Write with a common denominator. **(Section R.5)**

$\qquad\qquad = \dfrac{x\sqrt[3]{5} - \sqrt[3]{4}}{x^3}$ Subtract the numerators.

✔ Now Try Exercises 85 and 87.

In **Example 9(a),** we saw that the product

$$\left(\sqrt{7} - \sqrt{10}\right)\left(\sqrt{7} + \sqrt{10}\right) \quad \text{equals} \quad -3, \text{ a rational number.}$$

This suggests a way to rationalize a denominator that is a binomial in which one or both terms is a square root radical. The expressions $a - b$ and $a + b$ are **conjugates.**

EXAMPLE 12   **Rationalizing a Binomial Denominator**

Rationalize the denominator of $\dfrac{1}{1 - \sqrt{2}}$.

SOLUTION

$\dfrac{1}{1 - \sqrt{2}} = \dfrac{1\left(1 + \sqrt{2}\right)}{\left(1 - \sqrt{2}\right)\left(1 + \sqrt{2}\right)}$ $\boxed{\dfrac{1 + \sqrt{2}}{1 + \sqrt{2}} = 1}$ Multiply numerator and denominator by the conjugate of the denominator: $1 + \sqrt{2}$.

$\qquad\quad = \dfrac{1 + \sqrt{2}}{1 - 2}$ $(x - y)(x + y) = x^2 - y^2$

$\qquad\quad = -1 - \sqrt{2}$ $\dfrac{1 + \sqrt{2}}{-1} = -1 - \sqrt{2}$

✔ Now Try Exercise 93.

R.7   **Exercises**

Write each root using exponents and evaluate. **See Example 1.**

1. $\sqrt[3]{125}$ **2.** $\sqrt[3]{216}$ **3.** $\sqrt[4]{81}$

4. $\sqrt[4]{256}$ **5.** $\sqrt[3]{-125}$ **6.** $\sqrt[3]{-343}$

7. $\sqrt[4]{-81}$ **8.** $\sqrt[4]{-256}$ **9.** $\sqrt[5]{32}$

10. $\sqrt[7]{128}$ **11.** $-\sqrt[5]{-32}$ **12.** $-\sqrt[3]{-343}$

Concept Check In Exercises 13 and 14, match the rational exponent expression in Column I with the equivalent radical expression in Column II. Assume that x is not 0. **See Example 2.**

I   II

13. (a) $(-3x)^{1/3}$ **14. (a)** $-3x^{1/3}$ **A.** $\dfrac{3}{\sqrt[3]{x}}$ **B.** $-3\sqrt[3]{x}$

(b) $(-3x)^{-1/3}$ **(b)** $-3x^{-1/3}$ **C.** $\dfrac{1}{\sqrt[3]{3x}}$ **D.** $\dfrac{-3}{\sqrt[3]{x}}$

(c) $(3x)^{1/3}$ **(c)** $3x^{-1/3}$ **E.** $3\sqrt[3]{x}$ **F.** $\sqrt[3]{-3x}$

(d) $(3x)^{-1/3}$ **(d)** $3x^{1/3}$ **G.** $\sqrt[3]{3x}$ **H.** $\dfrac{1}{\sqrt[3]{-3x}}$

If the expression is in exponential form, write it in radical form. If it is in radical form, write it in exponential form. Assume all variables represent positive real numbers. **See Examples 2 and 3.**

15. $m^{2/3}$ **16.** $p^{5/4}$ **17.** $(2m + p)^{2/3}$ **18.** $(5r + 3t)^{4/7}$

19. $\sqrt[5]{k^2}$ **20.** $\sqrt[4]{z^5}$ **21.** $-3\sqrt{5p^3}$ **22.** $-m\sqrt{2y^5}$

Concept Check Answer each question.

23. For which of the following cases is $\sqrt{ab} = \sqrt{a} \cdot \sqrt{b}$ a true statement?

    A. a and b both positive   **B.** a and b both negative

24. For what positive integers n greater than or equal to 2 is $\sqrt[n]{a^n} = a$ always a true statement?

25. For what values of x is $\sqrt{9ax^2} = 3x\sqrt{a}$ a true statement? Assume $a \geq 0$.

26. Which of the following expressions is *not* simplified? Give the simplified form.

    A. $\sqrt[3]{2y}$ **B.** $\dfrac{\sqrt{5}}{2}$ **C.** $\sqrt[4]{m^3}$ **D.** $\sqrt{\dfrac{3}{4}}$

Simplify each expression. **See Example 4.**

27. $\sqrt[4]{x^4}$ **28.** $\sqrt[6]{x^6}$ **29.** $\sqrt{25k^4m^2}$

30. $\sqrt[4]{81p^{12}q^4}$ **31.** $\sqrt{(4x - y)^2}$ **32.** $\sqrt[4]{(5 + 2m)^4}$

Simplify each expression. Assume all variables represent positive real numbers. **See Examples 1, 4–6, and 8–11.**

33. $\sqrt[3]{81}$ **34.** $\sqrt[3]{250}$ **35.** $-\sqrt[4]{32}$

36. $-\sqrt[4]{243}$ **37.** $\sqrt{14} \cdot \sqrt{3pqr}$ **38.** $\sqrt{7} \cdot \sqrt{5xt}$

39. $\sqrt[3]{7x} \cdot \sqrt[3]{2y}$ **40.** $\sqrt[3]{9x} \cdot \sqrt[3]{4y}$ **41.** $-\sqrt{\dfrac{9}{25}}$

42. $-\sqrt{\dfrac{16}{49}}$ **43.** $-\sqrt[3]{\dfrac{5}{8}}$ **44.** $-\sqrt[4]{\dfrac{3}{16}}$

45. $\sqrt[4]{\dfrac{m}{n^4}}$ **46.** $\sqrt[6]{\dfrac{r}{s^6}}$ **47.** $3\sqrt[5]{-3125}$

48. $5\sqrt[3]{-343}$ **49.** $\sqrt[3]{16(-2)^4(2)^8}$ **50.** $\sqrt[3]{25(-3)^4(5)^3}$

51. $\sqrt{8x^5z^8}$ **52.** $\sqrt{24m^6n^5}$ **53.** $\sqrt[4]{x^4 + y^4}$

54. $\sqrt[3]{27 + a^3}$ **55.** $\sqrt{\dfrac{2}{3x}}$ **56.** $\sqrt{\dfrac{5}{3p}}$

57. $\sqrt{\dfrac{x^5 y^3}{z^2}}$ **58.** $\sqrt{\dfrac{g^3 h^5}{r^3}}$ **59.** $\sqrt[3]{\dfrac{8}{x^4}}$

60. $\sqrt[3]{\dfrac{9}{16p^4}}$ **61.** $\sqrt[4]{\dfrac{g^3 h^5}{9r^6}}$ **62.** $\sqrt[4]{\dfrac{32x^5}{y^5}}$

63. $\sqrt[8]{3^4}$ **64.** $\sqrt[9]{5^3}$ **65.** $\sqrt[3]{\sqrt{4}}$

66. $\sqrt[4]{\sqrt{25}}$ **67.** $\sqrt[4]{\sqrt[3]{2}}$ **68.** $\sqrt[5]{\sqrt[3]{9}}$

*Simplify each expression. Assume all variables represent positive real numbers. **See Examples 7, 9, and 11.***

69. $8\sqrt{2x} - \sqrt{8x} + \sqrt{72x}$ **70.** $4\sqrt{18k} - \sqrt{72k} + \sqrt{50k}$

71. $2\sqrt[3]{3} + 4\sqrt[3]{24} - \sqrt[3]{81}$ **72.** $\sqrt[3]{32} - 5\sqrt[3]{4} + 2\sqrt[3]{108}$

73. $\sqrt[4]{81x^6 y^3} - \sqrt[4]{16x^{10} y^3}$ **74.** $\sqrt[4]{256x^5 y^6} + \sqrt[4]{625x^9 y^2}$

75. $5\sqrt{6} + 2\sqrt{10}$ **76.** $3\sqrt{11} - 5\sqrt{13}$

77. $\left(\sqrt{2} + 3\right)\left(\sqrt{2} - 3\right)$ **78.** $\left(\sqrt{5} + \sqrt{2}\right)\left(\sqrt{5} - \sqrt{2}\right)$

79. $\left(\sqrt[3]{11} - 1\right)\left(\sqrt[3]{11^2} + \sqrt[3]{11} + 1\right)$ **80.** $\left(\sqrt[3]{7} + 3\right)\left(\sqrt[3]{7^2} - 3\sqrt[3]{7} + 9\right)$

81. $\left(\sqrt{3} + \sqrt{8}\right)^2$ **82.** $\left(\sqrt{5} + \sqrt{10}\right)^2$

83. $\left(3\sqrt{2} + \sqrt{3}\right)\left(2\sqrt{3} - \sqrt{2}\right)$ **84.** $\left(4\sqrt{5} + \sqrt{2}\right)\left(3\sqrt{2} - \sqrt{5}\right)$

85. $\dfrac{\sqrt[3]{mn} \cdot \sqrt[3]{m^2}}{\sqrt[3]{n^2}}$ **86.** $\dfrac{\sqrt[3]{8m^2 n^3} \cdot \sqrt[3]{2m^2}}{\sqrt[3]{32m^4 n^3}}$

87. $\sqrt[3]{\dfrac{2}{x^6}} - \sqrt[3]{\dfrac{5}{x^9}}$ **88.** $\sqrt[4]{\dfrac{7}{t^{12}}} + \sqrt[4]{\dfrac{9}{t^4}}$

89. $\dfrac{1}{\sqrt{2}} + \dfrac{3}{\sqrt{8}} + \dfrac{1}{\sqrt{32}}$ **90.** $\dfrac{2}{\sqrt{12}} - \dfrac{1}{\sqrt{27}} - \dfrac{5}{\sqrt{48}}$

91. $\dfrac{-4}{\sqrt[3]{3}} + \dfrac{1}{\sqrt[3]{24}} - \dfrac{2}{\sqrt[3]{81}}$ **92.** $\dfrac{5}{\sqrt[3]{2}} - \dfrac{2}{\sqrt[3]{16}} + \dfrac{1}{\sqrt[3]{54}}$

*Rationalize the denominator of each radical expression. Assume all variables represent nonnegative numbers and that no denominators are 0. **See Example 12.***

93. $\dfrac{\sqrt{3}}{\sqrt{5} + \sqrt{3}}$ **94.** $\dfrac{\sqrt{7}}{\sqrt{3} - \sqrt{7}}$ **95.** $\dfrac{\sqrt{7} - 1}{2\sqrt{7} + 4\sqrt{2}}$

96. $\dfrac{1 + \sqrt{3}}{3\sqrt{5} + 2\sqrt{3}}$ **97.** $\dfrac{p - 4}{\sqrt{p} + 2}$ **98.** $\dfrac{9 - r}{3 - \sqrt{r}}$

99. $\dfrac{3m}{2 + \sqrt{m + n}}$ **100.** $\dfrac{a}{\sqrt{a + b} - 1}$ **101.** $\dfrac{5\sqrt{x}}{2\sqrt{x} + \sqrt{y}}$

102. *Concept Check* What should the numerator and denominator of

$$\dfrac{1}{\sqrt[3]{3} - \sqrt[3]{5}}$$

be multiplied by in order to rationalize the denominator? Write this fraction with a rationalized denominator.

(Modeling) *Solve each problem.*

103. *Rowing Speed* Olympic rowing events have one-, two-, four-, or eight-person crews, with each person pulling a single oar. Increasing the size of the crew increases the speed of the boat. An analysis of Olympic rowing events concluded that the approximate speed, *s*, of the boat (in feet per second) was given by the formula

$$s = 15.18 \sqrt[9]{n},$$

where *n* is the number of oarsmen. Estimate the speed of a boat with a four-person crew. (*Source:* Townend, M. Stewart, *Mathematics in Sport,* Chichester, Ellis Horwood Limited.)

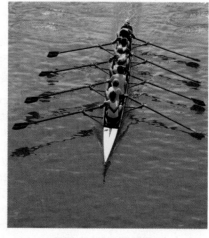

104. *Rowing Speed* **See Exercise 103.** Estimate the speed of a boat with an eight-person crew.

(Modeling) Windchill *The National Weather Service has used the formula*

$$\text{Windchill temperature} = 35.74 + 0.6215T - 35.75V^{0.16} + 0.4275TV^{0.16},$$

where T is the temperature in °F and V is the wind speed in miles per hour, to calculate windchill. (Source: National Oceanic and Atmospheric Administration, National Weather Service.) Use the formula to calculate the windchill to the nearest degree given the following conditions.

105. 10°F, 30 mph wind

106. 30°F, 15 mph wind

Concept Check *Simplify each expression mentally.*

107. $\sqrt[4]{8} \cdot \sqrt[4]{2}$

108. $\dfrac{\sqrt[3]{54}}{\sqrt[3]{2}}$

109. $\dfrac{\sqrt[5]{320}}{\sqrt[5]{10}}$

110. $\sqrt{0.1} \cdot \sqrt{40}$

111. $\dfrac{\sqrt[3]{15}}{\sqrt[3]{5}} \cdot \sqrt[3]{9}$

112. $\sqrt[6]{2} \cdot \sqrt[6]{4} \cdot \sqrt[6]{8}$

The screen in **Figure A** *seems to indicate that* π *and* $\sqrt[4]{\frac{2143}{22}}$ *are exactly equal, since the eight decimal values given by the calculator agree. However, as shown in* **Figure B** *using one more decimal place in the display, they differ in the ninth decimal place.* **The radical expression is a very good approximation for** π**, but it is still only an approximation.**

π	
	3.14159265
⁴√2143/22	
	3.14159265

π	
	3.141592654
⁴√2143/22	
	3.141592653

Figure A **Figure B**

Use your calculator to answer each question. Refer to the display for π *in* **Figure B**.

113. The Chinese of the fifth century used $\frac{355}{113}$ as an approximation for π. How many decimal places of accuracy does this fraction give?

114. A value for π that the Greeks used circa A.D. 150 is equivalent to $\frac{377}{120}$. In which decimal place does this value first differ from π?

115. The Hindu mathematician Bhaskara used $\frac{3927}{1250}$ as an approximation for π circa A.D. 1150. In which decimal place does this value first differ from π?

Chapter R Test Prep

Key Terms

R.1 set elements (members) infinite set finite set Venn diagram disjoint sets **R.2** number line coordinate system coordinate exponential expression multiplication property of zero	exponent base absolute value **R.3** algebraic expression term coefficient like terms polynomial polynomial in x degree of a term degree of a polynomial	trinomial binomial monomial descending order FOIL method **R.4** factoring factored form prime polynomial factored completely factoring by grouping **R.5** rational expression	domain of a rational expression lowest terms complex fraction **R.7** radicand index of a radical principal nth root like radicals unlike radicals rationalizing the denominator conjugates

New Symbols

$\{\ \}$	set braces	$\cup$	set union	
$\in$	is an element of	a^n	n factors of a	
$\notin$	is not an element of	$<$	is less than	
$\{x \mid x \textbf{ has property } p\}$	set-builder notation	$>$	is greater than	
U	universal set	$\leq$	is less than or equal to	
$\varnothing$, or $\{\ \}$	null (empty) set	$\geq$	is greater than or equal to	
$\subseteq$	is a subset of	$\lvert a \rvert$	absolute value of a	
$\not\subseteq$	is not a subset of	$\sqrt{}$	radical symbol	
A'	complement of a set A			
$\cap$	set intersection			

Quick Review

Concepts	Examples

R.1   Sets

Set Operations

For all sets A and B, with universal set U:

The **complement** of set A is the set A' of all elements in U that do not belong to set A.

$$A' = \{x \mid x \in U, x \notin A\}$$

The **intersection** of sets A and B, written $A \cap B$, is made up of all the elements belonging to both set A and set B.

$$A \cap B = \{x \mid x \in A \text{ and } x \in B\}$$

The **union** of sets A and B, written $A \cup B$, is made up of all the elements belonging to set A or to set B.

$$A \cup B = \{x \mid x \in A \text{ or } x \in B\}$$

Let $U = \{1, 2, 3, 4, 5, 6\}$, $A = \{1, 2, 3, 4\}$, and $B = \{3, 4, 6\}$.

$$A' = \{5, 6\}$$

$$A \cap B = \{3, 4\}$$

and

$$A \cup B = \{1, 2, 3, 4, 6\}$$

Concepts	Examples

R.2 Real Numbers and Their Properties

Sets of Numbers

Natural numbers
$$\{1, 2, 3, 4, \ldots\}$$

5, 17, 142

Whole numbers
$$\{0, 1, 2, 3, 4, \ldots\}$$

0, 27, 96

Integers
$$\{\ldots, -3, -2, -1, 0, 1, 2, 3, \ldots\}$$

$-24, 0, 19$

Rational numbers
$$\left\{ \tfrac{p}{q} \mid p \text{ and } q \text{ are integers and } q \neq 0 \right\}$$

$-\tfrac{3}{4}, -0.28, 0, 7, \tfrac{9}{16}, 0.66\overline{6}$

Irrational numbers
$$\{x \mid x \text{ is real but not rational}\}$$

$-\sqrt{15}, 0.101101110\ldots, \sqrt{2}, \pi$

Real numbers
$$\{x \mid x \text{ corresponds to a point on a number line}\}$$

$-46, 0.7, \pi, \sqrt{19}, \tfrac{8}{5}$

Properties of Real Numbers

For all real numbers a, b, and c:

Closure Properties
$$a + b \text{ is a real number.}$$
$$ab \text{ is a real number.}$$

$1 + \sqrt{2}$ is a real number.
$3\sqrt{7}$ is a real number.

Commutative Properties
$$a + b = b + a$$
$$ab = ba$$

$5 + 18 = 18 + 5$
$-4 \cdot 8 = 8 \cdot (-4)$

Associative Properties
$$(a + b) + c = a + (b + c)$$
$$(ab)c = a(bc)$$

$[6 + (-3)] + 5 = 6 + (-3 + 5)$
$(7 \cdot 6)20 = 7(6 \cdot 20)$

Identity Properties
There exists a unique real number 0 such that
$$a + 0 = a \quad \text{and} \quad 0 + a = a.$$

$145 + 0 = 145 \quad \text{and} \quad 0 + 145 = 145$

There exists a unique real number 1 such that
$$a \cdot 1 = a \quad \text{and} \quad 1 \cdot a = a.$$

$-60 \cdot 1 = -60 \quad \text{and} \quad 1 \cdot (-60) = -60$

Inverse Properties
There exists a unique real number $-a$ such that
$$a + (-a) = 0 \quad \text{and} \quad -a + a = 0.$$

$17 + (-17) = 0 \quad \text{and} \quad -17 + 17 = 0$

If $a \neq 0$, there exists a unique real number $\tfrac{1}{a}$ such that
$$a \cdot \frac{1}{a} = 1 \quad \text{and} \quad \frac{1}{a} \cdot a = 1.$$

$22 \cdot \dfrac{1}{22} = 1 \quad \text{and} \quad \dfrac{1}{22} \cdot 22 = 1$

Distributive Properties
$$a(b + c) = ab + ac$$
$$a(b - c) = ab - ac$$

$3(5 + 8) = 3 \cdot 5 + 3 \cdot 8$
$6(4 - 2) = 6 \cdot 4 - 6 \cdot 2$

(continued)

Concepts	Examples
Order $a > b$ if a is to the right of b on a number line. $a < b$ if a is to the left of b on a number line. **Absolute Value** $$\lvert a \rvert = \begin{cases} a & \text{if } a \geq 0 \\ -a & \text{if } a < 0 \end{cases}$$	$7 > -5$ $0 < 15$ $\lvert 3 \rvert = 3$ and $\lvert -3 \rvert = 3$

R.3 Polynomials

Special Products

Product of the Sum and Difference of Two Terms $$(x + y)(x - y) = x^2 - y^2$$	$(7 - x)(7 + x) = 7^2 - x^2 = 49 - x^2$
Square of a Binomial $$(x + y)^2 = x^2 + 2xy + y^2$$ $$(x - y)^2 = x^2 - 2xy + y^2$$	$(3a + b)^2 = (3a)^2 + 2(3a)(b) + b^2$ $\qquad = 9a^2 + 6ab + b^2$ $(2m - 5)^2 = (2m)^2 - 2(2m)(5) + 5^2$ $\qquad = 4m^2 - 20m + 25$

R.4 Factoring Polynomials

Factoring Patterns

Difference of Squares $$x^2 - y^2 = (x + y)(x - y)$$	$4t^2 - 9 = (2t + 3)(2t - 3)$
Perfect Square Trinomial $$x^2 + 2xy + y^2 = (x + y)^2$$ $$x^2 - 2xy + y^2 = (x - y)^2$$	$p^2 + 4pq + 4q^2 = (p + 2q)^2$ $9m^2 - 12mn + 4n^2 = (3m - 2n)^2$
Difference of Cubes $$x^3 - y^3 = (x - y)(x^2 + xy + y^2)$$	$r^3 - 8 = (r - 2)(r^2 + 2r + 4)$
Sum of Cubes $$x^3 + y^3 = (x + y)(x^2 - xy + y^2)$$	$27x^3 + 64 = (3x + 4)(9x^2 - 12x + 16)$

R.5 Rational Expressions

Operations
Let $\frac{a}{b}$ and $\frac{c}{d}$ $(b \neq 0, d \neq 0)$ represent fractions.

$$\frac{a}{b} \pm \frac{c}{d} = \frac{ad \pm bc}{bd}$$

$$\frac{a}{b} \cdot \frac{c}{d} = \frac{ac}{bd} \quad \text{and} \quad \frac{a}{b} \div \frac{c}{d} = \frac{ad}{bc} \quad (c \neq 0)$$

$$\frac{2}{x} + \frac{5}{y} = \frac{2y + 5x}{xy} \qquad \frac{x}{6} - \frac{2y}{5} = \frac{5x - 12y}{30}$$

$$\frac{3}{q} \cdot \frac{3}{2p} = \frac{9}{2pq} \qquad \frac{z}{4} \div \frac{z}{2t} = \frac{z}{4} \cdot \frac{2t}{z} = \frac{2zt}{4z} = \frac{t}{2}$$

Concepts	Examples

R.6 **Rational Exponents**

Rules for Exponents

Let r and s be rational numbers. The following results are valid for all positive numbers a and b.

$$a^r \cdot a^s = a^{r+s} \qquad (ab)^r = a^r b^r \qquad (a^r)^s = a^{rs}$$

$$\frac{a^r}{a^s} = a^{r-s} \qquad \left(\frac{a}{b}\right)^r = \frac{a^r}{b^r} \qquad a^{-r} = \frac{1}{a^r}$$

$$6^2 \cdot 6^3 = 6^5 \qquad (3x)^4 = 3^4 x^4 \qquad (m^2)^3 = m^6$$

$$\frac{p^5}{p^2} = p^3 \qquad \left(\frac{x}{3}\right)^2 = \frac{x^2}{3^2} \qquad 4^{-3} = \frac{1}{4^3}$$

R.7 **Radical Expressions**

Radical Notation

Suppose that a is a real number, n is a positive integer, and $a^{1/n}$ is defined.

$$\sqrt[n]{a} = a^{1/n}$$

Suppose that m is an integer, n is a positive integer, and a is a real number for which $\sqrt[n]{a}$ is defined.

$$a^{m/n} = \left(\sqrt[n]{a}\right)^m = \sqrt[n]{a^m}$$

$$\sqrt[4]{16} = 16^{1/4} = 2$$

$$8^{2/3} = \left(\sqrt[3]{8}\right)^2 = \sqrt[3]{8^2} = 4$$

Operations

Operations with radical expressions are performed like operations with polynomials.

$$\sqrt{8x} + \sqrt{32x} = 2\sqrt{2x} + 4\sqrt{2x}$$
$$= 6\sqrt{2x}$$
$$\left(\sqrt{5} - \sqrt{3}\right)\left(\sqrt{5} + \sqrt{3}\right) = 5 - 3$$
$$= 2$$
$$\left(\sqrt{2} + \sqrt{7}\right)\left(\sqrt{3} - \sqrt{6}\right)$$
$$= \sqrt{6} - 2\sqrt{3} + \sqrt{21} - \sqrt{42} \quad \text{FOIL; } \sqrt{12} = 2\sqrt{3}$$

Rationalize the denominator by multiplying numerator and denominator by a form of 1.

$$\frac{\sqrt{7y}}{\sqrt{5}} = \frac{\sqrt{7y}}{\sqrt{5}} \cdot \frac{\sqrt{5}}{\sqrt{5}}$$
$$= \frac{\sqrt{35y}}{5}$$

Chapter R | Review Exercises

1. Use set notation to list all the elements of the set $\{6, 8, 10, \ldots, 20\}$.

2. Is the set $\{x \mid x \text{ is a decimal between 0 and 1}\}$ finite or infinite?

3. *Concept Check* *True* or *false:* The set of negative integers and the set of whole numbers are disjoint sets.

4. *Concept Check* *True* or *false:* 9 is an element of {999}.

Determine whether each statement is true *or* false.

5. $1 \in \{6, 2, 5, 1\}$

6. $7 \notin \{1, 3, 5, 7\}$

7. $\{8, 11, 4\} = \{8, 11, 4, 0\}$

8. $\{0\} = \emptyset$

Let $A = \{1, 3, 4, 5, 7, 8\}$, $B = \{2, 4, 6, 8\}$, $C = \{1, 3, 5, 7\}$, $D = \{1, 2, 3\}$, $E = \{3, 7\}$, and $U = \{1, 2, 3, 4, 5, 6, 7, 8, 9, 10\}$.

Determine whether each statement is true *or* false.

9. $\emptyset \subseteq A$ **10.** $E \subseteq C$ **11.** $D \nsubseteq B$ **12.** $E \nsubseteq A$

Refer to the sets given for Exercises 9–12. Specify each set.

13. A' **14.** $B \cap A$ **15.** $B \cap E$

16. $C \cup E$ **17.** $D \cap \emptyset$ **18.** $B \cup \emptyset$

19. $(C \cap D) \cup B$ **20.** $(D' \cap U) \cup E$ **21.** $\emptyset'$

22. *Concept Check* *True or false:* For all sets A and B, $(A \cap B) \subseteq (A \cup B)$.

For Exercises 23 and 24, let set $K = \left\{ -12, -6, -0.9, -\sqrt{7}, -\sqrt{4}, 0, \frac{1}{8}, \frac{\pi}{4}, 6, \sqrt{11} \right\}$. *List all elements of K that belong to each set.*

23. Integers **24.** Rational numbers

For Exercises 25–28, choose all words from the following list that apply.

natural number	whole number	integer
rational number	irrational number	real number

25. $\dfrac{4\pi}{5}$ **26.** $\dfrac{\pi}{0}$ **27.** 0 **28.** $-\sqrt{36}$

Write each algebraic identity (true statement) as a complete English sentence without using the names of the variables. For instance, $z(x + y) = zx + zy$ can be stated as "The multiple of a sum is the sum of the multiples."

29. $\dfrac{1}{xy} = \dfrac{1}{x} \cdot \dfrac{1}{y}$ **30.** $a(b - c) = ab - ac$

31. $(ab)^n = a^n b^n$ **32.** $a^2 - b^2 = (a + b)(a - b)$

33. $\left(\dfrac{a}{b}\right)^n = \dfrac{a^n}{b^n}$ **34.** $|st| = |s| \cdot |t|$

Identify by name each property illustrated.

35. $8(5 + 9) = (5 + 9)8$ **36.** $4 \cdot 6 + 4 \cdot 12 = 4(6 + 12)$

37. $3 \cdot (4 \cdot 2) = (3 \cdot 4) \cdot 2$ **38.** $-8 + 8 = 0$

39. $(9 + p) + 0 = 9 + p$ **40.** $\dfrac{1}{\sqrt{2}} \cdot \dfrac{\sqrt{2}}{\sqrt{2}} = \dfrac{\sqrt{2}}{2}$

41. *Ages of College Undergraduates* The following table shows the age distribution of college students in 2008. In a random sample of 5000 such students, how many would you expect to be over 19?

Age	Percent
14–19	29
20–24	43
25–34	16
35 and older	12

Source: U.S. Census Bureau.

42. *Counting Marshmallows* In early 2011, there were media reports about students providing a correction to the following question posed on boxes of Swiss Miss Chocolate: *On average, how many mini-marshmallows are in one serving?*

$$3 + 2 \times 4 \div 2 - 3 \times 7 - 4 + 47 = \underline{\hspace{1.5cm}}$$

The company provided 92 as the answer. What is the *correct* calculation provided by the students? (*Source:* Swiss Miss Chocolate box.)

Simplify each expression.

43. $(-4 - 1)(-3 - 5) - 2^3$

44. $(6 - 9)(-2 - 7) \div (-4)$

45. $\left(-\dfrac{5}{9} - \dfrac{2}{3}\right) - \dfrac{5}{6}$

46. $\left(-\dfrac{2^3}{5} - \dfrac{3}{4}\right) - \left(-\dfrac{1}{2}\right)$

47. $\dfrac{6(-4) - 3^2(-2)^3}{-5[-2 - (-6)]}$

48. $\dfrac{(-7)(-3) - (-2^3)(-5)}{(-2^2 - 2)(-1 - 6)}$

Evaluate each expression for $a = -1$, $b = -2$, and $c = 4$.

49. $-c(2a - 5b)$

50. $(a - 2) \div 5 \cdot b + c$

51. $\dfrac{9a + 2b}{a + b + c}$

52. $\dfrac{3|b| - 4|c|}{|ac|}$

Perform the indicated operations.

53. $(3q^3 - 9q^2 + 6) + (4q^3 - 8q + 3)$

54. $2(3y^6 - 9y^2 + 2y) - (5y^6 - 4y)$

55. $(8y - 7)(2y + 7)$

56. $(2r + 11s)(4r - 9s)$

57. $(3k - 5m)^2$

58. $(4a - 3b)^2$

Perform each division.

59. $\dfrac{30m^3 - 9m^2 + 22m + 5}{5m + 1}$

60. $\dfrac{72r^2 + 59r + 12}{8r + 3}$

61. $\dfrac{3b^3 - 8b^2 + 12b - 30}{b^2 + 4}$

62. $\dfrac{5m^3 - 7m^2 + 14}{m^2 - 2}$

Factor as completely as possible.

63. $3(z - 4)^2 + 9(z - 4)^3$

64. $7z^2 - 9z^3 + z$

65. $z^2 - 6zk - 16k^2$

66. $r^2 + rp - 42p^2$

67. $48a^8 - 12a^7b - 90a^6b^2$

68. $6m^2 - 13m - 5$

69. $49m^8 - 9n^2$

70. $169y^4 - 1$

71. $6(3r - 1)^2 + (3r - 1) - 35$

72. $8y^3 - 1000z^6$

73. $xy + 2x - y - 2$

74. $15mp + 9mq - 10np - 6nq$

Factor each expression. (These expressions arise in calculus from a technique called the product rule that is used to determine the shape of a curve.)

75. $(3x - 4)^2 + (x - 5)(2)(3x - 4)(3)$

76. $(5 - 2x)(3)(7x - 8)^2(7) + (7x - 8)^3(-2)$

Perform the indicated operations.

77. $\dfrac{k^2 + k}{8k^3} \cdot \dfrac{4}{k^2 - 1}$

78. $\dfrac{3r^3 - 9r^2}{r^2 - 9} \div \dfrac{8r^3}{r + 3}$

79. $\dfrac{x^2 + x - 2}{x^2 + 5x + 6} \div \dfrac{x^2 + 3x - 4}{x^2 + 4x + 3}$

80. $\dfrac{27m^3 - n^3}{3m - n} \div \dfrac{9m^2 + 3mn + n^2}{9m^2 - n^2}$

81. $\dfrac{p^2 - 36q^2}{p^2 - 12pq + 36q^2} \cdot \dfrac{p^2 - 5pq - 6q^2}{p^2 + 2pq + q^2}$

82. $\dfrac{1}{4y} + \dfrac{8}{5y}$

83. $\dfrac{m}{4 - m} + \dfrac{3m}{m - 4}$

84. $\dfrac{3}{x^2 - 4x + 3} - \dfrac{2}{x^2 - 1}$

85. $\dfrac{p^{-1} + q^{-1}}{1 - (pq)^{-1}}$

86. $\dfrac{3 + \dfrac{2m}{m^2 - 4}}{\dfrac{5}{m - 2}}$

Simplify each expression. Write the answer with only positive exponents. Assume all variables represent positive real numbers.

87. $\left(-\dfrac{5}{4}\right)^{-2}$

88. $3^{-1} - 4^{-1}$

89. $(5z^3)(-2z^5)$

90. $(8p^2q^3)(-2p^5q^{-4})$

91. $(-6p^5w^4m^{12})^0$

92. $(-6x^2y^{-3}z^2)^{-2}$

93. $\dfrac{-8y^7p^{-2}}{y^{-4}p^{-3}}$

94. $\dfrac{a^{-6}(a^{-8})}{a^{-2}(a^{11})}$

95. $\dfrac{(p + q)^4(p + q)^{-3}}{(p + q)^6}$

96. $\dfrac{[p^2(m + n)^3]^{-2}}{p^{-2}(m + n)^{-5}}$

97. $(7r^{1/2})(2r^{3/4})(-r^{1/6})$

98. $(a^{3/4}b^{2/3})(a^{5/8}b^{-5/6})$

99. $\dfrac{y^{5/3} \cdot y^{-2}}{y^{-5/6}}$

100. $\left(\dfrac{25m^3n^5}{m^{-2}n^6}\right)^{-1/2}$

101. $\dfrac{(p^{15}q^{12})^{-4/3}}{(p^{24}q^{16})^{-3/4}}$

102. Simplify the product $-m^{3/4}(8m^{1/2} + 4m^{-3/2})$. Assume the variable represents a positive real number.

Simplify. Assume all variables represent positive real numbers.

103. $\sqrt{200}$

104. $\sqrt[3]{16}$

105. $\sqrt[4]{1250}$

106. $-\sqrt{\dfrac{16}{3}}$

107. $-\sqrt[3]{\dfrac{2}{5p^2}}$

108. $\sqrt{\dfrac{27y^8}{m^3}}$

109. $\sqrt[4]{\sqrt[3]{m}}$

110. $\dfrac{\sqrt[4]{8p^2q^5} \cdot \sqrt[4]{2p^3q}}{\sqrt[4]{p^5q^2}}$

111. $\left(\sqrt[3]{2} + 4\right)\left(\sqrt[3]{2^2} - 4\sqrt[3]{2} + 16\right)$

112. $\dfrac{3}{\sqrt{5}} - \dfrac{2}{\sqrt{45}} + \dfrac{6}{\sqrt{80}}$

113. $\sqrt{18m^3} - 3m\sqrt{32m} + 5\sqrt{m^3}$

114. $\dfrac{2}{7 - \sqrt{3}}$

115. $\dfrac{6}{3 - \sqrt{2}}$

116. $\dfrac{k}{\sqrt{k} - 3}$

*Concept Check Correct each **INCORRECT** statement by changing the right-hand side of the equation.*

117. $x(x^2 + 5) = x^3 + 5$

118. $-3^2 = 9$

119. $(m^2)^3 = m^5$

120. $(3x)(3y) = 3xy$

121. $\dfrac{\left(\dfrac{a}{b}\right)}{2} = \dfrac{2a}{b}$

122. $\dfrac{m}{r} \cdot \dfrac{n}{r} = \dfrac{mn}{r}$

123. $\dfrac{1}{(-2)^3} = 2^{-3}$

124. $(-5)^2 = -5^2$

125. $\left(\dfrac{8}{7} + \dfrac{a}{b}\right)^{-1} = \dfrac{7}{8} + \dfrac{b}{a}$

Chapter R   Test

Let $U = \{1, 2, 3, 4, 5, 6, 7, 8\}$, $A = \{1, 2, 3, 4, 5, 6\}$, $B = \{1, 3, 5\}$, $C = \{1, 6\}$, and $D = \{4\}$. *Tell whether each statement is* true *or* false.

1. $B' = \{2, 4, 6, 8\}$

2. $C \subseteq A$

3. $(B \cap C) \cup D = \{1, 3, 4, 5, 6\}$

4. $(A' \cup C) \cap B' = \{6, 7, 8\}$

5. Let $A = \left\{-13, -\frac{12}{4}, 0, \frac{3}{5}, \frac{\pi}{4}, 5.9, \sqrt{49}\right\}$. List the elements of A that belong to the given set.

    (a) Integers   **(b)** Rational numbers   **(c)** Real numbers

6. Evaluate the expression $\left|\dfrac{x^2 + 2yz}{3(x + z)}\right|$ for $x = -2$, $y = -4$, and $z = 5$.

7. Identify each property illustrated. Let a, b, and c represent any real numbers.

    (a) $a + (b + c) = (a + b) + c$ **(b)** $a + (c + b) = a + (b + c)$

    (c) $a(b + c) = ab + ac$ **(d)** $a + [b + (-b)] = a + 0$

8. *Passer Rating for NFL Quarterbacks* Approximate the quarterback rating (to the nearest tenth) of Drew Brees of the New Orleans Saints during the 2008 regular season. He attempted 514 passes, completed 363, had 4388 total yards, threw for 34 touchdowns, and had 11 interceptions. (*Source: World Almanac and Book of Facts.*)

$$\text{Rating} = \dfrac{\left(250 \cdot \dfrac{C}{A}\right) + \left(1000 \cdot \dfrac{T}{A}\right) + \left(12.5 \cdot \dfrac{Y}{A}\right) + 6.25 - \left(1250 \cdot \dfrac{I}{A}\right)}{3},$$

where A = attempted passes, C = completed passes, T = touchdown passes, Y = yards gained passing, and I = interceptions.

In addition to the weighting factors appearing in the formula, the four category ratios are limited to nonnegative values with the following maximums.

$$0.775 \text{ for } \dfrac{C}{A}, \quad 0.11875 \text{ for } \dfrac{T}{A}, \quad 12.5 \text{ for } \dfrac{Y}{A}, \quad 0.095 \text{ for } \dfrac{I}{A}$$

Perform the indicated operations.

9. $(x^2 - 3x + 2) - (x - 4x^2) + 3x(2x + 1)$

10. $(6r - 5)^2$

11. $(t + 2)(3t^2 - t + 4)$

12. $\dfrac{2x^3 - 11x^2 + 28}{x - 5}$

(Modeling) Adjusted Poverty Threshold The adjusted poverty threshold for a single person between the years 1995 and 2009 can be approximated by the polynomial

$$4.872x^2 + 170.2x + 7787,$$

where $x = 0$ corresponds to 1995, $x = 1$ corresponds to 1996, and so on, and the amount is in dollars. According to this model, what was the adjusted poverty threshold, to the nearest dollar, in each given year? *(Source: U.S. Census Bureau.)*

13. 2000　　　　　　　　　　　　　**14.** 2008

Factor completely.

15. $6x^2 - 17x + 7$　　　　　　　　**16.** $x^4 - 16$

17. $24m^3 - 14m^2 - 24m$　　　　　**18.** $x^3y^2 - 9x^3 - 8y^2 + 72$

19. $(a - b)^2 + 2(a - b)$　　　　　**20.** $1 - 27x^6$

Perform the indicated operations.

21. $\dfrac{5x^2 - 9x - 2}{30x^3 + 6x^2} \div \dfrac{x^4 - 3x^2 - 4}{2x^8 + 6x^7 + 4x^6}$　　　**22.** $\dfrac{x}{x^2 + 3x + 2} + \dfrac{2x}{2x^2 - x - 3}$

23. $\dfrac{a + b}{2a - 3} - \dfrac{a - b}{3 - 2a}$　　　　　**24.** $\dfrac{y - 2}{y - \dfrac{4}{y}}$

Simplify or evaluate as appropriate. Assume all variables represent positive real numbers.

25. $\sqrt{18x^5y^8}$　　　　　　　　　**26.** $\sqrt{32x} + \sqrt{2x} - \sqrt{18x}$

27. $\left(\sqrt{x} - \sqrt{y}\right)\left(\sqrt{x} + \sqrt{y}\right)$　　　　**28.** $\dfrac{14}{\sqrt{11} - \sqrt{7}}$

29. $\left(\dfrac{x^{-2}y^{-1/3}}{x^{-5/3}y^{-2/3}}\right)^3$　　　　　**30.** $\left(-\dfrac{64}{27}\right)^{-2/3}$

31. *Concept Check* *True* or *false:* For all real numbers x, $\sqrt{x^2} = x$.

32. *(Modeling) Period of a Pendulum* The period t, in seconds, of the swing of a pendulum is given by the equation

$$t = 2\pi\sqrt{\dfrac{L}{32}},$$

where L is the length of the pendulum in feet. Find the period of a pendulum 3.5 ft long. Use a calculator.

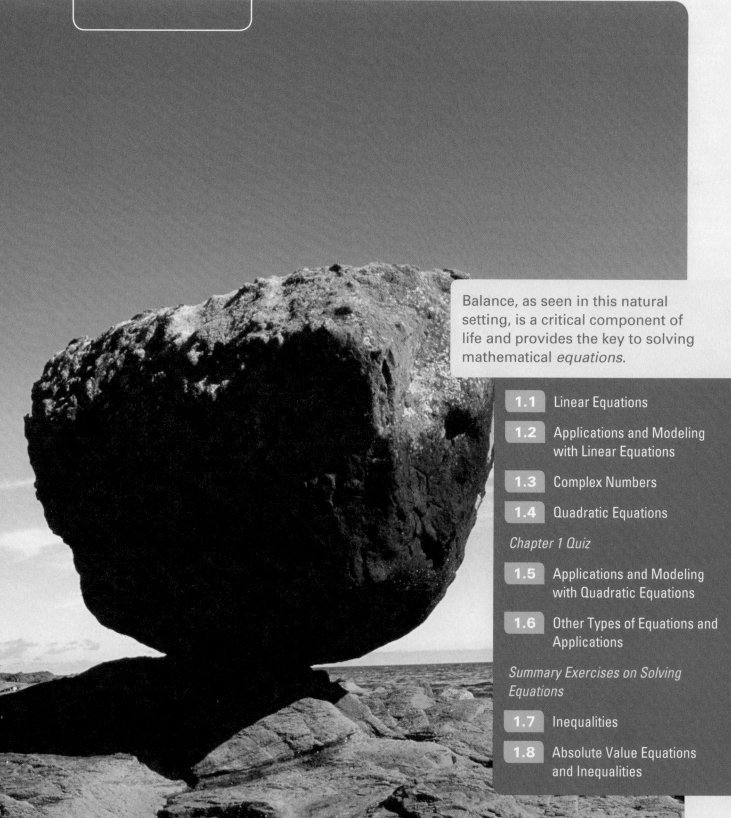

1 Equations and Inequalities

Balance, as seen in this natural setting, is a critical component of life and provides the key to solving mathematical *equations*.

1.1 Linear Equations

1.2 Applications and Modeling with Linear Equations

1.3 Complex Numbers

1.4 Quadratic Equations

Chapter 1 Quiz

1.5 Applications and Modeling with Quadratic Equations

1.6 Other Types of Equations and Applications

Summary Exercises on Solving Equations

1.7 Inequalities

1.8 Absolute Value Equations and Inequalities

1.1 Linear Equations

■ Basic Terminology of Equations
■ Solving Linear Equations
■ Identities, Conditional Equations, and Contradictions
■ Solving for a Specified Variable (Literal Equations)

Basic Terminology of Equations An **equation** is a statement that two expressions are equal.

$$x + 2 = 9, \quad 11x = 5x + 6x, \quad x^2 - 2x - 1 = 0 \quad \text{Equations}$$

To *solve* an equation means to find all numbers that make the equation a true statement. These numbers are the **solutions,** or **roots,** of the equation. A number that is a solution of an equation is said to *satisfy* the equation, and the solutions of an equation make up its **solution set.** Equations with the same solution set are **equivalent equations.** For example,

$$x = 4, \quad x + 1 = 5, \quad \text{and} \quad 6x + 3 = 27 \quad \text{are equivalent equations}$$

because they have the same solution set, $\{4\}$. However, the equations

$$x^2 = 9 \quad \text{and} \quad x = 3 \quad \text{are } not \text{ equivalent,}$$

since the first has solution set $\{-3, 3\}$ while the solution set of the second is $\{3\}$.

One way to solve an equation is to rewrite it as a series of simpler equivalent equations using the **addition and multiplication properties of equality.**

Addition and Multiplication Properties of Equality

Let a, b, and c represent real numbers.

$$\textbf{If } a = b \textbf{, then } a + c = b + c.$$

That is, the same number may be added to each side of an equation without changing the solution set.

$$\textbf{If } a = b \textbf{ and } c \neq 0 \textbf{, then } ac = bc.$$

That is, each side of an equation may be multiplied by the same nonzero number without changing the solution set. (Multiplying each side by zero leads to $0 = 0$.)

These properties can be extended: The same number may be subtracted from each side of an equation, and each side may be divided by the same nonzero number, without changing the solution set.

Solving Linear Equations We use the properties of equality to solve *linear equations.*

Linear Equation in One Variable

A **linear equation in one variable** is an equation that can be written in the form

$$ax + b = 0,$$

where a and b are real numbers with $a \neq 0$.

A linear equation is a **first-degree equation** since the greatest degree of the variable is 1.

$$3x + \sqrt{2} = 0, \quad \frac{3}{4}x = 12, \quad 0.5(x + 3) = 2x - 6 \qquad \text{Linear equations}$$

$$\sqrt{x} + 2 = 5, \quad \frac{1}{x} = -8, \quad x^2 + 3x + 0.2 = 0 \qquad \text{Nonlinear equations}$$

EXAMPLE 1 Solving a Linear Equation

Solve $3(2x - 4) = 7 - (x + 5)$.

SOLUTION

$$3(2x - 4) = 7 - (x + 5)$$ ⟵ Be careful with signs.

$$6x - 12 = 7 - x - 5 \qquad \text{Distributive property (Section R.2)}$$

$$6x - 12 = 2 - x \qquad \text{Combine like terms. (Section R.3)}$$

$$6x - 12 + x = 2 - x + x \qquad \text{Add } x \text{ to each side.}$$

$$7x - 12 = 2 \qquad \text{Combine like terms.}$$

$$7x - 12 + 12 = 2 + 12 \qquad \text{Add 12 to each side.}$$

$$7x = 14 \qquad \text{Combine like terms.}$$

$$\frac{7x}{7} = \frac{14}{7} \qquad \text{Divide each side by 7.}$$

$$x = 2$$

CHECK ⟵ A check of the solution is recommended.

$$3(2x - 4) = 7 - (x + 5) \qquad \text{Original equation}$$

$$3(2 \cdot 2 - 4) \overset{?}{=} 7 - (2 + 5) \qquad \text{Let } x = 2.$$

$$3(4 - 4) \overset{?}{=} 7 - (7) \qquad \text{Work inside the parentheses.}$$

$$0 = 0 \checkmark \qquad \text{True}$$

Since replacing x with 2 results in a true statement, 2 is a solution of the given equation. The solution set is $\{2\}$.

✔ *Now Try Exercise 11.*

EXAMPLE 2 Solving a Linear Equation with Fractions

Solve $\dfrac{2x + 4}{3} + \dfrac{1}{2}x = \dfrac{1}{4}x - \dfrac{7}{3}$.

SOLUTION

$$\frac{2x + 4}{3} + \frac{1}{2}x = \frac{1}{4}x - \frac{7}{3}$$

Distribute to *all* terms within the parentheses.

$$12\left(\frac{2x + 4}{3} + \frac{1}{2}x\right) = 12\left(\frac{1}{4}x - \frac{7}{3}\right) \qquad \begin{array}{l}\text{Multiply by 12, the LCD of the}\\\text{fractions. (Section R.5)}\end{array}$$

$$12\left(\frac{2x + 4}{3}\right) + 12\left(\frac{1}{2}x\right) = 12\left(\frac{1}{4}x\right) - 12\left(\frac{7}{3}\right) \qquad \text{Distributive property}$$

$$4(2x + 4) + 6x = 3x - 28 \qquad \text{Multiply.}$$

$$8x + 16 + 6x = 3x - 28 \qquad \text{Distributive property}$$

$$14x + 16 = 3x - 28 \qquad \text{Combine like terms.}$$

$$11x = -44 \qquad \text{Subtract } 3x. \text{ Subtract 16.}$$

$$x = -4 \qquad \text{Divide each side by 11.}$$

CHECK $\quad \dfrac{2x+4}{3} + \dfrac{1}{2}x = \dfrac{1}{4}x - \dfrac{7}{3}$ $\qquad$ Original equation

$$\dfrac{2(-4)+4}{3} + \dfrac{1}{2}(-4) \overset{?}{=} \dfrac{1}{4}(-4) - \dfrac{7}{3} \qquad \text{Let } x = -4.$$

$$\dfrac{-4}{3} + (-2) \overset{?}{=} -1 - \dfrac{7}{3} \qquad \text{Simplify.}$$

$$-\dfrac{10}{3} = -\dfrac{10}{3} \;\checkmark \qquad \text{True}$$

The solution set is $\{-4\}$.

✔ *Now Try Exercise 19.*

Identities, Conditional Equations, and Contradictions   An equation satisfied by every number that is a meaningful replacement for the variable is an **identity.**

$$3(x+1) = 3x + 3 \quad \text{Identity}$$

An equation that is satisfied by some numbers but not others is a **conditional equation.**

$$2x = 4 \quad \text{Conditional equation}$$

The equations in **Examples 1 and 2** are conditional equations. An equation that has no solution is a **contradiction.**

$$x = x + 1 \quad \text{Contradiction}$$

EXAMPLE 3   **Identifying Types of Equations**

Determine whether each equation is an *identity,* a *conditional equation,* or a *contradiction.* Give the solution set.

(a) $-2(x+4) + 3x = x - 8$ **(b)** $5x - 4 = 11$ **(c)** $3(3x - 1) = 9x + 7$

SOLUTION

(a) $-2(x + 4) + 3x = x - 8$

$\qquad -2x - 8 + 3x = x - 8 \qquad$ Distributive property

$\qquad\qquad\quad x - 8 = x - 8 \qquad$ Combine like terms.

$\qquad\qquad\qquad\;\; 0 = 0 \qquad\quad$ Subtract x. Add 8.

When a *true* statement such as $0 = 0$ results, the equation is an identity, and the solution set is **{all real numbers}.**

(b) $5x - 4 = 11$

$\qquad\quad 5x = 15 \qquad$ Add 4 to each side.

$\qquad\quad\; x = 3 \qquad$ Divide each side by 5.

This is a conditional equation, and its solution set is $\{3\}$.

(c) $3(3x - 1) = 9x + 7$

$\qquad 9x - 3 = 9x + 7 \qquad$ Distributive property

$\qquad\quad -3 = 7 \qquad\qquad$ Subtract $9x$.

When a *false* statement such as $-3 = 7$ results, the equation is a contradiction, and the solution set is the **empty set,** or **null set,** symbolized $\varnothing$.

✔ *Now Try Exercises 29, 31, and 35.*

Identifying Types of Linear Equations

1. If solving a linear equation leads to a true statement such as $0 = 0$, the equation is an **identity**. Its solution set is **{all real numbers}**. **(See Example 3(a).)**

2. If solving a linear equation leads to a single solution such as $x = 3$, the equation is **conditional**. Its solution set consists of a single element. **(See Example 3(b).)**

3. If solving a linear equation leads to a false statement such as $-3 = 7$, the equation is a **contradiction**. Its solution set is $\emptyset$. **(See Example 3(c).)**

Solving for a Specified Variable (Literal Equations)   A formula is an example of a **literal equation** (an equation involving letters).

EXAMPLE 4   Solving for a Specified Variable

Solve for the specified variable.

(a) $I = Prt$,   for t **(b)** $A - P = Prt$,   for P

(c) $3(2x - 5a) + 4b = 4x - 2$,   for x

SOLUTION

(a) This is the formula for **simple interest** I on a principal amount of P dollars at an annual interest rate r for t years. Here, treat t as if it were the only variable, and the other variables as if they were constants.

$$I = Prt \quad \boxed{\text{Goal: Isolate } t \text{ on one side.}}$$

$$\frac{I}{Pr} = \frac{Prt}{Pr} \quad \text{Divide each side by } Pr.$$

$$\frac{I}{Pr} = t, \quad \text{or} \quad t = \frac{I}{Pr}$$

(b) The formula $A = P(1 + rt)$, which can also be written $A - P = Prt$, gives the **future value**, or **maturity value**, A of P dollars invested for t years at annual simple interest rate r.

$$A - P = Prt \quad \boxed{\text{Goal: Isolate } P, \text{ the specified variable.}}$$

$$A = P + Prt \qquad \text{Transform so that all terms involving } P \text{ are on one side.}$$

$$A = P(1 + rt) \qquad \text{Factor out } P. \text{ (Section R.4)}$$

$\boxed{\text{Pay close attention to this step.}}$

$$\frac{A}{1 + rt} = P, \quad \text{or} \quad P = \frac{A}{1 + rt} \qquad \text{Divide by } 1 + rt.$$

(c) $3(2x - 5a) + 4b = 4x - 2$ Solve for x.

$6x - 15a + 4b = 4x - 2$ Distributive property

$6x - 4x = 15a - 4b - 2$ Isolate the x-terms on one side.

$2x = 15a - 4b - 2$ Combine like terms.

$$x = \frac{15a - 4b - 2}{2} \qquad \text{Divide each side by 2.}$$

✓ *Now Try Exercises 39, 47, and 49.*

EXAMPLE 5 **Applying the Simple Interest Formula**

Becky Brugman borrowed \$5240 for new furniture. She will pay it off in 11 months at an annual simple interest rate of 4.5%. How much interest will she pay?

SOLUTION Use the simple interest formula $I = Prt$.

$$I = Prt = 5240(0.045)\left(\frac{11}{12}\right) = \$216.15 \qquad \begin{array}{l} P = 5240, r = 0.045, \\ \text{and } t = \frac{11}{12} \text{ (year)} \end{array}$$

She will pay \$216.15 interest on her purchase.

☑ *Now Try Exercise 59.*

1.1 Exercises

Concept Check In Exercises 1–4, decide whether each statement is true *or* false.

1. The solution set of $2x + 5 = x - 3$ is $\{-8\}$.

2. The equation $5(x - 8) = 5x - 40$ is an example of an identity.

3. The equations $x^2 = 4$ and $x + 2 = 4$ are equivalent equations.

4. It is possible for a linear equation to have exactly two solutions.

5. Explain the difference between an identity and a conditional equation.

6. Make a complete list of the steps needed to solve a linear equation. (Some equations will not require every step.)

7. *Concept Check* Which one is not a linear equation?

    A. $5x + 7(x - 1) = -3x$ **B.** $9x^2 - 4x + 3 = 0$

    C. $7x + 8x = 13x$ **D.** $0.04x - 0.08x = 0.40$

8. In solving the equation $3(2x - 8) = 6x - 24$, a student obtains the result $0 = 0$ and gives the solution set $\{0\}$. Is this correct? Explain.

Solve each equation. ***See Examples 1 and 2.***

9. $5x + 4 = 3x - 4$ **10.** $9x + 11 = 7x + 1$

11. $6(3x - 1) = 8 - (10x - 14)$ **12.** $4(-2x + 1) = 6 - (2x - 4)$

13. $\dfrac{5}{6}x - 2x + \dfrac{4}{3} = \dfrac{5}{3}$ **14.** $\dfrac{7}{4} + \dfrac{1}{5}x - \dfrac{3}{2} = \dfrac{4}{5}x$

15. $3x + 5 - 5(x + 1) = 6x + 7$ **16.** $5(x + 3) + 4x - 3 = -(2x - 4) + 2$

17. $2[x - (4 + 2x) + 3] = 2x + 2$ **18.** $4[2x - (3 - x) + 5] = -6x - 28$

19. $\dfrac{1}{14}(3x - 2) = \dfrac{x + 10}{10}$ **20.** $\dfrac{1}{15}(2x + 5) = \dfrac{x + 2}{9}$

21. $0.2x - 0.5 = 0.1x + 7$ **22.** $0.01x + 3.1 = 2.03x - 2.96$

23. $-4(2x - 6) + 8x = 5x + 24 + x$ **24.** $-8(3x + 4) + 6x = 4(x - 8) + 4x$

25. $0.5x + \dfrac{4}{3}x = x + 10$ **26.** $\dfrac{2}{3}x + 0.25x = x + 2$

27. $0.08x + 0.06(x + 12) = 7.72$ **28.** $0.04(x - 12) + 0.06x = 1.52$

Determine whether each equation is an identity, *a* conditional equation, *or a* contradiction. *Give the solution set.* ***See Example 3.***

29. $4(2x + 7) = 2x + 22 + 3(2x + 2)$ **30.** $\dfrac{1}{2}(6x + 20) = x + 4 + 2(x + 3)$

31. $2(x - 8) = 3x - 16$

32. $-8(x + 5) = -8x - 5(x + 8)$

33. $0.3(x + 2) - 0.5(x + 2) = -0.2x - 0.4$

34. $-0.6(x - 5) + 0.8(x - 6) = 0.2x - 1.8$

35. $4(x + 7) = 2(x + 12) + 2(x + 1)$

36. $-6(2x + 1) - 3(x - 4) = -15x + 1$

37. A student claims that the equation $5x = 4x$ is a contradiction, since dividing both sides by x leads to $5 = 4$, a false statement. Explain why the student is incorrect.

38. If $k \neq 0$, is the equation $x + k = x$ a contradiction, a conditional equation, or an identity? Explain.

Solve each formula for the indicated variable. Assume that the denominator is not 0 if variables appear in the denominator. **See Examples 4(a) and (b).**

39. $V = lwh$,   for l (volume of a rectangular box)

40. $I = Prt$,   for P (simple interest)

41. $P = a + b + c$,   for c (perimeter of a triangle)

42. $P = 2l + 2w$,   for w (perimeter of a rectangle)

43. $\mathcal{A} = \dfrac{1}{2}h(B + b)$,   for B (area of a trapezoid)

44. $\mathcal{A} = \dfrac{1}{2}h(B + b)$,   for h (area of a trapezoid)

45. $S = 2\pi rh + 2\pi r^2$,   for h (surface area of a right circular cylinder)

46. $s = \dfrac{1}{2}gt^2$,   for g (distance traveled by a falling object)

47. $S = 2lw + 2wh + 2hl$,   for h (surface area of a rectangular box)

48. Refer to **Exercise 45.** Why is it not possible to solve this formula for r using the methods of this section?

Solve each equation for x. **See Example 4(c).**

49. $2(x - a) + b = 3x + a$

50. $5x - (2a + c) = 4(x + c)$

51. $ax + b = 3(x - a)$

52. $4a - ax = 3b + bx$

53. $\dfrac{x}{a - 1} = ax + 3$

54. $\dfrac{x - 1}{2a} = 2x - a$

55. $a^2x + 3x = 2a^2$

56. $ax + b^2 = bx - a^2$

57. $3x = (2x - 1)(m + 4)$

58. $-x = (5x + 3)(3k + 1)$

Work each problem. **See Example 5.**

59. *Simple Interest* Elmer Velasquez borrowed $3150 from his brother Julio to pay for books and tuition. He agreed to repay Julio in 6 months with simple annual interest at 4%.

(a) How much will the interest amount to?

(b) What amount must Elmer pay Julio at the end of the 6 months?

60. *Simple Interest* Levada Qualls borrows $30,900 from her bank to open a florist shop. She agrees to repay the money in 18 months with simple annual interest of 5.5%.

(a) How much must she pay the bank in 18 months?

(b) How much of the amount in part (a) is interest?

Celsius and Fahrenheit Temperatures *In the metric system of weights and measures, temperature is measured in degrees Celsius (°C) instead of degrees Fahrenheit (°F). To convert between the two systems, we use the equations*

$$C = \frac{5}{9}(F - 32) \quad \text{and} \quad F = \frac{9}{5}C + 32.$$

In each exercise, convert to the other system. Round answers to the nearest tenth of a degree if necessary.

61. 40°C   **62.** 200°C   **63.** 50°F

64. 77°F   **65.** 100°F   **66.** 350°F

Work each problem.

67. *Temperature of Venus* Venus is the hottest planet with a surface temperature of 867°F. What is this temperature in Celsius? (*Source: World Almanac and Book of Facts.*)

68. *Temperature at Soviet Antarctica Station* A record low temperature of −89.4°C was recorded at the Soviet Antarctica Station of Vostok on July 21, 1983. Find the corresponding Fahrenheit temperature. (*Source: World Almanac and Book of Facts.*)

69. *Temperature in Montreal* The average low temperature in Montreal, Canada, for the date January 30 is 7°F. What is the corresponding Celsius temperature to the nearest tenth of a degree? (*Source:* www.wunderground.com)

70. *Temperature in Haiti* The average annual temperature in Port-au-Prince, Haiti, is approximately 28.1°C. What is the corresponding Fahrenheit temperature to the nearest tenth of a degree? (*Source:* www.climatetemp.info/haiti)

1.2 Applications and Modeling with Linear Equations

- Solving Applied Problems
- Geometry Problems
- Motion Problems
- Mixture Problems
- Modeling with Linear Equations

Solving Applied Problems One of the main reasons for learning mathematics is to be able use it to solve application problems. While there is no one method that enables us to solve all types of applied problems, the following six steps provide a useful guide.

Solving an Applied Problem

Step 1 **Read** the problem carefully until you understand what is given and what is to be found.

Step 2 **Assign a variable** to represent the unknown value, using diagrams or tables as needed. Write down what the variable represents. If necessary, express any other unknown values in terms of the variable.

Step 3 **Write an equation** using the variable expression(s).

Step 4 **Solve** the equation.

Step 5 **State the answer** to the problem. Does it seem reasonable?

Step 6 **Check** the answer in the words of the original problem.

EXAMPLE 1   **Finding the Dimensions of a Square**

If the length of each side of a square is increased by 3 cm, the perimeter of the new square is 40 cm more than twice the length of each side of the original square. Find the dimensions of the original square.

SOLUTION

Step 1 **Read** the problem. We must find the length of each side of the original square.

Step 2 **Assign a variable.** Since the length of a side of the original square is to be found, let the variable represent this length.

Let x = the length of a side of the original square in centimeters.

The length of a side of the new square is 3 cm more than the length of a side of the old square.

Then $x + 3$ = the length of a side of the new square.

See **Figure 1.** Now write a variable expression for the perimeter of the new square. The perimeter of a square is 4 times the length of a side.

Thus, $4(x + 3)$ = the perimeter of the new square.

Original   Side is increased
square   by 3.

x and $x + 3$ are in centimeters.

Figure 1

Step 3 **Write an equation.** Translate the English sentence that follows into its equivalent algebraic equation.

The new perimeter	is	40	more than	twice the length of a side of the original square.
$4(x + 3)$	$=$	40	$+$	$2x$

Step 4 **Solve** the equation.

$$4x + 12 = 40 + 2x \quad \text{Distributive property (Section R.2)}$$

$$2x = 28 \quad \text{Subtract } 2x \text{ and } 12. \text{ (Section 1.1)}$$

$$x = 14 \quad \text{Divide by 2. (Section 1.1)}$$

Step 5 **State the answer.** Each side of the original square measures 14 cm.

Step 6 **Check.** Go back to the words of the original problem to see that all necessary conditions are satisfied. The length of a side of the new square would be $14 + 3 = 17$ cm. The perimeter of the new square would be $4(17) = 68$ cm. Twice the length of a side of the original square would be $2(14) = 28$ cm. Since $40 + 28 = 68$, the answer checks.

✔ *Now Try Exercise 13.*

LOOKING AHEAD TO CALCULUS

In calculus the concept of the **definite integral** is used to find the distance traveled by an object traveling at a *non-constant* velocity.

PROBLEM-SOLVING HINT   In a motion problem, the components *distance, rate,* and *time* are denoted by the letters d, r, and t, respectively. (The *rate* is also called the *speed* or *velocity*. Here, rate is understood to be constant.) These variables are related by the equation

$$d = rt, \quad \text{and its related forms} \quad r = \frac{d}{t} \quad \text{and} \quad t = \frac{d}{r}.$$

EXAMPLE 2 Solving a Motion Problem

Maria and Eduardo are traveling to a business conference. The trip takes 2 hr for Maria and 2.5 hr for Eduardo, since he lives 40 mi farther away. Eduardo travels 5 mph faster than Maria. Find their average rates.

SOLUTION

Step 1 **Read** the problem. We must find Maria's and Eduardo's average rates.

Step 2 **Assign a variable.** Since average rates are to be found, we let the variable represent one of these rates.

$$\text{Let} \qquad x = \text{Maria's rate.}$$

Because Eduardo travels 5 mph faster than Maria, we can express his average rate using the same variable.

$$\text{Then} \quad x + 5 = \text{Eduardo's rate.}$$

The table below organizes the information given in the problem. The expressions in the last column were found by multiplying the corresponding rates and times.

	r	t	d
Maria	x	2	$2x$
Eduardo	$x + 5$	2.5	$2.5(x + 5)$

Summarize the given information in a table.

Use $d = rt$.

Step 3 **Write an equation.** Eduardo's distance traveled exceeds Maria's distance by 40 mi. Translate this relationship into its algebraic form.

$$\underbrace{2.5(x + 5)}_{\text{Eduardo's distance}} \quad \underset{\text{is}}{=} \quad \underbrace{2x + 40}_{\substack{\text{40 more} \\ \text{than Maria's.}}}$$

Step 4 **Solve.**
$$2.5x + 12.5 = 2x + 40 \qquad \text{Distributive property}$$
$$0.5x = 27.5 \qquad \text{Subtract } 2x \text{ and } 12.5.$$
$$x = 55 \qquad \text{Divide by } 0.5.$$

Step 5 **State the answer.** Maria's rate of travel is 55 mph, and Eduardo's rate is

$$55 + 5 = 60 \text{ mph.}$$

Step 6 **Check.** The diagram below shows that the conditions of the problem are satisfied.

Distance traveled by Maria: $2(55) = 110$ mi

Distance traveled by Eduardo: $2.5(60) = 150$ mi

$150 - 110 = 40$

✔ *Now Try Exercise 19.*

Mixture Problems   Problems involving mixtures of two types of the same substance, salt solution, candy, and so on, often involve percent.

George Polya (1887–1985)

Polya, a native of Budapest, Hungary, wrote more than 250 papers and a number of books. He proposed a general outline for solving applied problems in his classic book, *How to Solve It*.

PROBLEM-SOLVING HINT   In mixture problems involving solutions, the rate (percent) of concentration is multiplied by the quantity to get the amount of pure substance present. ***The concentration of the final mixture must be between the concentrations of the two solutions making up the mixture.***

EXAMPLE 3 **Solving a Mixture Problem**

Lisa Harmon is a chemist. She needs a 20% solution of alcohol. She has a 15% solution on hand, as well as a 30% solution. How many liters of the 15% solution should she add to 3 L of the 30% solution to obtain her 20% solution?

SOLUTION

Step 1 **Read** the problem. We must find the required number of liters of 15% alcohol solution.

Step 2 **Assign a variable.**

Let x = the number of liters of 15% solution to be added.

Figure 2 and the table show what is happening in the problem. The numbers in the last column were found by multiplying the strengths and the numbers of liters.

Figure 2

Strength	Liters of Solution	Liters of Pure Alcohol	
15%	x	$0.15x$	←
30%	3	$0.30(3)$	← Sum must equal
20%	$x + 3$	$0.20(x + 3)$	←

Step 3 **Write an equation.** The number of liters of pure alcohol in the 15% solution plus the number of liters in the 30% solution must equal the number of liters in the final 20% solution.

$$\underbrace{0.15x}_{\text{Liters of pure alcohol in 15\%}} + \underbrace{0.30(3)}_{\text{Liters of pure alcohol in 30\%}} = \underbrace{0.20(x + 3)}_{\text{Liters of pure alcohol in 20\%}}$$

Step 4 **Solve.** $0.15x + 0.30(3) = 0.20(x + 3)$

$\qquad\qquad 0.15x + 0.90 = 0.20x + 0.60$ Distributive property

$\qquad\qquad\qquad\quad 0.30 = 0.05x$ Subtract 0.60 and 0.15x.

$\qquad\qquad\qquad\qquad\quad 6 = x$ Divide by 0.05.

Step 5 **State the answer.** Thus, 6 L of 15% solution should be mixed with 3 L of 30% solution, giving $6 + 3 = 9$ L of 20% solution.

Step 6 **Check.** The answer checks, since the amount of alcohol in the two solutions is equal to the amount of alcohol in the mixture.

$$0.15(6) + 0.9 = 0.9 + 0.9 = 1.8 \quad \text{Solutions}$$

$$0.20(6 + 3) = 0.20(9) = 1.8 \quad \text{Mixture}$$

✔ *Now Try Exercise 29.*

PROBLEM-SOLVING HINT   In mixed investment problems, multiply each principal by the interest rate and the time in years to find the amount of interest earned.

EXAMPLE 4   **Solving an Investment Problem**

An artist has sold a painting for $410,000. He needs some of the money in 6 months and the rest in 1 yr. He can get a Treasury bond for 6 months at 2.65% and one for a year at 2.91%. His broker tells him the two investments will earn a total of $8761. How much should be invested at each rate to obtain that amount of interest?

SOLUTION

Step 1 **Read** the problem. We must find the amount to be invested at each rate.

Step 2 **Assign a variable.**

Let $x =$ the dollar amount to be invested for 6 months at 2.65%.

$410,000 - x =$ the dollar amount to be invested for 1 yr at 2.91%.

Invested Amount	Interest Rate (%)	Time (in years)	Interest Earned
x	2.65	0.5	$x(0.0265)(0.5)$
$410,000 - x$	2.91	1	$(410,000 - x)(0.0291)(1)$

Summarize this information in a table using the formula $I = Prt$.

Step 3 **Write an equation.** The sum of the two interest amounts must equal the total interest earned.

$$\underbrace{0.5x(0.0265)}_{\text{Interest from 2.65\% investment}} + \underbrace{0.0291(410,000 - x)}_{\text{Interest from 2.91\% investment}} = \underbrace{8761}_{\text{Total interest}}$$

Step 4 **Solve.**

$$0.5x(0.0265) + 0.0291(410,000 - x) = 8761$$

$$0.01325x + 11,931 - 0.0291x = 8761 \qquad \text{Distributive property}$$

$$11,931 - 0.01585x = 8761 \qquad \text{Combine like terms.}$$

$$-0.01585x = -3170 \qquad \text{Subtract 11,931.}$$

$$x = 200,000 \qquad \text{Divide by } -0.01585.$$

Step 5 **State the answer.** The artist should invest $200,000 at 2.65% for 6 months and

$$\$410,000 - \$200,000 = \$210,000$$

at 2.91% for 1 yr to earn $8761 in interest.

Step 6 **Check.** The 6-month investment earns

$$\$200,000(0.0265)(0.5) = \$2650,$$

while the 1-yr investment earns

$$\$210,000(0.0291)(1) = \$6111.$$

The total amount of interest earned is

$$\$2650 + \$6111 = \$8761, \quad \text{as required.}$$

☑ *Now Try Exercise 35.*

Modeling with Linear Equations   A **mathematical model** is an equation (or inequality) that describes the relationship between two quantities. A **linear model** is a linear equation. The next example shows how a linear model is applied.

EXAMPLE 5 **Modeling Prevention of Indoor Pollutants**

If a vented range hood removes contaminants such as carbon monoxide and nitrogen dioxide from the air at a rate of F liters of air per second, then the percent P of contaminants that are also removed from the surrounding air can be modeled by the linear equation

$$P = 1.06F + 7.18, \quad \text{where } 10 \leq F \leq 75.$$

What flow F must a range hood have to remove 50% of the contaminants from the air? (*Source: Proceedings of the Third International Conference on Indoor Air Quality and Climate.*)

SOLUTION Replace P with 50 in the linear model, and solve for F.

$$P = 1.06F + 7.18 \quad \text{Given model}$$
$$50 = 1.06F + 7.18 \quad \text{Let } P = 50.$$
$$42.82 = 1.06F \quad \text{Subtract 7.18.}$$
$$F \approx 40.40 \quad \text{Divide by 1.06.}$$

Therefore, to remove 50% of the contaminants, the flow rate must be approximately 40.40 L of air per second.

✔ *Now Try Exercise 41.*

EXAMPLE 6 **Modeling Health Care Costs**

The projected per capita health care expenditures in the United States, where y is in dollars, and x is years after 2000, are given by the following linear equation.

$$y = 343x + 4512 \quad \text{Linear model}$$

(*Source:* Centers for Medicare and Medicaid Services.)

(a) What were the per capita health care expenditures in the year 2010?

(b) If this model continues to describe health care expenditures, when will the per capita expenditures reach $9200?

SOLUTION In part (a) we are given information to determine a value for x and asked to find the corresponding value of y, whereas in part (b) we are given a value for y and asked to find the corresponding value of x.

(a) The year 2010 is 10 yr after the year 2000. Let $x = 10$ and find the value of y.

$$y = 343x + 4512 \quad \text{Given model}$$
$$y = 343(10) + 4512 \quad \text{Let } x = 10.$$
$$y = 7942 \quad \text{Multiply and then add.}$$

In 2010, the estimated per capita health care expenditures were $7942.

(b) Let $y = 9200$ in the given model, and find the value of x.

$$9200 = 343x + 4512 \quad \text{Let } y = 9200.$$
$$4688 = 343x \quad \text{Subtract 4512.}$$
$$x \approx 13.7 \quad \text{Divide by 343.}$$

13 corresponds to 2000 + 13 = 2013.

The x-value of 13.7 indicates that per capita health care expenditures are projected to reach $9200 during the 13th year after 2000—that is, 2013.

✔ *Now Try Exercise 45.*

| 1.2 | **Exercises** |

Concept Check Exercises 1–8 should be done mentally. They will prepare you for some of the applications found in this exercise set.

1. If a train travels at 80 mph for 15 min, what is the distance traveled?

2. If 120 L of an acid solution is 75% acid, how much pure acid is there in the mixture?

3. If a person invests $500 at 2% simple interest for 4 yr, how much interest is earned?

4. If a jar of coins contains 40 half-dollars and 200 quarters, what is the monetary value of the coins?

5. *Acid Mixture* Suppose two acid solutions are mixed. One is 26% acid and the other is 34% acid. Which one of the following concentrations cannot possibly be the concentration of the mixture?

   A. 24%   **B.** 30%   **C.** 31%   **D.** 33%

6. *Sale Price* Suppose that a computer that originally sold for x dollars has been discounted 60%. Which one of the following expressions does not represent its sale price?

   A. $x - 0.60x$ **B.** $0.40x$ **C.** $\dfrac{4}{10}x$ **D.** $x - 0.60$

7. *Unknown Numbers* Consider the following problem.

   > One number is 3 less than 6 times a second number. Their sum is 46. Find the numbers.

   If x represents the second number, which equation is correct for solving this problem?

   A. $46 - (x + 3) = 6x$ **B.** $(3 - 6x) + x = 46$
   C. $46 - (3 - 6x) = x$ **D.** $(6x - 3) + x = 46$

8. *Unknown Numbers* Consider the following problem.

   > The difference between seven times a number and 9 is equal to five times the sum of the number and 2. Find the number.

   If x represents the number, which equation is correct for solving this problem?

   A. $7x - 9 = 5(x + 2)$ **B.** $9 - 7x = 5(x + 2)$
   C. $7x - 9 = 5x + 2$ **D.** $9 - 7x = 5x + 2$

Note: **Geometry formulas can be found on the back inside cover of this book.**

*Solve each problem. **See Example 1.***

9. *Perimeter of a Rectangle* The perimeter of a rectangle is 294 cm. The width is 57 cm. Find the length.

10. *Perimeter of a Storage Shed* Michael Gomski must build a rectangular storage shed. He wants the length to be 6 ft greater than the width, and the perimeter will be 44 ft. Find the length and the width of the shed.

11. *Dimensions of a Puzzle Piece* A puzzle piece in the shape of a triangle has perimeter 30 cm. Two sides of the triangle are each twice as long as the shortest side. Find the length of the shortest side. (Side lengths in the figure are in centimeters.)

12. *Dimensions of a Label* The length of a rectangular label is 2.5 cm less than twice the width. The perimeter is 40.6 cm. Find the width. (Side lengths in the figure are in centimeters.)

13. *Perimeter of a Plot of Land* The perimeter of a triangular plot of land is 2400 ft. The longest side is 200 ft less than twice the shortest. The middle side is 200 ft less than the longest side. Find the lengths of the three sides of the triangular plot.

14. *World Largest Ice Cream Cake* The world's largest ice cream cake, made at the Baxy ice cream factory in Beijing, China, on January 16, 2006, had length 5.9 ft greater than its width. Its perimeter was 51 ft. What were the length and width of this 8-ton cake? (*Sources:* www.chinadaily.com.cn, www.foodmall.org)

15. *World's Smallest Quran* The world's smallest Quran, published in Cairo, Egypt, in 1982 has width 0.42 cm shorter than its length. The book's perimeter is 5.96 cm. What are the width and length of this Quran, in centimeters? (*Source:* www.guinnessworldrecords.com)

16. *Cylinder Dimensions* A right circular cylinder has radius 6 in. and volume 144π in.^{3}. What is its height? (In the figure, h = height.)

h is in inches.

17. *Recycling Bin Dimensions* A recycling bin is in the shape of a rectangular box. Find the height of the box if its length is 18 ft, its width is 8 ft, and its surface area is 496 ft^2. (In the figure, h = height. Assume that the given surface area includes that of the top lid of the box.)

18. *Concept Check* Which one or more of the following cannot be a correct equation to solve a geometry problem, if x represents the length of a rectangle? (*Hint:* Solve each equation and consider the solution.)

A. $2x + 2(x - 1) = 14$ **B.** $-2x + 7(5 - x) = 52$

C. $5(x + 2) + 5x = 10$ **D.** $2x + 2(x - 3) = 22$

Solve each problem. **See Example 2.**

19. *Distance to an Appointment* Margaret drove to a business appointment at 50 mph. Her average speed on the return trip was 40 mph. The return trip took $\frac{1}{4}$ hr longer because of heavy traffic. How far did she travel to the appointment?

	r	t	d
Morning	50	x	
Afternoon	40	$x + \frac{1}{4}$	

20. *Distance between Cities* On a vacation, Elwyn averaged 50 mph traveling from Denver to Minneapolis. Returning by a different route that covered the same number of miles, he averaged 55 mph. What is the distance between the two cities if his total traveling time was 32 hr?

	r	t	d
Going	50	x	
Returning	55	$32 - x$	

21. *Distance to Work* David gets to work in 20 min when he drives his car. Riding his bike (by the same route) takes him 45 min. His average driving speed is 4.5 mph greater than his average speed on his bike. How far does he travel to work?

22. *Speed of a Plane* Two planes leave Los Angeles at the same time. One heads south to San Diego, while the other heads north to San Francisco. The San Diego plane flies 50 mph slower than the San Francisco plane. In $\frac{1}{2}$ hr, the planes are 275 mi apart. What are their speeds?

23. *Running Times* Mary and Janet are running in the Apple Hill Fun Run. Mary runs at 7 mph, Janet at 5 mph. If they start at the same time, how long will it be before they are 1.5 mi apart?

24. *Running Times* If the run in **Exercise 23** has a staggered start, and Janet starts first, with Mary starting 10 min later, how long will it be before Mary catches up with Janet?

25. *Track Event Speeds* At the 2008 Summer Olympics in Beijing, China, Usain Bolt (Jamaica) set a new Olympic and world record in the 100-m dash with a time of 9.69 sec. If this pace could be maintained for an entire 26-mi marathon, what would his time be? How would this time compare to the fastest time for a marathon, which is 2 hr, 3 min, 59 sec, also set in 2008? (*Hint:* 1 m ≈ 3.281 ft.) (*Source: World Almanac and Book of Facts.*)

26. *Track Event Speeds* On August 19, 2009, at the World Track and Field Championship in Berlin, Usain Bolt broke his own record in the 100-m dash with a time of 9.58 sec. Refer to **Exercise 25** and answer the questions using Bolt's 2009 time. (*Source:* www.sports.espn.go.com)

27. *Boat Speed* Callie took 20 min to drive her boat upstream to water-ski at her favorite spot. Coming back later in the day, at the same boat speed, took her 15 min. If the current in that part of the river is 5 km per hr, what was her boat speed?

28. *Wind Speed* Joe traveled against the wind in a small plane for 3 hr. The return trip with the wind took 2.8 hr. Find the speed of the wind if the speed of the plane in still air is 180 mph.

*Solve each problem. **See Example 3.***

29. *Acid Mixture* How many gallons of a 5% acid solution must be mixed with 5 gal of a 10% solution to obtain a 7% solution?

Strength	Gallons of Solution	Gallons of Pure Acid
5%	x	
10%	5	
7%	$x + 5$	

30. *Acid Mixture* Marin Caswell needs 10% hydrochloric acid for a chemistry experiment. How much 5% acid should she mix with 60 mL of 20% acid to get a 10% solution?

Strength	mL of Solution	mL of Pure Acid
5%	x	
20%	60	
10%	$x + 60$	

31. *Alcohol Mixture* Beau Glaser wishes to strengthen a mixture from 10% alcohol to 30% alcohol. How much pure alcohol should be added to 7 L of the 10% mixture?

32. *Alcohol Mixture* How many gallons of pure alcohol should be mixed with 20 gal of a 15% alcohol solution to obtain a mixture that is 25% alcohol?

33. *Saline Solution* How much water should be added to 8 mL of 6% saline solution to reduce the concentration to 4%?

34. *Acid Mixture* How much pure acid should be added to 18 L of 30% acid to increase the concentration to 50% acid?

Solve each problem. **See Example 4.**

35. *Real Estate Financing* Cody Westmoreland wishes to sell a piece of property for $240,000. He wants the money to be paid off in two ways—a short-term note at 2% interest and a long-term note at 2.5%. Find the amount of each note if the total annual interest paid is $5500.

Note Amount	Interest Rate (%)	Time (in years)	Interest Paid
x	2	1	$x(0.02)(1)$
$240{,}000 - x$	2.5	1	$(240{,}000 - x)(0.025)(1)$

36. *Buying and Selling Land* Roger bought two plots of land for a total of $120,000. When he sold the first plot, he made a profit of 15%. When he sold the second, he lost 10%. His total profit was $5500. How much did he pay for each piece of land?

37. *Retirement Planning* In planning her retirement, Janet Karrenbrock deposits some money at 2.5% interest, with twice as much deposited at 3%. Find the amount deposited at each rate if the total annual interest income is $850.

38. *Investing a Building Fund* A church building fund has invested some money in two ways: part of the money at 3% interest and four times as much at 2.75%. Find the amount invested at each rate if the total annual income from interest is $2800.

39. *Lottery Winnings* Linda won $200,000 in a state lottery. She first paid income tax of 30% on the winnings. She invested some of the rest at 1.5% and some at 4%, earning $4350 interest per year. How much did she invest at each rate?

40. *Cookbook Royalties* Becky Schantz earned $48,000 from royalties on her cookbook. She paid a 28% income tax on these royalties. The balance was invested in two ways, some of it at 3.25% interest and some at 1.75%. The investments produced $904.80 interest per year. Find the amount invested at each rate.

(Modeling) *Solve each problem.* **See Examples 5 and 6.**

41. *Warehouse Club Membership* Membership warehouse clubs offer shoppers low prices, along with rewards of cash back on club purchases. If the yearly fee for a warehouse club membership is $100 and the reward rate is 2% on club purchases for the year, then the linear equation

$$y = 100 - 0.02x$$

models the actual yearly cost of the membership y, in dollars. Here x represents the yearly amount of club purchases, also in dollars.

(a) Determine the actual yearly cost of the membership if club purchases for the year are $2400.

(b) What amount of club purchases would reduce the actual yearly cost of the membership to $50?

(c) How much would a member have to spend in yearly club purchases to reduce the yearly membership cost to $0?

42. *Warehouse Club Membership* Suppose that the yearly fee for a warehouse club membership is $50 and that the reward rate on club purchases for the year is 1.6%. Then the actual yearly cost of a membership *y*, in dollars, for an amount of yearly club purchases *x*, in dollars, can be modeled by the following linear equation.

$$y = 50 - 0.016x$$

(a) Determine the actual yearly cost of the membership if club purchases for the year are $1500.

(b) What amount of club purchases would reduce the actual yearly cost of the membership to $0?

(c) If club purchases for the year exceed $3125, how is the actual yearly membership cost affected?

43. *Indoor Air Pollution* Formaldehyde is an indoor air pollutant formerly found in plywood, foam insulation, and carpeting. When concentrations in the air reach 33 micrograms per cubic foot ($\mu g/ft^3$), eye irritation can occur. One square foot of new plywood could emit 140 μg per hr. (*Source:* A. Hines, *Indoor Air Quality & Control.*)

(a) A room has 100 ft^2 of new plywood flooring. Find a linear equation *F* that computes the amount of formaldehyde, in micrograms, emitted in *x* hours.

(b) The room contains 800 ft^3 of air and has no ventilation. Determine how long it would take for concentrations to reach 33 $\mu g/ft^3$.

44. *Classroom Ventilation* According to the American Society of Heating, Refrigerating and Air-Conditioning Engineers, Inc. (ASHRAE), a nonsmoking classroom should have a ventilation rate of 15 ft^3 per min for each person in the room.

(a) Write an equation that models the total ventilation *V* (in cubic feet per hour) necessary for a classroom with *x* students.

(b) A common unit of ventilation is air change per hour (ach). 1 ach is equivalent to exchanging all of the air in a room every hour. If *x* students are in a classroom having volume 15,000 ft^3, determine how many air exchanges per hour (*A*) are necessary to keep the room properly ventilated.

(c) Find the necessary number of ach (*A*) if the classroom has 40 students in it.

(d) In areas like bars and lounges that allow smoking, the ventilation rate should be increased to 50 ft^3 per min per person. Compared to classrooms, ventilation should be increased by what factor in heavy smoking areas?

45. *College Enrollments* The graph shows the projections in total enrollment at degree-granting institutions from fall 2009 to fall 2018.

Enrollments at Degree-Granting Institutions

Source: U.S. Department of Education, National Center for Education Statistics.

The following linear model provides the approximate enrollment, in millions, between the years 2009 and 2018, where *x* = 0 corresponds to 2009, *x* = 1 to 2010, and so on, and *y* is in millions of students.

$$y = 0.2309x + 18.35$$

(a) Use the model to determine projected enrollment for fall 2014.

(b) Use the model to determine the year in which enrollment is projected to reach 20 million.

(c) How do your answers to parts (a) and (b) compare to the corresponding values shown in the graph?

(d) The actual enrollment in fall 1997 was 14.5 million. The model here is based on data from 2009 to 2018. If you were to use the model for 1997, what would the projected enrollment be?

(e) Why do you think there is such a discrepancy between the actual value and the value based on the model in part (d)? Discuss the pitfalls of using the model to predict enrollment for years preceding 2009.

1.3 Complex Numbers

- Basic Concepts of Complex Numbers
- Operations on Complex Numbers

Basic Concepts of Complex Numbers The set of real numbers does not include all the numbers needed in algebra. For example, there is no real number solution of the equation

$$x^2 = -1,$$

since no real number, when squared, gives -1. To extend the real number system to include solutions of equations of this type, the number i is defined to have the following property.

The Imaginary Unit i

$$i = \sqrt{-1}, \quad \text{and therefore,} \quad i^2 = -1.$$

(Note that $-i$ is also a square root of -1.)

Square roots of negative numbers were not incorporated into an integrated number system until the 16th century. They were then used as solutions of equations and later (in the 18th century) in surveying. Today, such numbers are used extensively in science and engineering.

Complex numbers are formed by adding real numbers and multiples of i.

Complex Number

If a and b are real numbers, then any number of the form **$a + bi$** is a **complex number.** In the complex number $a + bi$, a is the **real part** and b is the **imaginary part.***

Two complex numbers $a + bi$ and $c + di$ are equal provided that their real parts are equal and their imaginary parts are equal; that is, they are equal if and only if $a = c$ and $b = d$.

---

* In some texts, the term bi is defined to be the imaginary part.

The calculator is in complex number mode. The top screen supports the definition of i. The bottom screen shows how the calculator returns the real and imaginary parts of the complex number $7 + 2i$.

Figure 3

Some graphing calculators, such as the TI 83/84 Plus, are capable of working with complex numbers, as seen in **Figure 3.** ■

For a complex number $a + bi$, if $b = 0$, then $a + bi = a$, which is a real number. Thus, the set of real numbers is a subset of the set of complex numbers. If $a = 0$ and $b \neq 0$, the complex number is said to be a **pure imaginary number.** For example, $3i$ is a pure imaginary number. A pure imaginary number, or a number such as $7 + 2i$ with $a \neq 0$ and $b \neq 0$, is a **nonreal complex number.** A complex number written in the form $a + bi$ (or $a + ib$) is in **standard form.** (The form $a + ib$ is used to write expressions such as $i\sqrt{5}$, since $\sqrt{5}i$ could be mistaken for $\sqrt{5i}$.)

The relationships among the subsets of the complex numbers are shown in **Figure 4.**

Figure 4

For a positive real number a, the expression $\sqrt{-a}$ is defined as follows.

The Expression $\sqrt{-a}$

If $a > 0$, then $\qquad\qquad \sqrt{-a} = i\sqrt{a}.$

EXAMPLE 1   Writing $\sqrt{-a}$ as $i\sqrt{a}$

Write as the product of a real number and i, using the definition of $\sqrt{-a}$.

(a) $\sqrt{-16}$ **(b)** $\sqrt{-70}$ **(c)** $\sqrt{-48}$

SOLUTION

(a) $\sqrt{-16} = i\sqrt{16} = 4i$ **(b)** $\sqrt{-70} = i\sqrt{70}$

(c) $\sqrt{-48} = i\sqrt{48} = i\sqrt{16 \cdot 3} = 4i\sqrt{3}$ Product rule for radicals **(Section R.7)**

☑ *Now Try Exercises 17, 19, and 21.*

Operations on Complex Numbers   Products or quotients with negative radicands are simplified by first rewriting $\sqrt{-a}$ as $i\sqrt{a}$ for a positive number a. Then the properties of real numbers and the fact that $i^2 = -1$ are applied.

CAUTION *When working with negative radicands, use the definition* $\sqrt{-a} = i\sqrt{a}$ *before using any of the other rules for radicals.* In particular, the rule $\sqrt{c} \cdot \sqrt{d} = \sqrt{cd}$ is valid only when c and d are *not* both negative. For example,

$$\sqrt{-4} \cdot \sqrt{-9} = 2i \cdot 3i = 6i^2 = -6 \quad \text{is correct,}$$

while $\quad \sqrt{-4} \cdot \sqrt{-9} = \sqrt{(-4)(-9)} = \sqrt{36} = 6 \quad$ is incorrect.

EXAMPLE 2 Finding Products and Quotients Involving $\sqrt{-a}$

Multiply or divide, as indicated. Simplify each answer.

(a) $\sqrt{-7} \cdot \sqrt{-7}$ **(b)** $\sqrt{-6} \cdot \sqrt{-10}$ **(c)** $\dfrac{\sqrt{-20}}{\sqrt{-2}}$ **(d)** $\dfrac{\sqrt{-48}}{\sqrt{24}}$

SOLUTION

(a) $\sqrt{-7} \cdot \sqrt{-7} = i\sqrt{7} \cdot i\sqrt{7}$ **(b)** $\sqrt{-6} \cdot \sqrt{-10} = i\sqrt{6} \cdot i\sqrt{10}$

First write all square roots in terms of *i*.
$$= i^2 \cdot \left(\sqrt{7}\right)^2 \qquad\qquad = i^2 \cdot \sqrt{60}$$
$$= -1 \cdot 7 \qquad\qquad\qquad = -1\sqrt{4 \cdot 15}$$
$$\qquad\qquad i^2 = -1 \qquad\qquad\qquad = -1 \cdot 2\sqrt{15}$$
$$= -7 \qquad\qquad\qquad\qquad = -2\sqrt{15}$$

(c) $\dfrac{\sqrt{-20}}{\sqrt{-2}} = \dfrac{i\sqrt{20}}{i\sqrt{2}} = \sqrt{\dfrac{20}{2}} = \sqrt{10}$ Quotient rule for radicals **(Section R.7)**

(d) $\dfrac{\sqrt{-48}}{\sqrt{24}} = \dfrac{i\sqrt{48}}{\sqrt{24}} = i\sqrt{\dfrac{48}{24}} = i\sqrt{2}$

✔ *Now Try Exercises 25, 27, 29, and 31.*

EXAMPLE 3 Simplifying a Quotient Involving $\sqrt{-a}$

Write $\dfrac{-8 + \sqrt{-128}}{4}$ in standard form $a + bi$.

SOLUTION

$$\frac{-8 + \sqrt{-128}}{4} = \frac{-8 + \sqrt{-64 \cdot 2}}{4}$$

$$= \frac{-8 + 8i\sqrt{2}}{4} \qquad \sqrt{-64} = 8i$$

Be sure to factor before simplifying.
$$= \frac{4\left(-2 + 2i\sqrt{2}\right)}{4} \qquad \text{Factor. (Section R.4)}$$

$$= -2 + 2i\sqrt{2} \qquad \text{Lowest terms (Section R.5)}$$

✔ *Now Try Exercise 37.*

With the definitions $i^2 = -1$ and $\sqrt{-a} = i\sqrt{a}$ for $a > 0$, all properties of real numbers are extended to complex numbers. As a result, complex numbers are added, subtracted, multiplied, and divided using real number properties and the definitions on the following pages.

Addition and Subtraction of Complex Numbers

For complex numbers $a + bi$ and $c + di$,

$$(a + bi) + (c + di) = (a + c) + (b + d)i$$

and $\qquad (a + bi) - (c + di) = (a - c) + (b - d)i.$

That is, to add or subtract complex numbers, add or subtract the real parts and add or subtract the imaginary parts.

EXAMPLE 4 Adding and Subtracting Complex Numbers

Find each sum or difference.

(a) $(3 - 4i) + (-2 + 6i)$ **(b)** $(-4 + 3i) - (6 - 7i)$

SOLUTION

Add real parts. Add imaginary parts.

(a) $(3 - 4i) + (-2 + 6i) = \left[3 + (-2)\right] + \left[-4 + 6\right]i$ Commutative, associative, distributive properties **(Section R.2)**

$$= 1 + 2i$$

(b) $(-4 + 3i) - (6 - 7i) = (-4 - 6) + \left[3 - (-7)\right]i$

$$= -10 + 10i$$

✔️ *Now Try Exercises 43 and 45.*

The product of two complex numbers is found by multiplying as though the numbers were binomials and using the fact that $i^2 = -1$, as follows.

$$(a + bi)(c + di) = ac + adi + bic + bidi \qquad \text{FOIL (Section R.3)}$$

$$= ac + adi + bci + bdi^2 \qquad \text{Associative property}$$

$$= ac + (ad + bc)i + bd(-1) \qquad \text{Distributive property; } i^2 = -1$$

$$= (ac - bd) + (ad + bc)i \qquad \text{Group like terms.}$$

Multiplication of Complex Numbers

For complex numbers $a + bi$ and $c + di$,

$$(a + bi)(c + di) = (ac - bd) + (ad + bc)i.$$

This definition is not practical in routine calculations. To find a given product, it is easier just to multiply as with binomials.

EXAMPLE 5 Multiplying Complex Numbers

Find each product.

(a) $(2 - 3i)(3 + 4i)$ **(b)** $(4 + 3i)^2$ **(c)** $(6 + 5i)(6 - 5i)$

SOLUTION

(a) $(2 - 3i)(3 + 4i) = 2(3) + 2(4i) - 3i(3) - 3i(4i)$ FOIL

$$= 6 + 8i - 9i - 12i^2 \qquad \text{Multiply.}$$

$$= 6 - i - 12(-1) \qquad \text{Combine like terms;} \\ i^2 = -1$$

$$= 18 - i \qquad \text{Standard form}$$

(b) $(4 + 3i)^2 = 4^2 + 2(4)(3i) + (3i)^2$ Square of a binomial **(Section R.3)**

> Remember to add twice the product of the two terms.

$$= 16 + 24i + 9i^2$$ Multiply.

$$= 16 + 24i + 9(-1)$$ $i^2 = -1$

$$= 7 + 24i$$ Standard form

(c) $(6 + 5i)(6 - 5i) = 6^2 - (5i)^2$ Product of the sum and difference of two terms **(Section R.3)**

$$= 36 - 25(-1)$$ Square 6 and 5; $i^2 = -1$.

$$= 36 + 25$$ Multiply.

$$= 61, \quad \text{or} \quad 61 + 0i$$ Standard form

✔ *Now Try Exercises 51, 55, and 59.*

```
(2-3i)(3+4i)
            18-i
(4+3i)²
            7+24i
(6+5i)(6-5i)
            61
```

This screen shows how the TI–83/84 Plus displays the results found in **Example 5.**

Example 5(c) showed that $(6 + 5i)(6 - 5i) = 61$. The numbers $6 + 5i$ and $6 - 5i$ differ only in the sign of their imaginary parts and are called **complex conjugates**. *The product of a complex number and its conjugate is always a real number.* This product is the sum of the squares of the real and imaginary parts.

Property of Complex Conjugates

For real numbers a and b,

$$(a + bi)(a - bi) = a^2 + b^2.$$

To find the quotient of two complex numbers in standard form, we multiply both the numerator and the denominator by the complex conjugate of the denominator.

EXAMPLE 6 Dividing Complex Numbers

Write each quotient in standard form $a + bi$.

(a) $\dfrac{3 + 2i}{5 - i}$ **(b)** $\dfrac{3}{i}$

SOLUTION

(a) $\dfrac{3 + 2i}{5 - i} = \dfrac{(3 + 2i)(5 + i)}{(5 - i)(5 + i)}$ Multiply by the complex conjugate of the denominator in both the numerator and the denominator.

$$= \frac{15 + 3i + 10i + 2i^2}{25 - i^2}$$ Multiply.

$$= \frac{13 + 13i}{26}$$ Combine like terms; $i^2 = -1$

$$= \frac{13}{26} + \frac{13i}{26}$$ $\frac{a + bi}{c} = \frac{a}{c} + \frac{bi}{c}$ **(Section R.5)**

$$= \frac{1}{2} + \frac{1}{2}i$$ Write in lowest terms and standard form.

CHECK $\left(\dfrac{1}{2} + \dfrac{1}{2}i\right)(5 - i) = 3 + 2i$ ✓ Quotient × Divisor = Dividend

This screen supports the results in **Example 6.**

Powers of i can be found on the TI–83/84 Plus calculator.

(b) $\dfrac{3}{i} = \dfrac{3(-i)}{i(-i)}$ $-i$ is the conjugate of i.

$\quad = \dfrac{-3i}{-i^2}$ Multiply.

$\quad = \dfrac{-3i}{1}$ $-i^2 = -(-1) = 1$

$\quad = -3i,$ or $0 - 3i$ Standard form

☑ *Now Try Exercises 69 and 75.*

Powers of i can be simplified using the facts

$$i^2 = -1 \quad \text{and} \quad i^4 = (i^2)^2 = (-1)^2 = 1.$$

Consider the following powers of i.

$i^1 = i$	$i^5 = i^4 \cdot i = 1 \cdot i = i$
$i^2 = -1$	$i^6 = i^4 \cdot i^2 = 1(-1) = -1$
$i^3 = i^2 \cdot i = (-1) \cdot i = -i$	$i^7 = i^4 \cdot i^3 = 1 \cdot (-i) = -i$
$i^4 = i^2 \cdot i^2 = (-1)(-1) = 1$	$i^8 = i^4 \cdot i^4 = 1 \cdot 1 = 1$ and so on.

Powers of i cycle through the same four outcomes $(i, -1, -i,$ and $1)$ since i^4 has the same multiplicative property as 1. Also, any power of i with an exponent that is a multiple of 4 has value 1. As with real numbers, $i^0 = 1$.

EXAMPLE 7   **Simplifying Powers of i**

Simplify each power of i.

(a) i^{15} **(b)** i^{-3}

SOLUTION

(a) Since $i^4 = 1$, write the given power as a product involving i^4.

$$i^{15} = i^{12} \cdot i^3 = (i^4)^3 \cdot i^3 = 1^3(-i) = -i$$

(b) Multiply i^{-3} by 1 in the form of i^4 to create the least positive exponent for i.

$$i^{-3} = i^{-3} \cdot 1 = i^{-3} \cdot i^4 = i \quad i^4 = 1$$

☑ *Now Try Exercises 85 and 93.*

1.3   Exercises

Concept Check   *Determine whether each statement is* true *or* false. *If it is false, tell why.*

1. Every real number is a complex number.

2. No real number is a pure imaginary number.

3. Every pure imaginary number is a complex number.

4. A number can be both real and complex.

5. There is no real number that is a complex number.

6. A complex number might not be a pure imaginary number.

Identify each number as real, complex, pure imaginary, *or* nonreal complex. *(More than one of these descriptions will apply.)*

7. -4 **8.** 0 **9.** $13i$ **10.** $-7i$ **11.** $5 + i$

12. $-6 - 2i$ **13.** π **14.** $\sqrt{24}$ **15.** $\sqrt{-25}$ **16.** $\sqrt{-36}$

Write each number as the product of a real number and i. **See Example 1.**

17. $\sqrt{-25}$ **18.** $\sqrt{-36}$ **19.** $\sqrt{-10}$ **20.** $\sqrt{-15}$

21. $\sqrt{-288}$ **22.** $\sqrt{-500}$ **23.** $-\sqrt{-18}$ **24.** $-\sqrt{-80}$

Multiply or divide, as indicated. Simplify each answer. **See Example 2.**

25. $\sqrt{-13} \cdot \sqrt{-13}$ **26.** $\sqrt{-17} \cdot \sqrt{-17}$ **27.** $\sqrt{-3} \cdot \sqrt{-8}$

28. $\sqrt{-5} \cdot \sqrt{-15}$ **29.** $\dfrac{\sqrt{-30}}{\sqrt{-10}}$ **30.** $\dfrac{\sqrt{-70}}{\sqrt{-7}}$

31. $\dfrac{\sqrt{-24}}{\sqrt{8}}$ **32.** $\dfrac{\sqrt{-54}}{\sqrt{27}}$ **33.** $\dfrac{\sqrt{-10}}{\sqrt{-40}}$

34. $\dfrac{\sqrt{-8}}{\sqrt{-72}}$ **35.** $\dfrac{\sqrt{-6} \cdot \sqrt{-2}}{\sqrt{3}}$ **36.** $\dfrac{\sqrt{-12} \cdot \sqrt{-6}}{\sqrt{8}}$

Write each number in standard form a + bi. **See Example 3.**

37. $\dfrac{-6 - \sqrt{-24}}{2}$ **38.** $\dfrac{-9 - \sqrt{-18}}{3}$ **39.** $\dfrac{10 + \sqrt{-200}}{5}$

40. $\dfrac{20 + \sqrt{-8}}{2}$ **41.** $\dfrac{-3 + \sqrt{-18}}{24}$ **42.** $\dfrac{-5 + \sqrt{-50}}{10}$

Find each sum or difference. Write the answer in standard form. **See Example 4.**

43. $(3 + 2i) + (9 - 3i)$ **44.** $(4 - i) + (8 + 5i)$

45. $(-2 + 4i) - (-4 + 4i)$ **46.** $(-3 + 2i) - (-4 + 2i)$

47. $(2 - 5i) - (3 + 4i) - (-2 + i)$ **48.** $(-4 - i) - (2 + 3i) + (-4 + 5i)$

49. $-i\sqrt{2} - 2 - \left(6 - 4i\sqrt{2}\right) - \left(5 - i\sqrt{2}\right)$

50. $3\sqrt{7} - \left(4\sqrt{7} - i\right) - 4i + \left(-2\sqrt{7} + 5i\right)$

Find each product. Write the answer in standard form. **See Example 5.**

51. $(2 + i)(3 - 2i)$ **52.** $(-2 + 3i)(4 - 2i)$ **53.** $(2 + 4i)(-1 + 3i)$

54. $(1 + 3i)(2 - 5i)$ **55.** $(3 - 2i)^2$ **56.** $(2 + i)^2$

57. $(3 + i)(3 - i)$ **58.** $(5 + i)(5 - i)$ **59.** $(-2 - 3i)(-2 + 3i)$

60. $(6 - 4i)(6 + 4i)$ **61.** $\left(\sqrt{6} + i\right)\left(\sqrt{6} - i\right)$ **62.** $\left(\sqrt{2} - 4i\right)\left(\sqrt{2} + 4i\right)$

63. $i(3 - 4i)(3 + 4i)$ **64.** $i(2 + 7i)(2 - 7i)$ **65.** $3i(2 - i)^2$

66. $-5i(4 - 3i)^2$ **67.** $(2 + i)(2 - i)(4 + 3i)$ **68.** $(3 - i)(3 + i)(2 - 6i)$

Find each quotient. Write the answer in standard form a + bi. **See Example 6.**

69. $\dfrac{6 + 2i}{1 + 2i}$ **70.** $\dfrac{14 + 5i}{3 + 2i}$ **71.** $\dfrac{2 - i}{2 + i}$ **72.** $\dfrac{4 - 3i}{4 + 3i}$

73. $\dfrac{1 - 3i}{1 + i}$ **74.** $\dfrac{-3 + 4i}{2 - i}$ **75.** $\dfrac{-5}{i}$ **76.** $\dfrac{-6}{i}$

77. $\dfrac{8}{-i}$ **78.** $\dfrac{12}{-i}$ **79.** $\dfrac{2}{3i}$ **80.** $\dfrac{5}{9i}$

(Modeling) Alternating Current Complex numbers are used to describe current I, voltage E, and impedance Z (the opposition to current). These three quantities are related by the equation

$$E = IZ, \quad \text{which is known as } \textbf{Ohm's Law.}$$

Thus, if any two of these quantities are known, the third can be found. In each exercise, solve the equation E = IZ for the remaining value.

81. $I = 5 + 7i, Z = 6 + 4i$ **82.** $I = 20 + 12i, Z = 10 - 5i$

83. $I = 10 + 4i, E = 88 + 128i$ **84.** $E = 57 + 67i, Z = 9 + 5i$

Simplify each power of i. ***See Example 7.***

85. i^{25} **86.** i^{29} **87.** i^{22} **88.** i^{26}

89. i^{23} **90.** i^{27} **91.** i^{32} **92.** i^{40}

93. i^{-13} **94.** i^{-14} **95.** $\dfrac{1}{i^{-11}}$ **96.** $\dfrac{1}{i^{-12}}$

97. Suppose that your friend, Kathy Strautz, tells you that she has discovered a method of simplifying a positive power of i. "Just divide the exponent by 2. Your answer is then the simplified form of i^2 raised to the quotient times i raised to the remainder." Explain why her method works.

98. Explain why the following method of simplifying i^{-42} works.

$$i^{-42} = \frac{1}{i^{42}} = \frac{1}{(i^2)^{21}} = \frac{1}{(-1)^{21}} = \frac{1}{-1} = -1$$

99. Show that $\dfrac{\sqrt{2}}{2} + \dfrac{\sqrt{2}}{2}i$ is a square root of i.

100. Show that $\dfrac{\sqrt{3}}{2} + \dfrac{1}{2}i$ is a cube root of i.

101. Show that $-2 + i$ is a solution of the equation $x^2 + 4x + 5 = 0$.

102. Show that $-3 + 4i$ is a solution of the equation $x^2 + 6x + 25 = 0$.

1.4 Quadratic Equations

- Solving a Quadratic Equation
- Completing the Square
- The Quadratic Formula
- Solving for a Specified Variable
- The Discriminant

A *quadratic equation* is defined as follows.

Quadratic Equation in One Variable

An equation that can be written in the form

$$ax^2 + bx + c = 0,$$

where a, b, and c are real numbers with $a \neq 0$, is a **quadratic equation.** The given form is called **standard form.**

A quadratic equation is a **second-degree equation**—that is, an equation with a squared variable term and no terms of greater degree.

$$x^2 = 25, \quad 4x^2 + 4x - 5 = 0, \quad 3x^2 = 4x - 8 \qquad \text{Quadratic equations}$$

Solving a Quadratic Equation The factoring method of solving a quadratic equation depends on the **zero-factor property.**

Zero-Factor Property

If a and b are complex numbers with $ab = 0$, then $a = 0$ or $b = 0$ or both equal zero.

EXAMPLE 1 **Using the Zero-Factor Property**

Solve $6x^2 + 7x = 3$.

SOLUTION

$\boxed{\text{Don't factor out } x \text{ here.}}$ — $6x^2 + 7x = 3$

$$6x^2 + 7x - 3 = 0 \qquad \text{Standard form}$$

$$(3x - 1)(2x + 3) = 0 \qquad \text{Factor. (Section R.4)}$$

$$3x - 1 = 0 \quad \text{or} \quad 2x + 3 = 0 \qquad \text{Zero-factor property}$$

$$3x = 1 \quad \text{or} \quad 2x = -3 \qquad \text{Solve each equation. (Section 1.1)}$$

$$x = \frac{1}{3} \quad \text{or} \quad x = -\frac{3}{2}$$

CHECK $\qquad\qquad 6x^2 + 7x = 3 \qquad$ Original equation

$$6\left(\frac{1}{3}\right)^2 + 7\left(\frac{1}{3}\right) \stackrel{?}{=} 3 \quad \text{Let } x = \tfrac{1}{3}. \quad \bigg| \quad 6\left(-\frac{3}{2}\right)^2 + 7\left(-\frac{3}{2}\right) \stackrel{?}{=} 3 \quad \text{Let } x = -\tfrac{3}{2}.$$

$$\frac{6}{9} + \frac{7}{3} \stackrel{?}{=} 3 \qquad\qquad\qquad\qquad \frac{54}{4} - \frac{21}{2} \stackrel{?}{=} 3$$

$$3 = 3 \checkmark \text{ True} \qquad\qquad\qquad\qquad 3 = 3 \checkmark \text{ True}$$

Both values check, since true statements result. The solution set is $\left\{\frac{1}{3}, -\frac{3}{2}\right\}$.

✔ *Now Try Exercise 15.*

A quadratic equation of the form $x^2 = k$ can also be solved by factoring.

$$x^2 = k$$

$$x^2 - k = 0 \qquad \text{Subtract } k.$$

$$\left(x - \sqrt{k}\right)\left(x + \sqrt{k}\right) = 0 \qquad \text{Factor.}$$

$$x - \sqrt{k} = 0 \quad \text{or} \quad x + \sqrt{k} = 0 \qquad \text{Zero-factor property}$$

$$x = \sqrt{k} \quad \text{or} \quad x = -\sqrt{k} \qquad \text{Solve each equation.}$$

This proves the **square root property.**

Square Root Property

If $x^2 = k$, then $x = \sqrt{k}$ or $x = -\sqrt{k}$.

That is, the solution set of $x^2 = k$ is

$$\left\{\sqrt{k}, -\sqrt{k}\right\}, \quad \textit{which may be abbreviated} \quad \left\{\pm\sqrt{k}\right\}.$$

Both solutions $\sqrt{k}$ and $-\sqrt{k}$ are real if $k > 0$, and both are pure imaginary if $k < 0$. If $k < 0$, then we write the solution set as

$$\left\{ \pm i\sqrt{|k|} \right\}.$$

If $k = 0$, then there is only one distinct solution, 0, sometimes called a **double solution.**

EXAMPLE 2 **Using the Square Root Property**

Solve each quadratic equation.

(a) $x^2 = 17$ **(b)** $x^2 = -25$ **(c)** $(x - 4)^2 = 12$

SOLUTION

(a) By the square root property, the solution set of $x^2 = 17$ is $\left\{ \pm\sqrt{17} \right\}$.

(b) Since $\sqrt{-1} = i$, the solution set of $x^2 = -25$ is $\left\{ \pm 5i \right\}$.

(c)
$$
\begin{aligned}
(x - 4)^2 &= 12 & \\
x - 4 &= \pm\sqrt{12} & \text{Generalized square root property} \\
x &= 4 \pm \sqrt{12} & \text{Add 4.} \\
x &= 4 \pm 2\sqrt{3} & \sqrt{12} = \sqrt{4 \cdot 3} = 2\sqrt{3} \text{ (Section R.7)}
\end{aligned}
$$

CHECK $(x - 4)^2 = 12$ Original equation

$\left(4 + 2\sqrt{3} - 4\right)^2 \stackrel{?}{=} 12$ Let $x = 4 + 2\sqrt{3}$.	$\left(4 - 2\sqrt{3} - 4\right)^2 \stackrel{?}{=} 12$ Let $x = 4 - 2\sqrt{3}$.
$\left(2\sqrt{3}\right)^2 \stackrel{?}{=} 12$	$\left(-2\sqrt{3}\right)^2 \stackrel{?}{=} 12$
$2^2 \cdot \left(\sqrt{3}\right)^2 \stackrel{?}{=} 12$	$(-2)^2 \cdot \left(\sqrt{3}\right)^2 \stackrel{?}{=} 12$
$12 = 12$ ✓ True	$12 = 12$ ✓ True

The solution set is $\left\{ 4 \pm 2\sqrt{3} \right\}$.

✔️ *Now Try Exercises 27, 29, and 31.*

Completing the Square   Any quadratic equation can be solved by the method of **completing the square,** summarized as follows.

Solving a Quadratic Equation by Completing the Square

To solve $ax^2 + bx + c = 0$, where $a \neq 0$, by completing the square, use these steps.

Step 1 If $a \neq 1$, divide both sides of the equation by a.

Step 2 Rewrite the equation so that the constant term is alone on one side of the equality symbol.

Step 3 Square half the coefficient of x, and add this square to each side of the equation.

Step 4 Factor the resulting trinomial as a perfect square and combine like terms on the other side.

Step 5 Use the square root property to complete the solution.

EXAMPLE 3 Using Completing the Square ($a = 1$)

Solve $x^2 - 4x - 14 = 0$.

SOLUTION $x^2 - 4x - 14 = 0$

Step 1 This step is not necessary since $a = 1$.

Step 2 $x^2 - 4x = 14$ Add 14 to each side.

Step 3 $x^2 - 4x + 4 = 14 + 4$ $\left[\frac{1}{2}(-4)\right]^2 = 4$; Add 4 to each side.

Step 4 $(x - 2)^2 = 18$ Factor. **(Section R.4)** Combine like terms.

Step 5 $x - 2 = \pm\sqrt{18}$ Square root property

Take *both* roots. $x = 2 \pm \sqrt{18}$ Add 2 to each side.

$x = 2 \pm 3\sqrt{2}$ Simplify the radical.

The solution set is $\left\{2 \pm 3\sqrt{2}\right\}$.

✔ *Now Try Exercise 41.*

EXAMPLE 4 Using Completing the Square ($a \neq 1$)

Solve $9x^2 - 12x + 9 = 0$.

SOLUTION $9x^2 - 12x + 9 = 0$

$x^2 - \frac{4}{3}x + 1 = 0$ Divide by 9. (Step 1)

$x^2 - \frac{4}{3}x = -1$ Subtract 1 from each side. (Step 2)

$x^2 - \frac{4}{3}x + \frac{4}{9} = -1 + \frac{4}{9}$ $\left[\frac{1}{2}\left(-\frac{4}{3}\right)\right]^2 = \frac{4}{9}$; Add $\frac{4}{9}$ to each side. (Step 3)

$\left(x - \frac{2}{3}\right)^2 = -\frac{5}{9}$ Factor. Combine like terms. (Step 4)

$x - \frac{2}{3} = \pm\sqrt{-\frac{5}{9}}$ Square root property (Step 5)

$x - \frac{2}{3} = \pm\frac{\sqrt{5}}{3}i$ $\sqrt{-\frac{5}{9}} = \frac{\sqrt{-5}}{\sqrt{9}} = \frac{i\sqrt{5}}{3}$, or $\frac{\sqrt{5}}{3}i$
(Section 1.3)

$x = \frac{2}{3} \pm \frac{\sqrt{5}}{3}i$ Add $\frac{2}{3}$ to each side.

The solution set is $\left\{\frac{2}{3} \pm \frac{\sqrt{5}}{3}i\right\}$.

✔ *Now Try Exercise 47.*

The Quadratic Formula   If we start with the equation, $ax^2 + bx + c = 0$, for $a > 0$, and complete the square to solve for x in terms of the constants a, b, and c, the result is a general formula for solving any quadratic equation.

$ax^2 + bx + c = 0$

$x^2 + \frac{b}{a}x + \frac{c}{a} = 0$ Divide each side by a. (Step 1)

$x^2 + \frac{b}{a}x = -\frac{c}{a}$ Subtract $\frac{c}{a}$ from each side. (Step 2)

Square half the coefficient of x: $\left[\frac{1}{2}\left(\frac{b}{a}\right)\right]^2 = \left(\frac{b}{2a}\right)^2 = \frac{b^2}{4a^2}.$

$$x^2 + \frac{b}{a}x + \frac{b^2}{4a^2} = -\frac{c}{a} + \frac{b^2}{4a^2}$$

Add $\frac{b^2}{4a^2}$ to each side. (Step 3)

$$\left(x + \frac{b}{2a}\right)^2 = \frac{b^2}{4a^2} + \frac{-c}{a}$$

Factor. Use the commutative property. (Step 4)

$$\left(x + \frac{b}{2a}\right)^2 = \frac{b^2}{4a^2} + \frac{-4ac}{4a^2}$$

Write fractions with a common denominator. **(Section R.5)**

$$\left(x + \frac{b}{2a}\right)^2 = \frac{b^2 - 4ac}{4a^2}$$

Add fractions. **(Section R.5)**

$$x + \frac{b}{2a} = \pm\sqrt{\frac{b^2 - 4ac}{4a^2}}$$

Square root property (Step 5)

$$x + \frac{b}{2a} = \frac{\pm\sqrt{b^2 - 4ac}}{2a}$$

Since $a > 0$, $\sqrt{4a^2} = 2a$. **(Section R.7)**

$$x = \frac{-b}{2a} \pm \frac{\sqrt{b^2 - 4ac}}{2a}$$

Subtract $\frac{b}{2a}$ from each side.

Quadratic Formula
This result is also ⟶ $x = \dfrac{-b \pm \sqrt{b^2 - 4ac}}{2a}$
true for $a < 0$.

Combine terms on the right.

> ### Quadratic Formula
>
> The solutions of the quadratic equation $ax^2 + bx + c = 0$, where $a \neq 0$, are given by the quadratic formula.
>
> $$x = \frac{-b \pm \sqrt{b^2 - 4ac}}{2a}$$

EXAMPLE 5 **Using the Quadratic Formula (Real Solutions)**

Solve $x^2 - 4x = -2$.

SOLUTION

$$x^2 - 4x + 2 = 0$$

Write in standard form. Here $a = 1$, $b = -4$, and $c = 2$.

$$x = \frac{-b \pm \sqrt{b^2 - 4ac}}{2a}$$

Quadratic formula

$$x = \frac{-(-4) \pm \sqrt{(-4)^2 - 4(1)(2)}}{2(1)}$$

Substitute $a = 1$, $b = -4$, and $c = 2$.

> The fraction bar extends *under* $-b$.

$$x = \frac{4 \pm \sqrt{16 - 8}}{2}$$

Simplify.

$$x = \frac{4 \pm 2\sqrt{2}}{2}$$

$\sqrt{16 - 8} = \sqrt{8} = \sqrt{4 \cdot 2} = 2\sqrt{2}$
(Section R.7)

$$x = \frac{2(2 \pm \sqrt{2})}{2}$$

Factor out 2 in the numerator.
(Section R.5)

> Factor first, then divide.

$$x = 2 \pm \sqrt{2}$$

Lowest terms

The solution set is $\left\{2 \pm \sqrt{2}\right\}$.

✔ *Now Try Exercise 53.*

CAUTION *Remember to extend the fraction bar in the quadratic formula under the* $-b$ *term in the numerator.*

Throughout this text, unless otherwise specified, we use the set of complex numbers as the domain when solving equations of degree 2 or greater.

EXAMPLE 6 Using the Quadratic Formula (Nonreal Complex Solutions)

Solve $2x^2 = x - 4$.

SOLUTION

$$2x^2 - x + 4 = 0 \qquad \text{Write in standard form.}$$

$$x = \frac{-(-1) \pm \sqrt{(-1)^2 - 4(2)(4)}}{2(2)} \qquad \begin{array}{l} \text{Quadratic formula with} \\ a = 2, b = -1, c = 4 \end{array}$$

> Use parentheses and substitute carefully to avoid errors.

$$x = \frac{1 \pm \sqrt{1 - 32}}{4}$$

$$x = \frac{1 \pm \sqrt{-31}}{4} \qquad \text{Simplify.}$$

$$x = \frac{1 \pm i\sqrt{31}}{4} \qquad \sqrt{-1} = i \text{ (Section 1.3)}$$

The solution set is $\left\{ \frac{1}{4} \pm \frac{\sqrt{31}}{4}i \right\}$.

✔ *Now Try Exercise 57.*

The equation $x^3 + 8 = 0$ is a **cubic equation** because the greatest degree of the terms is 3.

EXAMPLE 7 Solving a Cubic Equation

Solve $x^3 + 8 = 0$ using factoring and the quadratic formula.

SOLUTION $\qquad x^3 + 8 = 0$

$$(x + 2)(x^2 - 2x + 4) = 0 \qquad \begin{array}{l} \text{Factor as a sum} \\ \text{of cubes. (Section R.4)} \end{array}$$

$$x + 2 = 0 \quad \text{or} \quad x^2 - 2x + 4 = 0 \qquad \text{Zero-factor property}$$

$$x = -2 \quad \text{or} \quad x = \frac{-(-2) \pm \sqrt{(-2)^2 - 4(1)(4)}}{2(1)} \qquad \begin{array}{l} \text{Quadratic formula with} \\ a = 1, b = -2, c = 4 \end{array}$$

$$x = \frac{2 \pm \sqrt{-12}}{2} \qquad \text{Simplify.}$$

$$x = \frac{2 \pm 2i\sqrt{3}}{2} \qquad \text{Simplify the radical.}$$

$$x = \frac{2\left(1 \pm i\sqrt{3}\right)}{2} \qquad \begin{array}{l} \text{Factor out 2 in the} \\ \text{numerator.} \end{array}$$

$$x = 1 \pm i\sqrt{3} \qquad \text{Lowest terms}$$

The solution set is $\left\{ -2, 1 \pm i\sqrt{3} \right\}$.

✔ *Now Try Exercise 67.*

Solving for a Specified Variable To solve a quadratic equation for a specified variable, we usually apply the square root property or the quadratic formula.

EXAMPLE 8 **Solving for a Quadratic Variable in a Formula**

Solve for the specified variable. Use $\pm$ when taking square roots.

(a) $\mathcal{A} = \dfrac{\pi d^2}{4},$ for d **(b)** $rt^2 - st = k$ $(r \neq 0),$ for t

SOLUTION

(a) $\mathcal{A} = \dfrac{\pi d^2}{4}$ Goal: Isolate d, the specified variable.

$4\mathcal{A} = \pi d^2$ Multiply each side by 4.

$\dfrac{4\mathcal{A}}{\pi} = d^2$ Divide each side by π.

See the Note following this example.

$d = \pm\sqrt{\dfrac{4\mathcal{A}}{\pi}}$ Square root property

$d = \dfrac{\pm\sqrt{4\mathcal{A}}}{\sqrt{\pi}} \cdot \dfrac{\sqrt{\pi}}{\sqrt{\pi}}$ Multiply by $\dfrac{\sqrt{\pi}}{\sqrt{\pi}}$. (Section R.7)

$d = \dfrac{\pm\sqrt{4\mathcal{A}\pi}}{\pi}$ Multiply numerators. Multiply denominators.

$d = \dfrac{\pm 2\sqrt{\mathcal{A}\pi}}{\pi}$ Simplify the radical.

(b) Because $rt^2 - st = k$ has terms with t^2 and t, use the quadratic formula.

$rt^2 - st - k = 0$ Write in standard form.

$t = \dfrac{-b \pm \sqrt{b^2 - 4ac}}{2a}$ Quadratic formula

$t = \dfrac{-(-s) \pm \sqrt{(-s)^2 - 4(r)(-k)}}{2(r)}$ Here, $a = r$, $b = -s$, and $c = -k$.

$t = \dfrac{s \pm \sqrt{s^2 + 4rk}}{2r}$ Simplify.

✔ *Now Try Exercises 71 and 77.*

NOTE In **Example 8**, we took both positive and negative square roots. However, if the variable represents time or length in an application, we consider only the *positive* square root.

The Discriminant The quantity under the radical in the quadratic formula, $b^2 - 4ac$, is called the **discriminant**.

$$x = \dfrac{-b \pm \sqrt{\overbrace{b^2 - 4ac}}}{2a} \longleftarrow \text{Discriminant}$$

When the numbers a, b, and c are *integers* (but not necessarily otherwise), the value of the discriminant can be used to determine whether the solutions of a quadratic equation are rational, irrational, or nonreal complex numbers. The number and type of solutions based on the value of the discriminant are shown in the following table.

Discriminant	Number of Solutions	Type of Solutions	
Positive, perfect square	Two	Rational	
Positive, but not a perfect square	Two	Irrational	← As seen in Example 5
Zero	One (a double solution)	Rational	
Negative	Two	Nonreal complex	← As seen in Example 6

CAUTION *The restriction on a, b, and c is important.* For example,

$$x^2 - \sqrt{5}x - 1 = 0 \quad \text{has discriminant} \quad b^2 - 4ac = 5 + 4 = 9,$$

which would indicate two rational solutions *if the coefficients were integers.* By the quadratic formula, the two solutions $\dfrac{\sqrt{5} \pm 3}{2}$ are *irrational* numbers.

EXAMPLE 9   Using the Discriminant

Determine the number of distinct solutions, and tell whether they are *rational, irrational,* or *nonreal complex* numbers.

(a) $5x^2 + 2x - 4 = 0$ **(b)** $x^2 - 10x = -25$ **(c)** $2x^2 - x + 1 = 0$

SOLUTION

(a) For $5x^2 + 2x - 4 = 0$, use $a = 5$, $b = 2$, and $c = -4$.

$$b^2 - 4ac = 2^2 - 4(5)(-4) = 84 \leftarrow \text{Discriminant}$$

The discriminant 84 is positive and not a perfect square, so there are two distinct irrational solutions.

(b) First, write the equation in standard form as $x^2 - 10x + 25 = 0$. Thus, $a = 1$, $b = -10$, and $c = 25$.

$$b^2 - 4ac = (-10)^2 - 4(1)(25) = 0 \leftarrow \text{Discriminant}$$

There is one distinct rational solution, a double solution.

(c) For $2x^2 - x + 1 = 0$, use $a = 2$, $b = -1$, and $c = 1$.

$$b^2 - 4ac = (-1)^2 - 4(2)(1) = -7 \leftarrow \text{Discriminant}$$

There are two distinct nonreal complex solutions. (They are complex conjugates.)

✔ *Now Try Exercises 83, 85, and 89.*

1.4   Exercises

Concept Check Match the equation in Column I with its solution(s) in Column II.

I

1. $x^2 = 25$ **2.** $x^2 = -25$

3. $x^2 + 5 = 0$ **4.** $x^2 - 5 = 0$

5. $x^2 = -20$ **6.** $x^2 = 20$

7. $x - 5 = 0$ **8.** $x + 5 = 0$

II

A. $\pm 5i$ **B.** $\pm 2\sqrt{5}$

C. $\pm i\sqrt{5}$ **D.** 5

E. $\pm \sqrt{5}$ **F.** -5

G. ± 5 **H.** $\pm 2i\sqrt{5}$

Concept Check *Use Choices A–D to answer each question in Exercises 9–12.*

A. $3x^2 - 17x - 6 = 0$ **B.** $(2x + 5)^2 = 7$

C. $x^2 + x = 12$ **D.** $(3x - 1)(x - 7) = 0$

9. Which equation is set up for direct use of the zero-factor property? Solve it.

10. Which equation is set up for direct use of the square root property? Solve it.

11. Only one of the equations does not require Step 1 of the method for completing the square described in this section. Which one is it? Solve it.

12. Only one of the equations is set up so that the values of a, b, and c can be determined immediately. Which one is it? Solve it.

Solve each equation by the zero-factor property. ***See Example 1.***

13. $x^2 - 5x + 6 = 0$ **14.** $x^2 + 2x - 8 = 0$ **15.** $5x^2 - 3x - 2 = 0$

16. $2x^2 - x - 15 = 0$ **17.** $-4x^2 + x = -3$ **18.** $-6x^2 + 7x = -10$

19. $x^2 - 100 = 0$ **20.** $x^2 - 64 = 0$ **21.** $4x^2 - 4x + 1 = 0$

22. $9x^2 - 12x + 4 = 0$ **23.** $25x^2 + 30x + 9 = 0$ **24.** $36x^2 + 60x + 25 = 0$

Solve each equation by the square root property. ***See Example 2.***

25. $x^2 = 16$ **26.** $x^2 = 121$ **27.** $27 - x^2 = 0$

28. $48 - x^2 = 0$ **29.** $x^2 = -81$ **30.** $x^2 = -400$

31. $(3x - 1)^2 = 12$ **32.** $(4x + 1)^2 = 20$ **33.** $(x + 5)^2 = -3$

34. $(x - 4)^2 = -5$ **35.** $(5x - 3)^2 = -3$ **36.** $(-2x + 5)^2 = -8$

Solve each equation by completing the square. ***See Examples 3 and 4.***

37. $x^2 - 4x + 3 = 0$ **38.** $x^2 - 7x + 12 = 0$ **39.** $2x^2 - x - 28 = 0$

40. $4x^2 - 3x - 10 = 0$ **41.** $x^2 - 2x - 2 = 0$ **42.** $x^2 - 10x + 18 = 0$

43. $2x^2 + x = 10$ **44.** $3x^2 + 2x = 5$ **45.** $-2x^2 + 4x + 3 = 0$

46. $-3x^2 + 6x + 5 = 0$ **47.** $-4x^2 + 8x = 7$ **48.** $-3x^2 + 9x = 7$

49. *Concept Check* Francisco claimed that the equation $x^2 - 8x = 0$ cannot be solved by the quadratic formula since there is no value for c. Is he correct?

50. *Concept Check* Francesca, Francisco's twin sister, claimed that the equation $x^2 - 19 = 0$ cannot be solved by the quadratic formula since there is no value for b. Is she correct?

Solve each equation using the quadratic formula. ***See Examples 5 and 6.***

51. $x^2 - x - 1 = 0$ **52.** $x^2 - 3x - 2 = 0$ **53.** $x^2 - 6x = -7$

54. $x^2 - 4x = -1$ **55.** $x^2 = 2x - 5$ **56.** $x^2 = 2x - 10$

57. $-4x^2 = -12x + 11$ **58.** $-6x^2 = 3x + 2$ **59.** $\frac{1}{2}x^2 + \frac{1}{4}x - 3 = 0$

60. $\frac{2}{3}x^2 + \frac{1}{4}x = 3$ **61.** $0.2x^2 + 0.4x - 0.3 = 0$ **62.** $0.1x^2 - 0.1x = 0.3$

63. $(4x - 1)(x + 2) = 4x$ **64.** $(3x + 2)(x - 1) = 3x$ **65.** $(x - 9)(x - 1) = -16$

66. Explain why the following equations have the same solution set. (Do not actually solve.)

$$-2x^2 + 3x - 6 = 0 \quad \text{and} \quad 2x^2 - 3x + 6 = 0$$

Solve each cubic equation using factoring and the quadratic formula. ***See Example 7.***

67. $x^3 - 8 = 0$ **68.** $x^3 - 27 = 0$ **69.** $x^3 + 27 = 0$ **70.** $x^3 + 64 = 0$

Solve each equation for the indicated variable. Assume no denominators are 0. ***See Example 8.***

71. $s = \dfrac{1}{2}gt^2$, for t

72. $A = \pi r^2$, for r

73. $F = \dfrac{kMv^2}{r}$, for v

74. $E = \dfrac{e^2k}{2r}$, for e

75. $r = r_0 + \dfrac{1}{2}at^2$, for t

76. $s = s_0 + gt^2 + k$, for t

77. $h = -16t^2 + v_0t + s_0$, for t

78. $S = 2\pi rh + 2\pi r^2$, for r

*For each equation, **(a)** solve for x in terms of y, and **(b)** solve for y in terms of x.* ***See Example 8.***

79. $4x^2 - 2xy + 3y^2 = 2$

80. $3y^2 + 4xy - 9x^2 = -1$

81. $2x^2 + 4xy - 3y^2 = 2$

82. $5x^2 - 6xy + 2y^2 = 1$

Evaluate the discriminant for each equation. Then use it to predict the number of distinct solutions, and whether they are rational, irrational, *or* nonreal complex. *Do not solve the equation.* ***See Example 9.***

83. $x^2 - 8x + 16 = 0$ **84.** $x^2 + 4x + 4 = 0$ **85.** $3x^2 + 5x + 2 = 0$

86. $8x^2 = -14x - 3$ **87.** $4x^2 = -6x + 3$ **88.** $2x^2 + 4x + 1 = 0$

89. $9x^2 + 11x + 4 = 0$ **90.** $3x^2 = 4x - 5$ **91.** $8x^2 - 72 = 0$

92. Show that the discriminant for the equation

$$\sqrt{2}x^2 + 5x - 3\sqrt{2} = 0$$

is 49. If this equation is completely solved, it can be shown that the solution set is $\left\{-3\sqrt{2}, \frac{\sqrt{2}}{2}\right\}$. We have a discriminant that is positive and a perfect square, yet the two solutions are irrational. Does this contradict the discussion in this section? Explain.

93. Is it possible for the solution set of a quadratic equation with integer coefficients to consist of a single irrational number? Explain.

94. Is it possible for the solution set of a quadratic equation with real coefficients to consist of one real number and one nonreal complex number? Explain.

Find the values of a, b, and c for which the quadratic equation

$$ax^2 + bx + c = 0$$

has the given numbers as solutions. (Hint: Use the zero-factor property in reverse.)

95. $4, 5$ **96.** $-3, 2$ **97.** $1 + \sqrt{2}, 1 - \sqrt{2}$ **98.** $i, -i$

Chapter 1   **Quiz** (Sections 1.1–1.4)

1. Solve the linear equation $3(x - 5) + 2 = 1 - (4 + 2x)$.

2. Determine whether each equation is an *identity*, a *conditional equation*, or a *contradiction*. Give the solution set.

    (a) $4x - 5 = -2(3 - 2x) + 3$ **(b)** $5x - 9 = 5(-2 + x) + 1$

    (c) $5x - 4 = 3(6 - x)$

3. Solve the equation $ay + 2x = y + 5x$ for y. Assume $a \neq 1$.

4. *Earning Interest*   Johnny Ramistella deposits some money at 2.5% annual interest and twice as much at 3.0%. Find the amount deposited at each rate if his total annual interest income is $850.

5. *(Modeling) Minimum Hourly Wage*   One model for the minimum hourly wage in the United States for the period 1978–2009 is

$$y = 0.126x - 246.25,$$

where x represents the year and y represents the wage, in dollars. (*Source:* Bureau of Labor Statistics.) The actual 1999 minimum wage was $5.15. What does this model predict as the wage? What is the difference between the actual wage and the predicted wage?

6. Write $\frac{-4 + \sqrt{-24}}{8}$ in standard form $a + bi$.

7. Write the quotient $\frac{7 - 2i}{2 + 4i}$ in standard form $a + bi$.

Solve each equation.

8. $3x^2 - x = -1$ 9. $x^2 - 29 = 0$ 10. $A = \frac{1}{2}r^2\theta,$ for r

---

1.5   Applications and Modeling with Quadratic Equations

■ Geometry Problems

■ Using the Pythagorean Theorem

■ Height of a Projected Object

■ Modeling with Quadratic Equations

Geometry Problems   To solve these applications, we continue to use the problem-solving strategy discussed in **Section 1.2.**

EXAMPLE 1   **Solving a Problem Involving Volume**

A piece of machinery produces rectangular sheets of metal such that the length is three times the width. Equal-sized squares measuring 5 in. on a side can be cut from the corners so that the resulting piece of metal can be shaped into an open box by folding up the flaps. If specifications call for the volume of the box to be 1435 in.³, find the dimensions of the original piece of metal.

SOLUTION

Step 1   **Read** the problem. We must find the dimensions of the original piece of metal.

Step 2   **Assign a variable.** We know that the length is three times the width.

Let $x =$ the width (in inches) and thus, $3x =$ the length.

The box is formed by cutting $5 + 5 = 10$ in. from both the length and the width. See **Figure 5**. **Figure 6** indicates that the width of the bottom of the box is $x - 10$, the length of the bottom of the box is $3x - 10$, and the height is 5 in. (the length of the side of each cut-out square).

Figure 5

Figure 6

Step 3   **Write an equation.** The formula for volume of a box is $V = lwh$.

$$\underset{1435}{\text{Volume}} \quad \underset{=}{=} \quad \underset{(3x - 10)}{\text{length}} \times \underset{(x - 10)}{\text{width}} \times \underset{(5)}{\text{height}}$$

(Note that the dimensions of the box must be positive numbers, so $3x - 10$ and $x - 10$ must be greater than 0, which implies $x > \frac{10}{3}$ and $x > 10$. These are both satisfied when $x > 10$.)

Step 4 **Solve** the equation.

$$1435 = 15x^2 - 200x + 500 \qquad \text{Multiply. (Section R.3)}$$

$$0 = 15x^2 - 200x - 935 \qquad \text{Subtract 1435 from each side.}$$

$$0 = 3x^2 - 40x - 187 \qquad \text{Divide each side by 5.}$$

$$0 = (3x + 11)(x - 17) \qquad \text{Factor. (Section R.4)}$$

$$3x + 11 = 0 \quad \text{or} \quad x - 17 = 0 \qquad \text{Zero-factor property (Section 1.4)}$$

The width cannot be negative. $\longrightarrow$ $$x = -\frac{11}{3} \quad \text{or} \quad x = 17 \qquad \text{Solve each equation. (Section 1.1)}$$

Step 5 **State the answer.** Only 17 satisfies the restriction $x > 10$. Thus, the dimensions of the original piece should be 17 in. by $3(17) = 51$ in.

Step 6 **Check.** The length of the bottom of the box is $51 - 2(5) = 41$ in. The width is $17 - 2(5) = 7$ in. The height is 5 in. (the amount cut on each corner), so the volume of the box is

$$V = lwh = 41 \times 7 \times 5 = 1435 \text{ in.}^3, \quad \text{as required.}$$

✔ *Now Try Exercise 23.*

PROBLEM-SOLVING HINT   We may get a solution that does not satisfy the physical constraints of a problem. As seen in **Example 1,** a dimension of a box cannot be a negative number such as $-\frac{11}{3}$. Therefore, we reject this outcome.

Using the Pythagorean Theorem   **Example 2** requires the use of the **Pythagorean theorem** for right triangles. Recall that the **legs** of a right triangle form the right angle, and the **hypotenuse** is the side opposite the right angle.

Pythagorean Theorem

In a right triangle, the sum of the squares of the lengths of the legs is equal to the square of the length of the hypotenuse.

$$a^2 + b^2 = c^2$$

Hypotenuse
c
Leg *a*
Leg *b*

EXAMPLE 2   **Applying the Pythagorean Theorem**

A piece of property has the shape of a right triangle. The longer leg is 20 m longer than twice the length of the shorter leg. The hypotenuse is 10 m longer than the length of the longer leg. Find the lengths of the sides of the triangular lot.

SOLUTION

Step 1 **Read** the problem. We must find the lengths of the three sides.

Step 2 **Assign a variable.**

Let $x =$ the length of the shorter leg (in meters).

Then $2x + 20 =$ the length of the longer leg, and

$(2x + 20) + 10$, or $2x + 30 =$ the length of the hypotenuse.

See **Figure 7.**

x
2*x* + 30
2*x* + 20
x is in meters.

Figure 7

Step 3 **Write an equation.**

$$\underset{\downarrow}{a^2} \; + \; \underset{\downarrow}{b^2} \; = \; \underset{\downarrow}{c^2}$$

The hypotenuse is c.

$$x^2 \; + \; (2x + 20)^2 \; = \; (2x + 30)^2$$

Substitute into the Pythagorean theorem.

Step 4 **Solve** the equation.

Remember the middle terms.

$$x^2 + (4x^2 + 80x + 400) = 4x^2 + 120x + 900$$

Square the binomials. **(Section R.3)**

$$x^2 - 40x - 500 = 0$$

Standard form

$$(x - 50)(x + 10) = 0$$

Factor.

$$x - 50 = 0 \quad \text{or} \quad x + 10 = 0$$

Zero-factor property

$$x = 50 \quad \text{or} \qquad x = -10$$

Solve each equation.

Step 5 **State the answer.** Since x represents a length, -10 is not reasonable. The lengths of the sides of the triangular lot are 50 m, $2(50) + 20 = 120$ m, and $2(50) + 30 = 130$ m.

Step 6 **Check.** The lengths 50, 120, and 130 satisfy the words of the problem and also satisfy the Pythagorean theorem.

✔️ *Now Try Exercise 31.*

Galileo Galilei (1564–1642)

According to legend, Galileo dropped objects of different weights from the Leaning Tower of Pisa to disprove the Aristotelian view that heavier objects fall faster than lighter objects. He developed the formula $d = 16t^2$ for freely falling objects, where d is the distance in feet that an object falls (neglecting air resistance) in t seconds, regardless of weight.

Height of a Projected Object If air resistance is neglected, the height s (in feet) of an object projected directly upward from an initial height of s_0 feet, with initial velocity v_0 feet per second, is given by the following equation.

$$s = -16t^2 + v_0 t + s_0$$

Here t represents the number of seconds after the object is projected. The coefficient of t^2, -16, is a constant based on the gravitational force of Earth. This constant varies on other surfaces, such as the moon and other planets.

EXAMPLE 3 **Solving a Problem Involving Projectile Height**

If a projectile is launched vertically upward from the ground with an initial velocity of 100 ft per sec, neglecting air resistance, its height s (in feet) above the ground t seconds after projection is given by $s = -16t^2 + 100t$.

(a) After how many seconds will it be 50 ft above the ground?

(b) How long will it take for the projectile to return to the ground?

SOLUTION

(a) We must find value(s) of t so that height s is 50 ft.

$$50 = -16t^2 + 100t$$

Let $s = 50$ in the given equation.

$$0 = -16t^2 + 100t - 50$$

Standard form

$$0 = 8t^2 - 50t + 25$$

Divide by -2.

Substitute carefully.

$$t = \frac{-(-50) \pm \sqrt{(-50)^2 - 4(8)(25)}}{2(8)}$$

Quadratic formula **(Section 1.4)**

$$t = \frac{50 \pm \sqrt{1700}}{16}$$

Simplify.

$$t \approx 0.55 \quad \text{or} \quad t \approx 5.70$$

Use a calculator.

Both solutions are acceptable, since the projectile reaches 50 ft twice—once on its way up (after 0.55 sec) and once on its way down (after 5.70 sec).

LOOKING AHEAD TO CALCULUS

In calculus, you will need to be able to write an algebraic expression from the description in a problem like those in this section. Using calculus techniques, you will be asked to find the value of the variable that produces an optimum (a maximum or minimum) value of the expression.

(b) When the projectile returns to the ground, the height s will be 0 ft.

$$0 = -16t^2 + 100t \qquad \text{Let } s = 0.$$

$$0 = -4t(4t - 25) \qquad \text{Factor.}$$

$$-4t = 0 \quad \text{or} \quad 4t - 25 = 0 \qquad \text{Zero-factor property}$$

$$t = 0 \quad \text{or} \qquad t = 6.25 \qquad \text{Solve each equation.}$$

The first solution, 0, represents the time at which the projectile was on the ground prior to being launched, so it does not answer the question. The projectile will return to the ground 6.25 sec after it is launched.

☑ *Now Try Exercise 43.*

Modeling with Quadratic Equations

EXAMPLE 4 **Analyzing Trolley Ridership**

The I-Ride Trolley service carries passengers along the International Drive Resort Area of Orlando, Florida. The bar graph in **Figure 8** shows I-Ride Trolley ridership data in millions. The quadratic equation

$$y = -0.0072x^2 + 0.1081x + 1.619$$

models ridership from 2000 to 2010, where y represents ridership in millions and $x = 0$ represents 2000, $x = 1$ represents 2001, and so on.

Ridership

Source: I-Ride Trolley, International Drive Master Transit, www.itrolley.com

Figure 8

(a) Use the model to determine ridership in 2008. Compare the result to the actual ridership figure of 2.1 million.

(b) According to the model, in what year did ridership reach 1.9 million?

SOLUTION

(a) Since $x = 0$ represents the year 2000, use $x = 8$ to represent 2008.

$$y = -0.0072x^2 + 0.1081x + 1.619 \qquad \text{Given model}$$

$$y = -0.0072(8)^2 + 0.1081(8) + 1.619 \qquad \text{Let } x = 8.$$

$$y \approx 2.0 \text{ million} \qquad \text{Use a calculator.}$$

The prediction is about 0.1 million (that is, 100,000) less than the actual figure of 2.1 million.

(b) $1.9 = -0.0072x^2 + 0.1081x + 1.619 \qquad \text{Let } y = 1.9 \text{ in the model.}$

Solve this equation for *x.*

$$0 = -0.0072x^2 + 0.1081x - 0.281 \qquad \text{Standard form}$$

$$x = \frac{-0.1081 \pm \sqrt{(0.1081)^2 - 4(-0.0072)(-0.281)}}{2(-0.0072)} \qquad \text{Quadratic formula}$$

$$x \approx 3.3 \quad \text{or} \quad x \approx 11.7 \qquad \text{Use a calculator.}$$

The year 2003 corresponds to $x = 3.3$. Thus, according to the model, ridership reached 1.9 million in the year 2003. This outcome closely matches the bar graph and seems reasonable.

The year 2011 corresponds to $x = 11.7$. Round down to the year 2011 because 11.7 yr from 2000 occurs during 2011. There is no value on the bar graph to compare this to, because the last data value is for the year 2010. Always view results that are *beyond* the data in your model with skepticism, and realistically consider whether the model will continue as given. The model *predicts* that ridership will be 1.9 million again in the year 2011.

✔️ *Now Try Exercise 47.*

1.5 Exercises

Concept Check Answer each question.

1. *Area of a Parking Lot* For the rectangular parking area of the shopping center shown, which one of the following equations says that the area is 40,000 yd²?

x

$2x + 200$

A. $x(2x + 200) = 40{,}000$

B. $2x + 2(2x + 200) = 40{,}000$

C. $x + (2x + 200) = 40{,}000$

D. $x^2 + (2x + 200)^2 = 40{,}000^2$

2. *Diagonal of a Rectangle* If a rectangle is r feet long and s feet wide, which one of the following expressions is the length of its diagonal in terms of r and s?

A. $\sqrt{rs}$ **B.** $r + s$

C. $\sqrt{r^2 + s^2}$ **D.** $r^2 + s^2$

3. *Sides of a Right Triangle* To solve for the lengths of the right triangle sides, which equation is correct?

A. $x^2 = (2x - 2)^2 + (x + 4)^2$

B. $x^2 + (x + 4)^2 = (2x - 2)^2$

C. $x^2 = (2x - 2)^2 - (x + 4)^2$

D. $x^2 + (2x - 2)^2 = (x + 4)^2$

x $x + 4$ $2x - 2$

4. *Area of a Picture* The mat around the picture shown measures x inches across. Which one of the following equations says that the area of the picture itself is 600 in.²?

21 in.

34 in.

x in.

x in.

A. $2(34 - 2x) + 2(21 - 2x) = 600$

B. $(34 - 2x)(21 - 2x) = 600$

C. $(34 - x)(21 - x) = 600$

D. $x(34)(21) = 600$

To prepare for the applications that come later, work the following basic problems that lead to quadratic equations.

Unknown Numbers In Exercises 5–14, use the following facts.

If x represents an integer, then $x + 1$ represents the next consecutive integer.

If x represents an even integer, then $x + 2$ represents the next consecutive even integer.

If x represents an odd integer, then $x + 2$ represents the next consecutive odd integer.

5. Find two consecutive integers whose product is 56.

6. Find two consecutive integers whose product is 110.

7. Find two consecutive even integers whose product is 168.

8. Find two consecutive even integers whose product is 224.

9. Find two consecutive odd integers whose product is 63.

10. Find two consecutive odd integers whose product is 143.

11. The sum of the squares of two consecutive odd integers is 202. Find the integers.

12. The sum of the squares of two consecutive even integers is 52. Find the integers.

13. The difference of the squares of two positive consecutive even integers is 84. Find the integers.

14. The difference of the squares of two positive consecutive odd integers is 32. Find the integers.

15. *Dimensions of a Right Triangle* The lengths of the sides of a right triangle are consecutive even integers. Find these lengths. (*Hint:* Use the Pythagorean theorem.)

16. *Dimensions of a Right Triangle* The lengths of the sides of a right triangle are consecutive positive integers. Find these lengths. (*Hint:* Use the Pythagorean theorem.)

17. *Dimensions of a Square* The length of each side of a square is 3 in. more than the length of each side of a smaller square. The sum of the areas of the squares is 149 in.^{2}. Find the lengths of the sides of the two squares.

18. *Dimensions of a Square* The length of each side of a square is 5 in. more than the length of each side of a smaller square. The difference of the areas of the squares is 95 in.^{2}. Find the lengths of the sides of the two squares.

Solve each problem. See Example 1.

19. *Dimensions of a Parking Lot* A parking lot has a rectangular area of 40,000 yd^2. The length is 200 yd more than twice the width. Find the dimensions of the lot. (**See Exercise 1.**)

20. *Dimensions of a Garden* An ecology center wants to set up an experimental garden using 300 m of fencing to enclose a rectangular area of 5000 m^2. Find the dimensions of the garden.

$150 - x$

x

x is in meters.

21. *Dimensions of a Rug* Zachary Daniels wants to buy a rug for a room that is 12 ft wide and 15 ft long. He wants to leave a uniform strip of floor around the rug. He can afford to buy 108 ft^2 of carpeting. What dimensions should the rug have?

108 ft^2

12 ft

15 ft

22. *Width of a Flower Border* A landscape architect has included a rectangular flower bed measuring 9 ft by 5 ft in her plans for a new building. She wants to use two colors of flowers in the bed, one in the center and the other for a border of the same width on all four sides. If she has enough plants to cover 24 ft^2 for the border, how wide can the border be?

23. *Volume of a Box* A rectangular piece of metal is 10 in. longer than it is wide. Squares with sides 2 in. long are cut from the four corners, and the flaps are folded upward to form an open box. If the volume of the box is 832 in.^{3}, what were the original dimensions of the piece of metal?

24. *Volume of a Box* In **Exercise 23,** suppose that the piece of metal has length twice the width, and 4-in. squares are cut from the corners. If the volume of the box is 1536 in.^{3}, what were the original dimensions of the piece of metal?

25. *Manufacturing to Specifications* A manufacturing firm wants to package its product in a cylindrical container 3 ft high with surface area 8π ft^2. What should the radius of the circular top and bottom of the container be? (*Hint:* The surface area consists of the circular top and bottom and a rectangle that represents the side cut open vertically and unrolled.)

26. *Manufacturing to Specifications* In **Exercise 25,** what radius would produce a container with a volume of π times the radius? (*Hint:* The volume is the area of the circular base times the height.)

27. *Dimensions of a Square* What is the length of the side of a square if its area and perimeter are numerically equal?

28. *Dimensions of a Rectangle* A rectangle has an area that is numerically twice its perimeter. If the length is twice the width, what are its dimensions?

29. *Radius of a Can* A can of Blue Runner Red Kidney Beans has surface area 371 cm^2. Its height is 12 cm. What is the radius of the circular top? Round to the nearest hundredth.

30. *Dimensions of a Cereal Box* The volume of a 10-oz box of corn flakes is 180.4 in.^{3}. The width of the box is 3.2 in. less than the length, and its depth is 2.3 in. Find the length and width of the box to the nearest tenth.

*Solve each problem. **See Example 2.***

31. *Height of a Dock* A boat is being pulled into a dock with a rope attached to the boat at water level. When the boat is 12 ft from the dock, the length of the rope from the boat to the dock is 3 ft longer than twice the height of the dock above the water. Find the height of the dock.

32. *Height of a Kite* Grady is flying a kite on 50 ft of string. Its vertical distance from his hand is 10 ft more than the horizontal distance from his hand. Assuming that the string is being held 5 ft above ground level, find its horizontal distance from Grady and its vertical distance from the ground.

33. *Radius Covered by a Circular Lawn Sprinkler* A square lawn has area 800 ft². A sprinkler placed at the center of the lawn sprays water in a circular pattern as shown in the figure. What is the radius of the circle?

34. *Dimensions of a Solar Panel Frame* Molly has a solar panel with a width of 26 in. To get the proper inclination for her climate, she needs a right triangular support frame that has one leg twice as long as the other. To the nearest tenth of an inch, what dimensions should the frame have?

35. *Length of a Ladder* A building is 2 ft from a 9-ft fence that surrounds the property. A worker wants to wash a window in the building 13 ft from the ground. He plans to place a ladder over the fence so it rests against the building. (See the figure.) He decides he should place the ladder 8 ft from the fence for stability. To the nearest tenth of a foot, how long a ladder will he need?

36. *Range of Receivers* Tanner Jones and Sheldon Furst have received communications receivers for Christmas. If they leave from the same point at the same time, Tanner walking north at 2.5 mph and Sheldon walking east at 3 mph, how long will they be able to talk to each other if the range of the communications receivers is 4 mi? Round your answer to the nearest minute.

37. *Length of a Walkway* A nature conservancy group decides to construct a raised wooden walkway through a wetland area. To enclose the most interesting part of the wetlands, the walkway will have the shape of a right triangle with one leg 700 yd longer than the other and the hypotenuse 100 yd longer than the longer leg. Find the total length of the walkway.

38. *Broken Bamboo* Problems involving the Pythagorean theorem have appeared in mathematics for thousands of years. This one is taken from the ancient Chinese work, *Arithmetic in Nine Sections:*

> *There is a bamboo 10 ft high, the upper end of which, being broken, reaches the ground 3 ft from the stem. Find the height of the break.*

(Modeling) Solve each problem. **See Examples 3 and 4.**

Height of a Projectile A projectile is launched from ground level with an initial velocity of v_0 feet per second. Neglecting air resistance, its height in feet t seconds after launch is given by

$$s = -16t^2 + v_0 t.$$

In Exercises 39–42, find the time(s) that the projectile will **(a)** *reach a height of 80 ft and* **(b)** *return to the ground for the given value of* v_0. *Round answers to the nearest hundredth if necessary.*

39. $v_0 = 96$ **40.** $v_0 = 128$ **41.** $v_0 = 32$ **42.** $v_0 = 16$

43. *Height of a Projected Ball* An astronaut on the moon throws a baseball upward. The astronaut is 6 ft, 6 in. tall, and the initial velocity of the ball is 30 ft per sec. The height s of the ball in feet is given by the equation

$$s = -2.7t^2 + 30t + 6.5,$$

where t is the number of seconds after the ball was thrown.

(a) After how many seconds is the ball 12 ft above the moon's surface? Round to the nearest hundredth.

(b) How many seconds will it take for the ball to return to the surface? Round to the nearest hundredth.

44. The ball in **Exercise 43** will never reach a height of 100 ft. How can this be determined algebraically?

45. *Carbon Monoxide Exposure* Carbon monoxide (CO) combines with the hemoglobin of the blood to form carboxyhemoglobin (COHb), which reduces transport of oxygen to tissues. Smokers routinely have a 4% to 6% COHb level in their blood, which can cause symptoms such as blood flow alterations, visual impairment, and poorer vigilance ability. The quadratic model

$$T = 0.00787x^2 - 1.528x + 75.89$$

approximates the exposure time in hours necessary to reach this 4% to 6% level, where $50 \le x \le 100$ is the amount of carbon monoxide present in the air in parts per million (ppm). (*Source: Indoor Air Quality Environmental Information Handbook: Combustion Sources.*)

(a) A kerosene heater or a room full of smokers is capable of producing 50 ppm of carbon monoxide. How long would it take for a nonsmoking person to start feeling the above symptoms?

(b) Find the carbon monoxide concentration necessary for a person to reach the 4% to 6% COHb level in 3 hr. Round to the nearest tenth.

46. *Carbon Monoxide Exposure* Refer to **Exercise 45.** High concentrations of carbon monoxide (CO) can cause coma and death. The time required for a person to reach a COHb level capable of causing a coma can be approximated by the quadratic model

$$T = 0.0002x^2 - 0.316x + 127.9,$$

where T is the exposure time in hours necessary to reach this level and $500 \le x \le 800$ is the amount of carbon monoxide present in the air in parts per million (ppm). (*Source: Indoor Air Quality Environmental Information Handbook: Combustion Sources.*)

(a) What is the exposure time when $x = 600$ ppm?

(b) Estimate the concentration of CO necessary to produce a coma in 4 hr.

47. *Methane Gas Emissions* The table gives methane gas emissions from all sources in the United States, in millions of metric tons. The quadratic model

$$y = 1.493x^2 + 7.279x + 684.4$$

approximates the emissions for these years. In the model, x represents the number of years since 2004, so $x = 0$ represents 2004, $x = 1$ represents 2005, and so on.

Year	Millions of Metric Tons of Methane
2004	686.6
2005	691.8
2006	706.3
2007	722.7
2008	737.4

Source: U.S. Energy Information Administration.

(a) According to the model, what would be the emissions in 2010? Round to the nearest tenth of a million metric tons.

(b) Find the year beyond 2004 for which this model predicts that the emissions reached 700 million metric tons.

48. *Cost of Public Colleges* The average cost, in dollars, for tuition and fees for in-state students at four-year public colleges over the period 2000–2010 can be modeled by the equation

$$y = 3.026x^2 + 377.7x + 3449,$$

where $x = 0$ corresponds to 2000, $x = 1$ corresponds to 2001, and so on. Based on this model, for what year after 2000 was the average cost $7605? (*Source:* The College Board, *Annual Survey of Colleges.*)

49. *Internet Publishing and Broadcasting* Estimated revenue from Internet publishing and broadcasting in the United States during the years 2004 through 2008 can be modeled by the equation

$$y = 318.4x^2 + 1612x + 8596,$$

where $x = 0$ corresponds to the year 2004, $x = 1$ corresponds to 2005, and so on, and y is in millions of dollars. Approximate the revenue from Internet publishing and broadcasting in 2006 to the nearest million. (*Source:* U.S. Census Bureau.)

50. *Cable TV Subscribers* The number of U.S. households subscribing to cable TV for the period 2000 through 2010 can be modeled by the equation

$$y = -0.0746x^2 + 3.146x + 79.52,$$

where $x = 0$ corresponds to 2000, $x = 1$ corresponds to 2001, and so on, and y is in millions. Based on this model, approximately how many U.S. households, to the nearest tenth of a million, subscribed to cable TV in 2008? (*Source:* Nielsen Media Research.)

Relating Concepts

For individual or collaborative investigation *(Exercises 51–54)*

*If p units of an item are sold for x dollars per unit, the revenue is R = px. Use this idea to analyze the following problem, **working Exercises 51–54 in order.***

Number of Apartments Rented The manager of an 80-unit apartment complex knows from experience that at a rent of $300, all the units will be full. On the average, one additional unit will remain vacant for each $20 increase in rent over $300. Furthermore, the manager must keep at least 30 units rented due to other financial considerations. Currently, the revenue from the complex is $35,000. How many apartments are rented?

(continued)

51. Suppose that x represents the number of $20 increases over $300. Represent the number of apartment units that will be rented in terms of x.

52. Represent the rent per unit in terms of x.

53. Use the answers in **Exercises 51 and 52** to write an equation that defines the revenue generated when there are x increases of $20 over $300.

54. According to the problem, the revenue currently generated is $35,000. Substitute this value for revenue into the equation from **Exercise 53.** Solve for x to answer the question in the problem.

Solve each problem. (*See Relating Concepts Exercises 51–54.*)

55. *Number of Airline Passengers* The cost of a charter flight to Miami is $225 each for 75 passengers, with a refund of $5 per passenger for each passenger in excess of 75. How many passengers must take the flight to produce a revenue of $16,000?

56. *Number of Bus Passengers* A charter bus company charges a fare of $40 per person, plus $2 per person for each unsold seat on the bus. If the bus holds 100 passengers and x represents the number of unsold seats, how many passengers must ride the bus to produce revenue of $5950? (*Note:* Because of the company's commitment to efficient fuel use, the charter will not run unless filled to at least half-capacity.)

57. *Harvesting a Cherry Orchard* The manager of a cherry orchard wants to schedule the annual harvest. If the cherries are picked now, the average yield per tree will be 100 lb, and the cherries can be sold for 40 cents per pound. Past experience shows that the yield per tree will increase about 5 lb per week, while the price will decrease about 2 cents per pound per week. How many weeks should the manager wait to get an average revenue of $38.40 per tree?

58. *Recycling Aluminum Cans* A local group of scouts has been collecting old aluminum cans for recycling. The group has already collected 12,000 lb of cans, for which they could currently receive $4 per hundred pounds. The group can continue to collect cans at the rate of 400 lb per day. However, a glut in the old-can market has caused the recycling company to announce that it will lower its price, starting immediately, by $0.10 per hundred pounds per day. The scouts can make only one trip to the recycling center. How many days should they wait in order to get $490 for their cans?

1.6 Other Types of Equations and Applications

- Rational Equations
- Work Rate Problems
- Equations with Radicals
- Equations with Rational Exponents
- Equations Quadratic in Form

Rational Equations   A **rational equation** is an equation that has a rational expression for one or more terms. To solve a rational equation, multiply each side by the least common denominator (LCD) of the terms of the equation to eliminate fractions, and then solve the resulting equation.

A value of the variable that appears to be a solution after each side of a rational equation is multiplied by a variable expression (the LCD) is called a **proposed solution.** Because a rational expression is not defined when its denominator is 0, **proposed solutions for which any denominator equals 0 are excluded from the solution set.**

Be sure to check all proposed solutions in the original equation.

EXAMPLE 1 Solving Rational Equations That Lead to Linear Equations

Solve each equation.

(a) $\dfrac{3x-1}{3} - \dfrac{2x}{x-1} = x$

(b) $\dfrac{x}{x-2} = \dfrac{2}{x-2} + 2$

SOLUTION

(a) The least common denominator is $3(x-1)$, which is equal to 0 if $x = 1$. Therefore, 1 cannot possibly be a solution of this equation.

$$\frac{3x-1}{3} - \frac{2x}{x-1} = x$$

$$3(x-1)\left(\frac{3x-1}{3}\right) - 3(x-1)\left(\frac{2x}{x-1}\right) = 3(x-1)x \qquad \text{Multiply by the LCD.} \quad 3(x-1), \text{ where } x \ne 1.$$
$$\text{(Section R.5)}$$

$$(x-1)(3x-1) - 3(2x) = 3x(x-1) \qquad \text{Divide out common factors.}$$

$$3x^2 - 4x + 1 - 6x = 3x^2 - 3x \qquad \text{Multiply. (Section R.3)}$$

$$1 - 10x = -3x \qquad \text{Subtract } 3x^2. \text{ Combine like terms.}$$

$$1 = 7x \qquad \text{Solve the linear equation.} \\ \text{(Section 1.1)}$$

$$x = \frac{1}{7} \qquad \text{Proposed solution}$$

The proposed solution $\frac{1}{7}$ meets the requirement that $x \ne 1$ and does not cause any denominator to equal 0. Substitute to check for correct algebra.

$$\text{CHECK} \qquad \frac{3x-1}{3} - \frac{2x}{x-1} = x \qquad \text{Original equation}$$

$$\frac{3\left(\frac{1}{7}\right)-1}{3} - \frac{2\left(\frac{1}{7}\right)}{\frac{1}{7}-1} \overset{?}{=} \frac{1}{7} \qquad \text{Let } x = \frac{1}{7}.$$

$$-\frac{4}{21} - \left(-\frac{1}{3}\right) \overset{?}{=} \frac{1}{7} \qquad \text{Simplify the complex fractions.} \\ \text{(Section R.5)}$$

$$\frac{1}{7} = \frac{1}{7} \checkmark \qquad \text{True}$$

The solution set is $\left\{\frac{1}{7}\right\}$.

(b) $$\frac{x}{x-2} = \frac{2}{x-2} + 2$$

$$(x-2)\left(\frac{x}{x-2}\right) = (x-2)\left(\frac{2}{x-2}\right) + (x-2)2 \qquad \text{Multiply by the LCD,} \\ x-2, \text{ where } x \ne 2.$$

$$x = 2 + 2(x-2) \qquad \text{Divide out common factors.}$$

$$x = 2 + 2x - 4 \qquad \text{Distributive property} \\ \text{(Section R.2)}$$

$$-x = -2 \qquad \text{Solve the linear equation.}$$

$$x = 2 \qquad \text{Proposed solution}$$

The proposed solution is 2. However, the variable is restricted to real numbers except 2. If $x = 2$, then not only does it cause a zero denominator, but also multiplying by $x - 2$ in the first step is multiplying both sides by 0, which is not valid. Thus, the solution set is $\emptyset$.

✔ *Now Try Exercises 7 and 9.*

EXAMPLE 2 **Solving Rational Equations That Lead to Quadratic Equations**

Solve each equation.

(a) $\dfrac{3x+2}{x-2}+\dfrac{1}{x}=\dfrac{-2}{x^2-2x}$

(b) $\dfrac{-4x}{x-1}+\dfrac{4}{x+1}=\dfrac{-8}{x^2-1}$

SOLUTION

(a)

$$\frac{3x+2}{x-2}+\frac{1}{x}=\frac{-2}{x^2-2x}$$

$$\frac{3x+2}{x-2}+\frac{1}{x}=\frac{-2}{x(x-2)} \qquad \text{Factor the last denominator. (Section R.4)}$$

$$x(x-2)\left(\frac{3x+2}{x-2}\right)+x(x-2)\left(\frac{1}{x}\right)=x(x-2)\left(\frac{-2}{x(x-2)}\right) \qquad \begin{array}{l}\text{Multiply by}\\ x(x-2),\\ x\neq 0,2.\end{array}$$

$$x(3x+2)+(x-2)=-2 \qquad \text{Divide out common factors.}$$

$$3x^2+2x+x-2=-2 \qquad \text{Distributive property}$$

$$3x^2+3x=0 \qquad \begin{array}{l}\text{Standard form}\\ \text{(Section 1.4)}\end{array}$$

$$3x(x+1)=0 \qquad \text{Factor.}$$

Set *each* factor equal to 0. $\longrightarrow$ $3x=0 \quad \text{or} \quad x+1=0 \qquad \begin{array}{l}\text{Zero-factor property}\\ \text{(Section 1.4)}\end{array}$

$$x=0 \quad \text{or} \qquad x=-1 \qquad \text{Proposed solutions}$$

Because of the restriction $x\neq 0$, the only valid proposed solution is -1. Check -1 in the original equation. The solution set is $\{-1\}$.

(b)

$$\frac{-4x}{x-1}+\frac{4}{x+1}=\frac{-8}{x^2-1}$$

$$\frac{-4x}{x-1}+\frac{4}{x+1}=\frac{-8}{(x+1)(x-1)} \qquad \text{Factor.}$$

The restrictions on x are $x\neq\pm 1$. Multiply by the LCD, $(x+1)(x-1)$.

$$(x+1)(x-1)\left(\frac{-4x}{x-1}\right)+(x+1)(x-1)\left(\frac{4}{x+1}\right)=(x+1)(x-1)\left(\frac{-8}{(x+1)(x-1)}\right)$$

$$-4x(x+1)+4(x-1)=-8 \qquad \text{Divide out common factors.}$$

$$-4x^2-4x+4x-4=-8 \qquad \text{Distributive property}$$

$$-4x^2+4=0 \qquad \text{Standard form}$$

$$x^2-1=0 \qquad \text{Divide by } -4.$$

$$(x+1)(x-1)=0 \qquad \text{Factor.}$$

$$x+1=0 \quad \text{or} \quad x-1=0 \qquad \text{Zero-factor property}$$

$$x=-1 \quad \text{or} \qquad x=1 \qquad \text{Proposed solutions}$$

Neither proposed solution is valid, so the solution set is $\varnothing$.

☑ *Now Try Exercises 15 and 17.*

Work Rate Problems   If a job can be completed in 3 hr, then the rate of work is $\frac{1}{3}$ of the job per hr. After 1 hr the job would be $\frac{1}{3}$ complete, and after 2 hr the job would be $\frac{2}{3}$ complete. In 3 hr the job would be $\frac{3}{3}$ complete, meaning that 1 complete job had been accomplished.

PROBLEM-SOLVING HINT If a job can be completed in t units of time, then the rate of work, r, is $\frac{1}{t}$ of the job per unit time.

$$r = \frac{1}{t}$$

The amount of work completed, A, is found by multiplying the rate of work, r, and the amount of time worked, t. This formula is similar to the distance formula $d = rt$.

Amount of work completed = rate of work × amount of time worked

or $\qquad\qquad A = rt$

EXAMPLE 3 **Solving a Work Rate Problem**

One printer can do a job twice as fast as another. Working together, both printers can do the job in 2 hr. How long would it take each printer, working alone, to do the job?

SOLUTION

Step 1 **Read** the problem. We must find the time it would take each printer, working alone, to do the job.

Step 2 **Assign a variable.** Let x represent the number of hours it would take the faster printer, working alone, to do the job. The time for the slower printer to do the job alone is then $2x$ hours.

Therefore, $\qquad \dfrac{1}{x}$ = the rate of the faster printer (job per hour)

and $\qquad \dfrac{1}{2x}$ = the rate of the slower printer (job per hour).

The time for the printers to do the job together is 2 hr. Multiplying each rate by the time will give the fractional part of the job accomplished by each.

	Rate	Time	Part of the Job Accomplished
Faster Printer	$\frac{1}{x}$	2	$2\left(\frac{1}{x}\right) = \frac{2}{x}$
Slower Printer	$\frac{1}{2x}$	2	$2\left(\frac{1}{2x}\right) = \frac{1}{x}$

$A = rt$

Step 3 **Write an equation.** The sum of the two parts of the job accomplished is 1, since one whole job is done.

$$
\underbrace{\frac{2}{x}}_{\substack{\text{Part of the job}\\\text{done by the}\\\text{faster printer}}} + \underbrace{\frac{1}{x}}_{\substack{\text{Part of the job}\\\text{done by the}\\\text{slower printer}}} = \underbrace{1}_{\substack{\text{One whole}\\\text{job}}}
$$

Step 4 **Solve.**

$$x\left(\frac{2}{x} + \frac{1}{x}\right) = x(1) \qquad \text{Multiply each side by } x, \text{ where } x \neq 0.$$

$$x\left(\frac{2}{x}\right) + x\left(\frac{1}{x}\right) = x(1) \qquad \text{Distributive property}$$

$$2 + 1 = x \qquad \text{Multiply.}$$

$$3 = x \qquad \text{Add.}$$

Step 5 **State the answer.** The faster printer would take 3 hr to do the job alone, and the slower printer would take $2(3) = 6$ hr. Be sure to give *both* answers here.

Step 6 **Check.** The answer is reasonable, since the time working together (2 hr, as stated in the problem) is less than the time it would take the faster printer working alone (3 hr, as found in Step 4).

☑ *Now Try Exercise 29.*

NOTE   **Example 3** can also be solved by using the fact that the sum of the rates of the individual printers is equal to their rate working together. Since the printers can complete the job together in 2 hr, their combined rate is $\frac{1}{2}$ of the job per hr.

$$\frac{1}{x} + \frac{1}{2x} = \frac{1}{2}$$

$$2x\left(\frac{1}{x} + \frac{1}{2x}\right) = 2x\left(\frac{1}{2}\right) \qquad \text{Multiply each side by } 2x.$$

$$2 + 1 = x \qquad \text{Distributive property}$$

$$x = 3 \qquad \text{Same solution found earlier}$$

Equations with Radicals   To solve an equation such as

$$x - \sqrt{15 - 2x} = 0,$$

in which the variable appears in a radicand, we use the following **power property** to eliminate the radical.

Power Property

If P and Q are algebraic expressions, then every solution of the equation $P = Q$ is also a solution of the equation $P^n = Q^n$, for any positive integer n.

When the power property is used to solve equations, the new equation may have *more* solutions than the original equation. For example, the equation

$$x = -2 \quad \text{has solution set} \quad \{-2\}.$$

If we square each side of the equation $x = -2$, we obtain the new equation

$$x^2 = 4, \quad \text{which has solution set} \quad \{-2, 2\}.$$

Since the solution sets are not equal, the equations are not equivalent. *When we use the power property to solve an equation, it is essential to check all proposed solutions in the original equation.*

CAUTION   *Be very careful when using the power property.* It does *not* say that the equations $P = Q$ and $P^n = Q^n$ are equivalent. It says only that each solution of the original equation $P = Q$ is also a solution of the new equation $P^n = Q^n$.

> ### Solving an Equation Involving Radicals
>
> To solve an equation containing radicals, follow these steps.
>
> **Step 1** Isolate the radical on one side of the equation.
>
> **Step 2** Raise each side of the equation to a power that is the same as the index of the radical so that the radical is eliminated.
>
> *If the equation still contains a radical, repeat Steps 1 and 2.*
>
> **Step 3** Solve the resulting equation.
>
> **Step 4** Check each proposed solution in the *original* equation.

EXAMPLE 4 Solving an Equation Containing a Radical (Square Root)

Solve $x - \sqrt{15 - 2x} = 0$.

SOLUTION

$$x - \sqrt{15 - 2x} = 0$$

$$x = \sqrt{15 - 2x} \qquad \text{Isolate the radical. (Step 1)}$$

$$x^2 = \left(\sqrt{15 - 2x}\right)^2 \qquad \text{Square each side. (Step 2)}$$

$$x^2 = 15 - 2x \qquad \left(\sqrt{a}\right)^2 = a, \text{ for } a \geq 0. \text{ (Section R.7)}$$

$$x^2 + 2x - 15 = 0 \qquad \text{Solve the quadratic equation. (Step 3)}$$

$$(x + 5)(x - 3) = 0 \qquad \text{Factor.}$$

$$x + 5 = 0 \quad \text{or} \quad x - 3 = 0 \qquad \text{Zero-factor property}$$

$$x = -5 \quad \text{or} \qquad x = 3 \qquad \text{Proposed solutions}$$

CHECK $x - \sqrt{15 - 2x} = 0$ Original equation (Step 4)

$-5 - \sqrt{15 - 2(-5)} \stackrel{?}{=} 0$ Let $x = -5$. $\bigg|$ $3 - \sqrt{15 - 2(3)} \stackrel{?}{=} 0$ Let $x = 3$.

$-5 - \sqrt{25} \stackrel{?}{=} 0$ $\bigg|$ $3 - \sqrt{9} \stackrel{?}{=} 0$

$-5 - 5 \stackrel{?}{=} 0$ $\bigg|$ $3 - 3 \stackrel{?}{=} 0$

$-10 = 0$ False $\bigg|$ $0 = 0$ ✓ True

As the check shows, only 3 is a solution, so the solution set is $\{3\}$.

✔ *Now Try Exercise 35.*

EXAMPLE 5 Solving an Equation Containing Two Radicals

Solve $\sqrt{2x + 3} - \sqrt{x + 1} = 1$.

SOLUTION

$$\sqrt{2x + 3} - \sqrt{x + 1} = 1 \qquad \text{Isolate one of the radicals on one side of the equation.}$$

$$\sqrt{2x + 3} = 1 + \sqrt{x + 1} \qquad \text{Isolate } \sqrt{2x + 3}. \text{ (Step 1)}$$

$$\left(\sqrt{2x + 3}\right)^2 = \left(1 + \sqrt{x + 1}\right)^2 \qquad \text{Square each side. (Step 2)}$$

$$2x + 3 = 1 + 2\sqrt{x + 1} + (x + 1) \qquad \begin{array}{l}\textbf{Be careful:} \\ (a + b)^2 = a^2 + 2ab + b^2 \\ \textbf{(Section R.3)}\end{array}$$

Don't forget this term when squaring.

$$x + 1 = 2\sqrt{x + 1} \qquad \text{Isolate the remaining radical. (Step 1)}$$

$$(x + 1)^2 = \left(2\sqrt{x + 1}\right)^2 \qquad \text{Square again. (Step 2)}$$

$$x^2 + 2x + 1 = 4(x + 1) \quad \boxed{(ab)^2 = a^2b^2} \qquad \text{Apply the exponents.}$$

$$x^2 + 2x + 1 = 4x + 4 \qquad \text{Distributive property}$$

$$x^2 - 2x - 3 = 0 \qquad \text{Solve the quadratic equation. (Step 3)}$$

$$(x - 3)(x + 1) = 0 \qquad \text{Factor.}$$

$$x - 3 = 0 \quad \text{or} \quad x + 1 = 0 \qquad \text{Zero-factor property}$$

$$x = 3 \quad \text{or} \qquad x = -1 \qquad \text{Proposed solutions}$$

CHECK $\qquad\qquad \sqrt{2x + 3} - \sqrt{x + 1} = 1 \qquad$ Original equation (Step 4)

$$\sqrt{2(3) + 3} - \sqrt{3 + 1} \overset{?}{=} 1 \quad \text{Let } x = 3. \quad \Big| \quad \sqrt{2(-1) + 3} - \sqrt{-1 + 1} \overset{?}{=} 1 \quad \text{Let } x = -1.$$

$$\sqrt{9} - \sqrt{4} \overset{?}{=} 1 \qquad\qquad\qquad \sqrt{1} - \sqrt{0} \overset{?}{=} 1$$

$$3 - 2 \overset{?}{=} 1 \qquad\qquad\qquad\qquad 1 - 0 \overset{?}{=} 1$$

$$1 = 1 \; \checkmark \; \text{True} \qquad\qquad\qquad 1 = 1 \; \checkmark \; \text{True}$$

Both 3 and -1 are solutions of the original equation, so $\{-1, 3\}$ is the solution set.

☑ *Now Try Exercise 47.*

CAUTION Remember to isolate a radical in Step 1. It would be incorrect to square each term individually as the first step in **Example 5.**

EXAMPLE 6 **Solving an Equation Containing a Radical (Cube Root)**

Solve $\sqrt[3]{4x^2 - 4x + 1} - \sqrt[3]{x} = 0$.

SOLUTION

$$\sqrt[3]{4x^2 - 4x + 1} - \sqrt[3]{x} = 0$$

$$\sqrt[3]{4x^2 - 4x + 1} = \sqrt[3]{x} \qquad \text{Isolate a radical. (Step 1)}$$

$$\left(\sqrt[3]{4x^2 - 4x + 1}\right)^3 = \left(\sqrt[3]{x}\right)^3 \qquad \text{Cube each side. (Step 2)}$$

$$4x^2 - 4x + 1 = x \qquad \text{Apply the exponents.}$$

$$4x^2 - 5x + 1 = 0 \qquad \text{Solve the quadratic equation. (Step 3)}$$

$$(4x - 1)(x - 1) = 0 \qquad \text{Factor.}$$

$$4x - 1 = 0 \quad \text{or} \quad x - 1 = 0 \qquad \text{Zero-factor property}$$

$$x = \frac{1}{4} \quad \text{or} \qquad x = 1 \qquad \text{Proposed solutions}$$

CHECK $\qquad\qquad \sqrt[3]{4x^2 - 4x + 1} - \sqrt[3]{x} = 0 \qquad$ Original equation (Step 4)

$$\sqrt[3]{4\left(\tfrac{1}{4}\right)^2 - 4\left(\tfrac{1}{4}\right) + 1} - \sqrt[3]{\tfrac{1}{4}} \overset{?}{=} 0 \quad \text{Let } x = \tfrac{1}{4}. \; \Big| \; \sqrt[3]{4(1)^2 - 4(1) + 1} - \sqrt[3]{1} \overset{?}{=} 0 \quad \text{Let } x = 1.$$

$$\sqrt[3]{\tfrac{1}{4}} - \sqrt[3]{\tfrac{1}{4}} \overset{?}{=} 0 \qquad\qquad\qquad\qquad \sqrt[3]{1} - \sqrt[3]{1} \overset{?}{=} 0$$

$$0 = 0 \; \checkmark \; \text{True} \qquad\qquad\qquad\qquad 0 = 0 \; \checkmark \; \text{True}$$

Both are valid solutions, and the solution set is $\left\{\tfrac{1}{4}, 1\right\}$.

☑ *Now Try Exercise 59.*

Equations with Rational Exponents An **equation with a rational exponent** contains a variable, or variable expression, raised to an exponent that is a rational number. For example, the radical equation

$$\left(\sqrt[5]{x}\right)^3 = 27 \quad \text{can be written with a rational exponent as} \quad x^{3/5} = 27$$

and solved by raising each side to the reciprocal of the exponent, with care taken regarding signs as seen in **Example 7(b).**

EXAMPLE 7 **Solving Equations with Rational Exponents**

Solve each equation.

(a) $x^{3/5} = 27$ **(b)** $(x - 4)^{2/3} = 16$

SOLUTION

(a) $x^{3/5} = 27$

$\left(x^{3/5}\right)^{5/3} = 27^{5/3}$ Raise each side to the power $\frac{5}{3}$, the reciprocal of the exponent of x.

$x = 243$ $27^{5/3} = \left(\sqrt[3]{27}\right)^5 = 3^5 = 243$

CHECK   Let $x = 243$ in the original equation.

$$x^{3/5} = 243^{3/5} = \left(\sqrt[5]{243}\right)^3 = 27 \checkmark \text{ True}$$

The solution set is $\{243\}$.

(b) $(x - 4)^{2/3} = 16$ Raise each side to the power $\frac{3}{2}$. Insert $\pm$ since this involves an even root, as indicated by the 2 in the denominator.

$\left[(x - 4)^{2/3}\right]^{3/2} = \pm 16^{3/2}$

$x - 4 = \pm 64$ $\pm 16^{3/2} = \pm\left(\sqrt{16}\right)^3 = \pm 4^3 = \pm 64$

$x = 4 \pm 64$ Add 4 to each side.

$x = -60$ or $x = 68$ Proposed solutions

CHECK $(x - 4)^{2/3} = 16$ Original equation

$(-60 - 4)^{2/3} \overset{?}{=} 16$ Let $x = -60$. $\qquad$ $(68 - 4)^{2/3} \overset{?}{=} 16$ Let $x = 68$.

$(-64)^{2/3} \overset{?}{=} 16$ $\qquad\qquad\qquad\qquad$ $64^{2/3} \overset{?}{=} 16$

$16 = 16 \checkmark$ True $\qquad\qquad\qquad\qquad$ $16 = 16 \checkmark$ True

Both proposed solutions check, so the solution set is $\{-60, 68\}$.

✔ *Now Try Exercises 65 and 69.*

Equations Quadratic in Form The equation

$$(x + 1)^{2/3} - (x + 1)^{1/3} - 2 = 0$$

is not a quadratic equation in x. However, with the substitutions

$$u = (x + 1)^{1/3} \quad \text{and} \quad u^2 = \left[(x + 1)^{1/3}\right]^2 = (x + 1)^{2/3},$$

the equation becomes

$$u^2 - u - 2 = 0,$$

which is a quadratic equation in u. This quadratic equation can be solved to find u, and then $u = (x + 1)^{1/3}$ can be used to find the values of x, the solutions to the original equation.

Equation Quadratic in Form

An equation is said to be **quadratic in form** if it can be written as

$$au^2 + bu + c = 0,$$

where $a \neq 0$ and u is some algebraic expression.

EXAMPLE 8 **Solving Equations Quadratic in Form**

Solve each equation.

(a) $(x + 1)^{2/3} - (x + 1)^{1/3} - 2 = 0$ **(b)** $6x^{-2} + x^{-1} = 2$

SOLUTION

(a)
$$(x + 1)^{2/3} - (x + 1)^{1/3} - 2 = 0 \quad \text{Since } (x+1)^{2/3} = \left[(x+1)^{1/3}\right]^2,$$
$$u^2 - u - 2 = 0 \quad \text{let } u = (x+1)^{1/3}.$$
$$(u - 2)(u + 1) = 0 \quad \text{Factor.}$$

$u - 2 = 0$ or $u + 1 = 0$ Zero-factor property

$u = 2$ or $u = -1$ Solve each equation.

> **Don't forget this step.**

$(x + 1)^{1/3} = 2$ or $(x + 1)^{1/3} = -1$ Replace u with $(x + 1)^{1/3}$.

$\left[(x + 1)^{1/3}\right]^3 = 2^3$ or $\left[(x + 1)^{1/3}\right]^3 = (-1)^3$ Cube each side.

$x + 1 = 8$ or $x + 1 = -1$ Apply the exponents.

$x = 7$ or $x = -2$ Proposed solutions

CHECK $(x + 1)^{2/3} - (x + 1)^{1/3} - 2 = 0$ Original equation

$(7 + 1)^{2/3} - (7 + 1)^{1/3} - 2 \stackrel{?}{=} 0$ Let $x = 7$. $\quad$ $(-2 + 1)^{2/3} - (-2 + 1)^{1/3} - 2 \stackrel{?}{=} 0$ Let $x = -2$.

$8^{2/3} - 8^{1/3} - 2 \stackrel{?}{=} 0$ $\qquad\qquad\qquad (-1)^{2/3} - (-1)^{1/3} - 2 \stackrel{?}{=} 0$

$4 - 2 - 2 \stackrel{?}{=} 0$ $\qquad\qquad\qquad\qquad\qquad\quad 1 + 1 - 2 \stackrel{?}{=} 0$

$0 = 0$ ✓ True $\qquad\qquad\qquad\qquad\qquad\quad 0 = 0$ ✓ True

Both proposed solutions check, so the solution set is $\{-2, 7\}$.

(b) $\qquad\qquad 6x^{-2} + x^{-1} = 2$

$6x^{-2} + x^{-1} - 2 = 0$ Subtract 2 from each side.

$6u^2 + u - 2 = 0$ Let $u = x^{-1}$. Then $u^2 = x^{-2}$.

$(3u + 2)(2u - 1) = 0$ Factor.

$3u + 2 = 0$ or $2u - 1 = 0$ Zero-factor property

> **Don't stop here. Remember to substitute for u.**

$u = -\dfrac{2}{3}$ or $u = \dfrac{1}{2}$ Solve each equation.

$x^{-1} = -\dfrac{2}{3}$ or $x^{-1} = \dfrac{1}{2}$ Replace u with x^{-1}.

$x = -\dfrac{3}{2}$ or $x = 2$ x^{-1} is the reciprocal of x. **(Section R.6)**

Both proposed solutions check, so the solution set is $\left\{-\frac{3}{2}, 2\right\}$.

✔ *Now Try Exercises 83 and 87.*

CAUTION *When using a substitution variable in solving an equation that is quadratic in form, do not forget the step that gives the solution in terms of the original variable.*

EXAMPLE 9 **Solving an Equation Quadratic in Form**

Solve $12x^4 - 11x^2 + 2 = 0$.

SOLUTION

$$12x^4 - 11x^2 + 2 = 0$$

$$12(x^2)^2 - 11x^2 + 2 = 0 \qquad x^4 = (x^2)^2$$

$$12u^2 - 11u + 2 = 0 \qquad \text{Let } u = x^2. \text{ Then } u^2 = x^4.$$

$$(3u - 2)(4u - 1) = 0 \qquad \text{Solve the quadratic equation.}$$

$3u - 2 = 0$	or $\quad 4u - 1 = 0$	Zero-factor property
$u = \dfrac{2}{3}$	or $\qquad u = \dfrac{1}{4}$	Solve each equation.
$x^2 = \dfrac{2}{3}$	or $\qquad x^2 = \dfrac{1}{4}$	Replace u with x^2.
$x = \pm\sqrt{\dfrac{2}{3}}$	or $\qquad x = \pm\sqrt{\dfrac{1}{4}}$	Square root property (Section 1.4)
$x = \dfrac{\pm\sqrt{2}}{\sqrt{3}} \cdot \dfrac{\sqrt{3}}{\sqrt{3}}$	or $\qquad x = \pm\dfrac{1}{2}$	Simplify radicals. (Section R.7)
$x = \pm\dfrac{\sqrt{6}}{3}$		

Check that the solution set is $\left\{ \pm\dfrac{\sqrt{6}}{3}, \ \pm\dfrac{1}{2} \right\}$.

Now Try Exercise 77.

NOTE Some equations that are quadratic in form are simple enough to avoid using the substitution variable technique. For example, to solve

$$12x^4 - 11x^2 + 2 = 0, \qquad \text{Equation from Example 9}$$

we could factor $12x^4 - 11x^2 + 2$ directly as $(3x^2 - 2)(4x^2 - 1)$, set each factor equal to zero, and then solve the resulting two quadratic equations. *Which method to use is a matter of personal preference.*

1.6 Exercises

*Decide what values of the variable cannot possibly be solutions for each equation. Do not solve. **See Examples 1 and 2.***

1. $\dfrac{5}{2x+3} - \dfrac{1}{x-6} = 0$

2. $\dfrac{2}{x+1} + \dfrac{3}{5x-2} = 0$

3. $\dfrac{3}{x-2} + \dfrac{1}{x+1} = \dfrac{3}{x^2-x-2}$

4. $\dfrac{2}{x+3} - \dfrac{5}{x-1} = \dfrac{-5}{x^2+2x-3}$

5. $\dfrac{1}{4x} - \dfrac{2}{x} = 3$

6. $\dfrac{5}{2x} + \dfrac{2}{x} = 6$

Solve each equation. See Example 1.

7. $\dfrac{2x+5}{2} - \dfrac{3x}{x-2} = x$

8. $\dfrac{4x+3}{4} - \dfrac{2x}{x+1} = x$

9. $\dfrac{x}{x-3} = \dfrac{3}{x-3} + 3$

10. $\dfrac{x}{x-4} = \dfrac{4}{x-4} + 4$

11. $\dfrac{-2}{x-3} + \dfrac{3}{x+3} = \dfrac{-12}{x^2-9}$

12. $\dfrac{3}{x-2} + \dfrac{1}{x+2} = \dfrac{12}{x^2-4}$

13. $\dfrac{4}{x^2+x-6} - \dfrac{1}{x^2-4} = \dfrac{2}{x^2+5x+6}$

14. $\dfrac{3}{x^2+x-2} - \dfrac{1}{x^2-1} = \dfrac{7}{2x^2+6x+4}$

Solve each equation. See Example 2.

15. $\dfrac{2x+1}{x-2} + \dfrac{3}{x} = \dfrac{-6}{x^2-2x}$

16. $\dfrac{4x+3}{x+1} + \dfrac{2}{x} = \dfrac{1}{x^2+x}$

17. $\dfrac{x}{x-1} - \dfrac{1}{x+1} = \dfrac{2}{x^2-1}$

18. $\dfrac{-x}{x+1} - \dfrac{1}{x-1} = \dfrac{-2}{x^2-1}$

19. $\dfrac{5}{x^2} - \dfrac{43}{x} = 18$

20. $\dfrac{7}{x^2} + \dfrac{19}{x} = 6$

21. $2 = \dfrac{3}{2x-1} + \dfrac{-1}{(2x-1)^2}$

22. $6 = \dfrac{7}{2x-3} + \dfrac{3}{(2x-3)^2}$

23. $\dfrac{2x-5}{x} = \dfrac{x-2}{3}$

24. $\dfrac{x+4}{2x} = \dfrac{x-1}{3}$

25. $\dfrac{2x}{x-2} = 5 + \dfrac{4x^2}{x-2}$

26. $\dfrac{3x^2}{x-1} + 2 = \dfrac{x}{x-1}$

Solve each problem. See Example 3.

27. *Painting a House* (This problem appears in the 1994 movie *Little Big League*.) If Joe can paint a house in 3 hr, and Sam can paint the same house in 5 hr, how long does it take them to do it together?

28. *Painting a House* Repeat **Exercise 27,** but assume that Joe takes 6 hr working alone, and Sam takes 8 hr working alone.

29. *Pollution in a River* Two chemical plants are polluting a river. If plant A produces a predetermined maximum amount of pollutant twice as fast as plant B, and together they produce the maximum pollutant in 26 hr, how long will it take plant B alone?

	Rate	Time	Part of Job Accomplished
Pollution from A	$\frac{1}{x}$	26	$\frac{1}{x}(26)$
Pollution from B		26	

30. *Filling a Settling Pond* A sewage treatment plant has two inlet pipes to its settling pond. One pipe can fill the pond 3 times as fast as the other pipe, and together they can fill the pond in 12 hr. How long will it take the faster pipe to fill the pond alone?

31. *Filling a Pool* An inlet pipe can fill Blake's pool in 5 hr, while an outlet pipe can empty it in 8 hr. In his haste to surf the Internet, Blake left both pipes open. How long did it take to fill the pool?

32. *Filling a Pool* Suppose Blake discovered his error (see **Exercise 31**) after an hour-long surf. If he then closed the outlet pipe, how much more time would be needed to fill the pool?

33. *Filling a Sink* With both taps open, Robert can fill his kitchen sink in 5 min. When full, the sink drains in 10 min. How long will it take to fill the sink if Robert forgets to put in the stopper?

34. *Filling a Sink* If Robert (see **Exercise 33**) remembers to put in the stopper after 1 min, how much longer will it take to fill the sink?

Solve each equation. See Examples 4–6.

35. $x - \sqrt{2x + 3} = 0$

36. $x - \sqrt{3x + 18} = 0$

37. $\sqrt{3x + 7} = 3x + 5$

38. $\sqrt{4x + 13} = 2x - 1$

39. $\sqrt{4x + 5} - 6 = 2x - 11$

40. $\sqrt{6x + 7} - 9 = x - 7$

41. $\sqrt{4x} - x + 3 = 0$

42. $\sqrt{2x} - x + 4 = 0$

43. $\sqrt{x} - \sqrt{x - 5} = 1$

44. $\sqrt{x} - \sqrt{x - 12} = 2$

45. $\sqrt{x + 7} + 3 = \sqrt{x - 4}$

46. $\sqrt{x + 5} + 2 = \sqrt{x - 1}$

47. $\sqrt{2x + 5} - \sqrt{x + 2} = 1$

48. $\sqrt{4x + 1} - \sqrt{x - 1} = 2$

49. $\sqrt{3x} = \sqrt{5x + 1} - 1$

50. $\sqrt{2x} = \sqrt{3x + 12} - 2$

51. $\sqrt{x + 2} = 1 - \sqrt{3x + 7}$

52. $\sqrt{2x - 5} = 2 + \sqrt{x - 2}$

53. $\sqrt{2\sqrt{7x + 2}} = \sqrt{3x + 2}$

54. $\sqrt{3\sqrt{2x + 3}} = \sqrt{5x - 6}$

55. $3 - \sqrt{x} = \sqrt{2\sqrt{x} - 3}$

56. $\sqrt{x} + 2 = \sqrt{4 + 7\sqrt{x}}$

57. $\sqrt[3]{4x + 3} = \sqrt[3]{2x - 1}$

58. $\sqrt[3]{2x} = \sqrt[3]{5x + 2}$

59. $\sqrt[3]{5x^2 - 6x + 2} - \sqrt[3]{x} = 0$

60. $\sqrt[3]{3x^2 - 9x + 8} = \sqrt[3]{x}$

61. $\sqrt[4]{x - 15} = 2$

62. $\sqrt[4]{3x + 1} = 1$

63. $\sqrt[4]{x^2 + 2x} = \sqrt[4]{3}$

64. $\sqrt[4]{x^2 + 6x} = 2$

Solve each equation. See Example 7.

65. $x^{3/2} = 125$

66. $x^{5/4} = 32$

67. $(x^2 + 24x)^{1/4} = 3$

68. $(3x^2 + 52x)^{1/4} = 4$

69. $(x - 3)^{2/5} = 4$

70. $(x + 200)^{2/3} = 36$

71. $(2x + 5)^{1/3} - (6x - 1)^{1/3} = 0$

72. $(3x + 7)^{1/3} - (4x + 2)^{1/3} = 0$

73. $(2x - 1)^{2/3} = x^{1/3}$

74. $(x - 3)^{2/5} = (4x)^{1/5}$

75. $x^{2/3} = 2x^{1/3}$

76. $3x^{3/4} = x^{1/2}$

Solve each equation. See Examples 8 and 9.

77. $2x^4 - 7x^2 + 5 = 0$

78. $4x^4 - 8x^2 + 3 = 0$

79. $x^4 + 2x^2 - 15 = 0$

80. $3x^4 + 10x^2 - 25 = 0$

81. $(x - 1)^{2/3} + (x - 1)^{1/3} - 12 = 0$

82. $(2x - 1)^{2/3} + 2(2x - 1)^{1/3} - 3 = 0$

83. $(x + 1)^{2/5} - 3(x + 1)^{1/5} + 2 = 0$

84. $(x + 5)^{2/3} + (x + 5)^{1/3} - 20 = 0$

85. $6(x + 2)^4 - 11(x + 2)^2 = -4$

86. $8(x - 4)^4 - 10(x - 4)^2 = -3$

87. $10x^{-2} + 33x^{-1} - 7 = 0$

88. $7x^{-2} - 10x^{-1} - 8 = 0$

89. $x^{-2/3} + x^{-1/3} - 6 = 0$

90. $2x^{-2/5} - x^{-1/5} - 1 = 0$

91. $16x^{-4} - 65x^{-2} + 4 = 0$

92. $625x^{-4} - 125x^{-2} + 4 = 0$

Relating Concepts

For individual or collaborative investigation (Exercises 93–96)

In this section we introduced methods of solving equations quadratic in form by substitution and solving equations involving radicals by raising each side of the equation to a power. Suppose we wish to solve

$$x - \sqrt{x} - 12 = 0.$$

We can solve this equation using either of the two methods. **Work Exercises 93–96 in order** *to see how both methods apply.*

93. Let $u = \sqrt{x}$ and solve the equation by substitution. What is the value of u that does not lead to a solution of the equation?

94. Solve the equation by isolating $\sqrt{x}$ on one side and then squaring. What is the value of x that does not satisfy the equation?

95. Which one of the methods used in **Exercises 93 and 94** do you prefer? Why?

96. Solve $3x - 2\sqrt{x} - 8 = 0$ using one of the two methods described.

Solve each equation for the indicated variable. Assume all denominators are nonzero.

97. $d = k\sqrt{h}$, for h

98. $x^{2/3} + y^{2/3} = a^{2/3}$, for y

99. $m^{3/4} + n^{3/4} = 1$, for m

100. $\dfrac{1}{R} = \dfrac{1}{r_1} + \dfrac{1}{r_2}$, for R

101. $\dfrac{E}{e} = \dfrac{R+r}{r}$, for e

102. $a^2 + b^2 = c^2$, for b

Summary Exercises on Solving Equations

This section of miscellaneous equations provides practice in solving all the types introduced in this chapter so far. Solve each equation.

1. $4x - 3 = 2x + 3$

2. $5 - (6x + 3) = 2(2 - 2x)$

3. $x(x + 6) = 9$

4. $x^2 = 8x - 12$

5. $\sqrt{x + 2} + 5 = \sqrt{x + 15}$

6. $\dfrac{5}{x + 3} - \dfrac{6}{x - 2} = \dfrac{3}{x^2 + x - 6}$

7. $\dfrac{3x + 4}{3} - \dfrac{2x}{x - 3} = x$

8. $\dfrac{x}{2} + \dfrac{4}{3}x = x + 5$

9. $5 - \dfrac{2}{x} + \dfrac{1}{x^2} = 0$

10. $(2x + 1)^2 = 9$

11. $x^{-2/5} - 2x^{-1/5} - 15 = 0$

12. $\sqrt{x + 2} + 1 = \sqrt{2x + 6}$

13. $x^4 - 3x^2 - 4 = 0$

14. $1.2x + 0.3 = 0.7x - 0.9$

15. $\sqrt[6]{2x + 1} = \sqrt[6]{9}$

16. $3x^2 - 2x = -1$

17. $3[2x - (6 - 2x) + 1] = 5x$

18. $\sqrt{x + 1} = \sqrt{11 - \sqrt{x}}$

19. $(14 - 2x)^{2/3} = 4$

20. $2x^{-1} - x^{-2} = 1$

21. $\dfrac{3}{x - 3} = \dfrac{3}{x - 3}$

22. $a^2 + b^2 = c^2$, for a

1.7 Inequalities

- Linear Inequalities
- Three-Part Inequalities
- Quadratic Inequalities
- Rational Inequalities

An **inequality** says that one expression is greater than, greater than or equal to, less than, or less than or equal to another (**Section R.2**). As with equations, a value of the variable for which the inequality is true is a solution of the inequality, and the set of all solutions is the solution set of the inequality. Two inequalities with the same solution set are equivalent.

Inequalities are solved with the properties of inequality, which are similar to the properties of equality in **Section 1.1.**

Properties of Inequality

Let a, b, and c represent real numbers.

1. If $a < b$, then $a + c < b + c$.
2. If $a < b$ and if $c > 0$, then $ac < bc$.
3. If $a < b$ and if $c < 0$, then $ac > bc$.

Replacing $<$ with $>$, $\leq$, or $\geq$ results in similar properties. (Restrictions on c remain the same.)

NOTE Multiplication may be replaced by division in Properties 2 and 3. *Always remember to reverse the direction of the inequality symbol when multiplying or dividing by a negative number.*

Linear Inequalities The definition of a *linear inequality* is similar to the definition of a linear equation.

Linear Inequality in One Variable

A **linear inequality in one variable** is an inequality that can be written in the form

$$ax + b > 0,$$

where a and b are real numbers, with $a \neq 0$. (Any of the symbols $\geq$, $<$, and $\leq$ may also be used.)

EXAMPLE 1 Solving a Linear Inequality

Solve $-3x + 5 > -7$.

SOLUTION

$$-3x + 5 > -7$$

$$-3x + 5 - 5 > -7 - 5 \quad \text{Subtract 5.}$$

$$-3x > -12 \quad \text{Combine like terms.}$$

> Don't forget to reverse the inequality symbol here.

$$\frac{-3x}{-3} < \frac{-12}{-3} \quad \begin{array}{l}\text{Divide by } -3. \text{ Reverse the direction of the} \\ \text{inequality symbol when multiplying or dividing} \\ \text{by a negative number.}\end{array}$$

$$x < 4$$

Figure 9

Thus, the original inequality $-3x + 5 > -7$ is satisfied by any real number less than 4. The solution set can be written $\{x \mid x < 4\}$. A graph of the solution set is shown in **Figure 9**, where the parenthesis is used to show that 4 itself does not belong to the solution set. Note that testing values from the solution set in the original inequality will produce true statements, while testing values outside the solution set produces false statements.

The solution set of the inequality,

$$\{x \mid x < 4\}, \quad \text{Set builder notation (Section R.1)}$$

is an example of an **interval.** We use a simplified notation, called **interval notation,** to write intervals. With this notation, we write the interval as

$$(-\infty, 4). \quad \text{Interval notation}$$

The symbol $-\infty$ does not represent an actual number. Rather it is used to show that the interval includes all real numbers less than 4. The interval $(-\infty, 4)$ is an example of an **open interval,** since the endpoint, 4, is not part of the interval. A **closed interval** includes both endpoints. A square bracket is used to show that a number *is* part of the graph, and a parenthesis is used to indicate that a number *is not* part of the graph.

☑ *Now Try Exercise 13.*

In the table that follows, we assume that $a < b$.

Type of Interval	Set	Interval Notation	Graph
Open interval	$\{x \mid x > a\}$	(a, ∞)	
	$\{x \mid a < x < b\}$	(a, b)	
	$\{x \mid x < b\}$	$(-\infty, b)$	
Other intervals	$\{x \mid x \geq a\}$	$[a, \infty)$	
	$\{x \mid a < x \leq b\}$	$(a, b]$	
	$\{x \mid a \leq x < b\}$	$[a, b)$	
	$\{x \mid x \leq b\}$	$(-\infty, b]$	
Closed interval	$\{x \mid a \leq x \leq b\}$	$[a, b]$	
Disjoint interval	$\{x \mid x < a \text{ or } x > b\}$	$(-\infty, a) \cup (b, \infty)$	
All real numbers	$\{x \mid x \text{ is a real number}\}$	$(-\infty, \infty)$	

EXAMPLE 2 **Solving a Linear Inequality**

Solve $4 - 3x \leq 7 + 2x$. Give the solution set in interval notation.

SOLUTION
$$4 - 3x \leq 7 + 2x$$

$$4 - 3x - 4 \leq 7 + 2x - 4 \quad \text{Subtract 4.}$$

$$-3x \leq 3 + 2x \quad \text{Combine like terms.}$$

$$-3x - 2x \leq 3 + 2x - 2x \quad \text{Subtract } 2x.$$

$$-5x \leq 3 \qquad \text{Combine like terms.}$$

$$\frac{-5x}{-5} \geq \frac{3}{-5} \qquad \text{Divide by } -5. \text{ Reverse the direction of the inequality symbol.}$$

$$x \geq -\frac{3}{5}$$

In interval notation the solution set is $\left[-\frac{3}{5}, \infty\right)$. See **Figure 10** for the graph.

☑ *Now Try Exercise 15.*

Figure 10

A product will break even, or begin to produce a profit, only if the revenue from selling the product at least equals the cost of producing it. If R represents revenue and C is cost, then the **break-even point** is the point where $R = C$.

EXAMPLE 3 **Finding the Break-Even Point**

If the revenue and cost of a certain product are given by

$$R = 4x \quad \text{and} \quad C = 2x + 1000,$$

where x is the number of units produced and sold, at what production level does R *at least equal* C?

SOLUTION Set $R \geq C$ and solve for x.

$$R \geq C$$

At least equal to translates as ≥.

$$4x \geq 2x + 1000 \qquad \text{Substitute.}$$

$$2x \geq 1000 \qquad \text{Subtract } 2x.$$

$$x \geq 500 \qquad \text{Divide by 2.}$$

The break-even point is at $x = 500$. This product will at least break even if the number of units produced and sold is in the interval $[500, \infty)$.

☑ *Now Try Exercise 25.*

Three-Part Inequalities The inequality $-2 < 5 + 3x < 20$ says that $5 + 3x$ is *between* -2 and 20. This inequality is solved using an extension of the properties of inequality given earlier, working with all three expressions at the same time.

EXAMPLE 4 **Solving a Three-Part Inequality**

Solve $-2 < 5 + 3x < 20$.

SOLUTION

$$-2 < \quad 5 + 3x \quad < 20$$

$$-2 - 5 < 5 + 3x - 5 < 20 - 5 \qquad \text{Subtract 5 from each part.}$$

$$-7 < \quad 3x \quad < 15 \qquad \text{Combine like terms in each part.}$$

$$\frac{-7}{3} < \quad \frac{3x}{3} \quad < \frac{15}{3} \qquad \text{Divide each part by 3.}$$

$$-\frac{7}{3} < \quad x \quad < 5$$

Figure 11

The solution set, graphed in **Figure 11,** is the interval $\left(-\frac{7}{3}, 5\right)$.

☑ *Now Try Exercise 29.*

Quadratic Inequalities   Solving *quadratic inequalities* is more complicated than solving linear inequalities and depends on finding solutions of quadratic equations.

Quadratic Inequality

A **quadratic inequality** is an inequality that can be written in the form

$$ax^2 + bx + c < 0,$$

for real numbers a, b, and c, with $a \neq 0$. (The symbol $<$ can be replaced with $>$, $\leq$, or $\geq$.)

One method of solving a quadratic inequality involves finding the solutions of the corresponding quadratic equation and then testing values in the intervals on a number line determined by those solutions.

Solving a Quadratic Inequality

Step 1 Solve the corresponding quadratic equation.

Step 2 Identify the intervals determined by the solutions of the equation.

Step 3 Use a test value from each interval to determine which intervals form the solution set.

EXAMPLE 5   **Solving a Quadratic Inequality**

Solve $x^2 - x - 12 < 0$.

SOLUTION

Step 1 Find the values of x that satisfy $x^2 - x - 12 = 0$.

$$x^2 - x - 12 = 0 \qquad \text{Corresponding quadratic equation}$$

$$(x + 3)(x - 4) = 0 \qquad \text{Factor. (Section R.4)}$$

$$x + 3 = 0 \quad \text{or} \quad x - 4 = 0 \qquad \text{Zero-factor property (Section 1.4)}$$

$$x = -3 \quad \text{or} \qquad x = 4 \qquad \text{Solve each equation.}$$

Step 2 The two numbers -3 and 4 cause the expression $x^2 - x - 12$ to *equal* zero and can be used to divide the number line into three intervals, as shown in **Figure 12.** The expression $x^2 - x - 12$ will take on a value that is either *less than* zero or *greater than* zero on each of these intervals. Since we are looking for x-values that make the expression *less than* zero, use open circles at -3 and 4 to indicate that they are not included in the solution set.

Use open circles since the inequality symbol does not include *equality*. -3 and 4 do not satisfy the *inequality*.

Figure 12

Step 3 Choose a test value in each interval to see whether it satisfies the original inequality, $x^2 - x - 12 < 0$. If the test value makes the statement true, then the entire interval belongs to the solution set.

Interval	Test Value	Is $x^2 - x - 12 < 0$ True or False?
A: $(-\infty, -3)$	-4	$(-4)^2 - (-4) - 12 \overset{?}{<} 0$ $8 < 0$ False
B: $(-3, 4)$	0	$0^2 - 0 - 12 \overset{?}{<} 0$ $-12 < 0$ True
C: $(4, \infty)$	5	$5^2 - 5 - 12 \overset{?}{<} 0$ $8 < 0$ False

Since the values in Interval B make the inequality true, the solution set is $(-3, 4)$. See **Figure 13.**

☑ *Now Try Exercise 41.*

Figure 13

> **EXAMPLE 6**   **Solving a Quadratic Inequality**

Solve $2x^2 + 5x - 12 \geq 0$.

SOLUTION

Step 1 Find the values of x that satisfy $2x^2 + 5x - 12 = 0$.

$$2x^2 + 5x - 12 = 0 \qquad \text{Corresponding quadratic equation}$$

$$(2x - 3)(x + 4) = 0 \qquad \text{Factor.}$$

$$2x - 3 = 0 \quad \text{or} \quad x + 4 = 0 \qquad \text{Zero-factor property}$$

$$x = \frac{3}{2} \quad \text{or} \qquad x = -4 \quad \text{Solve each equation.}$$

Step 2 The values $\frac{3}{2}$ and -4 cause the inequality $2x^2 + 5x - 12$ to equal 0 and can be used to form the intervals $(-\infty, -4)$, $\left(-4, \frac{3}{2}\right)$, and $\left(\frac{3}{2}, \infty\right)$ on the number line, as seen in **Figure 14.**

Use closed circles since the inequality symbol includes *equality.* -4 and $\frac{3}{2}$ satisfy the *inequality.*

Figure 14

Step 3 Choose a test value from each interval.

Interval	Test Value	Is $2x^2 + 5x - 12 \geq 0$ True or False?
A: $(-\infty, -4)$	-5	$2(-5)^2 + 5(-5) - 12 \overset{?}{\geq} 0$ $13 \geq 0$ True
B: $\left(-4, \frac{3}{2}\right)$	0	$2(0)^2 + 5(0) - 12 \overset{?}{\geq} 0$ $-12 \geq 0$ False
C: $\left(\frac{3}{2}, \infty\right)$	2	$2(2)^2 + 5(2) - 12 \overset{?}{\geq} 0$ $6 \geq 0$ True

Figure 15

The values in Intervals A and C make the inequality true, so the solution set is the *union* (**Section R.1**) of the two intervals, written $(-\infty, -4] \cup \left[\frac{3}{2}, \infty\right)$. The graph of the solution set is shown in **Figure 15.**

☑ *Now Try Exercise 39.*

NOTE Inequalities that use the symbols $<$ and $>$ are **strict inequalities,** while $\leq$ and $\geq$ are used in **nonstrict inequalities.** The solutions of the equation in **Example 5** were not included in the solution set since the inequality was a *strict* inequality. In **Example 6,** the solutions of the equation *were* included in the solution set because of the nonstrict inequality.

EXAMPLE 7 **Finding Projectile Height**

If a projectile is launched from ground level with an initial velocity of 96 ft per sec, its height s in feet t seconds after launching is given by the following equation.

$$s = -16t^2 + 96t$$

When will the projectile be greater than 80 ft above ground level?

SOLUTION

$$-16t^2 + 96t > 80 \quad \text{Set } s \text{ greater than 80.}$$
$$-16t^2 + 96t - 80 > 0 \quad \text{Subtract 80.}$$

> Reverse the direction of the inequality symbol.

$$t^2 - 6t + 5 < 0 \quad \text{Divide by } -16.$$

Now solve the corresponding *equation.*

$$t^2 - 6t + 5 = 0$$
$$(t - 1)(t - 5) = 0 \quad \text{Factor.}$$
$$t - 1 = 0 \quad \text{or} \quad t - 5 = 0 \quad \text{Zero-factor property}$$
$$t = 1 \quad \text{or} \quad t = 5 \quad \text{Solve each equation.}$$

Use these values to determine the intervals $(-\infty, 1)$, $(1, 5)$, and $(5, \infty)$. Since we are solving a strict inequality, solutions of the equation $t^2 - 6t + 5 = 0$ are *not* included. See **Figure 16.**

Figure 16

Choose a test value from each interval and use the procedure of **Examples 5 and 6** to determine that values in Interval B, $(1, 5)$, satisfy the inequality. The projectile is greater than 80 ft above ground level between 1 and 5 sec after it is launched.

✔ *Now Try Exercise 95.*

Rational Inequalities Inequalities involving one or more rational expressions are **rational inequalities.**

$$\frac{5}{x + 4} \geq 1 \quad \text{and} \quad \frac{2x - 1}{3x + 4} < 5 \quad \text{Rational inequalities}$$

Solving a Rational Inequality

Step 1 Rewrite the inequality, if necessary, so that 0 is on one side and there is a single fraction on the other side.

Step 2 Determine the values that will cause either the numerator or the denominator of the rational expression to equal 0. These values determine the intervals on the number line to consider.

Step 3 Use a test value from each interval to determine which intervals form the solution set.

A value causing a denominator to equal zero will never be included in the solution set. If the inequality is strict, any value causing the numerator to equal zero will be excluded. If the inequality is nonstrict, any such value will be included.

CAUTION Solving a rational inequality such as

$$\frac{5}{x + 4} \geq 1$$

by multiplying each side by $x + 4$ to obtain $5 \geq x + 4$ requires considering *two cases,* since the sign of $x + 4$ depends on the value of x. If $x + 4$ is negative, then the inequality symbol must be reversed. The procedure described in the preceding box and used in the next two examples eliminates the need for considering separate cases.

EXAMPLE 8 Solving a Rational Inequality

Solve $\dfrac{5}{x + 4} \geq 1$.

SOLUTION

Step 1

$$\frac{5}{x + 4} - 1 \geq 0 \qquad \text{Subtract 1 so that 0 is on one side.}$$

Note the careful use of parentheses.

$$\frac{5}{x + 4} - \frac{x + 4}{x + 4} \geq 0 \qquad \begin{array}{l}\text{Use } x + 4 \text{ as the common} \\ \text{denominator.}\end{array}$$

$$\frac{5 - (x + 4)}{x + 4} \geq 0 \qquad \begin{array}{l}\text{Write as a single fraction.} \\ \text{(Section R.5)}\end{array}$$

$$\frac{1 - x}{x + 4} \geq 0 \qquad \begin{array}{l}\text{Combine like terms in the numerator,} \\ \text{being careful with signs.}\end{array}$$

Step 2 The quotient possibly changes sign only where x-values make the numerator or denominator 0. This occurs at

$$1 - x = 0 \quad \text{or} \quad x + 4 = 0$$

$$x = 1 \quad \text{or} \qquad x = -4.$$

These values form the intervals $(-\infty, -4)$, $(-4, 1)$, and $(1, \infty)$ on the number line, as seen in **Figure 17** on the next page.

Figure 17

Step 3 Choose test values.

Interval	Test Value	Is $\dfrac{5}{x+4} \geq 1$ True or False?
A: $(-\infty, -4)$	-5	$\dfrac{5}{-5+4} \stackrel{?}{\geq} 1$ $-5 \geq 1$ False
B: $(-4, 1)$	0	$\dfrac{5}{0+4} \stackrel{?}{\geq} 1$ $\dfrac{5}{4} \geq 1$ True
C: $(1, \infty)$	2	$\dfrac{5}{2+4} \stackrel{?}{\geq} 1$ $\dfrac{5}{6} \geq 1$ False

The values in the interval $(-4, 1)$ satisfy the original inequality. The value 1 makes the nonstrict inequality true, so it must be included in the solution set. Since -4 makes the denominator 0, it must be excluded. The solution set is $(-4, 1\,]$.

✔ *Now Try Exercise 73.*

CAUTION *Be careful with the endpoints of the intervals when solving rational inequalities.*

EXAMPLE 9 Solving a Rational Inequality

Solve $\dfrac{2x-1}{3x+4} < 5$.

SOLUTION

$$\dfrac{2x-1}{3x+4} - 5 < 0 \qquad \text{Subtract 5.}$$

$$\dfrac{2x-1}{3x+4} - \dfrac{5(3x+4)}{3x+4} < 0 \qquad \text{Common denominator is } 3x+4.$$

$$\dfrac{2x-1-5(3x+4)}{3x+4} < 0 \qquad \text{Write as a single fraction.}$$

Be careful with signs. $\quad \dfrac{2x-1-15x-20}{3x+4} < 0 \qquad \text{Distributive property \textbf{(Section R.2)}}$

$$\dfrac{-13x-21}{3x+4} < 0 \qquad \text{Combine like terms in the numerator.}$$

Set the numerator and denominator equal to 0 and solve the resulting equations to find the values of x where sign changes may occur.

$$-13x - 21 = 0 \qquad \text{or} \qquad 3x + 4 = 0$$

$$x = -\dfrac{21}{13} \qquad \text{or} \qquad x = -\dfrac{4}{3}$$

Use these values to form intervals on the number line, as seen in **Figure 18.**

Use an open circle at $-\frac{21}{13}$ because of the strict inequality, and use an open circle at $-\frac{4}{3}$ since it causes the denominator to equal 0.

Figure 18

Now choose test values from the intervals in **Figure 18.** Verify that

-2 from Interval A makes the inequality true.

-1.5 from Interval B makes the inequality false.

0 from Interval C makes the inequality true.

Because of the $<$ symbol, neither endpoint satisfies the inequality, so the solution set is

$$\left(-\infty, -\frac{21}{13}\right) \cup \left(-\frac{4}{3}, \infty\right).$$

✔ *Now Try Exercise 85.*

1.7 Exercises

Concept Check Match the inequality in each exercise in Column I with its equivalent interval notation in Column II.

I

1. $x < -6$

2. $x \leq 6$

3. $-2 < x \leq 6$

4. $x^2 \geq 0$

5. $x \geq -6$

6. $6 \leq x$

7. ![number line with bracket at -2 to parenthesis at 6]
$-2 \quad 0 \quad\quad 6$

8. ![number line from 0 parenthesis to 8]
$0 \quad\quad\quad 8$

9. ![number line parenthesis at -3 and 3]
$-3 \quad 0 \quad 3$

10. ![number line bracket at -6]
$-6 \quad\quad 0$

II

A. $(-2, 6]$

B. $[-2, 6)$

C. $(-\infty, -6]$

D. $[6, \infty)$

E. $(-\infty, -3) \cup (3, \infty)$

F. $(-\infty, -6)$

G. $(0, 8)$

H. $(-\infty, \infty)$

I. $[-6, \infty)$

J. $(-\infty, 6]$

11. Explain how to determine whether to use a parenthesis or a square bracket when graphing the solution set of a linear inequality.

12. *Concept Check* The three-part inequality $a < x < b$ means "a is less than x and x is less than b." Which one of the following inequalities is not satisfied by some real number x?

A. $-3 < x < 10$ **B.** $0 < x < 6$

C. $-3 < x < -1$ **D.** $-8 < x < -10$

Solve each inequality. Write each solution set in interval notation. ***See Examples 1 and 2.***

13. $-2x + 8 \leq 16$

14. $-3x - 8 \leq 7$

15. $-2x - 2 \leq 1 + x$

16. $-4x + 3 \geq -2 + x$

17. $3(x + 5) + 1 \geq 5 + 3x$

18. $6x - (2x + 3) \geq 4x - 5$

19. $8x - 3x + 2 < 2(x + 7)$

20. $2 - 4x + 5(x - 1) < -6(x - 2)$

21. $\dfrac{4x + 7}{-3} \leq 2x + 5$

22. $\dfrac{2x - 5}{-8} \leq 1 - x$

23. $\dfrac{1}{3}x + \dfrac{2}{5}x - \dfrac{1}{2}(x + 3) \leq \dfrac{1}{10}$

24. $-\dfrac{2}{3}x - \dfrac{1}{6}x + \dfrac{2}{3}(x + 1) \leq \dfrac{4}{3}$

Break-Even Interval *Find all intervals where each product will at least break even.* ***See Example 3.***

25. The cost to produce x units of picture frames is $C = 50x + 5000$, while the revenue is $R = 60x$.

26. The cost to produce x units of baseball caps is $C = 100x + 6000$, while the revenue is $R = 500x$.

27. The cost to produce x units of coffee cups is $C = 105x + 900$, while the revenue is $R = 85x$.

28. The cost to produce x units of briefcases is $C = 70x + 500$, while the revenue is $R = 60x$.

Solve each inequality. Write each solution set in interval notation. ***See Example 4.***

29. $-5 < 5 + 2x < 11$

30. $-7 < 2 + 3x < 5$

31. $10 \leq 2x + 4 \leq 16$

32. $-6 \leq 6x + 3 \leq 21$

33. $-11 > -3x + 1 > -17$

34. $2 > -6x + 3 > -3$

35. $-4 \leq \dfrac{x + 1}{2} \leq 5$

36. $-5 \leq \dfrac{x - 3}{3} \leq 1$

37. $-3 \leq \dfrac{x - 4}{-5} < 4$

38. $1 \leq \dfrac{4x - 5}{-2} < 9$

Solve each quadratic inequality. Write each solution set in interval notation. ***See Examples 5 and 6.***

39. $x^2 - x - 6 > 0$

40. $x^2 - 7x + 10 > 0$

41. $2x^2 - 9x \leq 18$

42. $3x^2 + x \leq 4$

43. $-x^2 - 4x - 6 \leq -3$

44. $-x^2 - 6x - 16 > -8$

45. $x(x - 1) \leq 6$

46. $x(x + 1) < 12$

47. $x^2 \leq 9$

48. $x^2 > 16$

49. $x^2 + 5x + 7 < 0$

50. $4x^2 + 3x + 1 \leq 0$

51. $x^2 - 2x \leq 1$

52. $x^2 + 4x > -1$

53. *Concept Check* Which one of the following inequalities has solution set $(-\infty, \infty)$?

    A. $(x - 3)^2 \geq 0$ **B.** $(5x - 6)^2 \leq 0$

    C. $(6x + 4)^2 > 0$ **D.** $(8x + 7)^2 < 0$

54. *Concept Check* Which one of the inequalities in **Exercise 53** has solution set $\varnothing$?

Relating Concepts

For individual or collaborative investigation *(Exercises 55–58)*

Inequalities that involve more than two factors, such as

$$(3x - 4)(x + 2)(x + 6) \leq 0,$$

*can be solved using an extension of the method shown in **Examples 5 and 6**. Work Exercises 55–58 in order, to see how the method is extended.*

55. Use the zero-factor property to solve $(3x - 4)(x + 2)(x + 6) = 0$.

56. Plot the three solutions in **Exercise 55** on a number line, using closed circles because of the nonstrict inequality, $\leq$.

57. The number line from **Exercise 56** should show four intervals formed by the three points. For each interval, choose a number from the interval and decide whether it satisfies the original inequality.

58. On a single number line, graph the intervals that satisfy the inequality, including endpoints. This is the graph of the solution set of the inequality. Write the solution set in interval notation.

*Use the technique described in **Relating Concepts Exercises 55–58** to solve each inequality. Write each solution set in interval notation.*

59. $(2x - 3)(x + 2)(x - 3) \geq 0$ **60.** $(x + 5)(3x - 4)(x + 2) \geq 0$

61. $4x - x^3 \geq 0$ **62.** $16x - x^3 \geq 0$

63. $(x + 1)^2 (x - 3) < 0$ **64.** $(x - 5)^2 (x + 1) < 0$

65. $x^3 + 4x^2 - 9x \geq 36$ **66.** $x^3 + 3x^2 - 16x \leq 48$

67. $x^2(x + 4)^2 \geq 0$ **68.** $-x^2(2x - 3)^2 \leq 0$

*Solve each rational inequality. Write each solution set in interval notation. **See Examples 8 and 9.***

69. $\dfrac{x - 3}{x + 5} \leq 0$ **70.** $\dfrac{x + 1}{x - 4} > 0$ **71.** $\dfrac{1 - x}{x + 2} < -1$

72. $\dfrac{6 - x}{x + 2} > 1$ **73.** $\dfrac{3}{x - 6} \leq 2$ **74.** $\dfrac{3}{x - 2} < 1$

75. $\dfrac{-4}{1 - x} < 5$ **76.** $\dfrac{-6}{3x - 5} \leq 2$ **77.** $\dfrac{10}{3 + 2x} \leq 5$

78. $\dfrac{1}{x + 2} \geq 3$ **79.** $\dfrac{7}{x + 2} \geq \dfrac{1}{x + 2}$ **80.** $\dfrac{5}{x + 1} > \dfrac{12}{x + 1}$

81. $\dfrac{3}{2x - 1} > \dfrac{-4}{x}$ **82.** $\dfrac{-5}{3x + 2} \geq \dfrac{5}{x}$ **83.** $\dfrac{4}{2 - x} \geq \dfrac{3}{1 - x}$

84. $\dfrac{4}{x + 1} < \dfrac{2}{x + 3}$ **85.** $\dfrac{x + 3}{x - 5} \leq 1$ **86.** $\dfrac{x + 2}{3 + 2x} \leq 5$

Solve each rational inequality. Write each solution set in interval notation.

87. $\dfrac{2x - 3}{x^2 + 1} \geq 0$ **88.** $\dfrac{9x - 8}{4x^2 + 25} < 0$ **89.** $\dfrac{(5 - 3x)^2}{(2x - 5)^3} > 0$

90. $\dfrac{(5x - 3)^3}{(25 - 8x)^2} \leq 0$ **91.** $\dfrac{(2x - 3)(3x + 8)}{(x - 6)^3} \geq 0$ **92.** $\dfrac{(9x - 11)(2x + 7)}{(3x - 8)^3} > 0$

(Modeling) Solve each problem. For Exercises 95 and 96, **see Example 7.**

93. *Box Office Receipts* U.S. movie box office receipts, in billions of dollars, are shown in 5-year increments from 1989 to 2009. (*Source:* www.boxofficemojo.com)

Year	Receipts
1989	5.033
1994	5.396
1999	7.413
2004	9.381
2009	10.595

These receipts R are reasonably approximated by the linear model

$$R = 0.3022x + 4.542,$$

where $x = 0$ corresponds to 1989, $x = 5$ corresponds to 1994, and so on. Using the model, calculate the year in which the receipts first exceed each amount.

(a) $6.5 billion   **(b)** $10 billion

94. *Recovery of Solid Waste* The percent W of municipal solid waste recovered is shown in the bar graph. The linear model

$$W = 0.68x + 30.6,$$

where $x = 1$ represents 2004, $x = 2$ represents 2005, and so on, fits the data reasonably well.

(a) Based on this model, when did the percent of waste recovered first exceed 32%?
(b) In what years was it between 33% and 34%?

Municipal Solid Waste Recovered

Source: U.S. Environmental Protection Agency.

95. *Height of a Projectile* A projectile is fired straight up from ground level. After t seconds, its height above the ground is s feet, where

$$s = -16t^2 + 220t.$$

For what time period is the projectile at least 624 ft above the ground?

96. *Height of a Baseball* A baseball is hit so that its height, s, in feet after t seconds is

$$s = -16t^2 + 44t + 4.$$

For what time period is the ball at least 32 ft above the ground?

97. *Velocity of an Object* Suppose the velocity, v, of an object is given by

$$v = 2t^2 - 5t - 12,$$

where t is time in seconds. (Here t can be positive or negative.) Find the intervals where the velocity is negative.

98. *Velocity of an Object* The velocity of an object, v, after t seconds is given by

$$v = 3t^2 - 18t + 24.$$

Find the interval where the velocity is negative.

99. A student attempted to solve the inequality

$$\frac{2x - 3}{x + 2} \le 0$$

by multiplying each side by $x + 2$ to get

$$2x - 3 \le 0$$

$$x \le \frac{3}{2}.$$

He wrote the solution set as $\left(-\infty, \frac{3}{2}\right]$. Is his solution correct? Explain.

100. A student solved the inequality $x^2 \le 144$ by taking the square root of each side to get $x \le 12$. She wrote the solution set as $(-\infty, 12]$. Is her solution correct? Explain.

1.8 Absolute Value Equations and Inequalities

- Basic Concepts
- Absolute Value Equations
- Absolute Value Inequalities
- Special Cases
- Absolute Value Models for Distance and Tolerance

Basic Concepts Recall from **Section R.2** that the **absolute value** of a number a, written $|a|$, gives the distance from a to 0 on a number line. By this definition, the equation $|x| = 3$ can be solved by finding all real numbers at a distance of 3 units from 0. As shown in **Figure 19,** two numbers satisfy this equation, 3 and -3, so the solution set is $\{-3, 3\}$.

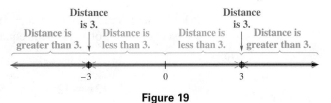

Figure 19

LOOKING AHEAD TO CALCULUS

The precise definition of a **limit** in calculus requires writing absolute value inequalities.

A standard problem in calculus is to find the "interval of convergence" of something called a **power series,** by solving an inequality of the form

$$|x - a| < r.$$

This inequality says that x can be any number within r units of a on the number line, so its solution set is indeed an interval—namely the interval $(a - r, a + r)$.

Similarly, $|x| < 3$ is satisfied by all real numbers whose distances from 0 are less than 3—that is, the interval

$$-3 < x < 3, \quad \text{or} \quad (-3, 3).$$

See **Figure 19.** Finally, $|x| > 3$ is satisfied by all real numbers whose distances from 0 are greater than 3. These numbers are less than -3 or greater than 3, so the solution set is

$$(-\infty, -3) \cup (3, \infty).$$

Notice in **Figure 19** that the union of the solution sets of $|x| = 3$, $|x| < 3$, and $|x| > 3$ is the set of real numbers.

These observations support the cases for solving absolute value equations and inequalities summarized in the table on the next page. If the equation or inequality fits the form of Case 1, 2, or 3, change it to its equivalent form and solve. The solution set and its graph will look similar to those shown.

For each equation or inequality in Cases 1–3 in the table, assume that $k > 0$.

Solving Absolute Value Equations and Inequalities

Absolute Value Equation or Inequality	Equivalent Form	Graph of the Solution Set	Solution Set
Case 1: $\lvert x \rvert = k$	$x = k$ or $x = -k$	$-k \qquad k$	$\{-k, k\}$
Case 2: $\lvert x \rvert < k$	$-k < x < k$	$-k \qquad k$	$(-k, k)$
Case 3: $\lvert x \rvert > k$	$x < -k$ or $x > k$	$-k \qquad k$	$(-\infty, -k) \cup (k, \infty)$

In Cases 2 and 3, the strict inequality may be replaced by its nonstrict form. Additionally, if an absolute value equation takes the form $\lvert a \rvert = \lvert b \rvert$, then a and b must be equal in value or opposite in value.

Thus, the equivalent form of $\lvert a \rvert = \lvert b \rvert$ is $a = b$ or $a = -b$.

Absolute Value Equations Because absolute value represents distance from 0 on a number line, solving an absolute value equation requires solving two possibilities, as shown in the examples that follow.

EXAMPLE 1 Solving Absolute Value Equations (Case 1 and the Special Case $\lvert a \rvert = \lvert b \rvert$)

Solve each equation.

(a) $\lvert 5 - 3x \rvert = 12$ **(b)** $\lvert 4x - 3 \rvert = \lvert x + 6 \rvert$

SOLUTION

(a) For the given expression $5 - 3x$ to have absolute value 12, it must represent either 12 or -12. This equation fits the form of Case 1.

$$\lvert 5 - 3x \rvert = 12$$

Don't forget this second possibility.

$$5 - 3x = 12 \quad \text{or} \quad 5 - 3x = -12 \quad \text{Case 1}$$
$$-3x = 7 \quad \text{or} \quad -3x = -17 \quad \text{Subtract 5.}$$
$$x = -\frac{7}{3} \quad \text{or} \quad x = \frac{17}{3} \quad \text{Divide by } -3.$$

Check the solutions $-\frac{7}{3}$ and $\frac{17}{3}$ by substituting them in the original absolute value equation. The solution set is $\left\{ -\frac{7}{3}, \frac{17}{3} \right\}$.

(b) If the absolute values of two expressions are equal, then those expressions are either equal in value or opposite in value.

$$\lvert 4x - 3 \rvert = \lvert x + 6 \rvert$$

$$4x - 3 = x + 6 \quad \text{or} \quad 4x - 3 = -(x + 6) \quad \text{Consider both possibilities.}$$
$$3x = 9 \quad \text{or} \quad 4x - 3 = -x - 6 \quad \text{Solve each linear equation.}$$
$$x = 3 \quad \text{or} \quad 5x = -3 \quad \text{(Section 1.1)}$$
$$x = -\frac{3}{5}$$

CHECK $\qquad |4x - 3| = |x + 6| \qquad$ Original equation

$$|4\left(-\tfrac{3}{5}\right) - 3| \overset{?}{=} |-\tfrac{3}{5} + 6| \quad \text{Let } x = -\tfrac{3}{5}. \qquad |4(3) - 3| \overset{?}{=} |3 + 6| \quad \text{Let } x = 3.$$

$$|-\tfrac{12}{5} - 3| \overset{?}{=} |-\tfrac{3}{5} + 6| \qquad\qquad\qquad\qquad |12 - 3| \overset{?}{=} |3 + 6|$$

$$|-\tfrac{27}{5}| = |\tfrac{27}{5}| \ \checkmark \quad \text{True} \qquad\qquad\qquad\quad |9| = |9| \ \checkmark \quad \text{True}$$

Both solutions check. The solution set is $\left\{-\tfrac{3}{5}, 3\right\}$.

✔ *Now Try Exercises 9 and 19.*

Absolute Value Inequalities

EXAMPLE 2 **Solving Absolute Value Inequalities (Cases 2 and 3)**

Solve each inequality.

(a) $|2x + 1| < 7$ $\qquad\qquad$ **(b)** $|2x + 1| > 7$

SOLUTION

(a) This inequality fits Case 2. If the absolute value of an expression is less than 7, then the value of the expression is *between* -7 and 7.

$$|2x + 1| < 7$$

$$-7 < 2x + 1 < 7 \quad \text{Case 2}$$

$$-8 < \quad 2x \quad < 6 \quad \text{Subtract 1 from each part. (Section 1.7)}$$

$$-4 < \quad x \quad < 3 \quad \text{Divide each part by 2.}$$

The final inequality gives the solution set $(-4, 3)$.

(b) This inequality fits Case 3. If the absolute value of an expression is greater than 7, then the value of the expression is either less than -7 *or* greater than 7.

$$|2x + 1| > 7$$

$$2x + 1 < -7 \quad \text{or} \quad 2x + 1 > 7 \quad \text{Case 3}$$

$$2x < -8 \quad \text{or} \quad\quad 2x > 6 \quad \text{Subtract 1 from each side.}$$

$$x < -4 \quad \text{or} \quad\quad\ x > 3 \quad \text{Divide each side by 2.}$$

The solution set is $(-\infty, -4) \cup (3, \infty)$.

✔ *Now Try Exercises 27 and 29.*

Cases 1, 2, and 3 require that the absolute value expression be isolated on one side of the equation or inequality.

EXAMPLE 3 **Solving an Absolute Value Inequality (Case 3)**

Solve $|2 - 7x| - 1 > 4$.

SOLUTION $\qquad\qquad |2 - 7x| - 1 > 4$

$$|2 - 7x| > 5 \qquad \text{Add 1 to each side.}$$

$$2 - 7x < -5 \quad \text{or} \quad 2 - 7x > 5 \qquad \text{Case 3}$$

$$-7x < -7 \quad \text{or} \quad\quad -7x > 3 \qquad \text{Subtract 2.}$$

$$x > 1 \quad \text{or} \quad\quad x < -\frac{3}{7} \qquad \begin{array}{l}\text{Divide by } -7. \text{ Reverse the} \\ \text{direction of each inequality.} \\ \text{(Section 1.7)}\end{array}$$

The solution set is $\left(-\infty, -\tfrac{3}{7}\right) \cup (1, \infty)$.

✔ *Now Try Exercise 51.*

> **Special Cases**   The three cases given in this section require the constant k to be positive. ***When $k \leq 0$, use the fact that the absolute value of any expression must be nonnegative, and consider the truth of the statement.***

EXAMPLE 4   **Solving Special Cases**

Solve each equation or inequality.

(a) $|2 - 5x| \geq -4$ **(b)** $|4x - 7| < -3$ **(c)** $|5x + 15| = 0$

SOLUTION

(a) Since the absolute value of a number is always nonnegative, the inequality

$$|2 - 5x| \geq -4 \text{ is always true.}$$

The solution set includes all real numbers, written $(-\infty, \infty)$.

(b) There is no number whose absolute value is less than -3 (or less than *any* negative number).

$$\text{The solution set of } |4x - 7| < -3 \text{ is } \emptyset.$$

(c) The absolute value of a number will be 0 only if that number is 0. Therefore, $|5x + 15| = 0$ is equivalent to

$$5x + 15 = 0, \quad \text{which has solution set} \quad \{-3\}.$$

   CHECK   Substitute -3 into the original equation.

$$|5x + 15| = 0 \qquad \text{Original equation}$$
$$|5(-3) + 15| \stackrel{?}{=} 0 \qquad \text{Let } x = -3.$$
$$0 = 0 \ \checkmark \ \text{True}$$

> ☑ *Now Try Exercises 55, 57, and 59.*

> **Absolute Value Models for Distance and Tolerance**   If a and b represent two real numbers, then the absolute value of their difference,
>
> $$\text{either} \quad |a - b| \quad \text{or} \quad |b - a|, \quad \text{(Section R.2)}$$
>
> represents the distance between them.

EXAMPLE 5   **Using Absolute Value Inequalities to Describe Distances**

Write each statement using an absolute value inequality.

(a) k is no less than 5 units from 8.   **(b)** n is within 0.001 unit of 6.

SOLUTION

(a) Since the distance from k to 8, written $|k - 8|$ or $|8 - k|$, is no less than 5, the distance is greater than or equal to 5. This can be written as

$$|k - 8| \geq 5, \quad \text{or, equivalently,} \quad |8 - k| \geq 5. \quad \text{Either form is acceptable.}$$

(b) This statement indicates that the distance between n and 6 is less than 0.001.

$$|n - 6| < 0.001, \quad \text{or, equivalently,} \quad |6 - n| < 0.001$$

> ☑ *Now Try Exercises 83 and 85.*

EXAMPLE 6 **Using Absolute Value to Model Tolerance**

In quality control and other applications, we often wish to keep the difference between two quantities within some predetermined amount, called the **tolerance.** Suppose $y = 2x + 1$ and we want y to be within 0.01 unit of 4. For what values of x will this be true?

SOLUTION

$\|y - 4\| < 0.01$	Write an absolute value inequality.
$\|2x + 1 - 4\| < 0.01$	Substitute $2x + 1$ for y.
$\|2x - 3\| < 0.01$	Combine like terms.
$-0.01 < 2x - 3 < 0.01$	Case 2
$2.99 < \quad 2x \quad < 3.01$	Add 3 to each part.
$1.495 < \quad x \quad < 1.505$	Divide each part by 2.

Reversing these steps shows that keeping x in the interval $(1.495, 1.505)$ ensures that the difference between y and 4 is within 0.01 unit.

☑ *Now Try Exercise 89.*

1.8 Exercises

Concept Check Match each equation or inequality in Column I with the graph of its solution set in Column II.

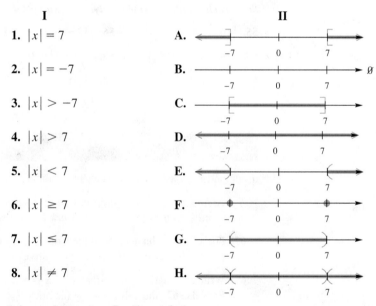

I

1. $|x| = 7$

2. $|x| = -7$

3. $|x| > -7$

4. $|x| > 7$

5. $|x| < 7$

6. $|x| \geq 7$

7. $|x| \leq 7$

8. $|x| \neq 7$

Solve each equation. See Example 1.

9. $|3x - 1| = 2$

10. $|4x + 2| = 5$

11. $|5 - 3x| = 3$

12. $|7 - 3x| = 3$

13. $\left|\dfrac{x - 4}{2}\right| = 5$

14. $\left|\dfrac{x + 2}{2}\right| = 7$

15. $\left|\dfrac{5}{x - 3}\right| = 10$

16. $\left|\dfrac{3}{2x - 1}\right| = 4$

17. $\left|\dfrac{6x + 1}{x - 1}\right| = 3$

18. $\left|\dfrac{2x + 3}{3x - 4}\right| = 1$

19. $|2x - 3| = |5x + 4|$

20. $|x + 1| = |1 - 3x|$

21. $|4 - 3x| = |2 - 3x|$

22. $|3 - 2x| = |5 - 2x|$

23. $|5x - 2| = |2 - 5x|$

24. The equation $|5x - 6| = 3x$ cannot have a negative solution. Why?

25. The equation $|7x + 3| = -5x$ cannot have a positive solution. Why?

26. *Concept Check* Determine the solution set of each equation by inspection.

    (a) $-|x| = |x|$ **(b)** $|-x| = |x|$ **(c)** $|x^2| = |x|$ **(d)** $-|x| = 9$

Solve each inequality. Give the solution set using interval notation. ***See Example 2.***

27. $|2x + 5| < 3$ **28.** $|3x - 4| < 2$ **29.** $|2x + 5| \geq 3$

30. $|3x - 4| \geq 2$ **31.** $\left|\dfrac{1}{2} - x\right| < 2$ **32.** $\left|\dfrac{3}{5} + x\right| < 1$

33. $4|x - 3| > 12$ **34.** $5|x + 1| > 10$ **35.** $|5 - 3x| > 7$

36. $|7 - 3x| > 4$ **37.** $|5 - 3x| \leq 7$ **38.** $|7 - 3x| \leq 4$

39. $\left|\dfrac{2}{3}x + \dfrac{1}{2}\right| \leq \dfrac{1}{6}$ **40.** $\left|\dfrac{5}{3} - \dfrac{1}{2}x\right| > \dfrac{2}{9}$ **41.** $|0.01x + 1| < 0.01$

42. Explain why the equation $|x| = \sqrt{x^2}$ has infinitely many solutions.

Solve each equation or inequality. ***See Examples 3 and 4.***

43. $|4x + 3| - 2 = -1$ **44.** $|8 - 3x| - 3 = -2$ **45.** $|6 - 2x| + 1 = 3$

46. $|4 - 4x| + 2 = 4$ **47.** $|3x + 1| - 1 < 2$ **48.** $|5x + 2| - 2 < 3$

49. $\left|5x + \dfrac{1}{2}\right| - 2 < 5$ **50.** $\left|2x + \dfrac{1}{3}\right| + 1 < 4$ **51.** $|10 - 4x| + 1 \geq 5$

52. $|12 - 6x| + 3 \geq 9$ **53.** $|3x - 7| + 1 < -2$ **54.** $|-5x + 7| - 4 < -6$

Solve each equation or inequality. ***See Example 4.***

55. $|10 - 4x| \geq -4$ **56.** $|12 - 9x| \geq -12$ **57.** $|6 - 3x| < -11$

58. $|18 - 3x| < -13$ **59.** $|8x + 5| = 0$ **60.** $|7 + 2x| = 0$

61. $|4.3x + 9.8| < 0$ **62.** $|1.5x - 14| < 0$ **63.** $|2x + 1| \leq 0$

64. $|3x + 2| \leq 0$ **65.** $|3x + 2| > 0$ **66.** $|4x + 3| > 0$

Relating Concepts

For individual or collaborative investigation *(Exercises 67–70)*

To see how to solve an equation that involves the absolute value of a quadratic polynomial, such as $|x^2 - x| = 6$, ***work Exercises 67–70 in order.***

67. For $x^2 - x$ to have an absolute value equal to 6, what are the two possible values that it may be? (*Hint:* One is positive and the other is negative.)

68. Write an equation stating that $x^2 - x$ is equal to the positive value you found in **Exercise 67,** and solve it using factoring.

69. Write an equation stating that $x^2 - x$ is equal to the negative value you found in **Exercise 67,** and solve it using the quadratic formula. (*Hint:* The solutions are not real numbers.)

70. Give the complete solution set of $|x^2 - x| = 6$, using the results from **Exercises 68 and 69.**

Use the method described in **Relating Concepts Exercises 67–70,** *if applicable, and properties of absolute value to solve each equation or inequality. (Hint: Exercises 77 and 78 can be solved by inspection.)*

71. $|3x^2 + x| = 14$ **72.** $|2x^2 - 3x| = 5$ **73.** $|4x^2 - 23x - 6| = 0$

74. $\left|6x^3 + 23x^2 + 7x\right| = 0$ **75.** $\left|x^2 + 1\right| - \left|2x\right| = 0$ **76.** $\left|\dfrac{x^2 + 2}{x}\right| - \dfrac{11}{3} = 0$

77. $\left|x^4 + 2x^2 + 1\right| < 0$ **78.** $\left|x^2 + 10\right| < 0$

79. $\left|\dfrac{x - 4}{3x + 1}\right| \geq 0$ **80.** $\left|\dfrac{9 - x}{7 + 8x}\right| \geq 0$

81. *Concept Check* Write an equation involving absolute value that says the distance between p and q is 2 units.

82. *Concept Check* Write an equation involving absolute value that says the distance between r and s is 6 units.

Write each statement as an absolute value equation or inequality. See Example 5.

83. m is no more than 2 units from 7. **84.** z is no less than 5 units from 4.

85. p is within 0.0001 unit of 9. **86.** k is within 0.0002 unit of 10.

87. r is no less than 1 unit from 29. **88.** q is no more than 8 units from 22.

89. *Tolerance* Suppose that $y = 5x + 1$ and we want y to be within 0.002 unit of 6. For what values of x will this be true?

90. *Tolerance* Repeat **Exercise 89**, but let $y = 10x + 2$.

(Modeling) *Solve each problem. See Example 6.*

91. *Weights of Babies* Dr. Tydings has found that, over the years, 95% of the babies he has delivered weighed x pounds, where

$$\left|x - 8.2\right| \leq 1.5.$$

What range of weights corresponds to this inequality?

92. *Temperatures on Mars* The temperatures on the surface of Mars in degrees Celsius approximately satisfy the inequality $\left|C + 84\right| \leq 56$. What range of temperatures corresponds to this inequality?

93. *Conversion of Methanol to Gasoline* The industrial process that is used to convert methanol to gasoline is carried out at a temperature range of 680°F to 780°F. Using F as the variable, write an absolute value inequality that corresponds to this range.

94. *Wind Power Extraction Tests* When a model kite was flown in crosswinds in tests to determine its limits of power extraction, it attained speeds of 98 to 148 ft per sec in winds of 16 to 26 ft per sec. Using x as the variable in each case, write absolute value inequalities that correspond to these ranges.

(Modeling) *Carbon Dioxide Emissions* When humans breathe, carbon dioxide is emitted. In one study, the emission rates of carbon dioxide by college students were measured during both lectures and exams. The average individual rate R_L (in grams per hour) during a lecture class satisfied the inequality

$$\left|R_L - 26.75\right| \leq 1.42,$$

whereas during an exam the rate R_E satisfied the inequality

$$\left|R_E - 38.75\right| \leq 2.17.$$

(Source: Wang, T. C., *ASHRAE Trans.,* 81 (Part 1), 32.)

Use this information in Exercises 95–96.

95. Find the range of values for R_L and R_E.

96. The class had 225 students. If T_L and T_E represent the total amounts of carbon dioxide in grams emitted during a one-hour lecture and exam, respectively, write inequalities that model the ranges for T_L and T_E.

Chapter 1 Test Prep

Key Terms

1.1
equation
solution or root
solution set
equivalent equations
linear equation in one
  variable
first-degree equation
identity
conditional equation
contradiction
simple interest
literal equation
future or maturity
  value

1.2
mathematical model
linear model
1.3
imaginary unit
complex number
real part
imaginary part
pure imaginary number
nonreal complex
  number
standard form
complex conjugate
1.4
quadratic equation
standard form
second-degree
  equation

double solution
cubic equation
discriminant
1.5
leg
hypotenuse
1.6
rational equation
proposed solution
equation quadratic
  in form
1.7
inequality
linear inequality in
  one variable

interval
interval notation
open interval
closed interval
break-even point
quadratic inequality
strict inequality
nonstrict inequality
rational inequality
1.8
tolerance

New Symbols

$\emptyset$	empty or null set
i	imaginary unit
∞	infinity

$$\left.\begin{array}{l} (a, b) \\ (-\infty, a] \\ [a, b) \end{array}\right\} \text{ interval notation}$$

$|a|$ absolute value of a

Quick Review

Concepts	Examples

1.1 Linear Equations

Addition and Multiplication Properties of Equality
Let a, b, and c represent real numbers.

$$\text{If } a = b, \text{ then } a + c = b + c.$$

$$\text{If } a = b \text{ and } c \neq 0, \text{ then } ac = bc.$$

Solve. $5(x + 3) = 3x + 7$

$$\begin{aligned} 5x + 15 &= 3x + 7 && \text{Distributive property} \\ 2x &= -8 && \text{Subtract } 3x. \text{ Subtract } 15. \\ x &= -4 && \text{Divide by 2.} \end{aligned}$$

Solution set: $\{-4\}$

1.2 Applications and Modeling with Linear Equations

Problem-Solving Steps
Step 1 Read the problem.

How many liters of 30% alcohol solution and 80% alcohol solution must be mixed to obtain 50 L of 50% alcohol solution?

Concepts	Examples

Step 2 Assign a variable.

Let $x =$ the number of liters of 30% solution.
$50 - x =$ the number of liters of 80% solution.
Summarize the information of the problem in a table.

Strength	Liters of Solution	Liters of Pure Alcohol
30%	x	$0.30x$
80%	$50 - x$	$0.80(50 - x)$
50%	50	$0.50(50)$

Step 3 Write an equation.

The equation is $0.30x + 0.80(50 - x) = 0.50(50)$.

Step 4 Solve the equation.

Solve the equation to obtain $x = 30$.

Step 5 State the answer.

Therefore, 30 L of the 30% solution and $50 - 30 = 20$ L of the 80% solution must be mixed.

Step 6 Check.

CHECK $\quad 0.30(30) + 0.80(50 - 30) \stackrel{?}{=} 0.50(50)$

$$25 = 25 \;\checkmark\; \text{True}$$

1.3 Complex Numbers

Definition of i

$$i = \sqrt{-1} \quad \text{and} \quad i^2 = -1$$

Definition of Complex Number (a and b real)

$$\underset{\substack{\uparrow \\ \text{Real} \\ \text{part}}}{a} + \underset{\substack{\uparrow \\ \text{Imaginary} \\ \text{part}}}{b i}$$

In the complex number $-6 + 2i$, the real part is -6 and the imaginary part is 2.

Definition of $\sqrt{-a}$

For $a > 0$, $\quad \sqrt{-a} = i\sqrt{a}.$

Simplify.

$$\sqrt{-4} = 2i$$
$$\sqrt{-12} = i\sqrt{12} = 2i\sqrt{3}$$

Adding and Subtracting Complex Numbers
Add or subtract the real parts, and add or subtract the imaginary parts.

$(2 + 3i) + (3 + i) - (2 - i)$

$\quad = (2 + 3 - 2) + (3 + 1 + 1)i$

$\quad = 3 + 5i$

Multiplying and Dividing Complex Numbers
Multiply complex numbers as with binomials, and use the fact that $i^2 = -1$.

$(6 + i)(3 - 2i) = 18 - 12i + 3i - 2i^2 \qquad$ FOIL

$\qquad = (18 + 2) + (-12 + 3)i \quad i^2 = -1$

$\qquad = 20 - 9i$

Divide complex numbers by multiplying the numerator and denominator by the complex conjugate of the denominator.

$\dfrac{3 + i}{1 + i} = \dfrac{(3 + i)(1 - i)}{(1 + i)(1 - i)} \quad$ Multiply by $\frac{1-i}{1-i}$.

$\qquad = \dfrac{3 - 3i + i - i^2}{1 - i^2} \quad$ Multiply.

$\qquad = \dfrac{4 - 2i}{2} \quad$ Combine like terms; $i^2 = -1$

$\qquad = \dfrac{2(2 - i)}{2} \quad$ Factor.

$\qquad = 2 - i \quad$ Divide out the common factor.

(continued)

Concepts	Examples

1.4 Quadratic Equations

Zero-Factor Property
If a and b are complex numbers with $ab = 0$, then $a = 0$ or $b = 0$ or both equal zero.

Solve. $6x^2 + x - 1 = 0$

$(3x - 1)(2x + 1) = 0$ Factor.

$3x - 1 = 0$ or $2x + 1 = 0$ Zero-factor property

$x = \dfrac{1}{3}$ or $x = -\dfrac{1}{2}$

Solution set: $\left\{ -\frac{1}{2}, \frac{1}{3} \right\}$

Square Root Property
The solution set of $x^2 = k$ is

$$\left\{ \sqrt{k}, -\sqrt{k} \right\}, \quad \text{abbreviated} \quad \left\{ \pm\sqrt{k} \right\}.$$

Solve. $x^2 = 12$

$$x = \pm\sqrt{12} = \pm 2\sqrt{3}$$

Solution set: $\left\{ \pm 2\sqrt{3} \right\}$

Quadratic Formula
The solutions of the quadratic equation $ax^2 + bx + c = 0$, where $a \neq 0$, are given by the quadratic formula.

$$x = \frac{-b \pm \sqrt{b^2 - 4ac}}{2a}$$

Solve. $x^2 + 2x + 3 = 0$

$$x = \frac{-2 \pm \sqrt{2^2 - 4(1)(3)}}{2(1)} \quad a = 1, b = 2, c = 3$$

$$= \frac{-2 \pm \sqrt{-8}}{2} \qquad \text{Simplify.}$$

$$= \frac{-2 \pm 2i\sqrt{2}}{2} \qquad \text{Simplify the radical.}$$

$$= \frac{2\left(-1 \pm i\sqrt{2}\right)}{2} \qquad \begin{array}{l}\text{Factor out 2 in the}\\\text{numerator.}\end{array}$$

$$= -1 \pm i\sqrt{2} \qquad \text{Lowest terms}$$

Solution set: $\left\{ -1 \pm i\sqrt{2} \right\}$

1.5 Applications and Modeling with Quadratic Equations

Pythagorean Theorem
In a right triangle, the sum of the squares of the lengths of legs a and b is equal to the square of the length of hypotenuse c.

$$a^2 + b^2 = c^2$$

In a right triangle, the shorter leg is 7 in. less than the longer leg, and the hypotenuse is 2 in. greater than the longer leg. What are the lengths of the sides?

Let $x = $ the length of the longer leg.

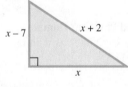

$$(x - 7)^2 + x^2 = (x + 2)^2$$
$$x^2 - 14x + 49 + x^2 = x^2 + 4x + 4$$
$$x^2 - 18x + 45 = 0$$
$$(x - 15)(x - 3) = 0$$
$$x = 15 \quad \text{or} \quad x = 3$$

Height of a Projected Object
The height s (in feet) of an object projected directly upward from an initial height of s_0 feet, with initial velocity v_0 feet per second, is

$$s = -16t^2 + v_0 t + s_0,$$

where t is the number of seconds after the object is projected.

The value 3 must be rejected. The lengths of the sides are 15 in., 8 in., and 17 in. Check to see that the conditions of the problem are satisfied.

Concepts	Examples

1.6 Other Types of Equations and Applications

Power Property

If P and Q are algebraic expressions, then every solution of the equation $P = Q$ is also a solution of the equation $P^n = Q^n$ for any positive integer n.

Quadratic in Form

An equation in the form $au^2 + bu + c = 0$, where u is an algebraic expression, can often be solved by using a substitution variable.

If the power property is applied, or if both sides of an equation are multiplied by a variable expression, check all proposed solutions.

Solve. $(x + 1)^{2/3} + (x + 1)^{1/3} - 6 = 0$

$$u^2 + u - 6 = 0$$

Let $u = (x + 1)^{1/3}$.

$$(u + 3)(u - 2) = 0$$

$u = -3$ or $u = 2$

$(x + 1)^{1/3} = -3$ or $(x + 1)^{1/3} = 2$

$x + 1 = -27$ or $x + 1 = 8$ Cube.

$x = -28$ or $x = 7$ Subtract 1.

Both solutions check. The solution set is $\{-28, 7\}$.

1.7 Inequalities

Properties of Inequality

Let a, b, and c represent real numbers.

1. **If $a < b$, then $a + c < b + c$.**

2. **If $a < b$ and if $c > 0$, then $ac < bc$.**

3. **If $a < b$ and if $c < 0$, then $ac > bc$.**

Solve. $-3(x + 4) + 2x < 6$

$$-3x - 12 + 2x < 6$$

$$-x < 18$$

$x > -18$ Multiply by -1.
Change $<$ to $>$.

Solution set: $(-18, \infty)$

Solving a Quadratic Inequality

Step 1 Solve the corresponding quadratic equation.

Solve. $x^2 + 6x \leq 7$

$x^2 + 6x - 7 = 0$ Corresponding equation

$(x + 7)(x - 1) = 0$ Factor.

$x = -7$ or $x = 1$ Zero-factor property

Step 2 Identify the intervals determined by the solutions of the equation.

Step 3 Use a test value from each interval to determine which intervals form the solution set.

The intervals formed are $(-\infty, -7)$, $(-7, 1)$, and $(1, \infty)$.

Test values show that values in the intervals $(-\infty, -7)$ and $(1, \infty)$ do not satisfy the original inequality, while those in $(-7, 1)$ do. Since the symbol $\leq$ includes equality, the endpoints are included.

Solution set: $[-7, 1]$

Solving a Rational Inequality

Step 1 Rewrite the inequality so that 0 is on one side and a single fraction is on the other.

Solve. $\dfrac{x}{x + 3} \geq \dfrac{5}{x + 3}$

$$\dfrac{x}{x + 3} - \dfrac{5}{x + 3} \geq 0$$

$$\dfrac{x - 5}{x + 3} \geq 0$$

The values -3 and 5 make either the numerator or the denominator 0. The intervals formed are

$$(-\infty, -3), \ (-3, 5), \ \text{and} \ (5, \infty).$$

Step 2 Find the values that make either the numerator or the denominator 0.

Step 3 Use a test value from each interval to determine which intervals form the solution set.

The value -3 must be excluded and 5 must be included. Test values show that values in the intervals $(-\infty, -3)$ and $(5, \infty)$ yield true statements.

Solution set: $(-\infty, -3) \cup [5, \infty)$

(continued)

Concepts	Examples

1.8 Absolute Value Equations and Inequalities

Solving Absolute Value Equations and Inequalities
For each equation or inequality in Cases 1-3, assume that $k > 0$.

Case 1: To solve $|x| = k$, use the equivalent form

$$x = k \quad \text{or} \quad x = -k.$$

Solve. $\qquad |5x - 2| = 3$

$$5x - 2 = 3 \quad \text{or} \quad 5x - 2 = -3$$
$$5x = 5 \quad \text{or} \qquad 5x = -1$$
$$x = 1 \quad \text{or} \qquad x = -\frac{1}{5}$$

Solution set: $\left\{ -\frac{1}{5}, 1 \right\}$

Case 2: To solve $|x| < k$, use the equivalent form

$$-k < x < k.$$

Solve. $\qquad |5x - 2| < 3$

$$-3 < 5x - 2 < 3$$
$$-1 < \quad 5x \quad < 5$$
$$-\frac{1}{5} < \quad x \quad < 1$$

Solution set: $\left(-\frac{1}{5}, 1 \right)$

Case 3: To solve $|x| > k$, use the equivalent form

$$x < -k \quad \text{or} \quad x > k.$$

Solve. $\qquad |5x - 2| \geq 3$

$$5x - 2 \leq -3 \quad \text{or} \quad 5x - 2 \geq 3$$
$$5x \leq -1 \quad \text{or} \qquad 5x \geq 5$$
$$x \leq -\frac{1}{5} \quad \text{or} \qquad x \geq 1$$

Solution set: $\left(-\infty, -\frac{1}{5} \right] \cup [1, \infty)$

Chapter 1 Review Exercises

Solve each equation.

1. $2x + 8 = 3x + 2$

2. $\frac{1}{6}x - \frac{1}{12}(x - 1) = \frac{1}{2}$

3. $5x - 2(x + 4) = 3(2x + 1)$

4. $9x - 11(k + p) = x(a - 1)$, for x

5. $A = \dfrac{24f}{B(p + 1)}$, for f (approximate annual interest rate)

6. *Concept Check* Which of the following cannot be a correct equation to solve a geometry problem, if x represents the measure of a side of a rectangle? (*Hint:* Solve the equations and consider the solutions.)

    A. $2x + 2(x + 2) = 20$ **B.** $2x + 2(5 + x) = -2$

    C. $8(x + 2) + 4x = 16$ **D.** $2x + 2(x - 3) = 10$

7. *Concept Check* If *x* represents the number of pennies in a jar in an applied problem, which of the following equations cannot be a correct equation for finding *x*? (*Hint:* Solve the equations and consider the solutions.)

A. $5x + 3 = 11$ **B.** $12x + 6 = -4$

C. $100x = 50(x + 3)$ **D.** $6(x + 4) = x + 24$

8. *Airline Carry-On Baggage Size* Carry-on rules for domestic economy-class travel differ from one airline to another, as shown in the table.

Airline	Size (linear inches)
AirTran	55
Alaska/Horizon	51
American	45
Delta	45
Southwest	50
United	45
USAirways	45

Source: Individual airline websites.

To determine the number of linear inches for a carry-on, add the length, width, and height of the bag.

(a) One Samsonite rolling bag measures 9 in. by 12 in. by 21 in. Are there any airlines that would not allow it as a carry-on?

(b) A Lark wheeled bag measures 10 in. by 14 in. by 22 in. On which airlines does it qualify as a carry-on?

Solve each problem.

9. *Dimensions of a Square* If the length of each side of a square is decreased by 4 in., the perimeter of the new square is 10 in. more than half the perimeter of the original square. What are the dimensions of the original square?

10. *Distance from a Library* Becky Anderson can ride her bike to the university library in 20 min. The trip home, which is all uphill, takes her 30 min. If her rate is 8 mph faster on her trip there than her trip home, how far does she live from the library?

11. *Alcohol Mixture* Alan wishes to strengthen a mixture that is 10% alcohol to one that is 30% alcohol. How much pure alcohol should he add to 12 L of the 10% mixture?

12. *Loan Interest Rates* A realtor borrowed $90,000 to develop some property. He was able to borrow part of the money at 5.5% interest and the rest at 6%. The annual interest on the two loans amounts to $5125. How much was borrowed at each rate?

13. *Speed of an Excursion Boat* An excursion boat travels upriver to a landing and then returns to its starting point. The trip upriver takes 1.2 hr, and the trip back takes 0.9 hr. If the average speed on the return trip is 5 mph faster than on the trip upriver, what is the boat's speed upriver?

14. *Toxic Waste* Two chemical plants are releasing toxic waste into a holding tank. Plant I releases waste twice as fast as Plant II. Together they fill the tank in 3 hr. How long would it take the slower plant to fill the tank working alone?

15. *(Modeling) Lead Intake* As directed by the "Safe Drinking Water Act" of December 1974, the EPA proposed a maximum lead level in public drinking water of 0.05 mg per liter. This standard assumed an individual consumption of two liters of water per day.

(a) If EPA guidelines are followed, write an equation that models the maximum amount of lead *A* ingested in *x* years. Assume that there are 365.25 days in a year.

(b) If the average life expectancy is 72 yr, find the EPA maximum lead intake from water over a lifetime.

16. *(Modeling) Online Retail Sales* Projected e-commerce sales (in billions of dollars) for the years 2007–2012 can be modeled by the equation

$$y = 32.13x + 172.59,$$

where $x = 0$ corresponds to 2007, $x = 1$ corresponds to 2008, and so on. Based on this model, what would you expect the retail e-commerce sales to be in 2012? (*Source:* Forrester Research, Inc.)

17. *(Modeling) Minimum Wage* Some values of the U.S. minimum hourly wage, in dollars, for selected years from 1956 to 2009 are shown in the table. The linear model

$$y = 0.1132x + 0.4609$$

approximates the minimum wage during this time period, where x is the number of years after 1956 and y is the minimum wage in dollars.

Year	Minimum Wage	Year	Minimum Wage
1956	1.00	1996	4.75
1963	1.25	1997	5.15
1975	2.10	2007	5.85
1981	3.35	2008	6.55
1990	3.80	2009	7.25

Source: Bureau of Labor Statistics.

(a) Use the model to approximate the minimum wage in 1990. How does it compare to the data in the table?

(b) Use the model to approximate the year in which the minimum wage was $5.85. How does your answer compare to the data in the table?

18. *(Modeling) Mobility of the U.S. Population* The U.S. population, in millions, is given in the table. The figures are for the midyear of each decade from the 1960s through the 2000s. If we let the midyear shown in the table represent each decade, we can use the population figures of people moving with the data given in the graph.

Year	Population
1965	194
1975	216
1985	238
1995	263
2005	296

Source: U.S. Census Bureau.

Less Mobile Americans

Percent — 19.7, 18.2, 17.9, 16.7, 15.8 for 1960s, 1970s, 1980s, 1990s, 2000s (Decade)

Source: U.S. Census Bureau.

(a) Find the number of Americans, to the nearest tenth of a million, moving in each year given in the table.

(b) The percents given in the graph decrease each decade, while the populations given in the table increase each decade. From your answers to part (a), is the number of Americans moving each decade increasing or decreasing? Explain.

Perform each operation and write answers in standard form.

19. $(6 - i) + (7 - 2i)$

20. $(-11 + 2i) - (8 - 7i)$

21. $15i - (3 + 2i) - 11$

22. $-6 + 4i - (8i - 2)$

23. $(5 - i)(3 + 4i)$

24. $(-8 + 2i)(-1 + i)$

25. $(5 - 11i)(5 + 11i)$

26. $(4 - 3i)^2$

27. $-5i(3 - i)^2$

28. $4i(2 + 5i)(2 - i)$

29. $\dfrac{-12 - i}{-2 - 5i}$

30. $\dfrac{-7 + i}{-1 - i}$

Find each power of i.

31. i^{11} **32.** i^{60} **33.** i^{1001} **34.** i^{110} **35.** i^{-27} **36.** $\dfrac{1}{i^{17}}$

Solve each equation.

37. $(x + 7)^2 = 5$

38. $(2 - 3x)^2 = 8$

39. $2x^2 + x - 15 = 0$

40. $12x^2 = 8x - 1$

41. $-2x^2 + 11x = -21$

42. $-x(3x + 2) = 5$

43. $(2x + 1)(x - 4) = x$

44. $\sqrt{2}x^2 - 4x + \sqrt{2} = 0$

45. $x^2 - \sqrt{5}\,x - 1 = 0$

46. $(x + 4)(x + 2) = 2x$

47. *Concept Check* Which one of the following equations has two real, distinct solutions? Do not actually solve.

    A. $(3x - 4)^2 = -9$ **B.** $(4 - 7x)^2 = 0$

    C. $(5x - 9)(5x - 9) = 0$ **D.** $(7x + 4)^2 = 11$

48. *Concept Check* Which equations in **Exercise 47** have only one distinct, real solution?

49. *Concept Check* Which one of the equations in **Exercise 47** has two nonreal complex solutions?

Evaluate the discriminant for each equation, and then use it to predict the number and type of solutions.

50. $8x^2 = -2x - 6$

51. $-6x^2 + 2x = -3$

52. $16x^2 + 3 = -26x$

53. $-8x^2 + 10x = 7$ **54.** $25x^2 + 110x + 121 = 0$ **55.** $x(9x + 6) = -1$

 56. Explain how the discriminant of $ax^2 + bx + c = 0\,(a \neq 0)$ is used to determine the number and type of solutions.

Solve each problem.

57. *(Modeling) Height of a Projectile* A projectile is fired straight up from ground level. After t seconds its height s, in feet above the ground, is given by

$$s = 220t - 16t^2.$$

At what times is the projectile exactly 750 ft above the ground?

58. *Dimensions of a Picture Frame* Zach Levy went into a frame-it-yourself shop. He wanted a frame 3 in. longer than it was wide. The frame he chose extended 1.5 in. beyond the picture on each side. Find the outside dimensions of the frame if the area of the unframed picture is 70 in.^{2}.

59. *Kitchen Flooring* Paula Story plans to replace the vinyl floor covering in her 10-ft by 12-ft kitchen. She wants to have a border of even width of a special material. She can afford only 21 ft^2 of this material. How wide a border can she have?

60. *(Modeling) Airplane Landing Speed* To determine the appropriate landing speed of a small airplane, the formula

$$D = 0.1s^2 - 3s + 22$$

is used, where s is the initial landing speed in feet per second and D is the length of the runway in feet. If the landing speed is too fast, the pilot may run out of runway. If the speed is too slow, the plane may stall. If the runway is 800 ft long, what is the appropriate landing speed? Round to the nearest tenth.

61. *(Modeling) U.S. Government Spending on Medical Care* The amount spent in billions of dollars by the U.S. government on medical care during the period 1990–2010 can be approximated by the equation

$$y = 1.717x^2 + 0.8179x + 167.3,$$

where $x = 0$ corresponds to 1990, $x = 1$ corresponds to 1991, and so on. According to this model, about how much was spent by the U.S. government on medical care in 2005? Round to the nearest tenth of a billion. (*Source: U.S. Office of Management and Budget.*)

62. *Dimensions of a Right Triangle* The lengths of the sides of a right triangle are such that the shortest side is 7 in. shorter than the middle side, while the longest side (the hypotenuse) is 1 in. longer than the middle side. Find the lengths of the sides.

Solve each equation.

63. $4x^4 + 3x^2 - 1 = 0$ **64.** $x^2 - 2x^4 = 0$ **65.** $\dfrac{2}{x} - \dfrac{4}{3x} = 8 + \dfrac{3}{x}$

66. $2 - \dfrac{5}{x} = \dfrac{3}{x^2}$ **67.** $\dfrac{10}{4x - 4} = \dfrac{1}{1 - x}$ **68.** $\dfrac{13}{x^2 + 10} = \dfrac{2}{x}$

69. $\dfrac{x}{x + 2} + \dfrac{1}{x} + 3 = \dfrac{2}{x^2 + 2x}$ **70.** $\dfrac{2}{x + 2} + \dfrac{1}{x + 4} = \dfrac{4}{x^2 + 6x + 8}$

71. $(2x + 3)^{2/3} + (2x + 3)^{1/3} - 6 = 0$ **72.** $(x + 3)^{-2/3} - 2(x + 3)^{-1/3} = 3$

73. $\sqrt{4x - 2} = \sqrt{3x + 1}$ **74.** $\sqrt{2x + 3} = x + 2$

75. $\sqrt{x + 2} - x = 2$ **76.** $\sqrt{x} - \sqrt{x + 3} = -1$

77. $\sqrt{x + 3} - \sqrt{3x + 10} = 1$ **78.** $\sqrt{5x - 15} - \sqrt{x + 1} = 2$

79. $\sqrt{x^2 + 3x} - 2 = 0$ **80.** $\sqrt[5]{2x} = \sqrt[5]{3x + 2}$

81. $\sqrt[3]{6x + 2} - \sqrt[3]{4x} = 0$ **82.** $(x - 2)^{2/3} = x^{1/3}$

Solve each inequality. Write each solution set using interval notation.

83. $-9x + 3 < 4x + 10$ **84.** $11x \geq 2(x - 4)$

85. $-5x - 4 \geq 3(2x - 5)$ **86.** $7x - 2(x - 3) \leq 5(2 - x)$

87. $5 \leq 2x - 3 \leq 7$ **88.** $-8 > 3x - 5 > -12$

89. $x^2 + 3x - 4 \leq 0$ **90.** $x^2 + 4x - 21 > 0$

91. $6x^2 - 11x < 10$ **92.** $x^2 - 3x \geq 5$

93. $x^3 - 16x \leq 0$ **94.** $2x^3 - 3x^2 - 5x < 0$

95. $\dfrac{3x + 6}{x - 5} > 0$ **96.** $\dfrac{x + 7}{2x + 1} - 1 \leq 0$

97. $\dfrac{3x-2}{x} - 4 > 0$

98. $\dfrac{5x+2}{x} < -1$

99. $\dfrac{3}{x-1} \le \dfrac{5}{x+3}$

100. $\dfrac{3}{x+2} > \dfrac{2}{x-4}$

(Modeling) *Solve each problem.*

101. *Ozone Concentration* Guideline levels for indoor ozone are less than 50 parts per billion (ppb). In a scientific study, a Purafil air filter was used to reduce an initial ozone concentration of 140 ppb. The filter removed 43% of the ozone. (*Source:* Parmar and Grosjean, *Removal of Air Pollutants from Museum Display Cases,* Getty Conservation Institute, Marina del Rey, CA.)

(a) What is the ozone concentration after the Purafil air filter is used?

(b) What is the maximum initial concentration of ozone that this filter will reduce to an acceptable level?

102. *Break-Even Interval* A company produces earbuds. The revenue from the sale of x units of these earbuds is $R = 8x$. The cost to produce x units of earbuds is $C = 3x + 1500$. In what interval will the company at least break even?

103. *Height of a Projectile* A projectile is launched upward from the ground. Its height s in feet above the ground after t seconds is given by

$$s = 320t - 16t^2.$$

(a) After how many seconds in the air will it hit the ground?

(b) During what time interval is the projectile more than 576 ft above the ground?

104. *Social Security* The total amount paid by the U.S. government to individuals for Social Security retirement and disability insurance benefits during the period 2000–2010 can be approximated by the linear model

$$y = 29.4x + 391.2,$$

where $x = 0$ corresponds to 2000, $x = 1$ corresponds to 2001, and so on. The variable y is in billions of dollars. Based on this model, during what year did the amount paid by the government first exceed \$600 billion? Round your answer to the nearest year. Compare your answer to the bar graph.

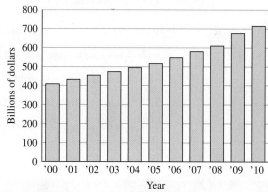

Social Security Retirement and Disability Insurance Benefits

Source: U.S. Office of Management and Budget.

105. Without actually solving the inequality, explain why 3 cannot be in the solution set of $\frac{14x+9}{x-3} < 0$.

106. Without actually solving the inequality, explain why -4 must be in the solution set of $\frac{x+4}{2x+1} \ge 0$.

Rewrite the numerical part of each statement as an inequality using the indicated variable.

107. *Teacher Retirement* A teacher in the Public School Retirement System of Missouri can retire at any age with at least 30 yr of credit. (*Source:* Public School Retirement System of Missouri.) Let c represent the number of years of credit that a retiree has earned.

108. *Honor Roll* To make the honor list for the 2010 fall semester, students at Missouri University of Science and Technology had to carry a minimum of 12 credit hours and have a grade point average of 3.2 or above. (*Source: Warren County Record.*) Let h represent the number of credit hours, and let g represent the grade point average of a student on the honor list.

109. *Blood Donation* Severe winter weather throughout much of the eastern half of the United States caused the cancellation of more than 14,000 blood and platelet donations through the American Red Cross. (*Source: O'Fallon Community News.*) Let x represent the number of donation cancellations.

110. *Blood Donation* Blood is perishable and has no substitute. Red blood cells have a shelf life of 42 days, and platelets have a shelf life of just 5 days. (*Source: O'Fallon Community News.*) Let d represent the number of days that red blood cells may be stored prior to use. Let p represent the number of days that platelets may be stored prior to use.

Solve each equation or inequality.

111. $|x + 4| = 7$

112. $|2 - x| - 3 = 0$

113. $\left| \dfrac{7}{2 - 3x} \right| - 9 = 0$

114. $\left| \dfrac{8x - 1}{3x + 2} \right| = 7$

115. $|5x - 1| = |2x + 3|$

116. $|x + 10| = |x - 11|$

117. $|2x + 9| \le 3$

118. $|8 - 5x| \ge 2$

119. $|7x - 3| > 4$

120. $\left| \dfrac{1}{2}x + \dfrac{2}{3} \right| < 3$

121. $|3x + 7| - 5 = 0$

122. $|7x + 8| - 6 > -3$

123. $|4x - 12| \ge -3$

124. $|7 - 2x| \le -9$

125. $|x^2 + 4x| \le 0$

126. $|x^2 + 4x| > 0$

Write as an absolute value equation or inequality.

127. k is 12 units from 6 on the number line.

128. p is at least 3 units from 1 on the number line.

129. t is no less than 0.01 unit from 5.

130. s is no more than 0.001 unit from 100.

Chapter 1 Test

Solve each equation.

1. $3(x - 4) - 5(x + 2) = 2 - (x + 24)$

2. $\dfrac{2}{3}x + \dfrac{1}{2}(x - 4) = x - 4$

3. $6x^2 - 11x - 7 = 0$

4. $(3x + 1)^2 = 8$

5. $3x^2 + 2x = -2$

6. $\dfrac{12}{x^2 - 9} = \dfrac{2}{x - 3} - \dfrac{3}{x + 3}$

7. $\dfrac{4x}{x - 2} + \dfrac{3}{x} = \dfrac{-6}{x^2 - 2x}$

8. $\sqrt{3x + 4} + 5 = 2x + 1$

9. $\sqrt{-2x + 3} + \sqrt{x + 3} = 3$

10. $\sqrt[3]{3x - 8} = \sqrt[3]{9x + 4}$

11. $x^4 - 17x^2 + 16 = 0$

12. $(x + 3)^{2/3} + (x + 3)^{1/3} - 6 = 0$

13. $|4x + 3| = 7$

14. $|2x + 1| = |5 - x|$

15. *Surface Area of a Rectangular Solid* The formula for the surface area of a rectangular solid is

$$S = 2HW + 2LW + 2LH,$$

where S, H, W, and L represent surface area, height, width, and length, respectively. Solve this formula for W.

16. Perform each operation. Give the answer in standard form.

(a) $(9 - 3i) - (4 + 5i)$

(b) $(4 + 3i)(-5 + 3i)$

(c) $(8 + 3i)^2$

(d) $\dfrac{3 + 19i}{1 + 3i}$

17. Simplify each power of i.

(a) i^{42}

(b) i^{-31}

(c) $\dfrac{1}{i^{19}}$

Solve each problem.

18. *(Modeling) Water Consumption for Snowmaking* Ski resorts require large amounts of water in order to make snow. Snowmass Ski Area in Colorado plans to pump between 1120 and 1900 gal of water per minute at least 12 hr per day from Snowmass Creek between mid-October and late December. (*Source:* York Snow Incorporated.)

(a) Determine an equation that will calculate the *minimum* amount of water A (in gallons) pumped after x days during mid-October to late December.

(b) Find the minimum amount of water pumped in 30 days.

(c) Suppose the water being pumped from Snowmass Creek was used to fill swimming pools. The average backyard swimming pool holds 20,000 gal of water. Determine an equation that will give the minimum number of pools P that could be filled after x days. How many pools could be filled each day?

(d) In how many days could a minimum of 1000 pools be filled?

19. *Dimensions of a Rectangle* The perimeter of a rectangle is 620 m. The length is 20 m less than twice the width. What are the length and width?

20. *Nut Mixture* To make a special mix, the owner of a fruit and nut stand wants to combine cashews that sell for $7.00 per lb with walnuts that sell for $5.50 per lb to obtain 35 lb of a mixture that sells for $6.50 per lb. How many pounds of each type of nut should be used in the mixture?

21. *Speed of a Plane* Mary Lynn left by plane to visit her mother in Louisiana, 420 km away. Fifteen minutes later, her mother left to meet her at the airport. She drove the 20 km to the airport at 40 km per hr, arriving just as the plane taxied in. What was the speed of the plane?

22. *(Modeling) Height of a Projectile* A projectile is launched straight up from ground level with an initial velocity of 96 ft per sec. Its height in feet, s, after t seconds is given by the equation

$$s = -16t^2 + 96t.$$

(a) At what time(s) will it reach a height of 80 ft?
(b) After how many seconds will it return to the ground?

23. *(Modeling) U.S. Airline Passengers* The number of fliers on scheduled flights of U.S. airline companies was approximately 548 million in 1995, 739 million in 2005, and 704 million in 2009. (*Source:* Air Transport Association of America.) Here are three possible models for these data, where $x = 0$ corresponds to the year 1995.

Year	A	B	C
0	562	547.8	544.5
10	688.8	738.6	726.9
14	739.5	703.9	708.5

A. $y = 12.68x + 562.0$
B. $y = -1.983x^2 + 38.91x + 547.8$
C. $y = -1.632x^2 + 34.56x + 544.5$

The table shows each equation evaluated at the years 1995, 2004, and 2009. Decide which equation most closely models the data for these years.

Solve each inequality. Give the answer using interval notation.

24. $-2(x - 1) - 12 < 2(x + 1)$

25. $-3 \leq \dfrac{1}{2}x + 2 \leq 3$

26. $2x^2 - x \geq 3$

27. $\dfrac{x + 1}{x - 3} < 5$

28. $|2x - 5| < 9$

29. $|2x + 1| - 11 \geq 0$

30. $|3x + 7| \leq 0$

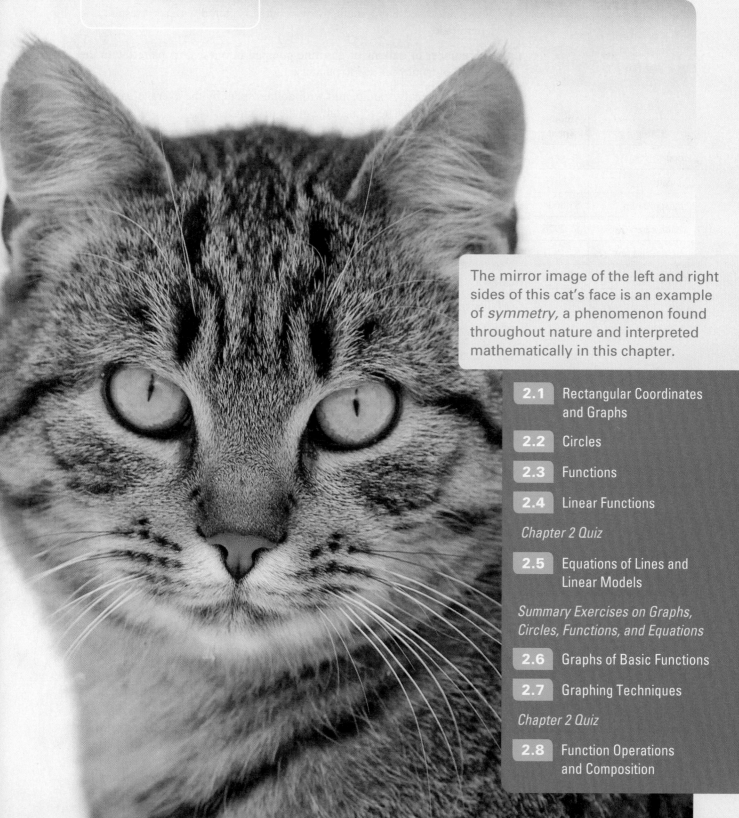

2 Graphs and Functions

The mirror image of the left and right sides of this cat's face is an example of *symmetry*, a phenomenon found throughout nature and interpreted mathematically in this chapter.

2.1 Rectangular Coordinates and Graphs

2.2 Circles

2.3 Functions

2.4 Linear Functions

Chapter 2 Quiz

2.5 Equations of Lines and Linear Models

Summary Exercises on Graphs, Circles, Functions, and Equations

2.6 Graphs of Basic Functions

2.7 Graphing Techniques

Chapter 2 Quiz

2.8 Function Operations and Composition

2.1 Rectangular Coordinates and Graphs

- Ordered Pairs
- The Rectangular Coordinate System
- The Distance Formula
- The Midpoint Formula
- Graphing Equations in Two Variables

Category	Amount Spent
food	$ 6443
housing	$17,109
transportation	$ 8604
health care	$ 2976
apparel and services	$ 1801
entertainment	$ 2835

Source: U.S. Bureau of Labor Statistics.

Ordered Pairs   The idea of pairing one quantity with another is often encountered in everyday life.

- A numerical score in a mathematics course is paired with a corresponding letter grade.

- The number of gallons of gasoline pumped into a tank is paired with the amount of money needed to purchase it.

- Expense categories are paired with dollars spent by the average American household in 2008. (See the table in the margin.)

Pairs of related quantities, such as a 96 determining a grade of A, 3 gallons of gasoline costing $10.50, and 2008 spending on food of $6443, can be expressed as *ordered pairs:* (96, A), (3, $10.50), (food, $6443). An **ordered pair** consists of two components, written inside parentheses.

EXAMPLE 1   **Writing Ordered Pairs**

Use the table to write ordered pairs to express the relationship between each category and the amount spent on it.

(a) housing   **(b)** entertainment

SOLUTION

(a) Use the data in the second row: (housing, $17,109).

(b) Use the data in the last row: (entertainment, $2835).

✔ *Now Try Exercise 9.*

In mathematics, we are most often interested in ordered pairs whose components are numbers. The ordered pairs (a, b) and (c, d) are equal provided that $a = c$ *and* $b = d$.

NOTE   Notation such as $(2, 4)$ was used in **Chapter 1** to show an interval on the number line, and the same notation is used to indicate an ordered pair of numbers. The intended use is usually clear from the context of the discussion.

The Rectangular Coordinate System   As mentioned in **Section R.2,** each real number corresponds to a point on a number line. This idea is extended to ordered pairs of real numbers by using two perpendicular number lines, one horizontal and one vertical, that intersect at their zero-points. This point of intersection is called the **origin.** The horizontal line is called the **x-axis,** and the vertical line is called the **y-axis.**

The x-axis and y-axis together make up a **rectangular coordinate system,** or **Cartesian coordinate system** (named for one of its coinventors, René Descartes. The other coinventor was Pierre de Fermat). The plane into which the coordinate system is introduced is the **coordinate plane,** or **xy-plane.** See **Figure 1.** The x-axis and y-axis divide the plane into four regions, or **quadrants,** labeled as shown. The points on the x-axis and y-axis belong to no quadrant.

Figure 1

Figure 2

Each point P in the xy-plane corresponds to a unique ordered pair (a, b) of real numbers. The point P corresponding to the ordered pair (a, b) often is written $P(a, b)$ as in **Figure 1** and referred to as "the point (a, b)." The numbers a and b are the **coordinates** of point P. To locate on the xy-plane the point corresponding to the ordered pair $(3, 4)$, for example, start at the origin, move 3 units in the positive x-direction, and then move 4 units in the positive y-direction. See **Figure 2**. Point A corresponds to the ordered pair $(3, 4)$.

The Distance Formula   Recall that the distance on a number line between points P and Q with coordinates x_1 and x_2 is

$$d(P, Q) = |x_1 - x_2| = |x_2 - x_1|.$$ Definition of distance **(Section R.2)**

By using the coordinates of their ordered pairs, we can extend this idea to find the distance between any two points in a plane.

Figure 3 shows the points $P(-4, 3)$ and $R(8, -2)$. We complete a right triangle as in the figure. This right triangle has its $90°$ angle at $Q(8, 3)$. The legs have lengths

$$d(P, Q) = |8 - (-4)| = 12$$

and

$$d(Q, R) = |3 - (-2)| = 5.$$

By the Pythagorean theorem, the hypotenuse has length

$$\sqrt{12^2 + 5^2} = \sqrt{144 + 25} = \sqrt{169} = 13.$$ (Section 1.5)

Thus, the distance between $(-4, 3)$ and $(8, -2)$ is 13.

Figure 3

$$d(P, R) = \sqrt{(x_2 - x_1)^2 + (y_2 - y_1)^2}$$

Figure 4

To obtain a general formula, let $P(x_1, y_1)$ and $R(x_2, y_2)$ be any two distinct points in a plane, as shown in **Figure 4**. Complete a triangle by locating point Q with coordinates (x_2, y_1). The Pythagorean theorem gives the distance between P and R.

$$d(P, R) = \sqrt{(x_2 - x_1)^2 + (y_2 - y_1)^2}$$

Absolute value bars are not necessary in this formula, since for all real numbers a and b,

$$|a - b|^2 = (a - b)^2.$$

The **distance formula** can be summarized as follows.

René Descartes (1596–1650)

The initial flash of *analytic geometry* may have come to Descartes as he was watching a fly crawling about on the ceiling near a corner of his room. It struck him that the path of the fly on the ceiling could be described if only one knew the relation connecting the fly's distances from two adjacent walls.

Source: An Introduction to the History of Mathematics by Howard Eves.

Distance Formula

Suppose that $P(x_1, y_1)$ and $R(x_2, y_2)$ are two points in a coordinate plane. The distance between P and R, written $d(P, R)$, is given by the following formula.

$$d(P, R) = \sqrt{(x_2 - x_1)^2 + (y_2 - y_1)^2}$$

LOOKING AHEAD TO CALCULUS

In analytic geometry and calculus, the distance formula is extended to two points in space. Points in space can be represented by **ordered triples.** The distance between the two points (x_1, y_1, z_1) and (x_2, y_2, z_2) is given by the following expression.

$$\sqrt{(x_2 - x_1)^2 + (y_2 - y_1)^2 + (z_2 - z_1)^2}$$

The distance formula can be stated in words.

> *The distance between two points in a coordinate plane is the square root of the sum of the square of the difference between their x-coordinates and the square of the difference between their y-coordinates.*

Although our derivation of the distance formula assumed that P and R are not on a horizontal or vertical line, the result is true for any two points.

EXAMPLE 2 **Using the Distance Formula**

Find the distance between $P(-8, 4)$ and $Q(3, -2)$.

SOLUTION Use the distance formula.

$$
\begin{aligned}
d(P, Q) &= \sqrt{(x_2 - x_1)^2 + (y_2 - y_1)^2} \qquad \text{Distance formula} \\
&= \sqrt{[3 - (-8)]^2 + (-2 - 4)^2} \qquad x_1 = -8,\, y_1 = 4,\, x_2 = 3,\, y_2 = -2 \\
&= \sqrt{11^2 + (-6)^2} \qquad \text{Be careful when subtracting a negative number.} \\
&= \sqrt{121 + 36} \\
&= \sqrt{157}
\end{aligned}
$$

✔ *Now Try Exercise 11(a).*

A statement of the form "If p, then q" is called a **conditional statement.** The related statement "If q, then p" is called its **converse.** In **Section 1.5** we studied the Pythagorean theorem. Its *converse* is also a true statement.

> *If the sides a, b, and c of a triangle satisfy $a^2 + b^2 = c^2$, then the triangle is a right triangle with legs having lengths a and b and hypotenuse having length c.*

We can use this fact to determine whether three points are the vertices of a right triangle.

EXAMPLE 3 **Applying the Distance Formula**

Are points $M(-2, 5)$, $N(12, 3)$, and $Q(10, -11)$ the vertices of a right triangle?

SOLUTION A triangle with the three given points as vertices, shown in **Figure 5,** is a right triangle if the square of the length of the longest side equals the sum of the squares of the lengths of the other two sides. Use the distance formula to find the length of each side of the triangle.

$$d(M, N) = \sqrt{[12 - (-2)]^2 + (3 - 5)^2} = \sqrt{196 + 4} = \sqrt{200}$$

$$d(M, Q) = \sqrt{[10 - (-2)]^2 + (-11 - 5)^2} = \sqrt{144 + 256} = \sqrt{400} = 20$$

$$d(N, Q) = \sqrt{(10 - 12)^2 + (-11 - 3)^2} = \sqrt{4 + 196} = \sqrt{200}$$

The longest side, of length 20 units, is chosen as the possible hypotenuse. Since

$$\left(\sqrt{200}\right)^2 + \left(\sqrt{200}\right)^2 = 400 = 20^2$$

is true, the triangle is a right triangle with hypotenuse joining M and Q.

✔ *Now Try Exercise 19.*

Figure 5

$$d(P, Q) + d(Q, R) = d(P, R)$$

Figure 6

Using a similar procedure, we can tell whether three points are **collinear** (that is, lying on a straight line). *Three points are collinear if the sum of the distances between two pairs of the points is equal to the distance between the remaining pair of points.* See **Figure 6**.

EXAMPLE 4   **Applying the Distance Formula**

Are the points $P(-1, 5)$, $Q(2, -4)$, and $R(4, -10)$ collinear?

SOLUTION

$$d(P, Q) = \sqrt{(-1 - 2)^2 + [5 - (-4)]^2} = \sqrt{9 + 81} = \sqrt{90} = 3\sqrt{10}$$

(Section R.7)

$$d(Q, R) = \sqrt{(2 - 4)^2 + [-4 - (-10)]^2} = \sqrt{4 + 36} = \sqrt{40} = 2\sqrt{10}$$

$$d(P, R) = \sqrt{(-1 - 4)^2 + [5 - (-10)]^2} = \sqrt{25 + 225} = \sqrt{250} = 5\sqrt{10}$$

Because $3\sqrt{10} + 2\sqrt{10} = 5\sqrt{10}$ is true, the three points are collinear.

☑ *Now Try Exercise 25.*

NOTE   In **Exercises 76–80** of **Section 2.5,** we examine another method of determining whether three points are collinear.

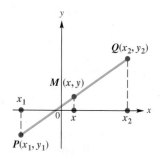

Figure 7

The Midpoint Formula   The midpoint of a line segment is equidistant from the endpoints of the segment. The **midpoint formula** is used to find the coordinates of the midpoint of a line segment. To develop the midpoint formula, let $P(x_1, y_1)$ and $Q(x_2, y_2)$ be any two distinct points in a plane. (Although **Figure 7** shows $x_1 < x_2$, no particular order is required.) Let $M(x, y)$ be the midpoint of the segment joining P and Q. Draw vertical lines from each of the three points to the x-axis, as shown in **Figure 7**.

Since $M(x, y)$ is the midpoint of the line segment joining P and Q, the distance between x and x_1 equals the distance between x and x_2.

$$x_2 - x = x - x_1$$

$$x_2 + x_1 = 2x \qquad \text{Add } x \text{ and } x_1. \text{ (Section 1.1)}$$

$$x = \frac{x_1 + x_2}{2} \qquad \text{Divide by 2 and rewrite.}$$

Similarly, the y-coordinate is $\dfrac{y_1 + y_2}{2}$, yielding the following formula.

Midpoint Formula

The coordinates of the midpoint M of the line segment with endpoints $P(x_1, y_1)$ and $Q(x_2, y_2)$ are given by the following.

$$M = \left(\frac{x_1 + x_2}{2}, \frac{y_1 + y_2}{2} \right)$$

*That is, the x-coordinate of the midpoint of a line segment is the **average of the x-coordinates** of the segment's endpoints, and the y-coordinate is the **average** of the y-coordinates of the segment's endpoints.*

EXAMPLE 5 **Using the Midpoint Formula**

Use the midpoint formula to do each of the following.

(a) Find the coordinates of the midpoint M of the segment with endpoints $(8, -4)$ and $(-6, 1)$.

(b) Find the coordinates of the other endpoint Q of a segment with one endpoint $P(-6, 12)$ and midpoint $M(8, -2)$.

SOLUTION

(a) The coordinates of M are found using the midpoint formula.

$$M = \left(\frac{8 + (-6)}{2}, \frac{-4 + 1}{2} \right) = \left(1, -\frac{3}{2} \right) \quad \text{Substitute in } \left(\frac{x_1 + x_2}{2}, \frac{y_1 + y_2}{2} \right).$$

The coordinates of midpoint M are $\left(1, -\frac{3}{2} \right)$.

(b) Let (x, y) represent the coordinates of Q. Use the midpoint formula twice.

x-value of P x-value of M	y-value of P y-value of M
$\dfrac{x + (-6)}{2} = 8$	$\dfrac{y + 12}{2} = -2$

Substitute carefully.

$$x - 6 = 16 \qquad\qquad y + 12 = -4$$
$$x = 22 \qquad\qquad\quad y = -16$$

The coordinates of endpoint Q are $(22, -16)$.

> ✔ *Now Try Exercises 11(b) and 31.*

EXAMPLE 6 **Applying the Midpoint Formula**

Figure 8 depicts how a graph might indicate the increase in the revenue generated by fast-food restaurants in the United States from $69.8 billion in 1990 to $164.8 billion in 2010. Use the midpoint formula and the two given points to estimate the revenue from fast-food restaurants in 2000, and compare it to the actual figure of $107.1 billion.

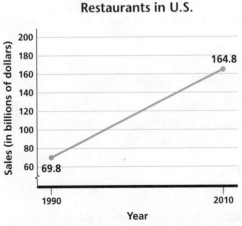

Revenue of Fast-Food Restaurants in U.S.

Source: National Restaurant Association.

Figure 8

SOLUTION The year 2000 lies halfway between 1990 and 2010, so we must find the coordinates of the midpoint M of the segment that has endpoints

$$(1990, 69.8) \quad \text{and} \quad (2010, 164.8).$$

(Here, the second component is in billions of dollars.)

$$M = \left(\frac{1990 + 2010}{2}, \frac{69.8 + 164.8}{2} \right) = (2000, 117.3) \quad \text{Use the midpoint formula.}$$

Thus, our estimate is $117.3 billion, which is greater than the actual figure of $107.1 billion. (This discrepancy is due to actual revenue with an average increase of $3.73 billion between 1990 and 2000, and then of $5.77 billion between 2000 and 2010. Graphs such as this can sometimes be misleading.)

☑ *Now Try Exercise 37.*

Graphing Equations in Two Variables Ordered pairs are used to express the solutions of equations in two variables. When an ordered pair represents the solution of an equation with the variables x and y, the x-value is written first. For example, we say that

$$(1, 2) \quad \text{is a solution of} \quad 2x - y = 0,$$

since substituting 1 for x and 2 for y in the equation gives a true statement.

$$2x - y = 0$$

$$2(1) - 2 \stackrel{?}{=} 0 \qquad \text{Let } x = 1 \text{ and } y = 2.$$

$$0 = 0 \checkmark \quad \text{True}$$

EXAMPLE 7 Finding Ordered-Pair Solutions of Equations

For each equation, find at least three ordered pairs that are solutions.

(a) $y = 4x - 1$ **(b)** $x = \sqrt{y - 1}$ **(c)** $y = x^2 - 4$

SOLUTION

(a) Choose any real number for x or y and substitute in the equation to get the corresponding value of the other variable. For example, let $x = -2$ and then let $y = 3$.

$$y = 4x - 1 \qquad\qquad\qquad y = 4x - 1$$

$$y = 4(-2) - 1 \quad \text{Let } x = -2. \qquad 3 = 4x - 1 \quad \text{Let } y = 3.$$

$$y = -8 - 1 \qquad \text{Multiply.} \qquad\qquad 4 = 4x \qquad \text{Add 1.}$$

$$y = -9 \qquad\quad \text{Subtract.} \qquad\qquad 1 = x \qquad \text{Divide by 4.}$$

This gives the ordered pairs $(-2, -9)$ and $(1, 3)$. Verify that the ordered pair $(0, -1)$ is also a solution.

(b)
$$x = \sqrt{y - 1} \qquad \text{Given equation}$$

$$1 = \sqrt{y - 1} \qquad \text{Let } x = 1.$$

$$1 = y - 1 \qquad \text{Square each side. (Section 1.6)}$$

$$2 = y \qquad\quad \text{Add 1.}$$

One ordered pair is $(1, 2)$. Verify that the ordered pairs $(0, 1)$ and $(2, 5)$ are also solutions of the equation.

(c) A table provides an organized method for determining ordered pairs. Here, we let x equal $-2, -1, 0, 1,$ and 2 in $y = x^2 - 4$ and determine the corresponding y-values.

x	y	
-2	0	$(-2)^2 - 4 = 4 - 4 = 0$
-1	-3	$(-1)^2 - 4 = 1 - 4 = -3$
0	-4	$0^2 - 4 = -4$
1	-3	$1^2 - 4 = -3$
2	0	$2^2 - 4 = 0$

Five ordered pairs are $(-2, 0)$, $(-1, -3)$, $(0, -4)$, $(1, -3)$, and $(2, 0)$.

✔️ *Now Try Exercises 43(a), 47(a), and 49(a).*

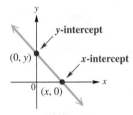

The **graph** of an equation is found by plotting ordered pairs that are solutions of the equation. The **intercepts** of the graph are good points to plot first. An **x-intercept** is a point where the graph intersects the x-axis. A **y-intercept** is a point where the graph intersects the y-axis. In other words, the x-intercept is represented by an ordered pair where $y = 0$, and the y-intercept is represented by an ordered pair where $x = 0$.

A general algebraic approach for graphing an equation using intercepts and point-plotting follows.

Graphing an Equation by Point Plotting

Step 1 Find the intercepts.

Step 2 Find as many additional ordered pairs as needed.

Step 3 Plot the ordered pairs from Steps 1 and 2.

Step 4 Join the points from Step 3 with a smooth line or curve.

EXAMPLE 8 **Graphing Equations**

Graph each of the equations here, from **Example 7.**

(a) $y = 4x - 1$ **(b)** $x = \sqrt{y - 1}$ **(c)** $y = x^2 - 4$

SOLUTION

(a) Step 1 Let $y = 0$ to find the x-intercept, and let $x = 0$ to find the y-intercept.

$$y = 4x - 1 \qquad\qquad y = 4x - 1$$
$$0 = 4x - 1 \quad \text{Let } y = 0. \qquad y = 4(0) - 1 \quad \text{Let } x = 0.$$
$$1 = 4x \qquad\qquad\qquad y = 0 - 1$$
$$\frac{1}{4} = x \qquad\qquad\qquad y = -1$$

The intercepts are $\left(\frac{1}{4}, 0\right)$ and $(0, -1)$.* Note that the y-intercept is one of the ordered pairs we found in **Example 7(a).**

---

*The intercepts are sometimes defined as numbers such as x-intercept $\frac{1}{4}$ and y-intercept -1. In this text, we define them as ordered pairs such as $\left(\frac{1}{4}, 0\right)$ and $(0, -1)$.

Step 2 We use the other ordered pairs found in **Example 7(a):**

$$(-2, -9) \quad \text{and} \quad (1, 3).$$

Step 3 Plot the four ordered pairs from Steps 1 and 2 as shown in **Figure 9.**

Step 4 Join the points plotted in Step 3 with a straight line. This line, also shown in **Figure 9,** is the graph of the equation $y = 4x - 1.$

Figure 9

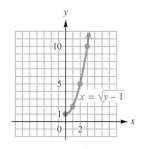

Figure 10

(b) For $x = \sqrt{y - 1}$, the y-intercept $(0, 1)$ was found in **Example 7(b).** Solve

$$x = \sqrt{0 - 1} \quad \text{Let } y = 0.$$

for the x-intercept. Since the quantity under the radical is negative, there is no x-intercept. In fact, $y - 1$ must be greater than or equal to 0, so y must be greater than or equal to 1.

We start by plotting the ordered pairs from **Example 7(b)** and then join the points with a smooth curve as in **Figure 10.** To confirm the direction the curve will take as x increases, we find another solution, $(3, 10)$. (Point plotting for graphs other than lines is often inefficient. We will examine other graphing methods later.)

(c) In **Example 7(c),** we made a table of five ordered pairs that satisfy the equation $y = x^2 - 4.$

$$(-2, 0), \quad (-1, -3), \quad (0, -4), \quad (1, -3), \quad (2, 0)$$
$$\quad\quad\uparrow \quad\quad\quad\quad\quad\quad\quad\quad \uparrow \quad\quad\quad\quad\quad\quad\quad \uparrow$$
$$x\text{-intercept} \quad\quad\quad\quad y\text{-intercept} \quad\quad\quad\quad x\text{-intercept}$$

Plotting the points and joining them with a smooth curve gives the graph in **Figure 11.** This curve is called a **parabola.**

✔️ *Now Try Exercises 43(b), 47(b), and 49(b).*

Figure 11

To graph an equation on a calculator, such as

$$y = 4x - 1, \quad \text{Equation from \textbf{Example 8(a)}}$$

we must first solve it for y (if necessary). Here the equation is already in the correct form, $y = 4x - 1$, so we enter $4x - 1$ for Y_1.

The intercepts can help determine an appropriate window, since we want them to appear in the graph. A good choice is often the **standard viewing window** for the TI-83/84 Plus, which has X minimum $= -10$, X maximum $= 10$, Y minimum $= -10$, Y maximum $= 10$, with X scale $= 1$ and Y scale $= 1$. (The X and Y scales determine the spacing of the tick marks.) Since the intercepts here are very close to the origin, we have chosen the X and Y minimum and maximum to be -3 and 3 instead. See **Figure 12.** ■

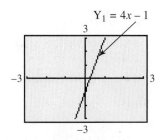

Figure 12

2.1 Exercises

Concept Check Decide whether each statement in Exercises 1–5 is true *or* false. *If the statement is false, tell why.*

1. The point $(-1, 3)$ lies in quadrant III of the rectangular coordinate system.

2. The distance from $P(x_1, y_1)$ to $Q(x_2, y_2)$ is given by
$$d(P, Q) = \sqrt{(x_1 - y_1)^2 + (x_2 - y_2)^2}.$$

3. The distance from the origin to the point (a, b) is $\sqrt{a^2 + b^2}$.

4. The midpoint of the segment joining (a, b) and $(3a, -3b)$ has coordinates $(2a, -b)$.

5. The graph of $y = 2x + 4$ has x-intercept $(-2, 0)$ and y-intercept $(0, 4)$.

6. In your own words, list the steps for graphing an equation.

In Exercises 7–10, give three ordered pairs from each table. **See Example 1.**

7.

x	y
2	−5
−1	7
3	−9
5	−17
6	−21

8.

x	y
3	3
−5	−21
8	18
4	6
0	−6

9. *Percent of High School Students Who Smoke*

Year	Percent
1997	36
1999	35
2001	29
2003	22
2005	23
2007	20

Source: Centers for Disease Control and Prevention.

10. *Number of U.S. Viewers of the Super Bowl*

Year	Viewers (millions)
1998	90.0
2000	88.5
2002	86.8
2004	89.8
2006	90.7
2008	97.4
2010	106.5

Source: www.tvbythenumbers.com

For the points P and Q, find **(a)** *the distance* $d(P, Q)$ *and* **(b)** *the coordinates of the midpoint M of the segment PQ.* **See Examples 2 and 5(a).**

11. $P(-5, -7)$, $Q(-13, 1)$

12. $P(-4, 3)$, $Q(2, -5)$

13. $P(8, 2)$, $Q(3, 5)$

14. $P(-8, 4)$, $Q(3, -5)$

15. $P(-6, -5)$, $Q(6, 10)$

16. $P(6, -2)$, $Q(4, 6)$

17. $P(3\sqrt{2}, 4\sqrt{5})$, $Q(\sqrt{2}, -\sqrt{5})$

18. $P(-\sqrt{7}, 8\sqrt{3})$, $Q(5\sqrt{7}, -\sqrt{3})$

Determine whether the three points are the vertices of a right triangle. **See Example 3.**

19. $(-6, -4)$, $(0, -2)$, $(-10, 8)$

20. $(-2, -8)$, $(0, -4)$, $(-4, -7)$

21. $(-4, 1)$, $(1, 4)$, $(-6, -1)$

22. $(-2, -5)$, $(1, 7)$, $(3, 15)$

23. $(-4, 3)$, $(2, 5)$, $(-1, -6)$

24. $(-7, 4)$, $(6, -2)$, $(0, -15)$

Determine whether the three points are collinear. See Example 4.

25. $(0, -7)$, $(-3, 5)$, $(2, -15)$

26. $(-1, 4)$, $(-2, -1)$, $(1, 14)$

27. $(0, 9)$, $(-3, -7)$, $(2, 19)$

28. $(-1, -3)$, $(-5, 12)$, $(1, -11)$

29. $(-7, 4)$, $(6, -2)$, $(-1, 1)$

30. $(-4, 3)$, $(2, 5)$, $(-1, 4)$

Find the coordinates of the other endpoint of each segment, given its midpoint and one endpoint. See Example 5(b).

31. midpoint $(5, 8)$, endpoint $(13, 10)$

32. midpoint $(-7, 6)$, endpoint $(-9, 9)$

33. midpoint $(12, 6)$, endpoint $(19, 16)$

34. midpoint $(-9, 8)$, endpoint $(-16, 9)$

35. midpoint (a, b), endpoint (p, q)

36. midpoint $\left(\frac{a + b}{2}, \frac{c + d}{2}\right)$, endpoint (b, d)

Solve each problem. See Example 6.

37. *Bachelor's Degree Attainment* The graph shows a straight line that approximates the percentage of Americans 25 years and older who had earned bachelor's degrees or higher for the years 1990–2008. Use the midpoint formula and the two given points to estimate the percent in 1999. Compare your answer with the actual percent of 25.2.

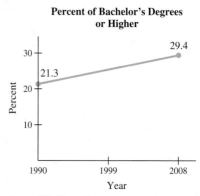

Percent of Bachelor's Degrees or Higher

Source: U.S. Census Bureau.

38. *Temporary Assistance for Needy Families (TANF)* The graph shows an idealized linear relationship for the average monthly payment per recipient to needy families in the TANF program. Based on this information, what was the average payment to families in 2004?

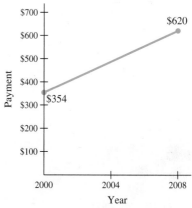

Average Monthly Payment to TANF Program Recipients

Source: U.S. Department of Health and Human Services.

39. *Poverty Level Income Cutoffs* The table lists how poverty level income cutoffs (in dollars) for a family of four have changed over time. Use the midpoint formula to approximate the poverty level cutoff in 2006 to the nearest dollar.

Year	Income (in dollars)
1980	8414
1990	13,359
2000	17,604
2004	19,307
2008	22,025

Source: U.S. Census Bureau.

40. *Public College Enrollment* Enrollments in public colleges for recent years are shown in the table. Assuming a linear relationship, estimate the enrollments for **(a)** 2002 and **(b)** 2006. Give answers to the nearest tenth of thousands if applicable.

Year	Enrollment (in thousands)
2000	11,753
2004	12,980
2008	13,972

Source: U.S. Census Bureau.

41. Show that if M is the midpoint of the line segment with endpoints $P(x_1, y_1)$ and $Q(x_2, y_2)$, then $d(P, M) + d(M, Q) = d(P, Q)$ and $d(P, M) = d(M, Q)$.

42. Write the distance formula $d = \sqrt{(x_2 - x_1)^2 + (y_2 - y_1)^2}$ using a rational exponent.

*For each equation, **(a)** give a table with at least three ordered pairs that are solutions, and **(b)** graph the equation. **See Examples 7 and 8.***

43. $y = \dfrac{1}{2}x - 2$ **44.** $y = -x + 3$ **45.** $2x + 3y = 5$

46. $3x - 2y = 6$ **47.** $y = x^2$ **48.** $y = x^2 + 2$

49. $y = \sqrt{x - 3}$ **50.** $y = \sqrt{x} - 3$ **51.** $y = |x - 2|$

52. $y = -|x + 4|$ **53.** $y = x^3$ **54.** $y = -x^3$

Concept Check Answer the following.

55. If a vertical line is drawn through the point $(4, 3)$, at what point will it intersect the x-axis?

56. If a horizontal line is drawn through the point $(4, 3)$, at what point will it intersect the y-axis?

57. If the point (a, b) is in the second quadrant, in what quadrant is $(a, -b)$? $(-a, b)$? $(-a, -b)$? (b, a)?

58. Show that the points $(-2, 2)$, $(13, 10)$, $(21, -5)$, and $(6, -13)$ are the vertices of a rhombus (all sides equal in length).

59. Are the points $A(1, 1)$, $B(5, 2)$, $C(3, 4)$, and $D(-1, 3)$ the vertices of a parallelogram (opposite sides equal in length)? of a rhombus (all sides equal in length)?

60. Find the coordinates of the points that divide the line segment joining $(4, 5)$ and $(10, 14)$ into three equal parts.

2.2 Circles

- Center-Radius Form
- General Form
- An Application

Center-Radius Form   By definition, a **circle** is the set of all points in a plane that lie a given distance from a given point. The given distance is the **radius** of the circle, and the given point is the **center.**

We can find the equation of a circle from its definition by using the distance formula. Suppose that the point (h, k) is the center and the circle has radius r, where $r > 0$. Let (x, y) represent any point on the circle. See **Figure 13.**

$$\sqrt{(x_2 - x_1)^2 + (y_2 - y_1)^2} = r \quad \text{Distance formula (Section 2.1)}$$

$$\sqrt{(x - h)^2 + (y - k)^2} = r \quad (h, k) = (x_1, y_1) \text{ and } (x, y) = (x_2, y_2)$$

$$(x - h)^2 + (y - k)^2 = r^2 \quad \text{Square each side. (Section 1.6)}$$

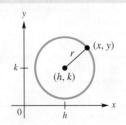

Figure 13

LOOKING AHEAD TO CALCULUS
The circle $x^2 + y^2 = 1$ is called the **unit circle.** It is important in interpreting the *trigonometric* or *circular* functions that appear in the study of calculus.

Center-Radius Form of the Equation of a Circle

A circle with center (h, k) and radius r has equation

$$(x - h)^2 + (y - k)^2 = r^2,$$

which is the **center-radius form** of the equation of the circle. As a special case, a circle with center $(0, 0)$ and radius r has the following equation.

$$x^2 + y^2 = r^2$$

EXAMPLE 1 **Finding the Center-Radius Form**

Find the center-radius form of the equation of each circle described.

(a) center at $(-3, 4)$, radius 6
(b) center at $(0, 0)$, radius 3

SOLUTION

(a) $\qquad (x - h)^2 + (y - k)^2 = r^2$ Center-radius form

$\qquad [x - (-3)]^2 + (y - 4)^2 = 6^2$ Substitute. Let $(h, k) = (-3, 4)$ and $r = 6$.

Watch signs here.

$\qquad (x + 3)^2 + (y - 4)^2 = 36$ Simplify.

(b) The center is the origin and $r = 3$.

$\qquad x^2 + y^2 = r^2$ Special case of the center-radius form

$\qquad x^2 + y^2 = 3^2$ Let $r = 3$.

$\qquad x^2 + y^2 = 9$ Apply the exponent.

✔ *Now Try Exercises 1(a) and 7(a).*

EXAMPLE 2 **Graphing Circles**

Graph each circle discussed in **Example 1.**

(a) $(x + 3)^2 + (y - 4)^2 = 36$
(b) $x^2 + y^2 = 9$

SOLUTION

(a) Writing the given equation in center-radius form

$$[x - (-3)]^2 + (y - 4)^2 = 6^2$$

gives $(-3, 4)$ as the center and 6 as the radius. See **Figure 14.**

Figure 14

Figure 15

(b) The graph with center $(0, 0)$ and radius 3 is shown in **Figure 15.**

✔ *Now Try Exercises 1(b) and 7(b).*

The circles graphed in **Figures 14 and 15** of **Example 2** can be generated on a graphing calculator by first solving for y and then entering two functions y_1 and y_2. See **Figures 16 and 17.** In both cases, the plot of y_1 yields the top half of the circle while that of y_2 yields the bottom half.

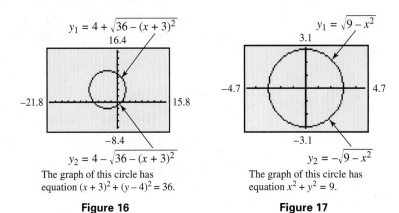

The graph of this circle has equation $(x + 3)^2 + (y - 4)^2 = 36$.

Figure 16

The graph of this circle has equation $x^2 + y^2 = 9$.

Figure 17

Because of the design of the viewing window, it is necessary to use a **square viewing window** to avoid distortion when graphing circles. Refer to your owner's manual to see how to do this. ■

General Form   Consider the center-radius form of the equation of a circle, and rewrite it so that the binomials are expanded and the right side equals 0.

Don't forget this term when squaring.		Remember this term.

$$(x - h)^2 + (y - k)^2 = r^2 \quad \text{Center-radius form}$$

$$x^2 - 2xh + h^2 + y^2 - 2yk + k^2 - r^2 = 0 \quad \text{Square each binomial, and subtract } r^2. \text{ (Section R.3)}$$

$$x^2 + y^2 + (-2h)x + (-2k)y + (h^2 + k^2 - r^2) = 0 \quad \text{Properties of real numbers (Section R.2)}$$

$$\underset{c}{\uparrow} \qquad \underset{d}{\uparrow} \qquad \underset{e}{\uparrow}$$

If $r > 0$, then the graph of this equation is a circle with center (h, k) and radius r, as seen earlier. This is the **general form of the equation of a circle.**

General Form of the Equation of a Circle

For some real numbers c, d, and e, the equation

$$x^2 + y^2 + cx + dy + e = 0$$

can have a graph that is a circle or a point, or is nonexistent.

Starting with an equation in this general form, we can complete the square to get an equation of the form

$$(x - h)^2 + (y - k)^2 = m, \quad \text{for some number } m.$$

There are three possibilities for the graph based on the value of m.

1. If $m > 0$, then $r^2 = m$, and the graph of the equation is a circle with radius $\sqrt{m}$.

2. If $m = 0$, then the graph of the equation is the single point (h, k).

3. If $m < 0$, then no points satisfy the equation, and the graph is nonexistent.

EXAMPLE 3 Finding the Center and Radius by Completing the Square

Show that $x^2 - 6x + y^2 + 10y + 25 = 0$ has a circle as its graph. Find the center and radius.

SOLUTION We complete the square twice, once for x and once for y. Begin by subtracting 25 from each side.

$$x^2 - 6x + y^2 + 10y + 25 = 0$$

$$(x^2 - 6x \quad) + (y^2 + 10y \quad) = -25$$

Think: $\left[\dfrac{1}{2}(-6)\right]^2 = (-3)^2 = 9$ and $\left[\dfrac{1}{2}(10)\right]^2 = 5^2 = 25$

Add 9 and 25 on the left to complete the two squares, and to compensate, add 9 and 25 on the right.

$$(x^2 - 6x + 9) + (y^2 + 10y + 25) = -25 + 9 + 25 \qquad \text{Complete the square.}$$

Add 9 and 25 on *both* sides. (Section 1.4)

$$(x - 3)^2 + (y + 5)^2 = 9 \qquad \text{Factor. (Section R.4)}$$

Add on the right.

$$(x - 3)^2 + (y - (-5))^2 = 3^2 \qquad \text{Center-radius form}$$

Since $3^2 = 9$ and $9 > 0$, the equation represents a circle with center at $(3, -5)$ and radius 3.

✔ *Now Try Exercise 19.*

EXAMPLE 4 Finding the Center and Radius by Completing the Square

Show that $2x^2 + 2y^2 - 6x + 10y = 1$ has a circle as its graph. Find the center and radius.

SOLUTION To complete the square, the coefficients of the x^2- and y^2-terms must be 1. In this case they are both 2, so begin by dividing each side by 2.

$$2x^2 + 2y^2 - 6x + 10y = 1$$

$$x^2 + y^2 - 3x + 5y = \dfrac{1}{2} \qquad \text{Divide by 2.}$$

$$(x^2 - 3x \quad) + (y^2 + 5y \quad) = \dfrac{1}{2} \qquad \begin{array}{l}\text{Rearrange and regroup terms}\\ \text{in anticipation of completing}\\ \text{the square.}\end{array}$$

$$\left(x^2 - 3x + \dfrac{9}{4}\right) + \left(y^2 + 5y + \dfrac{25}{4}\right) = \dfrac{1}{2} + \dfrac{9}{4} + \dfrac{25}{4} \qquad \begin{array}{l}\text{Complete the square for } both\\ x \text{ and } y; \left[\frac{1}{2}(-3)\right]^2 = \frac{9}{4} \text{ and}\\ \left[\frac{1}{2}(5)\right]^2 = \frac{25}{4}\end{array}$$

$$\left(x - \dfrac{3}{2}\right)^2 + \left(y + \dfrac{5}{2}\right)^2 = 9 \qquad \text{Factor and add.}$$

$$\left(x - \dfrac{3}{2}\right)^2 + \left(y - \left(-\dfrac{5}{2}\right)\right)^2 = 3^2 \qquad \text{Center-radius form}$$

The equation has a circle with center at $\left(\dfrac{3}{2}, -\dfrac{5}{2}\right)$ and radius 3 as its graph.

✔ *Now Try Exercise 23.*

EXAMPLE 5 **Determining Whether a Graph Is a Point or Nonexistent**

The graph of the equation $x^2 + 10x + y^2 - 4y + 33 = 0$ either is a point or is nonexistent. Which is it?

SOLUTION

$$x^2 + 10x + y^2 - 4y + 33 = 0$$

$$x^2 + 10x + y^2 - 4y = -33 \qquad \text{Subtract 33.}$$

Think: $\left[\dfrac{1}{2}(10)\right]^2 = 25$ and $\left[\dfrac{1}{2}(-4)\right]^2 = 4$ — Prepare to complete the square for both x and y.

$$(x^2 + 10x + 25) + (y^2 - 4y + 4) = -33 + 25 + 4 \qquad \text{Complete the square.}$$

$$(x + 5)^2 + (y - 2)^2 = -4 \qquad \text{Factor on the left, and add.}$$

Since $-4 < 0$, there are *no* ordered pairs (x, y), with x and y both real numbers, satisfying the equation. The graph of the given equation is nonexistent—it contains no points. (If the constant on the right side were 0, the graph would consist of the single point $(-5, 2)$.)

☑ *Now Try Exercise 25.*

An Application Seismologists can locate the epicenter of an earthquake by determining the intersection of three circles. The radii of these circles represent the distances from the epicenter to each of three receiving stations. The centers of the circles represent the receiving stations.

EXAMPLE 6 **Locating the Epicenter of an Earthquake**

Suppose receiving stations A, B, and C are located on a coordinate plane at the points $(1, 4)$, $(-3, -1)$, and $(5, 2)$. Let the distances from the earthquake epicenter to these stations be 2 units, 5 units, and 4 units, respectively. Where on the coordinate plane is the epicenter located?

SOLUTION Graph the three circles as shown in **Figure 18.** From the graph it appears that the epicenter is located at $(1, 2)$. To check this algebraically, determine the equation for each circle and substitute $x = 1$ and $y = 2$.

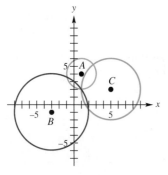

Figure 18

Station A:	Station B:	Station C:
$(x - 1)^2 + (y - 4)^2 = 4$	$(x + 3)^2 + (y + 1)^2 = 25$	$(x - 5)^2 + (y - 2)^2 = 16$
$(1 - 1)^2 + (2 - 4)^2 \overset{?}{=} 4$	$(1 + 3)^2 + (2 + 1)^2 \overset{?}{=} 25$	$(1 - 5)^2 + (2 - 2)^2 \overset{?}{=} 16$
$0 + 4 \overset{?}{=} 4$	$16 + 9 \overset{?}{=} 25$	$16 + 0 \overset{?}{=} 16$
$4 = 4$	$25 = 25$	$16 = 16$

The point $(1, 2)$ lies on all three graphs. Thus, we can conclude that the epicenter of the earthquake is at $(1, 2)$.

☑ *Now Try Exercise 41.*

2.2 Exercises

In Exercises 1–12, **(a)** *find the center-radius form of the equation of each circle, and* **(b)** *graph it.* **See Examples 1 and 2.**

1. center $(0, 0)$, radius 6   **2.** center $(0, 0)$, radius 9   **3.** center $(2, 0)$, radius 6

4. center $(3, 0)$, radius 3   **5.** center $(0, 4)$, radius 4   **6.** center $(0, -3)$, radius 7

7. center $(-2, 5)$, radius 4   **8.** center $(4, 3)$, radius 5

9. center $(5, -4)$, radius 7   **10.** center $(-3, -2)$, radius 6

11. center $(\sqrt{2}, \sqrt{2})$, radius $\sqrt{2}$ **12.** center $(-\sqrt{3}, -\sqrt{3})$, radius $\sqrt{3}$

Connecting Graphs with Equations *In Exercises 13–16, use each graph to determine the equation of the circle in* **(a)** *center-radius form and* **(b)** *general form.*

13.

14.

15.

16.

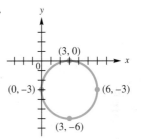

17. *Concept Check* Which one of the two screens is the correct graph of the circle with center $(-3, 5)$ and radius 4?

A.

B.

18. When the equation of a circle is written in the form

$$(x - h)^2 + (y - k)^2 = m,$$

how does the value of m indicate whether the graph is a circle, is a point, or is nonexistent?

Decide whether or not each equation has a circle as its graph. If it does, give the center and the radius. If it does not, describe the graph. **See Examples 3–5.**

19. $x^2 + y^2 + 6x + 8y + 9 = 0$ **20.** $x^2 + y^2 + 8x - 6y + 16 = 0$

21. $x^2 + y^2 - 4x + 12y = -4$ **22.** $x^2 + y^2 - 12x + 10y = -25$

23. $4x^2 + 4y^2 + 4x - 16y - 19 = 0$ **24.** $9x^2 + 9y^2 + 12x - 18y - 23 = 0$

25. $x^2 + y^2 + 2x - 6y + 14 = 0$ **26.** $x^2 + y^2 + 4x - 8y + 32 = 0$

27. $x^2 + y^2 - 6x - 6y + 18 = 0$ **28.** $x^2 + y^2 + 4x + 4y + 8 = 0$

29. $9x^2 + 9y^2 - 6x + 6y - 23 = 0$ **30.** $4x^2 + 4y^2 + 4x - 4y - 7 = 0$

Relating Concepts

For individual or collaborative investigation *(Exercises 31–36)*

The distance formula, the midpoint formula, and the center-radius form of the equation of a circle are closely related in the following problem.

A circle has a diameter with endpoints $(-1, 3)$ and $(5, -9)$. Find the center-radius form of the equation of this circle.

Work Exercises 31–36 in order *to see the relationships among these concepts.*

31. To find the center-radius form, we must find both the radius and the coordinates of the center. Find the coordinates of the center using the midpoint formula. (The center of the circle must be the midpoint of the diameter.)

32. There are several ways to find the radius of the circle. One way is to find the distance between the center and the point $(-1, 3)$. Use your result from **Exercise 31** and the distance formula to find the radius.

33. Another way to find the radius is to repeat **Exercise 32,** but use the point $(5, -9)$ rather than $(-1, 3)$. Do this to obtain the same answer you found in **Exercise 32.**

34. There is yet another way to find the radius. Because the radius is half the diameter, it can be found by finding half the length of the diameter. Using the endpoints of the diameter given in the problem, find the radius in this manner. You should once again obtain the same answer you found in **Exercise 32.**

35. Using the center found in **Exercise 31** and the radius found in **Exercises 32–34,** give the center-radius form of the equation of the circle.

36. Use the method described in **Exercises 31–35** to find the center-radius form of the equation of the circle with diameter having endpoints $(3, -5)$ and $(-7, 3)$.

---

*Find the center-radius form of the circle described or graphed. (**See Relating Concepts Exercises 31–36.**)*

37. a circle having a diameter with endpoints $(-1, 2)$ and $(11, 7)$

38. a circle having a diameter with endpoints $(5, 4)$ and $(-3, -2)$

39.

40.

Epicenter of an Earthquake   Solve each problem. To visualize the situation, use graph paper and a pair of compasses to carefully draw the graphs of the circles. **See Example 6.**

41. Suppose that receiving stations X, Y, and Z are located on a coordinate plane at the points $(7, 4)$, $(-9, -4)$, and $(-3, 9)$, respectively. The epicenter of an earthquake is determined to be 5 units from X, 13 units from Y, and 10 units from Z. Where on the coordinate plane is the epicenter located?

42. Suppose that receiving stations P, Q, and R are located on a coordinate plane at the points $(3, 1)$, $(5, -4)$, and $(-1, 4)$, respectively. The epicenter of an earthquake is determined to be $\sqrt{5}$ units from P, 6 units from Q, and $2\sqrt{10}$ units from R. Where on the coordinate plane is the epicenter located?

43. The locations of three receiving stations and the distances to the epicenter of an earthquake are contained in the following three equations: $(x - 2)^2 + (y - 1)^2 = 25$, $(x + 2)^2 + (y - 2)^2 = 16$, and $(x - 1)^2 + (y + 2)^2 = 9$. Determine the location of the epicenter.

44. The locations of three receiving stations and the distances to the epicenter of an earthquake are contained in the following three equations: $(x - 2)^2 + (y - 4)^2 = 25$, $(x - 1)^2 + (y + 3)^2 = 25$, and $(x + 3)^2 + (y + 6)^2 = 100$. Determine the location of the epicenter.

Concept Check   Work each of the following.

45. Find the center-radius form of the equation of a circle with center $(3, 2)$ and tangent to the x-axis. (*Hint:* A line **tangent** to a circle touches it at exactly one point.)

46. Find the equation of a circle with center at $(-4, 3)$, passing through the point $(5, 8)$. Write it in center-radius form.

47. Find all points (x, y) with $x = y$ that are 4 units from $(1, 3)$.

48. Find all points satisfying $x + y = 0$ that are 8 units from $(-2, 3)$.

49. Find the coordinates of all points whose distance from $(1, 0)$ is $\sqrt{10}$ and whose distance from $(5, 4)$ is $\sqrt{10}$.

50. Find the equation of the circle of least radius that contains the points $(1, 4)$ and $(-3, 2)$ within or on its boundary.

51. Find all values of y such that the distance between $(3, y)$ and $(-2, 9)$ is 12.

52. Suppose that a circle is tangent to both axes, is in the third quadrant, and has radius $\sqrt{2}$. Find the center-radius form of its equation.

53. Find the shortest distance from the origin to the graph of the circle with equation $x^2 - 16x + y^2 - 14y + 88 = 0$.

54. Find the coordinates of the points of intersection of the line $y = 1$ and the circle centered at $(3, 0)$ with radius 2.

55. Phlash Phelps is the morning radio personality on SiriusXM Satellite Radio's *Sixties on Six* Decades channel. Phlash is an expert on U.S. geography and loves traveling around the country to strange, out-of-the-way locations. The photo shows Curt Gilchrist (standing) and Phlash (seated) visiting a small Arizona settlement called *Nothing*. (Nothing is so small that it's not named on current maps.) The sign indicates that Nothing is 50 mi from Wickenburg, AZ, 75 mi from Kingman, AZ, 105 mi from Phoenix, AZ, and 180 mi from Las Vegas, NV. Discuss how the concepts of **Example 6** can be used to locate Nothing, AZ, on a map of Arizona and southern Nevada.

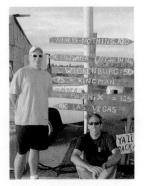

2.3 **Functions**

■ Relations and Functions
■ Domain and Range
■ Determining Whether Relations Are Functions
■ Function Notation
■ Increasing, Decreasing, and Constant Functions

Relations and Functions   Recall from **Section 2.1** how we described one quantity in terms of another.

- The letter grade you receive in a mathematics course depends on your numerical scores.
- The amount you pay (in dollars) for gas at the gas station depends on the number of gallons pumped.
- The dollars spent on household expenses depends on the category.

We used ordered pairs to represent these corresponding quantities. For example, (3, $10.50) indicates that you pay $10.50 for 3 gallons of gas. Since the amount you pay *depends* on the number of gallons pumped, the amount (in dollars) is called the *dependent variable,* and the number of gallons pumped is called the *independent variable.*

Generalizing, if the value of the second component y depends on the value of the first component x, then y is the **dependent variable** and x is the **independent variable.**

$$\underset{\substack{\downarrow \\ (x, \; y)}}{\overset{\text{Independent variable} \qquad \text{Dependent variable}}{}}$$

A set of ordered pairs such as $\{(3, 10.50), (8, 28.00), (10, 35.00)\}$ is called a *relation.* A special kind of relation called a *function* is very important in mathematics and its applications.

Relation and Function

A **relation** is a set of ordered pairs. A **function** is a relation in which, for each distinct value of the first component of the ordered pairs, there is *exactly one* value of the second component.

NOTE   The relation from the beginning of this section representing the number of gallons of gasoline and the corresponding cost is a function since each x-value is paired with exactly one y-value.

EXAMPLE 1   **Deciding Whether Relations Define Functions**

Decide whether each relation defines a function.

$$F = \{(1, 2), (-2, 4), (3, 4)\}$$
$$G = \{(1, 1), (1, 2), (1, 3), (2, 3)\}$$
$$H = \{(-4, 1), (-2, 1), (-2, 0)\}$$

SOLUTION   Relation F is a function, because for each different x-value there is exactly one y-value. We can show this correspondence as follows.

$$\{1, -2, 3\} \quad \text{x-values of F}$$
$$\downarrow \quad \downarrow \quad \downarrow$$
$$\{2, \quad 4, \quad 4\} \quad \text{y-values of F}$$

As the correspondence below shows, relation G is not a function because one first component corresponds to *more than one* second component.

$$\{1, 2\} \quad \text{\small{x-values of G}}$$

$$\{1, 2, 3\} \quad \text{\small{y-values of G}}$$

In relation H the last two ordered pairs have the same x-value paired with two different y-values (-2 is paired with both 1 and 0), so H is a relation but not a function. ***In a function, no two ordered pairs can have the same first component and different second components.***

Different y-values

$$H = \{(-4, 1), (-2, 1), (-2, 0)\} \quad \text{\small Not a function}$$

Same x-value

✔ *Now Try Exercises 1 and 3.*

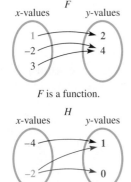

x-values F *y*-values

F is a function.

x-values H *y*-values

H is not a function.

Figure 19

Relations and functions can also be expressed as a correspondence or *mapping* from one set to another, as shown in **Figure 19** for function F and relation H from **Example 1.** The arrow from 1 to 2 indicates that the ordered pair $(1, 2)$ belongs to F—each first component is paired with exactly one second component. In the mapping for relation H, which is not a function, the first component -2 is paired with two different second components, 1 and 0.

Since relations and functions are sets of ordered pairs, we can represent them using tables and graphs. A table and graph for function F are shown in **Figure 20.**

x	y
1	2
-2	4
3	4

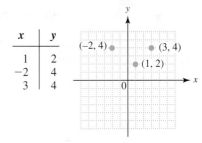

Graph of F

Figure 20

Finally, we can describe a relation or function using a rule that tells how to determine the dependent variable for a specific value of the independent variable. The rule may be given in words: for instance, "the dependent variable is twice the independent variable." Usually the rule is an equation, such as the one below.

Dependent variable $\longrightarrow y = 2x \longleftarrow$ Independent variable

NOTE Another way to think of a function relationship is to think of the independent variable as an input and the dependent variable as an output. This is illustrated by the input-output (function) machine for the function defined by $y = 2x$.

Function machine

In a function, there is exactly one value of the dependent variable, the second component, for each value of the independent variable, the first component.

On this particular day, an *input* of pumping 7.870 gallons of gasoline led to an *output* of $29.58 from the purchaser's wallet. This is an example of a function whose domain consists of numbers of gallons pumped, and whose range consists of amounts from the purchaser's wallet. Dividing the dollar amount by the number of gallons pumped gives the exact price of gasoline that day. Use your calculator to check this. Was this pump fair? (Later we will see that this price is an example of the slope *m* of a linear function of the form $y = mx$.)

Domain and Range For every relation there are two important sets of elements called the *domain* and *range*.

> ### Domain and Range
>
> In a relation consisting of ordered pairs (x, y), the set of all values of the independent variable (x) is the **domain.** The set of all values of the dependent variable (y) is the **range.**

EXAMPLE 2 Finding Domains and Ranges of Relations

Give the domain and range of each relation. Tell whether the relation defines a function.

(a) $\{(3, -1), (4, 2), (4, 5), (6, 8)\}$

(b)

(c)

x	y
-5	2
0	2
5	2

SOLUTION

(a) The domain is the set of *x*-values, $\{3, 4, 6\}$. The range is the set of *y*-values, $\{-1, 2, 5, 8\}$. This relation is not a function because the same *x*-value, 4, is paired with two different *y*-values, 2 and 5.

(b) The domain is $\{4, 6, 7, -3\}$ and the range is $\{100, 200, 300\}$. This mapping defines a function. Each *x*-value corresponds to exactly one *y*-value.

(c) This relation is a set of ordered pairs, so the domain is the set of *x*-values $\{-5, 0, 5\}$ and the range is the set of *y*-values $\{2\}$. The table defines a function because each different *x*-value corresponds to exactly one *y*-value (even though it is the same *y*-value).

☑ *Now Try Exercises 9, 11, and 13.*

EXAMPLE 3 Finding Domains and Ranges from Graphs

Give the domain and range of each relation.

(a)

(b)

(c)

(d)

SOLUTION

(a) The domain is the set of x-values, $\{-1, 0, 1, 4\}$. The range is the set of y-values, $\{-3, -1, 1, 2\}$.

(b) The x-values of the points on the graph include all numbers between -4 and 4, inclusive. The y-values include all numbers between -6 and 6, inclusive.

$$\text{The domain is } [-4, 4].$$ Use interval notation.

$$\text{The range is } [-6, 6].$$ **(Section 1.7)**

(c) The arrowheads indicate that the line extends indefinitely left and right, as well as up and down. Therefore, both the domain and the range include all real numbers, which is written $(-\infty, \infty)$.

(d) The arrowheads indicate that the graph extends indefinitely left and right, as well as upward. The domain is $(-\infty, \infty)$. Because there is a least y-value, -3, the range includes all numbers greater than or equal to -3, written $[-3, \infty)$.

> ✔ *Now Try Exercises 17 and 19.*

Since relations are often defined by equations, such as $y = 2x + 3$ and $y^2 = x$, we must sometimes determine the domain of a relation from its equation. In this book, we assume the following agreement on the domain of a relation.

Agreement on Domain

Unless specified otherwise, the domain of a relation is assumed to be all real numbers that produce real numbers when substituted for the independent variable.

To illustrate this agreement, since any real number can be used as a replacement for x in $y = 2x + 3$, the domain of this function is the set of all real numbers. As another example, the function defined by $y = \frac{1}{x}$ has all real numbers *except* 0 as domain, since y is undefined if $x = 0$.

> *In general, the domain of a function defined by an algebraic expression is all real numbers, except those numbers that lead to division by **0** or to an even root of a negative number.*

(There are also exceptions for logarithmic and trigonometric functions. They are covered in further treatment of precalculus mathematics.)

Determining Whether Relations Are Functions Since each value of x leads to only one value of y in a function, any vertical line must intersect the graph in at most one point. This is the **vertical line test** for a function.

Vertical Line Test

If every vertical line intersects the graph of a relation in no more than one point, then the relation is a function.

The graph in **Figure 21(a)** represents a function because each vertical line intersects the graph in no more than one point. The graph in **Figure 21(b)** is not the graph of a function since a vertical line intersects the graph in more than one point.

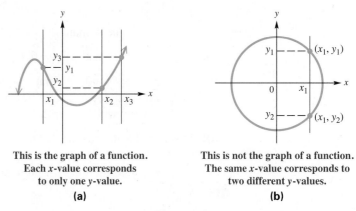

This is the graph of a function.
Each *x*-value corresponds
to only one *y*-value.
(a)

This is not the graph of a function.
The same *x*-value corresponds to
two different *y*-values.
(b)

Figure 21

EXAMPLE 4 Using the Vertical Line Test

Use the vertical line test to determine whether each relation graphed in **Example 3** is a function.

SOLUTION We repeat each graph from **Example 3,** this time with vertical lines drawn through the graphs.

(a)

(b)

(c)

(d)

- The graphs of the relations in parts (a), (c), and (d) pass the vertical line test, since every vertical line intersects each graph no more than once. Thus, these graphs represent functions.

- The graph of the relation in part (b) fails the vertical line test, since the same *x*-value corresponds to two different *y*-values. Therefore, it is not the graph of a function.

☑ *Now Try Exercises 17 and 19.*

The vertical line test is a simple method for identifying a function defined by a graph. Deciding whether a relation defined by an equation or an inequality is a function, as well as determining the domain and range, is more difficult. The next example gives some hints that may help.

EXAMPLE 5 | **Identifying Functions, Domains, and Ranges**

Decide whether each relation defines a function and give the domain and range.

(a) $y = x + 4$ **(b)** $y = \sqrt{2x - 1}$ **(c)** $y^2 = x$

(d) $y \leq x - 1$ **(e)** $y = \dfrac{5}{x - 1}$

SOLUTION

(a) In the defining equation (or rule), $y = x + 4$, y is always found by adding 4 to x. Thus, each value of x corresponds to just one value of y, and the relation defines a function. The variable x can represent any real number, so the domain is

$$\{x \mid x \text{ is a real number}\}, \quad \text{or} \quad (-\infty, \infty).$$

Since y is always 4 more than x, y also may be any real number, and so the range is $(-\infty, \infty)$.

(b) For any choice of x in the domain of $y = \sqrt{2x - 1}$, there is exactly one corresponding value for y (the radical is a nonnegative number), so this equation defines a function. Since the equation involves a square root, the quantity under the radical sign cannot be negative.

$$2x - 1 \geq 0 \quad \text{Solve the inequality. (Section 1.7)}$$

$$2x \geq 1 \quad \text{Add 1.}$$

$$x \geq \frac{1}{2} \quad \text{Divide by 2.}$$

The domain of the function is $\left[\frac{1}{2}, \infty\right)$. Because the radical must represent a nonnegative number, as x takes values greater than or equal to $\frac{1}{2}$, the range is $\{y \mid y \geq 0\}$, or $[0, \infty)$. See **Figure 22.**

Figure 22

(c) The ordered pairs $(16, 4)$ and $(16, -4)$ both satisfy the equation $y^2 = x$. Since one value of x, 16, corresponds to two values of y, 4 and -4, this equation does not define a function.

Because x is equal to the square of y, the values of x must always be nonnegative. The domain of the relation is $[0, \infty)$. Any real number can be squared, so the range of the relation is $(-\infty, \infty)$. See **Figure 23.**

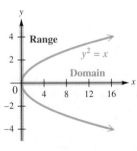

Figure 23

(d) By definition, y is a function of x if every value of x leads to exactly one value of y. Substituting a particular value of x, say 1, into $y \leq x - 1$ corresponds to many values of y. The ordered pairs

$$(1, 0), \quad (1, -1), \quad (1, -2), \quad (1, -3), \quad \text{and} \quad \text{so on}$$

all satisfy the inequality, so y is not a function of x here. Any number can be used for x or for y, so the domain and the range of this relation are both the set of real numbers, $(-\infty, \infty)$.

(e) Given any value of x in the domain of

$$y = \frac{5}{x - 1},$$

we find y by subtracting 1 from x, and then dividing the result into 5. This process produces exactly one value of y for each value in the domain, so this equation defines a function.

Figure 24

The domain of $y = \frac{5}{x-1}$ includes all real numbers except those that make the denominator 0. We find these numbers by setting the denominator equal to 0 and solving for x.

$$x - 1 = 0$$

$$x = 1 \quad \text{Add 1. (Section 1.1)}$$

Thus, the domain includes all real numbers except 1, written as the interval $(-\infty, 1) \cup (1, \infty)$. Values of y can be positive or negative, but never 0, because a fraction cannot equal 0 unless its numerator is 0. Therefore, the range is the interval $(-\infty, 0) \cup (0, \infty)$, as shown in **Figure 24.**

✔ *Now Try Exercises 27, 29, and 35.*

Variations of the Definition of Function

1. A **function** is a relation in which, for each distinct value of the first component of the ordered pairs, there is exactly one value of the second component.

2. A **function** is a set of ordered pairs in which no first component is repeated.

3. A **function** is a rule or correspondence that assigns exactly one range value to each distinct domain value.

LOOKING AHEAD TO CALCULUS

One of the most important concepts in calculus, that of the **limit of a function,** is defined using function notation:

$$\lim_{x \to a} f(x) = L$$

(read "the limit of $f(x)$ as x approaches a is equal to L") means that the values of $f(x)$ become as close as we wish to L when we choose values of x sufficiently close to a.

Function Notation When a function f is defined with a rule or an equation using x and y for the independent and dependent variables, we say, "y is a function of x" to emphasize that y *depends on x.* We use the notation

$$y = f(x),$$

called **function notation,** to express this and read $f(x)$ as **"f of x."** The letter f is the name given to this function.

For example, if $y = 3x - 5$, we can name the function f and write

$$f(x) = 3x - 5.$$

Note that $f(x)$ is just another name for the dependent variable y. For example, if $y = f(x) = 3x - 5$ and $x = 2$, then we find y, or $f(2)$, by replacing x with 2.

$$f(2) = 3 \cdot 2 - 5 \quad \text{Let } x = 2.$$

$$f(2) = 1 \qquad \text{Multiply, and then subtract.}$$

The statement "In the function f, if $x = 2$, then $y = 1$" represents the ordered pair $(2, 1)$ and is abbreviated with function notation as follows.

$$f(2) = 1$$

The symbol $f(2)$ is read "f of 2" or "f at 2."

These ideas can be illustrated as follows.

Name of the function

Defining expression

$$y \;=\; f(x) \;=\; 3x - 5$$

Value of the function   Name of the independent variable

CAUTION *The symbol $f(x)$ does not indicate "f times x,"* but represents the y-value for the indicated x-value. As just shown, $f(2)$ is the y-value that corresponds to the x-value 2.

EXAMPLE 6 Using Function Notation

Let $f(x) = -x^2 + 5x - 3$ and $g(x) = 2x + 3$. Find and simplify each of the following.

(a) $f(2)$ **(b)** $f(q)$ **(c)** $g(a + 1)$

SOLUTION

(a) $f(x) = -x^2 + 5x - 3$

$\quad f(2) = -2^2 + 5 \cdot 2 - 3$ Replace x with 2.

$\qquad = -4 + 10 - 3$ Apply the exponent and multiply. **(Section R.2)**

$\qquad = 3$ Add and subtract.

Thus, $f(2) = 3$, and the ordered pair $(2, 3)$ belongs to f.

(b) $f(x) = -x^2 + 5x - 3$

$\quad f(q) = -q^2 + 5q - 3$ Replace x with q.

(c) $\quad g(x) = 2x + 3$

$\quad g(a + 1) = 2(a + 1) + 3$ Replace x with $a + 1$.

$\qquad = 2a + 2 + 3$ Distributive property **(Section R.2)**

$\qquad = 2a + 5$ Add.

The replacement of one variable with another variable or expression, as in parts (b) and (c), is important in later courses.

✔ *Now Try Exercises 41, 49, and 55.*

Functions can be evaluated in a variety of ways, as shown in **Example 7.**

EXAMPLE 7 Using Function Notation

For each function, find $f(3)$.

(a) $f(x) = 3x - 7$ **(b)** $f = \{(-3, 5), (0, 3), (3, 1), (6, -1)\}$

(c)   **(d)**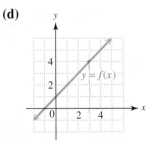

SOLUTION

(a) $f(x) = 3x - 7$

$\quad f(3) = 3(3) - 7$ Replace x with 3.

$\quad f(3) = 2$ Simplify.

Figure 25

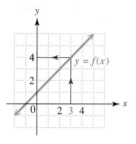

Figure 26

(b) For $f = \{(-3, 5), (0, 3), (3, 1), (6, -1)\}$, we want $f(3)$, the y-value of the ordered pair where $x = 3$. As indicated by the ordered pair $(3, 1)$, when $x = 3$, $y = 1$, so $f(3) = 1$.

(c) In the mapping, repeated in **Figure 25,** the domain element 3 is paired with 5 in the range, so $f(3) = 5$.

(d) To evaluate $f(3)$ using the graph, find 3 on the x-axis. See **Figure 26.** Then move up until the graph of f is reached. Moving horizontally to the y-axis gives 4 for the corresponding y-value. Thus, $f(3) = 4$.

☑ *Now Try Exercises 57, 59, and 61.*

If a function f is defined by an equation with x and y (and not with function notation), use the following steps to find $f(x)$.

Finding an Expression for $f(x)$

Consider an equation involving x and y. Assume that y can be expressed as a function f of x. To find an expression for $f(x)$, use the following steps.

Step 1 Solve the equation for y.

Step 2 Replace y with $f(x)$.

EXAMPLE 8 **Writing Equations Using Function Notation**

Assume that y is a function f of x. Rewrite each equation using function notation. Then find $f(-2)$ and $f(a)$.

(a) $y = x^2 + 1$ 　　　　　　　　　　(b) $x - 4y = 5$

SOLUTION

(a)
$$y = x^2 + 1 \quad \text{Equation is already solved for } y. \text{ (Step 1)}$$
$$f(x) = x^2 + 1 \quad \text{Let } y = f(x). \text{ (Step 2)}$$

Now find $f(-2)$ and $f(a)$.

$$f(-2) = (-2)^2 + 1 \quad \text{Let } x = -2. \qquad\qquad f(a) = a^2 + 1 \quad \text{Let } x = a.$$
$$f(-2) = 4 + 1$$
$$f(-2) = 5$$

(b)
$$x - 4y = 5 \qquad\qquad \text{Solve for } y. \text{ (Section 1.1)}$$
$$-4y = -x + 5 \qquad \text{(Step 1)}$$

$$y = \frac{x - 5}{4} \qquad\qquad \text{Multiply by } -1. \text{ Divide by 4.}$$

$$f(x) = \frac{1}{4}x - \frac{5}{4} \qquad \begin{array}{l} \text{Let } y = f(x). \text{ (Step 2)} \\ \frac{a-b}{c} = \frac{a}{c} - \frac{b}{c} \text{ (Section R.5)} \end{array}$$

Now find $f(-2)$ and $f(a)$.

$$f(-2) = \frac{1}{4}(-2) - \frac{5}{4} \quad \text{Let } x = -2. \qquad\qquad f(a) = \frac{1}{4}a - \frac{5}{4} \quad \text{Let } x = a.$$
$$f(-2) = -\frac{7}{4}$$

☑ *Now Try Exercises 63 and 67.*

Increasing, Decreasing, and Constant Functions   Informally speaking, a function *increases* on an open interval of its domain if its graph rises from left to right on the interval. It *decreases* on an open interval of its domain if its graph falls from left to right on the interval. It is *constant* on an open interval of its domain if its graph is horizontal on the interval.

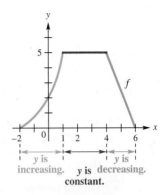

Figure 27

For example, consider **Figure 27.** The function increases on the open interval $(-2, 1)$ because the y-values continue to get larger for x-values in that interval. Similarly, the function is constant on the open interval $(1, 4)$ because the y-values are always 5 for all x-values there. Finally, the function decreases on the open interval $(4, 6)$ because there the y-values continuously get smaller. ***The intervals refer to the x-values where the y-values either increase, decrease, or are constant.***

The formal definitions of these concepts follow.

Increasing, Decreasing, and Constant Functions

Suppose that a function f is defined over an *open* interval I and x_1 and x_2 are in I.

(a) f **increases** on I if, whenever $x_1 < x_2$, $f(x_1) < f(x_2)$.

(b) f **decreases** on I if, whenever $x_1 < x_2$, $f(x_1) > f(x_2)$.

(c) f is **constant** on I if, for every x_1 and x_2, $f(x_1) = f(x_2)$.

Figure 28 illustrates these ideas.

Whenever $x_1 < x_2$, and $f(x_1) < f(x_2)$, f is increasing.

(a)

Whenever $x_1 < x_2$, and $f(x_1) > f(x_2)$, f is decreasing.

(b)

For every x_1 and x_2, if $f(x_1) = f(x_2)$, then f is constant.

(c)

Figure 28

NOTE   To decide whether a function is increasing, decreasing, or constant on an interval, ask yourself, ***"What does y do as x goes from left to right?" The concepts of increasing and decreasing functions apply to intervals of the domain, not to individual points.***

> **EXAMPLE 9** Determining Intervals over Which a Function Is Increasing, Decreasing, or Constant

Figure 29 shows the graph of a function. Determine the largest intervals over which the function is increasing, decreasing, or constant.

Figure 29

SOLUTION We should ask, "What is happening to the y-values as the x-values are getting larger?" Moving from left to right on the graph, we see the following:

- On the interval $(-\infty, 1)$, the y-values are *decreasing*.

- On the interval $(1, 3)$, the y-values are *increasing*.

- On the interval $(3, \infty)$, the y-values are *constant* (and equal to 6).

Therefore, the function is decreasing on $(-\infty, 1)$, increasing on $(1, 3)$, and constant on $(3, \infty)$.

✔ *Now Try Exercise 77.*

> **EXAMPLE 10** Interpreting a Graph

Figure 30 shows the relationship between the number of gallons, $g(t)$, of water in a small swimming pool and time in hours, t. By looking at this graph of the function, we can answer questions about the water level in the pool at various times. For example, at time 0 the pool is empty. The water level then increases, stays constant for a while, decreases, and then becomes constant again. Use the graph to respond to the following.

Figure 30

(a) What is the maximum number of gallons of water in the pool? When is the maximum water level first reached?

(b) For how long is the water level increasing? decreasing? constant?

(c) How many gallons of water are in the pool after 90 hr?

(d) Describe a series of events that could account for the water level changes shown in the graph.

SOLUTION

(a) The maximum range value is 3000, as indicated by the horizontal line segment for the hours 25 to 50. This maximum number of gallons, 3000, is first reached at $t = 25$ hr.

(b) The water level is increasing for $25 - 0 = 25$ hr and is decreasing for $75 - 50 = 25$ hr. It is constant for

$$(50 - 25) + (100 - 75) = 25 + 25, \quad \text{or} \quad 50 \text{ hr.}$$

(c) When $t = 90$, $y = g(90) = 2000$. There are 2000 gal after 90 hr.

(d) Looking at the graph in **Figure 30,** we might write the following description:

The pool is empty at the beginning and then is filled to a level of 3000 gal during the first 25 hr. For the next 25 hr, the water level remains the same. At 50 hr, the pool starts to be drained, and this draining lasts for 25 hr, until only 2000 gal remain. For the next 25 hr, the water level is unchanged.

Now Try Exercise 83.

2.3 Exercises

Decide whether each relation defines a function. **See Example 1.**

1. $\{(5, 1), (3, 2), (4, 9), (7, 8)\}$

2. $\{(8, 0), (5, 7), (9, 3), (3, 8)\}$

3. $\{(2, 4), (0, 2), (2, 6)\}$

4. $\{(9, -2), (-3, 5), (9, 1)\}$

5. $\{(-3, 1), (4, 1), (-2, 7)\}$

6. $\{(-12, 5), (-10, 3), (8, 3)\}$

7.

x	y
3	-4
7	-4
10	-4

8.

x	y
-4	$\sqrt{2}$
0	$\sqrt{2}$
4	$\sqrt{2}$

Decide whether each relation defines a function and give the domain and range. **See Examples 1–4.**

9. $\{(1, 1), (1, -1), (0, 0), (2, 4), (2, -4)\}$ **10.** $\{(2, 5), (3, 7), (3, 9), (5, 11)\}$

11.

12.

13.

x	y
0	0
-1	1
-2	2

14.

x	y
0	0
1	-1
2	-2

15. *Number of Visits to U.S. National Parks*

Year (x)	Number of Visits (y) (millions)
2005	63.5
2006	60.4
2007	62.3
2008	61.2

Source: National Park Service.

16. *Attendance at NCAA Women's College Basketball Games*

Season* (x)	Attendance (y)
2006	10,878,322
2007	11,120,822
2008	11,160,293
2009	11,134,738

Source: NCAA.

*Each season overlaps the starting year (given) with the following year.

17.

18.

19.

20.

21.

22.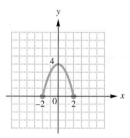

Decide whether each relation defines y as a function of x. Give the domain and range.
See Example 5.

23. $y = x^2$

24. $y = x^3$

25. $x = y^6$

26. $x = y^4$

27. $y = 2x - 5$

28. $y = -6x + 4$

29. $x + y < 3$

30. $x - y < 4$

31. $y = \sqrt{x}$

32. $y = -\sqrt{x}$

33. $xy = 2$

34. $xy = -6$

35. $y = \sqrt{4x + 1}$

36. $y = \sqrt{7 - 2x}$

37. $y = \dfrac{2}{x - 3}$

38. $y = \dfrac{-7}{x - 5}$

39. *Concept Check* Choose the correct answer: For function f, the notation $f(3)$ means

    A. the variable f times 3, or $3f$.

    B. the value of the dependent variable when the independent variable is 3.

    C. the value of the independent variable when the dependent variable is 3.

    D. f equals 3.

40. *Concept Check* Give an example of a function from everyday life. (*Hint:* Fill in the blanks: _____ depends on _____, so _____ is a function of _____.)

Let $f(x) = -3x + 4$ and $g(x) = -x^2 + 4x + 1$. Find and simplify each of the following.
See Example 6.

41. $f(0)$

42. $f(-3)$

43. $g(-2)$

44. $g(10)$

45. $f\left(\dfrac{1}{3}\right)$

46. $f\left(-\dfrac{7}{3}\right)$

47. $g\left(\dfrac{1}{2}\right)$

48. $g\left(-\dfrac{1}{4}\right)$

49. $f(p)$

50. $g(k)$

51. $f(-x)$

52. $g(-x)$

53. $f(x + 2)$

54. $f(a + 4)$

55. $f(2m - 3)$

56. $f(3t - 2)$

For each function, find (a) $f(2)$ and (b) $f(-1)$. See Example 7.

57. $f = \{(-1, 3), (4, 7), (0, 6), (2, 2)\}$

58. $f = \{(2, 5), (3, 9), (-1, 11), (5, 3)\}$

59. f

60. f

61.

62.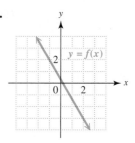

*An equation that defines y as a function of x is given. (**a**) Solve for y in terms of x and replace y with the function notation f(x). (**b**) Find f(3). See Example 8.*

63. $x + 3y = 12$ **64.** $x - 4y = 8$ **65.** $y + 2x^2 = 3 - x$

66. $y - 3x^2 = 2 + x$ **67.** $4x - 3y = 8$ **68.** $-2x + 5y = 9$

Concept Check Answer each question.

69. If $(3, 4)$ is on the graph of $y = f(x)$, which one of the following must be true: $f(3) = 4$ or $f(4) = 3$?

70. The figure shows a portion of the graph of $f(x) = x^2 + 3x + 1$ and a rectangle with its base on the x-axis and a vertex on the graph. What is the area of the rectangle? (*Hint:* $f(0.2)$ is the height.)

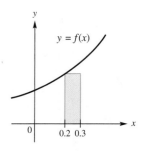

71. The graph of $Y_1 = f(X)$ is shown with a display at the bottom. What is $f(3)$?

72. The graph of $Y_1 = f(X)$ is shown with a display at the bottom. What is $f(-2)$?

*In Exercises 73–76, use the graph of $y = f(x)$ to find each function value: (**a**) $f(-2)$, (**b**) $f(0)$, (**c**) $f(1)$, and (**d**) $f(4)$. See Example 7(d).*

73.

74.

75.

76.

*Determine the largest intervals of the domain for which each function is **(a)** increasing, **(b)** decreasing, and **(c)** constant. **See Example 9.***

77.

78.

79.

80.

81.

82.

*Solve each problem. **See Example 10.***

83. *Electricity Usage* The graph shows the daily megawatts of electricity used on a record-breaking summer day in Sacramento, California.

(a) Is this the graph of a function?

(b) What is the domain?

(c) Estimate the number of megawatts used at 8 A.M.

(d) At what time was the most electricity used? the least electricity?

(e) Call this function f. What is $f(12)$? Interpret your answer.

(f) During what time intervals is electricity usage increasing? decreasing?

Electricity Use

Source: Sacramento Municipal Utility District.

84. *Height of a Ball* A ball is thrown straight up into the air. The function defined by $y = h(t)$ in the graph gives the height of the ball (in feet) at t seconds. (*Note:* The graph does *not* show the path of the ball. The ball is rising straight up and then falling straight down.)

(a) What is the height of the ball at 2 sec?

(b) When will the height be 192 ft?

(c) During what time intervals is the ball going up? down?

(d) How high does the ball go, and when does the ball reach its maximum height?

(e) After how many seconds does the ball hit the ground?

Height of a Thrown Ball

85. *Temperature* The graph shows temperatures on a given day in Bratenahl, Ohio.

(a) At what times during the day was the temperature over 55°?

(b) When was the temperature below 40°?

(c) Greenville, South Carolina, is 500 mi south of Bratenahl, Ohio, and its temperature is 7° higher all day long. At what time was the temperature in Greenville the same as the temperature at noon in Bratenahl?

(d) Use the graph to give a word description of the 24-hr period in Bratenahl.

Temperature in Bratenahl, Ohio

86. *Drug Levels in the Bloodstream* When a drug is taken orally, the amount of the drug in the bloodstream after t hours is given by the function defined by $y = f(t)$, as shown in the graph.

(a) How many units of the drug are in the bloodstream at 8 hr?

(b) During what time interval is the drug level in the bloodstream increasing? decreasing?

(c) When does the level of the drug in the bloodstream reach its maximum value, and how many units are in the bloodstream at that time?

(d) When the drug reaches its maximum level in the bloodstream, how many additional hours are required for the level to drop to 16 units?

(e) Use the graph to give a word description of the 12-hr period.

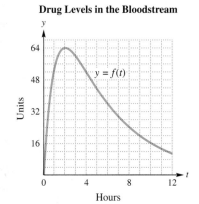

Drug Levels in the Bloodstream

2.4 Linear Functions

- Graphing Linear Functions
- Standard Form
  $Ax + By = C$
- Slope
- Average Rate of Change
- Linear Models

Graphing Linear Functions We begin our study of specific functions by looking at *linear functions*.

Linear Function

A function f is a **linear function** if

$$f(x) = ax + b,$$

for real numbers a and b. If $a \neq 0$, the domain and the range of a linear function are both $(-\infty, \infty)$.

In **Section 2.1,** we graphed lines by finding ordered pairs and plotting them. Although only two points are necessary to graph a linear function, we usually plot a third point as a check. The intercepts are often good points to choose for graphing lines.

> **EXAMPLE 1** **Graphing a Linear Function Using Intercepts**

Graph $f(x) = -2x + 6$. Give the domain and range.

SOLUTION The x-intercept is found by letting $f(x) = 0$ and solving for x.

$$f(x) = -2x + 6$$

$$0 = -2x + 6 \quad \text{Let } f(x) = 0.$$

$$x = 3 \qquad \text{Add } 2x \text{ and divide by } 2. \text{ (Section 1.1)}$$

We plot the x-intercept $(3, 0)$. The y-intercept is found by evaluating $f(0)$.

$$f(0) = -2(0) + 6 \quad \text{Let } x = 0.$$

$$f(0) = 6 \qquad \text{Simplify.}$$

Therefore, another point on the graph is the y-intercept $(0, 6)$. We plot this point and join the two points with a straight-line graph. We use the point $(2, 2)$ as a check. See **Figure 31.** The domain and the range are both $(-\infty, \infty)$.

The corresponding calculator graph is shown in **Figure 32.**

Figure 31   Figure 32

> ✔ *Now Try Exercise 9.*

If $a = 0$ in the definition of linear function, then the equation becomes $f(x) = b$. In this case, the domain is $(-\infty, \infty)$ and the range is $\{b\}$. A function of the form $\boldsymbol{f(x) = b}$ is a **constant function,** and its graph is a horizontal line.

> **EXAMPLE 2** **Graphing a Horizontal Line**

Graph $f(x) = -3$. Give the domain and range.

SOLUTION Since $f(x)$, or y, always equals -3, the value of y can never be 0. This means that the graph has no x-intercept. The only way a straight line can have no x-intercept is for it to be parallel to the x-axis, as shown in **Figure 33.** The domain of this linear function is $(-\infty, \infty)$ and the range is $\{-3\}$. **Figure 34** shows the calculator graph.

Figure 33   Figure 34

> ✔ *Now Try Exercise 13.*

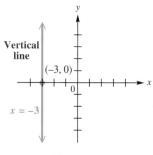

Figure 35

EXAMPLE 3 Graphing a Vertical Line

Graph $x = -3$. Give the domain and range of this relation.

SOLUTION Since x always equals -3, the value of x can never be 0, and the graph has no y-intercept. Using reasoning similar to that of **Example 2**, we find that this graph is parallel to the y-axis, as shown in **Figure 35.** The domain of this relation, which is *not* a function, is $\{-3\}$, while the range is $(-\infty, \infty)$.

✔ *Now Try Exercise 19.*

Standard Form $Ax + By = C$ Equations of lines are often written in the form $Ax + By = C$, which is called **standard form.**

NOTE The definition of "standard form" is, ironically, not standard from one text to another. Any linear equation can be written in infinitely many different, but equivalent, forms. For example, the equation $2x + 3y = 8$ can be written equivalently as

$$2x + 3y - 8 = 0, \quad 3y = 8 - 2x, \quad x + \frac{3}{2}y = 4, \quad 4x + 6y = 16,$$

and so on. In this text we will agree that if the coefficients and constant in a linear equation are rational numbers, then we will consider the standard form to be $Ax + By = C$, where $A \geq 0, A, B,$ and C are integers, and the greatest common factor of $A, B,$ and C is 1. If $A = 0$, then we choose $B > 0$. (If two or more integers have a greatest common factor of 1, they are said to be **relatively prime.**)

In **Example 8(a)** of **Section 2.1,** we graphed $y = 4x - 1$ (or, equivalently, $4x - y = 1$) by finding the x- and y-intercepts. The following example is similar.

EXAMPLE 4 Graphing $Ax + By = C$ $(C = 0)$

Graph $4x - 5y = 0$. Give the domain and range.

SOLUTION Find the intercepts.

$4(0) - 5y = 0$ Let $x = 0$.
$\qquad\qquad y = 0$ y-intercept is $(0, 0)$.

$4x - 5(0) = 0$ Let $y = 0$.
$\qquad\qquad x = 0$ x-intercept is $(0, 0)$.

The graph of this function has just one intercept: the origin $(0, 0)$. We need to find an additional point to graph the function by choosing a different value for x (or y).

$4(5) - 5y = 0$ We choose $x = 5$.
$\quad 20 - 5y = 0$ Multiply.
$\qquad\qquad 4 = y$ Add 5y. Divide by 5.

This leads to the ordered pair $(5, 4)$. Complete the graph using the two points $(0, 0)$ and $(5, 4)$, with a third point as a check. The domain and range are both $(-\infty, \infty)$. See **Figure 36.**

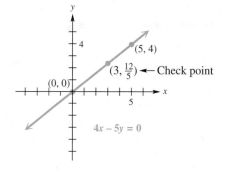

Figure 36

✔ *Now Try Exercise 17.*

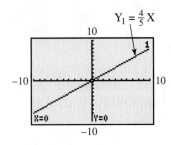

Figure 37

To use a graphing calculator to graph a linear function given in standard form, we must first solve the defining equation for y.

$$4x - 5y = 0 \qquad \text{Equation from **Example 4**}$$

$$y = \frac{4}{5}x \qquad \text{Subtract } 4x. \text{ Divide by } -5.$$

We graph $Y_1 = \frac{4}{5}X$. See **Figure 37**. ∎

Slope   An important characteristic of a straight line is its *slope,* a numerical measure of the steepness and orientation of the line. (Geometrically, this may be interpreted as the ratio of **rise** to **run.**) The slope of a highway (sometimes called the *grade*) is often given as a percent. For example, a 10% $\left(\text{or } \frac{10}{100} = \frac{1}{10}\right)$ slope means the highway rises 1 unit for every 10 horizontal units.

To find the slope of a line, start with two distinct points (x_1, y_1) and (x_2, y_2) on the line, as shown in **Figure 38**, where $x_1 \neq x_2$. As we move along the line from (x_1, y_1) to (x_2, y_2), the horizontal difference

$$\Delta x = x_2 - x_1$$

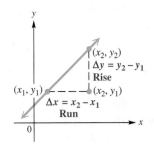

Figure 38

is called the **change in x,** denoted by Δx (read **"delta x"**), where Δ is the Greek letter **delta.** The vertical difference, called the **change in y,** can be written

$$\Delta y = y_2 - y_1.$$

The *slope* of a nonvertical line is defined as the quotient (ratio) of the change in y and the change in x, as follows.

Slope

The **slope** m of the line through the points (x_1, y_1) and (x_2, y_2) is given by the following.

$$m = \frac{\textbf{rise}}{\textbf{run}} = \frac{\Delta y}{\Delta x} = \frac{y_2 - y_1}{x_2 - x_1}, \quad \text{where } \Delta x \neq 0$$

That is, the slope of a line is the change in y divided by the corresponding change in x, where the change in x is not 0.

CAUTION   *When using the slope formula, it makes no difference which point is (x_1, y_1) or (x_2, y_2). However, be consistent.* Start with the x- and y-values of *one* point (either one) and subtract the corresponding values of the *other* point.

$$\textit{Use} \quad \frac{y_2 - y_1}{x_2 - x_1} \quad \text{or} \quad \frac{y_1 - y_2}{x_1 - x_2}, \quad \textit{not} \quad \frac{y_2 - y_1}{x_1 - x_2} \quad \text{or} \quad \frac{y_1 - y_2}{x_2 - x_1}.$$

Be sure to write the difference of the y-values in the numerator and the difference of the x-values in the denominator.

LOOKING AHEAD TO CALCULUS

The concept of slope of a line is extended in calculus to general curves. The **slope of a curve at a point** is understood to mean the slope of the line tangent to the curve at that point.

The line in the figure is tangent to the curve at point P.

The slope of a line can be found only if the line is nonvertical. This guarantees that $x_2 \neq x_1$ so that the denominator $x_2 - x_1 \neq 0$.

Undefined Slope

The slope of a vertical line is undefined.

EXAMPLE 5 Finding Slopes with the Slope Formula

Find the slope of the line through the given points.

(a) $(-4, 8), (2, -3)$ **(b)** $(2, 7), (2, -4)$ **(c)** $(5, -3), (-2, -3)$

SOLUTION

(a) Let $x_1 = -4$, $y_1 = 8$, and $x_2 = 2$, $y_2 = -3$. Then the slope is

$$m = \frac{\text{rise}}{\text{run}} = \frac{\Delta y}{\Delta x} \qquad \text{Definition of slope}$$

$$= \frac{-3 - 8}{2 - (-4)} \qquad \boxed{\text{Substitute carefully.}}$$

$$= \frac{-11}{6}, \quad \text{or} \quad -\frac{11}{6}.$$

We can also subtract in the opposite order, letting $x_1 = 2, y_1 = -3$ and $x_2 = -4, y_2 = 8$. The same slope results.

$$m = \frac{8 - (-3)}{-4 - 2} = \frac{11}{-6}, \quad \text{or} \quad -\frac{11}{6}$$

(b) If we attempt to use the formula with the points $(2, 7)$ and $(2, -4)$, we get a zero denominator.

$$m = \frac{-4 - 7}{2 - 2} = \frac{-11}{0} \quad \text{Undefined}$$

The formula is not valid here because $\Delta x = x_2 - x_1 = 2 - 2 = 0$. A sketch would show that the line through $(2, 7)$ and $(2, -4)$ is vertical. As mentioned above, the slope of a vertical line is undefined.

(c) For $(5, -3)$ and $(-2, -3)$, the slope is

$$m = \frac{-3 - (-3)}{-2 - 5} = \frac{0}{-7} = 0.$$

A sketch would show that the line through $(5, -3)$ and $(-2, -3)$ is horizontal.

☑ *Now Try Exercises 35, 41, and 43.*

The results in **Example 5(c)** suggest the following generalization.

LOOKING AHEAD TO CALCULUS

The **derivative** of a function provides a formula for determining the slope of a line tangent to a curve. If the slope is positive on a given interval, then the function is increasing there. If it is negative, then the function is decreasing. If it is 0, then the function is constant.

Zero Slope

The slope of a horizontal line is 0.

Theorems for similar triangles can be used to show that the slope of a line is independent of the choice of points on the line. *That is, slope is the same no matter which pair of distinct points on the line are used to find it.*

If the equation of a line is in the form $y = ax + b$, we can show that the slope of the line is a. To do this, we use function notation and the definition of slope. Let $f(x) = ax + b$, $x_1 = x$, and $x_2 = x + 1$.

$$m = \frac{f(x_2) - f(x_1)}{x_2 - x_1}$$

$$= \frac{[a(x + 1) + b] - (ax + b)}{(x + 1) - x} \qquad \text{Substitute in the slope formula.}$$

$$= \frac{ax + a + b - ax - b}{x + 1 - x} \qquad \text{Distributive property (Section R.2)}$$

$$= \frac{a}{1} \qquad \text{Combine like terms.}$$

$$= a \qquad \text{The slope is } a.$$

This discussion enables us to find the slope of the graph of any linear equation by solving for y and identifying the coefficient of x, which is that slope.

EXAMPLE 6 Finding the Slope from an Equation

Find the slope of the line $4x + 3y = 12$.

SOLUTION Solve the equation for y.

$$4x + 3y = 12$$

$$3y = -4x + 12 \qquad \text{Subtract } 4x.$$

$$y = -\frac{4}{3}x + 4 \qquad \text{Divide by 3.}$$

The slope is $-\frac{4}{3}$, the coefficient of x.

✔ *Now Try Exercise 49(a).*

Since the slope of a line is the ratio of vertical change (rise) to horizontal change (run), if we know the slope of a line and the coordinates of a point on the line, we can draw the graph of the line.

EXAMPLE 7 Graphing a Line Using a Point and the Slope

Graph the line passing through $(-1, 5)$ and having slope $-\frac{5}{3}$.

SOLUTION First locate the point $(-1, 5)$ as shown in **Figure 39**. Since the slope of this line is $\frac{-5}{3}$, a change of -5 units vertically (that is, 5 units *down*) corresponds to a change of 3 units horizontally (that is, 3 units to the *right*). This gives a second point, $(2, 0)$, which can then be used to complete the graph.

Because $\frac{-5}{3} = \frac{5}{-3}$, another point could be obtained by starting at $(-1, 5)$ and moving 5 units *up* and 3 units to the *left*. We would reach a different second point, $(-4, 10)$, but the graph would be the same. Confirm this in **Figure 39**.

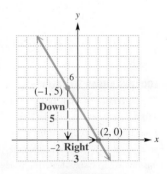

Figure 39

✔ *Now Try Exercise 53.*

Figure 40 shows lines with various slopes. *Notice the following important concepts:*

● A line with a positive slope rises from left to right. The corresponding linear function is increasing on its entire domain.

● A line with a negative slope falls from left to right. The corresponding linear function is decreasing on its entire domain.

● A line with 0 slope neither rises nor falls. The corresponding linear function is constant on its entire domain.

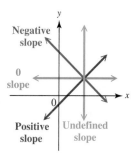

Figure 40

Average Rate of Change   We know that the slope of a line is the ratio of the vertical change in *y* to the horizontal change in *x*. ***Thus, slope gives the average rate of change in y per unit of change in x, where the value of y depends on the value of x.*** If *f* is a linear function defined on $[a, b]$, then we have the following.

$$\text{Average rate of change on } [a, b] = \frac{f(b) - f(a)}{b - a}$$

This is simply another way to write the slope formula, using function notation.

EXAMPLE 8   **Interpreting Slope as Average Rate of Change**

In 1980, the average monthly rate for basic cable TV in the United States was $7.69. In 2008, the average monthly rate was $46.13. Assume a linear relationship, and find the average rate of change in the monthly rate per year. Graph as a line segment and interpret the result. (*Source:* SNL Kagan.)

SOLUTION   To use the slope formula, we need two ordered pairs. Here, if *x* = 1980, then *y* = 7.69, and if *x* = 2008, then *y* = 46.13. This gives the ordered pairs (1980, 7.69) and (2008, 46.13). (Note that *y* is in dollars.)

$$\text{Average rate of change} = \frac{46.13 - 7.69}{2008 - 1980} = \frac{38.44}{28} \approx 1.37$$

The graph in **Figure 41** confirms that the line through the ordered pairs rises from left to right and therefore has positive slope. Thus, the monthly rate for basic cable *increased* by an average of about $1.37 each year from 1980 to 2008.

✔ *Now Try Exercise 75.*

Basic Cable TV Rates

Figure 41

Linear Models   In **Example 8,** we used the graph of a line to approximate real data. Recall from **Section 1.2** that this is called **mathematical modeling.** Points on the straight-line graph model, or approximate, the actual points that correspond to the data.

A **linear cost function** has the form

$$C(x) = mx + b, \qquad \text{Linear cost function}$$

where *x* represents the number of items produced, *m* represents the **cost** per item, and *b* represents the **fixed cost.** The fixed cost is constant for a particular product and does not change as more items are made. The value of *mx*, which increases as more items are produced, covers labor, materials, packaging, shipping, and so on.

The **revenue function** for selling a product depends on the price per item p and the number of items sold x. It is given by the following function.

$$R(x) = px \qquad \text{Revenue function}$$

Profit is found by subtracting cost from revenue and is described by the **profit function.**

$$P(x) = R(x) - C(x) \qquad \text{Profit function}$$

EXAMPLE 9 **Writing Linear Cost, Revenue, and Profit Functions**

Assume that the cost to produce an item is a linear function and all items produced are sold. The fixed cost is \$1500, the variable cost per item is \$100, and the item sells for \$125. Write linear functions to model

(a) cost, **(b)** revenue, and **(c)** profit.

(d) How many items must be sold for the company to make a profit?

SOLUTION

(a) Since the cost function is linear, it will have the following form.

$$C(x) = mx + b \qquad \text{Cost function}$$

$$C(x) = 100x + 1500 \qquad \text{Let } m = 100 \text{ and } b = 1500.$$

(b) The revenue function is defined by the product of 125 and x.

$$R(x) = px \qquad \text{Revenue function}$$

$$R(x) = 125x \qquad \text{Let } p = 125.$$

(c) The profit function is found by subtracting the cost function from the revenue function.

$$P(x) = R(x) - C(x)$$

> Use parentheses here.

$$= 125x - (100x + 1500)$$

$$= 125x - 100x - 1500 \qquad \text{Distributive property}$$

$$P(x) = 25x - 1500 \qquad \text{Combine like terms.}$$

ALGEBRAIC SOLUTION

(d) To make a profit, $P(x)$ must be positive.

$$P(x) = 25x - 1500 \qquad \text{Profit function from part (c)}$$

Set $P(x) > 0$ and solve.

$$P(x) > 0$$

$$25x - 1500 > 0 \qquad P(x) = 25x - 1500$$

$$25x > 1500 \qquad \text{Add 1500 to each side.}$$
$$\qquad\qquad\quad \text{(Section 1.7)}$$

$$x > 60 \qquad \text{Divide by 25.}$$

Since the number of items must be a whole number, at least 61 items must be sold for the company to make a profit.

GRAPHING CALCULATOR SOLUTION

(d) Define Y_1 as $25X - 1500$ and graph the line. Use the capability of your calculator to locate the x-intercept. See **Figure 42.** As the graph shows, y-values for x less than 60 are negative, and y-values for x greater than 60 are positive, so at least 61 items must be sold for the company to make a profit.

Figure 42

✔ *Now Try Exercise 87.*

2.4 Exercises

Concept Check Match the description in Column I with the correct response in Column II. *Some choices may not be used.*

I	II
1. a linear function whose graph has y-intercept $(0, 6)$	**A.** $f(x) = 5x$
2. a vertical line	**B.** $f(x) = 3x + 6$
3. a constant function	**C.** $f(x) = -8$
4. a linear function whose graph has x-intercept $(-2, 0)$ and y-intercept $(0, 4)$	**D.** $f(x) = x^2$
	E. $x + y = -6$
5. a linear function whose graph passes through the origin	**F.** $f(x) = 3x + 4$
	G. $2x - y = -4$
6. a function that is not linear	**H.** $x = 9$

Graph each linear function. Identify any constant functions. Give the domain and range. *See Examples 1 and 2.*

7. $f(x) = x - 4$

8. $f(x) = -x + 4$

9. $f(x) = \dfrac{1}{2}x - 6$

10. $f(x) = \dfrac{2}{3}x + 2$

11. $f(x) = 3x$

12. $f(x) = -2x$

13. $f(x) = -4$

14. $f(x) = 3$

Graph each line. Give the domain and range. *See Examples 3 and 4.*

15. $-4x + 3y = 12$

16. $2x + 5y = 10$

17. $3y - 4x = 0$

18. $3x + 2y = 0$

19. $x = 3$

20. $x = -4$

21. $2x + 4 = 0$

22. $-3x + 6 = 0$

23. $-x + 5 = 0$

24. $3 + x = 0$

Match each equation with the sketch that most closely resembles its graph. *See Examples 2 and 3.*

25. $y = 5$ **26.** $y = -5$ **27.** $x = 5$ **28.** $x = -5$

A.   **B.**   **C.**   **D.**

 Use a graphing calculator to graph each equation in the standard viewing window. *See Examples 1, 2, and 4.*

29. $y = 3x + 4$

30. $y = -2x + 3$

31. $3x + 4y = 6$

32. $-2x + 5y = 10$

33. *Concept Check* If a walkway rises 2.5 ft for every 10 ft on the horizontal, which of the following express its slope (or grade)? (There are several correct choices.)

A. 0.25 **B.** 4 **C.** $\dfrac{2.5}{10}$ **D.** 25%

E. $\dfrac{1}{4}$ **F.** $\dfrac{10}{2.5}$ **G.** 400% **H.** 2.5%

34. *Concept Check* If the pitch of a roof is $\frac{1}{4}$, how many feet in the horizontal direction correspond to a rise of 4 ft?

Find the slope of the line satisfying the given conditions. **See Example 5.**

35. through $(2, -1)$ and $(-3, -3)$

36. through $(-3, 4)$ and $(2, -8)$

37. through $(5, 8)$ and $(3, 12)$

38. through $(5, -3)$ and $(1, -7)$

39. through $(5, 9)$ and $(-2, 9)$

40. through $(-2, 4)$ and $(6, 4)$

41. horizontal, through $(5, 1)$

42. horizontal, through $(3, 5)$

43. vertical, through $(4, -7)$

44. vertical, through $(-8, 5)$

For each line, **(a)** *find the slope and* **(b)** *sketch the graph.* **See Examples 6 and 7.**

45. $y = 3x + 5$ **46.** $y = 2x - 4$ **47.** $2y = -3x$

48. $-4y = 5x$ **49.** $5x - 2y = 10$ **50.** $4x + 3y = 12$

Graph the line passing through the given point and having the indicated slope. Plot two points on the line. **See Example 7.**

51. through $(-1, 3)$, $m = \frac{3}{2}$

52. through $(-2, 8)$, $m = \frac{2}{5}$

53. through $(3, -4)$, $m = -\frac{1}{3}$

54. through $(-2, -3)$, $m = -\frac{3}{4}$

55. through $\left(-\frac{1}{2}, 4\right)$, $m = 0$

56. through $\left(\frac{3}{2}, 2\right)$, $m = 0$

57. through $\left(-\frac{5}{2}, 3\right)$, undefined slope

58. through $\left(\frac{9}{4}, 2\right)$, undefined slope

Concept Check For each given slope in Exercises 59–64, identify the line in A–F that could have this slope.

59. $\dfrac{1}{3}$ **60.** -3 **61.** 0 **62.** $-\dfrac{1}{3}$ **63.** 3 **64.** undefined

A.

B.

C.

D.

E.

F.

Concept Check   *Find and interpret the average rate of change illustrated in each graph.*

65.

66.

67.

68.

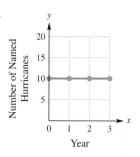

69. *Concept Check* For a constant function, is the average rate of change positive, negative, or zero? (Choose one.)

Solve each problem. ***See Example 8.***

70. *(Modeling) Olympic Times for 5000-Meter Run* The graph shows the winning times (in minutes) at the Olympic Games for the men's 5000-m run, together with a linear approximation of the data.

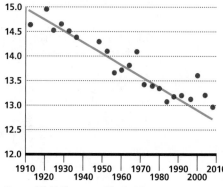

Olympic Times for 5000-Meter Run (in minutes)

Source: World Almanac and Book of Facts.

(a) An equation for the linear model, based on data from 1912–2008 (where *x* represents the year), is

$$y = -0.0193x + 51.73.$$

Determine the slope. (**See Example 6.**) What does the slope of this line represent? Why is the slope negative?

(b) Can you think of any reason why there are no data points for the years 1916, 1940, and 1944?

(c) The winning time for the 1996 Olympic Games was 13.13 min. What does the model predict? How far is the prediction from the actual value?

71. *(Modeling) U.S. Radio Stations* The graph shows the number of U.S. radio stations on the air, along with the graph of a linear function that models the data.

U.S. Radio Stations

Source: National Association of Broadcasters.

(a) Discuss the predictive accuracy of the linear function.

(b) Use the two data points $(1950, 2773)$ and $(2007, 13,977)$ to find the approximate slope of the line shown. Interpret this number.

72. *Cellular Telephone Subscribers* The table gives the number of cellular telephone subscribers in the U.S. (in thousands) from 2004 through 2009.

(a) Find the change in subscribers for 2004–2005, 2005–2006, and so on.

(b) Are the changes in successive years approximately the same? If the ordered pairs in the table were plotted, could an approximately straight line be drawn through them?

Year	Subscribers (in thousands)
2004	182,140
2005	207,896
2006	233,041
2007	255,396
2008	270,334
2009	285,646

Source: CTIA-The Wireless Association.

73. *Mobile Homes* The graph provides a good approximation of the number of mobile homes (in thousands) placed in use in the United States from 1999 through 2009.

Mobile Homes Placed in Use

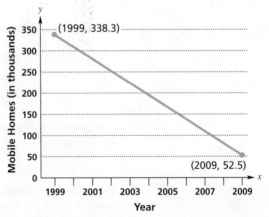

Source: U.S. Census Bureau.

    (a) Use the given ordered pairs to find the average rate of change in the number of mobile homes per year during this period.

    (b) Interpret what a negative slope means in this situation.

74. *Olympic Times for 5000-Meter Run* **Exercise 70** showed the winning times for the Olympic men's 5000-m run. Find and interpret the average rate of change for the following periods.

    (a) The winning time in 1912 was 14.6 min. In 1996 it was 13.1 min.

    (b) The winning time in 1912 was 14.6 min. In 2008 it was 13.0 min.

75. *High School Dropouts* In 1980, the number of high school dropouts in the United States was 5212 thousand. By 2008, this number had decreased to 3118 thousand. Find and interpret the average rate of change per year in the number of high school dropouts. (*Source:* U.S. Census Bureau.)

76. *Plasma Flat-Panel TV Sales* The total amount spent on plasma flat-panel TVs in the United States changed from \$5302 million in 2006 to \$2907 million in 2009. Find and interpret the average rate of change in sales, in millions of dollars per year. Round your answer to the nearest hundredth. *(Source: Consumer Electronics Association.)*

Relating Concepts

For individual or collaborative investigation *(Exercises 77–86)*

The table shows several points on the graph of a linear function. **Work Exercises 77–86 in order,** *to see connections between the slope formula, the distance formula, the midpoint formula, and linear functions.*

x	y
0	−6
1	−3
2	0
3	3
4	6
5	9
6	12

77. Use the first two points in the table to find the slope of the line.

78. Use the second and third points in the table to find the slope of the line.

79. Make a conjecture by filling in the blank: If we use any two points on a line to find its slope, we find that the slope is ——— in all cases.

80. Find the distance between the first two points in the table. (*Hint:* Use the distance formula.)

81. Find the distance between the second and fourth points in the table.

82. Find the distance between the first and fourth points in the table.

83. Add the results in **Exercises 80 and 81,** and compare the sum to the answer you found in **Exercise 82.** What do you notice?

84. Fill in the following blanks, basing your answers on your observations in **Exercises 80–83:** If points A, B, and C lie on a line in that order, then the distance between A and B added to the distance between ——— and ——— is equal to the distance between ——— and ———.

85. Find the midpoint of the segment joining $(0, -6)$ and $(6, 12)$. Compare your answer to the middle entry in the table. What do you notice?

86. If the table were set up to show an x-value of 4.5, what would be the corresponding y-value?

(Modeling) Cost, Revenue, and Profit Analysis A firm will break even (no profit and no loss) as long as revenue just equals cost. The value of x (the number of items produced and sold) where $C(x) = R(x)$ is called the **break-even point.** Assume that each of the following can be expressed as a linear function. Find

(a) the cost function, **(b)** the revenue function, and **(c)** the profit function.

(d) Find the break-even point and decide whether the product should be produced, given the restrictions on sales.

See Example 9.

	Fixed Cost	Variable Cost	Price of Item	
87.	$ 500	$ 10	$ 35	No more than 18 units can be sold.
88.	$2700	$150	$280	No more than 25 units can be sold.
89.	$1650	$400	$305	All units produced can be sold.
90.	$ 180	$ 11	$ 20	No more than 30 units can be sold.

(Modeling) Break-Even Point The manager of a small company that produces roof tile has determined that the total cost in dollars, $C(x)$, of producing x units of tile is given by

$$C(x) = 200x + 1000,$$

while the revenue in dollars, $R(x)$, from the sale of x units of tile is given by

$$R(x) = 240x.$$

91. Find the break-even point and the cost and revenue at the break-even point.

92. Suppose the variable cost is actually $220 per unit, instead of $200. How does this affect the break-even point? Is the manager better off or not?

Chapter 2 **Quiz** (Sections 2.1–2.4)

1. For $A(-4, 2)$ and $B(-8, -3)$, find $d(A, B)$, the distance between A and B.

2. *Two-Year College Enrollment* Enrollments in two-year colleges for selected years are shown in the table. Use the midpoint formula to estimate the enrollments for 2002 and 2006.

Year	Enrollment (in millions)
2000	5.95
2004	6.55
2008	6.97

Source: U.S. Center for Education Statistics.

3. Sketch the graph of $y = -x^2 + 4$ by plotting points.

4. Sketch the graph of $x^2 + y^2 = 16$.

5. Determine the radius and the coordinates of the center of the circle with equation

$$x^2 + y^2 - 4x + 8y + 3 = 0.$$

For Exercises 6–8, refer to the graph of $f(x) = |x + 3|$.

6. Find $f(-1)$.

7. Give the domain and the range of f.

8. Give the largest interval over which the function f is
(**a**) decreasing, (**b**) increasing, (**c**) constant.

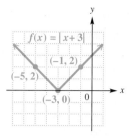

9. Find the slope of the line through the given points.

(**a**) $(1, 5)$ and $(5, 11)$ (**b**) $(-7, 4)$ and $(-1, 4)$ (**c**) $(6, 12)$ and $(6, -4)$

10. *Motor Vehicle Sales* The graph shows a straight line segment that approximates new motor vehicle sales in the United States from 2005 to 2009. Determine the average rate of change from 2005 to 2009, and interpret the results.

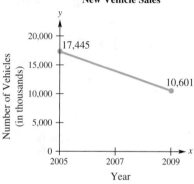

New Vehicle Sales

Source: U.S. Bureau of Economic Analysis.

2.5   Equations of Lines and Linear Models

- ■ **Point-Slope Form**
- ■ **Slope-Intercept Form**
- ■ **Vertical and Horizontal Lines**
- ■ **Parallel and Perpendicular Lines**
- ■ **Modeling Data**
- ■ **Solving Linear Equations in One Variable by Graphing**

Point-Slope Form   The graph of a linear function is a straight line. We now develop various forms for the equation of a line.

Figure 43 shows the line passing through the fixed point (x_1, y_1) having slope m. (Assuming that the line has a slope guarantees that it is not vertical.) Let (x, y) be any other point on the line. Since the line is not vertical, $x - x_1 \neq 0$. Now use the definition of slope.

$$m = \frac{y - y_1}{x - x_1} \qquad \text{Slope formula (Section 2.4)}$$

$$m(x - x_1) = y - y_1 \qquad \text{Multiply each side by } x - x_1.$$

or $\qquad y - y_1 = m(x - x_1) \qquad$ Interchange sides.

This result is the *point-slope form* of the equation of a line.

Figure 43

Point-Slope Form

The **point-slope form** of the equation of the line with slope m passing through the point (x_1, y_1) is

$$y - y_1 = m(x - x_1).$$

LOOKING AHEAD TO CALCULUS

A standard problem in calculus is to find the equation of the line tangent to a curve at a given point. The derivative (see *Looking Ahead to Calculus* in **Section 2.4**) is used to find the slope of the desired line, and then the slope and the given point are used in the point-slope form to solve the problem.

EXAMPLE 1 **Using the Point-Slope Form (Given a Point and the Slope)**

Write an equation of the line through $(-4, 1)$ having slope -3.

SOLUTION Here $x_1 = -4$, $y_1 = 1$, and $m = -3$.

$$y - y_1 = m(x - x_1) \qquad \text{Point-slope form}$$

$$y - 1 = -3[x - (-4)] \qquad x_1 = -4, y_1 = 1, m = -3$$

$$y - 1 = -3(x + 4) \qquad \boxed{\text{Be careful with signs.}}$$

$$y - 1 = -3x - 12 \qquad \text{Distributive property (Section R.2)}$$

$$y = -3x - 11 \qquad \text{Add 1. (Section 1.1)}$$

☑ *Now Try Exercise 23.*

EXAMPLE 2 **Using the Point-Slope Form (Given Two Points)**

Write an equation of the line through $(-3, 2)$ and $(2, -4)$. Write the result in standard form $Ax + By = C$.

SOLUTION Find the slope first.

$$m = \frac{-4 - 2}{2 - (-3)} = -\frac{6}{5} \qquad \text{Definition of slope}$$

The slope m is $-\frac{6}{5}$. Either $(-3, 2)$ or $(2, -4)$ can be used for (x_1, y_1). We choose $(-3, 2)$.

$$y - y_1 = m(x - x_1) \qquad \text{Point-slope form}$$

$$y - 2 = -\frac{6}{5}[x - (-3)] \qquad x_1 = -3, y_1 = 2, m = -\frac{6}{5}$$

$$5(y - 2) = -6(x + 3) \qquad \text{Multiply by 5.}$$

$$5y - 10 = -6x - 18 \qquad \text{Distributive property}$$

$$6x + 5y = -8 \qquad \text{Standard form (Section 2.4)}$$

Verify that we obtain the same equation if we use $(2, -4)$ instead of $(-3, 2)$ in the point-slope form.

☑ *Now Try Exercise 13.*

NOTE The lines in **Examples 1 and 2** both have negative slopes. Keep in mind that a slope of the form $-\frac{A}{B}$ may be interpreted as either $\frac{-A}{B}$ or $\frac{A}{-B}$.

Slope-Intercept Form As a special case of the point-slope form of the equation of a line, suppose that a line has y-intercept $(0, b)$. If the line has slope m, then using the point-slope form with $x_1 = 0$ and $y_1 = b$ gives the following.

$$y - y_1 = m(x - x_1) \qquad \text{Point-slope form}$$

$$y - b = m(x - 0) \qquad x_1 = 0, y_1 = b$$

$$y = mx + b \qquad \text{Solve for } y.$$

$$\underset{\text{Slope}}{\uparrow} \qquad \underset{y\text{-intercept is } (0, b).}{\uparrow}$$

Since this result shows the slope of the line and the y-value of the y-intercept, it is called the *slope-intercept form* of the equation of the line.

Slope-Intercept Form

The **slope-intercept form** of the equation of the line with slope m and y-intercept $(0, b)$ is

$$y = mx + b.$$

EXAMPLE 3 **Finding the Slope and y-Intercept from an Equation of a Line**

Find the slope and y-intercept of the line with equation $4x + 5y = -10$.

SOLUTION Write the equation in slope-intercept form.

$$4x + 5y = -10$$

$$5y = -4x - 10 \quad \text{Subtract } 4x.$$

$$y = -\frac{4}{5}x - 2 \quad \text{Divide by 5.}$$

$$\underset{m}{\uparrow} \qquad \underset{b}{\uparrow}$$

The slope is $-\frac{4}{5}$ and the y-intercept is $(0, -2)$.

Now Try Exercise 35.

NOTE Generalizing from **Example 3,** the slope m of the graph of

$$Ax + By = C$$

is $-\frac{A}{B}$, and the y-intercept is $\left(0, \frac{C}{B}\right)$.

EXAMPLE 4 **Using the Slope-Intercept Form (Given Two Points)**

Write an equation of the line through $(1, 1)$ and $(2, 4)$. Then graph the line using the slope-intercept form.

SOLUTION In **Example 2,** we used the *point-slope form* in a similar problem. Here we show an alternative method using the *slope-intercept form*. First, find the slope.

$$m = \frac{4 - 1}{2 - 1} = \frac{3}{1} = 3 \quad \text{Definition of slope}$$

Now substitute 3 for m in $y = mx + b$ and choose one of the given points, say $(1, 1)$, to find the value of b.

$$y = mx + b \quad \text{Slope-intercept form}$$

$$1 = 3(1) + b \quad m = 3, x = 1, y = 1$$

y-value of the y-intercept $\rightarrow b = -2 \quad$ Solve for b.

The slope-intercept form is

$$y = 3x - 2.$$

The graph is shown in **Figure 44.** We can plot $(0, -2)$ and then use the definition of slope to arrive at $(1, 1)$. Verify that $(2, 4)$ also lies on the line.

Now Try Exercise 13.

Figure 44

EXAMPLE 5   **Finding an Equation from a Graph**

Use the graph of the linear function f shown in **Figure 45** to complete the following.

(a) Find the slope, y-intercept, and x-intercept.

(b) Write the equation that defines f.

SOLUTION

(a) The line falls 1 unit each time the x-value increases by 3 units. Therefore, the slope is $\frac{-1}{3} = -\frac{1}{3}$. The graph intersects the y-axis at the point $(0, -1)$ and intersects the x-axis at the point $(-3, 0)$. Therefore, the y-intercept is $(0, -1)$ and the x-intercept is $(-3, 0)$.

Figure 45

(b) The slope is $m = -\frac{1}{3}$, and the y-value of the y-intercept is $b = -1$.

$$y = f(x) = mx + b \qquad \text{Slope-intercept form}$$

$$f(x) = -\frac{1}{3}x - 1 \qquad m = -\frac{1}{3}, b = -1$$

✔ *Now Try Exercise 39.*

(a)

(b)

Figure 46

Vertical and Horizontal Lines   We first saw graphs of vertical and horizontal lines in **Section 2.4.** The vertical line through the point (a, b) passes through all points of the form (a, y), for any value of y. Consequently, the equation of a vertical line through (a, b) is $x = a$. For example, the vertical line through $(-3, 1)$ has equation $x = -3$. See **Figure 46(a).** Since each point on the y-axis has x-coordinate 0, *the equation of the y-axis is $x = 0$.*

The horizontal line through the point (a, b) passes through all points of the form (x, b), for any value of x. Therefore, the equation of a horizontal line through (a, b) is $y = b$. For example, the horizontal line through $(1, -3)$ has equation $y = -3$. See **Figure 46(b).** Since each point on the x-axis has y-coordinate 0, *the equation of the x-axis is $y = 0$.*

Equations of Vertical and Horizontal Lines

An equation of the **vertical line** through the point (a, b) is $x = a.$

An equation of the **horizontal line** through the point (a, b) is $y = b.$

Parallel and Perpendicular Lines   Since two parallel lines are equally "steep," they should have the same slope. Also, two distinct lines with the same "steepness" are parallel. The following result summarizes this discussion. (The statement "p if and only if q" means "if p then q *and* if q then p.")

Parallel Lines

Two distinct nonvertical lines are parallel if and only if they have the same slope.

When two lines have slopes with a product of -1, the lines are perpendicular.

Perpendicular Lines

Two lines, neither of which is vertical, are perpendicular if and only if their slopes have a product of -1. Thus, the slopes of perpendicular lines, neither of which is vertical, are *negative reciprocals*.

For example, if the slope of a line is $-\frac{3}{4}$, the slope of any line perpendicular to it is $\frac{4}{3}$, since $-\frac{3}{4}\left(\frac{4}{3}\right) = -1$. (Numbers like $-\frac{3}{4}$ and $\frac{4}{3}$ are **negative reciprocals** of each other.) A proof of this result is outlined in **Exercises 69–75.**

NOTE Because a vertical line has *undefined* slope, it does not follow the *mathematical* rules for parallel and perpendicular lines. We intuitively know that all vertical lines are parallel and that a vertical line and a horizontal line are perpendicular.

EXAMPLE 6 **Finding Equations of Parallel and Perpendicular Lines**

Write the equation in both slope-intercept and standard form of the line that passes through the point $(3, 5)$ and satisfies the given condition.

(a) parallel to the line $2x + 5y = 4$

(b) perpendicular to the line $2x + 5y = 4$

SOLUTION

(a) Since we know that the point $(3, 5)$ is on the line, we need only find the slope to use the point-slope form. We find the slope by writing the equation of the given line in slope-intercept form. (That is, we solve for y.)

$$2x + 5y = 4$$

$$5y = -2x + 4 \quad \text{Subtract } 2x.$$

$$y = -\frac{2}{5}x + \frac{4}{5} \quad \text{Divide by 5.}$$

The slope is $-\frac{2}{5}$. Since the lines are parallel, $-\frac{2}{5}$ is also the slope of the line whose equation is to be found. Now substitute this slope and the given point $(3, 5)$ in the point-slope form.

$$y - y_1 = m(x - x_1) \quad \text{Point-slope form}$$

$$y - 5 = -\frac{2}{5}(x - 3) \quad m = -\frac{2}{5}, x_1 = 3, y_1 = 5$$

$$y - 5 = -\frac{2}{5}x + \frac{6}{5} \quad \text{Distributive property}$$

Slope-intercept form $\longrightarrow y = -\frac{2}{5}x + \frac{31}{5} \quad \text{Add } 5 = \frac{25}{5}.$

$$5y = -2x + 31 \quad \text{Multiply by 5.}$$

Standard form $\longrightarrow 2x + 5y = 31 \quad \text{Add } 2x.$

(b) There is no need to find the slope again, because in part (a) we found that the slope of the line $2x + 5y = 4$ is $-\frac{2}{5}$. The slope of any line perpendicular to it is $\frac{5}{2}$.

$$y - y_1 = m(x - x_1) \quad \text{Point-slope form}$$

$$y - 5 = \frac{5}{2}(x - 3) \quad m = \frac{5}{2}, x_1 = 3, y_1 = 5$$

$$y - 5 = \frac{5}{2}x - \frac{15}{2} \quad \text{Distributive property}$$

Slope-intercept form $\longrightarrow y = \frac{5}{2}x - \frac{5}{2} \quad$ Add $5 = \frac{10}{2}$.

$$2y = 5x - 5 \quad \text{Multiply by 2.}$$

Standard form $\longrightarrow 5x - 2y = 5 \quad$ Subtract $2y$, add 5, and rewrite.

☑ *Now Try Exercises 45 and 47.*

⊞ We can use a graphing calculator to support the results of **Example 6.** In **Figure 47(a),** we graph the equations of the parallel lines

$$y = -\frac{2}{5}x + \frac{4}{5} \quad \text{and} \quad y = -\frac{2}{5}x + \frac{31}{5}$$

from **Example 6(a).** The lines appear to be parallel, giving visual support for our result. We must use caution, however, when viewing such graphs, as the limited resolution of a graphing calculator screen may cause two lines to *appear* to be parallel even when they are not. For example, **Figure 47(b)** shows the graphs of the equations

$$y = 2x + 6 \quad \text{and} \quad y = 2.01x - 3$$

in the standard viewing window, and they appear to be parallel. This is not the case, however, because their slopes, 2 and 2.01, are different.

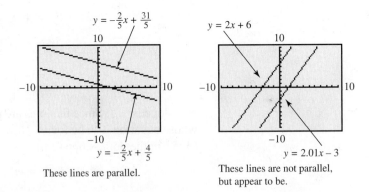

These lines are parallel.

(a)

These lines are not parallel, but appear to be.

(b)

Figure 47

To support the result of **Example 6(b),** we graph the equations of the perpendicular lines

$$y = -\frac{2}{5}x + \frac{4}{5} \quad \text{and} \quad y = \frac{5}{2}x - \frac{5}{2}.$$

If we use the standard viewing window, the lines do not appear to be perpendicular. See **Figure 48(a).** To obtain the correct perspective, we must use a square viewing window, as in **Figure 48(b).**

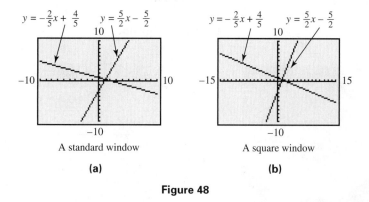

(a) A standard window

(b) A square window

Figure 48

A summary of the various forms of linear equations follows.

Equation	Description	When to Use
$y = mx + b$	**Slope-Intercept Form** Slope is m. y-intercept is $(0, b)$.	The slope and y-intercept can be easily identified and used to quickly graph the equation. This form can also be used to find the equation of a line given a point and the slope.
$y - y_1 = m(x - x_1)$	**Point-Slope Form** Slope is m. Line passes through (x_1, y_1).	This form is ideal for finding the equation of a line if the slope and a point on the line or two points on the line are known.
$Ax + By = C$	**Standard Form** (If the coefficients and constant are rational, then A, B, and C are expressed as relatively prime integers, with $A \geq 0$.) Slope is $-\frac{A}{B}$ $\quad (B \neq 0)$. x-intercept is $\left(\frac{C}{A}, 0\right)$ $\ (A \neq 0)$. y-intercept is $\left(0, \frac{C}{B}\right)$ $\ (B \neq 0)$.	The x- and y-intercepts can be found quickly and used to graph the equation. The slope must be calculated.
$y = b$	**Horizontal Line** Slope is 0. y-intercept is $(0, b)$.	If the graph intersects only the y-axis, then y is the only variable in the equation.
$x = a$	**Vertical Line** Slope is undefined. x-intercept is $(a, 0)$.	If the graph intersects only the x-axis, then x is the only variable in the equation.

Modeling Data   We can write equations of lines that mathematically describe, or model, real data if the data change at a fairly constant rate. In this case, the data fit a linear pattern, and the rate of change is the slope of the line.

> **EXAMPLE 7** **Finding an Equation of a Line That Models Data**

Average annual tuition and fees for in-state students at public four-year colleges are shown in the table for selected years and graphed as ordered pairs of points in **Figure 49,** where $x = 0$ represents 2005, $x = 1$ represents 2006, and so on, and y represents the cost in dollars. This graph of ordered pairs of data is called a **scatter diagram.**

Year	Cost (in dollars)
2005	5492
2006	5804
2007	6191
2008	6591
2009	7050
2010	7605

Source: *Trends in College Pricing 2010*, The College Board.

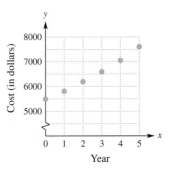

Figure 49

(a) Find an equation that models the data.

(b) Use the equation from part (a) to predict the cost of tuition and fees at public four-year colleges in 2012.

SOLUTION

(a) Since the points in **Figure 49** lie approximately on a straight line, we can write a linear equation that models the relationship between year x and cost y. We choose two data points, $(0, 5492)$ and $(5, 7605)$, to find the slope of the line.

$$m = \frac{7605 - 5492}{5 - 0} = \frac{2113}{5} = 422.6$$

The slope 422.6 indicates that the cost of tuition and fees for in-state students at public four-year colleges increased by about \$423 per year from 2005 to 2010. We use this slope, the y-intercept $(0, 5492)$, and the slope-intercept form to write an equation of the line.

$$y = mx + b \qquad \text{Slope-intercept form}$$

$$y = 422.6x + 5492 \qquad \text{Substitute.}$$

(b) The value $x = 7$ corresponds to the year 2012, so we substitute 7 for x.

$$y = 422.6x + 5492 \qquad \text{Model from part (a)}$$

$$y = 422.6(7) + 5492 \qquad \text{Let } x = 7.$$

$$y = 8450.2 \qquad \text{Multiply, and then add.}$$

The model predicts that average tuition and fees for in-state students at public four-year colleges in 2012 would be about \$8450.

☑ *Now Try Exercise 57(a) and (b).*

NOTE In **Example 7,** if we had chosen different data points, we would have gotten a slightly different equation.

Guidelines for Modeling

Step 1 Make a scatter diagram of the data.

Step 2 Find an equation that models the data. For a line, this involves selecting two data points and finding the equation of the line through them.

A technique from statistics called **linear regression** provides the line of "best fit." **Figure 50** shows how a TI-83/84 Plus calculator can accept the data points, calculate the equation of this line of best fit (in this case, $Y_1 = 420.1X + 5405$), and plot both the data points and the line on the same screen.

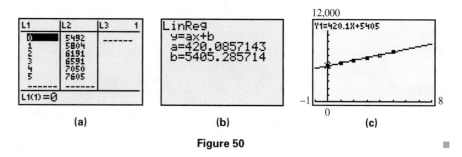

| (a) | (b) | (c) |

Figure 50

Solving Linear Equations in One Variable by Graphing Suppose that Y_1 and Y_2 are linear polynomials in x, and we want to solve the equation $Y_1 = Y_2$. Assuming that the equation has a unique solution, the equation can be found graphically as follows and is illustrated in **Example 8.**

1. Rewrite the equation as $Y_1 - Y_2 = 0$.

2. Graph the linear function $Y_3 = Y_1 - Y_2$.

3. Find the x-intercept of the graph of the function Y_3. This is the solution of $Y_1 = Y_2$.

EXAMPLE 8 **Solving an Equation with a Graphing Calculator**

Use a graphing calculator to solve $-2x - 4(2 - x) = 3x + 4$.

SOLUTION We write an equivalent equation with 0 on one side.

$$-2x - 4(2 - x) - 3x - 4 = 0 \quad \text{Subtract } 3x \text{ and } 4.$$

Then we graph $Y = -2X - 4(2 - X) - 3X - 4$ to find the x-intercept. The standard viewing window cannot be used because the x-intercept does not lie in the interval $[-10, 10]$.

As seen in **Figure 51,** the x-intercept of the graph is $(-12, 0)$, and thus the solution of the equation is -12. (-12 is the zero of the function Y.) The solution set is $\{-12\}$.

$$Y = -2X - 4(2 - X) - 3X - 4$$

Figure 51

✔ *Now Try Exercise 63.*

2.5 Exercises

Concept Check Match each equation in Exercises 1–4 to the correct graph in A–D.

1. $y = \frac{1}{4}x + 2$

2. $4x + 3y = 12$

3. $y - (-1) = \frac{3}{2}(x - 1)$

4. $y = 4$

A.

B.

C.

D.

In Exercises 5–26, write an equation for the line described. Give answers in standard form for Exercises 5–14 and in slope-intercept form (if possible) for Exercises 15–26. *See Examples 1–4.*

5. through $(1, 3)$, $m = -2$

6. through $(2, 4)$, $m = -1$

7. through $(-5, 4)$, $m = -\frac{3}{2}$

8. through $(-4, 3)$, $m = \frac{3}{4}$

9. through $(-8, 4)$, undefined slope

10. through $(5, 1)$, undefined slope

11. through $(5, -8)$, $m = 0$

12. through $(-3, 12)$, $m = 0$

13. through $(-1, 3)$ and $(3, 4)$

14. through $(2, 3)$ and $(-1, 2)$

15. x-intercept $(3, 0)$, y-intercept $(0, -2)$

16. x-intercept $(-4, 0)$, y-intercept $(0, 3)$

17. vertical, through $(-6, 4)$

18. vertical, through $(2, 7)$

19. horizontal, through $(-7, 4)$

20. horizontal, through $(-8, -2)$

21. $m = 5$, $b = 15$

22. $m = -2$, $b = 12$

23. through $(-2, 5)$ having slope -4

24. through $(4, -7)$ having slope -2

25. slope 0, y-intercept $\left(0, \frac{3}{2}\right)$

26. slope 0, y-intercept $\left(0, -\frac{5}{4}\right)$

27. *Concept Check* Fill in each blank with the appropriate response: The line $x + 2 = 0$ has x-intercept _____. It _____ have a y-intercept.
(does/does not)
The slope of this line is _____. The line $4y = 2$ has y-intercept
(0/undefined)
_____. It _____ have an x-intercept. The slope of
(does/does not)
this line is _____.
(0/undefined)

28. *Concept Check* Match each equation with the line that would most closely resemble its graph. (*Hint:* Consider the signs of m and b in the slope-intercept form.)

(a) $y = 3x + 2$ (b) $y = -3x + 2$ (c) $y = 3x - 2$ (d) $y = -3x - 2$

A.   **B.**   **C.**   **D.**

Give the slope and y-intercept of each line, and graph it. **See Example 3.**

29. $y = 3x - 1$ **30.** $y = -2x + 7$ **31.** $4x - y = 7$

32. $2x + 3y = 16$ **33.** $4y = -3x$ **34.** $2y = x$

35. $x + 2y = -4$ **36.** $x + 3y = -9$ **37.** $y - \dfrac{3}{2}x - 1 = 0$

38. *Concept Check* The table represents a linear function f.

(a) Find the slope of the line defined by $y = f(x)$.
(b) Find the y-intercept of the line.
(c) Find the equation for this line in slope-intercept form.

x	y
-2	-11
-1	-8
0	-5
1	-2
2	1
3	4

Connecting Graphs with Equations *The graph of a linear function f is shown.* **(a)** *Identify the slope, y-intercept, and x-intercept.* **(b)** *Write the equation that defines f.* **See Example 5.**

39.   **40.**   **41.**

42.   **43.**   **44.**

In Exercises 45–52, write an equation **(a)** *in standard form and* **(b)** *in slope-intercept form for the line described.* **See Example 6.**

45. through $(-1, 4)$, parallel to $x + 3y = 5$

46. through $(3, -2)$, parallel to $2x - y = 5$

47. through $(1, 6)$, perpendicular to $3x + 5y = 1$

48. through $(-2, 0)$, perpendicular to $8x - 3y = 7$

49. through $(4, 1)$, parallel to $y = -5$ **50.** through $(-2, -2)$, parallel to $y = 3$

51. through $(-5, 6)$, perpendicular to $x = -2$

52. through $(4, -4)$, perpendicular to $x = 4$

53. Find k so that the line through $(4, -1)$ and $(k, 2)$ is

    (a) parallel to $3y + 2x = 6$;   **(b)** perpendicular to $2y - 5x = 1$.

54. Find r so that the line through $(2, 6)$ and $(-4, r)$ is

    (a) parallel to $2x - 3y = 4$;   **(b)** perpendicular to $x + 2y = 1$.

(Modeling) Solve each problem. **See Example 7.**

55. *Annual Tuition and Fees* Use the data points $(0, 5492)$ and $(4, 7050)$ to find a linear equation that models the data shown in the table accompanying **Figure 49** in **Example 7.** Then use it to predict the average annual tuition and fees for in-state students at public four-year colleges in 2010. How does the result compare to the actual figure given in the table, $7605?

56. *Annual Tuition and Fees* Repeat **Exercise 55** using the data points for the years 2006 and 2008 to predict the average annual tuition and fees for 2010. How does the result compare to the actual figure given in the table, $7605?

57. *Cost of Private College Education* The table lists the average annual cost (in dollars) of tuition and fees at private four-year colleges for selected years.

    (a) Determine a linear function $f(x) = mx + b$ that models the data, where $x = 0$ represents 1996, $x = 1$ represents 1997, and so on. Use the points $(0, 12{,}881)$ and $(12, 22{,}449)$ to graph f and a scatter diagram of the data on the same coordinate axes. (You may wish to use a graphing calculator.) What does the slope of the graph of f indicate?

    (b) Use this function to approximate tuition and fees in 2007. Compare your approximation to the actual value of $21,979.

    (c) Use the linear regression feature of a graphing calculator to find the equation of the line of best fit.

Year	Tuition and Fees (in dollars)
1996	12,881
1998	13,973
2000	15,470
2002	16,826
2004	18,604
2006	20,517
2008	22,449

Source: The College Board, *Annual Survey of Colleges.*

58. *Distances and Velocities of Galaxies* The table lists the distances (in megaparsecs; 1 megaparsec = 3.085×10^{24} cm, and 1 megaparsec = 3.26 million light-years) and velocities (in kilometers per second) of four galaxies moving rapidly away from Earth.

Galaxy	Distance	Velocity
Virgo	15	1600
Ursa Minor	200	15,000
Corona Borealis	290	24,000
Bootes	520	40,000

Source: Acker, A., and C. Jaschek, *Astronomical Methods and Calculations,* John Wiley and Sons. Karttunen, H. (editor), *Fundamental Astronomy,* Springer-Verlag.

    (a) Plot the data using distances for the x-values and velocities for the y-values. What type of relationship seems to hold between the data?

    (b) Find a linear equation in the form $y = mx$ that models these data using the points $(520, 40{,}000)$ and $(0, 0)$. Graph your equation with the data on the same coordinate axes.

(c) The galaxy Hydra has a velocity of 60,000 km per sec. How far away is it according to the model in part (b)?

(d) The value of m is called the **Hubble constant.** The Hubble constant can be used to estimate the age of the universe A (in years) using the formula

$$A = \frac{9.5 \times 10^{11}}{m}.$$

Approximate A using your value of m.

(e) Astronomers currently place the value of the Hubble constant between 50 and 100. What is the range for the age of the universe A?

59. *Celsius and Fahrenheit Temperatures* When the Celsius temperature is $0°$, the corresponding Fahrenheit temperature is $32°$. When the Celsius temperature is $100°$, the corresponding Fahrenheit temperature is $212°$. Let C represent the Celsius temperature and F the Fahrenheit temperature.

(a) Express F as an exact linear function of C.

(b) Solve the equation in part (a) for C, thus expressing C as a function of F.

(c) For what temperature is $F = C$ a true statement?

60. *Water Pressure on a Diver* The pressure p of water on a diver's body is a linear function of the diver's depth, x. At the water's surface, the pressure is 1 atmosphere. At a depth of 100 ft, the pressure is about 3.92 atmospheres.

(a) Find the linear function that relates p to x.

(b) Compute the pressure at a depth of 10 fathoms (60 ft).

61. *Consumption Expenditures* In Keynesian macroeconomic theory, total consumption expenditure on goods and services, C, is assumed to be a linear function of national personal income, I. The table gives the values of C and I for 2004 and 2009 in the United States (in billions of dollars).

Year	2004	2009
Total consumption (C)	$8285	$10,089
National income (I)	$9937	$12,026

Source: U.S. Bureau of Economic Analysis.

(a) Find the formula for C as a function of I.

(b) The slope of the linear function is called the **marginal propensity to consume.** What is the marginal propensity to consume for the United States from 2004–2009?

Use a graphing calculator to solve each linear equation. See Example 8.

62. $2x + 7 - x = 4x - 2$

63. $7x - 2x + 4 - 5 = 3x + 1$

64. $3(2x + 1) - 2(x - 2) = 5$

65. $4x - 3(4 - 2x) = 2(x - 3) + 6x + 2$

66. The graph of $y = f(x)$ is shown in the standard viewing window. Which is the only value of x that could possibly be the solution of the equation $f(x) = 0$?

A. -15 **B.** 0 **C.** 5 **D.** 15

67. (a) Solve $-2(x - 5) = -x - 2$ using the methods of **Chapter 1.**

(b) Explain why the standard viewing window of a graphing calculator cannot graphically support the solution found in part (a). What minimum and maximum x-values would make it possible for the solution to be seen?

68. Using a graphing calculator, try to solve $-3(2x + 6) = -4x + 8 - 2x$. Explain what happens. What is the solution set?

Relating Concepts

For individual or collaborative investigation *(Exercises 69–75)*

In this section we state that two lines, neither of which is vertical, are perpendicular if and only if their slopes have a product of −1*. In Exercises 69–75, we outline a partial proof of this for the case where the two lines intersect at the origin.* **Work these exercises in order,** *and refer to the figure as needed.*

By the converse of the Pythagorean theorem, if

$$[d(O, P)]^2 + [d(O, Q)]^2 = [d(P, Q)]^2,$$

then triangle POQ is a right triangle with right angle at O.

69. Find an expression for the distance $d(O, P)$.

70. Find an expression for the distance $d(O, Q)$.

71. Find an expression for the distance $d(P, Q)$.

72. Use your results from **Exercises 69–71,** and substitute into the equation from the Pythagorean theorem. Simplify to show that this leads to the equation

$$-2m_1m_2x_1x_2 - 2x_1x_2 = 0.$$

73. Factor $-2x_1x_2$ from the final form of the equation in **Exercise 72.**

74. Use the property that if $ab = 0$ then $a = 0$ or $b = 0$ to solve the equation in **Exercise 73,** showing that $m_1m_2 = -1$.

75. State your conclusion based on **Exercises 69–74.**

76. Refer to **Example 4** in **Section 2.1,** and prove that the three points are collinear by taking them two at a time and showing that in all three cases, the slope is the same.

Determine whether the three points are collinear by using slopes as in **Exercise 76.** *(Note: These problems were first seen in* **Exercises 25–28 in Section 2.1.**)

77. $(-1, 4)$, $(-2, -1)$, $(1, 14)$ **78.** $(0, -7)$, $(-3, 5)$, $(2, -15)$

79. $(-1, -3)$, $(-5, 12)$, $(1, -11)$ **80.** $(0, 9)$, $(-3, -7)$, $(2, 19)$

Summary Exercises on Graphs, Circles, Functions, and Equations

These summary exercises provide practice with some of the concepts from **Sections 2.1–2.5.**

     For the points P and Q, find **(a)** *the distance* $d(P, Q)$, **(b)** *the coordinates of the midpoint of the segment PQ, and* **(c)** *an equation for the line through the two points. Write the equation in slope-intercept form if possible.*

1. $P(3, 5)$, $Q(2, -3)$ **2.** $P(-1, 0)$, $Q(4, -2)$

3. $P(-2, 2)$, $Q(3, 2)$ **4.** $P\left(2\sqrt{2}, \sqrt{2}\right)$, $Q\left(\sqrt{2}, 3\sqrt{2}\right)$

5. $P(5, -1),\ Q(5, 1)$

6. $P(1, 1),\ Q(-3, -3)$

7. $P(2\sqrt{3}, 3\sqrt{5}),\ Q(6\sqrt{3}, 3\sqrt{5})$

8. $P(0, -4),\ Q(3, 1)$

Write an equation for each of the following, and sketch the graph.

9. the line through $(-2, 1)$ and $(4, -1)$

10. the horizontal line through $(2, 3)$

11. the circle with center $(2, -1)$ and radius 3

12. the circle with center $(0, 2)$ and tangent to the *x*-axis

13. the line through $(3, -5)$ with slope $-\frac{5}{6}$

14. the line through the origin and perpendicular to the line $3x - 4y = 2$

15. the line through $(-3, 2)$ and parallel to the line $2x + 3y = 6$

16. the vertical line through $(-4, 3)$

Decide whether or not each equation has a circle as its graph. If it does, give the center and the radius.

17. $x^2 + y^2 - 4x + 2y = 4$

18. $x^2 + y^2 + 6x + 10y + 36 = 0$

19. $x^2 + y^2 - 12x + 20 = 0$

20. $x^2 + y^2 + 2x + 16y = -61$

21. $x^2 + y^2 - 2x + 10 = 0$

22. $x^2 + y^2 - 8y - 9 = 0$

23. Find the coordinates of the points of intersection of the line $y = 2$ and the circle with center at $(4, 5)$ and radius 4.

24. Find the shortest distance from the origin to the graph of the circle with equation

$$x^2 + y^2 - 10x - 24y + 144 = 0.$$

*For each relation, (**a**) find the domain and range, and (**b**) if the relation defines y as a function f of x, rewrite the relation using function notation and find $f(-2)$.*

25. $x - 4y = -6$

26. $y^2 - x = 5$

27. $(x + 2)^2 + y^2 = 25$

28. $x^2 - 2y = 3$

2.6 Graphs of Basic Functions

- Continuity
- The Identity, Squaring, and Cubing Functions
- The Square Root and Cube Root Functions
- The Absolute Value Function
- Piecewise-Defined Functions
- The Relation $x = y^2$

Continuity　Earlier in this chapter we graphed linear functions. The graph of a linear function, a straight line, may be drawn by hand over any interval of its domain without picking the pencil up from the paper. In mathematics we say that a function with this property is *continuous* over any interval. The formal definition of continuity requires concepts from calculus, but we can give an informal definition at the college algebra level.

Continuity (Informal Definition)

A function is **continuous** over an interval of its domain if its hand-drawn graph over that interval can be sketched without lifting the pencil from the paper.

The function is
discontinuous at $x = 2$.

Figure 52

If a function is not continuous at a *point,* then it has a *discontinuity* there. **Figure 52** shows the graph of a function with a discontinuity at the point where $x = 2$.

EXAMPLE 1 **Determining Intervals of Continuity**

Describe the intervals of continuity for each function in **Figure 53.**

(a)   (b)

Figure 53

SOLUTION The function in **Figure 53(a)** is continuous over its entire domain, $(-\infty, \infty)$. The function in **Figure 53(b)** has a point of discontinuity at $x = 3$. Thus, it is continuous over the intervals

$$(-\infty, 3) \quad \text{and} \quad (3, \infty).$$

✔ *Now Try Exercises 11 and 15.*

Graphs of the basic functions studied in college algebra can be sketched by careful point plotting or generated by a graphing calculator. As you become more familiar with these graphs, you should be able to provide quick rough sketches of them.

The Identity, Squaring, and Cubing Functions The **identity function** $f(x) = x$ pairs every real number with itself. See **Figure 54.**

Identity Function $f(x) = x$

Domain: $(-\infty, \infty)$ Range: $(-\infty, \infty)$

x	y
-2	-2
-1	-1
0	0
1	1
2	2

$f(x) = x$

Figure 54

- $f(x) = x$ is increasing on its entire domain, $(-\infty, \infty)$.
- It is continuous on its entire domain, $(-\infty, \infty)$.

LOOKING AHEAD TO CALCULUS
Many calculus theorems apply only to continuous functions.

The **squaring function** $f(x) = x^2$ pairs each real number with its square. Its graph is called a **parabola.** The point $(0, 0)$ at which the graph changes from decreasing to increasing is called the **vertex** of the parabola. See **Figure 55.** (For a parabola that opens downward, the vertex is the point at which the graph changes from increasing to decreasing.)

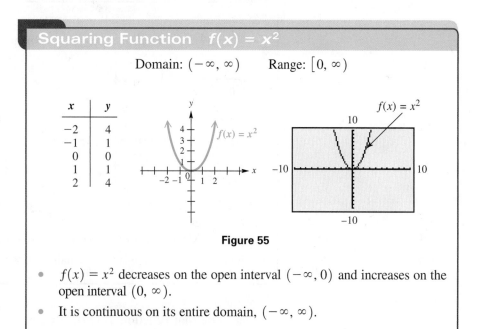

Squaring Function $f(x) = x^2$

Domain: $(-\infty, \infty)$ Range: $[0, \infty)$

x	y
-2	4
-1	1
0	0
1	1
2	4

Figure 55

- $f(x) = x^2$ decreases on the open interval $(-\infty, 0)$ and increases on the open interval $(0, \infty)$.
- It is continuous on its entire domain, $(-\infty, \infty)$.

The function $f(x) = x^3$ is called the **cubing function.** It pairs each real number with the cube of the number. See **Figure 56.** The point $(0, 0)$ at which the graph changes from "opening downward" to "opening upward" is an **inflection point.**

Cubing Function $f(x) = x^3$

Domain: $(-\infty, \infty)$ Range: $(-\infty, \infty)$

x	y
-2	-8
-1	-1
0	0
1	1
2	8

Figure 56

- $f(x) = x^3$ increases on its entire domain, $(-\infty, \infty)$.
- It is continuous on its entire domain, $(-\infty, \infty)$.

The Square Root and Cube Root Functions The function $f(x) = \sqrt{x}$ is called the **square root function.** It pairs each real number with its principal square root. See **Figure 57.** For the function value to be a real number, the domain must be restricted to $[0, \infty)$.

Square Root Function $f(x) = \sqrt{x}$

Domain: $[0, \infty)$ Range: $[0, \infty)$

x	y
0	0
1	1
4	2
9	3
16	4

Figure 57

- $f(x) = \sqrt{x}$ increases on the open interval $(0, \infty)$.
- It is continuous on its entire domain, $[0, \infty)$.

The **cube root function** $f(x) = \sqrt[3]{x}$ pairs each real number with its cube root. See **Figure 58.** The cube root function differs from the square root function in that *any* real number has a real number cube root. Thus, the domain is $(-\infty, \infty)$.

Cube Root Function $f(x) = \sqrt[3]{x}$

Domain: $(-\infty, \infty)$ Range: $(-\infty, \infty)$

x	y
-8	-2
-1	-1
0	0
1	1
8	2

Figure 58

- $f(x) = \sqrt[3]{x}$ increases on its entire domain, $(-\infty, \infty)$.
- It is continuous on its entire domain, $(-\infty, \infty)$.

The Absolute Value Function The **absolute value function,** $f(x) = |x|$, which pairs every real number with its absolute value, is graphed in **Figure 59** on the next page and is defined as follows.

$$f(x) = |x| = \begin{cases} x & \text{if } x \geq 0 \\ -x & \text{if } x < 0 \end{cases} \quad \text{Absolute value function}$$

That is, we use $|x| = x$ if x is positive or 0, and we use $|x| = -x$ if x is negative.

Absolute Value Function $f(x) = |x|$

Domain: $(-\infty, \infty)$ Range: $[0, \infty)$

x	y
-2	2
-1	1
0	0
1	1
2	2

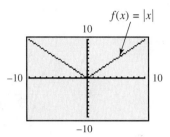

Figure 59

- $f(x) = |x|$ decreases on the open interval $(-\infty, 0)$ and increases on the open interval $(0, \infty)$.

- It is continuous on its entire domain, $(-\infty, \infty)$.

Piecewise-Defined Functions   The absolute value function is defined by different rules over different intervals of its domain. Such functions are called **piecewise-defined functions.**

EXAMPLE 2   **Graphing Piecewise-Defined Functions**

Graph each function.

(a) $f(x) = \begin{cases} -2x + 5 & \text{if } x \leq 2 \\ x + 1 & \text{if } x > 2 \end{cases}$ **(b)** $f(x) = \begin{cases} 2x + 3 & \text{if } x \leq 0 \\ -x^2 + 3 & \text{if } x > 0 \end{cases}$

ALGEBRAIC SOLUTION

(a) We graph each interval of the domain separately. If $x \leq 2$, the graph of $f(x) = -2x + 5$ has an endpoint at $x = 2$. We find the corresponding y-value by substituting 2 for x in $-2x + 5$ to get $y = 1$. To get another point on this part of the graph, we choose $x = 0$, so $y = 5$. We draw the graph through $(2, 1)$ and $(0, 5)$ as a partial line with endpoint $(2, 1)$.

We graph the function for $x > 2$ similarly, using $f(x) = x + 1$. This partial line has an open endpoint at $(2, 3)$. We use $y = x + 1$ to find another point with x-value greater than 2 to complete the graph. See **Figure 60.**

Figure 60

GRAPHING CALCULATOR SOLUTION

(a) By defining

$$Y_1 = (-2X + 5)(X \leq 2)$$

and

$$Y_2 = (X + 1)(X > 2)$$

we obtain the graph of f shown in **Figure 61.** Remember that inclusion or exclusion of endpoints is not readily apparent when we are observing a calculator graph. We must make this decision using our knowledge of inequalities.

$Y_2 = (X + 1)(X > 2)$

Figure 61

(b) First graph $f(x) = 2x + 3$ for $x \leq 0$. Then for $x > 0$, graph $f(x) = -x^2 + 3$. The two graphs meet at the point $(0, 3)$. See **Figure 62.**

Figure 62

(b) **Figure 63** shows an alternative method that can be used to obtain a calculator graph of this function. Here we entered the function as one expression.

$$Y_1 = (2X + 3)(X \leq 0) + (-X^2 + 3)(X > 0)$$

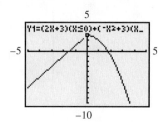

Figure 63

✔️ *Now Try Exercises 23 and 29.*

LOOKING AHEAD TO CALCULUS

The **greatest integer function** is used in calculus as a classic example of how the limit of a function may not exist at a particular value in its domain. For a limit to exist, the left- and right-hand limits must be equal. We can see from the graph of the greatest integer function that for an integer value such as 3, as x approaches 3 from the left, function values are all 2, while as x approaches 3 from the right, function values are all 3. Since the left- and right-hand limits are different, the limit as x approaches 3 does not exist.

Another piecewise-defined function is the *greatest integer function.*

$$f(x) = [\![x]\!]$$

The **greatest integer function** $f(x) = [\![x]\!]$ pairs every real number x with the greatest integer less than or equal to x.

For example, $[\![8.4]\!] = 8$, $[\![-5]\!] = -5$, $[\![\pi]\!] = 3$, and $[\![-6.9]\!] = -7$. In general, if $f(x) = [\![x]\!]$, then

$$\text{for } -2 \leq x < -1, \quad f(x) = -2,$$
$$\text{for } -1 \leq x < 0, \quad f(x) = -1,$$
$$\text{for } 0 \leq x < 1, \quad f(x) = 0,$$
$$\text{for } 1 \leq x < 2, \quad f(x) = 1,$$
$$\text{for } 2 \leq x < 3, \quad f(x) = 2, \quad \text{and so on.}$$

The graph of the greatest integer function is shown in **Figure 64.**

Greatest Integer Function $f(x) = [\![x]\!]$

Domain: $(-\infty, \infty)$

Range: $\{y \mid y \text{ is an integer}\} = \{\ldots, -3, -2, -1, 0, 1, 2, 3, \ldots\}$

x	y
-2	-2
-1.5	-2
-0.99	-1
0	0
0.001	0
3	3
3.99	3

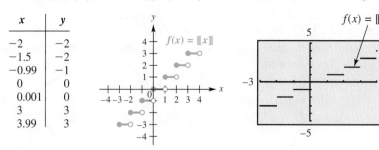

Figure 64

- $f(x) = [\![x]\!]$ is constant on the open intervals $\ldots, (-2, -1), (-1, 0), (0, 1), (1, 2), (2, 3), \ldots.$
- It is discontinuous at all integer values in its domain, $(-\infty, \infty)$.

EXAMPLE 3   **Graphing a Greatest Integer Function**

Graph $f(x) = [\![\frac{1}{2}x + 1]\!]$.

SOLUTION   If x is in the interval $[0, 2)$, then $y = 1$. For x in $[2, 4)$, $y = 2$, and so on. Some sample ordered pairs are given here.

x	0	$\frac{1}{2}$	1	$\frac{3}{2}$	2	3	4	-1	-2	-3
y	1	1	1	1	2	2	3	0	0	-1

The ordered pairs in the table suggest the graph shown in **Figure 65**. The domain is $(-\infty, \infty)$. The range is $\{\ldots, -2, -1, 0, 1, 2, \ldots\}$.

The dots indicate that the
graph continues indefinitely
in the same pattern.

Figure 65

✔ *Now Try Exercise 45.*

The greatest integer function is an example of a **step function,** a function with a graph that looks like a series of steps.

EXAMPLE 4   **Applying a Greatest Integer Function**

An express mail company charges $25 for a package weighing up to 2 lb. For each additional pound or fraction of a pound, there is an additional charge of $3. Let $y = D(x)$ represent the cost to send a package weighing x pounds. Graph $y = D(x)$ for x in the interval $(0, 6]$.

SOLUTION   For x in the interval $(0, 2]$, $y = 25$. For x in $(2, 3]$, $y = 25 + 3 = 28$. For x in $(3, 4]$, $y = 28 + 3 = 31$, and so on. The graph, which is that of a step function, is shown in **Figure 66**. In this case, the first step has a different width.

Figure 66

✔ *Now Try Exercise 47.*

The Relation $x = y^2$ *Recall that a function is a relation where every domain value is paired with one and only one range value.* Consider the relation defined by the equation $x = y^2$. Notice from the table of selected ordered pairs on the next page that this relation has two different y-values for each positive value of x.

If we plot the points from the table and join them with a smooth curve, we find that the graph of $x = y^2$ is a parabola opening to the right with vertex $(0, 0)$. See **Figure 67(a).** The domain is $[0, \infty)$ and the range is $(-\infty, \infty)$.

Selected Ordered Pairs
for $x = y^2$

x	y
0	0
1	±1
4	±2
9	±3

There are two different y-values for the same x-value.

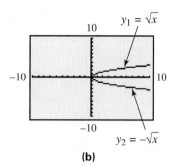

(a)   (b)

Figure 67

To use a calculator in function mode to graph the relation $x = y^2$, we graph the two functions $y_1 = \sqrt{x}$ (to generate the top half of the parabola) and $y_2 = -\sqrt{x}$ (to generate the bottom half). See **Figure 67(b).** ∎

2.6 Exercises

Concept Check For Exercises 1–10, refer to the following basic graphs.

A.

B.

C.

D.

E.

F.

G.

H.

I.

1. Which one is the graph of $y = x^2$? What is its domain?

2. Which one is the graph of $y = |x|$? What is the largest interval over which it is increasing?

3. Which one is the graph of $y = x^3$? What is its range?

4. Which one is not the graph of a function? What is its equation?

5. Which one is the identity function? What is its equation?

6. Which one is the graph of $y = [\![x]\!]$? What is the value of y when $x = 1.5$?

7. Which one is the graph of $y = \sqrt[3]{x}$? Is there any interval over which the function is decreasing?

8. Which one is the graph of $y = \sqrt{x}$? What is its domain?

9. Which one is discontinuous at many points? What is its range?

10. Which graphs of functions decrease over part of the domain and increase over the rest of the domain? What are the largest intervals on which they increase? decrease?

Determine the intervals of the domain over which each function is continuous. See Example 1.

11.

12.

13.

$(0, 3)$

14.

$(0, -1)$

15.

3

$(3, 1)$

-3 3

-3

16.

$(1, 2)$

For each piecewise-defined function, find (a) $f(-5)$, (b) $f(-1)$, (c) $f(0)$, and (d) $f(3)$. See Example 2.

17. $f(x) = \begin{cases} 2x & \text{if } x \leq -1 \\ x - 1 & \text{if } x > -1 \end{cases}$

18. $f(x) = \begin{cases} x - 2 & \text{if } x < 3 \\ 5 - x & \text{if } x \geq 3 \end{cases}$

19. $f(x) = \begin{cases} 2 + x & \text{if } x < -4 \\ -x & \text{if } -4 \leq x \leq 2 \\ 3x & \text{if } x > 2 \end{cases}$

20. $f(x) = \begin{cases} -2x & \text{if } x < -3 \\ 3x - 1 & \text{if } -3 \leq x \leq 2 \\ -4x & \text{if } x > 2 \end{cases}$

Graph each piecewise-defined function. See Example 2.

21. $f(x) = \begin{cases} x - 1 & \text{if } x \leq 3 \\ 2 & \text{if } x > 3 \end{cases}$

22. $f(x) = \begin{cases} 6 - x & \text{if } x \leq 3 \\ 3 & \text{if } x > 3 \end{cases}$

23. $f(x) = \begin{cases} 4 - x & \text{if } x < 2 \\ 1 + 2x & \text{if } x \geq 2 \end{cases}$

24. $f(x) = \begin{cases} 2x + 1 & \text{if } x \geq 0 \\ x & \text{if } x < 0 \end{cases}$

25. $f(x) = \begin{cases} -3 & \text{if } x \leq 1 \\ -1 & \text{if } x > 1 \end{cases}$

26. $f(x) = \begin{cases} -2 & \text{if } x \leq 1 \\ 2 & \text{if } x > 1 \end{cases}$

27. $f(x) = \begin{cases} 2 + x & \text{if } x < -4 \\ -x & \text{if } -4 \leq x \leq 5 \\ 3x & \text{if } x > 5 \end{cases}$

28. $f(x) = \begin{cases} -2x & \text{if } x < -3 \\ 3x - 1 & \text{if } -3 \leq x \leq 2 \\ -4x & \text{if } x > 2 \end{cases}$

29. $f(x) = \begin{cases} -\dfrac{1}{2}x^2 + 2 & \text{if } x \leq 2 \\ \dfrac{1}{2}x & \text{if } x > 2 \end{cases}$

30. $f(x) = \begin{cases} x^3 + 5 & \text{if } x \leq 0 \\ -x^2 & \text{if } x > 0 \end{cases}$

31. $f(x) = \begin{cases} 2x & \text{if } -5 \le x < -1 \\ -2 & \text{if } -1 \le x < 0 \\ x^2 - 2 & \text{if } 0 \le x \le 2 \end{cases}$ **32.** $f(x) = \begin{cases} 0.5x^2 & \text{if } -4 \le x \le -2 \\ x & \text{if } -2 < x < 2 \\ x^2 - 4 & \text{if } 2 \le x \le 4 \end{cases}$

33. $f(x) = \begin{cases} x^3 + 3 & \text{if } -2 \le x \le 0 \\ x + 3 & \text{if } 0 < x < 1 \\ 4 + x - x^2 & \text{if } 1 \le x \le 3 \end{cases}$ **34.** $f(x) = \begin{cases} -2x & \text{if } -3 \le x < -1 \\ x^2 + 1 & \text{if } -1 \le x \le 2 \\ \frac{1}{2}x^3 + 1 & \text{if } 2 < x \le 3 \end{cases}$

Connecting Graphs with Equations Give a rule for each piecewise-defined function. Also give the domain and range.

35. **36.** **37.**

38. **39.** **40.**

41. **42.**

Graph each function. Give the domain and range. ***See Example 3.***

43. $f(x) = [\![-x]\!]$ **44.** $f(x) = -[\![x]\!]$

45. $f(x) = [\![2x]\!]$ **46.** $g(x) = [\![2x - 1]\!]$

(Modeling) Solve each problem. ***See Example 4.***

47. *Postage Charges* Assume that postage rates are $0.44 for the first ounce, plus $0.20 for each additional ounce, and that each letter carries one $0.44 stamp and as many $0.20 stamps as necessary. Graph the function f that models the number of stamps on a letter weighing x ounces over the interval $(0, 5]$.

48. *Airport Parking Charges* The cost of parking a car at an airport hourly parking lot is $3 for the first half-hour and $2 for each additional half-hour or fraction of a half-hour. Graph the function f that models the cost of parking a car for x hours over the interval $(0, 2]$.

49. *Water in a Tank* Sketch a graph that depicts the amount of water in a 100-gal tank. The tank is initially empty and then filled at a rate of 5 gal per minute. Immediately after it is full, a pump is used to empty the tank at 2 gal per minute.

50. *Distance from Home* Sketch a graph showing the distance a person is from home after *x* hours if he or she drives on a straight road at 40 mph to a park 20 mi away, remains at the park for 2 hr, and then returns home at a speed of 20 mph.

51. *Pickup Truck Market Share* The light vehicle market share (in percent) in the United States for pickup trucks is shown in the graph. Let *x* = 0 represent 1995, *x* = 4 represent 1999, and so on.

    (a) Use the points on the graph to write equations for the line segments in the intervals $[0, 4]$ and $[4, 8]$.

    (b) Define this graph as a piecewise-defined function *f*.

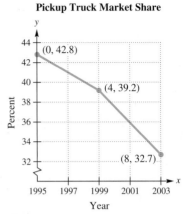

Pickup Truck Market Share

Source: Bureau of Transportation Statistics.

52. *Flow Rates* A water tank has an inlet pipe with a flow rate of 5 gal per minute and an outlet pipe with a flow rate of 3 gal per minute. A pipe can be either closed or completely open. The graph shows the number of gallons of water in the tank after *x* minutes. Use the concept of slope to interpret each piece of this graph.

Water in a Tank

53. *Swimming Pool Levels* The graph of $y = f(x)$ represents the amount of water in thousands of gallons remaining in a swimming pool after *x* days.

    (a) Estimate the initial and final amounts of water contained in the pool.

    (b) When did the amount of water in the pool remain constant?

    (c) Approximate $f(2)$ and $f(4)$.

    (d) At what rate was water being drained from the pool when $1 \le x \le 3$?

Water in a Swimming Pool

54. *Gasoline Usage* The graph shows the gallons of gasoline *y* in the gas tank of a car after *x* hours.

    (a) Estimate how much gasoline was in the gas tank when *x* = 3.

    (b) When did the car burn gasoline at the greatest rate?

Gasoline Use

55. *Lumber Costs* Lumber that is used to frame walls of houses is frequently sold in multiples of 2 ft. If the length of a board is not exactly a multiple of 2 ft, there is often no charge for the additional length. For example, if a board measures at least 8 ft, but less than 10 ft, then the consumer is charged for only 8 ft.

(a) Suppose that the cost of lumber is $0.80 every 2 ft. Find a formula for a function f that computes the cost of a board x feet long for $6 \le x \le 18$.

(b) Determine the costs of boards with lengths of 8.5 ft and 15.2 ft.

56. *Snow Depth* The snow depth in Michigan's Isle Royale National Park varies throughout the winter. In a typical winter, the snow depth in inches is approximated by the following function.

$$f(x) = \begin{cases} 6.5x & \text{if } 0 \le x \le 4 \\ -5.5x + 48 & \text{if } 4 < x \le 6 \\ -30x + 195 & \text{if } 6 < x \le 6.5 \end{cases}$$

Here, x represents the time in months with $x = 0$ representing the beginning of October, $x = 1$ representing the beginning of November, and so on.

(a) Graph $y = f(x)$.

(b) In what month is the snow deepest? What is the deepest snow depth?

(c) In what months does the snow begin and end?

2.7 Graphing Techniques

- Stretching and Shrinking
- Reflecting
- Symmetry
- Even and Odd Functions
- Translations

Graphing techniques presented in this section show how to graph functions that are defined by altering the equation of a basic function.

Stretching and Shrinking We begin by considering how the graphs of $y = af(x)$ and $y = f(ax)$ compare to the graph of $y = f(x)$, where $a > 0$.

EXAMPLE 1 Stretching or Shrinking a Graph

Graph each function.

(a) $g(x) = 2|x|$ (b) $h(x) = \dfrac{1}{2}|x|$ (c) $k(x) = |2x|$

SOLUTION

(a) Comparing the tables of values for $f(x) = |x|$ and $g(x) = 2|x|$ in **Figure 68,** we see that for corresponding x-values, the y-values of g are each twice those of f. The graph of $f(x) = |x|$ is *vertically stretched.* The graph of $g(x)$, shown in blue in **Figure 68,** is narrower than that of $f(x)$, shown in red for comparison.

x	$f(x) = |x|$	$g(x) = 2|x|$
-2	2	4
-1	1	2
0	0	0
1	1	2
2	2	4

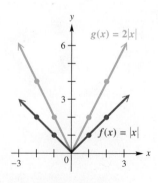

Figure 68

(b) The graph of $h(x) = \frac{1}{2}|x|$ is also the same general shape as that of $f(x)$, but here the coefficient $\frac{1}{2}$ is between 0 and 1 and causes a *vertical shrink*. The graph of $h(x)$ is wider than the graph of $f(x)$, as we see by comparing the tables of values. See **Figure 69.**

x	$f(x) = |x|$	$h(x) = \frac{1}{2}|x|$
-2	2	1
-1	1	$\frac{1}{2}$
0	0	0
1	1	$\frac{1}{2}$
2	2	1

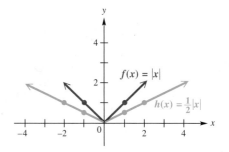

Figure 69

(c) Use Property 3 of absolute value $(|ab| = |a| \cdot |b|)$ to rewrite $|2x|$.

$$k(x) = |2x| = |2| \cdot |x| = 2|x| \quad \text{Property 3 (Section R.2)}$$

Therefore, the graph of $k(x) = |2x|$ is the same as the graph of $g(x) = 2|x|$ in part (a). This is a *horizontal shrink* of the graph of $f(x) = |x|$. See **Figure 68** on the previous page.

> ✔ *Now Try Exercises 7 and 9.*

Vertical stretching
$a > 1$

Vertical shrinking
$0 < a < 1$

Figure 70

Vertical Stretching or Shrinking of the Graph of a Function

Suppose that $a > 0$. If a point (x, y) lies on the graph of $y = f(x)$, then the point (x, ay) lies on the graph of $y = af(x)$.

(a) If $a > 1$, then the graph of $y = af(x)$ is a **vertical stretching** of the graph of $y = f(x)$.

(b) If $0 < a < 1$, then the graph of $y = af(x)$ is a **vertical shrinking** of the graph of $y = f(x)$.

Figure 70 shows graphical interpretations of vertical stretching and shrinking. Notice that in both cases, the x-intercepts of the graph remain the same but the y-intercepts *are* affected.

Graphs of functions can also be stretched and shrunk horizontally.

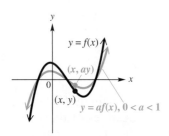

Horizontal stretching
$0 < a < 1$

Horizontal shrinking
$a > 1$

Figure 71

Horizontal Stretching or Shrinking of the Graph of a Function

Suppose that $a > 0$. If a point (x, y) lies on the graph of $y = f(x)$, then the point $\left(\frac{x}{a}, y\right)$ lies on the graph of $y = f(ax)$.

(a) If $0 < a < 1$, then the graph of $y = f(ax)$ is a **horizontal stretching** of the graph of $y = f(x)$.

(b) If $a > 1$, then the graph of $y = f(ax)$ is a **horizontal shrinking** of the graph of $y = f(x)$.

See **Figure 71** for graphical interpretations of horizontal stretching and shrinking. Notice that in both cases, the y-intercept remains the same but the x-intercepts *are* affected.

Reflecting   Forming the mirror image of a graph across a line is called **reflecting the graph across the line.**

EXAMPLE 2   **Reflecting a Graph across an Axis**

Graph each function.

(a) $g(x) = -\sqrt{x}$

(b) $h(x) = \sqrt{-x}$

SOLUTION

(a) The tables of values for $g(x) = -\sqrt{x}$ and $f(x) = \sqrt{x}$ are shown with their graphs in **Figure 72.** As the tables suggest, every y-value of the graph of $g(x) = -\sqrt{x}$ is the negative of the corresponding y-value of $f(x) = \sqrt{x}$. This has the effect of reflecting the graph across the x-axis.

x	$f(x) = \sqrt{x}$	$g(x) = -\sqrt{x}$
0	0	0
1	1	-1
4	2	-2

Figure 72

(b) The domain of $h(x) = \sqrt{-x}$ is $x \le 0$, while the domain of $f(x) = \sqrt{x}$ is $x \ge 0$. Choosing x-values for $h(x)$ that are negatives of those used for $f(x)$, we see that corresponding y-values are the same. The graph of h is a reflection of the graph of f across the y-axis. See **Figure 73.**

x	$f(x) = \sqrt{x}$	$h(x) = \sqrt{-x}$
-4	undefined	2
-1	undefined	1
0	0	0
1	1	undefined
4	2	undefined

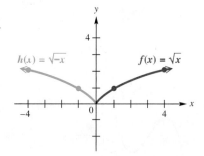

Figure 73

✔ *Now Try Exercises 17 and 23.*

The graphs in **Example 2** suggest the following generalizations.

Reflecting across an Axis

The graph of $y = -f(x)$ is the same as the graph of $y = f(x)$ reflected across the x-axis. (If a point (x, y) lies on the graph of $y = f(x)$, then $(x, -y)$ lies on this reflection.)

The graph of $y = f(-x)$ is the same as the graph of $y = f(x)$ reflected across the y-axis. (If a point (x, y) lies on the graph of $y = f(x)$, then $(-x, y)$ lies on this reflection.)

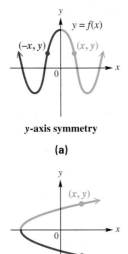

y-axis symmetry

(a)

x-axis symmetry

(b)

Figure 74

Symmetry The graph of f shown in **Figure 74(a)** is cut in half by the y-axis with each half the mirror image of the other half. Such a graph is *symmetric with respect to the y-axis*. **The point $(-x, y)$ is on the graph whenever the point (x, y) is on the graph.**

Similarly, if the graph in **Figure 74(b)** were folded in half along the x-axis, the portion at the top would exactly match the portion at the bottom. Such a graph is *symmetric with respect to the x-axis*. **The point $(x, -y)$ is on the graph whenever the point (x, y) is on the graph.**

Symmetry with Respect to an Axis

The graph of an equation is **symmetric with respect to the y-axis** if the replacement of x with $-x$ results in an equivalent equation.

The graph of an equation is **symmetric with respect to the x-axis** if the replacement of y with $-y$ results in an equivalent equation.

In **Section 2.6** we introduced graphs of basic functions. The squaring function and the absolute value function are examples of functions that are symmetric with respect to the y-axis.

EXAMPLE 3 **Testing for Symmetry with Respect to an Axis**

Test for symmetry with respect to the x-axis and the y-axis.

(a) $y = x^2 + 4$ **(b)** $x = y^2 - 3$ **(c)** $x^2 + y^2 = 16$ **(d)** $2x + y = 4$

SOLUTION

(a) In $y = x^2 + 4$, replace x with $-x$.

Use parentheses around $-x$.

$$y = x^2 + 4 \quad \longleftarrow$$
$$y = (-x)^2 + 4 \quad \text{Equivalent}$$
$$y = x^2 + 4 \quad \longleftarrow$$

The result is the same as the original equation, so the graph, shown in **Figure 75,** is symmetric with respect to the y-axis. Substituting $-y$ for y does not result in an equivalent equation, and thus the graph is *not* symmetric with respect to the x-axis.

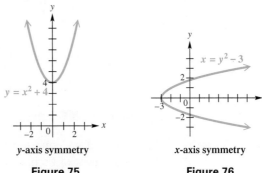

y-axis symmetry

Figure 75

x-axis symmetry

Figure 76

(b) In $x = y^2 - 3$, replace y with $-y$.

$$x = (-y)^2 - 3 = y^2 - 3 \quad \text{Same as the original equation}$$

The graph is symmetric with respect to the x-axis, as shown in **Figure 76.** It is *not* symmetric with respect to the y-axis.

LOOKING AHEAD TO CALCULUS
The tools of calculus enable us to find areas of regions in the plane. To find the area of the region below the graph of $y = x^2$, above the x-axis, bounded on the left by the line $x = -2$ and on the right by $x = 2$, draw a sketch of this region. Notice that due to the symmetry of the graph of $y = x^2$, the desired area is twice the area to the right of the y-axis. Thus, symmetry can be used to reduce the original problem to an easier one by simply finding the area to the right of the y-axis and then doubling the answer.

(c) Substitute $-x$ for x and then $-y$ for y in $x^2 + y^2 = 16$.

$$(-x)^2 + y^2 = 16 \quad \text{and} \quad x^2 + (-y)^2 = 16$$

Both simplify to the original equation,

$$x^2 + y^2 = 16.$$

The graph, a circle of radius 4 centered at the origin, is symmetric with respect to *both* axes. See **Figure 77.**

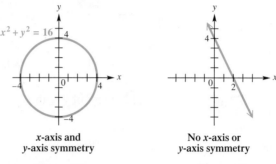

x-axis and y-axis symmetry	No x-axis or y-axis symmetry
Figure 77	**Figure 78**

(d) In $2x + y = 4$, replace x with $-x$ to get $-2x + y = 4$. Then replace y with $-y$ to get $2x - y = 4$. Neither case produces an equivalent equation, so this graph is not symmetric with respect to either axis. See **Figure 78.**

✔ *Now Try Exercise 35.*

Another kind of symmetry occurs when a graph can be rotated 180° about the origin, with the result coinciding exactly with the original graph. Symmetry of this type is called *symmetry with respect to the origin.* ***The point $(-x, -y)$ is on the graph whenever the point (x, y) is on the graph.*** See **Figure 79.**

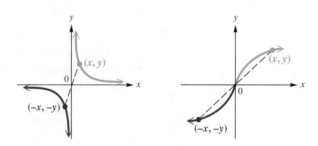

Origin symmetry

Figure 79

Symmetry with Respect to the Origin

The graph of an equation is **symmetric with respect to the origin** if the replacement of both x with $-x$ and y with $-y$ at the same time results in an equivalent equation.

The cubing and cube root functions are examples of functions whose graphs are symmetric with respect to the origin.

EXAMPLE 4   **Testing for Symmetry with Respect to the Origin**

Are the following graphs symmetric with respect to the origin?

(a) $x^2 + y^2 = 16$ **(b)** $y = x^3$

SOLUTION

(a) Replace x with $-x$ and y with $-y$.

> Use parentheses
> around $-x$ and $-y$.

$$x^2 + y^2 = 16 \longleftarrow$$
$$(-x)^2 + (-y)^2 = 16 \quad \text{Equivalent}$$
$$x^2 + y^2 = 16 \longleftarrow$$

The graph, which is the circle shown in **Figure 77** in **Example 3(c)**, is symmetric with respect to the origin.

(b) Replace x with $-x$ and y with $-y$.

$$y = x^3 \quad \longleftarrow$$
$$-y = (-x)^3 \quad \text{Equivalent}$$
$$-y = -x^3$$
$$y = x^3 \quad \longleftarrow$$

The graph, which is that of the cubing function, is symmetric with respect to the origin and is shown in **Figure 80**.

✔️ *Now Try Exercise 39.*

Notice the following important concepts regarding symmetry:

- A graph symmetric with respect to both the x- and y-axes is automatically symmetric with respect to the origin. (See **Figure 77.**)

- A graph symmetric with respect to the origin need *not* be symmetric with respect to either axis. (See **Figure 80.**)

- Of the three types of symmetry—with respect to the x-axis, with respect to the y-axis, and with respect to the origin—a graph possessing any two types must also exhibit the third type of symmetry.

The various tests for symmetry are summarized below.

Tests for Symmetry

	Symmetry with Respect to:		
	x-axis	**y-axis**	**Origin**
Equation is unchanged if:	y is replaced with $-y$	x is replaced with $-x$	x is replaced with $-x$ and y is replaced with $-y$
Example:			

Origin symmetry

Figure 80

Even and Odd Functions   The concepts of symmetry with respect to the y-axis and symmetry with respect to the origin are closely associated with the concepts of *even* and *odd functions*.

Even and Odd Functions

A function f is an **even function** if $f(-x) = f(x)$ for all x in the domain of f. (Its graph is symmetric with respect to the y-axis.)

A function f is an **odd function** if $f(-x) = -f(x)$ for all x in the domain of f. (Its graph is symmetric with respect to the origin.)

EXAMPLE 5   **Determining Whether Functions Are Even, Odd, or Neither**

Determine whether each function defined is *even, odd,* or *neither.*

(a) $f(x) = 8x^4 - 3x^2$ **(b)** $f(x) = 6x^3 - 9x$ **(c)** $f(x) = 3x^2 + 5x$

SOLUTION

(a) Replacing x with $-x$ gives the following.

$$f(x) = 8x^4 - 3x^2$$
$$f(-x) = 8(-x)^4 - 3(-x)^2 \quad \text{Replace } x \text{ with } -x.$$
$$f(-x) = 8x^4 - 3x^2 \qquad \text{Apply the exponents.}$$
$$f(-x) = f(x) \qquad\qquad 8x^4 - 3x^2 = f(x)$$

Since $f(-x) = f(x)$ for each x in the domain of the function, f is even.

(b) $f(x) = 6x^3 - 9x$
$$f(-x) = 6(-x)^3 - 9(-x) \quad \text{Replace } x \text{ with } -x.$$
$$f(-x) = -6x^3 + 9x \quad \boxed{\text{Be careful with signs.}}$$
$$f(-x) = -f(x) \qquad\qquad -6x^3 + 9x = -(6x^3 - 9x) = -f(x)$$

The function f is odd because $f(-x) = -f(x)$.

(c) $f(x) = 3x^2 + 5x$
$$f(-x) = 3(-x)^2 + 5(-x) \quad \text{Replace } x \text{ with } -x.$$
$$f(-x) = 3x^2 - 5x \qquad\qquad \text{Simplify.}$$

Since $f(-x) \neq f(x)$ and $f(-x) \neq -f(x)$, the function f is neither even nor odd.

✔ *Now Try Exercises 43, 45, and 47.*

NOTE   Consider a function defined by a polynomial in x.

- If the function has only *even* exponents on x (including the case of a constant where x^0 is understood to have the even exponent 0), it will *always* be an even function.

- Similarly, if only *odd* exponents appear on x, the function will be an odd function.

Translations   The next examples show the results of horizontal and vertical shifts, or **translations,** of the graph of $f(x) = |x|$.

EXAMPLE 6   **Translating a Graph Vertically**

Graph $g(x) = |x| - 4$.

SOLUTION   By comparing the table of values for $g(x) = |x| - 4$ and $f(x) = |x|$ shown with **Figure 81,** we see that for corresponding x-values, the y-values of g are each 4 *less* than those for f. Thus, the graph of $g(x) = |x| - 4$ is the same as that of $f(x) = |x|$, but translated 4 units down. See **Figure 81.** The lowest point is at $(0, -4)$. The graph is symmetric with respect to the y-axis and is therefore the graph of an even function.

x	$f(x) = |x|$	$g(x) = |x| - 4$
-4	4	0
-1	1	-3
0	0	-4
1	1	-3
4	4	0

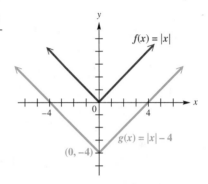

Figure 81

✔ *Now Try Exercise 57.*

The graphs in **Example 6** suggest the following generalization.

> **Vertical Translations**
>
> If a function g is defined by $g(x) = f(x) + c$, where c is a real number, then for every point (x, y) on the graph of f, there will be a corresponding point $(x, y + c)$ on the graph of g.
>
> The graph of g will be the same as the graph of f, but translated c units up if c is positive or $|c|$ units down if c is negative. The graph of g is called a **vertical translation** of the graph of f.

Figure 82 shows a graph of a function f and two vertical translations of f. **Figure 83** shows two vertical translations of $y_1 = x^2$ on a TI-83/84 Plus calculator screen.

Figure 82

y_2 is the graph of y_1 translated 2 units *up*. y_3 is that of y_1 translated 6 units *down*.

Figure 83

EXAMPLE 7 **Translating a Graph Horizontally**

Graph $g(x) = |x - 4|$.

SOLUTION   Comparing the tables of values given with **Figure 84** shows that for corresponding y-values, the x-values of g are each 4 *more* than those for f. The graph of $g(x) = |x - 4|$ is the same as that of $f(x) = |x|$, but translated 4 units to the right. The lowest point is at $(4, 0)$. As suggested by the graphs in **Figure 84,** this graph is symmetric with respect to the line $x = 4$.

x	$f(x) = \|x\|$	$g(x) = \|x - 4\|$
-2	2	6
0	0	4
2	2	2
4	4	0
6	6	2

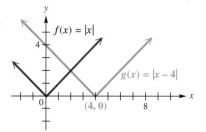

Figure 84

✔️ *Now Try Exercise 55.*

The graphs in **Example 7** suggest the following generalization.

Horizontal Translations

If a function g is defined by $g(x) = f(x - c)$, where c is a real number, then for every point (x, y) on the graph of f, there will be a corresponding point $(x + c, y)$ on the graph of g.

The graph of g will be the same as the graph of f, but translated c units to the right if c is positive or $|c|$ units to the left if c is negative. The graph of g is called a **horizontal translation** of the graph of f.

Figure 85 shows a graph of a function f and two horizontal translations of f. **Figure 86** shows two horizontal translations of $y_1 = x^2$ on a TI-83/84 Plus calculator screen.

$(c > 0)$ To Graph:	Shift the Graph of $y = f(x)$ by c Units:
$y = f(x) + c$	up
$y = f(x) - c$	down
$y = f(x + c)$	left
$y = f(x - c)$	right

Figure 85

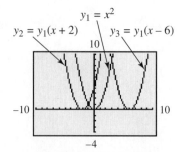

y_2 is the graph of y_1 translated 2 units to the *left.* y_3 is that of y_1 translated 6 units to the *right.*

Figure 86

Vertical and horizontal translations are summarized in the table in the margin, where f is a function, and c is a positive number.

CAUTION   *Be careful when translating graphs horizontally.* To determine the direction and magnitude of horizontal translations, find the value that would cause the expression in parentheses to equal 0.

For example, the graph of $y = (x - 5)^2$ would be translated 5 units to the *right* of $y = x^2$, because $x = +5$ would cause $x - 5$ to equal 0. On the other hand, the graph of $y = (x + 5)^2$ would be translated 5 units to the *left* of $y = x^2$, because $x = -5$ would cause $x + 5$ to equal 0.

EXAMPLE 8   Using More Than One Transformation

Graph each function.

(a) $f(x) = -|x + 3| + 1$ **(b)** $h(x) = |2x - 4|$ **(c)** $g(x) = -\dfrac{1}{2}x^2 + 4$

SOLUTION

(a) To graph $f(x) = -|x + 3| + 1$, the *lowest* point on the graph of $y = |x|$ is translated 3 units to the left and 1 unit up. The graph opens down because of the negative sign in front of the absolute value expression, making the lowest point now the highest point on the graph, as shown in **Figure 87.** The graph is symmetric with respect to the line $x = -3$.

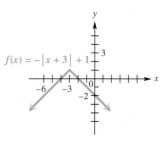

$f(x) = -|x + 3| + 1$

Figure 87

(b) To determine the horizontal translation, factor out 2.

$$h(x) = |2x - 4|$$

$$= |2(x - 2)| \qquad \text{Factor out 2.}$$

$$= |2| \cdot |x - 2| \quad |ab| = |a| \cdot |b| \text{ (Section R.2)}$$

$$= 2|x - 2| \qquad |2| = 2$$

The graph of h is the graph of $y = |x|$ translated 2 units to the right, and vertically stretched by a factor of 2. Horizontal shrinking gives the same appearance as vertical stretching for this function. See **Figure 88.**

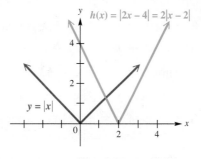

$h(x) = |2x - 4| = 2|x - 2|$

$y = |x|$

Figure 88

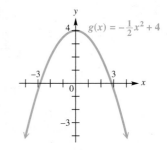

$g(x) = -\dfrac{1}{2}x^2 + 4$

Figure 89

(c) The graph of $g(x) = -\dfrac{1}{2}x^2 + 4$ has the same shape as that of $y = x^2$, but it is wider (that is, shrunken vertically), reflected across the x-axis because the coefficient $-\dfrac{1}{2}$ is negative, and then translated 4 units up. See **Figure 89.**

✔️ *Now Try Exercises 61, 63, and 71.*

EXAMPLE 9 **Graphing Translations of a Given Graph**

A graph of a function defined by $y = f(x)$ is shown in **Figure 90.** Use this graph to sketch each of the following graphs.

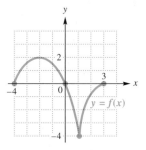

(a) $g(x) = f(x) + 3$ **(b)** $h(x) = f(x + 3)$

(c) $k(x) = f(x - 2) + 3$

Figure 90

SOLUTION In each part, pay close attention to how the plotted points in **Figure 90** are translated.

(a) The graph of $g(x) = f(x) + 3$ is the same as the graph in **Figure 90,** translated 3 units up. See **Figure 91(a).**

(b) To get the graph of $h(x) = f(x + 3)$, the graph of $y = f(x)$ must be translated 3 units to the left since $x + 3 = 0$ if $x = -3$. See **Figure 91(b).**

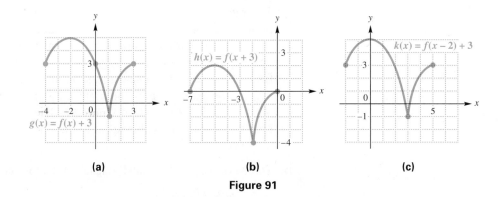

(a) (b) (c)

Figure 91

(c) The graph of $k(x) = f(x - 2) + 3$ will look like the graph of $f(x)$ translated 2 units to the right and 3 units up, as shown in **Figure 91(c).**

✔ *Now Try Exercise 77.*

Summary of Graphing Techniques

In the descriptions that follow, assume that $a > 0$, $h > 0$, and $k > 0$. In comparison with the graph of $y = f(x)$:

1. The graph of $y = f(x) + k$ is translated k units up.

2. The graph of $y = f(x) - k$ is translated k units down.

3. The graph of $y = f(x + h)$ is translated h units to the left.

4. The graph of $y = f(x - h)$ is translated h units to the right.

5. The graph of $y = af(x)$ is a vertical stretching of the graph of $y = f(x)$ if $a > 1$. It is a vertical shrinking if $0 < a < 1$.

6. The graph of $y = f(ax)$ is a horizontal stretching of the graph of $y = f(x)$ if $0 < a < 1$. It is a horizontal shrinking if $a > 1$.

7. The graph of $y = -f(x)$ is reflected across the x-axis.

8. The graph of $y = f(-x)$ is reflected across the y-axis.

2.7 Exercises

1. *Concept Check* Match each equation in Column I with a description of its graph from Column II as it relates to the graph of $y = x^2$.

I	II
(a) $y = (x - 7)^2$	**A.** a translation 7 units to the left
(b) $y = x^2 - 7$	**B.** a translation 7 units to the right
(c) $y = 7x^2$	**C.** a translation 7 units up
(d) $y = (x + 7)^2$	**D.** a translation 7 units down
(e) $y = x^2 + 7$	**E.** a vertical stretching by a factor of 7

2. *Concept Check* Match each equation in Column I with a description of its graph from Column II as it relates to the graph of $y = \sqrt[3]{x}$.

I	II
(a) $y = 4\sqrt[3]{x}$	**A.** a translation 4 units to the right
(b) $y = -\sqrt[3]{x}$	**B.** a translation 4 units down
(c) $y = \sqrt[3]{-x}$	**C.** a reflection across the x-axis
(d) $y = \sqrt[3]{x} - 4$	**D.** a reflection across the y-axis
(e) $y = \sqrt[3]{x - 4}$	**E.** a vertical stretching by a factor of 4

3. *Concept Check* Match each equation in parts (a)–(i) with the sketch of its graph.

(a) $y = x^2 + 2$ **(b)** $y = x^2 - 2$ **(c)** $y = (x + 2)^2$

(d) $y = (x - 2)^2$ **(e)** $y = 2x^2$ **(f)** $y = -x^2$

(g) $y = (x - 2)^2 + 1$ **(h)** $y = (x + 2)^2 + 1$ **(i)** $y = (x + 2)^2 - 1$

A.

B.

C.

D.

E.

F.

G.

H.

I.
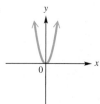

4. *Concept Check* Match each equation in parts (a)–(i) with the sketch of its graph.

(a) $y = \sqrt{x + 3}$ (b) $y = \sqrt{x} - 3$ (c) $y = \sqrt{x} + 3$

(d) $y = 3\sqrt{x}$ (e) $y = -\sqrt{x}$ (f) $y = \sqrt{x - 3}$

(g) $y = \sqrt{x - 3} + 2$ (h) $y = \sqrt{x + 3} + 2$ (i) $y = \sqrt{x - 3} - 2$

A.

B.

C.

D.

E.

F.

G.

H.

I.

5. *Concept Check* Match each equation in parts (a)–(i) with the sketch of its graph.

(a) $y = |x - 2|$ (b) $y = |x| - 2$ (c) $y = |x| + 2$

(d) $y = 2|x|$ (e) $y = -|x|$ (f) $y = |-x|$

(g) $y = -2|x|$ (h) $y = |x - 2| + 2$ (i) $y = |x + 2| - 2$

A.

B.

C.

D.

E.

F.

G.

H.

I.

6. (a) Suppose the equation $y = F(x)$ is changed to $y = c \cdot F(x)$, for some constant c. What is the effect on the graph of $y = F(x)$? Discuss the effect depending on whether $c > 0$ or $c < 0$, and on whether $|c| > 1$ or $|c| < 1$.

(b) Suppose $y = F(x)$ is changed to $y = F(x + h)$. How are the graphs of these equations related? Is the graph of $y = F(x) + h$ the same as the graph of $y = F(x + h)$? If not, how do they differ?

Graph each function. See Examples 1 and 2.

7. $y = 3|x|$ **8.** $y = 4|x|$ **9.** $y = \dfrac{2}{3}|x|$ **10.** $y = \dfrac{3}{4}|x|$

11. $y = 2x^2$ **12.** $y = 3x^2$ **13.** $y = \dfrac{1}{2}x^2$ **14.** $y = \dfrac{1}{3}x^2$

15. $y = -\dfrac{1}{2}x^2$ **16.** $y = -\dfrac{1}{3}x^2$ **17.** $y = -3|x|$ **18.** $y = -2|x|$

19. $y = \left| -\dfrac{1}{2}x \right|$ **20.** $y = \left| -\dfrac{1}{3}x \right|$ **21.** $y = \sqrt{4x}$

22. $y = \sqrt{9x}$ **23.** $y = -\sqrt{-x}$ **24.** $y = -|-x|$

Concept Check In Exercises 25–28, suppose the point $(8, 12)$ is on the graph of $y = f(x)$. Find a point on the graph of each function.

25. (a) $y = f(x + 4)$ **26. (a)** $y = \dfrac{1}{4}f(x)$ **27. (a)** $y = f(4x)$

     (b) $y = f(x) + 4$ **(b)** $y = 4f(x)$ **(b)** $y = f\left(\dfrac{1}{4}x\right)$

28. (a) the reflection of the graph of $y = f(x)$ across the x-axis

     (b) the reflection of the graph of $y = f(x)$ across the y-axis

Concept Check Plot each point, and then plot the points that are symmetric to the given point with respect to the (a) x-axis, (b) y-axis, and (c) origin.

29. $(5, -3)$ **30.** $(-6, 1)$ **31.** $(-4, -2)$ **32.** $(-8, 0)$

33. *Concept Check* The graph of $y = |x - 2|$ is symmetric with respect to a vertical line. What is the equation of that line?

34. *Concept Check* Repeat **Exercise 33** for $y = -|x + 1|$.

Without graphing, determine whether each equation has a graph that is symmetric with respect to the x-axis, the y-axis, the origin, or none of these. See Examples 3 and 4.

35. $y = x^2 + 5$ **36.** $y = 2x^4 - 3$ **37.** $x^2 + y^2 = 12$ **38.** $y^2 - x^2 = -6$

39. $y = -4x^3 + x$ **40.** $y = x^3 - x$ **41.** $y = x^2 - x + 8$ **42.** $y = x + 15$

Determine whether each function is even, odd, or neither. See Example 5.

43. $f(x) = -x^3 + 2x$ **44.** $f(x) = x^5 - 2x^3$

45. $f(x) = 0.5x^4 - 2x^2 + 6$ **46.** $f(x) = 0.75x^2 + |x| + 4$

47. $f(x) = x^3 - x + 9$ **48.** $f(x) = x^4 - 5x + 8$

Graph each function. See Examples 6–8.

49. $y = x^2 - 1$ **50.** $y = x^2 - 2$ **51.** $y = x^2 + 2$

52. $y = x^2 + 3$ **53.** $y = (x - 4)^2$ **54.** $y = (x - 2)^2$

55. $y = (x + 2)^2$ **56.** $y = (x + 3)^2$ **57.** $y = |x| - 1$

58. $y = |x + 3| + 2$ **59.** $y = -(x + 1)^3$ **60.** $y = -(x - 1)^3$

61. $y = 2x^2 - 1$ **62.** $y = 3x^2 - 2$ **63.** $f(x) = 2(x - 2)^2 - 4$

64. $f(x) = -3(x - 2)^2 + 1$ **65.** $f(x) = \sqrt{x + 2}$ **66.** $f(x) = \sqrt{x - 3}$

67. $f(x) = -\sqrt{x}$ **68.** $f(x) = \sqrt{x} - 2$ **69.** $f(x) = 2\sqrt{x} + 1$

70. $y = 3\sqrt{x} - 2$ **71.** $y = \dfrac{1}{2}x^3 - 4$ **72.** $y = \dfrac{1}{2}x^3 + 2$

73. $y = (x + 3)^3$ **74.** $y = (x - 2)^3$ **75.** $y = \dfrac{2}{3}(x - 2)^2$

76. *Concept Check* What is the relationship between the graphs of $f(x) = |x|$ and $g(x) = |-x|$?

*For Exercises 77 and 78, see **Example 9**.*

77. Given the graph of $y = g(x)$ in the figure, sketch the graph of each function, and explain how it is obtained from the graph of $y = g(x)$.

    (a) $y = g(-x)$

    (b) $y = g(x - 2)$

    (c) $y = -g(x) + 2$

78. Given the graph of $y = f(x)$ in the figure, sketch the graph of each function, and explain how it is obtained from the graph of $y = f(x)$.

    (a) $y = -f(x)$

    (b) $y = 2f(x)$

    (c) $y = f(-x)$

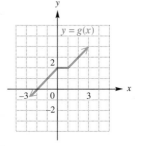

Connecting Graphs with Equations *Each of the following graphs is obtained from the graph of $f(x) = |x|$ or $g(x) = \sqrt{x}$ by applying several of the transformations discussed in this section. Describe the transformations and give the equation for the graph.*

79.

80.

81.

82.

83.

84.

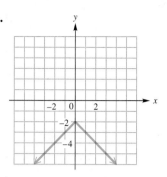

Concept Check Suppose that for a function f, f(3) = 6. For the given assumptions in Exercises 85–90, find another function value.

85. The graph of $y = f(x)$ is symmetric with respect to the origin.

86. The graph of $y = f(x)$ is symmetric with respect to the *y*-axis.

87. The graph of $y = f(x)$ is symmetric with respect to the line $x = 6$.

88. For all x, $f(-x) = f(x)$.

89. For all x, $f(-x) = -f(x)$.

90. f is an odd function.

91. Find the function $g(x) = ax + b$ whose graph can be obtained by translating the graph of $f(x) = 2x + 5$ up 2 units and to the left 3 units.

92. Find the function $g(x) = ax + b$ whose graph can be obtained by translating the graph of $f(x) = 3 - x$ down 2 units and to the right 3 units.

93. *Concept Check* Complete the left half of the graph of $y = f(x)$ in the figure for each condition.

(a) $f(-x) = f(x)$ **(b)** $f(-x) = -f(x)$

94. *Concept Check* Complete the right half of the graph of $y = f(x)$ in the figure for each condition.

(a) f is odd.   **(b)** f is even.

Chapter 2 **Quiz** (Sections 2.5–2.7)

1. For the line passing through the points $(-3, 5)$ and $(-1, 9)$, find the following:

(a) the slope-intercept form of its equation   **(b)** its *x*-intercept.

2. Find the slope-intercept form of the equation of the line passing through the point $(-6, 4)$ and perpendicular to the graph of $3x - 2y = 6$.

3. Suppose that P has coordinates $(-8, 5)$. Find the equation of the line through P that is

(a) vertical   **(b)** horizontal.

4. For each basic function graphed, give the name of the function, the domain, the range, and the largest intervals over which it is decreasing, increasing, or constant.

(a)

(b)

(c)

5. *(Modeling) Long-Distance Call Charges* A certain long-distance carrier provides service between Podunk and Nowheresville. If x represents the number of minutes for the call, where $x > 0$, then the function

$$f(x) = 0.40[\![x]\!] + 0.75$$

gives the total cost of the call in dollars. Find the cost of a 5.5-min call.

Graph each function.

6. $f(x) = \begin{cases} \sqrt{x} & \text{if } x \geq 0 \\ 2x + 3 & \text{if } x < 0 \end{cases}$ **7.** $f(x) = -x^3 + 1$ **8.** $f(x) = 2|x - 1| + 3$

9. *Connecting Graphs with Equations* The function graphed here is obtained by stretching, shrinking, reflecting, and/or translating the graph of $f(x) = \sqrt{x}$. Give the equation that defines this function.

10. Determine whether each function is *even, odd,* or *neither.*

(a) $f(x) = x^2 - 7$ **(b)** $f(x) = x^3 - x - 1$ **(c)** $f(x) = x^{101} - x^{99}$

2.8 Function Operations and Composition

- Arithmetic Operations on Functions
- The Difference Quotient
- Composition of Functions and Domain

Arithmetic Operations on Functions **Figure 92** shows the situation for a company that manufactures DVDs. The two lines are the graphs of the linear functions for revenue $R(x) = 168x$ and cost $C(x) = 118x + 800$, where x is the number of DVDs produced and sold, and x, $R(x)$, and $C(x)$ are given in thousands. When 30,000 (that is, 30 thousand) DVDs are produced and sold, profit is found as follows.

$$P(x) = R(x) - C(x)$$ Profit function **(Section 2.4)**

$$P(30) = R(30) - C(30)$$ Let $x = 30$.

$$= 5040 - 4340$$ $R(30) = 168(30); C(30) = 118(30) + 800$

$$P(30) = 700$$ Subtract.

Thus, the profit from the sale of 30,000 DVDs is $700,000.

The profit function is found by *subtracting* the cost function from the revenue function. New functions can be formed by using other operations as well.

Figure 92

Operations on Functions and Domains

Given two functions f and g, then for all values of x for which both $f(x)$ and $g(x)$ are defined, the functions $f + g$, $f - g$, fg, and $\frac{f}{g}$ are defined as follows.

$$(f + g)(x) = f(x) + g(x) \qquad \text{Sum}$$

$$(f - g)(x) = f(x) - g(x) \qquad \text{Difference}$$

$$(fg)(x) = f(x) \cdot g(x) \qquad \text{Product}$$

$$\left(\frac{f}{g}\right)(x) = \frac{f(x)}{g(x)}, \quad g(x) \neq 0 \quad \text{Quotient}$$

The **domains of $f + g$, $f - g$,** and **fg** include all real numbers in the intersection of the domains of f and g, while the **domain of** $\frac{f}{g}$ includes those real numbers in the intersection of the domains of f and g for which $g(x) \neq 0$.

NOTE The condition $g(x) \neq 0$ in the definition of the quotient means that the domain of $\left(\frac{f}{g}\right)(x)$ is restricted to all values of x for which $g(x)$ is not 0. The condition does not mean that $g(x)$ is a function that is never 0.

EXAMPLE 1 Using Operations on Functions

Let $f(x) = x^2 + 1$ and $g(x) = 3x + 5$. Find each of the following.

(a) $(f + g)(1)$ **(b)** $(f - g)(-3)$ **(c)** $(fg)(5)$ **(d)** $\left(\dfrac{f}{g}\right)(0)$

SOLUTION

(a) First determine $f(1) = 2$ and $g(1) = 8$. Then use the definition.

$$(f + g)(1) = f(1) + g(1) \qquad {\scriptstyle (f + g)(x) = f(x) + g(x)}$$
$$= 2 + 8 \qquad {\scriptstyle f(1) = 1^2 + 1;\, g(1) = 3(1) + 5}$$
$$= 10 \qquad {\scriptstyle \text{Add.}}$$

(b) $(f - g)(-3) = f(-3) - g(-3) \qquad {\scriptstyle (f - g)(x) = f(x) - g(x)}$
$$= 10 - (-4) \qquad {\scriptstyle f(-3) = (-3)^2 + 1;\, g(-3) = 3(-3) + 5}$$
$$= 14 \qquad {\scriptstyle \text{Subtract.}}$$

(c) $(fg)(5) = f(5) \cdot g(5)$
$$= (5^2 + 1)(3 \cdot 5 + 5)$$
$$= 26 \cdot 20$$
$$= 520$$

(d) $\left(\dfrac{f}{g}\right)(0) = \dfrac{f(0)}{g(0)} \qquad {\scriptstyle \left(\frac{f}{g}\right)(x) = \frac{f(x)}{g(x)}}$
$$= \dfrac{0^2 + 1}{3(0) + 5} \qquad {\scriptstyle f(x) = x^2 + 1 \atop g(x) = 3x + 5}$$
$$= \dfrac{1}{5} \qquad {\scriptstyle \text{Simplify.}}$$

✔ *Now Try Exercises 1, 3, 5, and 7.*

EXAMPLE 2 **Using Operations on Functions and Determining Domains**

Let $f(x) = 8x - 9$ and $g(x) = \sqrt{2x - 1}$. Find each function in (a)–(d).

(a) $(f + g)(x)$ **(b)** $(f - g)(x)$ **(c)** $(fg)(x)$ **(d)** $\left(\dfrac{f}{g}\right)(x)$

(e) Give the domains of the functions in parts (a)–(d).

SOLUTION

(a) $(f + g)(x) = f(x) + g(x)$

$= 8x - 9 + \sqrt{2x - 1}$

(b) $(f - g)(x) = f(x) - g(x)$

$= 8x - 9 - \sqrt{2x - 1}$

(c) $(fg)(x) = f(x) \cdot g(x)$

$= (8x - 9)\sqrt{2x - 1}$

(d) $\left(\dfrac{f}{g}\right)(x) = \dfrac{f(x)}{g(x)}$

$= \dfrac{8x - 9}{\sqrt{2x - 1}}$

(e) To find the domains of the functions in parts (a)–(d), we first find the domains of f and g. The domain of f is the set of all real numbers $(-\infty, \infty)$. Because g is defined by a square root radical, the radicand must be non-negative (that is, greater than or equal to 0).

$$g(x) = \sqrt{2x - 1} \quad \text{Rule for } g(x)$$

$$2x - 1 \geq 0 \quad \text{2x − 1 must be nonnegative.}$$

$$2x \geq 1 \quad \text{Add 1. (Section 1.7)}$$

$$x \geq \frac{1}{2} \quad \text{Divide by 2.}$$

Thus, the domain of g is $\left[\frac{1}{2}, \infty\right)$.

The domains of $f + g$, $f - g$, and fg are the intersection of the domains of f and g, which is

$$(-\infty, \infty) \cap \left[\frac{1}{2}, \infty\right) = \left[\frac{1}{2}, \infty\right). \quad \text{(Section R.1)}$$

The domain of $\frac{f}{g}$ includes those real numbers in the intersection above for which $g(x) = \sqrt{2x - 1} \neq 0$. That is, the domain of $\frac{f}{g}$ is $\left(\frac{1}{2}, \infty\right)$.

☑ *Now Try Exercises 9 and 13.*

EXAMPLE 3 **Evaluating Combinations of Functions**

If possible, use the given representations of functions f and g to evaluate

$$(f + g)(4), \quad (f - g)(-2), \quad (fg)(1), \quad \text{and} \quad \left(\frac{f}{g}\right)(0).$$

(a)

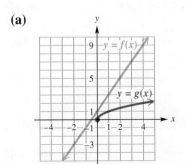

(b)

x	$f(x)$	$g(x)$
-2	-3	undefined
0	1	0
1	3	1
4	9	2

(c) $f(x) = 2x + 1$, $g(x) = \sqrt{x}$

SOLUTION

(a) From the figure, repeated in the margin, $f(4) = 9$ and $g(4) = 2$.

$$(f + g)(4) = f(4) + g(4) \quad (f + g)(x) = f(x) + g(x)$$

$$= 9 + 2 \qquad \text{Substitute.}$$

$$= 11 \qquad \text{Add.}$$

For $(f - g)(-2)$, although $f(-2) = -3$, $g(-2)$ is undefined because -2 is not in the domain of g. Thus $(f - g)(-2)$ is undefined. The domains of f and g include 1.

$$(fg)(1) = f(1) \cdot g(1) \quad (fg)(x) = f(x) \cdot g(x)$$

$$= 3 \cdot 1 \qquad \text{Substitute.}$$

$$= 3 \qquad \text{Multiply.}$$

The graph of g includes the origin, so $g(0) = 0$. Thus $\left(\frac{f}{g}\right)(0)$ is undefined.

x	$f(x)$	$g(x)$
-2	-3	undefined
0	1	0
1	3	1
4	9	2

(b) From the table, repeated in the margin, $f(4) = 9$ and $g(4) = 2$.

$$(f + g)(4) = f(4) + g(4) \quad (f + g)(x) = f(x) + g(x)$$

$$= 9 + 2 \qquad \text{Substitute.}$$

$$= 11 \qquad \text{Add.}$$

In the table, $g(-2)$ is undefined, and thus $(f - g)(-2)$ is also undefined.

$$(fg)(1) = f(1) \cdot g(1) \quad (fg)(x) = f(x) \cdot g(x)$$

$$= 3 \cdot 1 \qquad f(1) = 3 \text{ and } g(1) = 1$$

$$= 3 \qquad \text{Multiply.}$$

The quotient function value $\left(\frac{f}{g}\right)(0)$ is undefined since the denominator, $g(0)$, equals 0.

(c) Using $f(x) = 2x + 1$ and $g(x) = \sqrt{x}$, we can find $(f + g)(4)$ and $(fg)(1)$. Since -2 is not in the domain of g, $(f - g)(-2)$ is not defined.

$$(f + g)(4) = f(4) + g(4) \qquad\qquad (fg)(1) = f(1) \cdot g(1)$$

$$= (2 \cdot 4 + 1) + \sqrt{4} \qquad\qquad = (2 \cdot 1 + 1) \cdot \sqrt{1}$$

$$= 9 + 2 \qquad\qquad\qquad\qquad = 3(1)$$

$$= 11 \qquad\qquad\qquad\qquad\quad = 3$$

$\left(\frac{f}{g}\right)(0)$ is undefined since $g(0) = 0$.

✔ *Now Try Exercises 23 and 27.*

The Difference Quotient   Suppose the point P lies on the graph of $y = f(x)$ as in **Figure 93,** and suppose h is a positive number. If we let $(x, f(x))$ denote the coordinates of P and let $(x + h, f(x + h))$ denote the coordinates of Q, then the line joining P and Q has slope as follows.

$$m = \frac{f(x + h) - f(x)}{(x + h) - x} \qquad \text{Slope formula (Section 2.4)}$$

$$= \frac{f(x + h) - f(x)}{h}, \quad h \neq 0 \quad \text{Difference quotient}$$

This boldface expression is called the **difference quotient.**

Figure 93

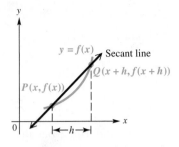

Figure 93 (repeated)

Figure 93 shows the graph of the line PQ (called a **secant line**). As h approaches 0, the slope of this secant line approaches the slope of the line tangent to the curve at P. Important applications of this idea are developed in calculus.

EXAMPLE 4 **Finding the Difference Quotient**

Let $f(x) = 2x^2 - 3x$. Find and simplify the expression for the difference quotient,

$$\frac{f(x + h) - f(x)}{h}.$$

SOLUTION We use a three-step process.

Step 1 Find the first term in the numerator, $f(x + h)$. Replace x in $f(x)$ with $x + h$.

$$f(x + h) = 2(x + h)^2 - 3(x + h)$$

Step 2 Find the entire numerator, $f(x + h) - f(x)$. From Step 1

$$\begin{aligned}
f(x + h) - f(x) &= \left[2(x + h)^2 - 3(x + h)\right] - (2x^2 - 3x) &&\text{Substitute.} \\
&= 2(x^2 + 2xh + h^2) - 3(x + h) - (2x^2 - 3x) \\
&&&\text{Square } x + h. \textbf{(Section R.3)} \\
&= 2x^2 + 4xh + 2h^2 - 3x - 3h - 2x^2 + 3x \\
&&&\text{Distributive property } \textbf{(Section R.2)} \\
&= 4xh + 2h^2 - 3h &&\text{Combine like terms.}
\end{aligned}$$

Remember this term when squaring $x + h$.

Step 3 Find the difference quotient by dividing by h.

$$\begin{aligned}
\frac{f(x + h) - f(x)}{h} &= \frac{4xh + 2h^2 - 3h}{h} &&\text{Substitute.} \\
&= \frac{h(4x + 2h - 3)}{h} &&\text{Factor out } h. \textbf{(Section R.4)} \\
&= 4x + 2h - 3 &&\text{Divide. } \textbf{(Section R.5)}
\end{aligned}$$

✔ *Now Try Exercises 35 and 45.*

CAUTION In **Example 4,** notice that the expression $f(x + h)$ is not equivalent to $f(x) + f(h)$.

$$f(x + h) = 2(x + h)^2 - 3(x + h) = 2x^2 + 4xh + 2h^2 - 3x - 3h$$

$$f(x) + f(h) = (2x^2 - 3x) + (2h^2 - 3h) = 2x^2 - 3x + 2h^2 - 3h$$

These expressions differ by $4xh$. In general, for a function f, $f(x + h)$ is *not* equivalent to $f(x) + f(h)$.

LOOKING AHEAD TO CALCULUS

The difference quotient is essential in the definition of the **derivative of a function** in calculus. The derivative provides a formula, in function form, for finding the slope of the tangent line to the graph of the function at a given point.

To illustrate, it is shown in calculus that the derivative of $f(x) = x^2 + 3$ is given by the function $f'(x) = 2x$. Now, $f'(0) = 2(0) = 0$, meaning that the slope of the tangent line to $f(x) = x^2 + 3$ at $x = 0$ is 0, which implies that the tangent line is horizontal. If you draw this tangent line, you will see that it is the line $y = 3$, which is indeed a horizontal line.

Composition of Functions and Domain The diagram in **Figure 94** shows a function f that assigns to each x in its domain a value $f(x)$. Then another function g assigns to each $f(x)$ in its domain a value $g(f(x))$. This two-step process takes an element x and produces a corresponding element $g(f(x))$.

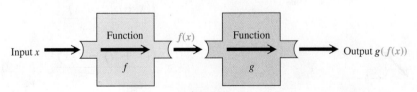

Figure 94

The function with y-values $g(f(x))$ is called the *composition* of functions g and f, which is written $g \circ f$ and read "$\boldsymbol{g}$ *of* $\boldsymbol{f}$."

Composition of Functions and Domain

If f and g are functions, then the **composite function,** or **composition,** of g and f is defined by

$$(g \circ f)(x) = g(f(x)).$$

The **domain of** $\boldsymbol{g \circ f}$ is the set of all numbers x in the domain of f such that $f(x)$ is in the domain of g.

As a real-life example of how composite functions occur, consider the following retail situation:

> *A* \$40 *pair of blue jeans is on sale for* 25% *off. If you purchase the jeans before noon, the retailer offers an additional* 10% *off. What is the final sale price of the blue jeans?*

You might be tempted to say that the jeans are 35% off and calculate $\$40(0.35) = \$14$, giving a final sale price of

$$\$40 - \$14 = \$26$$

for the jeans. ***This is not correct.*** To find the final sale price, we must first find the price after taking 25% off and then take an additional 10% off *that* price. See **Figure 95.**

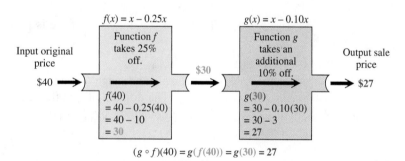

$$(g \circ f)(40) = g(f(40)) = g(30) = 27$$

Figure 95

EXAMPLE 5   Evaluating Composite Functions

Let $f(x) = 2x - 1$ and $g(x) = \frac{4}{x - 1}$.

(a) Find $(f \circ g)(2)$.   **(b)** Find $(g \circ f)(-3)$.

SOLUTION

(a) First find $g(2)$: $g(2) = \dfrac{4}{2 - 1} = \dfrac{4}{1} = 4$.

Now find $(f \circ g)(2)$.

$$
\begin{aligned}
(f \circ g)(2) &= f(g(2)) &&\text{Definition of composition} \\
&= f(4) &&\text{See above.} \\
&= 2(4) - 1 &&\text{Definition of } f \\
&= 7 &&\text{Simplify.}
\end{aligned}
$$

The screens show how a graphing calculator evaluates the expressions in **Example 5.**

(b) $(g \circ f)(-3) = g(f(-3))$ Definition of composition

$$= g[2(-3) - 1]$$ $f(-3) = 2(-3) - 1$

$$= g(-7)$$ $2(-3) - 1 = -7$

$$= \frac{4}{-7 - 1}$$ $g(x) = \frac{4}{x + 1}$

$$= \frac{4}{-8}, \quad \text{or} \quad -\frac{1}{2}$$ Simplify.

☑ *Now Try Exercise 47.*

EXAMPLE 6   **Determining Composite Functions and Their Domains**

Given that $f(x) = \sqrt{x}$ and $g(x) = 4x + 2$, find each of the following.

(a) $(f \circ g)(x)$ and its domain   **(b)** $(g \circ f)(x)$ and its domain

SOLUTION

(a) $(f \circ g)(x) = f(g(x)) = f(4x + 2) = \sqrt{4x + 2}$

The domain and range of g are both the set of all real numbers, $(-\infty, \infty)$. The domain of f is the set of all nonnegative real numbers, $[0, \infty)$. Thus, $g(x)$, which is defined as $4x + 2$, must be greater than or equal to zero.

> The radicand must be nonnegative.

$$4x + 2 \geq 0$$ Solve the inequality.

$$4x \geq -2$$ Subtract 2.

$$x \geq -\frac{1}{2}$$ Divide by 4.

Therefore, the domain of $f \circ g$ is $\left[-\frac{1}{2}, \infty\right)$.

(b) $(g \circ f)(x) = g(f(x)) = g(\sqrt{x}) = 4\sqrt{x} + 2$

The domain and range of f are both the set of all nonnegative real numbers, $[0, \infty)$. The domain of g is the set of all real numbers, $(-\infty, \infty)$. Therefore, the domain of $g \circ f$ is $[0, \infty)$.

☑ *Now Try Exercise 65.*

EXAMPLE 7   **Determining Composite Functions and Their Domains**

Given that $f(x) = \frac{6}{x - 3}$ and $g(x) = \frac{1}{x}$, find each of the following.

(a) $(f \circ g)(x)$ and its domain   **(b)** $(g \circ f)(x)$ and its domain

SOLUTION

(a) $(f \circ g)(x) = f(g(x)) = f\left(\frac{1}{x}\right)$ $g(x) = \frac{1}{x}$

$$= \frac{6}{\frac{1}{x} - 3}$$ $f(x) = \frac{6}{x - 3}$

$$= \frac{6x}{1 - 3x}$$ Multiply the numerator and denominator by x. **(Section R.5)**

The domain of g is all real numbers *except* 0, which makes $g(x)$ undefined. The domain of f is all real numbers *except* 3. The expression for $g(x)$, therefore, cannot equal 3. We determine the value that makes $g(x) = 3$ and *exclude* it from the domain of $f \circ g$.

LOOKING AHEAD TO CALCULUS

Finding the derivative of a function in calculus is called **differentiation.** To differentiate a composite function such as $h(x) = (3x + 2)^4$, we interpret $h(x)$ as $(f \circ g)(x)$, where $g(x) = 3x + 2$ and $f(x) = x^4$. The **chain rule** allows us to differentiate composite functions. Notice the use of the composition symbol and function notation in the following, which comes from the chain rule.

If $h(x) = (f \circ g)(x)$, then

$$h'(x) = f'(g(x)) \cdot g'(x).$$

$\dfrac{1}{x} = 3$ The solution must be excluded.

$1 = 3x$ Multiply by x.

$x = \dfrac{1}{3}$ Divide by 3.

Therefore, the domain of $f \circ g$ is the set of all real numbers *except* 0 and $\frac{1}{3}$, written in interval notation as

$$(-\infty, 0) \cup \left(0, \frac{1}{3}\right) \cup \left(\frac{1}{3}, \infty\right).$$

(b) $(g \circ f)(x) = g(f(x)) = g\left(\dfrac{6}{x - 3}\right)$

$= \dfrac{1}{\dfrac{6}{x - 3}}$ Note that this is meaningless if $x = 3$.

$= \dfrac{x - 3}{6}$ $\dfrac{1}{\frac{a}{b}} = 1 \div \dfrac{a}{b} = 1 \cdot \dfrac{b}{a} = \dfrac{b}{a}$

The domain of f is all real numbers *except* 3, and the domain of g is all real numbers *except* 0. The expression for $f(x)$, which is $\frac{6}{x-3}$, is never zero, since the numerator is the nonzero number 6. Therefore, the domain of $g \circ f$ is the set of all real numbers *except* 3, written

$$(-\infty, 3) \cup (3, \infty).$$

☑ *Now Try Exercise 77.*

NOTE   In a situation like **Example 7(b),** it often helps to consider the *unsimplified* form of the composition expression when determining the domain.

EXAMPLE 8   Showing That $(g \circ f)(x)$ Is Not Equivalent to $(f \circ g)(x)$

Let $f(x) = 4x + 1$ and $g(x) = 2x^2 + 5x$. Show that $(g \circ f)(x) \neq (f \circ g)(x)$. (This is sufficient to prove that this inequality is true in general.)

SOLUTION   First, find $(g \circ f)(x)$.

$(g \circ f)(x) = g(f(x)) = g(4x + 1)$ $f(x) = 4x + 1$

$= 2(4x + 1)^2 + 5(4x + 1)$ $g(x) = 2x^2 + 5x$

$= 2(16x^2 + 8x + 1) + 20x + 5$ Square $4x + 1$ and apply the distributive property.

$= 32x^2 + 16x + 2 + 20x + 5$ Distributive property

$(g \circ f)(x) = 32x^2 + 36x + 7$ Combine like terms.

Now, find $(f \circ g)(x)$.

$(f \circ g)(x) = f(g(x))$ By definition

$= f(2x^2 + 5x)$ $g(x) = 2x^2 + 5x$

$= 4(2x^2 + 5x) + 1$ $f(x) = 4x + 1$

$(f \circ g)(x) = 8x^2 + 20x + 1$ Distributive property

Thus, $(g \circ f)(x) \neq (f \circ g)(x)$.

☑ *Now Try Exercise 81.*

As **Example 8** shows, *it is not always true that $f \circ g = g \circ f$.* In fact, the composite functions $f \circ g$ and $g \circ f$ are equal only for a special class of functions, discussed in **Section 4.1.**

In calculus it is sometimes necessary to treat a function as a composition of two functions. The next example shows how this can be done.

EXAMPLE 9 **Finding Functions That Form a Given Composite**

Find functions f and g such that

$$(f \circ g)(x) = (x^2 - 5)^3 - 4(x^2 - 5) + 3.$$

SOLUTION Note the repeated quantity $x^2 - 5$. If we choose $g(x) = x^2 - 5$ and $f(x) = x^3 - 4x + 3$, then we have the following.

$$
\begin{aligned}
(f \circ g)(x) &= f(g(x)) && \text{By definition} \\
&= f(x^2 - 5) && g(x) = x^2 - 5 \\
&= (x^2 - 5)^3 - 4(x^2 - 5) + 3 && \text{Use the rule for } f.
\end{aligned}
$$

There are other pairs of functions f and g that also satisfy these conditions. Here is another such pair.

$$f(x) = (x - 5)^3 - 4(x - 5) + 3 \quad \text{and} \quad g(x) = x^2$$

☑ *Now Try Exercise 89.*

2.8 Exercises

Let $f(x) = x^2 + 3$ and $g(x) = -2x + 6$. Find each of the following. **See Example 1.**

1. $(f + g)(3)$ **2.** $(f + g)(-5)$ **3.** $(f - g)(-1)$ **4.** $(f - g)(4)$

5. $(fg)(4)$ **6.** $(fg)(-3)$ **7.** $\left(\dfrac{f}{g}\right)(-1)$ **8.** $\left(\dfrac{f}{g}\right)(5)$

For the pair of functions defined, find $(f + g)(x)$, $(f - g)(x)$, $(fg)(x)$, and $\left(\frac{f}{g}\right)(x)$. Give the domain of each. **See Example 2.**

9. $f(x) = 3x + 4, \; g(x) = 2x - 5$ **10.** $f(x) = 6 - 3x, \; g(x) = -4x + 1$

11. $f(x) = 2x^2 - 3x, \; g(x) = x^2 - x + 3$ **12.** $f(x) = 4x^2 + 2x, \; g(x) = x^2 - 3x + 2$

13. $f(x) = \sqrt{4x - 1}, \; g(x) = \dfrac{1}{x}$ **14.** $f(x) = \sqrt{5x - 4}, \; g(x) = -\dfrac{1}{x}$

Associate's Degrees Earned The graph shows the number of associate's degrees earned (in thousands) in the United States from 2000 through 2008. $M(x)$ gives the number of degrees earned by males, $F(x)$ gives the number earned by females, and $T(x)$ gives the total number for both groups. Use the graph in Exercises 15–18.

15. Estimate $M(2004)$ and $F(2004)$, and use your results to estimate $T(2004)$.

16. Estimate $M(2008)$ and $F(2008)$, and use your results to estimate $T(2008)$.

Associate's Degrees Earned

Source: U.S. National Center for Education Statistics.

17. Use the slopes of the line segments to decide in which period (2000–2004 or 2004–2008) the total number of associate's degrees earned increased more rapidly.

18. *Concept Check* Refer to the graph of Associate's Degrees Earned on the previous page. If $2000 \le k \le 2008$, $T(k) = r$, and $F(k) = s$, then $M(k) =$ _____ .

Science and Space/Technology Spending *The graph shows dollars (in billions) spent for general science and for space/other technologies in selected years. $G(x)$ represents the dollars spent for general science, and $S(x)$ represents the dollars spent for space and other technologies. $T(x)$ represents the total expenditures for these two categories. Use the graph in Exercises 19–22.*

Science and Space Spending

Source: U.S. Office of Management and Budget.

19. Estimate $(T - S)(2000)$. What does this function represent?

20. Estimate $(T - G)(2010)$. What does this function represent?

21. In which of the categories was spending almost static for several years? In which years did this occur?

22. In which period and which category does spending for $G(x)$ or $S(x)$ increase most?

Use the graph to evaluate each expression. See Example 3(a).

23. (a) $(f + g)(2)$ (b) $(f - g)(1)$ **24.** (a) $(f + g)(0)$ (b) $(f - g)(-1)$

(c) $(fg)(0)$ (d) $\left(\dfrac{f}{g}\right)(1)$ (c) $(fg)(1)$ (d) $\left(\dfrac{f}{g}\right)(2)$

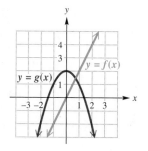

25. (a) $(f + g)(-1)$ (b) $(f - g)(-2)$ **26.** (a) $(f + g)(1)$ (b) $(f - g)(0)$

(c) $(fg)(0)$ (d) $\left(\dfrac{f}{g}\right)(2)$ (c) $(fg)(-1)$ (d) $\left(\dfrac{f}{g}\right)(1)$

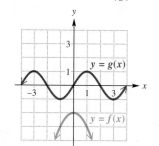

In Exercises 27 and 28, use the table to evaluate each expression in parts (a)–(d), if possible. **See Example 3(b).**

(a) $(f + g)(2)$ **(b)** $(f - g)(4)$ **(c)** $(fg)(-2)$ **(d)** $\left(\dfrac{f}{g}\right)(0)$

27.

x	$f(x)$	$g(x)$
-2	0	6
0	5	0
2	7	-2
4	10	5

28.

x	$f(x)$	$g(x)$
-2	-4	2
0	8	-1
2	5	4
4	0	0

29. Use the table in **Exercise 27** to complete the following table.

x	$(f + g)(x)$	$(f - g)(x)$	$(fg)(x)$	$\left(\frac{f}{g}\right)(x)$
-2				
0				
2				
4				

30. Use the table in **Exercise 28** to complete the following table.

x	$(f + g)(x)$	$(f - g)(x)$	$(fg)(x)$	$\left(\frac{f}{g}\right)(x)$
-2				
0				
2				
4				

31. How is the difference quotient related to slope?

32. Refer to **Figure 93.** How is the secant line PQ related to the tangent line to a curve at point P?

For each of the functions in Exercises 33–46, find **(a)** $f(x + h)$, **(b)** $f(x + h) - f(x)$, *and* **(c)** $\dfrac{f(x + h) - f(x)}{h}$. **See Example 4.**

33. $f(x) = 2 - x$ **34.** $f(x) = 1 - x$ **35.** $f(x) = 6x + 2$

36. $f(x) = 4x + 11$ **37.** $f(x) = -2x + 5$ **38.** $f(x) = -4x + 2$

39. $f(x) = \dfrac{1}{x}$ **40.** $f(x) = \dfrac{1}{x^2}$ **41.** $f(x) = x^2$

42. $f(x) = -x^2$ **43.** $f(x) = 1 - x^2$ **44.** $f(x) = 1 + 2x^2$

45. $f(x) = x^2 + 3x + 1$ **46.** $f(x) = x^2 - 4x + 2$

Let $f(x) = 2x - 3$ *and* $g(x) = -x + 3$. *Find each function value.* **See Example 5.**

47. $(f \circ g)(4)$ **48.** $(f \circ g)(2)$ **49.** $(f \circ g)(-2)$ **50.** $(g \circ f)(3)$

51. $(g \circ f)(0)$ **52.** $(g \circ f)(-2)$ **53.** $(f \circ f)(2)$ **54.** $(g \circ g)(-2)$

Concept Check *The tables give some selected ordered pairs for functions f and g.*

x	3	4	6
$f(x)$	1	3	9

x	2	7	1	9
$g(x)$	3	6	9	12

Find each of the following.

55. $(f \circ g)(2)$ **56.** $(f \circ g)(7)$ **57.** $(g \circ f)(3)$

58. $(g \circ f)(6)$ **59.** $(f \circ f)(4)$ **60.** $(g \circ g)(1)$

61. *Concept Check* Why can you not determine $(f \circ g)(1)$ given the information in the tables for **Exercises 55–60?**

62. *Concept Check* Extend the concept of composition of functions to evaluate $(g \circ (f \circ g))(7)$ using the tables for **Exercises 55–60.**

*Given functions f and g, find (**a**) $(f \circ g)(x)$ and its domain, and (**b**) $(g \circ f)(x)$ and its domain. See Examples 6 and 7.*

63. $f(x) = -6x + 9,$ $g(x) = 5x + 7$ **64.** $f(x) = 8x + 12,$ $g(x) = 3x - 1$

65. $f(x) = \sqrt{x},$ $g(x) = x + 3$ **66.** $f(x) = \sqrt{x},$ $g(x) = x - 1$

67. $f(x) = x^3,$ $g(x) = x^2 + 3x - 1$ **68.** $f(x) = x + 2,$ $g(x) = x^4 + x^2 - 4$

69. $f(x) = \sqrt{x - 1},$ $g(x) = 3x$ **70.** $f(x) = \sqrt{x - 2},$ $g(x) = 2x$

71. $f(x) = \dfrac{2}{x},$ $g(x) = x + 1$ **72.** $f(x) = \dfrac{4}{x},$ $g(x) = x + 4$

73. $f(x) = \sqrt{x + 2},$ $g(x) = -\dfrac{1}{x}$ **74.** $f(x) = \sqrt{x + 4},$ $g(x) = -\dfrac{2}{x}$

75. $f(x) = \sqrt{x},$ $g(x) = \dfrac{1}{x + 5}$ **76.** $f(x) = \sqrt{x},$ $g(x) = \dfrac{3}{x + 6}$

77. $f(x) = \dfrac{1}{x - 2},$ $g(x) = \dfrac{1}{x}$ **78.** $f(x) = \dfrac{1}{x + 4},$ $g(x) = -\dfrac{1}{x}$

79. *Concept Check* Fill in the missing entries in the table.

x	$f(x)$	$g(x)$	$g(f(x))$
1	3	2	7
2	1	5	
3	2		

80. *Concept Check* Suppose $f(x)$ is an odd function and $g(x)$ is an even function. Fill in the missing entries in the table.

x	-2	-1	0	1	2
$f(x)$			0	-2	
$g(x)$	0	2	1		
$(f \circ g)(x)$		1	-2		

81. Show that $(f \circ g)(x)$ is not equivalent to $(g \circ f)(x)$ for

$$f(x) = 3x - 2 \quad \text{and} \quad g(x) = 2x - 3.$$

82. Describe the steps required to find the composite function $f \circ g$, given

$$f(x) = 2x - 5 \quad \text{and} \quad g(x) = x^2 + 3.$$

For certain pairs of functions f and g, $(f \circ g)(x) = x$ and $(g \circ f)(x) = x$. Show that this is true for each pair in **Exercises 83–86.**

83. $f(x) = 4x + 2$, $g(x) = \frac{1}{4}(x - 2)$

84. $f(x) = -3x$, $g(x) = -\frac{1}{3}x$

85. $f(x) = \sqrt[3]{5x + 4}$, $g(x) = \frac{1}{5}x^3 - \frac{4}{5}$

86. $f(x) = \sqrt[3]{x + 1}$, $g(x) = x^3 - 1$

Find functions f and g such that $(f \circ g)(x) = h(x)$. (There are many possible ways to do this.) **See Example 9.**

87. $h(x) = (6x - 2)^2$

88. $h(x) = (11x^2 + 12x)^2$

89. $h(x) = \sqrt{x^2 - 1}$

90. $h(x) = (2x - 3)^3$

91. $h(x) = \sqrt{6x} + 12$

92. $h(x) = \sqrt[3]{2x + 3} - 4$

Solve each problem.

93. *Relationship of Measurement Units* The function defined by $f(x) = 12x$ computes the number of inches in x feet, and the function defined by $g(x) = 5280x$ computes the number of feet in x miles. What does $(f \circ g)(x)$ compute?

94. *Perimeter of a Square* The perimeter x of a square with side of length s is given by the formula $x = 4s$.

(a) Solve for s in terms of x.

(b) If y represents the area of this square, write y as a function of the perimeter x.

(c) Use the composite function of part (b) to find the area of a square with perimeter 6.

95. *Area of an Equilateral Triangle* The area of an equilateral triangle with sides of length x is given by the function defined by $\mathcal{A}(x) = \frac{\sqrt{3}}{4}x^2$.

(a) Find $\mathcal{A}(2x)$, the function representing the area of an equilateral triangle with sides of length twice the original length.

(b) Find the area of an equilateral triangle with side length 16. Use the formula $\mathcal{A}(2x)$ found in part (a).

96. *Software Author Royalties* A software author invests his royalties in two accounts for 1 yr.

(a) The first account pays 4% simple interest. If he invests x dollars in this account, write an expression for y_1 in terms of x, where y_1 represents the amount of interest earned.

(b) He invests in a second account $500 more than he invested in the first account. This second account pays 2.5% simple interest. Write an expression for y_2, where y_2 represents the amount of interest earned.

(c) What does $y_1 + y_2$ represent?

(d) How much interest will he receive if $250 is invested in the first account?

97. *Oil Leak* An oil well off the Gulf Coast is leaking, with the leak spreading oil over the water's surface as a circle. At any time t, in minutes, after the beginning of the leak, the radius of the circular oil slick on the surface is $r(t) = 4t$ feet. Let $\mathcal{A}(r) = \pi r^2$ represent the area of a circle of radius r.

(a) Find $(\mathcal{A} \circ r)(t)$.

(b) Interpret $(\mathcal{A} \circ r)(t)$.

(c) What is the area of the oil slick after 3 min?

98. *Emission of Pollutants* When a thermal inversion layer is over a city (as happens in Los Angeles), pollutants cannot rise vertically but are trapped below the layer and must disperse horizontally. Assume that a factory smokestack begins emitting a pollutant at 8 A.M. Assume that the pollutant disperses horizontally over a circular area. If t represents the time, in hours, since the factory began emitting pollutants ($t = 0$ represents 8 A.M.), assume that the radius of the circle of pollutants at time t is $r(t) = 2t$ miles. Let $\mathcal{A}(r) = \pi r^2$ represent the area of a circle of radius r.

(a) Find $(\mathcal{A} \circ r)(t)$.   **(b)** Interpret $(\mathcal{A} \circ r)(t)$.

(c) What is the area of the circular region covered by the layer at noon?

99. *(Modeling) Catering Cost* The cost to hire a caterer for a party depends on the number of guests attending. If 100 people attend, the cost per person will be $20. For each person less than 100, the cost will increase by $5. Assume that no more than 100 people will attend. Let x represent the number less than 100 who do not attend. For example, if 95 attend, $x = 5$.

(a) Write a function defined by $N(x)$ giving the number of guests.

(b) Write a function defined by $G(x)$ giving the cost per guest.

(c) Write a function defined by $N(x) \cdot G(x)$ for the total cost, $C(x)$.

(d) What is the total cost if 80 people attend?

100. *Area of a Square* The area of a square is x^2 square inches. Suppose that 3 in. is added to one dimension and 1 in. is subtracted from the other dimension. Express the area $\mathcal{A}(x)$ of the resulting rectangle as a product of two functions.

Chapter 2 Test Prep

Key Terms

2.1 ordered pair
origin
x-axis
y-axis
rectangular (Cartesian)
    coordinate system
coordinate plane
    (xy-plane)
quadrants
coordinates
collinear
graph of an equation
x-intercept
y-intercept

2.2 circle
radius
center of a circle

2.3 dependent variable
independent variable
relation
function
domain
range
function notation
increasing function
decreasing function
constant function

2.4 linear function

standard form
relatively prime
change in x
change in y
slope
average rate of change
linear cost function
cost
fixed cost
revenue function
profit function

2.5 point-slope form
slope-intercept form
scatter diagram

2.6 continuous function
parabola
vertex
piecewise-defined
    function
step function

2.7 symmetry
even function
odd function
vertical translation
horizontal translation

2.8 difference quotient
composite function
    (composition)

New Symbols

(a, b)	ordered pair	m	slope
$f(x)$	function f evaluated at x (read "f of x")	$[\![x]\!]$	the greatest integer less than or equal to x
Δx	change in x	$g \circ f$	composite function
Δy	change in y		

(continued)

Quick Review

Concepts	Examples

2.1 Rectangular Coordinates and Graphs

Distance Formula
Suppose $P(x_1, y_1)$ and $Q(x_2, y_2)$ are two points in a coordinate plane. Then the distance between P and Q, written $d(P, Q)$, is given by the following.

$$d(P, Q) = \sqrt{(x_2 - x_1)^2 + (y_2 - y_1)^2}$$

Find the distance between the points $P(-1, 4)$ and $Q(6, -3)$.

$$d(P, Q) = \sqrt{[6 - (-1)]^2 + (-3 - 4)^2}$$
$$= \sqrt{49 + 49}$$
$$= \sqrt{98}, \quad \text{or} \quad 7\sqrt{2}$$

Midpoint Formula
The coordinates of the midpoint M of the line segment with endpoints (x_1, y_1) and (x_2, y_2) are given by the following.

$$M = \left(\frac{x_1 + x_2}{2}, \frac{y_1 + y_2}{2}\right)$$

Find the coordinates of the midpoint M of the line segment with endpoints $(-1, 4)$ and $(6, -3)$.

$$M = \left(\frac{-1 + 6}{2}, \frac{4 + (-3)}{2}\right) = \left(\frac{5}{2}, \frac{1}{2}\right)$$

2.2 Circles

Center-Radius Form of the Equation of a Circle
The equation of a circle with center at (h, k) and radius r is given by the following.

$$(x - h)^2 + (y - k)^2 = r^2$$

Find the center-radius form of the equation of the circle with center at $(-2, 3)$ and radius 4.

$$[x - (-2)]^2 + (y - 3)^2 = 4^2$$
or
$$(x + 2)^2 + (y - 3)^2 = 16$$

General Form of the Equation of a Circle

$$x^2 + y^2 + cx + dy + e = 0$$

The general form of the equation of the preceding circle is

$$x^2 + y^2 + 4x - 6y - 3 = 0.$$

2.3 Functions

A **relation** is a set of ordered pairs. A **function** is a relation in which, for each value of the first component of the ordered pairs, there is *exactly one* value of the second component. The set of first components is the **domain,** and the set of second components is the **range.**

The relation $y = x^2$ defines a function, because each choice of a number for x corresponds to one and only one number for y. The domain is $(-\infty, \infty)$, and the range is $[0, \infty)$.

The relation $x = y^2$ does *not* define a function because a number x may correspond to two numbers for y. The domain is $[0, \infty)$, and the range is $(-\infty, \infty)$.

Vertical Line Test
If every vertical line intersects the graph of a relation in no more than one point, then the relation is a function.

Determine whether the graphs are functions.

By the vertical line test, graph A is the graph of a function, but graph B is not.

Concepts	Examples

Increasing, Decreasing, and Constant Functions

f is increasing. *f* is decreasing. *f* is **constant**.

Discuss the function in graph A on the preceding page in terms of whether it is increasing, decreasing, or constant.

The function in graph A is decreasing on the open interval $(-\infty, 0)$ and increasing on the open interval $(0, \infty)$.

2.4 Linear Functions

A function f is a **linear function** if, for real numbers a and b,

$$f(x) = ax + b.$$

The graph of a linear function is a line.

The equation

$$y = \frac{1}{2}x - 4$$

defines y as a linear function of x.

Definition of Slope

The slope m of the line through the points (x_1, y_1) and (x_2, y_2) is given by the following.

$$m = \frac{\text{rise}}{\text{run}} = \frac{\Delta y}{\Delta x} = \frac{y_2 - y_1}{x_2 - x_1}, \quad \text{where } \Delta x \neq 0$$

Find the slope of the line through the points $(2, 4)$ and $(-1, 7)$.

$$m = \frac{7 - 4}{-1 - 2} = \frac{3}{-3} = -1$$

2.5 Equations of Lines and Linear Models

Forms of Linear Equations

Equation	Description
$y = mx + b$	**Slope-Intercept Form** Slope is m. y-intercept is $(0, b)$.
$y - y_1 = m(x - x_1)$	**Point-Slope Form** Slope is m. Line passes through (x_1, y_1).
$Ax + By = C$	**Standard Form** (A, B, and C integers, $A \geq 0$) Slope is $-\frac{A}{B}$ $(B \neq 0)$. x-intercept is $\left(\frac{C}{A}, 0\right)$ $(A \neq 0)$. y-intercept is $\left(0, \frac{C}{B}\right)$ $(B \neq 0)$.
$y = b$	**Horizontal Line** Slope is 0. y-intercept is $(0, b)$.
$x = a$	**Vertical Line** Slope is undefined. x-intercept is $(a, 0)$.

Consider the following equation.

$$y = 3x + \frac{2}{3} \quad \text{Slope-intercept form}$$

The slope of the graph is $m = 3$ and the y-intercept is $\left(0, \frac{2}{3}\right)$.

Consider the following equation.

$$y - 3 = -2(x - 4) \quad \text{Point-slope form}$$

The slope is $m = -2$. The line passes through the point $(4, 3)$.

Consider the following equation.

$$4x + 5y = 7 \quad \text{Standard form with } A = 4, B = 5, C = 7$$

The slope is $m = -\frac{A}{B} = -\frac{4}{5}$.

The x-intercept is $\left(\frac{C}{A}, 0\right) = \left(\frac{7}{4}, 0\right)$.

The y-intercept is $\left(0, \frac{C}{B}\right) = \left(0, \frac{7}{5}\right)$.

Consider the following equations.

$$y = -6 \quad \text{Horizontal line}$$

The slope is 0. The y-intercept is $(0, -6)$.

$$x = 3 \quad \text{Vertical line}$$

The slope is undefined. The x-intercept is $(3, 0)$.

(continued)

Concepts	Examples

2.6 Graphs of Basic Functions

Basic Functions
Identity Function $f(x) = x$
Squaring Function $f(x) = x^2$
Cubing Function $f(x) = x^3$
Square Root Function $f(x) = \sqrt{x}$
Cube Root Function $f(x) = \sqrt[3]{x}$
Absolute Value Function $f(x) = |x|$
Greatest Integer Function $f(x) = [\![x]\!]$

Refer to the function boxes in **Section 2.6.** Graphs of the basic functions are also shown on the back inside cover.

2.7 Graphing Techniques

Stretching and Shrinking
If $a > 1$, then the graph of $y = af(x)$ is a **vertical stretching** of the graph of $y = f(x)$. If $0 < a < 1$, then the graph of $y = af(x)$ is a **vertical shrinking** of the graph of $y = f(x)$.

If $0 < a < 1$, then the graph of $y = f(ax)$ is a **horizontal stretching** of the graph of $y = f(x)$. If $a > 1$, then the graph of $y = f(ax)$ is a **horizontal shrinking** of the graph of $y = f(x)$.

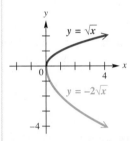

The graph of $y = -2\sqrt{x}$ is the graph of $y = \sqrt{x}$ stretched vertically by a factor of 2 and reflected across the x-axis.

Reflection Across an Axis
The graph of $y = -f(x)$ is the same as the graph of $y = f(x)$ reflected across the x-axis.

The graph of $y = f(-x)$ is the same as the graph of $y = f(x)$ reflected across the y-axis.

Symmetry
The graph of an equation is **symmetric with respect to the y-axis** if the replacement of x with $-x$ results in an equivalent equation.

The graph of an equation is **symmetric with respect to the x-axis** if the replacement of y with $-y$ results in an equivalent equation.

The graph of an equation is **symmetric with respect to the origin** if the replacement of both x with $-x$ and y with $-y$ at the same time results in an equivalent equation.

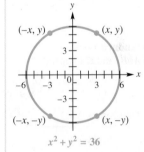

The graph of $x^2 + y^2 = 36$ is symmetric with respect to the y-axis, the x-axis, and the origin.

Translations
Let f be a function and c be a positive number.

To Graph:	Shift the Graph of $y = f(x)$ by c Units:
$y = f(x) + c$	up
$y = f(x) - c$	down
$y = f(x + c)$	left
$y = f(x - c)$	right

The graph of

$$y = (x - 1)^3 + 2$$

is the graph of $y = x^3$ translated 1 unit to the right and 2 units up.

Concepts	Examples

2.8 Function Operations and Composition

Operations on Functions
Given two functions f and g, then for all values of x for which both $f(x)$ and $g(x)$ are defined, the following operations are defined.

$$(f + g)(x) = f(x) + g(x) \quad \text{Sum}$$

$$(f - g)(x) = f(x) - g(x) \quad \text{Difference}$$

$$(fg)(x) = f(x) \cdot g(x) \quad \text{Product}$$

$$\left(\frac{f}{g}\right)(x) = \frac{f(x)}{g(x)}, \quad g(x) \neq 0 \quad \text{Quotient}$$

Let $f(x) = 2x - 4$ and $g(x) = \sqrt{x}$.

$$(f + g)(x) = 2x - 4 + \sqrt{x}$$

$$(f - g)(x) = 2x - 4 - \sqrt{x}$$ The domain is $[0, \infty)$.

$$(fg)(x) = (2x - 4)\sqrt{x}$$

$$\left(\frac{f}{g}\right)(x) = \frac{2x - 4}{\sqrt{x}}$$ The domain is $(0, \infty)$.

Difference Quotient
The line joining $P(x, f(x))$ and $Q(x + h, f(x + h))$ has slope

$$m = \frac{f(x + h) - f(x)}{h}, \quad h \neq 0.$$

Refer to **Example 4** in **Section 2.8.**

Composition of Functions
If f and g are functions, then the composite function, or composition, of g and f is defined by

$$(g \circ f)(x) = g(f(x)).$$

The domain of $g \circ f$ is the set of all x in the domain of f such that $f(x)$ is in the domain of g.

Using f and g as defined above,

$$(g \circ f)(x) = \sqrt{2x - 4}.$$

The domain is all x such that $2x - 4 \geq 0$. Solving this inequality gives the interval $[2, \infty)$.

Chapter 2 Review Exercises

Find the distance between each pair of points, and give the coordinates of the midpoint of the segment joining them.

1. $P(3, -1)$, $Q(-4, 5)$ 2. $M(-8, 2)$, $N(3, -7)$ 3. $A(-6, 3)$, $B(-6, 8)$

4. Are the points $(5, 7)$, $(3, 9)$, and $(6, 8)$ the vertices of a right triangle? If so, at what point is the right angle?

5. Determine the coordinates of B for segment AB, given that A has coordinates $(-6, 10)$ and the coordinates of its midpoint M are $(8, 2)$.

6. Use the distance formula to determine whether the points $(-2, -5)$, $(1, 7)$, and $(3, 15)$ are collinear.

Find the center-radius form of the equation for each circle described.

7. center $(-2, 3)$, radius 15   8. center $\left(\sqrt{5}, -\sqrt{7}\right)$, radius $\sqrt{3}$

9. center $(-8, 1)$, passing through $(0, 16)$ 10. center $(3, -6)$, tangent to the x-axis

Connecting Graphs with Equations Use each graph to determine the equation of the circle. Express in center-radius form.

11.

(3, 5)

12.

(−2, 3)

13.

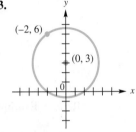

(−2, 6)

(0, 3)

14.

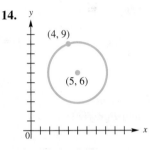

(4, 9)

(5, 6)

Find the center and radius of each circle.

15. $x^2 + y^2 - 4x + 6y + 12 = 0$

16. $x^2 + y^2 - 6x - 10y + 30 = 0$

17. $2x^2 + 2y^2 + 14x + 6y + 2 = 0$

18. $3x^2 + 3y^2 + 33x - 15y = 0$

19. Find all possible values of x so that the distance between $(x, -9)$ and $(3, -5)$ is 6.

For each graph, determine whether y is a function of x. Give the domain and range of each relation.

20.

21.

22.

23.

24.

25.

Determine whether each equation defines y as a function of x.

26. $x = \dfrac{1}{3}y^2$

27. $y = 6 - x^2$

28. $y = -\dfrac{4}{x}$

29. $y = \pm\sqrt{x - 2}$

Give the domain of each function.

30. $f(x) = -4 + |x|$

31. $f(x) = \dfrac{8 + x}{8 - x}$

32. $f(x) = \sqrt{6 - 3x}$

33. For the function graphed in **Exercise 22,** give the largest interval over which it is (a) increasing, (b) decreasing, and (c) constant.

34. Suppose the graph in **Exercise 24** is that of $y = f(x)$. What is $f(0)$?

Given $f(x) = -2x^2 + 3x - 6$, find each function value or expression.

35. $f(3)$ **36.** $f(-0.5)$ **37.** $f(k)$

Graph each equation.

38. $3x + 7y = 14$ **39.** $2x - 5y = 5$ **40.** $3y = x$

41. $2x + 5y = 20$ **42.** $x - 4y = 8$ **43.** $f(x) = x$

44. $f(x) = 3$ **45.** $x = -5$ **46.** $y + 2 = 0$

47. *Concept Check* The equation of the line that lies along the x-axis is ____.

Graph the line satisfying the given conditions.

48. through $(2, -4), m = \frac{3}{4}$ **49.** through $(0, 5), m = -\frac{2}{3}$

Find the slope for each line, provided that it has a slope.

50. through $(8, 7)$ and $\left(\frac{1}{2}, -2\right)$ **51.** through $(2, -2)$ and $(3, -4)$

52. through $(5, 6)$ and $(5, -2)$ **53.** through $(0, -7)$ and $(3, -7)$

54. $9x - 4y = 2$ **55.** $11x + 2y = 3$

56. $x - 5y = 0$ **57.** $x - 2 = 0$

58. *(Modeling) Job Market* The figure shows the number of jobs gained or lost in a recent period from September to May.

    (a) Is this the graph of a function?
    (b) In what month were the most jobs lost? the most gained?
    (c) What was the largest number of jobs lost? of jobs gained?
    (d) Do these data show an upward or a downward trend? If so, which is it?

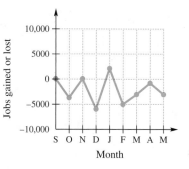

59. *(Modeling) Distance from Home* The graph depicts the distance y that a person driving a car on a straight road is from home after x hours. Interpret the graph. At what speeds did the car travel?

60. *Family Income* Family income in the United States has steadily increased for many years. In 1980 the median family income was about \$21,000 per year. In 2008 it was about \$61,500 per year. Find the average annual rate of change of median family income to the nearest dollar over that period. (*Source:* U.S. Census Bureau.)

61. *(Modeling) E-Filing Tax Returns* The percent of tax returns filed electronically in 2001 was 30.7%. In 2009, the figure was 67.2%. (*Source:* Internal Revenue Service.)

(a) Use the information given for the years 2001 and 2009, letting $x = 0$ represent 2001, $x = 8$ represent 2009, and y represent the percent of returns filed electronically, to find a linear equation that models the data. Write the equation in slope-intercept form. Interpret the slope of the graph of this equation.

(b) Use your equation from part (a) to approximate the percent of tax returns that were filed electronically in 2007.

For each line described, write the equation in (a) slope-intercept form, if possible, and (b) standard form.

62. through $(-2, 4)$ and $(1, 3)$

63. through $(3, -5)$ with slope -2

64. x-intercept $(-3, 0)$, y-intercept $(0, 5)$

65. through $(2, -1)$, parallel to $3x - y = 1$

66. through $(0, 5)$, perpendicular to $8x + 5y = 3$

67. through $(2, -10)$, perpendicular to a line with undefined slope

68. through $(3, -5)$, parallel to $y = 4$

69. through $(-7, 4)$, perpendicular to $y = 8$

Graph each function.

70. $f(x) = -|x|$

71. $f(x) = |x| - 3$

72. $f(x) = -\sqrt{x} - 2$

73. $f(x) = -(x + 1)^2 + 3$

74. $f(x) = 2\sqrt[3]{x + 1} - 2$

75. $f(x) = [\![x - 3]\!]$

76. $f(x) = \begin{cases} x^2 + 3 & \text{if } x < 2 \\ -x + 4 & \text{if } x \geq 2 \end{cases}$

77. $f(x) = \begin{cases} -4x + 2 & \text{if } x \leq 1 \\ 3x - 5 & \text{if } x > 1 \end{cases}$

78. $f(x) = \begin{cases} |x| & \text{if } x < 3 \\ 6 - x & \text{if } x \geq 3 \end{cases}$

79. *Concept Check* If x represents an integer, then what is the simplest form of the expression

$$[\![x]\!] + x?$$

Concept Check Decide whether each statement is true *or* false. *If false, tell why.*

80. The graph of a nonzero function cannot be symmetric with respect to the x-axis.

81. The graph of an even function is symmetric with respect to the y-axis.

82. The graph of an odd function is symmetric with respect to the origin.

83. If (a, b) is on the graph of an even function, so is $(a, -b)$.

84. If (a, b) is on the graph of an odd function, so is $(-a, b)$.

85. The constant function $f(x) = 0$ is both even and odd.

Decide whether each equation has a graph that is symmetric with respect to the x-axis, the y-axis, the origin, or none of these.

86. $5y^2 + 5x^2 = 30$

87. $x + y^2 = 10$

88. $y^3 = x + 4$

89. $x^2 = y^3$

90. $|y| = -x$

91. $6x + y = 4$

92. $|x| = |y|$

93. $y = 1$

94. $x^2 - y^2 = 0$

Describe how the graph of each function can be obtained from the graph of $f(x) = |x|$.

95. $g(x) = -|x|$

96. $h(x) = |x| - 2$

97. $k(x) = 2|x - 4|$

Let $f(x) = 3x - 4$. Find an equation for each reflection of the graph of $f(x)$.

98. across the *x*-axis   **99.** across the *y*-axis   **100.** across the origin

101. *Concept Check* The graph of a function *f* is shown in the figure. Sketch the graph of each function defined as follows.

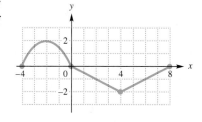

  (a) $y = f(x) + 3$

  (b) $y = f(x - 2)$

  (c) $y = f(x + 3) - 2$

  (d) $y = |f(x)|$

Let $f(x) = 3x^2 - 4$ and $g(x) = x^2 - 3x - 4$. Find each of the following.

102. $(f + g)(x)$ **103.** $(fg)(x)$ **104.** $(f - g)(4)$

105. $(f + g)(-4)$ **106.** $(f + g)(2k)$ **107.** $\left(\dfrac{f}{g}\right)(3)$

108. $\left(\dfrac{f}{g}\right)(-1)$ **109.** the domain of $(fg)(x)$ **110.** the domain of $\left(\dfrac{f}{g}\right)(x)$

111. Which of the following is *not* equivalent to $(f \circ g)(x)$ for $f(x) = \frac{1}{x}$ and $g(x) = x^2 + 1$? (*Hint:* There may be more than one.)

  A. $f(g(x))$ **B.** $\dfrac{1}{x^2 + 1}$ **C.** $\dfrac{1}{x^2}$ **D.** $(g \circ f)(x)$

For each function, find and simplify $\dfrac{f(x + h) - f(x)}{h}$, $h \neq 0$.

112. $f(x) = 2x + 9$ **113.** $f(x) = x^2 - 5x + 3$

Let $f(x) = \sqrt{x - 2}$ and $g(x) = x^2$. Find each of the following, if possible.

114. $(f \circ g)(x)$ **115.** $(g \circ f)(x)$ **116.** $(f \circ g)(-6)$

117. $(g \circ f)(3)$ **118.** $(g \circ f)(-1)$ **119.** the domain of $f \circ g$

Use the table to evaluate each expression, if possible.

120. $(f + g)(1)$

121. $(f - g)(3)$

122. $(fg)(-1)$

123. $\left(\dfrac{f}{g}\right)(0)$

124. $\left(\dfrac{f}{g}\right)(3)$

x	$f(x)$	$g(x)$
−1	3	−2
0	5	0
1	7	1
3	9	9

Tables for f and g are given. Use them to evaluate the expressions in Exercises 125 and 126.

125. $(g \circ f)(-2)$

126. $(f \circ g)(3)$

x	$f(x)$
−2	1
0	4
2	3
4	2

x	$g(x)$
1	2
2	4
3	−2
4	0

Concept Check *The graphs of two functions f and g are shown in the figures.*

127. Find $(f \circ g)(2)$.

128. Find $(g \circ f)(3)$.

Solve each problem.

129. *Relationship of Measurement Units* There are 36 in. in 1 yd, and there are 1760 yd in 1 mi. Express the number of inches in x miles by forming two functions and then considering their composition.

130. *(Modeling) Perimeter of a Rectangle* Suppose the length of a rectangle is twice its width. Let x represent the width of the rectangle. Write a formula for the perimeter P of the rectangle in terms of x alone. Then use $P(x)$ notation to describe it as a function. What type of function is this?

131. *(Modeling) Volume of a Sphere* The formula for the volume of a sphere is $V(r) = \frac{4}{3}\pi r^3$, where r represents the radius of the sphere. Construct a model function V representing the amount of volume gained when the radius r (in inches) of a sphere is increased by 3 in.

132. *(Modeling) Dimensions of a Cylinder* A cylindrical can makes the most efficient use of materials when its height is the same as the diameter of its top.

(a) Express the volume V of such a can as a function of the diameter d of its top.

(b) Express the surface area S of such a can as a function of the diameter d of its top. (*Hint:* The curved side is made from a rectangle whose length is the circumference of the top of the can.)

Chapter 2 Test

1. Match the set described in Column I with the correct interval notation from Column II. Choices in Column II may be used once, more than once, or not at all.

I	II
(a) Domain of $f(x) = \sqrt{x+3}$	**A.** $[-3, \infty)$
(b) Range of $f(x) = \sqrt{x-3}$	**B.** $[3, \infty)$
(c) Domain of $f(x) = x^2 - 3$	**C.** $(-\infty, \infty)$
(d) Range of $f(x) = x^2 + 3$	
(e) Domain of $f(x) = \sqrt[3]{x-3}$	**D.** $[0, \infty)$
(f) Range of $f(x) = \sqrt[3]{x+3}$	**E.** $(-\infty, 3)$
(g) Domain of $f(x) = |x| - 3$	**F.** $(-\infty, 3]$
(h) Range of $f(x) = |x+3|$	**G.** $(3, \infty)$
(i) Domain of $x = y^2$	**H.** $(-\infty, 0]$
(j) Range of $x = y^2$	

The graph shows the line that passes through the points $(-2, 1)$ *and* $(3, 4)$. *Refer to it to answer Exercises 2–6.*

2. What is the slope of the line?

3. What is the distance between the two points shown?

4. What are the coordinates of the midpoint of the segment joining the two points?

5. Find the standard form of the equation of the line.

6. Write the linear function $f(x) = ax + b$ that has this line as its graph.

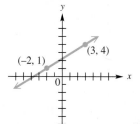

7. *Connecting Graphs with Equations* Use each graph to determine the equation of the circle. Express in center-radius form.

   (a)   **(b)**

8. Graph the circle with equation $x^2 + y^2 + 4x - 10y + 13 = 0$.

9. In each case, determine whether y is a function of x. Give the domain and range. If it is a function, give the intervals where it is increasing, decreasing, or constant.

   (a)   **(b)**

10. Suppose point A has coordinates $(5, -3)$.

    (a) What is the equation of the vertical line through A?

    (b) What is the equation of the horizontal line through A?

11. Find the slope-intercept form of the equation of the line passing through $(2, 3)$ and

    (a) parallel to the graph of $y = -3x + 2$;

    (b) perpendicular to the graph of $y = -3x + 2$.

12. Consider the graph of the function shown here. Give the interval(s) over which the function is

    (a) increasing,

    (b) decreasing,

    (c) constant,

    (d) continuous.

    (e) What is the domain of this function?

    (f) What is the range of this function?

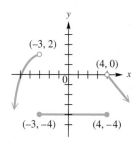

Graph each function.

13. $y = |x - 2| - 1$

14. $f(x) = [\![x + 1]\!]$

15. $f(x) = \begin{cases} 3 & \text{if } x < -2 \\ 2 - \frac{1}{2}x & \text{if } x \geq -2 \end{cases}$

16. The graph of $y = f(x)$ is shown here. Sketch the graph of each of the following. Use ordered pairs to indicate three points on the graph.

(a) $y = f(x) + 2$ (b) $y = f(x + 2)$

(c) $y = -f(x)$ (d) $y = f(-x)$

(e) $y = 2f(x)$

17. Describe how the graph of $y = -2\sqrt{x + 2} - 3$ can be obtained from the graph of $y = \sqrt{x}$.

18. Determine whether the graph of $3x^2 - 2y^2 = 3$ is symmetric with respect to

(a) the x-axis,   (b) the y-axis,   and   (c) the origin.

In Exercises 19 and 20, given $f(x) = 2x^2 - 3x + 2$ and $g(x) = -2x + 1$, find each of the following. Simplify the expressions when possible.

19. (a) $(f - g)(x)$ **(b)** $\left(\dfrac{f}{g}\right)(x)$

     (c) the domain of $\dfrac{f}{g}$ **(d)** $\dfrac{f(x + h) - f(x)}{h}$ $(h \neq 0)$

20. (a) $(f + g)(1)$ **(b)** $(fg)(2)$ **(c)** $(f \circ g)(0)$

In Exercises 21 and 22, let $f(x) = \sqrt{x + 1}$ and $g(x) = 2x - 7$.

21. Find $(f \circ g)(x)$ and give its domain.

22. Find $(g \circ f)(x)$ and give its domain.

23. *(Modeling) Cell Phone Charges* A cell phone service provides communication between two cities. If x represents the number of minutes for the call, where $x > 0$, then the function

$$f(x) = 0.75[\![x]\!] + 1.50$$

gives the total cost of the call in dollars. Find the cost of a 7.5-min call.

24. *(Modeling) Cost, Revenue, and Profit Analysis* Dotty McGinnis starts up a small business manufacturing bobble-head figures of famous soccer players. Her initial cost is $3300. Each figure costs $4.50 to manufacture.

(a) Write a cost function C, where x represents the number of figures manufactured.

(b) Find the revenue function R if each figure in part (a) sells for $10.50.

(c) Give the profit function P.

(d) How many figures must be produced and sold for Dotty to earn a profit?

3 Polynomial and Rational Functions

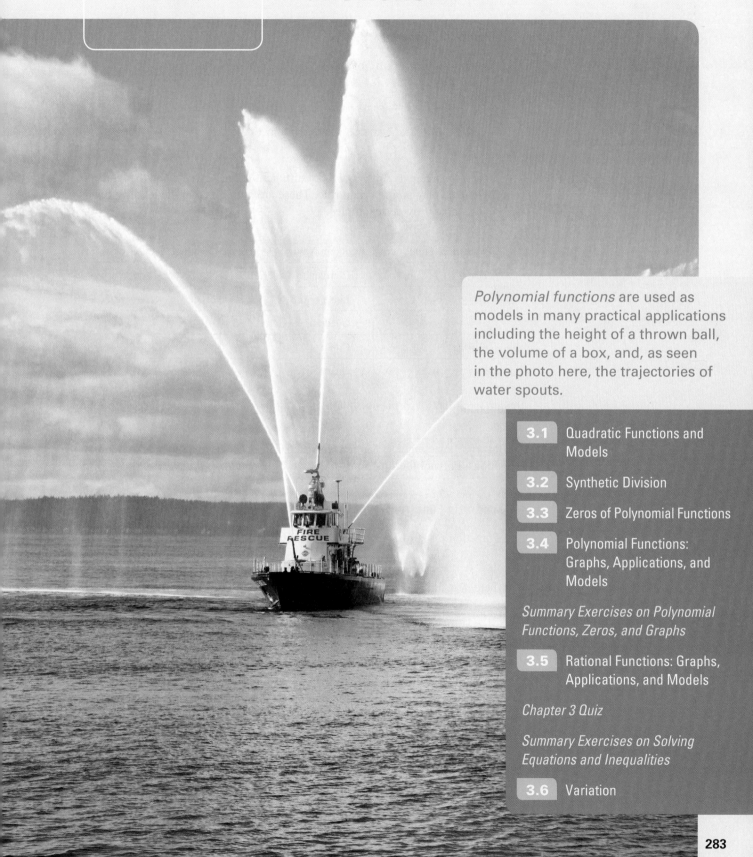

Polynomial functions are used as models in many practical applications including the height of a thrown ball, the volume of a box, and, as seen in the photo here, the trajectories of water spouts.

3.1 Quadratic Functions and Models

3.2 Synthetic Division

3.3 Zeros of Polynomial Functions

3.4 Polynomial Functions: Graphs, Applications, and Models

Summary Exercises on Polynomial Functions, Zeros, and Graphs

3.5 Rational Functions: Graphs, Applications, and Models

Chapter 3 Quiz

Summary Exercises on Solving Equations and Inequalities

3.6 Variation

3.1 Quadratic Functions and Models

- Quadratic Functions
- Graphing Techniques
- Completing the Square
- The Vertex Formula
- Quadratic Models

LOOKING AHEAD TO CALCULUS

In calculus, polynomial functions are used to approximate more complicated functions. For example, the trigonometric function $\sin x$ is approximated by the polynomial

$$x - \frac{x^3}{6} + \frac{x^5}{120} - \frac{x^7}{5040}.$$

A *polynomial function* is defined as follows.

Polynomial Function

A **polynomial function** f of degree n, where n is a nonnegative integer, is given by

$$f(x) = a_n x^n + a_{n-1} x^{n-1} + \cdots + a_1 x + a_0,$$

where $a_n, a_{n-1}, \ldots, a_1,$ and a_0 are real numbers, with $a_n \neq 0$.

When we are analyzing a polynomial function, the degree n and the **leading coefficient** a_n play an important role. These are both given in the **leading term** $a_n x^n$. The table provides examples.

Polynomial Function	Function Name	Degree n	Leading Coefficient a_n
$f(x) = 2$	Constant	0	2
$f(x) = 5x - 1$	Linear	1	5
$f(x) = 4x^2 - x + 1$	Quadratic	2	4
$f(x) = 2x^3 - \frac{1}{2}x + 5$	Cubic	3	2
$f(x) = x^4 + \sqrt{2}x^3 - 3x^2$	Quartic	4	1

The function $f(x) = 0$ is the **zero polynomial** and has no degree.

Quadratic Functions Earlier we discussed constant and linear polynomial functions. Polynomial functions of degree 2 are *quadratic functions*.

Quadratic Function

A function f is a **quadratic function** if

$$f(x) = ax^2 + bx + c,$$

where a, b, and c are real numbers, with $a \neq 0$.

The simplest quadratic function is

$$f(x) = x^2, \quad \text{Squaring function (Section 2.6)}$$

as shown in **Figure 1.** This graph is a **parabola.** Every quadratic function defined over the real numbers has a graph that is a parabola.

The domain of $f(x) = x^2$ is $(-\infty, \infty)$, and the range is $[0, \infty)$. The lowest point on the graph occurs at the origin $(0, 0)$. Thus, the function decreases on the interval $(-\infty, 0)$ and increases on the interval $(0, \infty)$. (Remember that these intervals indicate x-values.)

x	$f(x)$
-2	4
-1	1
0	0
1	1
2	4

Figure 1

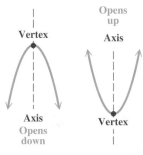

Figure 2

Parabolas are symmetric with respect to a line (the y-axis in **Figure 1**). This line is the **axis of symmetry,** or **axis,** of the parabola. The point where the axis intersects the parabola is the **vertex** of the parabola. As **Figure 2** shows, the vertex of a parabola that opens down is the highest point of the graph, and the vertex of a parabola that opens up is the lowest point of the graph.

Graphing Techniques The graphing techniques of **Section 2.7** may be applied to the graph of $f(x) = x^2$ to give the graph of *any* quadratic function. Compared to the basic graph of $f(x) = x^2$, the graph of $F(x) = a(x - h)^2 + k$ has the following characteristics.

$$F(x) = a(x - h)^2 + k$$

- Opens up if $a > 0$
- Opens down if $a < 0$
- Vertically stretched (narrower) if $|a| > 1$
- Vertically shrunk (wider) if $0 < |a| < 1$

Horizontal shift:
- h units right if $h > 0$
- $|h|$ units left if $h < 0$

Vertical shift:
- k units up if $k > 0$
- $|k|$ units down if $k < 0$

EXAMPLE 1 **Graphing Quadratic Functions**

Graph each function. Give the domain and range.

(a) $f(x) = x^2 - 4x - 2$ (by plotting points as in **Section 2.1**)

(b) $g(x) = -\frac{1}{2}x^2$ (and compare to $y = x^2$ and $y = \frac{1}{2}x^2$)

(c) $F(x) = -\frac{1}{2}(x - 4)^2 + 3$ (and compare to the graph in part (b))

SOLUTION

(a) See the table with **Figure 3**. The domain of $f(x) = x^2 - 4x - 2$ is $(-\infty, \infty)$, the range is $[-6, \infty)$, the vertex is $(2, -6)$, and the axis has equation $x = 2$. **Figure 4** shows how a graphing calculator displays this graph.

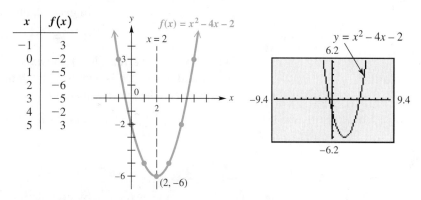

Figure 3   **Figure 4**

(b) Think of $g(x) = -\frac{1}{2}x^2$ as $g(x) = -\left(\frac{1}{2}x^2\right)$. The graph of $y = \frac{1}{2}x^2$ is a wider version of the graph of $y = x^2$, and the graph of $g(x) = -\left(\frac{1}{2}x^2\right)$ is a reflection of the graph of $y = \frac{1}{2}x^2$ across the x-axis. See **Figure 5** on the next page. The vertex is $(0, 0)$, and the axis of the parabola is the line $x = 0$ (the y-axis). The domain is $(-\infty, \infty)$, and the range is $(-\infty, 0]$.

Calculator graphs are shown in **Figure 6.** These may be used to verify the accuracy of the graphs drawn in **Figure 5.**

Figure 5

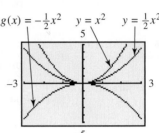

Figure 6

(c) Notice that $F(x) = -\frac{1}{2}(x - 4)^2 + 3$ is related to $g(x) = -\frac{1}{2}x^2$ from part (b). The graph of $F(x)$ is the graph of $g(x)$ translated 4 units to the right and 3 units up. See **Figure 7.** The vertex is $(4, 3)$, which is also shown in the calculator graph in **Figure 8,** and the axis of the parabola is the line $x = 4$. The domain is $(-\infty, \infty)$, and the range is $(-\infty, 3]$.

Figure 7

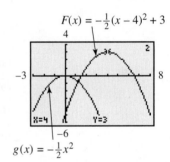

Figure 8

✔ *Now Try Exercises 9 and 11.*

Completing the Square In general, the graph of the quadratic function

$$f(x) = a(x - h)^2 + k$$

is a parabola with **vertex (h, k)** and **axis $x = h$.** The parabola opens up if a is positive and down if a is negative. With these facts in mind, we *complete the square* to graph the general quadratic function

$$f(x) = ax^2 + bx + c.$$

EXAMPLE 2 **Graphing a Parabola by Completing the Square ($a = 1$)**

Graph $f(x) = x^2 - 6x + 7$ by completing the square and locating the vertex. Find the largest intervals over which the function is increasing or decreasing.

SOLUTION We express $x^2 - 6x + 7$ in the form $(x - h)^2 + k$ by completing the square. In preparation for this, we first write

$$f(x) = (x^2 - 6x \qquad) + 7. \qquad \text{Prepare to complete the square. (Section 1.4)}$$

We must add a number inside the parentheses to get a perfect square trinomial. Find this number by taking half the coefficient of x and squaring the result.

$f(x) = x^2 - 6x + 7$
$= (x-3)^2 - 2$

This screen shows that the vertex of the graph in **Figure 9** is $(3, -2)$. Because it is the *lowest* point on the graph, we direct the calculator to find the *minimum*.

$$\left[\frac{1}{2}(-6)\right]^2 = (-3)^2 = 9 \qquad \text{Take half the coefficient of } x. \text{ Square the result.}$$

$f(x) = (x^2 - 6x + 9 - 9) + 7$ Add and subtract 9.   This is the same as adding 0.

$f(x) = (x^2 - 6x + 9) - 9 + 7$ Regroup terms.

$f(x) = (x-3)^2 - 2$ Factor and simplify. **(Section R.4)**

The vertex of the parabola is $(3, -2)$, and the axis is the line $x = 3$. Now we find additional ordered pairs that satisfy the equation, as shown in the table, and plot and connect these points to obtain the graph in **Figure 9.**

x	y	
0	7	← y-intercept
1	2	
5	2	← Find by using symmetry about the axis.
6	7	

$f(x) = x^2 - 6x + 7$
$f(x) = (x-3)^2 - 2$

Figure 9

The domain of this function is $(-\infty, \infty)$, and the range is $[-2, \infty)$. Since the lowest point on the graph is the vertex $(3, -2)$, the function is decreasing on $(-\infty, 3)$ and increasing on $(3, \infty)$.

☑ *Now Try Exercise 21.*

NOTE In **Example 2** we added and subtracted 9 *on the same side* of the equation to complete the square. This differs from adding the same number to *each side of the equation,* as when we completed the square in **Section 1.4.** Since we want $f(x)$ (or y) alone on one side of the equation, we adjusted that step in the process of completing the square.

EXAMPLE 3   **Graphing a Parabola by Completing the Square ($a \neq 1$)**

Graph $f(x) = -3x^2 - 2x + 1$ by completing the square and locating the vertex. Identify the intercepts of the graph.

SOLUTION   To complete the square, the coefficient of x^2 must be 1.

$f(x) = -3\left(x^2 + \frac{2}{3}x \qquad\right) + 1$ Factor -3 from the first two terms.

$f(x) = -3\left(x^2 + \frac{2}{3}x + \frac{1}{9} - \frac{1}{9}\right) + 1$ $\left[\frac{1}{2}\left(\frac{2}{3}\right)\right]^2 = \left(\frac{1}{3}\right)^2 = \frac{1}{9}$, so add and subtract $\frac{1}{9}$.

$f(x) = -3\left(x^2 + \frac{2}{3}x + \frac{1}{9}\right) - 3\left(-\frac{1}{9}\right) + 1$ Distributive property **(Section R.2)**   Be careful here.

$f(x) = -3\left(x + \frac{1}{3}\right)^2 + \frac{4}{3}$ Factor and simplify.

The vertex is $\left(-\frac{1}{3}, \frac{4}{3}\right)$. The intercepts are good additional points to find. The y-intercept is found by evaluating $f(0)$.

$f(0) = -3(0)^2 - 2(0) + 1 = 1$ ← The y-intercept is $(0, 1)$.

The *x*-intercepts are found by setting $f(x)$ equal to 0 and solving for *x*.

$$0 = -3x^2 - 2x + 1 \qquad \text{Set } f(x) = 0.$$

$$0 = 3x^2 + 2x - 1 \qquad \text{Multiply by } -1.$$

$$0 = (3x - 1)(x + 1) \qquad \text{Factor.}$$

$$x = \frac{1}{3} \quad \text{or} \quad x = -1 \qquad \begin{array}{l}\text{Zero-factor property}\\\text{(Section 1.4)}\end{array}$$

Therefore, the *x*-intercepts are $\left(\frac{1}{3}, 0\right)$ and $(-1, 0)$. The graph is shown in **Figure 10**.

$f(x) = -3x^2 - 2x + 1$
$= -3\left(x + \frac{1}{3}\right)^2 + \frac{4}{3}$

This screen gives the vertex of the graph in **Figure 10** as $(-0.\overline{3}, 1.\overline{3}) = \left(-\frac{1}{3}, \frac{4}{3}\right)$. We want the highest point on the graph, so we direct the calculator to find the *maximum*.

Figure 10

✔ *Now Try Exercise 23.*

NOTE It is possible to reverse the process of **Example 3** and write the quadratic function from its graph if the vertex and any other point on the graph are known. Since quadratic functions take the form

$$f(x) = a(x - h)^2 + k,$$

substitute the *x*- and *y*-values of the vertex, $\left(-\frac{1}{3}, \frac{4}{3}\right)$, for *h* and *k*, respectively.

$$f(x) = a\left[x - \left(-\frac{1}{3}\right)\right]^2 + \frac{4}{3} \qquad \text{Let } h = -\frac{1}{3} \text{ and } k = \frac{4}{3}.$$

$$f(x) = a\left(x + \frac{1}{3}\right)^2 + \frac{4}{3} \qquad \text{Simplify.}$$

Now find the value of *a* by substituting the *x*- and *y*-coordinates of any other point on the graph, say $(0, 1)$, into this equation and solving for *a*.

$$1 = a\left(0 + \frac{1}{3}\right)^2 + \frac{4}{3} \qquad \text{Let } x = 0 \text{ and } y = 1.$$

$$1 = a\left(\frac{1}{9}\right) + \frac{4}{3} \qquad \text{Square.}$$

$$-\frac{1}{3} = \frac{1}{9}a \qquad \text{Subtract } \frac{4}{3}.$$

$$a = -3 \qquad \text{Multiply by 9. Interchange sides.}$$

Verify in **Example 3** that the vertex form of the function is

$$f(x) = -3\left(x + \frac{1}{3}\right)^2 + \frac{4}{3}.$$

In the Exercise set, problems of this type are labeled *Connecting Graphs with Equations*.

LOOKING AHEAD TO CALCULUS

An important concept in calculus is the **definite integral.** If the graph of f lies above the x-axis, the symbol

$$\int_a^b f(x)\,dx$$

represents the area of the region above the x-axis and below the graph of f from $x = a$ to $x = b$. For example, in **Figure 10** with

$$f(x) = -3x^2 - 2x + 1,$$

$a = -1$, and $b = \frac{1}{3}$, calculus provides the tools for determining that the area enclosed by the parabola and the x-axis is $\frac{32}{27}$ (square units).

The Vertex Formula   We can generalize the earlier work to obtain a formula for the vertex of a parabola.

$$f(x) = ax^2 + bx + c \qquad \text{General quadratic form}$$

$$= a\left(x^2 + \frac{b}{a}x \quad\right) + c \qquad \text{Factor } a \text{ from the first two terms.}$$

$$= a\left(x^2 + \frac{b}{a}x + \frac{b^2}{4a^2}\right) + c - a\left(\frac{b^2}{4a^2}\right) \qquad \text{Add } \left[\frac{1}{2}\left(\frac{b}{a}\right)\right]^2 = \frac{b^2}{4a^2} \text{ inside the parentheses. Subtract } a\left(\frac{b^2}{4a^2}\right) \text{ outside the parentheses.}$$

$$= a\left(x + \frac{b}{2a}\right)^2 + c - \frac{b^2}{4a} \qquad \text{Factor and simplify.}$$

$$f(x) = a\left[x - \left(-\frac{b}{2a}\right)\right]^2 + \frac{4ac - b^2}{4a} \qquad \begin{array}{l}\text{Vertex form of}\\ f(x) = a(x - h)^2 + k\end{array}$$

$$\underbrace{\phantom{x - \left(-\frac{b}{2a}\right)}}_{h} \qquad \underbrace{\phantom{\frac{4ac-b^2}{4a}}}_{k}$$

Thus, the vertex (h, k) can be expressed in terms of a, b, and c. **It is not necessary to memorize the expression for k, since it is equal to $f(h) = f\left(-\frac{b}{2a}\right)$.**

The following statements summarize this discussion.

Graph of a Quadratic Function

The quadratic function $f(x) = ax^2 + bx + c$ can be written as

$$y = f(x) = a(x - h)^2 + k, \quad a \neq 0,$$

where $\qquad h = -\dfrac{b}{2a} \quad$ and $\quad k = f(h)$.   Vertex formula

The graph of f has the following characteristics.

1. It is a parabola with vertex (h, k) and the vertical line $x = h$ as axis.
2. It opens up if $a > 0$ and down if $a < 0$.
3. It is wider than the graph of $y = x^2$ if $|a| < 1$ and narrower if $|a| > 1$.
4. The y-intercept is $(0, f(0)) = (0, c)$.
5. The x-intercepts are found by solving the equation $ax^2 + bx + c = 0$.
   - If $b^2 - 4ac > 0$, the x-intercepts are $\left(\dfrac{-b \pm \sqrt{b^2 - 4ac}}{2a}, 0\right)$.
   - If $b^2 - 4ac = 0$, the x-intercept is $\left(-\dfrac{b}{2a}, 0\right)$.
   - If $b^2 - 4ac < 0$, there are no x-intercepts.

EXAMPLE 4   **Using the Vertex Formula**

Find the axis and vertex of the parabola having equation $f(x) = 2x^2 + 4x + 5$.

SOLUTION   The axis of the parabola is the vertical line

$$x = h = -\frac{b}{2a} = -\frac{4}{2(2)} = -1. \qquad \begin{array}{l}\text{Use the vertex formula.}\\ \text{Here } a = 2 \text{ and } b = 4.\end{array}$$

The vertex is $(-1, f(-1))$. Since

$$f(-1) = 2(-1)^2 + 4(-1) + 5 = 3,$$

the vertex is $(-1, 3)$.

☑ *Now Try Exercise 21(a).*

Quadratic Models Since the vertex of a vertical parabola is the highest or lowest point on the graph, equations of the form

$$y = ax^2 + bx + c$$

are important in certain problems where we must find the maximum or minimum value of some quantity.

- When $a < 0$, the y-coordinate of the vertex gives the maximum value of y.

- When $a > 0$, the y-coordinate of the vertex gives the minimum value of y.

The x-coordinate of the vertex tells *where* the maximum or minimum value occurs.

If air resistance is neglected, the height s (in feet) of an object projected directly upward from an initial height s_0 feet with initial velocity v_0 feet per second is

$$s(t) = -16t^2 + v_0t + s_0, \quad \text{(Section 1.5)}$$

where t is the number of seconds after the object is projected. The coefficient of t^2 (that is, -16) is a constant based on the gravitational force of Earth. This constant is different on other surfaces, such as the moon and the other planets.

EXAMPLE 5 **Solving a Problem Involving Projectile Motion**

A ball is projected directly upward from an initial height of 100 ft with an initial velocity of 80 ft per sec.

(a) Give the function that describes the height of the ball in terms of time t.

(b) After how many seconds does the ball reach its maximum height? What is this maximum height?

(c) For what interval of time is the height of the ball greater than 160 ft?

(d) After how many seconds will the ball hit the ground?

ALGEBRAIC SOLUTION

(a) Use the projectile height function with $v_0 = 80$ and $s_0 = 100$.

$$s(t) = -16t^2 + v_0t + s_0$$

$$s(t) = -16t^2 + 80t + 100$$

(b) Since the coefficient of t^2 is -16, the graph of the projectile function is a parabola that opens downward. Find the coordinates of the vertex to determine the maximum height and when it occurs. Let $a = -16$ and $b = 80$ in the vertex formula.

$$t = -\frac{b}{2a} = -\frac{80}{2(-16)} = 2.5$$

$$s(t) = -16t^2 + 80t + 100$$

$$s(2.5) = -16(2.5)^2 + 80(2.5) + 100$$

$$s(2.5) = 200$$

Therefore, after 2.5 sec the ball reaches its maximum height of 200 ft.

GRAPHING CALCULATOR SOLUTION

(a) Use the projectile height function as in the algebraic solution.

$$s(t) = -16t^2 + 80t + 100$$

(b) Using the capabilities of the calculator, we see in **Figure 11** that the vertex coordinates are indeed $(2.5, 200)$.

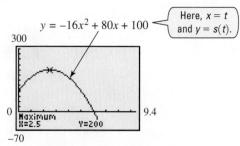

Figure 11

Do not misinterpret the graph in **Figure 11**. *It does not show the path followed by the ball. It defines height as a function of time.*

(c) We must solve the quadratic *inequality*

$$-16t^2 + 80t + 100 > 160.$$

$$-16t^2 + 80t - 60 > 0 \qquad \text{Subtract 160.}$$

$$4t^2 - 20t + 15 < 0 \qquad \text{(Section 1.7)}$$
   Divide by -4; reverse the inequality symbol.

Use the quadratic formula to find the solutions of $4t^2 - 20t + 15 = 0$.

$$t = \frac{-(-20) \pm \sqrt{(-20)^2 - 4(4)(15)}}{2(4)}$$
   (Section 1.4)

$$t = \frac{5 - \sqrt{10}}{2} \approx 0.92 \quad \text{or} \quad t = \frac{5 + \sqrt{10}}{2} \approx 4.08$$

These numbers divide the number line into three intervals: $(-\infty, 0.92)$, $(0.92, 4.08)$, and $(4.08, \infty)$. Using a test value from each interval shows that $(0.92, 4.08)$ satisfies the *inequality*. The ball is more than 160 ft above the ground between 0.92 sec and 4.08 sec.

(d) The height is 0 when the ball hits the ground. We use the quadratic formula to find the *positive* solution of

$$-16t^2 + 80t + 100 = 0.$$

Here, $a = -16$, $b = 80$, and $c = 100$.

$$t = \frac{-80 \pm \sqrt{80^2 - 4(-16)(100)}}{2(-16)}$$

$$t \approx \underset{\text{Reject}}{\cancel{-1.04}} \quad \text{or} \quad t \approx 6.04$$

The ball hits the ground after about 6.04 sec.

(c) If we graph

$$y_1 = -16x^2 + 80x + 100 \quad \text{and} \quad y_2 = 160,$$

as shown in **Figures 12 and 13,** and locate the two points of intersection, we find that the x-coordinates for these points are approximately 0.92 and 4.08. Therefore, between 0.92 sec and 4.08 sec, y_1 is greater than y_2, and the ball is more than 160 ft above the ground.

Figure 12   **Figure 13**

(d) **Figure 14** shows that the positive x-intercept of the graph of $y = -16x^2 + 80x + 100$ is approximately $(6.04, 0)$, which means that the ball hits the ground after about 6.04 sec.

Figure 14

☑ *Now Try Exercise 53.*

▱▱ **EXAMPLE 6**   **Modeling the Number of Hospital Outpatient Visits**

The number of hospital outpatient visits (in millions) for selected years is shown in the table.

Year	Visits	Year	Visits
95	483.2	102	640.5
96	505.5	103	648.6
97	520.6	104	662.1
98	545.5	105	673.7
99	573.5	106	690.4
100	592.7	107	693.5
101	612.0	108	710.0

Source: American Hospital Association.

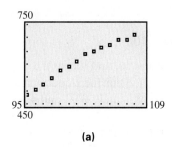

(a)

$f(x) = -0.6861x^2 + 157.0x - 8249$

(b)

(c)

Figure 15

In the table on the preceding page, 95 represents 1995, 100 represents 2000, and so on, and the number of outpatient visits is given in millions.

(a) Prepare a scatter diagram, and determine a quadratic model for these data.

(b) Use the model from part (a) to predict the number of visits in 2012.

SOLUTION

(a) In **Section 2.5** we used linear regression to determine linear equations that modeled data. With a graphing calculator, we can use **quadratic regression** to find quadratic equations that model data.

The scatter diagram in **Figure 15(a)** suggests that a quadratic function with a negative value of a (so the graph opens down) would be a reasonable model for the data. Using quadratic regression, the quadratic function

$$f(x) = -0.6861x^2 + 157.0x - 8249$$

approximates the data well. See **Figure 15(b).** The quadratic regression values of a, b, and c are displayed in **Figure 15(c).**

(b) The year 2012 corresponds to $x = 112$. The model predicts that there will be 729 million visits in 2012.

$$f(x) = -0.6861x^2 + 157.0x - 8249$$
$$f(112) = -0.6861(112)^2 + 157.0(112) - 8249$$
$$\approx 729 \text{ million}$$

☑ *Now Try Exercise 65.*

3.1 Exercises

*In Exercises 1–4, you are given an equation and the graph of a quadratic function. Do each of the following. **See Examples 1–4.***

(a) *Give the domain and range.*   **(b)** *Give the coordinates of the vertex.*
(c) *Give the equation of the axis.*   **(d)** *Find the y-intercept.*
(e) *Find the x-intercepts.*

1.

$f(x) = (x + 3)^2 - 4$

2.

$f(x) = (x - 5)^2 - 4$

3.

$f(x) = -2(x + 3)^2 + 2$

4.

$f(x) = -3(x - 2)^2 + 1$

Concept Check *Calculator graphs of the functions in Exercises 5–8 are shown in Figures A–D. Match each function with its graph without actually entering it into your calculator. Then, after you have completed the exercises, check your answers with your calculator. Use the standard viewing window.*

5. $f(x) = (x - 4)^2 - 3$

6. $f(x) = -(x - 4)^2 + 3$

7. $f(x) = (x + 4)^2 - 3$

8. $f(x) = -(x + 4)^2 + 3$

A.

B.

C.

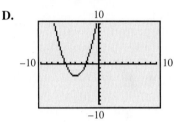

D.

9. Graph the following on the same coordinate system.

(a) $y = x^2$ (b) $y = 3x^2$ (c) $y = \dfrac{1}{3}x^2$

(d) How does the coefficient of x^2 affect the shape of the graph?

10. Graph the following on the same coordinate system.

(a) $y = x^2$ (b) $y = x^2 - 2$ (c) $y = x^2 + 2$

(d) How do the graphs in parts (b) and (c) differ from the graph of $y = x^2$?

11. Graph the following on the same coordinate system.

(a) $y = (x - 2)^2$ (b) $y = (x + 1)^2$ (c) $y = (x + 3)^2$

(d) How do these graphs differ from the graph of $y = x^2$?

12. *Concept Check* Match each equation with the description of the parabola that is its graph.

(a) $y = (x + 4)^2 + 2$ **A.** vertex $(-2, 4)$, opens up

(b) $y = (x + 2)^2 + 4$ **B.** vertex $(-2, 4)$, opens down

(c) $y = -(x + 4)^2 + 2$ **C.** vertex $(-4, 2)$, opens up

(d) $y = -(x + 2)^2 + 4$ **D.** vertex $(-4, 2)$, opens down

Graph each quadratic function. Give the (a) vertex, (b) axis, (c) domain, and (d) range. Then determine (e) the largest interval of the domain for which the function is increasing and (f) the largest interval for which the function is decreasing. **See Examples 1–4.**

13. $f(x) = (x - 2)^2$

14. $f(x) = (x + 4)^2$

15. $f(x) = (x + 3)^2 - 4$

16. $f(x) = (x - 5)^2 - 4$

17. $f(x) = -\dfrac{1}{2}(x + 1)^2 - 3$

18. $f(x) = -3(x - 2)^2 + 1$

19. $f(x) = x^2 - 2x + 3$

20. $f(x) = x^2 + 6x + 5$

21. $f(x) = x^2 - 10x + 21$

22. $f(x) = 2x^2 - 4x + 5$

23. $f(x) = -2x^2 - 12x - 16$

24. $f(x) = -3x^2 + 24x - 46$

25. $f(x) = -\dfrac{1}{2}x^2 - 3x - \dfrac{1}{2}$

26. $f(x) = \dfrac{2}{3}x^2 - \dfrac{8}{3}x + \dfrac{5}{3}$

Concept Check *The figure shows the graph of a quadratic function* $y = f(x)$. *Use it to work Exercises 27–30.*

27. What is the minimum value of $f(x)$?

28. For what value of x is $f(x)$ as small as possible?

29. How many real solutions are there to the equation $f(x) = 1$?

30. How many real solutions are there to the equation $f(x) = 4$?

Concept Check *Several possible graphs of the quadratic function*

$$f(x) = ax^2 + bx + c$$

*are shown below. For the restrictions on a, b, and c given in Exercises 31–36, select the corresponding graph from choices A–F. (Hint: Use the discriminant. See **Section 1.4** if necessary.)*

31. $a < 0;\ b^2 - 4ac = 0$ **32.** $a > 0;\ b^2 - 4ac < 0$

33. $a < 0;\ b^2 - 4ac < 0$ **34.** $a < 0;\ b^2 - 4ac > 0$

35. $a > 0;\ b^2 - 4ac > 0$ **36.** $a > 0;\ b^2 - 4ac = 0$

A.   **B.**   **C.**

D.   **E.**   **F.**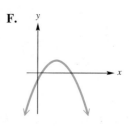

Connecting Graphs with Equations *In Exercises 37–40, find a quadratic function f whose graph matches the one in the figure. (Hint: See the Note following **Example 3**.)*

37.

38.

39.

40.

(Modeling) In each scatter diagram, tell whether a linear or a quadratic model is appropriate for the data. If linear, tell whether the slope should be positive or negative. If quadratic, tell whether the leading coefficient of x^2 should be positive or negative.

41. number of shopping centers as a function of time

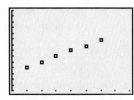

42. growth in science centers/museums as a function of time

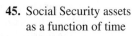

43. value of U.S. salmon catch as a function of time

44. height of an object projected upward as a function of time

45. Social Security assets as a function of time

46. newborns with AIDS as a function of time

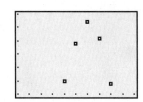

Solve each problem.

47. *Sum and Product of Two Numbers* Find two numbers whose sum is 20 and whose product is the maximum possible value. (*Hint: Let x be one number. Then* $20 - x$ *is the other number. Form a quadratic function by multiplying them, and then find the maximum value of the function.*)

48. *Sum and Product of Two Numbers* Find two numbers whose sum is 32 and whose product is the maximum possible value.

49. *Minimum Cost* Brigette Cole has a taco stand. She has found that her daily costs are approximated by

$$C(x) = x^2 - 40x + 610,$$

where $C(x)$ is the cost, in dollars, to sell x units of tacos. Find the number of units of tacos she should sell to minimize her costs. What is the minimum cost?

50. *Maximum Revenue* The revenue of a charter bus company depends on the number of unsold seats. If the revenue in dollars, $R(x)$, is given by

$$R(x) = -x^2 + 50x + 5000,$$

where x is the number of unsold seats, find the number of unsold seats that produce maximum revenue. What is the maximum revenue?

51. *Maximum Number of Mosquitos* The number of mosquitos, $M(x)$, in millions, in a certain area of Florida depends on the June rainfall, x, in inches. The function

$$M(x) = 10x - x^2$$

models this phenomenon. Find the amount of rainfall that will maximize the number of mosquitos. What is the maximum number of mosquitos?

52. *Height of an Object* If an object is projected upward from ground level with an initial velocity of 32 ft per sec, then its height in feet after t seconds is given by

$$s(t) = -16t^2 + 32t.$$

Find the number of seconds it will take to reach its maximum height. What is this maximum height?

(Modeling) Solve each problem. ***See Example 5.***

53. *Height of a Toy Rocket* A toy rocket (not internally powered) is launched straight up from the top of a building 50 ft tall at an initial velocity of 200 ft per sec.

    (a) Give the function that describes the height of the rocket in terms of time t.

    (b) Determine the time at which the rocket reaches its maximum height and the maximum height in feet.

    (c) For what time interval will the rocket be more than 300 ft above ground level?

    (d) After how many seconds will it hit the ground?

54. *Height of a Projected Rock* A rock is projected directly upward from ground level with an initial velocity of 90 ft per sec.

    (a) Give the function that describes the height of the rock in terms of time t.

    (b) Determine the time at which the rock reaches its maximum height and the maximum height in feet.

    (c) For what time interval will the rock be more than 120 ft above ground level?

    (d) After how many seconds will it return to the ground?

55. *Area of a Parking Lot* One campus of Houston Community College has plans to construct a rectangular parking lot on land bordered on one side by a highway. There are 640 ft of fencing available to fence the other three sides. Let x represent the length of each of the two parallel sides of fencing.

    (a) Express the length of the remaining side to be fenced in terms of x.

    (b) What are the restrictions on x?

    (c) Determine a function $\mathscr{A}$ that represents the area of the parking lot in terms of x.

    (d) Determine the values of x that will give an area between 30,000 and 40,000 ft^2.

    (e) What dimensions will give a maximum area, and what will this area be?

56. *Area of a Rectangular Region* A farmer wishes to enclose a rectangular region bordering a river with fencing, as shown in the diagram. Suppose that x represents the length of each of the three parallel pieces of fencing. She has 600 ft of fencing available.

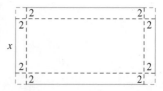

    (a) What is the length of the remaining piece of fencing in terms of x?

    (b) Determine a function $\mathscr{A}$ that represents the total area of the enclosed region. Give any restrictions on x.

    (c) What dimensions for the total enclosed region would give an area of 22,500 ft^2?

    (d) What is the maximum area that can be enclosed?

57. *Volume of a Box* A piece of cardboard is twice as long as it is wide. It is to be made into a box with an open top by cutting 2-in. squares from each corner and folding up the sides. Let x represent the width (in inches) of the original piece of cardboard.

    (a) Represent the length of the original piece of cardboard in terms of x.

    (b) What will be the dimensions of the bottom rectangular base of the box? Give the restrictions on x.

    (c) Determine a function V that represents the volume of the box in terms of x.

    (d) For what dimensions of the bottom of the box will the volume be 320 in.^{3}?

    (e) Find the values of x (to the nearest tenth of an inch) if such a box is to have a volume between 400 and 500 in.^{3}.

58. *Volume of a Box* A piece of sheet metal is 2.5 times as long as it is wide. It is to be made into a box with an open top by cutting 3-in. squares from each corner and folding up the sides. Let x represent the width (in inches) of the original piece.

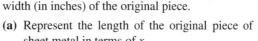

(a) Represent the length of the original piece of sheet metal in terms of x.

(b) What are the restrictions on x?

(c) Determine a function V that represents the volume of the box in terms of x.

(d) For what values of x (that is, original widths) will the volume of the box be between 600 and 800 in.^{3}? Give values to the nearest tenth of an inch.

59. *Shooting a Free Throw* If a person shoots a free throw from a position 8 ft above the floor, then the path of the ball may be modeled by the parabola

$$y = \frac{-16x^2}{0.434v^2} + 1.15x + 8,$$

where v is the initial velocity of the ball in feet per second, as illustrated in the figure. (*Source:* Rist, C., "The Physics of Foul Shots," *Discover*.)

(a) If the basketball hoop is 10 ft high and located 15 ft away, what initial velocity v should the basketball have?

(b) What is the maximum height of the basketball?

60. *Shooting a Free Throw* See **Exercise 59.** If a person shoots a free throw from an underhand position 3 ft above the floor, the path of the ball may be modeled by

$$y = \frac{-16x^2}{0.117v^2} + 2.75x + 3.$$

Repeat parts (a) and (b) from **Exercise 59.** Then compare the paths for the overhand shot and the underhand shot.

(Modeling) Solve each problem. See Example 6.

61. *Births to Unmarried Women* The percent of births to unmarried women in the United States from 1996 to 2007 are shown in the table. The data are modeled by the quadratic function

$$f(x) = 0.0773x^2 - 0.2115x + 32.64,$$

where $x = 0$ corresponds to 1996 and $f(x)$ is the percent. If this model continues to apply, what would it predict for the percent of these births in 2012?

Year	Percent	Year	Percent
1996	32.4	2002	34.0
1997	32.4	2003	34.6
1998	32.8	2004	35.8
1999	33.0	2005	36.9
2000	33.2	2006	38.5
2001	33.5	2007	39.7

Source: National Center for Health Statistics.

62. *Concentration of Atmospheric CO_2* The quadratic function

$$f(x) = 0.0098x^2 + 0.9010x + 316.8$$

models the worldwide atmospheric concentration of carbon dioxide in parts per million (ppm) over the period 1960–2009, where $x = 0$ represents the year 1960. If this model continues to hold, what will be the atmospheric CO_2 concentration in 2020? (*Source:* U.S. Department of Energy.)

63. *Spending on Shoes and Clothing* The total amount spent by Americans on shoes and clothing from 2000 to 2009 can be modeled by

$$f(x) = -4.979x^2 + 71.73x + 97.29,$$

where $x = 0$ represents 2000 and $f(x)$ is in billions of dollars. Based on this model, in what year did spending on shoes and clothing reach a maximum? (*Source:* Bureau of Economic Analysis.)

64. *Accident Rate* According to data from the National Highway Traffic Safety Administration, the accident rate as a function of the age of the driver in years x can be approximated by the function

$$f(x) = 0.0232x^2 - 2.28x + 60.0,$$

for $16 \le x \le 85$. Find both the age at which the accident rate is a minimum and the minimum rate.

65. *AIDS Cases in the United States* The table* lists the total (cumulative) number of AIDS cases diagnosed in the United States through 2007. For example, a total of 361,509 AIDS cases were diagnosed through 1993.

(a) Plot the data. Let $x = 0$ correspond to the year 1990.

(b) Would a linear or a quadratic function model the data better? Explain.

(c) Find a quadratic function defined by $f(x) = ax^2 + bx + c$ that models the data.

(d) Plot the data together with f on the same coordinate plane. How well does f model the number of AIDS cases?

(e) Use f to estimate the total number of AIDS cases diagnosed in the years 2009 and 2010.

(f) According to the model, how many new cases were diagnosed in the year 2010?

Year	AIDS Cases	Year	AIDS Cases
1990	193,245	1999	718,676
1991	248,023	2000	759,434
1992	315,329	2001	801,302
1993	361,509	2002	844,047
1994	441,406	2003	888,279
1995	515,586	2004	932,387
1996	584,394	2005	978,056
1997	632,249	2006	982,498
1998	673,572	2007	1,018,428

66. *AIDS Deaths in the United States* The table* lists the total (cumulative) number of known deaths caused by AIDS in the United States up to 2007.

(a) Plot the data. Let $x = 0$ correspond to the year 1990.

(b) Would a linear or a quadratic function model the data better? Explain.

(c) Find a quadratic function defined by $g(x) = ax^2 + bx + c$ that models the data.

(d) Plot the data together with g on the same coordinate plane. How well does g model the number of AIDS cases?

Year	AIDS Deaths	Year	AIDS Deaths
1990	119,821	1999	419,234
1991	154,567	2000	436,373
1992	191,508	2001	454,099
1993	220,592	2002	471,417
1994	269,992	2003	489,437
1995	320,692	2004	507,536
1996	359,892	2005	524,547
1997	381,738	2006	565,927
1998	400,743	2007	583,298

Sources: "Facts and Figures," Joint United Nations Programme on HIV/AIDS (UNAIDS), February, 1999; U.S. Census Bureau, 2007; and Centers for Disease Control and Prevention.

(e) Use g to estimate the total number of AIDS deaths in the year 2010.

(f) Consider the last two entries in the table for the years 2006 and 2007. Is it safe to assume that the quadratic model given will continue for 2008 and beyond?

67. *Foreign-Born Americans* The table lists the percent of the U.S. population that was foreign-born for selected years.

Year	Percent
1930	11.6
1940	8.8
1950	6.9
1960	5.4
1970	4.7
1980	6.2
1990	8.0
2000	10.4
2008	12.2

Source: U.S. Census Bureau.

(a) Plot the data. Let $x = 0$ correspond to the year 1930, $x = 10$ correspond to 1940, and so on.

(b) Find a quadratic function $f(x) = a(x - h)^2 + k$ that models the data. Use $(40, 4.7)$ as the vertex and $(20, 6.9)$ as the other point to determine a.

(c) Plot the data together with f in the same window. How well does f model the percent of the U.S. population that is foreign-born?

(d) Use the quadratic regression feature of a graphing calculator to determine the quadratic function g that provides the best fit for the data.

(e) Use the functions f and g to predict the percent of the U.S. population in 2015 that will be foreign-born.

68. *Automobile Stopping Distance* Selected values of the stopping distance y, in feet, of a car traveling x miles per hour are given in the table.

(a) Plot the data.

(b) The quadratic function

$$f(x) = 0.056057x^2 + 1.06657x$$

is one model that has been used to approximate stopping distances. Find and interpret $f(45)$.

(c) How well does f model the car's stopping distance?

Speed (in mph)	Stopping Distance (in feet)
20	46
30	87
40	140
50	240
60	282
70	371

Source: National Safety Institute Student Workbook.

Concept Check *Work Exercises 69–74.*

69. Find a value of c so that $y = x^2 - 10x + c$ has exactly one x-intercept.

70. For what values of a does $y = ax^2 - 8x + 4$ have no x-intercepts?

71. Define the quadratic function f having x-intercepts $(2, 0)$ and $(5, 0)$ and y-intercept $(0, 5)$.

72. Define the quadratic function f having x-intercepts $(1, 0)$ and $(-2, 0)$ and y-intercept $(0, 4)$.

73. The distance between the two points $P(x_1, y_1)$ and $R(x_2, y_2)$ is

$$d(P, R) = \sqrt{(x_1 - x_2)^2 + (y_1 - y_2)^2}. \quad \text{Distance formula (Section 2.1)}$$

Find the closest point on the line $y = 2x$ to the point $(1, 7)$. (*Hint:* Every point on $y = 2x$ has the form $(x, 2x)$, and the closest point has the minimum distance.)

74. A quadratic equation $f(x) = 0$ has a solution $x = 2$. Its graph has vertex $(5, 3)$. What is the other solution of the equation?

Relating Concepts

For individual or collaborative investigation *(Exercises 75–78)*

In **Section 1.7** we solved quadratic inequalities such as

$$x^2 + 2x - 8 < 0,$$

by solving the related quadratic equation

$$x^2 + 2x - 8 = 0,$$

identifying intervals determined by the solutions of this equation, and then using a test value from each interval to determine which intervals form the solution set. **Work Exercises 75–78 in order** to learn a graphical method of solving inequalities.

75. Graph $f(x) = x^2 + 2x - 8$.

76. The real solutions of $x^2 + 2x - 8 = 0$ are the x-values of the x-intercepts of the graph in **Exercise 75.** These are values of x for which $f(x) = 0$. What are the x-intercepts? What is the solution set of this equation?

77. The real solutions of $x^2 + 2x - 8 < 0$ are the x-values for which the graph in **Exercise 75** lies *below* the x-axis. These are values of x for which $f(x) < 0$. What interval of x-values represents the solution set of this inequality?

78. The real solutions of $x^2 + 2x - 8 > 0$ are the x-values for which the graph in **Exercise 75** lies *above* the x-axis. These are values of x for which $f(x) > 0$. What intervals of x-values represent the solution set of this inequality?

Use the technique described in **Relating Concepts Exercises 75–78** to solve each inequality. Write the solution set in interval notation.

79. $x^2 - x - 6 < 0$ **80.** $x^2 - 9x + 20 < 0$ **81.** $2x^2 - 9x \geq 18$

82. $3x^2 + x \geq 4$ **83.** $-x^2 + 4x + 1 \geq 0$ **84.** $-x^2 + 2x + 6 > 0$

3.2 Synthetic Division

- Synthetic Division
- Evaluating Polynomial Functions Using the Remainder Theorem
- Testing Potential Zeros

A division problem can be written using multiplication, even when the division involves polynomials. The **division algorithm** illustrates this.

Division Algorithm

Let $f(x)$ and $g(x)$ be polynomials with $g(x)$ of lesser degree than $f(x)$ and $g(x)$ of degree 1 or more. There exist unique polynomials $q(x)$ and $r(x)$ such that

$$f(x) = g(x) \cdot q(x) + r(x),$$

where either $r(x) = 0$ or the degree of $r(x)$ is less than the degree of $g(x)$.

For instance, we saw in **Example 11** of **Section R.3** that

$$\frac{3x^3 - 2x^2 - 150}{x^2 - 4} = 3x - 2 + \frac{12x - 158}{x^2 - 4}. \quad \text{(Section R.3)}$$

We can express this result using the division algorithm.

$$\underbrace{3x^3 - 2x^2 - 150}_{f(x)} = \underbrace{(x^2 - 4)}_{g(x)}\underbrace{(3x - 2)}_{q(x)} + \underbrace{12x - 158}_{r(x)}.$$

$$\underset{\substack{\text{Dividend} \\ \text{(original polynomial)}}}{} = \text{Divisor} \cdot \text{Quotient} + \text{Remainder}$$

Synthetic Division   When a given polynomial in x is divided by a first-degree binomial of the form $x - k$, a shortcut method called **synthetic division** may be used. The example on the left below is simplified by omitting all variables and writing only coefficients, with 0 used to represent the coefficient of any missing terms. Since the coefficient of x in the divisor is always 1 in these divisions, it too can be omitted. These omissions simplify the problem, as shown on the right.

$$
\begin{array}{r}
3x^2 + 10x + 40 \\
x - 4 \overline{\smash{)}3x^3 - 2x^2 + 0x - 150} \\
\underline{3x^3 - 12x^2} \\
10x^2 + 0x \\
\underline{10x^2 - 40x} \\
40x - 150 \\
\underline{40x - 160} \\
10
\end{array}
\qquad
\begin{array}{r}
3 \quad 10 \quad 40 \\
-4 \overline{\smash{)}3 \;\; -2 \quad\;\; 0 \; -150} \\
\underline{3 \; -12} \\
10 \quad 0 \\
\underline{10 \; - 40} \\
40 \; -150 \\
\underline{40 \; -160} \\
10
\end{array}
$$

The numbers in color that are repetitions of the numbers directly above them can also be omitted, as shown on the left below.

$$
\begin{array}{r}
3 \quad 10 \quad 40 \\
-4 \overline{\smash{)}3 \;\; -2 \quad\;\; 0 \; -150} \\
\underline{-12} \\
10 \quad\;\; 0 \\
\underline{-40} \\
40 \; -150 \\
\underline{-160} \\
10
\end{array}
\qquad
\begin{array}{r}
3 \quad 10 \quad 40 \\
-4 \overline{\smash{)}3 \;\; -2 \quad\;\; 0 \; -150} \\
\underline{-12} \\
10 \\
\underline{-40} \\
40 \\
\underline{-160} \\
10
\end{array}
$$

The numbers in color are again repetitions of those directly above them. They may be omitted, as shown on the right above.

The entire process can now be condensed vertically. The top row of numbers can be omitted since it duplicates the bottom row if the 3 is brought down.

$$
\begin{array}{r}
-4 \overline{\smash{)}3 \quad\; -2 \quad\;\; 0 \; -150} \\
\underline{-12 \; -40 \; -160} \\
3 \quad\;\; 10 \quad 40 \quad\;\; 10
\end{array}
$$

The rest of the bottom row is obtained by subtracting -12, -40, and -160 from the corresponding terms above them.

To simplify the arithmetic, we replace subtraction in the second row by addition and compensate by changing the -4 at the upper left to its additive inverse, 4.

$$
\begin{array}{r}
\text{Additive} \;\longrightarrow\; 4 \overline{\smash{)}3 \quad\;\; -2 \quad\;\; 0 \; -150} \\
\text{inverse} \qquad\qquad\qquad 12 \quad 40 \quad 160 \;\;\longleftarrow\; \text{Signs changed} \\
3 \quad\;\; 10 \quad 40 \quad\;\; 10 \\
\downarrow \quad\;\; \downarrow \quad\;\; \downarrow \qquad\;\; \downarrow \\
10 \;\;\longleftarrow\; \text{Remainder} \\
\text{Quotient} \;\longrightarrow\; 3x^2 + 10x + 40 + \dfrac{}{x - 4}
\end{array}
$$

Synthetic division provides an efficient process for dividing a polynomial by a binomial of the form $x - k$. Begin by writing the coefficients of the polynomial in decreasing powers of the variable, using 0 as the coefficient of any missing powers. The number k is written to the left in the same row. In the example above, $x - k$ is $x - 4$, so k is 4. The answer is found on the bottom row with the remainder farthest to the right and the coefficients of the quotient on the left when written in order of decreasing degree.

CAUTION *To avoid errors, use* **0** *as the coefficient for any missing terms, including a missing constant, when setting up the division.*

EXAMPLE 1 **Using Synthetic Division**

Use synthetic division to divide.

$$\frac{5x^3 - 6x^2 - 28x - 2}{x + 2}$$

SOLUTION Express $x + 2$ in the form $x - k$ by writing it as $x - (-2)$.

$$-2)\overline{5 \quad -6 \quad -28 \quad -2} \leftarrow \text{Coefficients of the polynomial}$$

x + 2 leads to −2.

Bring down the 5, and multiply: $-2(5) = -10$.

$$
\begin{array}{r}
-2)\overline{5 \quad -6 \quad -28 \quad -2} \\
\underline{\downarrow \quad -10 } \\
5
\end{array}
$$

Add -6 and -10 to obtain -16. Multiply: $-2(-16) = 32$.

$$
\begin{array}{r}
-2)\overline{5 \quad -6 \quad -28 \quad -2} \\
\underline{\quad -10 \quad 32 } \\
5 \quad -16
\end{array}
$$

Add -28 and 32, obtaining 4. Finally, $-2(4) = -8$.

$$
\begin{array}{r}
-2)\overline{5 \quad -6 \quad -28 \quad -2} \\
\underline{\quad -10 \quad 32 \quad -8} \\
5 \quad -16 \quad 4
\end{array}
$$

Add columns. Be careful with signs.

Add -2 and -8 to obtain -10.

$$
\begin{array}{r}
-2)\overline{5 \quad -6 \quad -28 \quad -2} \\
\underline{\quad -10 \quad 32 \quad -8} \\
\underbrace{5 \quad -16 \quad 4}_{\text{Quotient}} \quad -10 \leftarrow \text{Remainder}
\end{array}
$$

Since the divisor $x - k$ has degree 1, the degree of the quotient will always be one less than the degree of the polynomial to be divided.

$$\frac{5x^3 - 6x^2 - 28x - 2}{x + 2} = 5x^2 - 16x + 4 + \frac{-10}{x + 2}$$

Remember to add $\frac{\text{remainder}}{\text{divisor}}$.

✔ *Now Try Exercise 9.*

The result of the division in **Example 1** can be written as

$$5x^3 - 6x^2 - 28x - 2 = (x + 2)(5x^2 - 16x + 4) + (-10)$$

by multiplying each side by the denominator $x + 2$. The theorem that follows is a generalization of this product form.

Special Case of the Division Algorithm

For any polynomial $f(x)$ and any complex number k, there exists a [unique] polynomial $q(x)$ and number r such that the following holds.

$$f(x) = (x - k)q(x) + r$$

The mathematical statement

$$\underbrace{5x^3 - 6x^2 - 28x - 2}_{f(x)} = \underbrace{(x + 2)}_{= (x - k) \cdot}\underbrace{(5x^2 - 16x + 4)}_{q(x)} + \underbrace{(-10)}_{+ \quad r}$$

illustrates this connection. This form of the division algorithm is useful in developing the *remainder theorem*.

Evaluating Polynomial Functions Using the Remainder Theorem Suppose that $f(x)$ is written as $f(x) = (x - k)q(x) + r$. This equality is true for all complex values of x, so it is true for $x = k$. Replace x with k.

$$f(k) = (k - k)q(k) + r, \quad \text{or} \quad f(k) = r$$

This proves the following **remainder theorem,** which gives a new method of evaluating polynomial functions.

Remainder Theorem

If the polynomial $f(x)$ is divided by $x - k$, then the remainder is equal to $f(k)$.

In **Example 1,** when $f(x) = 5x^3 - 6x^2 - 28x - 2$ was divided by $x + 2$, or $x - (-2)$, the remainder was -10. Substitute -2 for x in $f(x)$.

$$f(-2) = 5(-2)^3 - 6(-2)^2 - 28(-2) - 2$$
$$= -40 - 24 + 56 - 2$$
$$= -10$$

> Use parentheses around substituted values to avoid errors.

As shown below, an alternative way to find the value of a polynomial is to use synthetic division. By the remainder theorem, instead of replacing x by -2 to find $f(-2)$, divide $f(x)$ by $x + 2$ using synthetic division as in **Example 1.** Then $f(-2)$ is the remainder, -10.

$$
\begin{array}{r|rrrr}
-2 & 5 & -6 & -28 & -2 \\
   &   & -10 & 32 & -8 \\
\hline
   & 5 & -16 & 4 & -10 \leftarrow f(-2)
\end{array}
$$

EXAMPLE 2 Applying the Remainder Theorem

Let $f(x) = -x^4 + 3x^2 - 4x - 5$. Use the remainder theorem to find $f(-3)$.

SOLUTION Use synthetic division with $k = -3$.

$$
\begin{array}{r|rrrrr}
-3 & -1 & 0 & 3 & -4 & -5 \\
   &   & 3 & -9 & 18 & -42 \\
\hline
   & -1 & 3 & -6 & 14 & -47 \leftarrow \text{Remainder}
\end{array}
$$

> $f(-3)$ is equal to the remainder when dividing by $x + 3$.

By this result, $f(-3) = -47$.

☑ *Now Try Exercise 31.*

Testing Potential Zeros   A **zero** of a polynomial function $f(x)$ is a number k such that $f(k) = 0$. ***The real number zeros are the x-intercepts of the graph of the function.***

The remainder theorem gives a quick way to decide whether a number k is a zero of a polynomial function defined by $f(x)$, as follows.

1. Use synthetic division to find $f(k)$.

2. If the remainder is 0, then $f(k) = 0$ and k is a zero of $f(x)$. If the remainder is not 0, then k is not a zero of $f(x)$.

A zero of $f(x)$ is a **root,** or **solution,** of the equation $f(x) = 0$.

EXAMPLE 3   **Deciding Whether a Number Is a Zero**

Decide whether the given number k is a zero of $f(x)$.

(a) $f(x) = x^3 - 4x^2 + 9x - 6;\quad k = 1$

(b) $f(x) = x^4 + x^2 - 3x + 1;\quad k = -1$

(c) $f(x) = x^4 - 2x^3 + 4x^2 + 2x - 5;\quad k = 1 + 2i$

SOLUTION

(a) Use synthetic division to decide whether 1 is a zero of $f(x) = x^3 - 4x^2 + 9x - 6$.

$$\text{Proposed zero} \longrightarrow 1\overline{)\begin{array}{rrrr} 1 & -4 & 9 & -6 \\ & 1 & -3 & 6 \\ \hline 1 & -3 & 6 & 0 \end{array}} \longleftarrow f(x) = x^3 - 4x^2 + 9x - 6$$

$\longleftarrow$ Remainder

$f(x) = x^3 - 4x^2 + 9x - 6$

Since the remainder is 0, $f(1) = 0$, and 1 is a zero of the given polynomial function. Therefore, the graph of $f(x) = x^3 - 4x^2 + 9x - 6$ includes the x-intercept $(1, 0)$. The graph in **Figure 16** supports this.

Figure 16

(b) For $f(x) = x^4 + x^2 - 3x + 1$, remember to use 0 as coefficient for the missing x^3-term in the synthetic division.

$$\text{Proposed zero} \longrightarrow -1\overline{)\begin{array}{rrrrr} 1 & 0 & 1 & -3 & 1 \\ & -1 & 1 & -2 & 5 \\ \hline 1 & -1 & 2 & -5 & 6 \end{array}} \longleftarrow \text{Remainder}$$

The remainder is not 0, so -1 is not a zero of $f(x) = x^4 + x^2 - 3x + 1$. In fact, $f(-1) = 6$, indicating that $(-1, 6)$ is on the graph of $f(x)$. The graph in **Figure 17** supports this.

Figure 17

(c) Use synthetic division and operations with complex numbers to determine whether $1 + 2i$ is a zero of $f(x) = x^4 - 2x^3 + 4x^2 + 2x - 5$.

$$1 + 2i\overline{)\begin{array}{rrrrr} 1 & -2 & 4 & 2 & -5 \\ & 1 + 2i & -5 & -1 - 2i & 5 \\ \hline 1 & -1 + 2i & -1 & 1 - 2i & 0 \end{array}} \longleftarrow \text{Remainder}$$

$(1 + 2i)(-1 + 2i)$
$= -1 + 4i^2$
$= -5$

$i^2 = -1$ (Section 1.3)

Since the remainder is 0, $1 + 2i$ is a zero of the given polynomial function. Notice that $1 + 2i$ is *not* a real number zero. Therefore, it cannot appear as an x-intercept on the graph of $f(x)$.

✔ *Now Try Exercises 43 and 55.*

3.2 Exercises

Use synthetic division to perform each division. See Example 1.

1. $\dfrac{x^3 + 3x^2 + 11x + 9}{x + 1}$

2. $\dfrac{x^3 + 7x^2 + 13x + 6}{x + 2}$

3. $\dfrac{5x^4 + 5x^3 + 2x^2 - x - 3}{x + 1}$

4. $\dfrac{2x^4 - x^3 - 7x^2 + 7x - 10}{x - 2}$

5. $\dfrac{x^4 + 4x^3 + 2x^2 + 9x + 4}{x + 4}$

6. $\dfrac{x^4 + 5x^3 + 4x^2 - 3x + 9}{x + 3}$

7. $\dfrac{x^5 + 3x^4 + 2x^3 + 2x^2 + 3x + 1}{x + 2}$

8. $\dfrac{x^6 - 3x^4 + 2x^3 - 6x^2 - 5x + 3}{x + 2}$

9. $\dfrac{-9x^3 + 8x^2 - 7x + 2}{x - 2}$

10. $\dfrac{-11x^4 + 2x^3 - 8x^2 - 4}{x + 1}$

11. $\dfrac{\frac{1}{3}x^3 - \frac{2}{9}x^2 + \frac{2}{27}x - \frac{1}{81}}{x - \frac{1}{3}}$

12. $\dfrac{x^3 + x^2 + \frac{1}{2}x + \frac{1}{8}}{x + \frac{1}{2}}$

13. $\dfrac{x^4 - 3x^3 - 4x^2 + 12x}{x - 2}$

14. $\dfrac{x^4 - x^3 - 5x^2 - 3x}{x + 1}$

15. $\dfrac{x^3 - 1}{x - 1}$

16. $\dfrac{x^4 - 1}{x - 1}$

17. $\dfrac{x^5 + 1}{x + 1}$

18. $\dfrac{x^7 + 1}{x + 1}$

Express $f(x)$ in the form $f(x) = (x - k)q(x) + r$ for the given value of k.

19. $f(x) = 2x^3 + x^2 + x - 8; \ k = -1$

20. $f(x) = 2x^3 + 3x^2 - 16x + 10; \ k = -4$

21. $f(x) = x^3 + 4x^2 + 5x + 2; \ k = -2$

22. $f(x) = -x^3 + x^2 + 3x - 2; \ k = 2$

23. $f(x) = 4x^4 - 3x^3 - 20x^2 - x; \ k = 3$

24. $f(x) = 2x^4 + x^3 - 15x^2 + 3x; \ k = -3$

25. $f(x) = 3x^4 + 4x^3 - 10x^2 + 15; \ k = -1$

26. $f(x) = -5x^4 + x^3 + 2x^2 + 3x + 1; \ k = 1$

For each polynomial function, use the remainder theorem and synthetic division to find $f(k)$. See Example 2.

27. $f(x) = x^2 + 5x + 6; \ k = -2$

28. $f(x) = x^2 - 4x - 5; \ k = 5$

29. $f(x) = 2x^2 - 3x - 3; \ k = 2$

30. $f(x) = -x^3 + 8x^2 + 63; \ k = 4$

31. $f(x) = x^3 - 4x^2 + 2x + 1; \ k = -1$

32. $f(x) = 2x^3 - 3x^2 - 5x + 4; \ k = 2$

33. $f(x) = 2x^5 - 10x^3 - 19x^2 - 50; \ k = 3$

34. $f(x) = x^4 + 6x^3 + 9x^2 + 3x - 3; \ k = 4$

35. $f(x) = 6x^4 + x^3 - 8x^2 + 5x + 6; \ k = \frac{1}{2}$

36. $f(x) = 6x^3 - 31x^2 - 15x; \ k = -\frac{1}{2}$

37. $f(x) = x^2 - 5x + 1; \ k = 2 + i$

38. $f(x) = x^2 - x + 3; \ k = 3 - 2i$

39. $f(x) = x^2 + 4; \ k = 2i$

40. $f(x) = 2x^2 + 10; \ k = i\sqrt{5}$

Use synthetic division to decide whether the given number k is a zero of the given polynomial function. If it is not, give the value of $f(k)$. ***See Examples 2 and 3.***

41. $f(x) = x^2 + 2x - 8; \ k = 2$

42. $f(x) = x^2 + 4x - 5; \ k = -5$

43. $f(x) = x^3 - 3x^2 + 4x - 4; \ k = 2$

44. $f(x) = x^3 + 2x^2 - x + 6; \ k = -3$

45. $f(x) = 2x^3 - 6x^2 - 9x + 4; \ k = 1$

46. $f(x) = 2x^3 + 9x^2 - 16x + 12; \ k = 1$

47. $f(x) = x^3 + 7x^2 + 10x; \ k = 0$

48. $f(x) = 2x^3 - 3x^2 - 5x; \ k = 0$

49. $f(x) = 4x^4 + x^2 + 17x + 3; \ k = -\frac{3}{2}$

50. $f(x) = 3x^4 + 13x^3 - 10x + 8; \ k = -\frac{4}{3}$

51. $f(x) = 5x^4 + 2x^3 - x + 3; \ k = \frac{2}{5}$

52. $f(x) = 16x^4 + 3x^2 - 2; \ k = \frac{1}{2}$

53. $f(x) = x^2 - 2x + 2; \ k = 1 - i$

54. $f(x) = x^2 - 4x + 5; \ k = 2 - i$

55. $f(x) = x^2 + 3x + 4; \ k = 2 + i$

56. $f(x) = x^2 - 3x + 5; \ k = 1 - 2i$

57. $f(x) = x^3 + 3x^2 - x + 1; \ k = 1 + i$

58. $f(x) = 2x^3 - x^2 + 3x - 5; \ k = 2 - i$

Relating Concepts

For individual or collaborative investigation *(Exercises 59–68)*

The remainder theorem indicates that when a polynomial $f(x)$ is divided by $x - k$, the remainder is equal to $f(k)$. For

$$f(x) = x^3 - 2x^2 - x + 2,$$

use the remainder theorem to find each of the following. Then determine the coordinates of the corresponding point on the graph of $f(x)$.

59. $f(-2)$ **60.** $f(-1)$ **61.** $f\left(-\dfrac{1}{2}\right)$ **62.** $f(0)$

63. $f(1)$ **64.** $f\left(\dfrac{3}{2}\right)$ **65.** $f(2)$ **66.** $f(3)$

 67. Use the results from **Exercises 59–66** to plot eight points on the graph of $f(x)$. Connect these points with a smooth curve. Describe a method for graphing polynomial functions using the remainder theorem.

68. Use the method described in **Exercises 59–67** to sketch the graph of $f(x) = -x^3 - x^2 + 2x$.

3.3   Zeros of Polynomial Functions

■ Factor Theorem

■ Rational Zeros Theorem

■ Number of Zeros

■ Conjugate Zeros Theorem

■ Finding Zeros of a Polynomial Function

■ Descartes' Rule of Signs

Factor Theorem   Consider the polynomial function

$$f(x) = x^2 + x - 2,$$

which is written in factored form as

$$f(x) = (x - 1)(x + 2). \quad \text{(Section R.4)}$$

For this function, $f(1) = 0$ and $f(-2) = 0$, and thus 1 and -2 are zeros of $f(x)$. Notice the special relationship between each linear factor and its corresponding zero. The **factor theorem** summarizes this relationship.

Factor Theorem

For any polynomial function $f(x)$, $x - k$ is a factor of the polynomial if and only if $f(k) = 0$.

EXAMPLE 1 **Deciding Whether $x - k$ Is a Factor**

Determine whether $x - 1$ is a factor of each polynomial.

(a) $f(x) = 2x^4 + 3x^2 - 5x + 7$

(b) $f(x) = 3x^5 - 2x^4 + x^3 - 8x^2 + 5x + 1$

SOLUTION

(a) By the factor theorem, $x - 1$ will be a factor if $f(1) = 0$. Use synthetic division and the remainder theorem to decide.

$$
\begin{array}{r|rrrrr}
1) & 2 & 0 & 3 & -5 & 7 \quad \text{(Section 3.2)} \\
   &   & 2 & 2 & 5 & 0 \\
\hline
   & 2 & 2 & 5 & 0 & 7 \longleftarrow f(1) = 7
\end{array}
$$

Use a zero coefficient for the missing term.

The remainder is 7 and not 0, so $x - 1$ is not a factor of $2x^4 + 3x^2 - 5x + 7$.

(b)

$$
\begin{array}{r|rrrrrr}
1) & 3 & -2 & 1 & -8 & 5 & 1 \\
   &   & 3 & 1 & 2 & -6 & -1 \\
\hline
   & 3 & 1 & 2 & -6 & -1 & 0 \longleftarrow f(1) = 0
\end{array}
$$

Because the remainder is 0, $x - 1$ is a factor. Additionally, we can determine from the coefficients in the bottom row that the other factor is

$$3x^4 + x^3 + 2x^2 - 6x - 1.$$

Thus, we can express the polynomial in factored form.

$$f(x) = (x - 1)(3x^4 + x^3 + 2x^2 - 6x - 1)$$

☑ *Now Try Exercises 5 and 7.*

We can use the factor theorem to factor a polynomial of greater degree into linear factors of the form $ax - b$.

EXAMPLE 2 **Factoring a Polynomial Given a Zero**

Factor $f(x) = 6x^3 + 19x^2 + 2x - 3$ into linear factors if -3 is a zero of f.

SOLUTION Since -3 is a zero of f, $x - (-3) = x + 3$ is a factor.

$$
\begin{array}{r|rrrr}
-3) & 6 & 19 & 2 & -3 \\
    &   & -18 & -3 & 3 \\
\hline
    & 6 & 1 & -1 & 0
\end{array}
$$

Use synthetic division to divide $f(x)$ by $x + 3$.

The quotient is $6x^2 + x - 1$, which is the factor that accompanies $x + 3$.

$$f(x) = (x + 3)(6x^2 + x - 1)$$

$$f(x) = (x + 3)(2x + 1)(3x - 1) \quad \text{Factor } 6x^2 + x - 1.$$

These factors are all linear.

☑ *Now Try Exercise 17.*

LOOKING AHEAD TO CALCULUS

Finding the derivative of a polynomial function is one of the basic skills required in a first calculus course. For the functions

$$f(x) = x^4 - x^2 + 5x - 4,$$
$$g(x) = -x^6 + x^2 - 3x + 4,$$

and $h(x) = 3x^3 - x^2 + 2x - 4,$

the derivatives are

$$f'(x) = 4x^3 - 2x + 5,$$
$$g'(x) = -6x^5 + 2x - 3,$$

and $h'(x) = 9x^2 - 2x + 2.$

Notice the use of the "prime" notation. For example, the derivative of $f(x)$ is denoted $f'(x)$.

Look for the pattern among the exponents and the coefficients. Using this pattern, what is the derivative of

$$F(x) = 4x^4 - 3x^3 + 6x - 4?$$

The answer is at the bottom of the page.

Rational Zeros Theorem The **rational zeros theorem** gives a method to determine all possible candidates for rational zeros of a polynomial function with integer coefficients.

Rational Zeros Theorem

If $\frac{p}{q}$ is a rational number written in lowest terms, and if $\frac{p}{q}$ is a zero of f, a polynomial function with integer coefficients, then p is a factor of the constant term, and q is a factor of the leading coefficient.

Proof $f\left(\frac{p}{q}\right) = 0$ since $\frac{p}{q}$ is a zero of $f(x)$.

$$a_n\left(\frac{p}{q}\right)^n + a_{n-1}\left(\frac{p}{q}\right)^{n-1} + \cdots + a_1\left(\frac{p}{q}\right) + a_0 = 0 \qquad \text{Definition of zero of } f$$

$$a_n\left(\frac{p^n}{q^n}\right) + a_{n-1}\left(\frac{p^{n-1}}{q^{n-1}}\right) + \cdots + a_1\left(\frac{p}{q}\right) + a_0 = 0 \qquad \text{Power rule for exponents (Section R.3)}$$

$$a_n p^n + a_{n-1} p^{n-1} q + \cdots + a_1 pq^{n-1} = -a_0 q^n \qquad \text{Multiply by } q^n \text{ and subtract } a_0 q^n.$$

$$p(a_n p^{n-1} + a_{n-1} p^{n-2} q + \cdots + a_1 q^{n-1}) = -a_0 q^n \qquad \text{Factor out } p.$$

This result shows that $-a_0 q^n$ equals the product of the two factors p and $(a_n p^{n-1} + \cdots + a_1 q^{n-1})$. For this reason, p must be a factor of $-a_0 q^n$. Since it was assumed that $\frac{p}{q}$ is written in lowest terms, p and q have no common factor other than 1, so p is not a factor of q^n. Thus, p must be a factor of a_0. In a similar way, it can be shown that q is a factor of a_n.

EXAMPLE 3 **Using the Rational Zeros Theorem**

Consider the polynomial function.

$$f(x) = 6x^4 + 7x^3 - 12x^2 - 3x + 2$$

(a) List all possible rational zeros.

(b) Find all rational zeros and factor $f(x)$ into linear factors.

SOLUTION

(a) For a rational number $\frac{p}{q}$ to be a zero, p must be a factor of $a_0 = 2$, and q must be a factor of $a_4 = 6$. Thus, p can be ± 1 or ± 2, and q can be ± 1, ± 2, ± 3, or ± 6. The possible rational zeros, $\frac{p}{q}$, are ± 1, ± 2, $\pm \frac{1}{2}$, $\pm \frac{1}{3}$, $\pm \frac{1}{6}$, and $\pm \frac{2}{3}$.

(b) Use the remainder theorem to show that 1 is a zero.

Use "trial and error" to find zeros.

$$
\begin{array}{r|rrrrr}
1) & 6 & 7 & -12 & -3 & 2 \\
  &   & 6 & 13 & 1 & -2 \\
\hline
  & 6 & 13 & 1 & -2 & 0 \leftarrow f(1) = 0
\end{array}
$$

The 0 remainder shows that 1 is a zero. The quotient is $6x^3 + 13x^2 + x - 2$.

$$f(x) = (x - 1)(6x^3 + 13x^2 + x - 2) \qquad \text{Begin factoring } f(x).$$

Now, use the quotient polynomial and synthetic division to find that -2 is a zero.

$$
\begin{array}{r|rrrr}
-2) & 6 & 13 & 1 & -2 \\
   &   & -12 & -2 & 2 \\
\hline
   & 6 & 1 & -1 & 0 \leftarrow f(-2) = 0
\end{array}
$$

Answer to Looking Ahead to Calculus:

$F'(x) = 16x^3 - 9x^2 + 6$

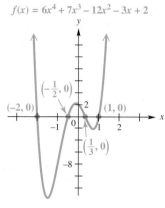

$f(x) = 6x^4 + 7x^3 - 12x^2 - 3x + 2$

Figure 18

The new quotient polynomial is $6x^2 + x - 1$. Therefore, $f(x)$ can now be completely factored as follows.

$$f(x) = (x - 1)(x + 2)(6x^2 + x - 1)$$
$$f(x) = (x - 1)(x + 2)(3x - 1)(2x + 1)$$

Setting $3x - 1 = 0$ and $2x + 1 = 0$ yields the zeros $\frac{1}{3}$ and $-\frac{1}{2}$. In summary, the rational zeros are 1, -2, $\frac{1}{3}$, and $-\frac{1}{2}$, and they are the x-values of the x-intercepts on the graph of $f(x)$ in **Figure 18**. The linear factorization of $f(x)$ is

$$f(x) = 6x^4 + 7x^3 - 12x^2 - 3x + 2$$
$$f(x) = (x - 1)(x + 2)(3x - 1)(2x + 1).$$

> Check by multiplying these factors.

✔ *Now Try Exercise 35.*

NOTE   In **Example 3**, once we obtained the quadratic factor

$$6x^2 + x - 1,$$

we were able to complete the work by factoring it directly. Had it not been easily factorable, we could have used the quadratic formula to find the other two zeros (and factors).

CAUTION   *The rational zeros theorem gives only possible rational zeros. It does not tell us whether these rational numbers are actual zeros.* We must rely on other methods to determine whether or not they are indeed zeros. Furthermore, the polynomial must have integer coefficients.

To apply the rational zeros theorem to a polynomial with fractional coefficients, multiply through by the least common denominator of all the fractions. For example, any rational zeros of $p(x)$ defined below will also be rational zeros of $q(x)$.

$$p(x) = x^4 - \frac{1}{6}x^3 + \frac{2}{3}x^2 - \frac{1}{6}x - \frac{1}{3}$$

$$q(x) = 6x^4 - x^3 + 4x^2 - x - 2 \qquad \text{Multiply the terms of } p(x) \text{ by 6.}$$

Number of Zeros   The **fundamental theorem of algebra** says that every function defined by a polynomial of degree 1 or more has a zero, which means that every such polynomial can be factored.

**Carl Friedrich Gauss
(1777–1855)**

The **fundamental theorem of algebra** was first proved by Carl Friedrich Gauss in his doctoral thesis in 1799, when he was 22 years old.

Fundamental Theorem of Algebra

Every function defined by a polynomial of degree 1 or more has at least one complex zero.

From the fundamental theorem, if $f(x)$ is of degree 1 or more, then there is some number k_1 such that $f(k_1) = 0$. By the factor theorem,

$$f(x) = (x - k_1)q_1(x), \quad \text{for some polynomial } q_1(x).$$

If $q_1(x)$ is of degree 1 or more, the fundamental theorem and the factor theorem can be used to factor $q_1(x)$ in the same way. There is some number k_2 such that $q_1(k_2) = 0$, so

$$q_1(x) = (x - k_2)q_2(x)$$

and $$f(x) = (x - k_1)(x - k_2)q_2(x).$$

Assuming that $f(x)$ has degree n and repeating this process n times gives

$$f(x) = a(x - k_1)(x - k_2)\cdots(x - k_n). \quad \text{\textit{a} is the leading coefficient.}$$

Each of these factors leads to a zero of $f(x)$, so $f(x)$ has the n zeros k_1, k_2, k_3, ..., k_n. This result suggests the **number of zeros theorem.**

Number of Zeros Theorem

A function defined by a polynomial of degree n has *at most* n distinct zeros.

For example, a polynomial function of degree 3 has *at most* three distinct zeros but can have as few as one zero. Consider the following polynomial.

$$f(x) = x^3 + 3x^2 + 3x + 1$$

$$f(x) = (x + 1)^3$$

The function f is of degree 3 but has only one zero, -1. Actually, the zero -1 occurs *three* times, since there are three factors of $x + 1$. The number of times a zero occurs is referred to as the **multiplicity of the zero.**

EXAMPLE 4   Finding a Polynomial Function That Satisfies Given Conditions (Real Zeros)

Find a function f defined by a polynomial of degree 3 that satisfies the given conditions.

(a) Zeros of -1, 2, and 4; $f(1) = 3$

(b) -2 is a zero of multiplicity 3; $f(-1) = 4$

SOLUTION

(a) These three zeros give $x - (-1) = x + 1$, $x - 2$, and $x - 4$ as factors of $f(x)$. Since $f(x)$ is to be of degree 3, these are the only possible factors by the number of zeros theorem. Therefore, $f(x)$ has the form

$$f(x) = a(x + 1)(x - 2)(x - 4), \quad \text{for some real number } a.$$

To find a, use the fact that $f(1) = 3$.

$$f(1) = a(1 + 1)(1 - 2)(1 - 4) \quad \text{Let } x = 1.$$

$$3 = a(2)(-1)(-3) \quad\quad f(1) = 3$$

$$3 = 6a \quad\quad \text{Multiply.}$$

$$a = \frac{1}{2} \quad\quad \text{Divide by 6.}$$

Thus, $$f(x) = \frac{1}{2}(x + 1)(x - 2)(x - 4),$$

or, $$f(x) = \frac{1}{2}x^3 - \frac{5}{2}x^2 + x + 4. \quad \text{Multiply.}$$

(b) The polynomial function $f(x)$ has the following form.

$$f(x) = a(x + 2)(x + 2)(x + 2) \quad \text{Factor theorem}$$

$$f(x) = a(x + 2)^3$$

To find a, use the fact that $f(-1) = 4$.

$$f(-1) = a(-1 + 2)^3 \quad \text{Let } x = -1.$$

$$4 = a(1)^3 \quad f(-1) = 4$$

Remember:
$(x + 2)^3 \neq x^3 + 2^3$

$$a = 4 \quad \text{Solve for } a.$$

Thus, $f(x) = 4(x + 2)^3,$

or, $f(x) = 4x^3 + 24x^2 + 48x + 32. \quad \text{Multiply.}$

✔ *Now Try Exercises 49 and 53.*

NOTE In **Example 4(a)**, we cannot clear the denominators in $f(x)$ by multiplying each side by 2 because the result would equal $2 \cdot f(x)$, not $f(x)$.

Conjugate Zeros Theorem The following properties of complex conjugates are needed to prove the **conjugate zeros theorem**. We use a simplified notation for conjugates here. If $z = a + bi$, then the conjugate of z is written $\bar{z}$, where $\bar{z} = a - bi$. For example, if $z = -5 + 2i$, then $\bar{z} = -5 - 2i$. The proofs of the first two of these properties are left for **Exercises 113 and 114**.

Properties of Conjugates

For any complex numbers c and d, the following properties hold.

$$\overline{c + d} = \bar{c} + \bar{d}, \quad \overline{c \cdot d} = \bar{c} \cdot \bar{d}, \quad \text{and} \quad \overline{c^n} = (\bar{c})^n$$

The remainder theorem can be used to show that both $2 + i$ and $2 - i$ are zeros of $f(x) = x^3 - x^2 - 7x + 15$. In general, if z is a zero of a polynomial function with *real* coefficients, then so is $\bar{z}$.

Conjugate Zeros Theorem

If $f(x)$ defines a polynomial function *having only real coefficients* and if $z = a + bi$ is a zero of $f(x)$, where a and b are real numbers, then

$$\bar{z} = a - bi \text{ is also a zero of } f(x).$$

Proof Start with the polynomial function

$$f(x) = a_n x^n + a_{n-1} x^{n-1} + \cdots + a_1 x + a_0,$$

where all coefficients are real numbers. If the complex number z is a zero of $f(x)$, then we have the following.

$$f(z) = a_n z^n + a_{n-1} z^{n-1} + \cdots + a_1 z + a_0 = 0$$

$$a_n z^n + a_{n-1} z^{n-1} + \cdots + a_1 z + a_0 = 0 \qquad \text{From the preceding discussion}$$

Take the conjugate of both sides of this equation.

$$\overline{a_n z^n + a_{n-1} z^{n-1} + \cdots + a_1 z + a_0} = \overline{0}$$

$$\overline{a_n z^n} + \overline{a_{n-1} z^{n-1}} + \cdots + \overline{a_1 z} + \overline{a_0} = \overline{0} \qquad \text{Use generalizations of the properties}$$
$$\overline{c + d} = \overline{c} + \overline{d} \text{ and } \overline{c \cdot d} = \overline{c} \cdot \overline{d}.$$

$$\overline{a_n}\, \overline{z}^n + \overline{a_{n-1}}\, \overline{z}^{n-1} + \cdots + \overline{a_1}\, \overline{z} + \overline{a_0} = \overline{0}$$

$$a_n(\overline{z})^n + a_{n-1}(\overline{z})^{n-1} + \cdots + a_1(\overline{z}) + a_0 = 0 \qquad \text{Use the property } \overline{c^n} = (\overline{c})^n \text{ and}$$
$$\text{the fact that for any real number } a,$$

$$f(\overline{z}) = 0 \qquad \overline{a} = a.$$

Hence $\overline{z}$ is also a zero of $f(x)$, which completes the proof.

CAUTION   *When the conjugate zeros theorem is applied, it is essential that the polynomial have only real coefficients.* For example,

$$f(x) = x - (1 + i)$$

has $1 + i$ as a zero, but the conjugate $1 - i$ is not a zero.

> **EXAMPLE 5**   **Finding a Polynomial Function That Satisfies Given Conditions (Complex Zeros)**

Find a polynomial function of least degree having only real coefficients and zeros 3 and $2 + i$.

SOLUTION   The complex number $2 - i$ must also be a zero, so the polynomial has at least three zeros: 3, $2 + i$, and $2 - i$. For the polynomial to be of least degree, these must be the only zeros. By the factor theorem there must be three factors: $x - 3$, $x - (2 + i)$, and $x - (2 - i)$.

$$f(x) = (x - 3)\big[x - (2 + i)\big]\big[x - (2 - i)\big] \qquad \text{Factor theorem}$$

$$f(x) = (x - 3)(x - 2 - i)(x - 2 + i) \qquad \text{Distribute negative signs.}$$

$$f(x) = (x - 3)(x^2 - 4x + 5) \qquad \text{Multiply and combine like terms; } i^2 = -1.$$
$$\text{(Section 1.3)}$$

$$f(x) = x^3 - 7x^2 + 17x - 15 \qquad \text{Multiply again.}$$

Any nonzero multiple of $x^3 - 7x^2 + 17x - 15$ also satisfies the given conditions on zeros. The information on zeros given in the problem is not sufficient to give a specific value for the leading coefficient.

☑ *Now Try Exercise 65.*

Finding Zeros of a Polynomial Function   The theorem on conjugate zeros helps predict the number of real zeros of polynomial functions with real coefficients.

- A polynomial function with real coefficients of odd degree n, where $n \geq 1$, must have at least one real zero (since zeros of the form $a + bi$, where $b \neq 0$, occur in conjugate pairs).

- A polynomial function with real coefficients of even degree n may have no real zeros.

EXAMPLE 6 **Finding All Zeros Given One Zero**

Find all zeros of $f(x) = x^4 - 7x^3 + 18x^2 - 22x + 12$, given that $1 - i$ is a zero.

SOLUTION Since the polynomial function has only real coefficients and since $1 - i$ is a zero, by the conjugate zeros theorem $1 + i$ is also a zero. To find the remaining zeros, first use synthetic division to divide the original polynomial by $x - (1 - i)$.

$$(1 - i)(-6 - i) = -6 - i + 6i + i^2 = -7 + 5i$$

$$\begin{array}{r|rrrrr} 1 - i) & 1 & -7 & 18 & -22 & 12 \\ & & 1 - i & -7 + 5i & 16 - 6i & -12 \\ \hline & 1 & -6 - i & 11 + 5i & -6 - 6i & 0 \end{array}$$

By the factor theorem, since $x = 1 - i$ is a zero of $f(x)$, $x - (1 - i)$ is a factor, and $f(x)$ can be written as follows.

$$f(x) = [x - (1 - i)][x^3 + (-6 - i)x^2 + (11 + 5i)x + (-6 - 6i)]$$

We know that $x = 1 + i$ is also a zero of $f(x)$. Continue to use synthetic division and divide the quotient polynomial above by $x - (1 + i)$.

$$\begin{array}{r|rrrr} 1 + i) & 1 & -6 - i & 11 + 5i & -6 - 6i \\ & & 1 + i & -5 - 5i & 6 + 6i \\ \hline & 1 & -5 & 6 & 0 \end{array}$$

Now $f(x)$ can be written in the following factored form.

$$f(x) = [x - (1 - i)][x - (1 + i)](x^2 - 5x + 6)$$
$$f(x) = [x - (1 - i)][x - (1 + i)](x - 2)(x - 3)$$

The remaining zeros are 2 and 3. The four zeros are $1 - i$, $1 + i$, 2, and 3.

✔ *Now Try Exercise 31.*

NOTE In **Example 6,** if we had been unable to factor $x^2 - 5x + 6$ into linear factors, we would have used the quadratic formula to solve the equation $x^2 - 5x + 6 = 0$ to find the remaining two zeros of the function.

Descartes' Rule of Signs The following rule helps to determine the number of positive and negative real zeros of a polynomial function. A **variation in sign** is a change from positive to negative or from negative to positive in successive terms of the polynomial when they are written in order of descending powers of the variable. *Missing terms (those with 0 coefficients) are counted as no change in sign and can be ignored.*

Descartes' Rule of Signs

Let $f(x)$ define a polynomial function with real coefficients and a nonzero constant term, with terms in descending powers of x.

(a) The number of positive real zeros of f either equals the number of variations in sign occurring in the coefficients of $f(x)$, or is less than the number of variations by a positive even integer.

(b) The number of negative real zeros of f either equals the number of variations in sign occurring in the coefficients of $f(-x)$, or is less than the number of variations by a positive even integer.

EXAMPLE 7   **Applying Descartes' Rule of Signs**

Determine the different possibilities for the numbers of positive, negative, and nonreal complex zeros of

$$f(x) = x^4 - 6x^3 + 8x^2 + 2x - 1.$$

SOLUTION   We first consider the possible number of positive zeros by observing that $f(x)$ has three variations in signs.

$$f(x) = +x^4 - 6x^3 + 8x^2 + 2x - 1$$

Thus, by Descartes' rule of signs, $f(x)$ has either three or one (since $3 - 2 = 1$) positive real zeros.

For negative zeros, consider the variations in signs for $f(-x)$.

$$f(-x) = (-x)^4 - 6(-x)^3 + 8(-x)^2 + 2(-x) - 1$$
$$= x^4 + 6x^3 + 8x^2 - 2x - 1$$

Since there is only one variation in sign, $f(x)$ has exactly one negative real zero.

Because $f(x)$ is a fourth-degree polynomial function, it must have four complex zeros, some of which may be repeated. Descartes' rule of signs has indicated that exactly one of these zeros is a negative real number.

- One possible combination of the zeros is one negative real zero, three positive real zeros, and no nonreal complex zeros.

- Another possible combination of the zeros is one negative real zero, one positive real zero, and two nonreal complex zeros.

By the conjugate zeros theorem, any possible nonreal complex zeros must occur in conjugate pairs since $f(x)$ has real coefficients. The table below summarizes these possibilities.

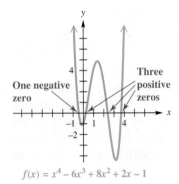

$f(x) = x^4 - 6x^3 + 8x^2 + 2x - 1$

Figure 19

Possible Number of Zeros		
Positive	**Negative**	**Nonreal Complex**
3	1	0
1	1	2

The graph of $f(x)$ in **Figure 19** verifies the correct combination of three positive real zeros with one negative real zero, as seen in the first row of the table.*

☑ *Now Try Exercise 75.*

> **NOTE**   *Descartes' rule of signs does not identify the multiplicity of the zeros of a function.* For example, if it indicates that a function $f(x)$ has exactly two positive real zeros, then $f(x)$ may have two distinct positive real zeros or one positive real zero of multiplicity 2.

---

* The authors would like to thank Mary Hill of College of DuPage for her input into **Example 7**.

3.3 Exercises

Concept Check *Decide whether each statement is* true *or* false. *If false, tell why.*

1. Since $x - 1$ is a factor of $f(x) = x^6 - x^4 + 2x^2 - 2$, we can conclude that $f(1) = 0$.

2. Since $f(1) = 0$ for $f(x) = x^6 - x^4 + 2x^2 - 2$, we can conclude that $x - 1$ is a factor of $f(x)$.

3. For $f(x) = (x + 2)^4(x - 3)$, 2 is a zero of multiplicity 4.

4. Since $2 + 3i$ is a zero of $f(x) = x^2 - 4x + 13$, we can conclude that $2 - 3i$ is also a zero.

Use the factor theorem and synthetic division to decide whether the second polynomial is a factor of the first. *See Example 1.*

5. $x^3 - 5x^2 + 3x + 1$; $x - 1$ 6. $x^3 + 6x^2 - 2x - 7$; $x + 1$

7. $2x^4 + 5x^3 - 8x^2 + 3x + 13$; $x + 1$ 8. $-3x^4 + x^3 - 5x^2 + 2x + 4$; $x - 1$

9. $-x^3 + 3x - 2$; $x + 2$ 10. $-2x^3 + x^2 - 63$; $x + 3$

11. $4x^2 + 2x + 54$; $x - 4$ 12. $5x^2 - 14x + 10$; $x + 2$

13. $x^3 + 2x^2 + 3$; $x - 1$ 14. $2x^3 + x + 2$; $x + 1$

15. $2x^4 + 5x^3 - 2x^2 + 5x + 6$; $x + 3$ 16. $5x^4 + 16x^3 - 15x^2 + 8x + 16$; $x + 4$

Factor $f(x)$ *into linear factors given that* k *is a zero of* $f(x)$. *See Example 2.*

17. $f(x) = 2x^3 - 3x^2 - 17x + 30$; $k = 2$

18. $f(x) = 2x^3 - 3x^2 - 5x + 6$; $k = 1$

19. $f(x) = 6x^3 + 13x^2 - 14x + 3$; $k = -3$

20. $f(x) = 6x^3 + 17x^2 - 63x + 10$; $k = -5$

21. $f(x) = 6x^3 + 25x^2 + 3x - 4$; $k = -4$

22. $f(x) = 8x^3 + 50x^2 + 47x - 15$; $k = -5$

23. $f(x) = x^3 + (7 - 3i)x^2 + (12 - 21i)x - 36i$; $k = 3i$

24. $f(x) = 2x^3 + (3 + 2i)x^2 + (1 + 3i)x + i$; $k = -i$

25. $f(x) = 2x^3 + (3 - 2i)x^2 + (-8 - 5i)x + (3 + 3i)$; $k = 1 + i$

26. $f(x) = 6x^3 + (19 - 6i)x^2 + (16 - 7i)x + (4 - 2i)$; $k = -2 + i$

27. $f(x) = x^4 + 2x^3 - 7x^2 - 20x - 12$; $k = -2$ (multiplicity 2)

28. $f(x) = 2x^4 + x^3 - 9x^2 - 13x - 5$; $k = -1$ (multiplicity 3)

For each polynomial function, one zero is given. Find all others. *See Examples 2 and 6.*

29. $f(x) = x^3 - x^2 - 4x - 6$; 3   30. $f(x) = x^3 + 4x^2 - 5$; 1

31. $f(x) = x^3 - 7x^2 + 17x - 15$; $2 - i$ 32. $f(x) = 4x^3 + 6x^2 - 2x - 1$; $\frac{1}{2}$

33. $f(x) = x^4 + 5x^2 + 4$; $-i$ 34. $f(x) = x^4 + 26x^2 + 25$; i

For each polynomial function, *(a)* *list all possible rational zeros,* *(b)* *find all rational zeros, and* *(c)* *factor* $f(x)$. *See Example 3.*

35. $f(x) = x^3 - 2x^2 - 13x - 10$ 36. $f(x) = x^3 + 5x^2 + 2x - 8$

37. $f(x) = x^3 + 6x^2 - x - 30$ 38. $f(x) = x^3 - x^2 - 10x - 8$

39. $f(x) = 6x^3 + 17x^2 - 31x - 12$

40. $f(x) = 15x^3 + 61x^2 + 2x - 8$

41. $f(x) = 24x^3 + 40x^2 - 2x - 12$

42. $f(x) = 24x^3 + 80x^2 + 82x + 24$

For each polynomial function, find all zeros and their multiplicities.

43. $f(x) = (x - 2)^3(x^2 - 7)$

44. $f(x) = (x + 1)^2(x - 1)^3(x^2 - 10)$

45. $f(x) = 3x(x - 2)(x + 3)(x^2 - 1)$

46. $f(x) = 5x^2(x^2 - 16)(x + 5)$

47. $f(x) = (x^2 + x - 2)^5(x - 1 + \sqrt{3})^2$

48. $f(x) = (2x^2 - 7x + 3)^3(x - 2 - \sqrt{5})$

*Find a polynomial function $f(x)$ of degree 3 with real coefficients that satisfies the given conditions. **See Example 4.***

49. Zeros of -3, 1, and 4; $f(2) = 30$

50. Zeros of 1, -1, and 0; $f(2) = 3$

51. Zeros of -2, 1, and 0; $f(-1) = -1$

52. Zeros of 2, -3, and 5; $f(3) = 6$

53. Zero of -3 having multiplicity 3; $f(3) = 36$

54. Zero of 2 and zero of 4 having multiplicity 2; $f(1) = -18$

55. Zero of 0 and zero of 1 having multiplicity 2; $f(2) = 10$

56. Zero of -4 and zero of 0 having multiplicity 2; $f(-1) = -6$

*Find a polynomial function $f(x)$ of least degree having only real coefficients with zeros as given. **See Examples 4–6.***

57. $5 + i$ and $5 - i$

58. $7 - 2i$ and $7 + 2i$

59. $0, i,$ and $1 + i$

60. $0, -i,$ and $2 + i$

61. $1 + \sqrt{2}, 1 - \sqrt{2},$ and 1

62. $1 - \sqrt{3}, 1 + \sqrt{3},$ and 1

63. $2 - i, 3,$ and -1

64. $3 + 2i, -1,$ and 2

65. 2 and $3 + i$

66. -1 and $4 - 2i$

67. $1 - \sqrt{2}, 1 + \sqrt{2},$ and $1 - i$

68. $2 + \sqrt{3}, 2 - \sqrt{3},$ and $2 + 3i$

69. $2 - i$ and $6 - 3i$

70. $5 + i$ and $4 - i$

71. $4, 1 - 2i,$ and $3 + 4i$

72. $-1, 5 - i,$ and $1 + 4i$

73. $1 + 2i$ and 2 (multiplicity 2)

74. $2 + i$ and -3 (multiplicity 2)

*Use Descartes' rule of signs to determine the different possibilities for the numbers of positive, negative, and nonreal complex zeros for each function. **See Example 7.***

75. $f(x) = 2x^3 - 4x^2 + 2x + 7$

76. $f(x) = x^3 + 2x^2 + x - 10$

77. $f(x) = 4x^3 - x^2 + 2x - 7$

78. $f(x) = 3x^3 + 6x^2 + x + 7$

79. $f(x) = 5x^4 + 3x^2 + 2x - 9$

80. $f(x) = 3x^4 + 2x^3 - 8x^2 - 10x - 1$

81. $f(x) = -8x^4 + 3x^3 - 6x^2 + 5x - 7$

82. $f(x) = 6x^4 + 2x^3 + 9x^2 + x + 5$

83. $f(x) = x^5 + 3x^4 - x^3 + 2x + 3$

84. $f(x) = 2x^5 - x^4 + x^3 - x^2 + x + 5$

85. $f(x) = 7x^5 + 6x^4 + 2x^3 + 9x^2 + x + 5$

86. $f(x) = -2x^5 + 10x^4 - 6x^3 + 8x^2 - x + 1$

87. $f(x) = 2x^5 - 7x^3 + 6x + 8$

88. $f(x) = 11x^5 - x^3 + 7x - 5$

89. $f(x) = 5x^6 - 6x^5 + 7x^3 - 4x^2 + x + 2$ **90.** $f(x) = 9x^6 - 7x^4 + 8x^2 + x + 6$

Find all complex zeros of each polynomial function. Give exact values. List multiple zeros as necessary.*

91. $f(x) = x^4 + 2x^3 - 3x^2 + 24x - 180$

92. $f(x) = x^3 - x^2 - 8x + 12$

93. $f(x) = x^4 + x^3 - 9x^2 + 11x - 4$

94. $f(x) = x^3 - 14x + 8$

95. $f(x) = 2x^5 + 11x^4 + 16x^3 + 15x^2 + 36x$

96. $f(x) = 3x^3 - 9x^2 - 31x + 5$

97. $f(x) = x^5 - 6x^4 + 14x^3 - 20x^2 + 24x - 16$

98. $f(x) = 9x^4 + 30x^3 + 241x^2 + 720x + 600$

99. $f(x) = 2x^4 - x^3 + 7x^2 - 4x - 4$

100. $f(x) = 32x^4 - 188x^3 + 261x^2 + 54x - 27$

101. $f(x) = 5x^3 - 9x^2 + 28x + 6$

102. $f(x) = 4x^3 + 3x^2 + 8x + 6$

103. $f(x) = x^4 + 29x^2 + 100$

104. $f(x) = x^4 + 4x^3 + 6x^2 + 4x + 1$

105. $f(x) = x^4 + 2x^2 + 1$

106. $f(x) = x^4 - 8x^3 + 24x^2 - 32x + 16$

107. $f(x) = x^4 - 6x^3 + 7x^2$

108. $f(x) = 4x^4 - 65x^2 + 16$

109. $f(x) = x^4 - 8x^3 + 29x^2 - 66x + 72$

110. $f(x) = 12x^4 - 43x^3 + 50x^2 + 38x - 12$

111. $f(x) = x^6 - 9x^4 - 16x^2 + 144$

112. $f(x) = x^6 - x^5 - 26x^4 + 44x^3 + 91x^2 - 139x + 30$

If c and d are complex numbers, prove each statement. (Hint: Let $c = a + bi$ and $d = m + ni$ and form all the conjugates, the sums, and the products.)

113. $\overline{c + d} = \overline{c} + \overline{d}$

114. $\overline{c \cdot d} = \overline{c} \cdot \overline{d}$

115. $\overline{a} = a$ for any real number a

116. $\overline{c^2} = (\overline{c})^2$

In 1545, a method of solving a cubic equation of the form

$$x^3 + mx = n,$$

developed by Niccolo Tartaglia, was published in the Ars Magna, *a work by Girolamo Cardano. The formula for finding the one real solution of the equation is*

$$x = \sqrt[3]{\frac{n}{2} + \sqrt{\left(\frac{n}{2}\right)^2 + \left(\frac{m}{3}\right)^3}} - \sqrt[3]{\frac{-n}{2} + \sqrt{\left(\frac{n}{2}\right)^2 + \left(\frac{m}{3}\right)^3}}.$$

(Source: Gullberg, J., *Mathematics from the Birth of Numbers,* W.W. Norton & Company.) *Use the formula to solve each equation for the one real solution.*

117. $x^3 + 9x = 26$

118. $x^3 + 15x = 124$

---

* The authors would like to thank Aileen Solomon of Trident Technical College for preparing and suggesting the inclusion of **Exercises 91–104.**

3.4 Polynomial Functions: Graphs, Applications, and Models

■ Graphs of $f(x) = ax^n$

■ Graphs of General Polynomial Functions

■ Behavior at Zeros

■ Turning Points and End Behavior

■ Graphing Techniques

■ Intermediate Value and Boundedness Theorems

■ Approximating Real Zeros

■ Polynomial Models

Graphs of $f(x) = ax^n$ We can now graph polynomial functions of degree 3 or more with real number domains (since we will be graphing in the real number plane). We begin by inspecting the graphs of several functions of the form

$$f(x) = ax^n, \quad \text{with } a = 1,$$

and noticing their behavior. The identity function $f(x) = x$, the squaring function $f(x) = x^2$, and the cubing function $f(x) = x^3$ were first graphed in **Section 2.6** using a general point-plotting method of graphing.

Each function in **Figure 20** has odd degree and is an odd function exhibiting symmetry about the origin. Each has domain $(-\infty, \infty)$ and range $(-\infty, \infty)$ and is continuous on its entire domain $(-\infty, \infty)$. Additionally, these odd functions are increasing on their entire domain $(-\infty, \infty)$, appearing as though they fall to the left and rise to the right.

Figure 20

Each function in **Figure 21** has even degree and is an even function exhibiting symmetry about the y-axis. Each has domain $(-\infty, \infty)$ but restricted range $[0, \infty)$. These even functions are also continuous on their entire domain $(-\infty, \infty)$. However, they are decreasing on $(-\infty, 0)$ and increasing on $(0, \infty)$, appearing as though they rise both to the left and to the right.

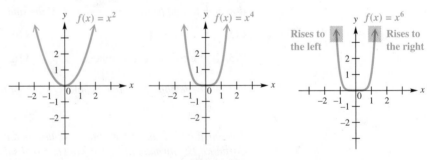

Figure 21

The behaviors that we see in the graphs of these basic polynomial functions also apply to more complicated polynomial functions.

Graphs of General Polynomial Functions As with quadratic functions, the absolute value of a in $f(x) = ax^n$ determines the width of the graph.

- When $|a| > 1$, the graph is stretched vertically, making it narrower.

- When $0 < |a| < 1$, the graph is shrunk or compressed vertically, making it wider.

Compared to the graph of $f(x) = ax^n$, the following also hold true.

- The graph of $\boldsymbol{f(x) = -ax^n}$ is reflected across the x-axis.

- The graph of $\boldsymbol{f(x) = ax^n + k}$ is translated (shifted) k units up if $k > 0$ and $|k|$ units down if $k < 0$.

- The graph of $\boldsymbol{f(x) = a(x - h)^n}$ is translated h units to the right if $h > 0$ and $|h|$ units to the left if $h < 0$.

- The graph of $\boldsymbol{f(x) = a(x - h)^n + k}$ shows a combination of these translations.

EXAMPLE 1   **Examining Vertical and Horizontal Translations**

Graph each function. Determine the largest intervals of the domain for which each function is increasing or decreasing.

(a) $f(x) = x^5 - 2$ **(b)** $f(x) = (x + 1)^6$ **(c)** $f(x) = -2(x - 1)^3 + 3$

SOLUTION

(a) The graph of $f(x) = x^5 - 2$ will be the same as that of $f(x) = x^5$, but translated 2 units down. See **Figure 22.** This function is increasing on its entire domain $(-\infty, \infty)$.

(b) In $f(x) = (x + 1)^6$, function f has a graph like that of $f(x) = x^6$, but since $x + 1 = x - (-1)$, it is translated 1 unit to the left. See **Figure 23.** This function is decreasing on $(-\infty, -1)$ and increasing on $(-1, \infty)$.

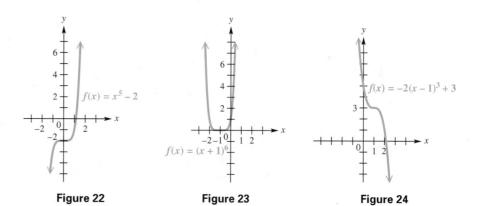

Figure 22   Figure 23   Figure 24

(c) The negative sign in -2 causes the graph of

$$f(x) = -2(x - 1)^3 + 3$$

to be reflected across the x-axis when compared with the graph of $f(x) = x^3$. Because $|-2| > 1$, the graph is stretched vertically when compared to the graph of $f(x) = x^3$. As shown in **Figure 24,** the graph is also translated 1 unit to the right and 3 units up. This function is decreasing on its entire domain $(-\infty, \infty)$.

Now Try Exercises 13, 15, and 19.

Unless otherwise restricted, the domain of a polynomial function is the set of all real numbers. Polynomial functions are smooth, continuous curves on the interval $(-\infty, \infty)$. The range of a polynomial function of odd degree is also the set of all real numbers.

Typical graphs of polynomial functions of odd degree are shown in **Figure 25.** These graphs suggest that for every polynomial function f of odd degree there is at least one real value of x that makes $f(x) = 0$. The real zeros are located at the x-intercepts of the graph and can be determined by inspecting the factored form of each polynomial.

Odd Degree

$$f(x) = 2x^3 + 8x^2 + 2x - 12$$
$$= 2(x - 1)(x + 2)(x + 3)$$

(a)

$$f(x) = -x^3 + 2x^2 - x + 2$$
$$= -(x - 2)(x - i)(x + i)$$

(b)

$$f(x) = x^5 + 4x^4 + x^3 - 10x^2 - 4x + 8$$
$$= (x - 1)^2(x + 2)^3$$

(c)

Figure 25

A polynomial function of even degree has a range of the form $(-\infty, k]$ or $[k, \infty)$, for some real number k. **Figure 26** shows two typical graphs.

Even Degree

$$f(x) = x^4 - 5x^2 + 4$$
$$= (x - 1)(x + 1)(x - 2)(x + 2)$$

(a)

The graph crosses the x-axis at $(-2, 0)$. The graph touches the x-axis and then turns at $(-1, 0)$. The graph crosses the x-axis and changes shape at $(1, 0)$.

$$f(x) = -x^6 - x^5 + 4x^4 + 2x^3 - 5x^2 - x + 2$$
$$= -(x + 2)(x + 1)^2(x - 1)^3$$

(b)

Figure 26

Behavior at Zeros   Take a close look at the graph in **Figure 26(b).** Here we see a sixth-degree polynomial function with three distinct zeros, yet the behavior of the graph at each zero is different. This behavior depends on the multiplicity of the zero as determined by the exponent on the corresponding factor. The factored form of the polynomial in **Figure 26(b)** is $f(x) = -(x + 2)^1(x + 1)^2(x - 1)^3$.

- $(x + 2)$ is a factor of multiplicity 1. Therefore, the graph crosses the x-axis at $(-2, 0)$.

- $(x + 1)$ is a factor of multiplicity 2. Therefore, the graph is tangent to the x-axis at $(-1, 0)$. This means that it touches the x-axis, then turns and changes behavior from decreasing to increasing similar to that of the squaring function $f(x) = x^2$ at its zero.

- $(x - 1)$ is a factor of multiplicity 3. Therefore, the graph crosses the x-axis **and** is tangent to the x-axis at $(1, 0)$. This causes a change in concavity, or shape, at this x-intercept with behavior similar to that of the cubing function $f(x) = x^3$ at its zero.

See **Figure 27** for a generalization of the behavior of the graphs of polynomial functions at their zeros.

The graph crosses the x-axis at $(c, 0)$ if c is a zero of multiplicity 1.

(a)

The graph is tangent to the x-axis at $(c, 0)$ if c is a zero of even multiplicity. The graph bounces, or turns, at c.

(b)

The graph crosses **and** is tangent to the x-axis at $(c, 0)$ if c is a zero of odd multiplicity greater than 1. The graph wiggles at c.

(c)

Figure 27

Turning Points and End Behavior   The graphs in **Figures 25 and 26** show that polynomial functions often have **turning points** where the function changes from increasing to decreasing or from decreasing to increasing.

Turning Points

A polynomial function of degree n has at most $n - 1$ turning points, with at least one turning point between each pair of successive zeros.

The **end behavior** of a polynomial graph is determined by the **dominating term**—that is, the term of greatest degree. A polynomial of the form

$$f(x) = a_n x^n + a_{n-1} x^{n-1} + \cdots + a_0$$

has the same end behavior as $f(x) = a_n x^n$. For example,

$$f(x) = 2x^3 + 8x^2 + 2x - 12$$

has the same end behavior as $f(x) = 2x^3$. It is large and positive for large positive values of x and large and negative for negative values of x with large absolute value. That is, it rises to the right and falls to the left.

Figure 25(a) shows that as x increases without bound, y does also. For the same graph, as x decreases without bound, y does also.

$$\text{As} \quad x \to \infty, \quad y \to \infty \quad \text{and as} \quad x \to -\infty, \quad y \to -\infty.$$

LOOKING AHEAD TO CALCULUS

To find the x-coordinates of the two turning points of the graph of

$$f(x) = 2x^3 + 8x^2 + 2x - 12$$

we can use the "maximum" and "minimum" capabilities of a graphing calculator and determine that, to the nearest thousandth, they are -0.131 and -2.535. In calculus, their exact values can be found by determining the zeros of the derivative function of $f(x)$,

$$f'(x) = 6x^2 + 16x + 2,$$

because the turning points occur precisely where the tangent line has 0 slope. Using the quadratic formula would show that the zeros are

$$\frac{-4 \pm \sqrt{13}}{3},$$

which agree with the calculator approximations.

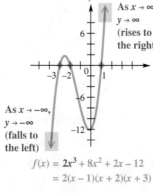

$$f(x) = 2x^3 + 8x^2 + 2x - 12$$
$$= 2(x - 1)(x + 2)(x + 3)$$

Figure 25(a) (repeated)

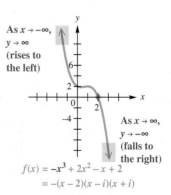

$$f(x) = -x^3 + 2x^2 - x + 2$$
$$= -(x - 2)(x - i)(x + i)$$

Figure 25(b) (repeated)

The graph in **Figure 25(b)** has the same end behavior as $f(x) = -x^3$.

$$\text{As} \quad x \to \infty, \quad y \to -\infty \quad \text{and as} \quad x \to -\infty, \quad y \to \infty.$$

The graph of a polynomial function with a dominating term of even degree will show end behavior in the same direction. See **Figure 26**.

End Behavior of Graphs of Polynomial Functions

Suppose that ax^n is the dominating term of a polynomial function f of **odd degree.**

1. If $a > 0$, then as $x \to \infty$, $f(x) \to \infty$, and as $x \to -\infty$, $f(x) \to -\infty$. Therefore, the end behavior of the graph is of the type shown in **Figure 28(a).** We symbolize it as $\nearrow$.

2. If $a < 0$, then as $x \to \infty$, $f(x) \to -\infty$, and as $x \to -\infty$, $f(x) \to \infty$. Therefore, the end behavior of the graph is of the type shown in **Figure 28(b).** We symbolize it as $\searrow$.

$a > 0$ n odd	$a < 0$ n odd	$a > 0$ n even	$a < 0$ n even
(a)	**(b)**	**(a)**	**(b)**
Figure 28		**Figure 29**	

Suppose that ax^n is the dominating term of a polynomial function f of **even degree.**

1. If $a > 0$, then as $|x| \to \infty$, $f(x) \to \infty$. Therefore, the end behavior of the graph is of the type shown in **Figure 29(a).** We symbolize it as $\smile$.

2. If $a < 0$, then as $|x| \to \infty$, $f(x) \to -\infty$. Therefore, the end behavior of the graph is of the type shown in **Figure 29(b).** We symbolize it as $\frown$.

EXAMPLE 2 **Determining End Behavior**

The graphs of the functions defined as follows are shown in A–D.

$$f(x) = x^4 - x^2 + 5x - 4, \qquad g(x) = -x^6 + x^2 - 3x - 4,$$
$$h(x) = 3x^3 - x^2 + 2x - 4, \quad \text{and} \quad k(x) = -x^7 + x - 4$$

Based on the previous discussion of end behavior, match each function with its graph.

A. **B.** **C.** **D.**

SOLUTION

Function f is of even degree with positive leading coefficient. Its graph is in C.

Function g is of even degree with negative leading coefficient. Its graph is in A.

Function h has odd degree and a dominating term with positive coefficient. Its graph is in B.

Function k has odd degree and a dominating term with negative coefficient. Its graph is in D.

✔ *Now Try Exercises 21, 23, 25, and 27.*

Graphing Techniques We have discussed several characteristics of the graphs of polynomial functions that are useful for graphing the function by hand. A **comprehensive graph** of a polynomial function $f(x)$ will show the following characteristics.

- all x-intercepts (locations of zeros) and the behavior of the graph there

- the y-intercept

- the sign of $f(x)$ within the intervals formed by the x-intercepts

- enough of the domain to show the end behavior

In **Examples 3 and 4,** we sketch the graphs of two polynomial functions by hand. While there are several ways to approach this process, we use the following general guidelines.

Graphing a Polynomial Function

Let $f(x) = a_n x^n + a_{n-1} x^{n-1} + \cdots + a_1 x + a_0$, with $a_n \neq 0$, be a polynomial function of degree n. To sketch its graph, follow these steps.

Step 1 Find the real zeros of f. Plot the corresponding x-intercepts.

Step 2 Find $f(0) = a_0$. Plot the corresponding y-intercept.

Step 3 Use end behavior, whether the graph crosses, bounces on, or wiggles through the x-axis at the x-intercepts, and selected points as necessary to complete the graph.

EXAMPLE 3 **Graphing a Polynomial Function**

Graph $f(x) = 2x^3 + 5x^2 - x - 6$.

SOLUTION

Step 1 The possible rational zeros are ± 1, ± 2, ± 3, ± 6, $\pm \frac{1}{2}$, and $\pm \frac{3}{2}$. Use synthetic division to show that 1 is a zero.

$$\begin{array}{r} 1)\overline{\begin{array}{cccc} 2 & 5 & -1 & -6 \end{array}} \qquad \text{(Section 3.2)} \\ \begin{array}{cccc} & 2 & 7 & 6 \end{array} \\ \hline \begin{array}{cccc} 2 & 7 & 6 & 0 \end{array} \leftarrow f(1) = 0 \end{array}$$

Thus,

$$f(x) = (x - 1)(2x^2 + 7x + 6)$$

$$f(x) = (x - 1)(2x + 3)(x + 2). \quad \text{Factor. (Section R.4)}$$

Set each linear factor equal to 0, and then solve for x to find zeros. The three zeros of f are 1, $-\frac{3}{2}$, and -2. See **Figure 30.**

Step 2 $f(0) = -6$, so plot $(0, -6)$. See **Figure 30.**

Step 3 Since the dominating term of $f(x)$ is $2x^3$, the graph will have end behavior similar to that of $f(x) = x^3$. It will rise to the right and fall to the left as $\nearrow$. See **Figure 30.** Each zero of $f(x)$ occurs with multiplicity 1, meaning that the graph of $f(x)$ will cross the x-axis at each of its zeros. Because the graph of a polynomial function has no breaks, gaps, or sudden jumps, we now have sufficient information to sketch the graph of $f(x)$.

Figure 30

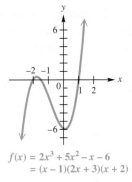

$f(x) = 2x^3 + 5x^2 - x - 6$
$= (x - 1)(2x + 3)(x + 2)$

Figure 31

Begin sketching at either end of the graph with the appropriate end behavior, and draw a smooth curve that crosses the x-axis at each zero, has a turning point between successive zeros, and passes through the y-intercept as shown in **Figure 31.**

Additional points may be used to verify whether the graph is above or below the x-axis between the zeros and to add detail to the sketch of the graph. The zeros divide the x-axis into four intervals:

$$(-\infty, -2), \quad \left(-2, -\frac{3}{2}\right), \quad \left(-\frac{3}{2}, 1\right) \quad \text{and} \quad (1, \infty).$$

Select an x-value as a test point in each interval, and substitute it into the equation for $f(x)$ to determine additional points on the graph. A typical selection of test points and the results of the tests are shown in the table.

Interval	Test Point x	Value of $f(x)$	Sign of $f(x)$	Graph Above or Below x-Axis
$(-\infty, -2)$	-3	-12	Negative	Below
$\left(-2, -\frac{3}{2}\right)$	$-\frac{7}{4}$	$\frac{11}{32}$	Positive	Above
$\left(-\frac{3}{2}, 1\right)$	0	-6	Negative	Below
$(1, \infty)$	2	28	Positive	Above

The sketch could be improved by plotting the points found in each interval in the table.

✔ *Now Try Exercise 29.*

EXAMPLE 4   **Graphing a Polynomial Function**

Graph $f(x) = -(x - 1)(x - 3)(x + 2)^2$.

SOLUTION

Step 1 Since the polynomial is given in factored form, the zeros can be determined by inspection. They are 1, 3, and -2. Plot the corresponding x-intercepts of the graph of $f(x)$. See **Figure 32.**

Step 2
$$f(0) = -(0 - 1)(0 - 3)(0 + 2)^2 \qquad \text{Find } f(0).$$
$$= -(-1)(-3)(2)^2, \quad \text{or} \quad -12$$

Plot the y-intercept $(0, -12)$ on the y-axis. See **Figure 32.**

Step 3 The dominating term of $f(x)$ can be found by multiplying the factors and identifying the term of greatest degree. Here it is $-(x)(x)(x)^2 = -x^4$, indicating that the end behavior of the graph is ⌢⌣. Because 1 and 3 are zeros of multiplicity 1, the graph will cross the x-axis at these zeros. The graph of $f(x)$ will touch the x-axis at -2 and then turn and change direction because it is a zero of even multiplicity.

Begin at either end of the graph with the appropriate end behavior and draw a smooth curve that crosses the x-axis at 1 and 3 and that touches the x-axis at -2, then turns and changes direction. The graph will also pass through the y-intercept $(0, -12)$. See **Figure 32.** Using test points within intervals formed by the x-intercepts is a good way to add detail to the graph and verify the accuracy of the sketch. A typical selection of test points is $(-3, -24)$, $(-1, -8)$, $(2, 16)$, and $(4, -108)$.

✔ *Now Try Exercise 33.*

$f(x) = -(x - 1)(x - 3)(x + 2)^2$

Figure 32

NOTE It is possible to reverse the process of **Example 4** and write the polynomial function from its graph if the zeros and any other point on the graph are known. Suppose that you are asked to find a polynomial function of least possible degree having the graph shown in **Figure 32** on the previous page. Because the graph crosses the x-axis at 1 and 3 and bounces at -2, we know that the factored form of the function is as follows.

Multiplicity
one

Multiplicity
two

$$f(x) = a(x - 1)^1(x - 3)^1(x + 2)^2$$

Now find the value of a by substituting the x- and y-values of any other point on the graph, say $(0, -12)$, into this function and solving for a.

$f(x) = a(x - 1)(x - 3)(x + 2)^2$

$-12 = a(0 - 1)(0 - 3)(0 + 2)^2$ Let $x = 0$ and $y = -12$.

$-12 = a(12)$ Simplify.

$a = -1$ Divide by 12. Interchange sides.

Verify in **Example 4** that the polynomial function is

$$f(x) = -(x - 1)(x - 3)(x + 2)^2.$$

In the Exercise set, problems of this type are labeled *Connecting Graphs with Equations*.

We emphasize the important relationships among the following concepts.

- the x-intercepts of the graph of $y = f(x)$

- the zeros of the function f

- the solutions of the equation $f(x) = 0$

- the factors of $f(x)$

For example, the graph of the function in **Example 3,**

$$f(x) = 2x^3 + 5x^2 - x - 6$$

$$f(x) = (x - 1)(2x + 3)(x + 2), \quad \text{Factored form}$$

has x-intercepts $(1, 0)$, $\left(-\frac{3}{2}, 0\right)$, and $(-2, 0)$, as shown in **Figure 31** on the previous page. Since 1, $-\frac{3}{2}$, and -2 are the x-values where the function is 0, they are the zeros of f. Also, 1, $-\frac{3}{2}$, and -2 are the solutions of the polynomial equation

$$2x^3 + 5x^2 - x - 6 = 0.$$

This discussion is summarized as follows.

---

x-Intercepts, Zeros, Solutions, and Factors

If f is a polynomial function and $(a, 0)$ is an x-intercept of the graph of $y = f(x)$, then

a is a zero of f, a is a solution of $f(x) = 0$,

and $x - a$ is a factor of $f(x)$.

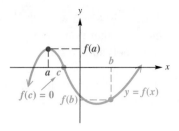

Figure 33

As **Examples 3 and 4** show, one key to graphing a polynomial function is locating its zeros. In the special case where the potential zeros are rational numbers, the zeros are found by the rational zeros theorem (**Section 3.3**).

Occasionally, irrational zeros can be found by inspection. For instance, $f(x) = x^3 - 2$ has the irrational zero $\sqrt[3]{2}$. The next two theorems presented in this section apply to the zeros of every polynomial function with real coefficients. The first theorem uses the fact that graphs of polynomial functions are continuous curves. The proof requires advanced methods, so it is not given here. **Figure 33** illustrates the theorem.

> ### Intermediate Value Theorem for Polynomials
>
> If $f(x)$ is a polynomial function with only real coefficients, and if for real numbers a and b, the values $f(a)$ and $f(b)$ are opposite in sign, then there exists at least one real zero between a and b.

This theorem helps identify intervals where zeros of polynomial functions are located. If $f(a)$ and $f(b)$ are opposite in sign, then 0 is between $f(a)$ and $f(b)$, and so there must be a number c between a and b where $f(c) = 0$.

EXAMPLE 5 **Locating a Zero**

Use synthetic division and a graph to show that $f(x) = x^3 - 2x^2 - x + 1$ has a real zero between 2 and 3.

ALGEBRAIC SOLUTION

Use synthetic division to find $f(2)$ and $f(3)$.

$$
\begin{array}{r}
2)\overline{1 \quad -2 \quad -1 \quad 1} \\
\phantom{2)\overline{1}} 2 \quad 0 \quad -2 \\
\hline
1 \quad 0 \quad -1 \quad -1 = f(2)
\end{array}
$$

$$
\begin{array}{r}
3)\overline{1 \quad -2 \quad -1 \quad 1} \\
\phantom{3)\overline{1}} 3 \quad 3 \quad 6 \\
\hline
1 \quad 1 \quad 2 \quad 7 = f(3)
\end{array}
$$

Since $f(2)$ is negative and $f(3)$ is positive, by the intermediate value theorem there must be a real zero between 2 and 3.

GRAPHING CALCULATOR SOLUTION

The graphing calculator screen in **Figure 34** indicates that this zero is approximately 2.2469796. (Notice that there are two other zeros as well.)

Figure 34

☑ *Now Try Exercise 49.*

CAUTION *Be careful how you interpret the intermediate value theorem.* If $f(a)$ and $f(b)$ are **not** opposite in sign, it does not necessarily mean that there is no zero between a and b. In **Figure 35,** $f(a)$ and $f(b)$ are both negative, but -3 and -1, which are between a and b, are zeros of $f(x)$.

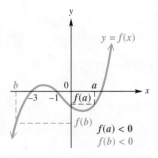

Figure 35

The intermediate value theorem for polynomials helps limit the search for real zeros to smaller and smaller intervals. In **Example 5,** we used the theorem to verify that there is a real zero between 2 and 3. We can use the theorem repeatedly to locate the zero more accurately.

The **boundedness theorem** shows how the bottom row of a synthetic division is used to place upper and lower bounds on possible real zeros of a polynomial function.

Boundedness Theorem

Let $f(x)$ be a polynomial function of degree $n \geq 1$ with real coefficients and with a positive leading coefficient. Suppose $f(x)$ is divided synthetically by $x - c$.

(a) If $c > 0$ and all numbers in the bottom row of the synthetic division are nonnegative, then $f(x)$ has no zero greater than c.

(b) If $c < 0$ and the numbers in the bottom row of the synthetic division alternate in sign (with 0 considered positive or negative, as needed), then $f(x)$ has no zero less than c.

Proof We outline the proof of part (a). The proof for part (b) is similar. By the division algorithm, if $f(x)$ is divided by $x - c$, then for some $q(x)$ and r,

$$f(x) = (x - c)q(x) + r,$$

where all coefficients of $q(x)$ are nonnegative, $r \geq 0$, and $c > 0$. If $x > c$, then $x - c > 0$. Since $q(x) > 0$ and $r \geq 0$,

$$f(x) = (x - c)q(x) + r > 0.$$

This means that $f(x)$ will never be 0 for $x > c$.

EXAMPLE 6 **Using the Boundedness Theorem**

Show that the real zeros of $f(x) = 2x^4 - 5x^3 + 3x + 1$ satisfy these conditions.

(a) No real zero is greater than 3.   **(b)** No real zero is less than -1.

SOLUTION

(a) Since $f(x)$ has real coefficients and the leading coefficient, 2, is positive, use the boundedness theorem. Divide $f(x)$ synthetically by $x - 3$.

$$
\begin{array}{r|rrrrr}
3) & 2 & -5 & 0 & 3 & 1 \\
   &   & 6 & 3 & 9 & 36 \\
\hline
   & 2 & 1 & 3 & 12 & 37 \\
\end{array}
\longleftarrow \text{All are nonnegative.}
$$

Since $3 > 0$ and all numbers in the last row of the synthetic division are nonnegative, $f(x)$ has no real zero greater than 3.

(b) We use the boundedness theorem again and divide $f(x)$ synthetically by $x - (-1)$, or $x + 1$.

$$
\begin{array}{r|rrrrr}
-1) & 2 & -5 & 0 & 3 & 1 \\
    &   & -2 & 7 & -7 & 4 \\
\hline
    & 2 & -7 & 7 & -4 & 5 \\
\end{array}
\longleftarrow \text{These numbers alternate in sign.}
$$

Here $-1 < 0$ and the numbers in the last row alternate in sign, so $f(x)$ has no real zero less than -1.

✔️ *Now Try Exercises 57 and 59.*

> **Approximating Real Zeros** We can approximate the irrational real zeros of a polynomial function using a graphing calculator.

EXAMPLE 7 **Approximating Real Zeros of a Polynomial Function**

Approximate the real zeros of $f(x) = x^4 - 6x^3 + 8x^2 + 2x - 1$.

SOLUTION The dominating term is x^4, so the graph will have end behavior similar to the graph of $f(x) = x^4$, which is positive for all values of x with large absolute values. That is, the end behavior is up at the left and the right, $\smile$. There are at most four real zeros, since the polynomial is fourth-degree.

Since $f(0) = -1$, the y-intercept is $(0, -1)$. Because the end behavior is positive on the left and the right, by the intermediate value theorem f has at least one real zero on either side of $x = 0$. To approximate the zeros, we use a graphing calculator. The graph in **Figure 36** shows that there are four real zeros, and the table indicates that they are between

$$-1 \text{ and } 0, \quad 0 \text{ and } 1, \quad 2 \text{ and } 3, \quad \text{and} \quad 3 \text{ and } 4$$

because there is a sign change in $f(x) = Y_1$ in each case.

$$f(x) = x^4 - 6x^3 + 8x^2 + 2x - 1$$

Figure 36   **Figure 37**

Using the capability of the calculator, we can find the zeros to a great degree of accuracy. **Figure 37** shows that the negative zero is approximately -0.4142136. Similarly, we find that the other three zeros are approximately

$$0.26794919, \quad 2.4142136, \quad \text{and} \quad 3.7320508.$$

☑ *Now Try Exercise 77.*

> **Polynomial Models**

EXAMPLE 8 **Examining a Polynomial Model**

The table shows the number of transactions, in millions, by users of bank debit cards for selected years.

(a) Using $x = 0$ to represent 1995, $x = 3$ to represent 1998, and so on, use the regression feature of a calculator to determine the quadratic function that best fits the data. Plot the data and the graph.

(b) Repeat part (a) for a cubic function (degree 3).

(c) Repeat part (a) for a quartic function (degree 4).

(d) The **correlation coefficient,** R, is a measure of the strength of the relationship between two variables. The values of R and R^2 are used to determine how well a regression model fits a set of data. The closer the value of R^2 is to 1, the better the fit. Compare R^2 for the three functions found in parts (a)–(c) to decide which function best fits the data.

Year	Transactions (in millions)
1995	829
1998	3765
2000	5290
2004	14,106
2008	28,464
2011*	39,049

Source: Statistical Abstract of the United States.
*Projected

SOLUTION

(a) The best-fitting quadratic function for the data is

$$y = 127.7x^2 + 377.3x + 868.7.$$

The regression coordinates screen and the graph are shown in **Figure 38.**

Figure 38

(b) The best-fitting cubic function is shown in **Figure 39** and is

$$y = -2.412x^3 + 184.7x^2 + 50.58x + 1125.$$

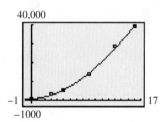

Figure 39

(c) The best-fitting quartic function is shown in **Figure 40** and is

$$y = -1.496x^4 + 45.07x^3 - 274.5x^2 + 1367x + 859.0.$$

Figure 40

(d) With the statistical diagnostics turned on, the value of R^2 is displayed with the regression results on the TI 83/84 Plus each time that a regression model is executed. By inspecting the R^2 value for each model above, we see that the quartic function provides the best fit since it has the largest R^2 value of 0.9999026772.

☑ *Now Try Exercise 105.*

NOTE In **Example 8(d),** we selected the quartic function as the best model based on the comparison of R^2 values of the models. In practice, however, the best choice of a model should also depend on the set of data being analyzed as well as analysis of its trends and attributes.

3.4 Exercises

Concept Check *Comprehensive graphs of four polynomial functions are shown in A–D. They represent the graphs of functions defined by these four equations, but not necessarily in the order listed.*

$$y = x^3 - 3x^2 - 6x + 8 \qquad y = x^4 + 7x^3 - 5x^2 - 75x$$

$$y = -x^3 + 9x^2 - 27x + 17 \qquad y = -x^5 + 36x^3 - 22x^2 - 147x - 90$$

Apply the concepts of this section to work Exercises 1–8.

A.

B.

C.

D.

1. Which one of the graphs is that of $y = x^3 - 3x^2 - 6x + 8$?

2. Which one of the graphs is that of $y = x^4 + 7x^3 - 5x^2 - 75x$?

3. How many real zeros does the function graphed in C have?

4. Which one of C and D is the graph of $y = -x^3 + 9x^2 - 27x + 17$?

5. Which of the graphs cannot be that of a cubic polynomial function?

6. Which one of the graphs is that of a function whose range is *not* $(-\infty, \infty)$?

7. The function $f(x) = x^4 + 7x^3 - 5x^2 - 75x$ has the graph shown in B. Use the graph to factor the polynomial.

8. The function $f(x) = -x^5 + 36x^3 - 22x^2 - 147x - 90$ has the graph shown in D. Use the graph to factor the polynomial.

Sketch the graph of each polynomial function. Determine the largest intervals of the domain for which each function is (a) increasing or (b) decreasing. See Example 1.

9. $f(x) = 2x^4$

10. $f(x) = \dfrac{1}{4}x^6$

11. $f(x) = -\dfrac{2}{3}x^5$

12. $f(x) = -\dfrac{5}{4}x^5$

13. $f(x) = \dfrac{1}{2}x^3 + 1$

14. $f(x) = -x^4 + 2$

15. $f(x) = -(x + 1)^3 + 1$

16. $f(x) = (x + 2)^3 - 1$

17. $f(x) = (x - 1)^4 + 2$

18. $f(x) = \dfrac{1}{3}(x + 3)^4 - 3$

19. $f(x) = \dfrac{1}{2}(x - 2)^2 + 4$

20. $f(x) = \dfrac{1}{3}(x + 1)^3 - 3$

Use an end behavior diagram, ⌣, ⌢, ↘, *or* ↗, *to describe the end behavior of the graph of each polynomial function.* ***See Example 2.***

21. $f(x) = 5x^5 + 2x^3 - 3x + 4$

22. $f(x) = -x^3 - 4x^2 + 2x - 1$

23. $f(x) = -4x^3 + 3x^2 - 1$

24. $f(x) = 4x^7 - x^5 + x^3 - 1$

25. $f(x) = 9x^6 - 3x^4 + x^2 - 2$

26. $f(x) = 10x^6 - x^5 + 2x - 2$

27. $f(x) = 3 + 2x - 4x^2 - 5x^{10}$

28. $f(x) = 7 + 2x - 5x^2 - 10x^4$

Graph each polynomial function. Factor first if the expression is not in factored form. ***See Examples 3 and 4.***

29. $f(x) = x^3 + 5x^2 + 2x - 8$

30. $f(x) = x^3 + 3x^2 - 13x - 15$

31. $f(x) = 2x(x - 3)(x + 2)$

32. $f(x) = x(x + 1)(x - 1)$

33. $f(x) = x^2(x - 2)(x + 3)^2$

34. $f(x) = x^2(x - 5)(x + 3)(x - 1)$

35. $f(x) = (3x - 1)(x + 2)^2$

36. $f(x) = (4x + 3)(x + 2)^2$

37. $f(x) = x^3 + 5x^2 - x - 5$

38. $f(x) = x^3 + x^2 - 36x - 36$

39. $f(x) = x^3 - x^2 - 2x$

40. $f(x) = 3x^4 + 5x^3 - 2x^2$

41. $f(x) = 2x^3(x^2 - 4)(x - 1)$

42. $f(x) = x^2(x - 3)^3(x + 1)$

43. $f(x) = 2x^3 - 5x^2 - x + 6$

44. $f(x) = 2x^4 + x^3 - 6x^2 - 7x - 2$

45. $f(x) = 3x^4 - 7x^3 - 6x^2 + 12x + 8$

46. $f(x) = x^4 + 3x^3 - 3x^2 - 11x - 6$

Use the intermediate value theorem for polynomials to show that each polynomial function has a real zero between the numbers given. ***See Example 5.***

47. $f(x) = 2x^2 - 7x + 4$; 2 and 3

48. $f(x) = 3x^2 - x - 4$; 1 and 2

49. $f(x) = 2x^3 - 5x^2 - 5x + 7$; 0 and 1

50. $f(x) = 2x^3 - 9x^2 + x + 20$; 2 and 2.5

51. $f(x) = 2x^4 - 4x^2 + 4x - 8$; 1 and 2

52. $f(x) = x^4 - 4x^3 - x + 3$; 0.5 and 1

53. $f(x) = x^4 + x^3 - 6x^2 - 20x - 16$; 3.2 and 3.3

54. $f(x) = x^4 - 2x^3 - 2x^2 - 18x + 5$; 3.7 and 3.8

55. $f(x) = x^4 - 4x^3 - 20x^2 + 32x + 12$; -1 and 0

56. $f(x) = x^5 + 2x^4 + x^3 + 3$; -1.8 and -1.7

Show that the real zeros of each polynomial function satisfy the given conditions. ***See Example 6.***

57. $f(x) = x^4 - x^3 + 3x^2 - 8x + 8$; no real zero greater than 2

58. $f(x) = 2x^5 - x^4 + 2x^3 - 2x^2 + 4x - 4$; no real zero greater than 1

59. $f(x) = x^4 + x^3 - x^2 + 3$; no real zero less than -2

60. $f(x) = x^5 + 2x^3 - 2x^2 + 5x + 5$; no real zero less than -1

61. $f(x) = 3x^4 + 2x^3 - 4x^2 + x - 1$; no real zero greater than 1

62. $f(x) = 3x^4 + 2x^3 - 4x^2 + x - 1$; no real zero less than -2

63. $f(x) = x^5 - 3x^3 + x + 2$; no real zero greater than 2

64. $f(x) = x^5 - 3x^3 + x + 2$; no real zero less than -3

Connecting Graphs with Equations In Exercises 65–70, find a polynomial function $f(x)$ of least possible degree having the graph shown. (Hint: See the Note following *Example 4.*)

65.

66.

67.

68.

69.

70.

Use a graphing calculator to graph the function in the viewing window specified. Compare the graph to the one shown in the answer section of this text. Then use the graph to find $f(1.25)$.

71. $f(x) = 2x(x - 3)(x + 2)$; window: $[-3, 4]$ by $[-20, 12]$
Compare to **Exercise 31.**

72. $f(x) = x^2(x - 2)(x + 3)^2$; window: $[-4, 3]$ by $[-24, 4]$
Compare to **Exercise 33.**

73. $f(x) = (3x - 1)(x + 2)^2$; window: $[-4, 2]$ by $[-15, 15]$
Compare to **Exercise 35.**

74. $f(x) = x^3 + 5x^2 - x - 5$; window: $[-6, 2]$ by $[-30, 30]$
Compare to **Exercise 37.**

Use a graphing calculator to approximate the real zero discussed in each specified exercise. *See Example 7.*

75. Exercise 47

76. Exercise 49

77. Exercise 51

78. Exercise 50

For the given polynomial function, approximate each zero as a decimal to the nearest tenth. **See Example 7.**

79. $f(x) = x^3 + 3x^2 - 2x - 6$

80. $f(x) = x^3 - 3x + 3$

81. $f(x) = -2x^4 - x^2 + x + 5$

82. $f(x) = -x^4 + 2x^3 + 3x^2 + 6$

Use a graphing calculator to find the coordinates of the turning points of the graph of each polynomial function in the given domain interval. Give answers to the nearest hundredth.

83. $f(x) = x^3 + 4x^2 - 8x - 8; \ [-3.8, -3]$

84. $f(x) = x^3 + 4x^2 - 8x - 8; \ [0.3, 1]$

85. $f(x) = 2x^3 - 5x^2 - x + 1; \ [-1, 0]$

86. $f(x) = 2x^3 - 5x^2 - x + 1; \ [1.4, 2]$

87. $f(x) = x^4 - 7x^3 + 13x^2 + 6x - 28; \ [-1, 0]$

88. $f(x) = x^3 - x + 3; \ [-1, 0]$

Solve each problem.

89. *(Modeling) Social Security Numbers* Your Social Security number (SSN) is unique, and with it you can construct your own personal Social Security polynomial. Let the polynomial function be defined as follows, where a_i represents the ith digit in your SSN:

$$\text{SSN}(x) = (x - a_1)(x + a_2)(x - a_3)(x + a_4)(x - a_5) \cdot$$
$$(x + a_6)(x - a_7)(x + a_8)(x - a_9).$$

For example, if the SSN is 539-58-0954, the polynomial function is

$$\text{SSN}(x) = (x - 5)(x + 3)(x - 9)(x + 5)(x - 8)(x + 0)(x - 9)(x + 5)(x - 4).$$

A comprehensive graph of this function is shown in **Figure A.** In **Figure B,** we show a screen obtained by zooming in on the positive zeros, as the comprehensive graph does not show the local behavior well in this region. Use a graphing calculator to graph your own "personal polynomial."

Figure A

Figure B

90. A comprehensive graph of $f(x) = x^4 - 7x^3 + 18x^2 - 22x + 12$ is shown in the two screens, along with displays of the two real zeros. Find the two remaining non-real complex zeros.

Relating Concepts

For individual or collaborative investigation (Exercises 91–96)

For any function $y = f(x)$, the following hold true.

(a) The real solutions of $f(x) = 0$ are the x-values of x-intercepts of the graph.

(b) The real solutions of $f(x) < 0$ are the x-values for which the graph lies *below* the x-axis.

(c) The real solutions of $f(x) > 0$ are the x-values for which the graph lies *above* the x-axis.

In Exercises 91–96, a polynomial function $f(x)$ is given in both expanded and factored forms. Graph the function, and solve the equations and inequalities. Give multiplicities of solutions when applicable.

91. $f(x) = x^3 - 3x^2 - 6x + 8$
$= (x - 4)(x - 1)(x + 2)$

    (a) $f(x) = 0$ **(b)** $f(x) < 0$

    (c) $f(x) > 0$

92. $f(x) = x^3 + 4x^2 - 11x - 30$
$= (x - 3)(x + 2)(x + 5)$

    (a) $f(x) = 0$ **(b)** $f(x) < 0$

    (c) $f(x) > 0$

93. $f(x) = 2x^4 - 9x^3 - 5x^2 + 57x - 45$
$= (x - 3)^2(2x + 5)(x - 1)$

    (a) $f(x) = 0$ **(b)** $f(x) < 0$

    (c) $f(x) > 0$

94. $f(x) = 4x^4 + 27x^3 - 42x^2$
$- 445x - 300$
$= (x + 5)^2(4x + 3)(x - 4)$

    (a) $f(x) = 0$ **(b)** $f(x) < 0$

    (c) $f(x) > 0$

95. $f(x) = -x^4 - 4x^3 + 3x^2 + 18x$
$= x(2 - x)(x + 3)^2$

    (a) $f(x) = 0$ **(b)** $f(x) \geq 0$

    (c) $f(x) \leq 0$

96. $f(x) = -x^4 + 2x^3 + 8x^2$
$= x^2(4 - x)(x + 2)$

    (a) $f(x) = 0$ **(b)** $f(x) \geq 0$

    (c) $f(x) \leq 0$

(Modeling) Exercises 97–104 are geometric in nature and lead to polynomial models. Solve each problem.

97. *Volume of a Box* A rectangular piece of cardboard measuring 12 in. by 18 in. is to be made into a box with an open top by cutting equal-size squares from each corner and folding up the sides. Let x represent the length of a side of each such square in inches.

    (a) Give the restrictions on x.

    (b) Determine a function V that gives the volume of the box as a function of x.

    (c) For what value of x will the volume be a maximum? What is this maximum volume? (*Hint:* Use the function of a graphing calculator that allows you to determine a maximum point within a given interval.)

    (d) For what values of x will the volume be greater than 80 in.^{3}?

98. *Construction of a Rain Gutter* A piece of rectangular sheet metal is 20 in. wide. It is to be made into a rain gutter by turning up the edges to form parallel sides. Let x represent the length of each of the parallel sides.

    (a) Give the restrictions on x.

    (b) Determine a function $\mathcal{A}$ that gives the area of a cross section of the gutter.

    (c) For what value of x will $\mathcal{A}$ be a maximum (and thus maximize the amount of water that the gutter will hold)? What is this maximum area?

    (d) For what values of x will the area of a cross section be less than 40 in.^{2}?

99. *Sides of a Right Triangle* A certain right triangle has area 84 in.^{2}. One leg of the triangle measures 1 in. less than the hypotenuse. Let x represent the length of the hypotenuse.

(a) Express the length of the leg mentioned above in terms of x. Give the domain of x.

(b) Express the length of the other leg in terms of x.

(c) Write an equation based on the information determined thus far. Square both sides and then write the equation with one side as a polynomial with integer coefficients, in descending powers, and the other side equal to 0.

(d) Solve the equation in part (c) graphically. Find the lengths of the three sides of the triangle.

100. *Area of a Rectangle* Find the value of x in the figure that will maximize the area of rectangle *ABCD*.

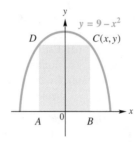

101. *Butane Gas Storage* A storage tank for butane gas is to be built in the shape of a right circular cylinder of altitude 12 ft, with a half sphere attached to each end. If x represents the radius of each half sphere, what radius should be used to cause the volume of the tank to be 144π ft^3?

102. *Volume of a Box* A standard piece of notebook paper measuring 8.5 in. by 11 in. is to be made into a box with an open top by cutting equal-size squares from each corner and folding up the sides. Let x represent the length of a side of each such square in inches. Use the table feature of your graphing calculator to do the following.

(a) Find the maximum volume of the box.

(b) Determine when the volume of the box will be greater than 40 in.^{3}.

103. *Floating Ball* The polynomial function

$$f(x) = \frac{\pi}{3}x^3 - 5\pi x^2 + \frac{500\pi d}{3}$$

can be used to find the depth that a ball 10 cm in diameter sinks in water. The constant d is the density of the ball, where the density of water is 1. The smallest *positive* zero of $f(x)$ equals the depth that the ball sinks. Approximate this depth for each material and interpret the results.

(a) A wooden ball with $d = 0.8$

(b) A solid aluminum ball with $d = 2.7$

(c) A spherical water balloon with $d = 1$

104. *Floating Ball* Refer to **Exercise 103.** If a ball has a 20-cm diameter, then the function becomes

$$f(x) = \frac{\pi}{3}x^3 - 10\pi x^2 + \frac{4000\pi d}{3}.$$

This function can be used to determine the depth that the ball sinks in water. Find the depth that this size ball sinks when $d = 0.6$.

⊞ *(Modeling) Solve each problem involving a polynomial function model.* ***See Example 8.***

105. *Highway Design* To allow enough distance for cars to pass on two-lane highways, engineers calculate minimum sight distances between curves and hills. The table shows the minimum sight distance y in feet for a car traveling at x miles per hour.

x (in mph)	20	30	40	50	60	65	70
y (in feet)	810	1090	1480	1840	2140	2310	2490

Source: Haefner, L., *Introduction to Transportation Systems,* Holt, Rinehart and Winston.

(a) Make a scatter diagram of the data.

(b) Use the regression feature of a calculator to find the best-fitting linear function for the data. Graph the function with the data.

(c) Repeat part (b) for a cubic function.

(d) Estimate the minimum sight distance for a car traveling 43 mph using the functions from parts (b) and (c).

(e) By comparing the graphs of the functions in parts (b) and (c) with the data, decide which function best fits the given data.

106. *Water Pollution* Copper in high doses can be lethal to aquatic life. The table lists copper concentrations in freshwater mussels after 45 days at various distances downstream from an electroplating plant. The concentration C is measured in micrograms of copper per gram of mussel x kilometers downstream.

x	5	21	37	53	59
C	20	13	9	6	5

Source: Foster, R., and J. Bates, "Use of mussels to monitor point source industrial discharges," *Environ. Sci. Technol.*; Mason, C., *Biology of Freshwater Pollution,* John Wiley & Sons.

(a) Make a scatter diagram of the data.

(b) Use the regression feature of a calculator to find the best-fitting quadratic function for the data. Graph the function with the data.

(c) Repeat part (b) for a cubic function.

(d) By comparing graphs of the functions in parts (b) and (c) with the data, decide which function best fits the given data.

(e) Concentrations above 10 are lethal to mussels. Find the values of x (using the cubic function) for which this is the case.

107. *Government Spending on Health Research and Training* The table lists the annual amount (in billions of dollars) spent by the federal government on health research and training programs over an 8-yr period.

Year	Amount	Year	Amount
2002	21.4	2006	28.8
2003	24.0	2007	29.3
2004	27.1	2008	29.9
2005	28.1	2009	30.6

Source: U.S. Office of Management and Budget.

Which one of the following provides the best model for these data, where x represents the year?

A. $f(x) = 0.5(x - 2002)^2 + 21.4$

B. $g(x) = (x - 2002) + 21.4$

C. $h(x) = 3.5\sqrt{x - 2002} + 21.4$

D. $k(x) = (x - 2002)^3 + 21.4$

108. *Swing of a Pendulum* Grandfather clocks use pendulums to keep accurate time. The relationship between the length of a pendulum L and the time T for one complete oscillation can be expressed by the equation $L = kT^n$, where k is a constant and n is a positive integer to be determined. The data in the table were taken for different lengths of pendulums.

L (ft)	T (sec)	L (ft)	T (sec)
1.0	1.11	3.0	1.92
1.5	1.36	3.5	2.08
2.0	1.57	4.0	2.22
2.5	1.76		

(a) As the length of the pendulum increases, what happens to T?

(b) Discuss how n and k could be found.

(c) Use the data to approximate k and determine the best value for n.

(d) Using the values of k and n from part (c), predict T for a pendulum having length 5 ft.

(e) If the length L of a pendulum doubles, what happens to the period T?

Summary Exercises on Polynomial Functions, Zeros, and Graphs

We use all of the theorems for finding complex zeros of polynomial functions in the next example.

EXAMPLE **Finding All Zeros of a Polynomial Function**

Find all zeros of $f(x) = x^4 - 3x^3 + 6x^2 - 12x + 8$.

SOLUTION We consider the number of positive zeros by observing the variations in signs for $f(x)$.

$$f(x) = +x^4 - 3x^3 + 6x^2 - 12x + 8 \quad \text{(Section 3.3)}$$

$$\ \ 1 \quad\ \ 2 \quad\ \ 3 \quad\ \ 4$$

Since $f(x)$ has four sign changes, we can use Descartes' rule of signs to determine that there are four, two, or zero positive real zeros. For negative zeros, we consider the variations in signs for $f(-x)$.

$$f(-x) = (-x)^4 - 3(-x)^3 + 6(-x)^2 - 12(-x) + 8$$

$$f(-x) = x^4 + 3x^3 + 6x^2 + 12x + 8$$

Since $f(-x)$ has no sign changes, there are no negative real zeros. Because the function is of degree 4, it has a maximum of four zeros with possibilities summarized as follows.

Positive	Negative	Nonreal Complex
4	0	0
2	0	2
0	0	4

We can now use the rational zeros theorem to determine that the possible rational zeros are ± 1, ± 2, ± 4, and ± 8. Based on Descartes' rule of signs, we discard the negative rational zeros from this list and try to find a positive rational zero. Suppose we start by using synthetic division to check 4.

$$\text{Proposed zero} \longrightarrow \begin{array}{r} 4\overline{)1 \quad -3 \quad 6 \quad -12 \quad 8} \\ \underline{4 \quad 4 \quad 40 \quad 112} \\ 1 \quad 1 \quad 10 \quad 28 \quad 120 \end{array} \longleftarrow f(4) = 120 \qquad \text{(Section 3.2)}$$

We find that 4 is not a zero. However, $4 > 0$, and the numbers in the bottom row of the synthetic division are nonnegative. Thus, the boundedness theorem indicates that there are no zeros greater than 4. We can discard 8 as a possible rational zero and use synthetic division to show that 1 and 2 are zeros.

$$\begin{array}{r} 1\overline{)1 \quad -3 \quad 6 \quad -12 \quad 8} \\ \underline{1 \quad -2 \quad 4 \quad -8} \\ 2\overline{)1 \quad -2 \quad 4 \quad -8 \quad 0} \longleftarrow f(1) = 0 \\ \underline{2 \quad 0 \quad 8} \\ 1 \quad 0 \quad 4 \quad 0 \qquad \longleftarrow f(2) = 0 \end{array}$$

The polynomial now factors as

$$f(x) = (x - 1)(x - 2)(x^2 + 4).$$

We find the remaining two zeros using algebra to solve for x in the quadratic factor of the following equation.

$$(x - 1)(x - 2)(x^2 + 4) = 0$$

$$x - 1 = 0 \quad \text{or} \quad x - 2 = 0 \quad \text{or} \quad x^2 + 4 = 0 \qquad \text{Zero-factor property}$$
$$\text{(Section 1.4)}$$

$$x = 1 \quad \text{or} \qquad x = 2 \quad \text{or} \qquad x^2 = -4$$

$$x = \pm 2i \qquad \text{Square root property}$$
$$\text{(Section 1.4)}$$

The linear factored form of the polynomial is

$$f(x) = (x - 1)(x - 2)(x - 2i)(x + 2i),$$

and the corresponding zeros are $1, 2, 2i,$ and $-2i$.

☑ *Now Try Exercise 3.*

For each polynomial function, do the following in order.

(a) Use Descartes' rule of signs to determine the different possibilities for the numbers of positive, negative, and nonreal complex zeros.

(b) Use the rational zeros theorem to determine the possible rational zeros.

(c) Use synthetic division with the boundedness theorem where appropriate and/or factoring to find the rational zeros, if any.

(d) Find all other complex zeros (both real and nonreal), if any.

1. $f(x) = 6x^3 - 41x^2 + 26x + 24$

2. $f(x) = 2x^3 - 5x^2 - 4x + 3$

3. $f(x) = 3x^4 - 5x^3 + 14x^2 - 20x + 8$

4. $f(x) = 2x^4 - 3x^3 + 16x^2 - 27x - 18$

5. $f(x) = 6x^4 - 5x^3 - 11x^2 + 10x - 2$

6. $f(x) = 5x^4 + 8x^3 - 19x^2 - 24x + 12$

7. $f(x) = x^5 - 6x^4 + 16x^3 - 24x^2 + 16x$ (*Hint:* Factor out x first.)

8. $f(x) = 2x^4 + 8x^3 - 7x^2 - 42x - 9$

9. $f(x) = 8x^4 + 8x^3 - x - 1$ (*Hint:* Factor the polynomial.)

10. $f(x) = 2x^5 + 5x^4 - 9x^3 - 11x^2 + 19x - 6$

For each polynomial function, do the following in order.

(a) Use Descartes' rule of signs to determine the different possibilities for the numbers of positive, negative, and nonreal complex zeros.

(b) Use the rational zeros theorem to determine the possible rational zeros.

(c) Find the rational zeros, if any.

(d) Find all other real zeros, if any.

(e) Find any other complex zeros (that is, zeros that are not real), if any.

(f) Find the *x*-intercepts of the graph, if any.

(g) Find the *y*-intercept of the graph.

(h) Use synthetic division to find $f(4)$, and give the coordinates of the corresponding point on the graph.

(i) Determine the end behavior of the graph.

(j) Sketch the graph.

11. $f(x) = x^4 + 3x^3 - 3x^2 - 11x - 6$

12. $f(x) = -2x^5 + 5x^4 + 34x^3 - 30x^2 - 84x + 45$

13. $f(x) = 2x^5 - 10x^4 + x^3 - 5x^2 - x + 5$

14. $f(x) = 3x^4 - 4x^3 - 22x^2 + 15x + 18$

15. $f(x) = -2x^4 - x^3 + x + 2$

16. $f(x) = 4x^5 + 8x^4 + 9x^3 + 27x^2 + 27x$ (*Hint:* Factor out *x* first.)

17. $f(x) = 3x^4 - 14x^2 - 5$ (*Hint:* Factor the polynomial.)

18. $f(x) = -x^5 - x^4 + 10x^3 + 10x^2 - 9x - 9$

19. $f(x) = -3x^4 + 22x^3 - 55x^2 + 52x - 12$

20. For the polynomial functions in **Exercises 11–19** that have irrational zeros, find approximations to the nearest thousandth.

3.5 Rational Functions: Graphs, Applications, and Models

- The Reciprocal Function $f(x) = \frac{1}{x}$
- The Function $f(x) = \frac{1}{x^2}$
- Asymptotes
- Steps for Graphing Rational Functions
- Rational Function Models

A rational expression is a fraction that is the quotient of two polynomials. A function defined by a rational expression is a *rational function.*

Rational Function

A function f of the form

$$f(x) = \frac{p(x)}{q(x)},$$

where $p(x)$ and $q(x)$ are polynomials, with $q(x) \neq 0$, is a **rational function.**

Some examples of rational functions are

$$f(x) = \frac{1}{x}, \quad f(x) = \frac{x + 1}{2x^2 + 5x - 3}, \quad \text{and} \quad f(x) = \frac{3x^2 - 3x - 6}{x^2 + 8x + 16}.$$

Rational functions

Since any values of *x* such that $q(x) = 0$ are excluded from the domain of a rational function, this type of function often has a **discontinuous graph**—that is, a graph that has one or more breaks in it. (See **Chapter 2.**)

The Reciprocal Function $f(x) = \frac{1}{x}$ The simplest rational function with a variable denominator is the **reciprocal function.**

$$f(x) = \frac{1}{x} \quad \text{Reciprocal function}$$

The domain of this function is the set of all real numbers except 0. The number 0 cannot be used as a value of x, but it is helpful to find values of $f(x)$ for some values of x very close to 0. We use the table feature of a graphing calculator to do this. The tables in **Figure 41** suggest that $|f(x)|$ increases without bound as x gets closer and closer to 0, which is written in symbols as

$$|f(x)| \to \infty \quad \text{as} \quad x \to 0.$$

(The symbol $x \to 0$ means that x approaches 0, without necessarily ever being equal to 0.) Since x cannot equal 0, the graph of $f(x) = \frac{1}{x}$ will never intersect the vertical line $x = 0$. This line is a **vertical asymptote.**

As X approaches ∞, $Y_1 = \frac{1}{X}$ approaches 0 through positive values.

As X approaches $-\infty$, $Y_1 = \frac{1}{X}$ approaches 0 through negative values.

Figure 42

As X approaches 0 from the left, $Y_1 = \frac{1}{X}$ approaches $-\infty$. ($-1E-6$ means -1×10^{-6}.)

As X approaches 0 from the right, $Y_1 = \frac{1}{X}$ approaches ∞.

Figure 41

As $|x|$ increases without bound, the values of $f(x) = \frac{1}{x}$ get closer and closer to 0, as shown in the tables in **Figure 42.** Letting $|x|$ increase without bound (written $|x| \to \infty$) causes the graph of $f(x) = \frac{1}{x}$ to move closer and closer to the horizontal line $y = 0$. This line is a **horizontal asymptote.**

The graph and important features of $f(x) = \frac{1}{x}$ are summarized in the following box and shown in **Figure 43.**

Reciprocal Function $f(x) = \dfrac{1}{x}$

Domain: $(-\infty, 0) \cup (0, \infty)$ Range: $(-\infty, 0) \cup (0, \infty)$

x	y
-2	$-\frac{1}{2}$
-1	-1
$-\frac{1}{2}$	-2
0	undefined
$\frac{1}{2}$	2
1	1
2	$\frac{1}{2}$

Figure 43

- $f(x) = \frac{1}{x}$ decreases on the intervals $(-\infty, 0)$ and $(0, \infty)$.
- It is discontinuous at $x = 0$.
- The y-axis is a vertical asymptote, and the x-axis is a horizontal asymptote.
- It is an odd function, and its graph is symmetric with respect to the origin.

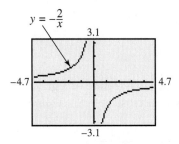

$y = -\frac{2}{x}$

The graph in **Figure 44** is shown here using a **decimal window.** In older models of graphing calculators, using a nondecimal window *may* produce an extraneous vertical line that is not part of the graph.

The graph of $y = \frac{1}{x}$ can be translated and reflected in the same way as other basic graphs in **Chapter 2.**

EXAMPLE 1 Graphing a Rational Function

Graph $y = -\frac{2}{x}$. Give the domain and range and the largest intervals of the domain for which the function is increasing or decreasing.

SOLUTION The expression $-\frac{2}{x}$ can be written as $-2\left(\frac{1}{x}\right)$ or $2\left(\frac{1}{-x}\right)$, indicating that the graph may be obtained by stretching the graph of $y = \frac{1}{x}$ vertically by a factor of 2 and reflecting it across either the x-axis or y-axis. The x- and y-axes remain the horizontal and vertical asymptotes. The domain and range are both still $(-\infty, 0) \cup (0, \infty)$. See **Figure 44.**

The graph shows that $f(x)$ is increasing on both sides of its vertical asymptote. Thus, it is increasing on

$$(-\infty, 0) \quad \text{and on} \quad (0, \infty).$$

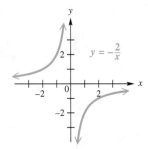

Figure 44

✔ Now Try Exercise 17.

EXAMPLE 2 Graphing a Rational Function

Graph $f(x) = \frac{2}{x + 1}$. Give the domain and range and the largest intervals of the domain for which the function is increasing or decreasing.

ALGEBRAIC SOLUTION

The expression $\frac{2}{x + 1}$ can be written as $2\left(\frac{1}{x + 1}\right)$, indicating that the graph may be obtained by shifting the graph of $y = \frac{1}{x}$ to the left 1 unit and stretching it vertically by a factor of 2. The graph is shown in **Figure 45.**

The horizontal shift affects the domain, which is now $(-\infty, -1) \cup (-1, \infty)$. The line $x = -1$ is the vertical asymptote, and the line $y = 0$ (the x-axis) remains the horizontal asymptote. The range is still $(-\infty, 0) \cup (0, \infty)$. The graph shows that $f(x)$ is decreasing on both sides of its vertical asymptote. Thus, it is decreasing on $(-\infty, -1)$ and on $(-1, \infty)$.

Figure 45

GRAPHING CALCULATOR SOLUTION

*When entering this rational function into the function editor, make sure that the numerator is **2** and the denominator is the entire expression $(x + 1)$.* The graph of this function has a vertical asymptote at $x = -1$ and a horizontal asymptote at $y = 0$, so it is reasonable to choose a viewing window that contains both asymptotes as well as enough of the graph to determine its basic characteristics. Notice in **Figure 46** that the asymptotes do not actually appear on the calculator screen.

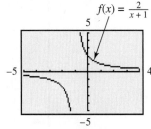

Figure 46

We may need to experiment with various modes and windows to obtain an accurate calculator graph of a rational function.

✔ Now Try Exercise 19.

The Function $f(x) = \frac{1}{x^2}$ The rational function

$$f(x) = \frac{1}{x^2}$$ Rational function

also has domain $(-\infty, 0) \cup (0, \infty)$. We can use the table feature of a graphing calculator to examine values of $f(x)$ for some x-values close to 0. See **Figure 47.**

X	Y1
1	1
10	.01
100	1E-4
1000	1E-6
10000	1E-8
100000	1E-10
1E6	1E-12

Y₁☐1/X²

As X approaches ∞, $Y_1 = \dfrac{1}{X^2}$ approaches 0 through positive values.

X	Y1
-1	1
-.1	100
-.01	10000
-.001	1E6
-1E-4	1E8
-1E-5	1E10
-1E-6	1E12

Y₁☐1/X²

As X approaches 0 from the left, $Y_1 = \dfrac{1}{X^2}$ approaches ∞.

X	Y1
1	1
.1	100
.01	10000
.001	1E6
1E-4	1E8
1E-5	1E10
1E-6	1E12

Y₁☐1/X²

As X approaches 0 from the right, $Y_1 = \dfrac{1}{X^2}$ approaches ∞.

Figure 47

X	Y1
-1	1
-10	.01
-100	1E-4
-1000	1E-6
-10000	1E-8
-1E5	1E-10
-1E6	1E-12

Y₁☐1/X²

As X approaches $-\infty$, $Y_1 = \dfrac{1}{X^2}$ approaches 0 through positive values.

Figure 48

The tables suggest that $f(x)$ increases without bound as x gets closer and closer to 0. Notice that as x approaches 0 from *either* side, function values are all positive and there is symmetry with respect to the y-axis. Thus, $f(x) \to \infty$ as $x \to 0$. The y-axis ($x = 0$) is the vertical asymptote.

As $|x|$ increases without bound, $f(x)$ approaches 0, as suggested by the tables in **Figure 48.** Again, function values are all positive. The x-axis is the horizontal asymptote of the graph.

The graph and important features of $f(x) = \frac{1}{x^2}$ are summarized in the following box and shown in **Figure 49.**

Rational Function $f(x) = \dfrac{1}{x^2}$

Domain: $(-\infty, 0) \cup (0, \infty)$ Range: $(0, \infty)$

x	y
± 3	$\frac{1}{9}$
± 2	$\frac{1}{4}$
± 1	1
$\pm \frac{1}{2}$	4
$\pm \frac{1}{4}$	16
0	undefined

Figure 49

- $f(x) = \frac{1}{x^2}$ increases on the interval $(-\infty, 0)$ and decreases on the interval $(0, \infty)$.
- It is discontinuous at $x = 0$.
- The y-axis is a vertical asymptote, and the x-axis is a horizontal asymptote.
- It is an even function, and its graph is symmetric with respect to the y-axis.

EXAMPLE 3   **Graphing a Rational Function**

Graph $y = \dfrac{1}{(x+2)^2} - 1$. Give the domain and range and the largest intervals of the domain for which the function is increasing or decreasing.

SOLUTION   The equation $y = \dfrac{1}{(x+2)^2} - 1$ is equivalent to

$$y = f(x+2) - 1, \quad \text{where} \quad f(x) = \frac{1}{x^2}.$$

This indicates that the graph will be shifted 2 units to the left and 1 unit down. The horizontal shift affects the domain, now $(-\infty, -2) \cup (-2, \infty)$, while the vertical shift affects the range, now $(-1, \infty)$.

The vertical asymptote has equation $x = -2$, and the horizontal asymptote has equation $y = -1$. A traditional graph is shown in **Figure 50,** with a calculator graph in **Figure 51.** Both graphs show that this function is increasing on $(-\infty, -2)$ and decreasing on $(-2, \infty)$.

This is the graph of $y = \frac{1}{x^2}$ shifted 2 units to the left and 1 unit down.

Figure 50

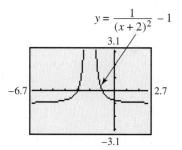

With older model calculators, a choice of window other than the one here *may* produce an undesired vertical line at $x = -2$.

Figure 51

✔ *Now Try Exercise 27.*

Asymptotes   The preceding examples suggest the following definitions of vertical and horizontal asymptotes.

Asymptotes

Let $p(x)$ and $q(x)$ define polynomials. Consider the rational function $f(x) = \dfrac{p(x)}{q(x)}$, written in lowest terms, and real numbers a and b.

1. If $|f(x)| \to \infty$ as $x \to a$, then the line $x = a$ is a **vertical asymptote.**

2. If $f(x) \to b$ as $|x| \to \infty$, then the line $y = b$ is a **horizontal asymptote.**

Locating asymptotes is important when graphing rational functions. We find *vertical asymptotes* by determining the values of x that make the denominator equal to 0. To find *horizontal asymptotes* (and, in some cases, *oblique asymptotes*), we must consider what happens to $f(x)$ as $|x| \to \infty$. These asymptotes determine the end behavior of the graph.

LOOKING AHEAD TO CALCULUS

The rational function defined by

$$f(x) = \frac{2}{x+1}$$

in **Example 2** has a vertical asymptote at $x = -1$. In calculus, the behavior of the graph of this function for values close to -1 is described using **one-sided limits**. As x approaches -1 from the *left*, the function values decrease without bound. This is written

$$\lim_{x \to -1^-} f(x) = -\infty.$$

As x approaches -1 from the *right*, the function values increase without bound. This is written

$$\lim_{x \to -1^+} f(x) = \infty.$$

Determining Asymptotes

To find the asymptotes of a rational function defined by a rational expression in *lowest terms,* use the following procedures.

1. **Vertical Asymptotes**
   Find any vertical asymptotes by setting the denominator equal to 0 and solving for x. If a is a zero of the denominator, then **the line $x = a$ is a vertical asymptote.**

2. **Other Asymptotes**
   Determine any other asymptotes by considering three possibilities:

   (a) If the numerator has lesser degree than the denominator, then there is a **horizontal asymptote $y = 0$** (the x-axis).

   (b) If the numerator and denominator have the same degree, and the function is of the form

   $$f(x) = \frac{a_n x^n + \cdots + a_0}{b_n x^n + \cdots + b_0}, \quad \text{where} \quad a_n, b_n \neq 0,$$

   then the **horizontal asymptote has equation $y = \frac{a_n}{b_n}$.**

   (c) If the numerator is of degree exactly one more than the denominator, then there will be an **oblique (slanted) asymptote.** To find it, divide the numerator by the denominator and disregard the remainder. Set the rest of the quotient equal to y to obtain the equation of the asymptote.

NOTE   The graph of a rational function may have more than one vertical asymptote, or it may have none at all. *The graph cannot intersect any vertical asymptote. There can be at most one other (nonvertical) asymptote, and the graph may intersect that asymptote,* as we shall see in **Example 7.**

EXAMPLE 4   **Finding Asymptotes of Rational Functions**

For each rational function f, find all asymptotes.

(a) $f(x) = \dfrac{x+1}{(2x-1)(x+3)}$ **(b)** $f(x) = \dfrac{2x+1}{x-3}$ **(c)** $f(x) = \dfrac{x^2+1}{x-2}$

SOLUTION

(a) *To find the vertical asymptotes, set the denominator equal to 0 and solve.*

$$(2x-1)(x+3) = 0$$

$2x - 1 = 0$ or $x + 3 = 0$ Zero-factor property **(Section 1.4)**

$x = \dfrac{1}{2}$ or $x = -3$ Solve each equation. **(Section 1.1)**

The equations of the vertical asymptotes are $x = \frac{1}{2}$ and $x = -3$.

To find the equation of the horizontal asymptote, divide each term by the greatest power of x in the expression. First, multiply the factors in the denominator.

$$f(x) = \frac{x+1}{(2x-1)(x+3)} = \frac{x+1}{2x^2 + 5x - 3}$$

Now divide each term in the numerator and denominator by x^2 since 2 is the greatest power of x.

$$f(x) = \dfrac{\dfrac{x}{x^2} + \dfrac{1}{x^2}}{\dfrac{2x^2}{x^2} + \dfrac{5x}{x^2} - \dfrac{3}{x^2}} = \dfrac{\dfrac{1}{x} + \dfrac{1}{x^2}}{2 + \dfrac{5}{x} - \dfrac{3}{x^2}} \quad \boxed{\begin{array}{l}\text{Stop here. Leave}\\\text{the expression in}\\\text{complex form.}\end{array}}$$

As $|x|$ increases without bound, the quotients $\frac{1}{x}$, $\frac{1}{x^2}$, $\frac{5}{x}$, and $\frac{3}{x^2}$ all approach 0, and the value of $f(x)$ approaches

$$\frac{0 + 0}{2 + 0 - 0} = \frac{0}{2}, \quad \text{or} \quad 0.$$

The line $y = 0$ (that is, the x-axis) is therefore the horizontal asymptote. This supports procedure 2(a) of determining asymptotes on the previous page.

(b) Set the denominator $x - 3$ equal to 0 to find that the vertical asymptote has equation $x = 3$. To find the horizontal asymptote, divide each term in the rational expression by x since the greatest power of x in the expression is 1.

$$f(x) = \frac{2x + 1}{x - 3} = \dfrac{\dfrac{2x}{x} + \dfrac{1}{x}}{\dfrac{x}{x} - \dfrac{3}{x}} = \dfrac{2 + \dfrac{1}{x}}{1 - \dfrac{3}{x}}$$

As $|x|$ increases without bound, both $\frac{1}{x}$ and $\frac{3}{x}$ approach 0, and $f(x)$ approaches

$$\frac{2 + 0}{1 - 0} = \frac{2}{1}, \quad \text{or} \quad 2,$$

so the line $y = 2$ is the horizontal asymptote. This supports procedure 2(b) of determining asymptotes on the previous page.

(c) Setting the denominator $x - 2$ equal to 0 shows that the vertical asymptote has equation $x = 2$. If we divide by the greatest power of x as before (x^2 in this case), we see that there is no horizontal asymptote because

$$f(x) = \frac{x^2 + 1}{x - 2} = \dfrac{\dfrac{x^2}{x^2} + \dfrac{1}{x^2}}{\dfrac{x}{x^2} - \dfrac{2}{x^2}} = \dfrac{1 + \dfrac{1}{x^2}}{\dfrac{1}{x} - \dfrac{2}{x^2}}$$

does not approach any real number as $|x| \to \infty$, since $\frac{1 + 0}{0 - 0} = \frac{1}{0}$ is undefined. This happens whenever the degree of the numerator is greater than the degree of the denominator.

In such cases, divide the denominator into the numerator to write the expression in another form. We use synthetic division, as shown in the margin. The result enables us to write the function as follows.

$$f(x) = x + 2 + \frac{5}{x - 2}$$

For very large values of $|x|$, $\frac{5}{x-2}$ is close to 0, and the graph approaches the line $y = x + 2$. This line is an **oblique asymptote** (slanted, neither vertical nor horizontal) for the graph of the function. This supports procedure 2(c) of determining asymptotes.

$$\begin{array}{r} 2\,\overline{)\,1 \quad 0 \quad 1} \\ \underline{2 \quad 4} \\ 1 \quad 2 \quad 5 \end{array} \quad \text{(Section 3.2)}$$

☑️ *Now Try Exercises 37, 39, and 41.*

Steps for Graphing Rational Functions A comprehensive graph of a rational function will show the following characteristics.

- all x- and y-intercepts

- all asymptotes: vertical, horizontal, and/or oblique

- the point at which the graph intersects its nonvertical asymptote (if there is any such point)

- the behavior of the function on each domain interval determined by the vertical asymptotes and x-intercepts

Graphing a Rational Function

Let $f(x) = \dfrac{p(x)}{q(x)}$ define a function where $p(x)$ and $q(x)$ are polynomials and the rational expression is written in lowest terms. To sketch its graph, follow these steps.

Step 1 Find any vertical asymptotes.

Step 2 Find any horizontal or oblique asymptotes.

Step 3 Find the y-intercept by evaluating $f(0)$.

Step 4 Find the x-intercepts, if any, by solving $f(x) = 0$. (These will correspond to the zeros of the numerator, $p(x)$.)

Step 5 Determine whether the graph will intersect its nonvertical asymptote $y = b$ or $y = mx + b$ by solving $f(x) = b$ or $f(x) = mx + b$.

Step 6 Plot selected points, as necessary. Choose an x-value in each domain interval determined by the vertical asymptotes and x-intercepts.

Step 7 Complete the sketch.

EXAMPLE 5 **Graphing a Rational Function with the x-Axis as Horizontal Asymptote**

Graph $f(x) = \dfrac{x + 1}{2x^2 + 5x - 3}$.

SOLUTION

Step 1 Since $2x^2 + 5x - 3 = (2x - 1)(x + 3)$, from **Example 4(a),** the vertical asymptotes have equations $x = \frac{1}{2}$ and $x = -3$.

Step 2 As shown in **Example 4(a),** the horizontal asymptote is the x-axis.

Step 3 The y-intercept is $\left(0, -\frac{1}{3}\right)$, as justified below.

$$f(0) = \frac{0 + 1}{2(0)^2 + 5(0) - 3} = -\frac{1}{3} \qquad \text{The } y\text{-intercept corresponds to the quotient of the constant terms.}$$

Step 4 The x-intercept is found by solving $f(x) = 0$.

$$\frac{x + 1}{2x^2 + 5x - 3} = 0 \qquad \text{Set } f(x) = 0.$$

$$x + 1 = 0 \qquad \text{If a rational expression is equal to 0, then its numerator must equal 0.}$$

$$x = -1 \qquad \text{The } x\text{-intercept is } (-1, 0).$$

Step 5 To determine whether the graph intersects its horizontal asymptote, solve this equation.

$$f(x) = 0 \longleftarrow y\text{-value of horizontal asymptote}$$

Since the horizontal asymptote is the x-axis, the solution of the equation $f(x) = 0$ was found in Step 4. The graph intersects its horizontal asymptote at $(-1, 0)$.

Step 6 Plot a point in each of the intervals determined by the x-intercepts and vertical asymptotes, $(-\infty, -3)$, $(-3, -1)$, $\left(-1, \frac{1}{2}\right)$ and $\left(\frac{1}{2}, \infty\right)$, to get an idea of how the graph behaves in each interval.

$$f(x) = \frac{x+1}{2x^2 + 5x - 3}$$

Figure 52

Interval	Test Point x	Value of $f(x)$	Sign of $f(x)$	Graph Above or Below x-Axis
$(-\infty, -3)$	-4	$-\frac{1}{3}$	Negative	Below
$(-3, -1)$	-2	$\frac{1}{5}$	Positive	Above
$\left(-1, \frac{1}{2}\right)$	0	$-\frac{1}{3}$	Negative	Below
$\left(\frac{1}{2}, \infty\right)$	2	$\frac{1}{5}$	Positive	Above

Step 7 Complete the sketch as shown in **Figure 52.** This function is decreasing on each interval of its domain—that is, on $(-\infty, -3)$, $\left(-3, \frac{1}{2}\right)$ and $\left(\frac{1}{2}, \infty\right)$.

✔ Now Try Exercise 67.

EXAMPLE 6 **Graphing a Rational Function That Does Not Intersect Its Horizontal Asymptote**

Graph $f(x) = \dfrac{2x+1}{x-3}$.

SOLUTION

Steps 1 and 2 As determined in **Example 4(b),** the equation of the vertical asymptote is $x = 3$. The horizontal asymptote has equation $y = 2$.

Step 3 $f(0) = -\frac{1}{3}$, so the y-intercept is $\left(0, -\frac{1}{3}\right)$.

Step 4 Solve $f(x) = 0$ to find any x-intercepts.

$$\frac{2x+1}{x-3} = 0 \qquad \text{Set } f(x) = 0.$$

$$2x + 1 = 0 \qquad \text{If a rational expression is equal to 0, then its numerator must equal 0.}$$

$$x = -\frac{1}{2} \qquad \text{The } x\text{-intercept is } \left(-\frac{1}{2}, 0\right).$$

Step 5 The graph does not intersect its horizontal asymptote since $f(x) = 2$ has no solution.

$$\frac{2x+1}{x-3} = 2 \qquad \text{Set } f(x) = 2.$$

$$2x + 1 = 2x - 6 \qquad \text{Multiply each side by } x - 3.$$

> A false statement results. The solution set is ∅.

$$1 = -6 \qquad \text{Subtract } 2x.$$

Steps 6 and 7 The points $(-4, 1)$, $\left(1, -\frac{3}{2}\right)$, and $\left(6, \frac{13}{3}\right)$ are on the graph and can be used to complete the sketch of this function, which decreases on every interval of its domain. See **Figure 53.**

Figure 53

✔ Now Try Exercise 63.

LOOKING AHEAD TO CALCULUS

The rational function

$$f(x) = \frac{2x + 1}{x - 3},$$

seen in **Example 6,** has horizontal asymptote $y = 2$. In calculus, the behavior of the graph of this function as x approaches $-\infty$ and as x approaches ∞ is described using **limits at infinity.** As x approaches $-\infty$, $f(x)$ approaches 2. This is written

$$\lim_{x \to -\infty} f(x) = 2.$$

As x approaches ∞, $f(x)$ approaches 2. This is written

$$\lim_{x \to \infty} f(x) = 2.$$

EXAMPLE 7 **Graphing a Rational Function That Intersects Its Horizontal Asymptote**

Graph $f(x) = \dfrac{3x^2 - 3x - 6}{x^2 + 8x + 16}$.

SOLUTION

Step 1 To find the vertical asymptote(s), solve $x^2 + 8x + 16 = 0$.

$$x^2 + 8x + 16 = 0 \qquad \text{Set the denominator equal to 0.}$$

$$(x + 4)^2 = 0 \qquad \text{Factor. (Section R.4)}$$

$$x = -4 \qquad \text{Zero-factor property}$$

Since the numerator is not 0 when $x = -4$, the vertical asymptote has equation $x = -4$.

Step 2 We divide all terms by x^2 and consider the behavior of each term as $|x|$ increases without bound to get the equation of the horizontal asymptote,

$$y = \frac{3}{1}, \quad \begin{array}{l} \leftarrow \text{ Leading coefficient of numerator} \\ \leftarrow \text{ Leading coefficient of denominator} \end{array} \quad \text{or} \quad y = 3.$$

Step 3 Because $f(0) = -\frac{3}{8}$, the y-intercept is $\left(0, -\frac{3}{8}\right)$.

Step 4 To find the x-intercept(s), if any, we solve $f(x) = 0$.

$$\frac{3x^2 - 3x - 6}{x^2 + 8x + 16} = 0 \qquad \text{Set } f(x) = 0.$$

$$3x^2 - 3x - 6 = 0 \qquad \text{Set the numerator equal to 0.}$$

$$x^2 - x - 2 = 0 \qquad \text{Divide by 3.}$$

$$(x - 2)(x + 1) = 0 \qquad \text{Factor. (Section R.4)}$$

$$x = 2 \quad \text{or} \quad x = -1 \qquad \text{Zero-factor property}$$

The x-intercepts are $(-1, 0)$ and $(2, 0)$.

Step 5 We set $f(x) = 3$ and solve to locate the point where the graph intersects the horizontal asymptote.

$$\frac{3x^2 - 3x - 6}{x^2 + 8x + 16} = 3 \qquad \text{Set } f(x) = 3.$$

$$3x^2 - 3x - 6 = 3x^2 + 24x + 48 \qquad \text{Multiply each side by } x^2 + 8x + 16.$$

$$-3x - 6 = 24x + 48 \qquad \text{Subtract } 3x^2.$$

$$-27x = 54 \qquad \text{Subtract } 24x \text{ and add 6.}$$

$$x = -2 \qquad \text{Divide by } -27.$$

The graph intersects its horizontal asymptote at $(-2, 3)$.

Steps 6 and 7 Some other points that lie on the graph are $(-10, 9)$, $\left(-8, 13\frac{1}{8}\right)$, and $\left(5, \frac{2}{3}\right)$. These are used to complete the graph, as shown in **Figure 54.**

✔️ *Now Try Exercise 83.*

$$f(x) = \frac{3x^2 - 3x - 6}{x^2 + 8x + 16}$$

Figure 54

Notice the behavior of the graph of the function in **Figure 54** near the line $x = -4$. As $x \to -4$ from either side, $f(x) \to \infty$. If we examine the behavior of the graph of the function in **Figure 53** (on the previous page) near the line $x = 3$, we find that $f(x) \to -\infty$ as x approaches 3 from the left, while $f(x) \to \infty$ as x approaches 3 from the right.

LOOKING AHEAD TO CALCULUS
Different types of discontinuity are discussed in calculus. The function in **Example 9,**

$$f(x) = \frac{x^2 - 4}{x - 2},$$

is said to have a **removable discontinuity** at $x = 2$, since the discontinuity can be removed by redefining f at 2. The function in **Example 8,**

$$f(x) = \frac{x^2 + 1}{x - 2},$$

has **infinite discontinuity** at $x = 2$, as indicated by the vertical asymptote there. The greatest integer function, discussed in **Section 2.6,** has **jump discontinuities** because the function values "jump" from one value to another for integer domain values.

The behavior of the graph of a rational function near a vertical asymptote $x = a$ partially depends on the exponent on $x - a$ in the denominator.

Behavior of Graphs of Rational Functions near Vertical Asymptotes

Suppose that $f(x)$ is a rational expression in lowest terms. If n is the largest positive integer such that $(x - a)^n$ is a factor of the denominator of $f(x)$, the graph will behave in the manner illustrated.

In **Section 3.4** we observed that the behavior of the graph of a polynomial function near its zeros is dependent on the multiplicity of the zero. The same statement can be made for rational functions. Suppose that $f(x)$ is defined by a rational expression in lowest terms. If n is the greatest positive integer such that $(x - c)^n$ is a factor of the numerator of $f(x)$, the graph will behave in the manner illustrated.

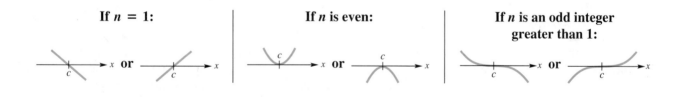

NOTE   Suppose that we are asked to reverse the process of **Example 7** and find the equation of a rational function having the graph shown in **Figure 54** on the previous page. Since the graph crosses the x-axis at its x-intercepts $(-1, 0)$ and $(2, 0)$, the numerator must have factors $(x + 1)$ and $(x - 2)$, each of degree 1.

The graph's behavior at its vertical asymptote $x = -4$ suggests that there is a factor of $(x + 4)$ of even degree in the denominator. The horizontal asymptote at $y = 3$ indicates that the numerator and denominator have the same degree (both 2) and that the ratio of leading coefficients is 3.

Verify in **Example 7** that the rational function is as follows.

$$f(x) = \frac{3(x + 1)(x - 2)}{(x + 4)^2},$$

or $\qquad f(x) = \dfrac{3x^2 - 3x - 6}{x^2 + 8x + 16}$ Multiply the factors in the numerator.
Square in the denominator.

In the Exercise set, problems of this type are labeled *Connecting Graphs with Equations*.

| EXAMPLE 8 | Graphing a Rational Function with an Oblique Asymptote |

Graph $f(x) = \dfrac{x^2 + 1}{x - 2}$.

SOLUTION As shown in **Example 4,** the vertical asymptote has equation $x = 2$, and the graph has an oblique asymptote with equation $y = x + 2$. The y-intercept is $\left(0, -\frac{1}{2}\right)$, and the graph has no x-intercepts since the numerator, $x^2 + 1$, has no real zeros. The graph does not intersect its oblique asymptote because the following has no solution.

$$\frac{x^2 + 1}{x - 2} = x + 2$$

$$x^2 + 1 = x^2 - 4 \quad \text{Multiply each side by } x - 2.$$

$$1 = -4 \quad \text{False}$$

Using the y-intercept, asymptotes, the points $\left(4, \frac{17}{2}\right)$ and $\left(-1, -\frac{2}{3}\right)$, and the general behavior of the graph near its asymptotes leads to the graph in **Figure 55.**

☑ *Now Try Exercise 87.*

Figure 55

A rational function that is not in lowest terms often has a "hole," or **point of discontinuity,** in its graph.

| EXAMPLE 9 | Graphing a Rational Function Defined by an Expression That Is Not in Lowest Terms |

Graph $f(x) = \dfrac{x^2 - 4}{x - 2}$.

ALGEBRAIC SOLUTION

The domain of this function cannot include 2. The expression $\frac{x^2 - 4}{x - 2}$ should be written in lowest terms.

$$f(x) = \frac{x^2 - 4}{x - 2} \quad \boxed{\text{Factor and then divide.}}$$

$$= \frac{(x + 2)(x - 2)}{x - 2} \quad \text{Factor.}$$

$$= x + 2, \quad x \neq 2$$

Therefore, the graph of this function will be the same as the graph of $y = x + 2$ (a straight line), with the exception of the point with x-value 2. A "hole" appears in the graph at $(2, 4)$. See **Figure 56.**

Figure 56

GRAPHING CALCULATOR SOLUTION

If we set the window of a graphing calculator so that an x-value of 2 is displayed, then we can see that the calculator cannot determine a value for y. We define

$$Y_1 = \frac{X^2 - 4}{X - 2}$$

and graph it in such a window, as in **Figure 57.** The error message in the table further supports the existence of a discontinuity at $X = 2$. (For the table, $Y_2 = X + 2$.)

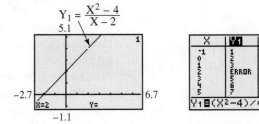

Figure 57

Notice the visible discontinuity at $X = 2$ in the graph. The window was chosen so the "hole" would be visible. This requires a decimal viewing window or a window with x-values centered at 2. Other window choices may not show this discontinuity.

☑ *Now Try Exercise 91.*

Rational Function Models   Rational functions have a variety of applications.

EXAMPLE 10   **Modeling Traffic Intensity with a Rational Function**

Vehicles arrive randomly at a parking ramp at an average rate of 2.6 vehicles per minute. The parking attendant can admit 3.2 vehicles per minute. However, since arrivals are random, lines form at various times. (*Source:* Mannering, F. and W. Kilareski, *Principles of Highway Engineering and Traffic Analysis,* 2nd ed., John Wiley & Sons.)

(a) The **traffic intensity** x is defined as the ratio of the average arrival rate to the average admittance rate. Determine x for this parking ramp.

(b) The average number of vehicles waiting in line to enter the ramp is given by

$$f(x) = \frac{x^2}{2(1 - x)},$$

where $0 \le x < 1$ is the traffic intensity. Graph $f(x)$ and compute $f(0.8125)$ for this parking ramp.

(c) What happens to the number of vehicles waiting as the traffic intensity approaches 1?

SOLUTION

(a) The average arrival rate is 2.6 vehicles per minute and the average admittance rate is 3.2 vehicles per minute, so

$$x = \frac{2.6}{3.2} = 0.8125.$$

(b) A calculator graph of f is shown in **Figure 58.**

$$f(0.8125) = \frac{0.8125^2}{2(1 - 0.8125)} \approx 1.76 \text{ vehicles}$$

(c) From the graph we see that as x approaches 1, $y = f(x)$ gets very large. Thus, the average number of waiting vehicles gets very large. This is what we would expect.

✔ *Now Try Exercise 113.*

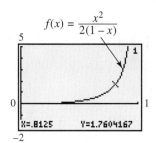

$f(x) = \dfrac{x^2}{2(1 - x)}$

Figure 58

3.5   Exercises

Concept Check   Provide a short answer to each question.

1. What is the domain of $f(x) = \dfrac{1}{x}$? What is its range?

2. What is the domain of $f(x) = \dfrac{1}{x^2}$? What is its range?

3. What is the largest interval over which $f(x) = \dfrac{1}{x}$ increases? decreases? is constant?

4. What is the largest interval over which $f(x) = \dfrac{1}{x^2}$ increases? decreases? is constant?

5. What is the equation of the vertical asymptote of the graph of $y = \dfrac{1}{x-3} + 2$? of the horizontal asymptote?

6. What is the equation of the vertical asymptote of the graph of $y = \dfrac{1}{(x+2)^2} - 4$? of the horizontal asymptote?

7. Is $f(x) = \dfrac{1}{x^2}$ an even or an odd function? What symmetry does its graph exhibit?

8. Is $f(x) = \dfrac{1}{x}$ an even or an odd function? What symmetry does its graph exhibit?

Concept Check   *Use the graphs of the rational functions in choices A–D to answer each question in Exercises 9–16. There may be more than one correct choice.*

A.

B.

C.

D.

9. Which choices have domain $(-\infty, 3) \cup (3, \infty)$?

10. Which choices have range $(-\infty, 3) \cup (3, \infty)$?

11. Which choices have range $(-\infty, 0) \cup (0, \infty)$?

12. Which choices have range $(0, \infty)$?

13. If f represents the function, only one choice has a single solution to the equation $f(x) = 3$. Which one is it?

14. Which choices have domain $(-\infty, 0) \cup (0, 3) \cup (3, \infty)$?

15. Which choices have the x-axis as a horizontal asymptote?

16. Which choices are symmetric with respect to a vertical line?

Explain how the graph of each function can be obtained from the graph of $y = \frac{1}{x}$ or $y = \frac{1}{x^2}$. Then graph f and give the (a) domain and (b) range. Determine the intervals of the domain for which the function is (c) increasing or (d) decreasing. See Examples 1–3.

17. $f(x) = \dfrac{2}{x}$ **18.** $f(x) = -\dfrac{3}{x}$ **19.** $f(x) = \dfrac{1}{x+2}$

20. $f(x) = \dfrac{1}{x-3}$ **21.** $f(x) = \dfrac{1}{x} + 1$ **22.** $f(x) = \dfrac{1}{x} - 2$

23. $f(x) = -\dfrac{2}{x^2}$ **24.** $f(x) = \dfrac{1}{x^2} + 3$ **25.** $f(x) = \dfrac{1}{(x-3)^2}$

26. $f(x) = \dfrac{-2}{(x-3)^2}$ **27.** $f(x) = \dfrac{-1}{(x+2)^2} - 3$ **28.** $f(x) = \dfrac{-1}{(x-4)^2} + 2$

Concept Check Match the rational function in Column I with the appropriate description in Column II. Choices in Column II can be used only once.

I

29. $f(x) = \dfrac{x+7}{x+1}$

30. $f(x) = \dfrac{x+10}{x+2}$

31. $f(x) = \dfrac{1}{x+4}$

32. $f(x) = \dfrac{-3}{x^2}$

33. $f(x) = \dfrac{x^2-16}{x+4}$

34. $f(x) = \dfrac{4x+3}{x-7}$

35. $f(x) = \dfrac{x^2+3x+4}{x-5}$

36. $f(x) = \dfrac{x+3}{x-6}$

II

A. The x-intercept is $(-3, 0)$.

B. The y-intercept is $(0, 5)$.

C. The horizontal asymptote is $y = 4$.

D. The vertical asymptote is $x = -1$.

E. There is a "hole" in its graph at $x = -4$.

F. The graph has an oblique asymptote.

G. The x-axis is its horizontal asymptote, and the y-axis is not its vertical asymptote.

H. The x-axis is its horizontal asymptote, and the y-axis is its vertical asymptote.

Give the equations of any vertical, horizontal, or oblique asymptotes for the graph of each rational function. **See Example 4.**

37. $f(x) = \dfrac{3}{x-5}$ **38.** $f(x) = \dfrac{-6}{x+9}$ **39.** $f(x) = \dfrac{4-3x}{2x+1}$

40. $f(x) = \dfrac{2x+6}{x-4}$ **41.** $f(x) = \dfrac{x^2-1}{x+3}$ **42.** $f(x) = \dfrac{x^2+4}{x-1}$

43. $f(x) = \dfrac{x^2-2x-3}{2x^2-x-10}$ **44.** $f(x) = \dfrac{3x^2-6x-24}{5x^2-26x+5}$

45. $f(x) = \dfrac{x^2+1}{x^2+9}$ **46.** $f(x) = \dfrac{4x^2+25}{x^2+9}$

47. *Concept Check* Let f be the function whose graph is obtained by translating the graph of $y = \frac{1}{x}$ to the right 3 units and up 2 units.

(a) Write an equation for $f(x)$ as a quotient of two polynomials.

(b) Determine the zero(s) of f.

(c) Identify the asymptotes of the graph of $f(x)$.

48. *Concept Check* Repeat **Exercise 47** with f the function whose graph is obtained by translating the graph of $y = -\frac{1}{x^2}$ to the left 3 units and up 1 unit.

49. *Concept Check* After the numerator is divided by the denominator,

$$f(x) = \dfrac{x^5 + x^4 + x^2 + 1}{x^4 + 1} \quad \text{becomes} \quad f(x) = x + 1 + \dfrac{x^2 - x}{x^4 + 1}.$$

(a) What is the oblique asymptote of the graph of the function?

(b) Where does the graph of the function intersect its asymptote?

(c) As $x \to \infty$, does the graph of the function approach its asymptote from above or below?

50. *Concept Check* Choices A–D below show the four ways in which the graph of a rational function can approach the vertical line $x = 2$ as an asymptote. Identify the graph of each rational function defined in parts (a)–(d).

(a) $f(x) = \dfrac{1}{(x-2)^2}$ **(b)** $f(x) = \dfrac{1}{x-2}$ **(c)** $f(x) = \dfrac{-1}{x-2}$ **(d)** $f(x) = \dfrac{-1}{(x-2)^2}$

A.

B.

C.

D.

51. *Concept Check* Which function has a graph that does not have a vertical asymptote?

A. $f(x) = \dfrac{1}{x^2 + 2}$ **B.** $f(x) = \dfrac{1}{x^2 - 2}$ **C.** $f(x) = \dfrac{3}{x^2}$ **D.** $f(x) = \dfrac{2x + 1}{x - 8}$

52. *Concept Check* Which function has a graph that does not have a horizontal asymptote?

A. $f(x) = \dfrac{2x - 7}{x + 3}$ **B.** $f(x) = \dfrac{3x}{x^2 - 9}$

C. $f(x) = \dfrac{x^2 - 9}{x + 3}$ **D.** $f(x) = \dfrac{x + 5}{(x + 2)(x - 3)}$

Identify any vertical, horizontal, or oblique asymptotes in the graph of $y = f(x)$. State the domain of f.

53.

54.

55.

56.

57.

58.

59.

60.

Sketch the graph of each rational function. See Examples 5–9.

61. $f(x) = \dfrac{x+1}{x-4}$

62. $f(x) = \dfrac{x-5}{x+3}$

63. $f(x) = \dfrac{x+2}{x-3}$

64. $f(x) = \dfrac{x-3}{x+4}$

65. $f(x) = \dfrac{4-2x}{8-x}$

66. $f(x) = \dfrac{6-3x}{4-x}$

67. $f(x) = \dfrac{3x}{x^2-x-2}$

68. $f(x) = \dfrac{2x+1}{x^2+6x+8}$

69. $f(x) = \dfrac{5x}{x^2-1}$

70. $f(x) = \dfrac{x}{4-x^2}$

71. $f(x) = \dfrac{(x+6)(x-2)}{(x+3)(x-4)}$

72. $f(x) = \dfrac{(x+3)(x-5)}{(x+1)(x-4)}$

73. $f(x) = \dfrac{3x^2+3x-6}{x^2-x-12}$

74. $f(x) = \dfrac{4x^2+4x-24}{x^2-3x-10}$

75. $f(x) = \dfrac{9x^2-1}{x^2-4}$

76. $f(x) = \dfrac{16x^2-9}{x^2-9}$

77. $f(x) = \dfrac{(x-3)(x+1)}{(x-1)^2}$

78. $f(x) = \dfrac{x(x-2)}{(x+3)^2}$

79. $f(x) = \dfrac{x}{x^2-9}$

80. $f(x) = \dfrac{-5}{2x+4}$

81. $f(x) = \dfrac{1}{x^2+1}$

82. $f(x) = \dfrac{(x-5)(x-2)}{x^2+9}$

83. $f(x) = \dfrac{(x+4)^2}{(x-1)(x+5)}$

84. $f(x) = \dfrac{(x+1)^2}{(x+2)(x-3)}$

85. $f(x) = \dfrac{20+6x-2x^2}{8+6x-2x^2}$

86. $f(x) = \dfrac{18+6x-4x^2}{4+6x+2x^2}$

87. $f(x) = \dfrac{x^2+1}{x+3}$

88. $f(x) = \dfrac{2x^2+3}{x-4}$

89. $f(x) = \dfrac{x^2+2x}{2x-1}$

90. $f(x) = \dfrac{x^2-x}{x+2}$

91. $f(x) = \dfrac{x^2-9}{x+3}$

92. $f(x) = \dfrac{x^2-16}{x+4}$

93. $f(x) = \dfrac{2x^2-5x-2}{x-2}$

94. $f(x) = \dfrac{x^2-5}{x-3}$

95. $f(x) = \dfrac{x^2-1}{x^2-4x+3}$

96. $f(x) = \dfrac{x^2-4}{x^2+3x+2}$

97. $f(x) = \dfrac{(x^2-9)(2+x)}{(x^2-4)(3+x)}$

98. $f(x) = \dfrac{(x^2-16)(3+x)}{(x^2-9)(4+x)}$

99. $f(x) = \dfrac{x^4-20x^2+64}{x^4-10x^2+9}$

100. $f(x) = \dfrac{x^4-5x^2+4}{x^4-24x^2+108}$

Connecting Graphs with Equations *Find an equation for each rational function graph.* *(Hint: See the note preceding* **Example 8.**)

101.

102.

103.

104.

105.

106.

Concept Check *In Exercises 107 and 108, find a possible equation for the function with a graph having the given features.*

107. x-intercepts: $(-1, 0)$ and $(3, 0)$
y-intercept: $(0, -3)$
vertical asymptote: $x = 1$
horizontal asymptote: $y = 1$

108. x-intercepts: $(1, 0)$ and $(3, 0)$
y-intercept: none
vertical asymptotes: $x = 0$ and $x = 2$
horizontal asymptote: $y = 1$

Use a graphing calculator to graph the rational function in the specified exercise. Then use the graph to find $f(1.25)$.

109. Exercise 61   **110. Exercise 67**   **111. Exercise 89**   **112. Exercise 91**

(Modeling) *Solve each problem.* **See Example 10.**

113. *Traffic Intensity* Let the average number of vehicles arriving at the gate of an amusement park per minute be equal to k, and let the average number of vehicles admitted by the park attendants be equal to r. Then, the average waiting time T (in minutes) for each vehicle arriving at the park is given by the rational function

$$T(r) = \frac{2r - k}{2r^2 - 2kr},$$

where $r > k$. (*Source:* Mannering, F., and W. Kilareski, *Principles of Highway Engineering and Traffic Analysis,* 2nd ed., John Wiley & Sons.)

(a) It is known from experience that on Saturday afternoon $k = 25$. Use graphing to estimate the admittance rate r that is necessary to keep the average waiting time T for each vehicle to 30 sec.

(b) If one park attendant can serve 5.3 vehicles per minute, how many park attendants will be needed to keep the average wait to 30 sec?

114. *Waiting in Line* **Queuing theory** (also known as **waiting-line theory**) investigates the problem of providing adequate service economically to customers waiting in line. Suppose customers arrive at a fast-food service window at the rate of 9 people per hour. With reasonable assumptions, the average time (in hours) that a customer will wait in line before being served is modeled by

$$f(x) = \frac{9}{x(x - 9)},$$

where x is the average number of people served per hour. A graph of $f(x)$ for $x > 9$ is shown in the figure on the next page.

(a) Why is the function meaningless if the average number of people served per hour is less than 9?

Suppose the average time to serve a customer is 5 min.

(b) How many customers can be served in an hour?

(c) How many minutes will a customer have to wait in line (on the average)?

(d) Suppose we want to halve the average waiting time to 7.5 min $\left(\frac{1}{8}\,\text{hr}\right)$. How fast must an employee work to serve a customer (on the average)? (*Hint:* Let $f(x) = \frac{1}{8}$ and solve the equation for x. Convert your answer to minutes.) How might this reduction in serving time be accomplished?

Average Waiting Time

115. *Braking Distance* The rational function

$$d(x) = \frac{8710x^2 - 69,400x + 470,000}{1.08x^2 - 324x + 82,200}$$

can be used to accurately model the braking distance for automobiles traveling at x miles per hour, where $20 \le x \le 70$. (*Source:* Mannering, F., and W. Kilareski, *Principles of Highway Engineering and Traffic Analysis,* 2nd ed., John Wiley & Sons.)

(a) Use graphing to estimate x when $d(x) = 300$.
(b) Complete the table for each value of x.
(c) If a car doubles its speed, does the braking distance double or more than double? Explain.
(d) Suppose that the automobile braking distance doubled whenever the speed doubled. What type of relationship would exist between the braking distance and the speed?

x	$d(x)$	x	$d(x)$
20		50	
25		55	
30		60	
35		65	
40		70	
45			

116. *Braking Distance* The **grade** x of a hill is a measure of its steepness. For example, if a road rises 10 ft for every 100 ft of horizontal distance, then it has an uphill grade of

$$x = \frac{10}{100}, \quad \text{or} \quad 10\%.$$

Grades are typically kept quite small—usually less than 10%. The braking distance D for a car traveling at 50 mph on a wet, uphill grade is given by

$$D(x) = \frac{2500}{30(0.3 + x)}.$$

(*Source:* Haefner, L., *Introduction to Transportation Systems,* Holt, Rinehart and Winston.)

100 ft

10 ft

(a) Evaluate $D(0.05)$ and interpret the result.
(b) Describe what happens to braking distance as the hill becomes steeper. Does this agree with your driving experience?
(c) Estimate the grade associated with a braking distance of 220 ft.

117. *Tax Revenue* Economist Arthur Laffer has been a center of controversy because of his **Laffer curve,** an idealized version of which is shown here. According to this curve, increasing a tax rate, say from x_1 percent to x_2 percent on the graph, can actually lead to a decrease in government revenue. All economists agree on the endpoints, 0 revenue at tax rates of both 0% and 100%, but there is much disagreement on the location of the rate x_1 that produces maximum revenue. Suppose an economist studying the Laffer curve produces the rational function

$$R(x) = \frac{80x - 8000}{x - 110},$$

where $R(x)$ is government revenue in tens of millions of dollars for a tax rate of x percent, with the function valid for $55 \leq x \leq 100$. Find the revenue for the following tax rates.

  (a) 55%   **(b)** 60%   **(c)** 70%   **(d)** 90%   **(e)** 100%
  (f) Graph R in the window $[0, 100]$ by $[0, 80]$.

118. *Tax Revenue* **See Exercise 117.** Suppose an economist determines that

$$R(x) = \frac{60x - 6000}{x - 120},$$

where $y = R(x)$ is government revenue in tens of millions of dollars for a tax rate of x percent, with $y = R(x)$ valid for $50 \leq x \leq 100$. Find the revenue for each tax rate.

  (a) 50%   **(b)** 60%   **(c)** 80%   **(d)** 100%
  (e) Graph R in the window $[0, 100]$ by $[0, 50]$.

Relating Concepts

For individual or collaborative investigation *(Exercises 119–128)*

Consider the following "monster" rational function.

$$f(x) = \frac{x^4 - 3x^3 - 21x^2 + 43x + 60}{x^4 - 6x^3 + x^2 + 24x - 20}$$

Analyzing this function will synthesize many of the concepts of this and earlier chapters. **Work Exercises 119–128 in order.**

119. Find the equation of the horizontal asymptote.

120. Given that -4 and -1 are zeros of the numerator, factor the numerator completely.

121. (a) Given that 1 and 2 are zeros of the denominator, factor the denominator completely.
  (b) Write the entire quotient for f so that the numerator and the denominator are in factored form.

122. (a) What is the common factor in the numerator and the denominator?
  (b) For what value of x will there be a point of discontinuity (i.e., a "hole")?

123. What are the x-intercepts of the graph of f?

124. What is the y-intercept of the graph of f?

125. Find the equations of the vertical asymptotes.

(continued)

126. Determine the point or points of intersection of the graph of f with its horizontal asymptote.

127. Sketch the graph of f.

128. Use the graph of f to solve each inequality.

    (a) $f(x) < 0$ **(b)** $f(x) > 0$

Chapter 3   Quiz (Sections 3.1–3.5)

1. Graph each quadratic function. Give the vertex, axis, domain, range, and the largest intervals of the domain over which each function is increasing or decreasing.

    (a) $f(x) = -2(x + 3)^2 - 1$ **(b)** $f(x) = 2x^2 - 8x + 3$

2. *Height of a Projected Object* A ball is projected directly upward from an initial height of 200 ft with an initial velocity of 64 ft per sec.

    (a) Use the projectile height function $s(t) = -16t^2 + v_0 t + s_0$ to describe the height of the ball in terms of time t.

    (b) For what interval of time is the height of the ball greater than 240 ft?

Use synthetic division to decide whether the given number k is a zero of the polynomial function. If it is not, give the value of $f(k)$.

3. $f(x) = 2x^4 + x^3 - 3x + 4$; $k = 2$ **4.** $f(x) = x^2 - 4x + 5$; $k = 2 + i$

5. Find a polynomial of least degree having only real coefficients with zeros -2, 3, and $3 - i$.

Graph each polynomial function. Factor first if the polynomial is not in factored form.

6. $f(x) = x(x - 2)^3(x + 2)^2$

7. $f(x) = 2x^4 - 9x^3 - 5x^2 + 57x - 45$

8. $f(x) = -4x^5 + 16x^4 + 13x^3 - 76x^2 - 3x + 18$

Sketch the graph of each rational function.

9. $f(x) = \dfrac{3x + 1}{x^2 + 7x + 10}$ **10.** $f(x) = \dfrac{x^2 + 2x + 1}{x - 1}$

Summary Exercises on Solving Equations and Inequalities

In **Section 1.7** we solved rational inequalities by rewriting them so that 0 is on one side. We then determined the values that cause either the numerator or denominator to equal 0, and by using a test value from each interval determined by these values, we identified the solution set.

We now solve rational inequalities by inspecting the graph of a related function. The graphs can be obtained by using technology or by using the steps for graphing a rational function given in **Section 3.5.**

EXAMPLE **Solving a Rational Inequality**

Solve the inequality.

$$\frac{1-x}{x+4} \geq 0$$

SOLUTION Graph the related rational function

$$y = \frac{1-x}{x+4}.$$

The real solutions of $\frac{1-x}{x+4} \geq 0$ are the x-values for which the graph lies above or on the x-axis. This is true for all x to the right of the vertical asymptote at $x = -4$, up to and including the x-value of the x-intercept at $(1, 0)$. Therefore, the solution set of the inequality is $(-4, 1]$.

By inspecting the graph of the related function, we can also determine that the solution set of $\frac{1-x}{x+4} < 0$ is $(-\infty, -4) \cup (1, \infty)$ and that the solution set of the equation $\frac{1-x}{x+4} = 0$ is $\{1\}$, because the x-intercept is $(1, 0)$. This graphical method may be used to solve other equations and inequalities including those defined by polynomials.

☑ *Now Try Exercise 19.*

In Exercises 1–8, solve each equation.

1. $\dfrac{5x + 8}{-2} = 2x - 10$

2. $\dfrac{1}{5}x + 0.25x = \dfrac{1}{2}x - 1$

3. $(x - 5)^{-4} - 13(x - 5)^{-2} = -36$

4. $x = 13\sqrt{x} - 40$

5. $\sqrt{2x - 5} - \sqrt{x - 3} = 1$

6. $3 = \sqrt{x + 2} + \sqrt{x - 1}$

7. $x^{2/3} + \dfrac{1}{2} = \dfrac{3}{4}$

8. $27 - (x - 4)^{3/2} = 0$

In Exercises 9–12, use the graph of the function to solve each equation or inequality.

9. **(a)** $f(x) > 0$ **(b)** $f(x) \leq 0$

10. **(a)** $f(x) < 0$ **(b)** $f(x) > 0$

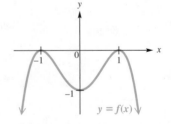

11. **(a)** $f(x) = 0$ **(b)** $f(x) > 0$

12. **(a)** $f(x) = 0$ **(b)** $f(x) \leq 0$

*In Exercises 13–22, sketch the graph of an appropriate function and then use the graph to solve each equation or inequality. (Note: First identify the expression as an equation or an inequality. When appropriate, use the steps for graphing polynomial functions from **Section 3.4** or rational functions from **Section 3.5**.)*

13. $25x^2 - 20x + 4 > 0$

14. $3x + 4 \leq x^2$

15. $x^4 - 2x^3 - 3x^2 + 4x + 4 = 0$

16. $x^3 + 5x^2 + 3x - 9 \geq 0$

17. $-x^4 - x^3 + 12x^2 = 0$

18. $-4x^4 + 13x^2 - 3 > 0$

19. $\dfrac{2x^2 - 13x + 15}{x^2 - 3x} \geq 0$

20. $\dfrac{x^2 + 3x - 1}{x + 1} > 3$

21. $\dfrac{x - 1}{(x - 3)^2} = 0$

22. $\dfrac{x}{x^2 - 4} > 0$

3.6 Variation

- **Direct Variation**
- **Inverse Variation**
- **Combined and Joint Variation**

To apply mathematics we often need to express relationships between quantities. For example,

- In chemistry, the ideal gas law describes how temperature, pressure, and volume are related.

- In physics, various formulas in optics describe the relationship between the focal length of a lens and the size of an image.

This section introduces some special applications of polynomial and rational functions.

Direct Variation   When one quantity is a constant multiple of another quantity, the two quantities are said to *vary directly*. For example, if you work for an hourly wage of $10, then

$$[\text{pay}] = 10[\text{hours worked}].$$

Doubling the hours doubles the pay. Tripling the hours triples the pay, and so on. This is stated more precisely as follows.

Direct Variation

y **varies directly** as *x*, or *y* is **directly proportional** to *x*, if there exists a nonzero real number *k*, called the **constant of variation**, such that for all *x*,

$$y = kx.$$

The direct variation equation $y = kx$ defines a linear function, where the constant of variation *k* is the slope of the line. For $k > 0$,

- As the value of *x increases,* the value of *y increases.*

- As the value of *x decreases,* the value of *y decreases.*

When used to describe a direct variation relationship, the phrase "directly proportional" is sometimes abbreviated to just "proportional."

The steps involved in solving a variation problem are summarized here.

Solving Variation Problems

Step 1 Write the general relationship among the variables as an equation. Use the constant k.

Step 2 Substitute given values of the variables and find the value of k.

Step 3 Substitute this value of k into the equation from Step 1, obtaining a specific formula.

Step 4 Substitute the remaining values and solve for the required unknown.

EXAMPLE 1   Solving a Direct Variation Problem

The area of a rectangle varies directly as its length. If the area is 50 m² when the length is 10 m, find the area when the length is 25 m. (See **Figure 59.**)

SOLUTION

Step 1 Since the area varies directly as the length,

$$\mathscr{A} = kL,$$

where $\mathscr{A}$ represents the area of the rectangle, L is the length, and k is a nonzero constant.

$\mathscr{A} = 50\ m^2$

10 m

$\mathscr{A} = ?$

25 m

Figure 59

Step 2 Since $\mathscr{A} = 50$ when $L = 10$, the equation $\mathscr{A} = kL$ becomes

$$50 = 10k \qquad \text{Substitute for } \mathscr{A} \text{ and } L.$$
$$k = 5. \qquad \text{Divide by 10. Interchange sides.}$$

Step 3 Using this value of k, we can express the relationship between the area and the length as

$$\mathscr{A} = 5L. \qquad \text{Direct variation equation}$$

Step 4 To find the area when the length is 25, we replace L with 25.

$$\mathscr{A} = 5L$$
$$\mathscr{A} = 5(25) \qquad \text{Substitute for } L.$$
$$\mathscr{A} = 125 \qquad \text{Multiply.}$$

The area of the rectangle is 125 m² when the length is 25 m.

✔ *Now Try Exercise 21.*

Sometimes y varies as a power of x. If n is a positive integer greater than or equal to 2, then y is a greater-power polynomial function of x.

Direct Variation as *n*th Power

Let n be a positive real number. Then y **varies directly as the *n*th power** of x, or y is **directly proportional to the *n*th power** of x, if for all x there exists a nonzero real number k such that

$$y = kx^n.$$

For example, the area of a square of side x is given by the formula $\mathcal{A} = x^2$, so the area varies directly as the square of the length of a side. Here $k = 1$.

Inverse Variation   Another type of variation is *inverse variation*. With inverse variation, where $k > 0$, as the value of one variable increases, the value of the other decreases. This relationship can be expressed as a rational function.

Inverse Variation as *n*th Power

Let n be a positive real number. Then y **varies inversely as the *n*th power** of x, or y is **inversely proportional to the *n*th power** of x, if for all x there exists a nonzero real number k such that

$$y = \frac{k}{x^n}.$$

If $n = 1$, then $y = \frac{k}{x}$, and y **varies inversely** as x.

EXAMPLE 2   **Solving an Inverse Variation Problem**

In a certain manufacturing process, the cost of producing a single item varies inversely as the square of the number of items produced. If 100 items are produced, each costs \$2. Find the cost per item if 400 items are produced.

SOLUTION

Step 1   Let x represent the number of items produced and y represent the cost per item. Then, for some nonzero constant k,

$$y = \frac{k}{x^2}. \qquad \text{\textit{y} varies inversely as the square of \textit{x}.}$$

Step 2 $2 = \dfrac{k}{100^2}$ Substitute; $y = 2$ when $x = 100$.

                   $k = 20{,}000$ Solve for k. **(Section 1.1)**

Step 3   The relationship between x and y is $y = \dfrac{20{,}000}{x^2}$.

Step 4   When 400 items are produced, the cost per item is

$$y = \frac{20{,}000}{x^2} = \frac{20{,}000}{400^2} = 0.125, \quad \text{or} \quad 12.5 \text{ cents.}$$

                                            ✔ *Now Try Exercise 31.*

Combined and Joint Variation   In **combined variation,** one variable depends on more than one other variable. Specifically, when a variable depends on the *product* of two or more other variables, it is referred to as *joint variation*.

Joint Variation

Let m and n be real numbers. Then y **varies jointly** as the *n*th power of x and the *m*th power of z if for all x and z, there exists a nonzero real number k such that

$$y = kx^n z^m.$$

CAUTION Note that *and* in the expression "*y* varies jointly as *x* and *z*" translates as the product $y = kxz$. **The word "and" does not indicate addition here.**

Solving a Joint Variation Problem

The area of a triangle varies jointly as the lengths of the base and the height. A triangle with base 10 ft and height 4 ft has area 20 ft². Find the area of a triangle with base 3 ft and height 8 ft. (See **Figure 60.**)

SOLUTION

Step 1 Let $\mathcal{A}$ represent the area, b the base, and h the height of the triangle. Then, for some number k,

$$\mathcal{A} = kbh. \quad \mathcal{A} \text{ varies jointly as } b \text{ and } h.$$

Step 2 Since $\mathcal{A}$ is 20 when b is 10 and h is 4, substitute and solve for k.

$$20 = k(10)(4)$$

$$\frac{1}{2} = k$$

Step 3 The relationship among the variables is the familiar formula for the area of a triangle,

$$\mathcal{A} = \frac{1}{2}bh.$$

Step 4 To find $\mathcal{A}$ when $b = 3$ ft and $h = 8$ ft, substitute into the formula.

$$\mathcal{A} = \frac{1}{2}(3)(8) = 12 \text{ ft}^2$$

✔ *Now Try Exercise 33.*

Area = 20 ft²

$\mathcal{A} = ?$

Figure 60

Solving a Combined Variation Problem

The number of vibrations per second (the pitch) of a steel guitar string varies directly as the square root of the tension and inversely as the length of the string. If the number of vibrations per second is 50 when the tension is 225 newtons and the length is 0.60 m, find the number of vibrations per second when the tension is 196 newtons and the length is 0.65 m.

SOLUTION Let n represent the number of vibrations per second, T represent the tension, and L represent the length of the string. Then, from the information in the problem, write the variation equation (Step 1).

$$n = \frac{k\sqrt{T}}{L} \quad \begin{array}{l}n \text{ varies directly as the square root of} \\ T \text{ and inversely as } L.\end{array}$$

Substitute the given values for n, T, and L to find k (Step 2).

$$50 = \frac{k\sqrt{225}}{0.60} \quad \text{Let } n = 50, T = 225, L = 0.60.$$

$$30 = k\sqrt{225} \quad \text{Multiply by 0.60.}$$

$$30 = 15k \quad \sqrt{225} = 15$$

$$k = 2 \quad \text{Divide by 15. Interchange sides.}$$

Substitute for k to find the relationship among the variables (Step 3).

$$n = \frac{2\sqrt{T}}{L}$$

Now use the second set of values for T and L to find n (Step 4).

$$n = \frac{2\sqrt{196}}{0.65} \approx 43 \quad \text{Let } T = 196, L = 0.65.$$

The number of vibrations per second is approximately 43.

✔️ *Now Try Exercise 37.*

3.6 Exercises

Concept Check Write each formula as an English phrase using the word varies *or pro-portional.*

1. $C = 2\pi r$, where C is the circumference of a circle of radius r

2. $d = \frac{1}{5}s$, where d is the approximate distance (in miles) from a storm, and s is the number of seconds between seeing lightning and hearing thunder

3. $r = \frac{d}{t}$, where r is the speed when traveling d miles in t hours

4. $d = \dfrac{1}{4\pi nr^2}$, where d is the distance a gas atom of radius r travels between collisions, and n is the number of atoms per unit volume

5. $s = kx^3$, where s is the strength of a muscle that has length x

6. $f = \dfrac{mv^2}{r}$, where f is the centripetal force of an object of mass m moving along a circle of radius r at velocity v

Concept Check Match each statement in Exercises 7–10 with its corresponding graph in choices A–D. In each case, $k > 0$.

7. y varies directly as x. $(y = kx)$ **8.** y varies inversely as x. $\left(y = \frac{k}{x}\right)$

9. y varies directly as the second power of x. $(y = kx^2)$ **10.** x varies directly as the second power of y. $(x = ky^2)$

A.   **B.**   **C.**   **D.**

*Solve each variation problem. **See Examples 1–4.***

11. If y varies directly as x, and $y = 20$ when $x = 4$, find y when $x = -6$.

12. If y varies directly as x, and $y = 9$ when $x = 30$, find y when $x = 40$.

13. If m varies jointly as x and y, and $m = 10$ when $x = 2$ and $y = 14$, find m when $x = 21$ and $y = 8$.

14. If m varies jointly as z and p, and $m = 10$ when $z = 2$ and $p = 7.5$, find m when $z = 6$ and $p = 9$.

15. If y varies inversely as x, and $y = 10$ when $x = 3$, find y when $x = 20$.

16. If y varies inversely as x, and $y = 20$ when $x = \frac{1}{4}$, find y when $x = 15$.

17. Suppose r varies directly as the square of m, and inversely as s. If $r = 12$ when $m = 6$ and $s = 4$, find r when $m = 6$ and $s = 20$.

18. Suppose p varies directly as the square of z, and inversely as r. If $p = \frac{32}{5}$ when $z = 4$ and $r = 10$, find p when $z = 3$ and $r = 32$.

19. Let a be directly proportional to m and n^2, and inversely proportional to y^3. If $a = 9$ when $m = 4$, $n = 9$, and $y = 3$, find a when $m = 6$, $n = 2$, and $y = 5$.

20. If y varies directly as x, and inversely as m^2 and r^2, and $y = \frac{5}{3}$ when $x = 1$, $m = 2$, and $r = 3$, find y when $x = 3$, $m = 1$, and $r = 8$.

*Solve each problem. **See Examples 1–4.***

21. *Circumference of a Circle* The circumference of a circle varies directly as the radius. A circle with radius 7 in. has circumference 43.96 in. Find the circumference of the circle if the radius changes to 11 in.

22. *Pressure Exerted by a Liquid* The pressure exerted by a certain liquid at a given point varies directly as the depth of the point beneath the surface of the liquid. The pressure at 10 ft is 50 pounds per square inch (psi). What is the pressure at 15 ft?

23. *Resistance of a Wire* The resistance in ohms of a platinum wire temperature sensor varies directly as the temperature in kelvins (K). If the resistance is 646 ohms at a temperature of 190 K, find the resistance at a temperature of 250 K.

24. *Weight on the Moon* The weight of an object on Earth is directly proportional to the weight of that same object on the moon. A 200-lb astronaut would weigh 32 lb on the moon. How much would a 50-lb dog weigh on the moon?

25. *Distance to the Horizon* The distance that a person can see to the horizon on a clear day from a point above the surface of Earth varies directly as the square root of the height at that point. If a person 144 m above the surface of Earth can see 18 km to the horizon, how far can a person see to the horizon from a point 64 m above the surface?

26. *Water Emptied by a Pipe* The amount of water emptied by a pipe varies directly as the square of the diameter of the pipe. For a certain constant water flow, a pipe emptying into a canal will allow 200 gal of water to escape in an hour. The diameter of the pipe is 6 in. How much water would a 12-in. pipe empty into the canal in an hour, assuming the same water flow?

27. *Hooke's Law for a Spring* Hooke's law for an elastic spring states that the distance a spring stretches varies directly as the force applied. If a force of 15 lb stretches a certain spring 8 in., how much will a force of 30 lb stretch the spring?

8 in.

15 lb

28. *Current in a Circuit* The current in a simple electrical circuit varies inversely as the resistance. If the current is 50 amps when the resistance is 10 ohms, find the current if the resistance is 5 ohms.

29. *Speed of a Pulley* The speed of a pulley varies inversely as its diameter. One kind of pulley, with diameter 3 in., turns at 150 revolutions per minute. Find the speed of a similar pulley with diameter 5 in.

30. *Weight of an Object* The weight of an object varies inversely as the square of its distance from the center of Earth. If an object 8000 mi from the center of Earth weighs 90 lb, find its weight when it is 12,000 mi from the center of Earth.

31. *Current Flow* In electric current flow, it is found that the resistance offered by a fixed length of wire of a given material varies inversely as the square of the diameter of the wire. If a wire 0.01 in. in diameter has a resistance of 0.4 ohm, what is the resistance of a wire of the same length and material with diameter 0.03 in., to the nearest ten-thousandth of an ohm?

32. *Illumination* The illumination produced by a light source varies inversely as the square of the distance from the source. The illumination of a light source at 5 m is 70 candela. What is the illumination 12 m from the source?

33. *Simple Interest* Simple interest varies jointly as principal and time. If $1000 invested for 2 yr earned $70, find the amount of interest earned by $5000 for 5 yr.

34. *Volume of a Gas* The volume of a gas varies inversely as the pressure and directly as the temperature in kelvins (K). If a certain gas occupies a volume of 1.3 L at 300 K and a pressure of 18 newtons, find the volume at 340 K and a pressure of 24 newtons.

35. *Force of Wind* The force of the wind blowing on a vertical surface varies jointly as the area of the surface and the square of the velocity. If a wind of 40 mph exerts a force of 50 lb on a surface of $\frac{1}{2}$ ft^2, how much force will a wind of 80 mph place on a surface of 2 ft^2?

36. *Volume of a Cylinder* The volume of a right circular cylinder is jointly proportional to the square of the radius of the circular base and to the height. If the volume is 300 cm^3 when the height is 10.62 cm and the radius is 3 cm, find the volume to the nearest tenth of a cylinder with radius 4 cm and height 15.92 cm.

37. *Sports Arena Construction* The roof of a new sports arena rests on round concrete pillars. The maximum load a cylindrical column of circular cross section can hold varies directly as the fourth power of the diameter and inversely as the square of the height. The arena has 9-m-tall columns that are 1 m in diameter and will support a load of 8 metric tons. How many metric tons will be supported by a column 12 m high and $\frac{2}{3}$ m in diameter?

Load = 8 metric tons

38. *Sports Arena Construction* The sports arena in **Exercise 37** requires a horizontal beam 16 m long, 24 cm wide, and 8 cm high. The maximum load of a horizontal beam that is supported at both ends varies directly as the width of the beam and the square of its height and inversely as the length between supports. If a beam of the same material 8 m long, 12 cm wide, and 15 cm high can support a maximum of 400 kg, what is the maximum load the beam in the arena will support?

39. *Period of a Pendulum* The period of a pendulum varies directly as the square root of the length of the pendulum and inversely as the square root of the acceleration due to gravity. Find the period when the length is 121 cm and the acceleration due to gravity is 980 cm per second squared, if the period is 6π seconds when the length is 289 cm and the acceleration due to gravity is 980 cm per second squared.

40. *Long-Distance Phone Calls* The number of long-distance phone calls between two cities in a certain time period varies directly as the populations p_1 and p_2 of the cities and inversely as the distance between them. If 10,000 calls are made between two cities 500 mi apart, having populations of 50,000 and 125,000, find the number of calls between two cities 800 mi apart, having populations of 20,000 and 80,000.

41. *Body Mass Index* The federal government has developed the **body mass index** (BMI) to determine ideal weights. A person's BMI is directly proportional to his or her weight in pounds and inversely proportional to the square of his or her height in inches. (A BMI of 19 to 25 corresponds to a healthy weight.) A 6-foot-tall person weighing 177 lb has BMI 24. Find the BMI (to the nearest whole number) of a person whose weight is 130 lb and whose height is 66 in.

42. *Poiseuille's Law* According to Poiseuille's law, the resistance to flow of a blood vessel, R, is directly proportional to the length, l, and inversely proportional to the fourth power of the radius, r. If $R = 25$ when $l = 12$ and $r = 0.2$, find R to the nearest hundredth as r increases to 0.3, while l is unchanged.

43. *Stefan-Boltzmann Law* The Stefan-Boltzmann law says that the radiation of heat R from an object is directly proportional to the fourth power of the kelvin temperature of the object. For a certain object, $R = 213.73$ at room temperature (293 K). Find R to the nearest hundredth if the temperature increases to 335 K.

44. *Nuclear Bomb Detonation* Suppose the effects of detonating a nuclear bomb will be felt over a distance from the point of detonation that is directly proportional to the cube root of the yield of the bomb. Suppose a 100-kiloton bomb has certain effects to a radius of 3 km from the point of detonation. Find the distance to the nearest tenth that the effects would be felt for a 1500-kiloton bomb.

45. *Malnutrition Measure* A measure of malnutrition, called the **pelidisi,** varies directly as the cube root of a person's weight in grams and inversely as the person's sitting height in centimeters. A person with a pelidisi below 100 is considered to be undernourished, while a pelidisi greater than 100 indicates overfeeding. A person who weighs 48,820 g with a sitting height of 78.7 cm has a pelidisi of 100. Find the pelidisi (to the nearest whole number) of a person whose weight is 54,430 g and whose sitting height is 88.9 cm. Is this individual undernourished or overfed?

Weight: 48,820 g   Weight: 54,430 g

46. *Photography* Variation occurs in a formula from photography. In

$$L = \frac{25F^2}{st},$$

the luminance, L, varies directly as the square of the F-stop, F, and inversely as the product of the film ASA number, s, and the shutter speed, t.

(a) What would an appropriate F-stop be for 200 ASA film and a shutter speed of $\frac{1}{250}$ sec when 500 footcandles of light is available?

(b) If 125 footcandles of light is available and an F-stop of 2 is used with 200 ASA film, what shutter speed should be used?

Concept Check *Work each problem.*

47. For $k > 0$, if y varies directly as x, then when x increases, y _____, and when x decreases, y _____.

48. For $k > 0$, if y varies inversely as x, then when x increases, y _____, and when x decreases, y _____.

49. What happens to y if y varies inversely as x, and x is doubled?

50. What happens to y if y varies directly as x, and x is halved?

51. Suppose y is directly proportional to x, and x is replaced by $\frac{1}{3}x$. What happens to y?

52. Suppose y is inversely proportional to x, and x is tripled. What happens to y?

53. Suppose p varies directly as r^3 and inversely as t^2. If r is halved and t is doubled, what happens to p?

54. Suppose m varies directly as p^2 and q^4. If p doubles and q triples, what happens to m?

Chapter 3 Test Prep

Key Terms

3.1 polynomial function
leading coefficient
leading term
zero polynomial
quadratic function
parabola
axis (axis of
symmetry)
vertex

quadratic regression
3.2 synthetic division
zero of a polynomial
function
root (or solution) of an
equation
3.3 multiplicity of a zero
3.4 turning points
end behavior

dominating term
3.5 rational function
discontinuous graph
vertical asymptote
horizontal
asymptote
oblique asymptote
point of
discontinuity

3.6 varies directly
(directly
proportional to)
constant of variation
varies inversely
(inversely
proportional to)
combined variation
varies jointly

New Symbols

$\bar{z}$ conjugate of $z = a + bi$

      end behavior diagrams

$|f(x)| \to \infty$ absolute value of $f(x)$ increases
without bound

$x \to a$ x approaches a

Quick Review

Concepts

Examples

| 3.1 | **Quadratic Functions and Models** |

1. The graph of

$$f(x) = a(x - h)^2 + k, \quad a \neq 0,$$

is a parabola with vertex at (h, k) and the vertical line $x = h$ as axis.

2. The graph opens up if a is positive and down if a is negative.

3. The graph is wider than the graph of $f(x) = x^2$ if $|a| < 1$ and narrower if $|a| > 1$.

Vertex Formula
The vertex of the graph of $f(x) = ax^2 + bx + c$, $a \neq 0$, may be found by completing the square. The vertex has coordinates

$$\left(-\frac{b}{2a}, f\left(-\frac{b}{2a} \right) \right).$$

Graphing a Quadratic Function $f(x) = ax^2 + bx + c$

Step 1 Find the vertex either by using the vertex formula or by completing the square.

Step 2 Find the y-intercept by evaluating $f(0)$.

Step 3 Find any x-intercepts by solving $f(x) = 0$.

Step 4 Find and plot any additional points as needed, using symmetry about the axis.

The graph opens up if $a > 0$ and down if $a < 0$.

Graph $f(x) = -(x + 3)^2 + 1$.

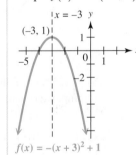

$f(x) = -(x + 3)^2 + 1$

The graph opens down since $a < 0$. It is the graph of $y = -x^2$ shifted 3 units left and 1 unit up, so the vertex is $(-3, 1)$, with axis $x = -3$. The domain is $(-\infty, \infty)$, and the range is $(-\infty, 1]$. The function is increasing on $(-\infty, -3)$ and decreasing on $(-3, \infty)$.

Graph $f(x) = x^2 + 4x + 3$. The vertex of the graph is

$$\left(-\frac{b}{2a}, f\left(-\frac{b}{2a} \right) \right) = (-2, -1). \quad a = 1, b = 4, c = 3$$

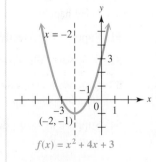

$f(x) = x^2 + 4x + 3$

The graph opens up since $a > 0$. Since $f(0) = 3$, the y-intercept is $(0, 3)$. The solutions of $x^2 + 4x + 3 = 0$ are -1 and -3, which determine the x-intercepts. The domain is $(-\infty, \infty)$, and the range is $[-1, \infty)$. The function is decreasing on $(-\infty, -2)$ and increasing on $(-2, \infty)$.

Concepts	Examples

3.2 Synthetic Division

Synthetic division is a shortcut method for dividing a polynomial by a binomial of the form $x - k$.

Use synthetic division to divide

$$f(x) = 2x^3 - 3x + 2 \quad \text{by} \quad x - 1,$$

and write the result as $f(x) = g(x) \cdot q(x) + r(x)$.

$$
\begin{array}{r|rrrr}
1) & 2 & 0 & -3 & 2 \\
   &   & 2 & 2 & -1 \\
\hline
   & 2 & 2 & -1 & 1
\end{array}
$$

$$\underbrace{\text{Coefficients of}}_{\text{the quotient}} \quad \underbrace{\text{Remainder}}$$

$$2x^3 - 3x + 2 = \underbrace{(x - 1)}_{g(x)}\underbrace{(2x^2 + 2x - 1)}_{q(x)} + \underbrace{1}_{r(x)}$$

Remainder Theorem

If the polynomial $f(x)$ is divided by $x - k$, the remainder is $f(k)$.

By the result above, for $f(x) = 2x^3 - 3x + 2$,

$$f(1) = 1.$$

3.3 Zeros of Polynomial Functions

Factor Theorem

For any polynomial function $f(x)$, $x - k$ is a factor of the polynomial if and only if $f(k) = 0$.

For the polynomial functions

$$f(x) = x^3 + x + 2 \quad \text{and} \quad g(x) = x^3 - 1,$$

$f(-1) = 0$. Therefore, $x - (-1)$, or $x + 1$, is a factor of $f(x)$. Also, since $x - 1$ is a factor of $g(x)$, $g(1) = 0$.

Rational Zeros Theorem

If $\frac{p}{q}$ is a rational number written in lowest terms, and if $\frac{p}{q}$ is a zero of f, a polynomial function with integer coefficients, then p is a factor of the constant term, and q is a factor of the leading coefficient.

The only rational numbers that can possibly be zeros of

$$f(x) = 2x^3 - 9x^2 - 4x - 5$$

are ± 1, ± 5, $\pm\frac{1}{2}$, and $\pm\frac{5}{2}$. By synthetic division, it can be shown that the only rational zero of $f(x)$ is 5.

$$
\begin{array}{r|rrrr}
5) & 2 & -9 & -4 & -5 \\
   &   & 10 & 5 & 5 \\
\hline
   & 2 & 1 & 1 & 0 \leftarrow f(5)
\end{array}
$$

Fundamental Theorem of Algebra

Every function defined by a polynomial of degree 1 or more has at least one complex zero.

$f(x) = x^3 + x + 2$ has at least one and at most three distinct zeros.

Number of Zeros Theorem

A function defined by a polynomial of degree n has at most n distinct zeros.

Conjugate Zeros Theorem

If $f(x)$ defines a polynomial function *having only real coefficients* and if $a + bi$ is a zero of $f(x)$, where a and b are real numbers, then the conjugate $a - bi$ is also a zero of $f(x)$.

Since $1 + 2i$ is a zero of

$$f(x) = x^3 - 5x^2 + 11x - 15,$$

its conjugate $1 - 2i$ is a zero as well.

(continued)

Concepts	Examples
Descartes' Rule of Signs See the discussion in **Section 3.3**.	For $f(x) = 3x^3 - 2x^2 + x - 4$, there are three sign changes, so there will be three or one positive real zeros. Since $f(-x) = -3x^3 - 2x^2 - x - 4$ has no sign changes, there will be no negative real zeros. The table shows the possibilities for the numbers of positive, negative, and nonreal complex zeros.

Positive	Negative	Nonreal Complex
3	0	0
1	0	2

3.4 Polynomial Functions: Graphs, Applications, and Models

Graphing Using Translations

The graph of the function

$$f(x) = a(x - h)^n + k$$

can be found by considering the effects of the constants a, h, and k on the graph of $y = x^n$.

Graph $f(x) = -(x + 2)^4 + 1$.

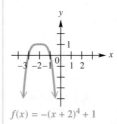

$f(x) = -(x + 2)^4 + 1$

The negative sign causes the graph to be reflected across the x-axis compared to the graph of $y = x^4$. The graph is translated 2 units to the left and 1 unit up. The graph shows that this function is increasing on $(-\infty, -2)$ and decreasing on $(-2, \infty)$.

Multiplicity of a Zero

The behavior of the graph of a polynomial function $f(x)$ near a zero depends on the multiplicity of the zero. If $(x - c)^n$ is a factor of $f(x)$, the graph will behave in the following manner.

- For $n = 1$, the graph will cross the x-axis at $(c, 0)$.

- For n even, the graph will bounce, or turn, at $(c, 0)$.

- For n an odd integer greater than 1, the graph will wiggle through the x-axis at $(c, 0)$.

Determine the behavior of

$$f(x) = (x - 1)(x - 3)^2(x + 1)^3$$

near its zeros, and graph.

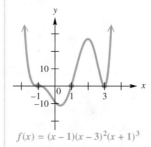

$f(x) = (x - 1)(x - 3)^2(x + 1)^3$

The graph will cross the x-axis at $x = 1$, bounce at $x = 3$, and wiggle through the x-axis at $x = -1$. Since the dominating term is x^6, the end behavior is $\cup\cup$. Since $f(0) = -9$, the y-intercept is $(0, -9)$.

Turning Points

A polynomial function of degree n has at most $n - 1$ turning points.

The graph of

$$f(x) = 4x^5 - 2x^3 + 3x^2 + x - 10$$

has at most four turning points (since $5 - 1 = 4$).

Concepts	Examples

End Behavior

The end behavior of the graph of a polynomial function $f(x)$ is determined by the dominating term, or term of greatest degree. If ax^n is the dominating term of $f(x)$, then the end behavior is as follows.

$a > 0$ n odd $a < 0$ n odd $a > 0$ n even $a < 0$ n even

The end behavior of

$$f(x) = 3x^5 + 2x^2 + 7$$

is .

The end behavior of

$$f(x) = -x^4 - 3x^3 + 2x - 9$$

is .

Graphing Polynomial Functions

To graph a polynomial function f, first find the x-intercepts and y-intercepts.

Then use end behavior, whether the graph crosses, bounces on, or wiggles through the x-axis at the x-intercepts, and selected points as necessary to complete the graph.

Graph $f(x) = (x + 2)(x - 1)(x + 3)$.

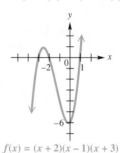

$f(x) = (x + 2)(x - 1)(x + 3)$

The zeros of f are $-2, 1$, and -3. Since $f(0) = 2(-1)(3) = -6$, the y-intercept is $(0, -6)$. The dominating term is $x(x)(x)$ or x^3, so the end behavior is . Begin at either end of the graph with the correct end behavior, and draw a smooth curve that crosses the x-axis at each zero, has a turning point between successive zeros, and passes through the y-intercept.

Intermediate Value Theorem for Polynomials

If $f(x)$ is a polynomial function with only *real coefficients,* and if for real numbers a and b, the values of $f(a)$ and $f(b)$ are opposite in sign, then there exists at least one real zero between a and b.

For the polynomial function

$$f(x) = -x^4 + 2x^3 + 3x^2 + 6,$$

$$f(3.1) = 2.0599 \text{ and } f(3.2) = -2.6016.$$

Since $f(3.1) > 0$ and $f(3.2) < 0$, there exists at least one real zero between 3.1 and 3.2.

Boundedness Theorem

Let $f(x)$ be a polynomial function of degree $n \geq 1$ with *real coefficients* and with a *positive* leading coefficient. Suppose $f(x)$ is divided synthetically by $x - c$.

(a) If $c > 0$ and all numbers in the bottom row of the synthetic division are nonnegative, then $f(x)$ has no zero greater than c.

(b) If $c < 0$ and the numbers in the bottom row of the synthetic division alternate in sign (with 0 considered positive or negative, as needed), then $f(x)$ has no zero less than c.

Show that $f(x) = x^3 - x^2 - 8x + 12$ has no zero greater than 4 and no zero less than -4.

$$
\begin{array}{r|rrrr}
4 & 1 & -1 & -8 & 12 \\
  &   & 4 & 12 & 16 \\
\hline
  & 1 & 3 & 4 & 28
\end{array}
$$
$\longleftarrow$ All signs positive

$$
\begin{array}{r|rrrr}
-4 & 1 & -1 & -8 & 12 \\
   &   & -4 & 20 & -48 \\
\hline
   & 1 & -5 & 12 & -36
\end{array}
$$
$\longleftarrow$ Alternating signs

(continued)

Concepts	Examples

3.5 Rational Functions: Graphs, Applications, and Models

Graphing Rational Functions

To graph a rational function in lowest terms, find the asymptotes and intercepts. Determine whether the graph intersects a nonvertical asymptote. Plot a few points, as necessary, to complete the sketch.

Graph $f(x) = \dfrac{x^2 - 1}{(x + 3)(x - 2)}$.

Point of Discontinuity

If a rational function is not written in lowest terms, there may be a "hole" in the graph instead of an asymptote.

Graph $f(x) = \dfrac{x^2 - 1}{x + 1}$.

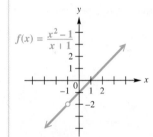

$$f(x) = \frac{x^2 - 1}{x + 1}$$
$$= \frac{(x + 1)(x - 1)}{x + 1}$$
$$= x - 1, \quad x \neq -1$$

3.6 Variation

Direct Variation

y varies directly as the nth power of x if for all x there exists a nonzero real number k such that

$$y = kx^n.$$

The area of a circle varies directly as the square of the radius.

$$\mathcal{A} = kr^2 \quad (k = \pi)$$

Inverse Variation

y varies inversely as the nth power of x if for all x there exists a nonzero real number k such that

$$y = \frac{k}{x^n}.$$

Pressure of a gas varies inversely as volume.

$$P = \frac{k}{V}$$

Joint Variation

For real numbers m and n, y varies jointly as the nth power of x and the mth power of z if for all x and z there exists a nonzero real number k such that

$$y = kx^n z^m.$$

The area of a triangle varies jointly as its base and its height.

$$\mathcal{A} = kbh \quad \left(k = \tfrac{1}{2}\right)$$

Chapter 3 | Review Exercises

Graph. Give the vertex, axis, x-intercepts, y-intercept, domain, range, and the largest intervals of the domain for which each function is increasing or decreasing.

1. $f(x) = 3(x + 4)^2 - 5$

2. $f(x) = -\dfrac{2}{3}(x - 6)^2 + 7$

3. $f(x) = -3x^2 - 12x - 1$

4. $f(x) = 4x^2 - 4x + 3$

Concept Check *In Exercises 5–8, consider the function*

$$f(x) = a(x - h)^2 + k, \quad \text{for } a > 0.$$

5. What are the coordinates of the lowest point of its graph?

6. What is the *y*-intercept of its graph?

7. Under what conditions will its graph have one or more *x*-intercepts? For these conditions, express the *x*-intercept(s) in terms of *a*, *h*, and *k*.

8. If *a* is positive, what is the least value of $ax^2 + bx + c$ in terms of *a*, *b*, and *c*?

(Modeling) *Solve each problem.*

9. *Area of a Rectangle* Use a quadratic function to find the dimensions of the rectangular region of maximum area that can be enclosed with 180 m of fencing, if no fencing is needed along one side of the region.

10. *Height of a Projectile* A projectile is fired vertically upward, and its height $s(t)$ in feet after *t* seconds is given by the function

$$s(t) = -16t^2 + 800t + 600.$$

(a) From what height was the projectile fired?

(b) After how many seconds will it reach its maximum height?

(c) What is the maximum height it will reach?

(d) Between what two times (in seconds, to the nearest tenth) will it be more than 5000 ft above the ground?

(e) After how many seconds, to the nearest tenth, will the projectile hit the ground?

11. *Food Bank Volunteers* During the course of a year, the number of volunteers available to run a food bank each month is modeled by $V(x)$, where

$$V(x) = 2x^2 - 32x + 150$$

between the months of January and August. Here *x* is time in months, with $x = 1$ representing January. From August to December, $V(x)$ is modeled by

$$V(x) = 31x - 226.$$

Find the number of volunteers in each of the following months.

(a) January   **(b)** May   **(c)** August   **(d)** October   **(e)** December

(f) Sketch a graph of $y = V(x)$ for January through December. In what month are the fewest volunteers available?

12. *Concentration of Atmospheric CO_2* In 1990, the International Panel on Climate Change (IPCC) stated that if current trends of burning fossil fuel and deforestation were to continue, then future amounts of atmospheric carbon dioxide in parts per million (ppm) would increase, as shown in the table.

Year	Carbon Dioxide
1990	353
2000	375
2075	590
2175	1090
2275	2000

Source: IPCC.

(a) Let $x = 0$ represent 1990, $x = 10$ represent 2000, and so on. Find a function of the form

$$f(x) = a(x - h)^2 + k$$

that models the data. Use $(0, 353)$ as the vertex and $(285, 2000)$ as another point to determine a.

(b) Use the function to predict the amount of carbon dioxide in 2300.

Consider the function $f(x) = -2.64x^2 + 5.47x + 3.54$ *for Exercises 13–16.*

13. Use the discriminant to explain how to determine the number of x-intercepts the graph of $f(x)$ will have before graphing it on your calculator.

14. Graph the function in the standard viewing window of your calculator, and use the calculator to solve the equation $f(x) = 0$. Express solutions as approximations to the nearest hundredth.

15. Use your answer to **Exercise 14** and the graph of f to solve
   (a) $f(x) > 0$,   and   (b) $f(x) < 0$.

16. Use the capabilities of your calculator to find the coordinates of the vertex of the graph. Express coordinates to the nearest hundredth.

Use synthetic division to perform each division.

17. $\dfrac{x^3 + x^2 - 11x - 10}{x - 3}$

18. $\dfrac{3x^3 + 8x^2 + 5x + 10}{x + 2}$

19. $\dfrac{2x^3 - x + 6}{x + 4}$

20. $\dfrac{3x^3 + 6x^2 - 8x + 3}{x + 3}$

Express $f(x)$ in the form $f(x) = (x - k)q(x) + r$ for the given value of k.

21. $5x^3 - 3x^2 + 2x - 6$; $k = 2$

22. $-3x^3 + 5x - 6$; $k = -1$

Use synthetic division to find $f(2)$.

23. $f(x) = -x^3 + 5x^2 - 7x + 1$

24. $f(x) = 2x^3 - 3x^2 + 7x - 12$

25. $f(x) = 5x^4 - 12x^2 + 2x - 8$

26. $f(x) = x^5 + 4x^2 - 2x - 4$

Use synthetic division to determine whether k is a zero of the function.

27. $f(x) = x^3 + 2x^2 + 3x + 2$; $k = -1$ **28.** $f(x) = 2x^3 + 5x^2 + 30$; $k = -4$

29. *Concept Check* If $f(x)$ is a polynomial function with real coefficients, and if $7 + 2i$ is a zero of the function, then what other complex number must also be a zero?

30. *Concept Check* Suppose the polynomial function f has a zero at $x = -3$. Which of the following statements *must* be true?

   A. $(3, 0)$ is an x-intercept of the graph of f.

   B. $(0, 3)$ is a y-intercept of the graph of f.

   C. $x - 3$ is a factor of $f(x)$.

   D. $f(-3) = 0$

Find a polynomial function with real coefficients and least degree having the given zeros.

31. $-1, 4, 7$

32. $8, 2, 3$

33. $\sqrt{3}, -\sqrt{3}, 2, 3$

34. $-2 + \sqrt{5}, -2 - \sqrt{5}, -2, 1$

35. $2, 4, -i$

36. $0, 5, 1 + 2i$

Find all rational zeros of each function.

37. $f(x) = 2x^3 - 9x^2 - 6x + 5$

38. $f(x) = 8x^4 - 14x^3 - 29x^2 - 4x + 3$

In Exercises 39–41, show that the polynomial function has a real zero as described in parts (a) and (b). In Exercises 39 and 40, then work part (c).

39. $f(x) = 3x^3 - 8x^2 + x + 2$

(a) between -1 and 0

(b) between 2 and 3

(c) Find the zero in part (b) to three decimal places.

40. $f(x) = 4x^3 - 37x^2 + 50x + 60$

(a) between 2 and 3

(b) between 7 and 8

(c) Find the zero in part (b) to three decimal places.

41. $f(x) = 6x^4 + 13x^3 - 11x^2 - 3x + 5$

(a) no zero greater than 1

(b) no zero less than -3

Solve each problem.

42. Use Descartes' rule of signs to determine the different possibilities for the numbers of positive, negative, and nonreal complex zeros of

$$f(x) = x^3 + 3x^2 - 4x - 2.$$

43. Is $x + 1$ a factor of $f(x) = x^3 + 2x^2 + 3x + 2$?

44. Find a polynomial function f with real coefficients of degree 4 with 3, 1, and $-1 + 3i$ as zeros, and $f(2) = -36$.

45. Find a polynomial function f of degree 3 with -2, 1, and 4 as zeros, and $f(2) = 16$.

46. Find all zeros of $f(x) = x^4 - 3x^3 - 8x^2 + 22x - 24$, given that $1 + i$ is a zero.

47. Find all zeros of $f(x) = 2x^4 - x^3 + 7x^2 - 4x - 4$, given that 1 and $-2i$ are zeros.

48. Find a value of k such that $x - 4$ is a factor of

$$f(x) = x^3 - 2x^2 + kx + 4.$$

49. Find a value of k such that when the polynomial $x^3 - 3x^2 + kx - 4$ is divided by $x - 2$, the remainder is 5.

50. *Concept Check* Give an example of a fourth-degree polynomial function having exactly two distinct real zeros, and then sketch its graph.

51. *Concept Check* Give an example of a cubic polynomial function having exactly one real zero, and then sketch its graph.

52. Give the maximum number of turning points of the graph of each function.

(a) $f(x) = x^5 - 9x^2$ (b) $f(x) = 4x^3 - 6x^2 + 2$

53. *Concept Check* Suppose the leading term of a polynomial function is $10x^7$. What can you conclude about each of the following features of the graph of the function?

(a) domain   (b) range   (c) end behavior   (d) number of zeros
(e) number of turning points

54. *Concept Check* Repeat **Exercise 53** for a polynomial function with leading term $-9x^6$.

Sketch the graph of each polynomial function.

55. $f(x) = (x - 2)^2(x + 3)$ **56.** $f(x) = -2x^3 + 7x^2 - 2x - 3$

57. $f(x) = 2x^3 + x^2 - x$ **58.** $f(x) = x^4 - 3x^2 + 2$

59. $f(x) = x^4 + x^3 - 3x^2 - 4x - 4$ **60.** $f(x) = -2x^4 + 7x^3 - 4x^2 - 4x$

Concept Check For each polynomial function, identify its graph from choices A–F.

61. $f(x) = (x - 2)^2(x - 5)$

62. $f(x) = -(x - 2)^2(x - 5)$

63. $f(x) = (x - 2)^2(x - 5)^2$

64. $f(x) = (x - 2)(x - 5)$

65. $f(x) = -(x - 2)(x - 5)$

66. $f(x) = -(x - 2)^2(x - 5)^2$

A.

B.

C.

D.

E.

F.

Use a graphing calculator to graph each polynomial function in the viewing window specified. Then determine the real zeros to as many decimal places as the calculator will provide.

67. $f(x) = x^3 - 8x^2 + 2x + 5$; window: $[-10, 10]$ by $[-60, 60]$

68. $f(x) = x^4 - 4x^3 - 5x^2 + 14x - 15$; window: $[-10, 10]$ by $[-60, 60]$

Solve each problem.

69. *(Modeling) Medicare Beneficiary Spending* Out-of-pocket spending projections for a typical Medicare beneficiary as a share of his or her income are given in the table. Let $x = 0$ represent 1990, so $x = 8$ represents 1998. Use a graphing calculator to do the following.

Year	Percent of Income
1998	18.6
2000	19.3
2005	21.7
2010	24.7
2015	27.5
2020	28.3
2025	28.6

Source: Urban Institute's Analysis of 1998 Medicare Trustees' Report.

(a) Graph the data points.

(b) Find a quadratic function to model the data.

(c) Find a cubic function to model the data.

(d) Graph each function in the same viewing window as the data points.

(e) Compare the two functions. Which is a better fit for the data?

70. *Dimensions of a Cube* After a 2-in. slice is cut off the top of a cube, the resulting solid has a volume of 32 in.^{3}. Find the dimensions of the original cube.

71. *Dimensions of a Box* The width of a rectangular box is three times its height, and its length is 11 in. more than its height. Find the dimensions of the box if its volume is 720 in.^{3}.

72. The function $f(x) = \frac{1}{x}$ is negative at $x = -1$ and positive at $x = 1$ but has no zero between -1 and 1. Explain why this does not contradict the intermediate value theorem.

Graph each rational function.

73. $f(x) = \dfrac{4}{x-1}$

74. $f(x) = \dfrac{4x-2}{3x+1}$

75. $f(x) = \dfrac{6x}{x^2+x-2}$

76. $f(x) = \dfrac{2x}{x^2-1}$

77. $f(x) = \dfrac{x^2+4}{x+2}$

78. $f(x) = \dfrac{x^2-1}{x}$

79. $f(x) = \dfrac{-2}{x^2+1}$

80. $f(x) = \dfrac{4x^2-9}{2x+3}$

81. *Concept Check*

(a) Sketch the graph of a function that does not intersect its horizontal asymptote $y = 1$, has the line $x = 3$ as a vertical asymptote, and has x-intercepts $(2, 0)$ and $(4, 0)$.

(b) Find an equation for a possible corresponding rational function.

82. *Concept Check*

(a) Sketch the graph of a function that is never negative and has the lines $x = -1$ and $x = 1$ as vertical asymptotes, the x-axis as a horizontal asymptote, and $(0, 0)$ as an x-intercept.

(b) Find an equation for a possible corresponding rational function.

83. *Connecting Graphs with Equations* Find a rational function $f(x)$ having the graph shown.

(Modeling) *Solve each problem.*

84. *Antique-Car Competition* Antique-car owners often enter their cars in a **concours d'elegance** in which a maximum of 100 points can be awarded to a particular car. Points are awarded for the general attractiveness of the car. The function

$$C(x) = \frac{10x}{49(101-x)}$$

models the cost, in thousands of dollars, of restoring a car so that it will win x points.

(a) Graph the function in the window $[0, 101]$ by $[0, 10]$.

(b) How much would an owner expect to pay to restore a car in order to earn 95 points?

85. *Environmental Pollution* In situations involving environmental pollution, a cost-benefit model expresses cost as a function of the percentage of pollutant removed from the environment. Suppose a cost-benefit model is expressed as

$$C(x) = \frac{6.7x}{100 - x},$$

where $C(x)$ is cost in thousands of dollars of removing x percent of a pollutant.

(a) Graph the function in the window $[0, 100]$ by $[0, 100]$.
(b) How much would it cost to remove 95% of the pollutant?

Solve each variation problem.

86. If x varies directly as y, and $x = 12$ when $y = 4$, find x when $y = 12$.

87. If x varies directly as y, and $x = 20$ when $y = 14$, find y when $x = 50$.

88. If z varies inversely as w, and $z = 10$ when $w = \frac{1}{2}$, find z when $w = 10$.

89. If t varies inversely as s, and $t = 3$ when $s = 5$, find s when $t = 20$.

90. p varies jointly as q and r^2, and $p = 100$ when $q = 2$ and $r = 3$. Find p when $q = 5$ and $r = 2$.

91. f varies jointly as g^2 and h, and $f = 50$ when $g = 5$ and $h = 4$. Find f when $g = 3$ and $h = 6$.

92. *Pressure in a Liquid* The pressure on a point in a liquid is directly proportional to the distance from the surface to the point. In a certain liquid, the pressure at a depth of 4 m is 60 kg per m^2. Find the pressure at a depth of 10 m.

93. *Skidding Car* The force needed to keep a car from skidding on a curve varies inversely as the radius r of the curve and jointly as the weight of the car and the square of the speed. It takes 3000 lb of force to keep a 2000-lb car from skidding on a curve of radius 500 ft at 30 mph. What force will keep the same car from skidding on a curve of radius 800 ft at 60 mph?

94. *Power of a Windmill* The power a windmill obtains from the wind varies directly as the cube of the wind velocity. If a wind of 10 km per hr produces 10,000 units of power, how much power is produced by a wind of 15 km per hr?

Chapter 3 Test

1. Sketch the graph of the quadratic function $f(x) = -2x^2 + 6x - 3$. Give the intercepts, vertex, axis, domain, range, and the largest intervals of the domain for which the function is increasing or decreasing.

2. *(Modeling) Height of a Projectile* A small rocket is fired directly upward, and its height s in feet after t seconds is given by the function

$$s(t) = -16t^2 + 88t + 48.$$

(a) Determine the time at which the rocket reaches its maximum height.
(b) Determine the maximum height.
(c) Between what two times (in seconds, to the nearest tenth) will the rocket be more than 100 ft above ground level?
(d) After how many seconds will the rocket return to the ground?

Use synthetic division to perform each division.

3. $\dfrac{3x^3 + 4x^2 - 9x + 6}{x + 2}$

4. $\dfrac{2x^3 - 11x^2 + 25}{x - 5}$

5. Use synthetic division to determine $f(5)$, if $f(x) = 2x^3 - 9x^2 + 4x + 8$.

6. Use the factor theorem to determine whether the polynomial $x - 3$ is a factor of

$$6x^4 - 11x^3 - 35x^2 + 34x + 24.$$

If it is, what is the other factor? If it is not, explain why.

7. Find all zeros of $f(x)$, given that $f(x) = x^3 + 8x^2 + 25x + 26$ and -2 is one zero.

8. Find a fourth degree polynomial function $f(x)$ having only real coefficients, -1, 2, and i as zeros, and $f(3) = 80$.

9. Why can't the polynomial function $f(x) = x^4 + 8x^2 + 12$ have any real zeros?

10. Consider the function defined by $f(x) = x^3 - 5x^2 + 2x + 7$.

    (a) Use the intermediate value theorem to show that f has a zero between 1 and 2.

    (b) Use Descartes' rule of signs to determine the different possibilities for the numbers of positive, negative, and nonreal complex zeros.

    (c) Use a graphing calculator to find all real zeros to as many decimal places as the calculator will give.

11. Graph the functions $f(x) = x^4$ and $g(x) = -2(x + 5)^4 + 3$ on the same axes. How can the graph of g be obtained by a transformation of the graph of f?

12. Use end behavior to determine which one of the following graphs is that of $f(x) = -x^7 + x - 4$.

A.

B.

C.

D.

Graph each polynomial function.

13. $f(x) = x^3 - 5x^2 + 3x + 9$

14. $f(x) = 2x^2(x - 2)^2$

15. $f(x) = -x^3 - 4x^2 + 11x + 30$

16. *Connecting Graphs with Equations* Find a cubic polynomial function having the graph shown.

17. *(Modeling) Oil Pressure* The pressure of oil in a reservoir tends to drop with time. Engineers found that the change in pressure is modeled by

$$f(t) = 1.06t^3 - 24.6t^2 + 180t,$$

for t (in years) in the interval $[0, 15]$.

(a) What was the change after 2 yr?

(b) For what time periods, to the nearest tenth of a year, is the amount of change in pressure increasing? decreasing? Use a graph to decide.

Graph each rational function.

18. $f(x) = \dfrac{3x - 1}{x - 2}$

19. $f(x) = \dfrac{x^2 - 1}{x^2 - 9}$

20. Consider the rational function $f(x) = \dfrac{2x^2 + x - 6}{x - 1}$.

(a) Determine the equation of the oblique asymptote.

(b) Determine the x-intercepts.

(c) Determine the y-intercept.

(d) Determine the equation of the vertical asymptote.

(e) Sketch the graph.

21. If y varies directly as the square root of x, and $y = 12$ when $x = 4$, find y when $x = 100$.

22. *Weight On and Above Earth* The weight w of an object varies inversely as the square of the distance d between the object and the center of Earth. If a man weighs 90 kg on the surface of Earth, how much would he weigh 800 km above the surface? (*Hint:* The radius of Earth is about 6400 km.)

4

Inverse, Exponential, and Logarithmic Functions

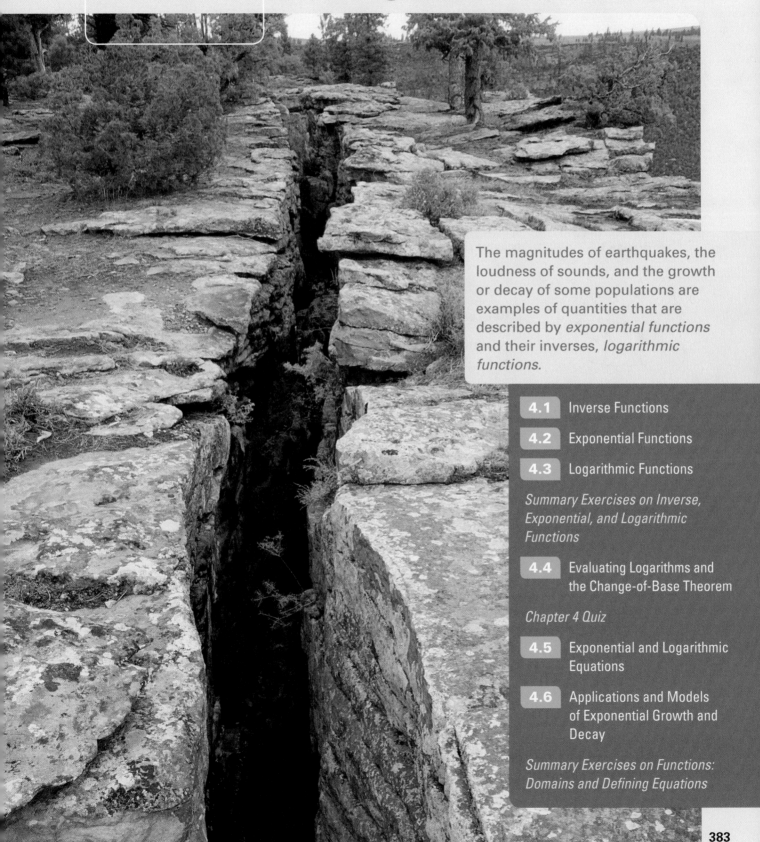

The magnitudes of earthquakes, the loudness of sounds, and the growth or decay of some populations are examples of quantities that are described by *exponential functions* and their inverses, *logarithmic functions*.

4.1 Inverse Functions

4.2 Exponential Functions

4.3 Logarithmic Functions

Summary Exercises on Inverse, Exponential, and Logarithmic Functions

4.4 Evaluating Logarithms and the Change-of-Base Theorem

Chapter 4 Quiz

4.5 Exponential and Logarithmic Equations

4.6 Applications and Models of Exponential Growth and Decay

Summary Exercises on Functions: Domains and Defining Equations

4.1 Inverse Functions

- One-to-One Functions
- Inverse Functions
- Equations of Inverses
- An Application of Inverse Functions to Cryptography

One-to-One Functions   Suppose we define the function

$$F = \{(-2, 2), (-1, 1), (0, 0), (1, 3), (2, 5)\}.$$

(Notice that we have defined F so that each *second* component is used only once.) We can form another set of ordered pairs from F by interchanging the x- and y-values of each pair in F. We call this set G, so

$$G = \{(2, -2), (1, -1), (0, 0), (3, 1), (5, 2)\}.$$

To show that these two sets are related, G is called the *inverse* of F. For a function f to have an inverse, f must be a *one-to-one function*.

> *In a one-to-one function, each x-value corresponds to only one y-value, and each y-value corresponds to only one x-value.*

The function f shown in **Figure 1** is not one-to-one because the y-value 7 corresponds to *two* x-values, 2 and 3. That is, the ordered pairs $(2, 7)$ and $(3, 7)$ both belong to the function. The function f in **Figure 2** is one-to-one.

Not One-to-One

Figure 1

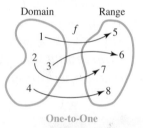

One-to-One

Figure 2

One-to-One Function

A function f is a **one-to-one function** if, for elements a and b in the domain of f,

$$a \neq b \quad \text{implies} \quad f(a) \neq f(b).$$

Using the concept of the *contrapositive* from the study of logic, the last line in the preceding box is equivalent to

$$f(a) = f(b) \quad \text{implies} \quad a = b.$$

We use this statement to decide whether a function f is one-to-one in the next example.

EXAMPLE 1   Deciding Whether Functions Are One-to-One

Decide whether each function is one-to-one.

(a) $f(x) = -4x + 12$ **(b)** $f(x) = \sqrt{25 - x^2}$

SOLUTION

(a) We must show that $f(a) = f(b)$ leads to the result $a = b$.

$$f(a) = f(b)$$

$$-4a + 12 = -4b + 12 \quad f(x) = -4x + 12$$

$$-4a = -4b \quad \text{Subtract 12. (Section 1.1)}$$

$$a = b \quad \text{Divide by } -4.$$

By the definition, $f(x) = -4x + 12$ is one-to-one.

(b) For the function $f(x) = \sqrt{25 - x^2}$, if we choose $a = 3$ and $b = -3$, then $3 \neq -3$, but

$$f(3) = \sqrt{25 - 3^2} = \sqrt{25 - 9} = \sqrt{16} = 4$$

and $\qquad f(-3) = \sqrt{25 - (-3)^2} = \sqrt{25 - 9} = 4.$

Here, even though $3 \neq -3$, $f(3) = f(-3) = 4$. By the definition, f is *not* a one-to-one function.

> ✔ *Now Try Exercises 9 and 11.*

As illustrated in **Example 1(b),** a way to show that a function is *not* one-to-one is to produce a pair of different domain elements that lead to the same function value. There is also a useful graphical test, the **horizontal line test,** that tells whether or not a function is one-to-one.

Figure 3

> ### Horizontal Line Test
>
> A function is one-to-one if every horizontal line intersects the graph of the function at most once.

NOTE In **Example 1(b),** the graph of the function is a semicircle, as shown in **Figure 3.** Because there is at least one horizontal line that intersects the graph in more than one point, this function is not one-to-one.

EXAMPLE 2 **Using the Horizontal Line Test**

Determine whether each graph is the graph of a one-to-one function.

(a)

(b)

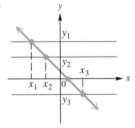

SOLUTION

(a) Each point where the horizontal line intersects the graph has the same value of y but a different value of x. Since more than one (here three) different values of x lead to the same value of y, the function is not one-to-one.

(b) Since every horizontal line will intersect the graph at exactly one point, this function is one-to-one.

> ✔ *Now Try Exercises 3 and 5.*

Notice that the function graphed in **Example 2(b)** decreases on its entire domain. *In general, a function that is either increasing or decreasing on its entire domain, such as $f(x) = -x$, $g(x) = x^3$, and $h(x) = \frac{1}{x}$, must be one-to-one.*

In summary, there are four ways to decide whether a function is one-to-one.

Tests to Determine Whether a Function Is One-to-One

1. Show that $f(a) = f(b)$ implies $a = b$. This means that f is one-to-one. **(Example 1(a))**
2. In a one-to-one function, every y-value corresponds to no more than one x-value. To show that a function is not one-to-one, find at least two x-values that produce the same y-value. **(Example 1(b))**
3. Sketch the graph and use the horizontal line test. **(Example 2)**
4. If the function either increases or decreases on its entire domain, then it is one-to-one. A sketch is helpful here, too. **(Example 2(b))**

Inverse Functions   Consider the functions

$$f(x) = 8x + 5 \quad \text{and} \quad g(x) = \frac{1}{8}x - \frac{5}{8}.$$

Let us choose an arbitrary element from the domain of f, say 10. Evaluate $f(10)$.

$$f(10) = 8 \cdot 10 + 5 \qquad \text{Let } x = 10.$$

$$f(10) = 85 \qquad \text{Multiply and then add.}$$

Now, we evaluate $g(85)$.

$$g(85) = \frac{1}{8}(85) - \frac{5}{8} \quad \text{Let } x = 85.$$

$$= \frac{85}{8} - \frac{5}{8} \qquad \text{Multiply.}$$

$$= \frac{80}{8} \qquad \text{Subtract.}$$

$$g(85) = 10 \qquad \text{Divide.}$$

Starting with 10, we "applied" function f and then "applied" function g to the result, which returned the number 10. See **Figure 4.**

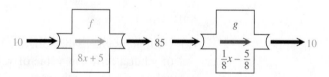

Figure 4

As further examples, check that

$$f(3) = 29 \quad \text{and} \quad g(29) = 3,$$

$$f(-5) = -35 \quad \text{and} \quad g(-35) = -5,$$

$$g(2) = -\frac{3}{8} \quad \text{and} \quad f\left(-\frac{3}{8}\right) = 2.$$

In particular, for this pair of functions,

$$f(g(2)) = 2 \quad \text{and} \quad g(f(2)) = 2.$$

In fact, for *any* value of x,

$$f(g(x)) = x \quad \text{and} \quad g(f(x)) = x.$$

Using the notation for composition introduced in **Section 2.8,** these two equations can be written as follows.

$$(f \circ g)(x) = x \quad \text{and} \quad (g \circ f)(x) = x \qquad \text{The result is the identity function.}$$

Because the compositions of f and g yield the *identity* function, they are *inverses* of each other.

Inverse Function

Let f be a one-to-one function. Then g is the **inverse function** of f if

$$(f \circ g)(x) = x \quad \text{for every } x \text{ in the domain of } g,$$

and $\quad(g \circ f)(x) = x \quad$ for every x in the domain of f.

The condition that f is one-to-one in the definition of inverse function is essential. Otherwise, g will not define a function.

EXAMPLE 3 **Deciding Whether Two Functions Are Inverses**

Let functions f and g be defined by

$$f(x) = x^3 - 1 \quad \text{and} \quad g(x) = \sqrt[3]{x + 1},$$

respectively. Is g the inverse function of f?

SOLUTION   As shown in **Figure 5,** the horizontal line test applied to the graph indicates that f is one-to-one, so the function does have an inverse. Since it is one-to-one, we now find $(f \circ g)(x)$ and $(g \circ f)(x)$.

Figure 5

$$(f \circ g)(x) = f(g(x)) = \left(\sqrt[3]{x + 1}\right)^3 - 1 \quad \text{(Section 2.8)}$$

$$= x + 1 - 1$$

$$= x$$

$$(g \circ f)(x) = g(f(x)) = \sqrt[3]{(x^3 - 1) + 1}$$

$$= \sqrt[3]{x^3}$$

$$= x$$

Since $(f \circ g)(x) = x$ and $(g \circ f)(x) = x$, function g is the inverse of function f.

✔ *Now Try Exercise 41.*

A special notation is used for inverse functions: If g is the inverse of a function f, then g is written as f^{-1} (read "*f*-inverse"). In **Example 3,**

$$f(x) = x^3 - 1 \quad \text{has inverse} \quad f^{-1}(x) = \sqrt[3]{x + 1}.$$

CAUTION *Do not confuse the -1 in f^{-1} with a negative exponent.* The symbol $f^{-1}(x)$ does not represent $\frac{1}{f(x)}$. It represents the inverse function of f.

By the definition of inverse function, the domain of f is the range of f^{-1}, and the range of f is the domain of f^{-1}. See **Figure 6.**

Figure 6

Year	Number of Unhealthy Days
2004	7
2005	32
2006	8
2007	24
2008	14
2009	13

Source: Illinois Environmental Protection Agency.

EXAMPLE 4 **Finding Inverses of One-to-One Functions**

Find the inverse of each function that is one-to-one.

(a) $F = \{(-2, 1), (-1, 0), (0, 1), (1, 2), (2, 2)\}$

(b) $G = \{(3, 1), (0, 2), (2, 3), (4, 0)\}$

(c) The table in the margin shows the number of days in Illinois that were unhealthy for sensitive groups for selected years using the Air Quality Index (AQI). Let f be the function defined in the table, with the years forming the domain and the numbers of unhealthy days forming the range.

SOLUTION

(a) Each x-value in F corresponds to just one y-value. However, the y-value 2 corresponds to two x-values, 1 and 2. Also, the y-value 1 corresponds to both -2 and 0. Because at least one y-value corresponds to more than one x-value, F is not one-to-one and does not have an inverse.

(b) Every x-value in G corresponds to only one y-value, and every y-value corresponds to only one x-value, so G is a one-to-one function. The inverse function is found by interchanging the x- and y-values in each ordered pair.

$$G^{-1} = \{(1, 3), (2, 0), (3, 2), (0, 4)\}$$

Notice how the domain and range of G become the range and domain, respectively, of G^{-1}.

(c) Each x-value in f corresponds to only one y-value, and each y-value corresponds to only one x-value, so f is a one-to-one function. The inverse function is found by interchanging the x- and y-values in the table.

$$f^{-1}(x) = \{(7, 2004), (32, 2005), (8, 2006), (24, 2007), (14, 2008), (13, 2009)\}$$

The domain and range of f become the range and domain of f^{-1}.

✔ *Now Try Exercises 37, 51, and 53.*

Equations of Inverses The inverse of a one-to-one function is found by interchanging the x- and y-values of each of its ordered pairs. The equation of the inverse of a function defined by $y = f(x)$ is found in the same way, as given in the box on the next page.

> ### Finding the Equation of the Inverse of $y = f(x)$
>
> For a one-to-one function f defined by an equation $y = f(x)$, find the defining equation of the inverse as follows. (If necessary, replace $f(x)$ with y first. Any restrictions on x and y should be considered.)
>
> **Step 1** Interchange x and y.
>
> **Step 2** Solve for y.
>
> **Step 3** Replace y with $f^{-1}(x)$.

EXAMPLE 5 **Finding Equations of Inverses**

Decide whether each equation defines a one-to-one function. If so, find the equation of the inverse.

(a) $f(x) = 2x + 5$ **(b)** $y = x^2 + 2$ **(c)** $f(x) = (x - 2)^3$

SOLUTION

(a) The graph of $y = 2x + 5$ is a nonhorizontal line, so by the horizontal line test, f is a one-to-one function. To find the equation of the inverse, follow the steps in the preceding box, first replacing $f(x)$ with y.

$$f(x) = 2x + 5$$

$$y = 2x + 5 \qquad \text{Let } y = f(x).$$

$$x = 2y + 5 \qquad \text{Interchange } x \text{ and } y. \text{ (Step 1)}$$

$$x - 5 = 2y \qquad \text{Subtract 5.}$$

$$y = \frac{x - 5}{2} \qquad \begin{matrix} \text{Divide by 2.} \\ \text{Rewrite.} \end{matrix} \quad \left. \begin{matrix} \text{Solve for } y. \\ \text{(Step 2)} \end{matrix} \right.$$

$$f^{-1}(x) = \frac{1}{2}x - \frac{5}{2} \qquad \begin{matrix} \text{Replace } y \text{ with } f^{-1}(x). \text{ (Step 3)} \\ \frac{a - b}{c} = \left(\frac{1}{c}\right)a - \frac{b}{c} \end{matrix}$$

Thus, $f^{-1}(x) = \frac{x - 5}{2} = \frac{1}{2}x - \frac{5}{2}$ is a linear function. In the function defined by $y = 2x + 5$, the value of y is found by starting with a value of x, multiplying by 2, and adding 5.

The form $f^{-1}(x) = \frac{x - 5}{2}$ for the equation of the inverse has us *subtract* 5 and then *divide* by 2. This shows how an inverse is used to "undo" what a function does to the variable x.

(b) The equation $y = x^2 + 2$ has a parabola opening up as its graph, so some horizontal lines will intersect the graph at two points. For example, both $x = 3$ and $x = -3$ correspond to $y = 11$. Because of the presence of the x^2-term, there are many pairs of x-values that correspond to the same y-value. This means that the function defined by $y = x^2 + 2$ is not one-to-one and does not have an inverse.

The steps for finding the equation of an inverse lead to the following.

$$y = x^2 + 2$$

$$x = y^2 + 2 \qquad \text{Interchange } x \text{ and } y.$$

$$x - 2 = y^2 \qquad \text{Solve for } y.$$

$$\pm\sqrt{x - 2} = y \qquad \text{Square root property (Section 1.4)}$$

> Remember
> *both* roots.

The last step shows that there are two y-values for each choice of x greater than 2, so the given function is not one-to-one and cannot have an inverse.

Figure 7

(c) **Figure 7** shows that the horizontal line test assures us that this horizontal translation of the graph of the cubing function is one-to-one.

$$f(x) = (x - 2)^3 \qquad \text{Given function}$$

$$y = (x - 2)^3 \qquad \text{Replace } f(x) \text{ with } y.$$

$$x = (y - 2)^3 \qquad \text{Interchange } x \text{ and } y.$$

$$\sqrt[3]{x} = \sqrt[3]{(y - 2)^3} \qquad \text{Take the cube root on each side. (Section 1.6)}$$

$$\sqrt[3]{x} = y - 2 \qquad \sqrt[3]{a^3} = a \text{ (Section R.7)}$$

$$\sqrt[3]{x} + 2 = y \qquad \text{Solve for } y \text{ by adding 2.}$$

$$f^{-1}(x) = \sqrt[3]{x} + 2 \qquad \text{Replace } y \text{ with } f^{-1}(x). \text{ Rewrite.}$$

☑ *Now Try Exercises 59(a), 63(a), and 65(a).*

EXAMPLE 6 **Finding the Equation of the Inverse of a Rational Function**

The rational function

$$f(x) = \frac{2x + 3}{x - 4}, \quad x \neq 4 \quad \text{(Section 3.5)}$$

is a one-to-one function. Find its inverse.

SOLUTION

$$f(x) = \frac{2x + 3}{x - 4}, \quad x \neq 4$$

$$y = \frac{2x + 3}{x - 4} \qquad \text{Replace } f(x) \text{ with } y.$$

$$x = \frac{2y + 3}{y - 4}, \quad y \neq 4 \qquad \begin{array}{l}\text{Interchange } x \text{ and } y.\\ \text{(Step 1)}\end{array}$$

$$x(y - 4) = 2y + 3 \qquad \text{Multiply by } y - 4.$$

Pay close attention here. → $$xy - 4x = 2y + 3 \qquad \text{Distributive property}$$

$$xy - 2y = 4x + 3 \qquad \text{Add } 4x \text{ and } -2y.$$

$$y(x - 2) = 4x + 3 \qquad \begin{array}{l}\text{Factor out } y.\\ \text{(Section R.4)}\end{array}$$

$$y = \frac{4x + 3}{x - 2}, \quad x \neq 2 \quad \text{Divide by } x - 2.$$

Solve for *y*. (Step 2)

In the final line, we give the condition $x \neq 2$. (Note that 2 was not in the *range* of f, so it is not in the domain of f^{-1}.)

$$f^{-1}(x) = \frac{4x + 3}{x - 2}, \quad x \neq 2 \qquad \text{Replace } y \text{ with } f^{-1}(x). \text{ (Step 3)}$$

☑ *Now Try Exercise 71(a).*

One way to graph the inverse of a function f whose equation is known follows.

Step 1 Find some ordered pairs that are on the graph of f.

Step 2 Interchange x and y to get ordered pairs that are on the graph of f^{-1}.

Step 3 Plot those points, and sketch the graph of f^{-1} through them.

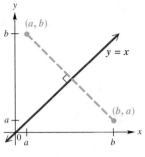

Figure 8

Another way is to select points on the graph of f and use symmetry to find corresponding points on the graph of f^{-1}.

For example, suppose the point (a, b) shown in **Figure 8** is on the graph of a one-to-one function f. Then the point (b, a) is on the graph of f^{-1}. The line segment connecting (a, b) and (b, a) is perpendicular to, and cut in half by, the line $y = x$. The points (a, b) and (b, a) are "mirror images" of each other with respect to $y = x$. **Thus, we can find the graph of f^{-1} from the graph of f by locating the mirror image of each point in f with respect to the line $y = x$.**

EXAMPLE 7 Graphing f^{-1} Given the Graph of f

In each set of axes in **Figure 9**, the graph of a one-to-one function f is shown in blue. Graph f^{-1} in red.

SOLUTION In **Figure 9**, the graphs of two functions f shown in blue are given with their inverses shown in red. In each case, the graph of f^{-1} is a reflection of the graph of f with respect to the line $y = x$.

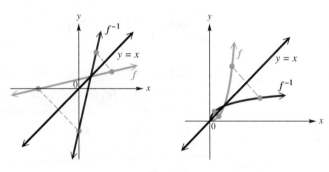

Figure 9

☑ *Now Try Exercises 77 and 81.*

EXAMPLE 8 Finding the Inverse of a Function with a Restricted Domain

Let $f(x) = \sqrt{x + 5}, \quad x \geq -5$. Find $f^{-1}(x)$.

SOLUTION First, notice that the domain of f is restricted to the interval $[-5, \infty)$. Function f is one-to-one because it is increasing on its entire domain and, thus, has an inverse function. Now we find the equation of the inverse.

$$f(x) = \sqrt{x + 5}, \quad x \geq -5$$

$$y = \sqrt{x + 5}, \quad x \geq -5 \qquad \text{Replace } f(x) \text{ with } y.$$

$$x = \sqrt{y + 5}, \quad y \geq -5 \qquad \text{Interchange } x \text{ and } y.$$

$$x^2 = \left(\sqrt{y + 5}\right)^2 \qquad \text{Square each side. (Section 1.6)}$$

$$x^2 = y + 5 \qquad \qquad \left(\sqrt{a}\right)^2 = a \text{ for } a \geq 0 \text{ (Section R.7)}$$

$$y = x^2 - 5 \qquad \qquad \text{Solve for } y.$$

However, we cannot define $f^{-1}(x)$ as $x^2 - 5$. The domain of f is $[-5, \infty)$, and its range is $[0, \infty)$. The range of f is the domain of f^{-1}, so f^{-1} must be defined as

$$f^{-1}(x) = x^2 - 5, \quad x \geq 0.$$

As a check, the range of f^{-1}, $[-5, \infty)$, is the domain of f.

Graphs of f and f^{-1} are shown in **Figures 10** and **11.** The line $y = x$ is included on the graphs to show that the graphs of f and f^{-1} are mirror images with respect to this line.

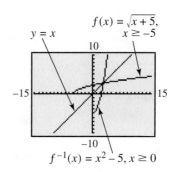

Figure 10   **Figure 11**

☑ *Now Try Exercise 75.*

Important Facts about Inverses

1. If f is one-to-one, then f^{-1} exists.

2. The domain of f is the range of f^{-1}, and the range of f is the domain of f^{-1}.

3. If the point (a, b) lies on the graph of f, then (b, a) lies on the graph of f^{-1}. The graphs of f and f^{-1} are reflections of each other across the line $y = x$.

4. To find the equation for f^{-1}, replace $f(x)$ with y, interchange x and y, and solve for y. This gives $f^{-1}(x)$.

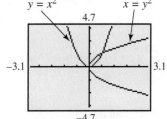

Despite the fact that $y = x^2$ is not one-to-one, the calculator will draw its "inverse," $x = y^2$.

Figure 12

☒☐ Some graphing calculators have the capability of "drawing" the reflection of a graph across the line $y = x$. This feature does not require that the function be one-to-one, however, so the resulting figure may not be the graph of a function. See **Figure 12.** *Again, it is necessary to understand the mathematics to interpret results correctly.* ■

An Application of Inverse Functions to Cryptography   A one-to-one function and its inverse can be used to make information secure. The function is used to encode a message, and its inverse is used to decode the coded message. In practice, complicated functions are used. We illustrate the process with a simple function in **Example 9.**

EXAMPLE 9   **Using Functions to Encode and Decode a Message**

Use the one-to-one function $f(x) = 3x + 1$ and the following numerical values assigned to each letter of the alphabet to encode and decode the message BE MY FACEBOOK FRIEND.

A	1	**H**	8	**O**	15	**V**	22
B	2	**I**	9	**P**	16	**W**	23
C	3	**J**	10	**Q**	17	**X**	24
D	4	**K**	11	**R**	18	**Y**	25
E	5	**L**	12	**S**	19	**Z**	26
F	6	**M**	13	**T**	20		
G	7	**N**	14	**U**	21		

A	1	N	14
B	2	O	15
C	3	P	16
D	4	Q	17
E	5	R	18
F	6	S	19
G	7	T	20
H	8	U	21
I	9	V	22
J	10	W	23
K	11	X	24
L	12	Y	25
M	13	Z	26

SOLUTION The message BE MY FACEBOOK FRIEND would be encoded as

$$7 \quad 16 \quad 40 \quad 76 \quad 19 \quad 4 \quad 10 \quad 16 \quad 7$$
$$46 \quad 46 \quad 34 \quad 19 \quad 55 \quad 28 \quad 16 \quad 43 \quad 13$$

because

B corresponds to 2   and $f(2) = 3(2) + 1 = 7$,

E corresponds to 5   and $f(5) = 3(5) + 1 = 16$,   and so on.

Using the inverse $f^{-1}(x) = \frac{1}{3}x - \frac{1}{3}$ to decode yields

$$f^{-1}(7) = \frac{1}{3}(7) - \frac{1}{3} = 2, \quad \text{which corresponds to B,}$$

$$f^{-1}(16) = \frac{1}{3}(16) - \frac{1}{3} = 5, \quad \text{which corresponds to E,} \quad \text{and so on.}$$

✔ *Now Try Exercise 97.*

4.1 Exercises

Concept Check In Exercises 1 and 2, answer the question and then write a short explanation.

1. The table shows the number of registered passenger cars in the United States for the years 2004–2008.

   If this correspondence is considered to be a function that pairs each year with the number of registered passenger cars, is it one-to-one? If not, explain why.

Year	Registered Passenger Cars (in thousands)
2004	136,431
2005	136,568
2006	135,400
2007	135,933
2008	137,080

Source: U.S. Federal Highway Administration.

2. The table shows the number of hazardous waste sites on the National Priority List for the Superfund program in 2008 for seven states in the top ten.

   If this correspondence is considered to be a function that pairs each state with its number of hazardous waste sites, is it one-to-one? If not, explain why.

State	Number of Sites
New Jersey	116
California	97
Pennsylvania	96
New York	86
Florida	52
Illinois	49
Texas	49

Source: U.S. Environmental Protection Agency.

Decide whether each function as graphed or defined is one-to-one. ***See Examples 1 and 2.***

3.

4.

5.

6.

7.

8.

9. $y = 2x - 8$

10. $y = 4x + 20$

11. $y = \sqrt{36 - x^2}$

12. $y = -\sqrt{100 - x^2}$

13. $y = 2x^3 - 1$

14. $y = 3x^3 - 6$

15. $y = \dfrac{-1}{x + 2}$

16. $y = \dfrac{4}{x - 8}$

17. $y = 2(x + 1)^2 - 6$

18. $y = -3(x - 6)^2 + 8$

19. $y = \sqrt[3]{x + 1} - 3$

20. $y = -\sqrt[3]{x + 2} - 8$

21. Explain why a constant function, such as $f(x) = 3$, defined over the set of real numbers, cannot be one-to-one.

22. (a) Explain why a polynomial function of even degree cannot have an inverse.
  (b) Explain why a polynomial function of odd degree *may* not be one-to-one.

Concept Check Answer each of the following.

23. For a function to have an inverse, it must be _____.

24. If two functions f and g are inverses, then $(f \circ g)(x) =$ _____, and _____ $= x$.

25. The domain of f is equal to the _____ of f^{-1}, and the range of f is equal to the _____ of f^{-1}.

26. If the point (a, b) lies on the graph of f, and f has an inverse, then the point _____ lies on the graph of f^{-1}.

27. *True or false*: If $f(x) = x^2$, then $f^{-1}(x) = \sqrt{x}$.

28. If a function f has an inverse, then the graph of f^{-1} may be obtained by reflecting the graph of f across the line with equation _____.

29. If a function f has an inverse and $f(-3) = 6$, then $f^{-1}(6) =$ _____.

30. If $f(-4) = 16$ and $f(4) = 16$, then f _____ have an inverse because
        (does/does not)
_____.

Concept Check In Exercises 31–36, an everyday activity is described. Keeping in mind that an inverse operation "undoes" what an operation does, describe each inverse activity.

31. tying your shoelaces

32. starting a car

33. entering a room

34. climbing the stairs

35. screwing in a light bulb

36. filling a cup

Decide whether the given functions are inverses. **See Example 4.**

37.

x	f(x)		x	g(x)
3	−4		−4	3
2	−6		−6	2
5	8		8	5
1	9		9	1
4	3		3	4

38.

x	f(x)		x	g(x)
−2	−8		8	−2
−1	−1		1	−1
0	0		0	0
1	1		−1	1
2	8		−8	2

39. $f = \{(2, 5), (3, 5), (4, 5)\}; \quad g = \{(5, 2)\}$

40. $f = \{(1, 1), (3, 3), (5, 5)\}; \quad g = \{(1, 1), (3, 3), (5, 5)\}$

Use the definition of inverses to determine whether f and g are inverses. **See Example 3.**

41. $f(x) = 2x + 4, \quad g(x) = \dfrac{1}{2}x - 2$

42. $f(x) = 3x + 9, \quad g(x) = \dfrac{1}{3}x - 3$

43. $f(x) = -3x + 12, \quad g(x) = -\dfrac{1}{3}x - 12$

44. $f(x) = -4x + 2, \quad g(x) = -\dfrac{1}{4}x - 2$

45. $f(x) = \dfrac{x + 1}{x - 2}, \quad g(x) = \dfrac{2x + 1}{x - 1}$

46. $f(x) = \dfrac{x - 3}{x + 4}, \quad g(x) = \dfrac{4x + 3}{1 - x}$

47. $f(x) = \dfrac{2}{x + 6}, \quad g(x) = \dfrac{6x + 2}{x}$

48. $f(x) = \dfrac{-1}{x + 1}, \quad g(x) = \dfrac{1 - x}{x}$

49. $f(x) = x^2 + 3, \quad x \geq 0; \quad g(x) = \sqrt{x - 3}, \quad x \geq 3$

50. $f(x) = \sqrt{x + 8}, \quad x \geq -8; \quad g(x) = x^2 - 8, \quad x \geq 0$

If the function is one-to-one, find its inverse. **See Example 4.**

51. $\{(-3, 6), (2, 1), (5, 8)\}$

52. $\left\{(3, -1), (5, 0), (0, 5), \left(4, \dfrac{2}{3}\right)\right\}$

53. $\{(1, -3), (2, -7), (4, -3), (5, -5)\}$

54. $\{(6, -8), (3, -4), (0, -8), (5, -4)\}$

Decide whether each pair of functions graphed are inverses. **See Example 7.**

55.

56.

57.

58.

*For each function as defined that is one-to-one, (**a**) write an equation for the inverse function in the form $y = f^{-1}(x)$, (**b**) graph f and f^{-1} on the same axes, and (**c**) give the domain and the range of f and f^{-1}. If the function is not one-to-one, say so. See Examples 5–8.*

59. $y = 3x - 4$ **60.** $y = 4x - 5$ **61.** $f(x) = -4x + 3$

62. $f(x) = -6x - 8$ **63.** $f(x) = x^3 + 1$ **64.** $f(x) = -x^3 - 2$

65. $y = x^2 + 8$ **66.** $y = -x^2 + 2$ **67.** $y = \dfrac{1}{x}, \quad x \neq 0$

68. $y = \dfrac{4}{x}, \quad x \neq 0$ **69.** $f(x) = \dfrac{1}{x - 3}, \quad x \neq 3$ **70.** $f(x) = \dfrac{1}{x + 2}, \quad x \neq -2$

71. $f(x) = \dfrac{x + 1}{x - 3}, \quad x \neq 3$ **72.** $f(x) = \dfrac{x + 2}{x - 1}, \quad x \neq 1$

73. $f(x) = \dfrac{2x + 6}{x - 3}, \quad x \neq 3$ **74.** $f(x) = \dfrac{-3x + 12}{x - 6}, \quad x \neq 6$

75. $f(x) = \sqrt{6 + x}, \quad x \geq -6$ **76.** $f(x) = -\sqrt{x^2 - 16}, \quad x \geq 4$

Graph the inverse of each one-to-one function. See Example 7.

77.

78.

79.

80.

81.

82.

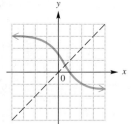

Concept Check The graph of a function f is shown in the figure. Use the graph to find each value.

83. $f^{-1}(4)$ **84.** $f^{-1}(2)$

85. $f^{-1}(0)$ **86.** $f^{-1}(-2)$

87. $f^{-1}(-3)$ **88.** $f^{-1}(-4)$

Concept Check Answer each of the following.

89. Suppose $f(x)$ is the number of cars that can be built for x dollars. What does $f^{-1}(1000)$ represent?

90. Suppose $f(r)$ is the volume (in cubic inches) of a sphere of radius r inches. What does $f^{-1}(5)$ represent?

91. If a line has slope a, what is the slope of its reflection across the line $y = x$?

92. For a one-to-one function f, find $(f^{-1} \circ f)(2)$, where $f(2) = 3$.

Use a graphing calculator to graph each function defined as follows, using the given viewing window. Use the graph to decide which functions are one-to-one. If a function is one-to-one, give the equation of its inverse.

93. $f(x) = 6x^3 + 11x^2 - 6$;
$[-3, 2]$ by $[-10, 10]$

94. $f(x) = x^4 - 5x^2$;
$[-3, 3]$ by $[-8, 8]$

95. $f(x) = \dfrac{x - 5}{x + 3}, \quad x \neq -3$;
$[-8, 8]$ by $[-6, 8]$

96. $f(x) = \dfrac{-x}{x - 4}, \quad x \neq 4$;
$[-1, 8]$ by $[-6, 6]$

*Use the alphabet coding assignment given in **Example 9** for Exercises 97–100.*

97. The function $f(x) = 3x - 2$ was used to encode a message as

37 25 19 61 13 34 22 1 55 1 52 52 25 64 13 10.

Find the inverse function and determine the message.

98. The function $f(x) = 2x - 9$ was used to encode a message as

−5 9 5 5 9 27 15 29 −1 21 19 31 −3 27 41.

Find the inverse function and determine the message.

99. Encode the message SEND HELP, using the one-to-one function $f(x) = x^3 - 1$. Give the inverse function that the decoder will need when the message is received.

100. Encode the message SAILOR BEWARE, using the one-to-one function $f(x) = (x + 1)^3$. Give the inverse function that the decoder will need when the message is received.

4.2 Exponential Functions

■ Exponents and Properties
■ Exponential Functions
■ Exponential Equations
■ Compound Interest
■ The Number *e* and Continuous Compounding
■ Exponential Models

Exponents and Properties   Recall the definition of $a^{m/n}$: If a is a real number, m is an integer, n is a positive integer, and $\sqrt[n]{a}$ is a real number, then

$$a^{m/n} = \left(\sqrt[n]{a}\right)^m. \quad \text{(Section R.7)}$$

For example,

$$16^{3/4} = \left(\sqrt[4]{16}\right)^3 = 2^3 = 8,$$

$$27^{-1/3} = \frac{1}{27^{1/3}} = \frac{1}{\sqrt[3]{27}} = \frac{1}{3}, \quad \text{and} \quad 64^{-1/2} = \frac{1}{64^{1/2}} = \frac{1}{\sqrt{64}} = \frac{1}{8}.$$

In this section we extend the definition of a^r to include all real (not just rational) values of the exponent r. For example, $2^{\sqrt{3}}$ might be evaluated by *approximating* the exponent $\sqrt{3}$ with the rational numbers 1.7, 1.73, 1.732, and so on. Since these decimals approach the value of $\sqrt{3}$ more and more closely, it seems reasonable that $2^{\sqrt{3}}$ should be approximated more and more closely by the numbers $2^{1.7}$, $2^{1.73}$, $2^{1.732}$, and so on. $\left(\text{Recall, for example, that } 2^{1.7} = 2^{17/10} = \left(\sqrt[10]{2}\right)^{17}.\right)$ To show that this assumption is reasonable, **Figure 13** gives graphs of the function $f(x) = 2^x$ with three different domains.

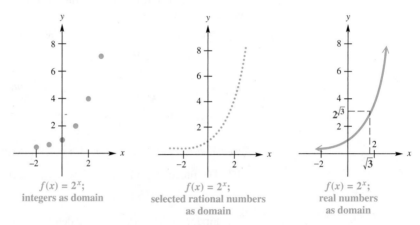

$f(x) = 2^x$;
integers as domain

$f(x) = 2^x$;
selected rational numbers
as domain

$f(x) = 2^x$;
real numbers
as domain

Figure 13

Using this interpretation of real exponents, all rules and theorems for exponents are valid for all real number exponents, not just rational ones. In addition to the rules for exponents presented earlier, we use several new properties in this chapter. These properties are generalized below. Proofs of the properties are not given here, because they require more advanced mathematics.

Additional Properties of Exponents

For any real number $a > 0$, $a \neq 1$, the following statements are true.

(a) a^x **is a unique real number for all real numbers x.**

(b) $a^b = a^c$ **if and only if $b = c$.**

(c) **If $a > 1$ and $m < n$, then $a^m < a^n$.**

(d) **If $0 < a < 1$ and $m < n$, then $a^m > a^n$.**

Properties (a) and (b) require $a > 0$ so that a^x is always defined. For example, $(-6)^x$ is not a real number if $x = \frac{1}{2}$. This means that a^x will always be positive, since a must be positive. In property (a), a cannot equal 1 because $1^x = 1$ for every real number value of x, so each value of x leads to the same real number, 1. For property (b) to hold, a must not equal 1 since, for example, $1^4 = 1^5$, even though $4 \neq 5$.

Properties (c) and (d) say that *when $a > 1$, increasing the exponent on "a" leads to a greater number, but when $0 < a < 1$, increasing the exponent on "a" leads to a lesser number.*

| EXAMPLE 1 | **Evaluating an Exponential Expression** |

If $f(x) = 2^x$, find each of the following.

(a) $f(-1)$ **(b)** $f(3)$ **(c)** $f\left(\dfrac{5}{2}\right)$ **(d)** $f(4.92)$

SOLUTION

(a) $f(-1) = 2^{-1} = \dfrac{1}{2}$ Replace x with -1.

(b) $f(3) = 2^3 = 8$

(c) $f\left(\dfrac{5}{2}\right) = 2^{5/2} = (2^5)^{1/2} = 32^{1/2} = \sqrt{32} = \sqrt{16 \cdot 2} = 4\sqrt{2}$ (Section R.7)

(d) $f(4.92) = 2^{4.92} \approx 30.2738447$ Use a calculator.

✔ *Now Try Exercises 3, 9, and 13.*

Exponential Functions   We now define a function $f(x) = a^x$ whose domain is the set of all real numbers. Notice how the independent variable x appears in the exponent in this function. In earlier chapters the independent variable did not appear in exponents.

Exponential Function

If $a > 0$ and $a \neq 1$, then

$$f(x) = a^x$$

defines the **exponential function with base a.**

NOTE   *We do not allow 1 as the base for an exponential function.* If $a = 1$, the function becomes the constant function $f(x) = 1$, which is not an exponential function.

    Figure 13 showed the graph of $f(x) = 2^x$ with three different domains. We repeat the final graph (with real numbers as domain) here.

- The y-value of the y-intercept is $y = 2^0 = 1$.

- Since $2^x > 0$ for all x and $2^x \to 0$ as $x \to -\infty$, the x-axis is a horizontal asymptote.

- As the graph suggests, the domain of the function is $(-\infty, \infty)$ and the range is $(0, \infty)$.

- The function is increasing on its entire domain, and it therefore is one-to-one.

    These observations from **Figure 13** lead to the following generalizations about the graphs of exponential functions.

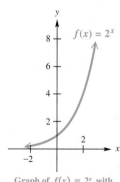

Graph of $f(x) = 2^x$ with domain $(-\infty, \infty)$

Figure 13 (repeated)

Exponential Function $f(x) = a^x$

Domain: $(-\infty, \infty)$ Range: $(0, \infty)$

For $f(x) = 2^x$:

x	$f(x)$
-2	$\frac{1}{4}$
-1	$\frac{1}{2}$
0	1
1	2
2	4
3	8

This is the general behavior seen on a calculator graph for **any base a, for $a > 1$.**

Figure 14

- $f(x) = a^x$, **for $a > 1$,** is increasing and continuous on its entire domain, $(-\infty, \infty)$.
- The x-axis is a horizontal asymptote as $x \to -\infty$.
- The graph passes through the points $\left(-1, \frac{1}{a}\right)$, $(0, 1)$, and $(1, a)$.

For $f(x) = \left(\frac{1}{2}\right)^x$:

x	$f(x)$
-3	8
-2	4
-1	2
0	1
1	$\frac{1}{2}$
2	$\frac{1}{4}$

This is the general behavior seen on a calculator graph for **any base a, for $0 < a < 1$.**

Figure 15

- $f(x) = a^x$, **for $0 < a < 1$,** is decreasing and continuous on its entire domain, $(-\infty, \infty)$.
- The x-axis is a horizontal asymptote as $x \to \infty$.
- The graph passes through the points $\left(-1, \frac{1}{a}\right)$, $(0, 1)$, and $(1, a)$.

From **Section 2.7,** the graph of $y = f(-x)$ is the graph of $y = f(x)$ reflected across the y-axis. Thus, we have the following.

$$\text{If } f(x) = 2^x, \quad \text{then} \quad f(-x) = 2^{-x} = 2^{-1 \cdot x} = (2^{-1})^x = \left(\frac{1}{2}\right)^x.$$

This is supported by the graphs in **Figures 14 and 15.**

The graph of $f(x) = 2^x$ is typical of graphs of $f(x) = a^x$ where $a > 1$. For larger values of a, the graphs rise more steeply, but the general shape is similar to the graph in **Figure 14.** When $0 < a < 1$, the graph decreases in a manner similar to the graph of $f(x) = \left(\frac{1}{2}\right)^x$. In **Figure 16** on the next page, the graphs of several typical exponential functions illustrate these facts.

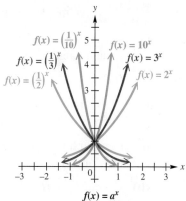

$$f(x) = a^x$$
Domain: $(-\infty, \infty)$; Range: $(0, \infty)$
- When $a > 1$, the function is increasing.
- When $0 < a < 1$, the function is decreasing.
- In every case, the x-axis is a horizontal asymptote.

Figure 16

In summary, the graph of a function of the form $f(x) = a^x$ has the following features.

Characteristics of the Graph of $f(x) = a^x$

1. The points $\left(-1, \frac{1}{a}\right)$, $(0, 1)$, and $(1, a)$ are on the graph.
2. If $a > 1$, then f is an increasing function. If $0 < a < 1$, then f is a decreasing function.
3. The x-axis is a horizontal asymptote.
4. The domain is $(-\infty, \infty)$, and the range is $(0, \infty)$.

EXAMPLE 2 **Graphing an Exponential Function**

Graph $f(x) = 5^x$. Give the domain and range.

SOLUTION The y-intercept is $(0, 1)$, and the x-axis is a horizontal asymptote. Plot a few ordered pairs, and draw a smooth curve through them as shown in **Figure 17.** Like the function $f(x) = 2^x$, this function also has domain $(-\infty, \infty)$ and range $(0, \infty)$ and is one-to-one. The function is increasing on its entire domain.

x	$f(x)$
-1	0.2
0	1
0.5	≈ 2.2
1	5
1.5	≈ 11.2
2	25

$f(x) = 5^x$

Figure 17

✔ *Now Try Exercise 17.*

EXAMPLE 3 **Graphing Reflections and Translations**

Graph each function. Show the graph of $y = 2^x$ for comparison. Give the domain and range.

(a) $f(x) = -2^x$ **(b)** $f(x) = 2^{x+3}$ **(c)** $f(x) = 2^{x-2} - 1$

SOLUTION In each graph, we show in particular how the point $(0, 1)$ on the graph of $y = 2^x$ has been translated.

(a) The graph of $f(x) = -2^x$ is that of $f(x) = 2^x$ reflected across the x-axis. The domain is $(-\infty, \infty)$, and the range is $(-\infty, 0)$. See **Figure 18.**

(b) The graph of $f(x) = 2^{x+3}$ is the graph of $f(x) = 2^x$ translated 3 units to the left, as shown in **Figure 19.** The domain is $(-\infty, \infty)$, and the range is $(0, \infty)$.

(c) The graph of $f(x) = 2^{x-2} - 1$ is that of $f(x) = 2^x$ translated 2 units to the right and 1 unit down. See **Figure 20.** The domain is $(-\infty, \infty)$, and the range is $(-1, \infty)$.

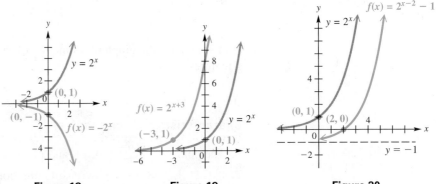

Figure 18   Figure 19   Figure 20

✔ *Now Try Exercises 29, 31, and 37.*

Exponential Equations Because the graph of $y = a^x$ is that of a one-to-one function, to solve $a^{x_1} = a^{x_2}$, we need only show that $x_1 = x_2$. Property (b) given earlier in this section is used to solve **exponential equations,** which are equations with variables as exponents.

EXAMPLE 4 **Solving an Exponential Equation**

Solve $\left(\frac{1}{3}\right)^x = 81$.

SOLUTION *Write each side of the equation using a common base.*

$$\left(\frac{1}{3}\right)^x = 81$$

$(3^{-1})^x = 81$ Definition of negative exponent **(Section R.6)**

$3^{-x} = 81$ $(a^m)^n = a^{mn}$ **(Section R.3)**

$3^{-x} = 3^4$ Write 81 as a power of 3. **(Section R.2)**

$-x = 4$ Set exponents equal (Property (b)).

$x = -4$ Multiply by -1. **(Section 1.1)**

The solution set of the original equation is $\{-4\}$.

By extending the method of solving linear equations graphically from **Section 2.5,** we can support the result in **Example 4.**

✔ *Now Try Exercise 63.*

EXAMPLE 5 Solving an Exponential Equation

Solve $2^{x+4} = 8^{x-6}$.

SOLUTION Write each side of the equation using a common base.

$$2^{x+4} = 8^{x-6}$$

$$2^{x+4} = \left(2^3\right)^{x-6} \qquad \text{Write 8 as a power of 2.}$$

$$2^{x+4} = 2^{3x-18} \qquad (a^m)^n = a^{mn}$$

$$x + 4 = 3x - 18 \qquad \text{Set exponents equal (Property (b)).}$$

$$-2x = -22 \qquad \text{Subtract } 3x \text{ and 4. (Section 1.1)}$$

$$x = 11 \qquad \text{Divide by } -2.$$

Check by substituting 11 for x in the original equation. The solution set is $\{11\}$.

✔ Now Try Exercise 71.

Later in this chapter, we describe a general method for solving exponential equations where the approach used in **Examples 4 and 5** is not possible. For instance, the above method could not be used to solve an equation like

$$7^x = 12,$$

since it is not easy to express both sides as exponential expressions with the same base.

In **Example 6,** we review a solution process first seen in **Section 1.6.**

EXAMPLE 6 Solving an Equation with a Fractional Exponent

Solve $x^{4/3} = 81$.

SOLUTION Notice that the variable is in the base rather than in the exponent.

$$x^{4/3} = 81$$

$$\left(\sqrt[3]{x}\right)^4 = 81 \qquad \text{Radical notation for } a^{m/n} \text{ (Section R.7)}$$

$$\sqrt[3]{x} = \pm 3 \qquad \begin{array}{l}\text{Take fourth roots on each side.}\\ \text{Remember to use } \pm. \text{ (Section 1.6)}\end{array}$$

$$x = \pm 27 \qquad \text{Cube each side.}$$

Check *both* solutions in the original equation. Both check, so the solution set is $\{\pm 27\}$.

Alternative Method There may be more than one way to solve an exponential equation, as shown here.

$$x^{4/3} = 81$$

$$\left(x^{4/3}\right)^3 = 81^3 \qquad \text{Cube each side.}$$

$$x^4 = \left(3^4\right)^3 \qquad \text{Write 81 as } 3^4.$$

$$x^4 = 3^{12} \qquad (a^m)^n = a^{mn}$$

$$x = \pm \sqrt[4]{3^{12}} \qquad \text{Take fourth roots on each side.}$$

$$x = \pm 3^3 \qquad \text{Simplify the radical.}$$

$$x = \pm 27 \qquad \text{Apply the exponent.}$$

The same solution set, $\{\pm 27\}$, results.

✔ Now Try Exercise 79.

Compound Interest   Recall the formula for simple interest, $I = Prt$, where P is principal (amount deposited), r is annual rate of interest expressed as a decimal, and t is time in years that the principal earns interest. Suppose $t = 1$ yr. Then at the end of the year, the amount has grown to

$$P + Pr = P(1 + r),$$

the original principal plus interest. If this balance earns interest at the same interest rate for another year, the balance at the end of *that* year will be

$$[P(1 + r)] + [P(1 + r)]r = [P(1 + r)](1 + r) \quad \text{Factor.}$$
$$= P(1 + r)^2. \quad \quad a \cdot a = a^2$$

After the third year, this will grow to

$$[P(1 + r)^2] + [P(1 + r)^2]r = [P(1 + r)^2](1 + r) \quad \text{Factor.}$$
$$= P(1 + r)^3. \quad \quad a^2 \cdot a = a^3$$

Continuing in this way produces a formula for interest compounded annually.

$$A = P(1 + r)^t$$

The general formula for compound interest can be derived in the same way.

Compound Interest

If P dollars are deposited in an account paying an annual rate of interest r compounded (paid) n times per year, then after t years the account will contain A dollars, according to the following formula.

$$A = P\left(1 + \frac{r}{n}\right)^{tn}$$

EXAMPLE 7   **Using the Compound Interest Formula**

Suppose $1000 is deposited in an account paying 4% interest per year compounded quarterly (four times per year).

(a) Find the amount in the account after 10 yr with no withdrawals.

(b) How much interest is earned over the 10-yr period?

SOLUTION

(a) $A = P\left(1 + \dfrac{r}{n}\right)^{tn}$ Compound interest formula

$A = 1000\left(1 + \dfrac{0.04}{4}\right)^{10(4)}$ Let $P = 1000$, $r = 0.04$, $n = 4$, and $t = 10$.

$A = 1000(1 + 0.01)^{40}$ Simplify.

$A = 1488.86$ Round to the nearest cent.

Thus, $1488.86 is in the account after 10 yr.

(b) The interest earned for that period is

$$\$1488.86 - \$1000 = \$488.86.$$

✔ Now Try Exercise 87(a).

In the formula for compound interest

$$A = P\left(1 + \frac{r}{n}\right)^{tn},$$

A is sometimes called the **future value** and P the **present value**. A is also called the **compound amount** and is the balance *after* interest has been earned.

EXAMPLE 8 **Finding Present Value**

Becky Anderson must pay a lump sum of $6000 in 5 yr.

(a) What amount deposited today (present value) at 3.1% compounded annually will grow to $6000 in 5 yr?

(b) If only $5000 is available to deposit now, what annual interest rate is necessary for the money to increase to $6000 in 5 yr?

SOLUTION

(a)

$$A = P\left(1 + \frac{r}{n}\right)^{tn} \qquad \text{Compound interest formula}$$

$$6000 = P\left(1 + \frac{0.031}{1}\right)^{5(1)} \qquad \text{Let } A = 6000, \ r = 0.031, n = 1, \text{ and } t = 5.$$

$$6000 = P(1.031)^5 \qquad \text{Simplify.}$$

$$P = \frac{6000}{(1.031)^5} \qquad \text{Divide by } (1.031)^5 \text{ to solve for } P. \ \textbf{(Section 1.1)}$$

$$P \approx 5150.60 \qquad \text{Use a calculator.}$$

If Becky leaves $5150.60 for 5 yr in an account paying 3.1% compounded annually, she will have $6000 when she needs it. Thus, $5150.60 is the present value of $6000 if interest of 3.1% is compounded annually for 5 yr.

(b)

$$A = P\left(1 + \frac{r}{n}\right)^{tn} \qquad \text{Compound interest formula}$$

$$6000 = 5000(1 + r)^5 \qquad \text{Let } A = 6000, P = 5000, n = 1, \text{ and } t = 5.$$

$$\frac{6}{5} = (1 + r)^5 \qquad \text{Divide by 5000.}$$

$$\left(\frac{6}{5}\right)^{1/5} = 1 + r \qquad \text{Take the fifth root on each side.}$$

$$\left(\frac{6}{5}\right)^{1/5} - 1 = r \qquad \text{Subtract 1.}$$

$$r \approx 0.0371 \qquad \text{Use a calculator.}$$

An interest rate of 3.71% will produce enough interest to increase the $5000 to $6000 by the end of 5 yr.

✔️ *Now Try Exercises 89 and 93.*

CAUTION When performing the computations in problems like those in **Examples 7 and 8,** do not round off during intermediate steps. Keep all calculator digits and round at the end of the process.

n	$\left(1 + \dfrac{1}{n}\right)^n$ (rounded)
1	2
2	2.25
5	2.48832
10	2.59374
100	2.70481
1000	2.71692
10,000	2.71815
1,000,000	2.71828

The Number e and Continuous Compounding The more often interest is compounded within a given time period, the more interest will be earned. Surprisingly, however, there is a limit on the amount of interest, no matter how often it is compounded.

Suppose that $1 is invested at 100% interest per year, compounded n times per year. Then the interest rate (in decimal form) is 1.00, and the interest rate per period is $\frac{1}{n}$. According to the formula (with $P = 1$), the compound amount at the end of 1 yr will be

$$A = \left(1 + \frac{1}{n}\right)^n.$$

A calculator gives the results in the margin for various values of n. The table suggests that as n increases, the value of $\left(1 + \frac{1}{n}\right)^n$ gets closer and closer to some fixed number. This is indeed the case. This fixed number is called e. *(Note that in mathematics, e is a real number and not a variable.)*

Figure 21

Value of e

$$e \approx 2.718281828459045$$

Figure 21 shows graphs of the functions

$$y = 2^x, \quad y = 3^x, \quad \text{and} \quad y = e^x.$$

Because $2 < e < 3$, the graph of $y = e^x$ lies "between" the other two graphs.

As mentioned above, the amount of interest earned increases with the frequency of compounding, but the value of the expression $\left(1 + \frac{1}{n}\right)^n$ approaches e as n gets larger. Consequently, the formula for compound interest approaches a limit as well, called the compound amount from **continuous compounding.**

Continuous Compounding

If P dollars are deposited at a rate of interest r compounded continuously for t years, the compound amount A in dollars on deposit is given by the following formula.

$$A = Pe^{rt}$$

EXAMPLE 9 **Solving a Continuous Compounding Problem**

Suppose $5000 is deposited in an account paying 3% interest compounded continuously for 5 yr. Find the total amount on deposit at the end of 5 yr.

SOLUTION

$A = Pe^{rt}$	Continuous compounding formula
$\quad = 5000e^{0.03(5)}$	Let $P = 5000$, $r = 0.03$, and $t = 5$.
$\quad = 5000e^{0.15}$	Multiply exponents.
$A \approx 5809.17 \quad \text{or} \quad \$5809.17$	Use a calculator.

Check that daily compounding would have produced a compound amount about $0.03 less.

✔ Now Try Exercise 87(b).

EXAMPLE 10 **Comparing Interest Earned as Compounding Is More Frequent**

In **Example 7,** we found that $1000 invested at 4% compounded quarterly for 10 yr grew to $1488.86. Compare this same investment compounded annually, semiannually, monthly, daily, and continuously.

SOLUTION Substitute 0.04 for r, 10 for t, and the appropriate number of compounding periods for n into

$$A = P\left(1 + \frac{r}{n}\right)^{tn} \qquad \text{Compound interest formula}$$

and also into

$$A = Pe^{rt}. \qquad \text{Continuous compounding formula}$$

The results for amounts of $1 and $1000 are given in the table.

Compounded	$1	$1000
Annually	$(1 + 0.04)^{10} \approx 1.48024$	$1480.24
Semiannually	$\left(1 + \dfrac{0.04}{2}\right)^{10(2)} \approx 1.48595$	$1485.95
Quarterly	$\left(1 + \dfrac{0.04}{4}\right)^{10(4)} \approx 1.48886$	$1488.86
Monthly	$\left(1 + \dfrac{0.04}{12}\right)^{10(12)} \approx 1.49083$	$1490.83
Daily	$\left(1 + \dfrac{0.04}{365}\right)^{10(365)} \approx 1.49179$	$1491.79
Continuously	$e^{10(0.04)} \approx 1.49182$	$1491.82

Comparing the results, we notice the following.

- Compounding semiannually rather than annually increases the value of the account after 10 yr by $5.71.

- Quarterly compounding grows to $2.91 more than semiannual compounding after 10 yr.

- Daily compounding yields only $0.96 more than monthly compounding.

- Continuous compounding yields only $0.03 more than daily compounding.

Each increase in compounding frequency earns less additional interest.

✔ *Now Try Exercise 95.*

LOOKING AHEAD TO CALCULUS

In calculus, the derivative allows us to determine the slope of a tangent line to the graph of a function. For the function $f(x) = e^x$, the derivative is the function f itself: $f'(x) = e^x$. Therefore, in calculus the exponential function with base e is much easier to work with than exponential functions having other bases.

Exponential Models The number e is important as the base of an exponential function in many practical applications. In situations involving growth or decay of a quantity, the amount or number present at time t often can be closely modeled by a function of the form

$$y = y_0 e^{kt},$$

where y_0 is the amount or number present at time $t = 0$ and k is a constant.

Example 11 on the next page illustrates exponential growth. Further examples of exponential growth and decay are given in **Section 4.6.**

EXAMPLE 11 Using Data to Model Exponential Growth

Data from recent past years indicate that future amounts of carbon dioxide in the atmosphere may grow according to the table. Amounts are given in parts per million.

(a) Make a scatter diagram of the data. Do the carbon dioxide levels appear to grow exponentially?

(b) One model for the data is the function

$$y = 0.001942e^{0.00609x},$$

where x is the year and $1990 \leq x \leq 2275$. Use a graph of this model to estimate when future levels of carbon dioxide will double and triple over the preindustrial level of 280 ppm.

Year	Carbon Dioxide (ppm)
1990	353
2000	375
2075	590
2175	1090
2275	2000

Source: International Panel on Climate Change (IPCC).

(a)

SOLUTION

(a) We show a calculator graph for the data in **Figure 22(a).** The data appear to resemble the graph of an increasing exponential function.

(b) A graph of $y = 0.001942e^{0.00609x}$ in **Figure 22(b)** shows that it is very close to the data points. We graph $y = 2 \cdot 280 = 560$ in **Figure 23(a)** and $y = 3 \cdot 280 = 840$ in **Figure 23(b)** on the same coordinate axes as the given function, and we use the calculator to find the intersection points.

$y = 0.001942e^{0.00609x}$

(b)

Figure 22

(a)   (b)

Figure 23

The graph of the function intersects the horizontal lines at approximately 2064.4 and 2130.9. According to this model, carbon dioxide levels will have doubled by 2064 and tripled by 2131.

☑ *Now Try Exercise 97.*

(a)

(b)

Figure 24

Graphing calculators are capable of fitting exponential curves to scatter diagrams like the one found in **Example 11. Figure 24(a)** shows how the TI-83/84 Plus displays another (different) equation for the atmospheric carbon dioxide example, approximated as follows.

$$y = 0.001923(1.006109)^x$$

Notice that this calculator form differs from the model in **Example 11. Figure 24(b)** shows the data points and the graph of this exponential regression equation. ∎

4.2 Exercises

For $f(x) = 3^x$ and $g(x) = \left(\frac{1}{4}\right)^x$, find each of the following. In Exercises 13–16, round the answer to the nearest thousandth. **See Example 1.**

1. $f(2)$ **2.** $f(3)$ **3.** $f(-2)$ **4.** $f(-3)$

5. $g(2)$ **6.** $g(3)$ **7.** $g(-2)$ **8.** $g(-3)$

9. $f\left(\frac{3}{2}\right)$ **10.** $f\left(-\frac{5}{2}\right)$ **11.** $g\left(\frac{3}{2}\right)$ **12.** $g\left(-\frac{5}{2}\right)$

13. $f(2.34)$ **14.** $f(-1.68)$ **15.** $g(-1.68)$ **16.** $g(2.34)$

Graph each function. **See Example 2.**

17. $f(x) = 3^x$ **18.** $f(x) = 4^x$ **19.** $f(x) = \left(\frac{1}{3}\right)^x$

20. $f(x) = \left(\frac{1}{4}\right)^x$ **21.** $f(x) = \left(\frac{3}{2}\right)^x$ **22.** $f(x) = \left(\frac{5}{3}\right)^x$

23. $f(x) = \left(\frac{1}{10}\right)^{-x}$ **24.** $f(x) = \left(\frac{1}{6}\right)^{-x}$ **25.** $f(x) = 4^{-x}$

26. $f(x) = 10^{-x}$ **27.** $f(x) = 2^{|x|}$ **28.** $f(x) = 2^{-|x|}$

Sketch the graph of $f(x) = 2^x$. Then refer to it and use the techniques of **Chapter 2** to graph each function as defined. **See Example 3.**

29. $f(x) = 2^x + 1$ **30.** $f(x) = 2^x - 4$ **31.** $f(x) = 2^{x+1}$ **32.** $f(x) = 2^{x-4}$

33. $f(x) = -2^{x+2}$ **34.** $f(x) = -2^{x-3}$ **35.** $f(x) = 2^{-x}$ **36.** $f(x) = -2^{-x}$

37. $f(x) = 2^{x-1} + 2$ **38.** $f(x) = 2^{x+3} + 1$ **39.** $f(x) = 2^{x+2} - 4$ **40.** $f(x) = 2^{x-3} - 1$

Sketch the graph of $f(x) = \left(\frac{1}{3}\right)^x$. Then refer to it and use the techniques of **Chapter 2** to graph each function as defined. **See Example 3.**

41. $f(x) = \left(\frac{1}{3}\right)^x - 2$ **42.** $f(x) = \left(\frac{1}{3}\right)^x + 4$ **43.** $f(x) = \left(\frac{1}{3}\right)^{x+2}$

44. $f(x) = \left(\frac{1}{3}\right)^{x-4}$ **45.** $f(x) = \left(\frac{1}{3}\right)^{-x+1}$ **46.** $f(x) = \left(\frac{1}{3}\right)^{-x-2}$

47. $f(x) = \left(\frac{1}{3}\right)^{-x}$ **48.** $f(x) = -\left(\frac{1}{3}\right)^{-x}$ **49.** $f(x) = \left(\frac{1}{3}\right)^{x-2} + 2$

50. $f(x) = \left(\frac{1}{3}\right)^{x-1} + 3$ **51.** $f(x) = \left(\frac{1}{3}\right)^{x+2} - 1$ **52.** $f(x) = \left(\frac{1}{3}\right)^{x+3} - 2$

53. *Concept Check* Fill in the blank: The graph of $f(x) = a^{-x}$ is the same as that of $g(x) = (\underline{\hspace{0.5in}})^x$.

54. *Concept Check* Fill in the blanks: If $a > 1$, then the graph of $f(x) = a^x$ $\underline{\hspace{1.2in}}$ from left to right. If $0 < a < 1$, then the graph of $g(x) = a^x$
(rises/falls)
$\underline{\hspace{1.2in}}$ from left to right.
(rises/falls)

Connecting Graphs with Equations Write an equation for the graph given. Each represents an exponential function f, with base 2 or 3, translated and/or reflected.

55.

56.

57.

58.

59.

60.

Solve each equation. ***See Examples 4–6.***

61. $4^x = 2$

62. $125^x = 5$

63. $\left(\dfrac{5}{2}\right)^x = \dfrac{4}{25}$

64. $\left(\dfrac{2}{3}\right)^x = \dfrac{9}{4}$

65. $2^{3-2x} = 8$

66. $5^{2+2x} = 25$

67. $e^{4x-1} = (e^2)^x$

68. $e^{3-x} = (e^3)^{-x}$

69. $27^{4x} = 9^{x+1}$

70. $32^{2x} = 16^{x-1}$

71. $4^{x-2} = 2^{3x+3}$

72. $2^{6-3x} = 8^{x+1}$

73. $\left(\dfrac{1}{e}\right)^{-x} = \left(\dfrac{1}{e^2}\right)^{x+1}$

74. $e^{x-1} = \left(\dfrac{1}{e^4}\right)^{x+1}$

75. $\left(\sqrt{2}\right)^{x+4} = 4^x$

76. $\left(\sqrt[3]{5}\right)^{-x} = \left(\dfrac{1}{5}\right)^{x+2}$

77. $\dfrac{1}{27} = x^{-3}$

78. $\dfrac{1}{32} = x^{-5}$

79. $x^{2/3} = 4$

80. $x^{2/5} = 16$

81. $x^{5/2} = 32$

82. $x^{3/2} = 27$

83. $x^{-6} = \dfrac{1}{64}$

84. $x^{-4} = \dfrac{1}{256}$

85. $x^{5/3} = -243$

86. $x^{7/5} = -128$

Solve each problem involving compound interest. ***See Examples 7–9.***

87. *Future Value* Find the future value and interest earned if $8906.54 is invested for 9 yr at 5% compounded

    (a) semiannually

    (b) continuously.

88. *Future Value* Find the future value and interest earned if $56,780 is invested at 5.3% compounded

(a) quarterly for 23 quarters   **(b)** continuously for 15 yr.

89. *Present Value* Find the present value that will grow to $25,000 if interest is 6% compounded quarterly for 11 quarters.

90. *Present Value* Find the present value that will grow to $45,000 if interest is 3.6% compounded monthly for 1 yr.

91. *Present Value* Find the present value that will grow to $5000 if interest is 3.5% compounded quarterly for 10 yr.

92. *Interest Rate* Find the required annual interest rate to the nearest tenth of a percent for $65,000 to grow to $65,325 if interest is compounded monthly for 6 months.

93. *Interest Rate* Find the required annual interest rate to the nearest tenth of a percent for $1200 to grow to $1500 if interest is compounded quarterly for 5 yr.

94. *Interest Rate* Find the required annual interest rate to the nearest tenth of a percent for $5000 to grow to $8400 if interest is compounded quarterly for 8 yr.

Solve each problem. See Example 10.

95. *Comparing Loans* Bank A is lending money at 6.4% interest compounded annually. The rate at Bank B is 6.3% compounded monthly, and the rate at Bank C is 6.35% compounded quarterly. At which bank will you pay the *least* interest?

96. *Future Value* Suppose $10,000 is invested at an annual rate of 5% for 10 yr. Find the future value if interest is compounded as follows.

(a) annually   **(b)** quarterly   **(c)** monthly   **(d)** daily (365 days)

(Modeling) *Solve each problem. See Example 11.*

97. *Atmospheric Pressure* The atmospheric pressure (in millibars) at a given altitude (in meters) is shown in the table.

Altitude	Pressure	Altitude	Pressure
0	1013	6000	472
1000	899	7000	411
2000	795	8000	357
3000	701	9000	308
4000	617	10,000	265
5000	541		

Source: Miller, A. and J. Thompson, *Elements of Meteorology,* Fourth Edition, Charles E. Merrill Publishing Company, Columbus, Ohio.

(a) Use a graphing calculator to make a scatter diagram of the data for atmospheric pressure P at altitude x.

(b) Would a linear or an exponential function fit the data better?

(c) The function

$$P(x) = 1013e^{-0.0001341x}$$

approximates the data. Use a graphing calculator to graph P and the data on the same coordinate axes.

(d) Use P to predict the pressures at 1500 m and 11,000 m, and compare them to the actual values of 846 millibars and 227 millibars, respectively.

98. *World Population Growth* Since 2000, world population in millions closely fits the exponential function

$$y = 6084e^{0.0120x},$$

where x is the number of years since 2000. (*Source:* U.S. Census Bureau.)

(a) The world population was about 6853 million in 2010. How closely does the function approximate this value?

(b) Use this model to predict the population in 2020.

(c) Use this model to predict the population in 2030.

(d) Explain why this model may not be accurate for 2030.

99. *Deer Population* The exponential growth of the deer population in Massachusetts can be calculated using the model

$$f(x) = 50{,}000(1 + 0.06)^x,$$

where 50,000 is the initial deer population and 0.06 is the rate of growth. $f(x)$ is the total population after x years have passed.

(a) Predict the total population after 4 yr.

(b) If the initial population was 30,000 and the growth rate was 0.12, approximately how many deer would be present after 3 yr?

(c) How many additional deer can we expect in 5 yr if the initial population is 45,000 and the current growth rate is 0.08?

100. *Employee Training* A person learning certain skills involving repetition tends to learn quickly at first. Then learning tapers off and skill acquisition approaches some upper limit. Suppose the number of symbols per minute that a person using a keyboard can type is given by

$$f(t) = 250 - 120(2.8)^{-0.5t},$$

where t is the number of months the operator has been in training. Find each value.

(a) $f(2)$ **(b)** $f(4)$ **(c)** $f(10)$

(d) What happens to the number of symbols per minute after several months of training?

Use a graphing calculator to find the solution set of each equation. Approximate the solution(s) to the nearest tenth.

101. $5e^{3x} = 75$ **102.** $6^{-x} = 1 - x$ **103.** $3x + 2 = 4^x$ **104.** $x = 2^x$

105. A function of the form $f(x) = x^r$, where r is a constant, is called a **power function.** Discuss the difference between an exponential function and a power function.

106. *Concept Check* If $f(x) = a^x$ and $f(3) = 27$, find each value of $f(x)$.

(a) $f(1)$ **(b)** $f(-1)$ **(c)** $f(2)$ **(d)** $f(0)$

Concept Check *Give an equation of the form $f(x) = a^x$ to define the exponential function whose graph contains the given point.*

107. $(3, 8)$ **108.** $(3, 125)$ **109.** $(-3, 64)$ **110.** $(-2, 36)$

Concept Check *Use properties of exponents to write each function in the form $f(t) = ka^t$, where k is a constant. (Hint: Recall that $a^{x+y} = a^x \cdot a^y$.)*

111. $f(t) = 3^{2t+3}$ **112.** $f(t) = 2^{3t+2}$ **113.** $f(t) = \left(\dfrac{1}{3}\right)^{1-2t}$ **114.** $f(t) = \left(\dfrac{1}{2}\right)^{1-2t}$

In calculus, it is shown that

$$e^x = 1 + x + \frac{x^2}{2 \cdot 1} + \frac{x^3}{3 \cdot 2 \cdot 1} + \frac{x^4}{4 \cdot 3 \cdot 2 \cdot 1} + \frac{x^5}{5 \cdot 4 \cdot 3 \cdot 2 \cdot 1} + \cdots.$$

By using more terms, one can obtain a more accurate approximation for e^x.

115. Use the terms shown, and replace x with 1 to approximate $e^1 = e$ to three decimal places. Check your result with a calculator.

116. Use the terms shown, and replace x with -0.05 to approximate $e^{-0.05}$ to four decimal places. Check your result with a calculator.

Relating Concepts

For individual or collaborative investigation *(Exercises 117–122)*

Assume $f(x) = a^x$, where $a > 1$. **Work these exercises in order.**

117. Is f a one-to-one function? If so, based on **Section 4.1**, what kind of related function exists for f?

118. If f has an inverse function f^{-1}, sketch f and f^{-1} on the same set of axes.

119. If f^{-1} exists, find an equation for $y = f^{-1}(x)$ using the method described in **Section 4.1**. You need not solve for y.

120. If $a = 10$, what is the equation for $y = f^{-1}(x)$? (You need not solve for y.)

121. If $a = e$, what is the equation for $y = f^{-1}(x)$? (You need not solve for y.)

122. If the point (p, q) is on the graph of f, then the point _____ is on the graph of f^{-1}.

4.3　Logarithmic Functions

- Logarithms
- Logarithmic Equations
- Logarithmic Functions
- Properties of Logarithms

Logarithms　　The previous section dealt with exponential functions of the form $y = a^x$ for all positive values of a, where $a \neq 1$. The horizontal line test shows that exponential functions are one-to-one and thus have inverse functions. The equation defining the inverse of a function is found by interchanging x and y in the equation that defines the function. Starting with $y = a^x$ and interchanging x and y yields

$$x = a^y.$$

Here y is the exponent to which a must be raised in order to obtain x. We call this exponent a **logarithm,** symbolized by the abbreviation "**log.**" The expression $\log_a x$ represents the logarithm in this discussion. The number a is called the **base** of the logarithm, and x is called the **argument** of the expression. It is read "**logarithm with base a of x**," or "**logarithm of x with base a**," or "**base a logarithm of x**."

Logarithm

For all real numbers y and all positive numbers a and x, where $a \neq 1$,

$$y = \log_a x \quad \text{is equivalent to} \quad x = a^y.$$

The expression $\log_a x$ represents the exponent to which the base a must be raised in order to obtain x.

EXAMPLE 1 **Writing Equivalent Logarithmic and Exponential Forms**

The table shows several pairs of equivalent statements, written in both logarithmic and exponential forms.

SOLUTION

Logarithmic Form	Exponential Form
$\log_2 8 = 3$	$2^3 = 8$
$\log_{1/2} 16 = -4$	$\left(\frac{1}{2}\right)^{-4} = 16$
$\log_{10} 100{,}000 = 5$	$10^5 = 100{,}000$
$\log_3 \frac{1}{81} = -4$	$3^{-4} = \frac{1}{81}$
$\log_5 5 = 1$	$5^1 = 5$
$\log_{3/4} 1 = 0$	$\left(\frac{3}{4}\right)^0 = 1$

To remember the relationships among a, x, and y in the two equivalent forms $y = \log_a x$ and $x = a^y$, refer to these diagrams.

A logarithm is an exponent.

Exponent
↓
Logarithmic form: $y = \log_a x$
↑
Base

Exponent
↓
Exponential form: $a^y = x$
↑
Base

☑ *Now Try Exercises 3, 5, 7, and 9.*

Logarithmic Equations   The definition of logarithm can be used to solve a **logarithmic equation,** which is an equation with a logarithm in at least one term. Many logarithmic equations can be solved by first writing the equation in exponential form.

EXAMPLE 2 **Solving Logarithmic Equations**

Solve each equation.

(a) $\log_x \dfrac{8}{27} = 3$ **(b)** $\log_4 x = \dfrac{5}{2}$ **(c)** $\log_{49} \sqrt[3]{7} = x$

SOLUTION

(a) $\log_x \dfrac{8}{27} = 3$

$x^3 = \dfrac{8}{27}$ Write in exponential form.

$x^3 = \left(\dfrac{2}{3}\right)^3$ $\frac{8}{27} = \left(\frac{2}{3}\right)^3$ **(Section R.2)**

$x = \dfrac{2}{3}$ Take cube roots. **(Section 1.6)**

CHECK $\log_x \dfrac{8}{27} = 3$ Original equation

$\log_{2/3} \dfrac{8}{27} \overset{?}{=} 3$ Let $x = \frac{2}{3}$.

$\left(\dfrac{2}{3}\right)^3 \overset{?}{=} \dfrac{8}{27}$ Write in exponential form.

$\dfrac{8}{27} = \dfrac{8}{27}$ ✓   True

The solution set is $\left\{\dfrac{2}{3}\right\}$.

(b)
$$\log_4 x = \frac{5}{2}$$

$4^{5/2} = x$ Write in exponential form.

$(4^{1/2})^5 = x$ $a^{mn} = (a^m)^n$ **(Section R.3)**

$2^5 = x$ $4^{1/2} = (2^2)^{1/2} = 2$

$32 = x$ Apply the exponent.

A check shows that the solution set is $\{32\}$.

(c)
$$\log_{49} \sqrt[3]{7} = x$$

$49^x = \sqrt[3]{7}$ Write in exponential form.

$(7^2)^x = 7^{1/3}$ Write with the same base.

$7^{2x} = 7^{1/3}$ Power rule for exponents

$2x = \dfrac{1}{3}$ Set exponents equal.

$x = \dfrac{1}{6}$ Divide by 2.

A check shows that the solution set is $\left\{\frac{1}{6}\right\}$.

✔ *Now Try Exercises 15, 27, and 29.*

Logarithmic Functions   We define the logarithmic function with base a as follows.

Logarithmic Function

If $a > 0$, $a \neq 1$, and $x > 0$, then

$$f(x) = \log_a x$$

defines the **logarithmic function with base a.**

Exponential and logarithmic functions are inverses of each other. The graph of $y = 2^x$ is shown in red in **Figure 25(a).** The graph of its inverse is found by reflecting the graph of $y = 2^x$ across the line $y = x$. The graph of the inverse function, defined by $y = \log_2 x$, shown in blue, has the y-axis as a vertical asymptote. **Figure 25(b)** shows a calculator graph of the two functions.

x	2^x		x	$\log_2 x$
-2	0.25		0.25	-2
-1	0.5		0.5	-1
0	1		1	0
1	2		2	1
2	4		4	2

The graph of $y = \log_2 x$ can be obtained by drawing the inverse of $y = 2^x$.

(a)   (b)

Figure 25

Since the domain of an exponential function is the set of all real numbers, the range of a logarithmic function also will be the set of all real numbers. In the same way, both the range of an exponential function and the domain of a logarithmic function are the set of all positive real numbers.

Thus, logarithms can be found for positive numbers only.

Logarithmic Function $f(x) = \log_a x$

Domain: $(0, \infty)$ Range: $(-\infty, \infty)$

For $f(x) = \log_2 x$:

x	$f(x)$
$\frac{1}{4}$	-2
$\frac{1}{2}$	-1
1	0
2	1
4	2
8	3

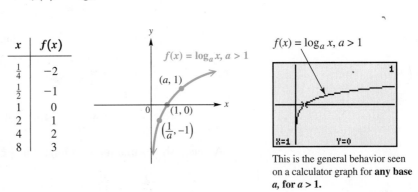

This is the general behavior seen on a calculator graph for **any base a, for $a > 1$.**

Figure 26

- $f(x) = \log_a x$, **for $a > 1$,** is increasing and continuous on its entire domain, $(0, \infty)$.
- The y-axis is a vertical asymptote as $x \to 0$ from the right.
- The graph passes through the points $\left(\frac{1}{a}, -1\right)$, $(1, 0)$, and $(a, 1)$.

For $f(x) = \log_{1/2} x$:

x	$f(x)$
$\frac{1}{4}$	2
$\frac{1}{2}$	1
1	0
2	-1
4	-2
8	-3

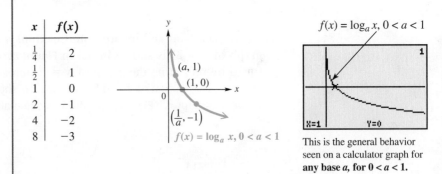

This is the general behavior seen on a calculator graph for **any base a, for $0 < a < 1$.**

Figure 27

- $f(x) = \log_a x$, **for $0 < a < 1$,** is decreasing and continuous on its entire domain, $(0, \infty)$.
- The y-axis is a vertical asymptote as $x \to 0$ from the right.
- The graph passes through the points $\left(\frac{1}{a}, -1\right)$, $(1, 0)$, and $(a, 1)$.

⊟ Calculator graphs of logarithmic functions do not, in general, give an accurate picture of the behavior of the graphs near the vertical asymptotes. While it may seem as if the graph has an endpoint, this is not the case. The resolution of the calculator screen is not precise enough to indicate that the graph approaches the vertical asymptote as the value of x gets closer to it. Do not draw incorrect conclusions just because the calculator does not show this behavior. ■

The graphs in **Figures 26 and 27** and the information with them suggest the following generalizations about the graphs of logarithmic functions of the form $f(x) = \log_a x$.

Characteristics of the Graph of $f(x) = \log_a x$

1. The points $\left(\frac{1}{a}, -1\right)$, $(1, 0)$, and $(a, 1)$ are on the graph.
2. If $a > 1$, then f is an increasing function. If $0 < a < 1$, then f is a decreasing function.
3. The y-axis is a vertical asymptote.
4. The domain is $(0, \infty)$, and the range is $(-\infty, \infty)$.

EXAMPLE 3 Graphing Logarithmic Functions

Graph each function.

(a) $f(x) = \log_{1/2} x$ **(b)** $f(x) = \log_3 x$

SOLUTION

(a) One approach is to first graph $y = \left(\frac{1}{2}\right)^x$, which defines the inverse function of f, by plotting points. Some ordered pairs are given in the table with the graph shown in red in **Figure 28**. The graph of $f(x) = \log_{1/2} x$ is the reflection of the graph of $y = \left(\frac{1}{2}\right)^x$ across the line $y = x$. The ordered pairs for $y = \log_{1/2} x$ are found by interchanging the x- and y-values in the ordered pairs for $y = \left(\frac{1}{2}\right)^x$. See the graph in blue in **Figure 28**.

x	$y = \left(\frac{1}{2}\right)^x$
-2	4
-1	2
0	1
1	$\frac{1}{2}$
2	$\frac{1}{4}$
4	$\frac{1}{16}$

x	$f(x) = \log_{1/2} x$
4	-2
2	-1
1	0
$\frac{1}{2}$	1
$\frac{1}{4}$	2
$\frac{1}{16}$	4

Figure 28

x	$f(x) = \log_3 x$
$\frac{1}{3}$	-1
1	0
3	1
9	2

Think: $x = 3^y$

Figure 29

(b) Another way to graph a logarithmic function is to write $f(x) = y = \log_3 x$ in exponential form as $x = 3^y$, and then select y-values and calculate corresponding x-values. Several selected ordered pairs are shown in the table for the graph in **Figure 29**.

✔ *Now Try Exercise 51.*

CAUTION If you write a logarithmic function in exponential form to graph, as in **Example 3(b),** start *first* with y-values to calculate corresponding x-values. *Be careful to write the values in the ordered pairs in the correct order.*

More general logarithmic functions can be obtained by forming the composition of $f(x) = \log_a x$ with a function $g(x)$. For example, if $f(x) = \log_2 x$ and $g(x) = x - 1$, then

$$(f \circ g)(x) = f(g(x)) = \log_2(x - 1). \quad \text{(Section 2.8)}$$

The next example shows how to graph such functions.

EXAMPLE 4 **Graphing Translated Logarithmic Functions**

Graph each function. Give the domain and range.

(a) $f(x) = \log_2(x - 1)$ **(b)** $f(x) = (\log_3 x) - 1$

(c) $f(x) = \log_4(x + 2) + 1$

SOLUTION

(a) The graph of $f(x) = \log_2(x - 1)$ is the graph of $f(x) = \log_2 x$ translated 1 unit to the right. The vertical asymptote has equation $x = 1$. Since logarithms can be found only for positive numbers, we solve $x - 1 > 0$ to find the domain, $(1, \infty)$. To determine ordered pairs to plot, use the equivalent exponential form of the equation $y = \log_2(x - 1)$.

$$y = \log_2(x - 1)$$

$$x - 1 = 2^y \qquad \text{Write in exponential form.}$$

$$x = 2^y + 1 \qquad \text{Add 1.}$$

We first choose values for y and then calculate each of the corresponding x-values. The range is $(-\infty, \infty)$. See **Figure 30.**

Figure 30   Figure 31

(b) The function $f(x) = (\log_3 x) - 1$ has the same graph as $g(x) = \log_3 x$ translated 1 unit down. We find ordered pairs to plot by writing the equation $y = (\log_3 x) - 1$ in exponential form.

$$y = (\log_3 x) - 1$$

$$y + 1 = \log_3 x \qquad \text{Add 1.}$$

$$x = 3^{y+1} \qquad \text{Write in exponential form.}$$

Again, choose y-values and calculate the corresponding x-values. The graph is shown in **Figure 31.** The domain is $(0, \infty)$, and the range is $(-\infty, \infty)$.

(c) The graph of $f(x) = \log_4(x + 2) + 1$ is obtained by shifting the graph of $y = \log_4 x$ to the left 2 units and up 1 unit. The domain is found by solving $x + 2 > 0$, which yields $(-2, \infty)$. The vertical asymptote has been shifted to the left 2 units as well, and it has equation $x = -2$. The range is unaffected by the vertical shift and remains $(-\infty, \infty)$. See **Figure 32.**

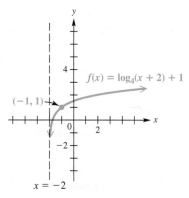

Figure 32

✓ Now Try Exercises 39, 43, and 55.

NOTE If we are given a graph such as the one in **Figure 31** and are asked to find its equation, we could reason as follows: The point $(1, 0)$ on the basic logarithmic graph has been shifted *down* 1 unit, and the point $(3, 0)$ on the given graph is 1 unit lower than $(3, 1)$, which is on the graph of $y = \log_3 x$. Thus, the equation will be

$$y = (\log_3 x) - 1.$$

Properties of Logarithms The properties of logarithms enable us to change the form of logarithmic statements so that products can be converted to sums, quotients can be converted to differences, and powers can be converted to products.

Properties of Logarithms

For $x > 0$, $y > 0$, $a > 0$, $a \neq 1$, and any real number r, the following properties hold.

Property	Description
Product Property $\log_a xy = \log_a x + \log_a y$	The logarithm of the product of two numbers is equal to the sum of the logarithms of the numbers.
Quotient Property $\log_a \dfrac{x}{y} = \log_a x - \log_a y$	The logarithm of the quotient of two numbers is equal to the difference between the logarithms of the numbers.
Power Property $\log_a x^r = r \log_a x$	The logarithm of a number raised to a power is equal to the exponent multiplied by the logarithm of the number.
Logarithm of 1 $\log_a 1 = 0$	The base a logarithm of 1 is 0.
Base a Logarithm of a $\log_a a = 1$	The base a logarithm of a is 1.

LOOKING AHEAD TO CALCULUS

A technique called **logarithmic differentiation,** which uses the properties of logarithms, can often be used to differentiate complicated functions.

Proof To prove the product property, let $m = \log_a x$ and $n = \log_a y$.

$$\log_a x = m \quad \text{means} \quad a^m = x$$
$$\log_a y = n \quad \text{means} \quad a^n = y$$

Now consider the product xy.

$$xy = a^m \cdot a^n \qquad \text{Substitute.}$$

$$xy = a^{m+n} \qquad \text{Product rule for exponents (Section R.3)}$$

$$\log_a xy = m + n \qquad \text{Write in logarithmic form.}$$

$$\log_a xy = \log_a x + \log_a y \quad \text{Substitute.}$$

The last statement is the result we wished to prove. The quotient and power properties are proved similarly. (See **Exercises 103 and 104.**)

EXAMPLE 5 **Using the Properties of Logarithms**

Rewrite each expression. Assume all variables represent positive real numbers, with $a \neq 1$ and $b \neq 1$.

(a) $\log_6(7 \cdot 9)$ **(b)** $\log_9 \dfrac{15}{7}$ **(c)** $\log_5 \sqrt{8}$

(d) $\log_a \dfrac{mnq}{p^2 t^4}$ **(e)** $\log_a \sqrt[3]{m^2}$ **(f)** $\log_b \sqrt[n]{\dfrac{x^3 y^5}{z^m}}$

SOLUTION

(a) $\log_6(7 \cdot 9) = \log_6 7 + \log_6 9$ Product property

(b) $\log_9 \dfrac{15}{7} = \log_9 15 - \log_9 7$ Quotient property

(c) $\log_5 \sqrt{8} = \log_5(8^{1/2}) = \dfrac{1}{2} \log_5 8$ Power property

> Use parentheses to avoid errors.

(d) $\log_a \dfrac{mnq}{p^2 t^4} = \log_a m + \log_a n + \log_a q - (\log_a p^2 + \log_a t^4)$

     Product and quotient properties

$$= \log_a m + \log_a n + \log_a q - (2 \log_a p + 4 \log_a t)$$

     Power property

$$= \log_a m + \log_a n + \log_a q - 2 \log_a p - 4 \log_a t$$

> Be careful with signs.

(e) $\log_a \sqrt[3]{m^2} = \log_a m^{2/3} = \dfrac{2}{3} \log_a m$

(f) $\log_b \sqrt[n]{\dfrac{x^3 y^5}{z^m}} = \log_b \left(\dfrac{x^3 y^5}{z^m}\right)^{1/n}$ $\sqrt[n]{a} = a^{1/n}$ (Section R.7)

$$= \dfrac{1}{n} \log_b \dfrac{x^3 y^5}{z^m} \qquad \text{Power property}$$

$$= \dfrac{1}{n} (\log_b x^3 + \log_b y^5 - \log_b z^m) \qquad \text{Product and quotient properties}$$

$$= \dfrac{1}{n} (3 \log_b x + 5 \log_b y - m \log_b z) \qquad \text{Power property}$$

$$= \dfrac{3}{n} \log_b x + \dfrac{5}{n} \log_b y - \dfrac{m}{n} \log_b z \qquad \begin{array}{l}\text{Distributive property} \\ \text{(Section R.2)}\end{array}$$

✔ *Now Try Exercises 63, 65, 69, and 73.*

EXAMPLE 6 **Using the Properties of Logarithms**

Write each expression as a single logarithm with coefficient 1. Assume all variables represent positive real numbers, with $a \neq 1$ and $b \neq 1$.

(a) $\log_3(x + 2) + \log_3 x - \log_3 2$

(b) $2 \log_a m - 3 \log_a n$

(c) $\dfrac{1}{2} \log_b m + \dfrac{3}{2} \log_b 2n - \log_b m^2 n$

SOLUTION

(a) $\log_3(x + 2) + \log_3 x - \log_3 2 = \log_3 \dfrac{(x + 2)x}{2}$ Product and quotient properties

(b) $2 \log_a m - 3 \log_a n = \log_a m^2 - \log_a n^3$ Power property

$$= \log_a \frac{m^2}{n^3} \qquad \text{Quotient property}$$

(c) $\dfrac{1}{2} \log_b m + \dfrac{3}{2} \log_b 2n - \log_b m^2 n$

$$= \log_b m^{1/2} + \log_b (2n)^{3/2} - \log_b m^2 n \qquad \text{Power property}$$

$$= \log_b \frac{m^{1/2}(2n)^{3/2}}{m^2 n} \quad \boxed{\text{Use parentheses on } 2n.} \qquad \text{Product and quotient properties}$$

$$= \log_b \frac{2^{3/2} n^{1/2}}{m^{3/2}} \qquad \begin{array}{l}\text{Rules for exponents} \\ \text{(Sections R.3 and R.6)}\end{array}$$

$$= \log_b \left(\frac{2^3 n}{m^3}\right)^{1/2} \qquad \text{Rules for exponents}$$

$$= \log_b \sqrt{\frac{8n}{m^3}} \qquad \text{Definition of } a^{1/n}$$

✔ *Now Try Exercises 75, 79, and 81.*

CAUTION *There is no property of logarithms to rewrite a logarithm of a sum or difference.* That is why, in **Example 6(a)**, $\log_3(x + 2)$ was not written as $\log_3 x + \log_3 2$. The distributive property does not apply in a situation like this because $\log_3(x + y)$ is one term. The abbreviation "log" is a function name, *not* a factor.

The next example uses = symbols for values of logarithms. These are actually approximations.

EXAMPLE 7 **Using the Properties of Logarithms with Numerical Values**

Assume that $\log_{10} 2 = 0.3010$. Find each logarithm.

(a) $\log_{10} 4$

(b) $\log_{10} 5$

SOLUTION

(a) $\log_{10} 4 = \log_{10} 2^2$

$$= 2 \log_{10} 2$$

$$= 2(0.3010)$$

$$= 0.6020$$

(b) $\log_{10} 5 = \log_{10} \dfrac{10}{2}$

$$= \log_{10} 10 - \log_{10} 2$$

$$= 1 - 0.3010$$

$$= 0.6990$$

✔ *Now Try Exercises 85 and 87.*

Napier's Rods

The search for ways to make calculations easier has been a long, ongoing process. Machines built by Charles Babbage and Blaise Pascal, a system of "rods" used by John Napier, and slide rules were the forerunners of today's calculators and computers. The invention of logarithms by John Napier in the 16th century was a great breakthrough in the search for easier calculation methods.

Source: IBM Corporate Archives.

Recall that for inverse functions f and g, $(f \circ g)(x) = (g \circ f)(x) = x$. We can use this property with exponential and logarithmic functions to state two more properties. If $f(x) = a^x$ and $g(x) = \log_a x$, then

$$(f \circ g)(x) = a^{\log_a x} \quad \text{and} \quad (g \circ f)(x) = \log_a(a^x).$$

Theorem on Inverses

For $a > 0$, $a \neq 1$, the following properties hold.

$$a^{\log_a x} = x \text{ (for } x > 0\text{)} \quad \text{and} \quad \log_a a^x = x$$

The following are examples of applications of this theorem.

$$7^{\log_7 10} = 10, \quad \log_5 5^3 = 3, \quad \text{and} \quad \log_r r^{k+1} = k + 1$$

The second statement in the theorem will be useful in **Sections 4.5 and 4.6** when we solve other logarithmic and exponential equations.

4.3 Exercises

Concept Check In Exercises 1 and 2, match the logarithm in Column I with its value in Column II. Remember that $\log_a x$ is the exponent to which a must be raised in order to obtain x.

	I		II			I		II
1.	**(a)** $\log_2 16$	**A.**	0		**2.**	**(a)** $\log_3 81$	**A.**	-2
	(b) $\log_3 1$	**B.**	$\dfrac{1}{2}$			**(b)** $\log_3 \dfrac{1}{3}$	**B.**	-1
	(c) $\log_{10} 0.1$	**C.**	4			**(c)** $\log_{10} 0.01$	**C.**	0
	(d) $\log_2 \sqrt{2}$	**D.**	-3			**(d)** $\log_6 \sqrt{6}$	**D.**	$\dfrac{1}{2}$
	(e) $\log_e \dfrac{1}{e^2}$	**E.**	-1			**(e)** $\log_e 1$	**E.**	$\dfrac{9}{2}$
	(f) $\log_{1/2} 8$	**F.**	-2			**(f)** $\log_3 27^{3/2}$	**F.**	4

If the statement is in exponential form, write it in an equivalent logarithmic form. If the statement is in logarithmic form, write it in exponential form. **See Example 1.**

3. $3^4 = 81$ **4.** $2^5 = 32$ **5.** $\left(\dfrac{2}{3}\right)^{-3} = \dfrac{27}{8}$ **6.** $10^{-4} = 0.0001$

7. $\log_6 36 = 2$ **8.** $\log_5 5 = 1$ **9.** $\log_{\sqrt{3}} 81 = 8$ **10.** $\log_4 \dfrac{1}{64} = -3$

11. Explain why logarithms of negative numbers are not defined.

12. *Concept Check* Why is $\log_a 1$ always equal to 0 for any valid base a?

Solve each logarithmic equation. **See Example 2.**

13. $x = \log_5 \dfrac{1}{625}$ **14.** $x = \log_3 \dfrac{1}{81}$ **15.** $\log_x \dfrac{1}{32} = 5$

16. $\log_x \dfrac{27}{64} = 3$ **17.** $x = \log_8 \sqrt[4]{8}$ **18.** $x = \log_7 \sqrt[5]{7}$

19. $x = 3^{\log_3 8}$ **20.** $x = 12^{\log_{12} 5}$ **21.** $x = 2^{\log_2 9}$

22. $x = 8^{\log_8 11}$ **23.** $\log_x 25 = -2$ **24.** $\log_x 16 = -2$

25. $\log_4 x = 3$ **26.** $\log_2 x = 3$ **27.** $x = \log_4 \sqrt[3]{16}$

28. $x = \log_5 \sqrt[4]{25}$ **29.** $\log_9 x = \dfrac{5}{2}$ **30.** $\log_4 x = \dfrac{7}{2}$

31. $\log_{1/2}(x + 3) = -4$ **32.** $\log_{1/3}(x + 6) = -2$

33. $\log_{(x+3)} 6 = 1$ **34.** $\log_{(x-4)} 19 = 1$

35. $3x - 15 = \log_x 1 \quad (x > 0, x \neq 1)$ **36.** $4x - 24 = \log_x 1 \quad (x > 0, x \neq 1)$

37. Compare the summary of characteristics of the graph of $f(x) = \log_a x$ with the similar summary about the graph of $f(x) = a^x$ in **Section 4.2.** Make a list of characteristics that reinforce the idea that these are inverse functions.

38. *Concept Check* The calculator graph of $Y = \log_2 X$ shows the values of the ordered pair with $X = 5$. What does the value of Y represent?

*Sketch the graph of $f(x) = \log_2 x$. Then refer to it and use the techniques of **Chapter 2** to graph each function. Give the domain and range. See **Example 4.***

39. $f(x) = (\log_2 x) + 3$ **40.** $f(x) = \log_2(x + 3)$ **41.** $f(x) = |\log_2(x + 3)|$

*Sketch the graph of $f(x) = \log_{1/2} x$. Then refer to it and use the techniques of **Chapter 2** to graph each function. Give the domain and range. See **Example 4.***

42. $f(x) = (\log_{1/2} x) - 2$ **43.** $f(x) = \log_{1/2}(x - 2)$ **44.** $f(x) = |\log_{1/2}(x - 2)|$

Concept Check In Exercises 45–50, match the function with its graph from choices A–F.

45. $f(x) = \log_2 x$ **46.** $f(x) = \log_2 2x$ **47.** $f(x) = \log_2 \dfrac{1}{x}$

48. $f(x) = \log_2\left(\dfrac{1}{2}x\right)$ **49.** $f(x) = \log_2(x - 1)$ **50.** $f(x) = \log_2(-x)$

A.

B.

C.

D.

E.

F.

Graph each function. ***See Examples 3 and 4.***

51. $f(x) = \log_5 x$ **52.** $f(x) = \log_{10} x$ **53.** $f(x) = \log_{1/2}(1 - x)$

54. $f(x) = \log_{1/3}(3 - x)$ **55.** $f(x) = \log_3(x - 1) + 2$ **56.** $f(x) = \log_2(x + 2) - 3$

Connecting Graphs with Equations *In Exercises 57–62, write an equation for the graph given. Each is a logarithmic function f with base 2 or 3, translated and/or reflected.* ***See the Note following Example 4.***

57.

58.

59.

60.

61.

62.

Use the properties of logarithms to rewrite each expression. Simplify the result if possible. Assume all variables represent positive real numbers. ***See Example 5.***

63. $\log_2 \dfrac{6x}{y}$ **64.** $\log_3 \dfrac{4p}{q}$ **65.** $\log_5 \dfrac{5\sqrt{7}}{3}$

66. $\log_2 \dfrac{2\sqrt{3}}{5}$ **67.** $\log_4(2x + 5y)$ **68.** $\log_6(7m + 3q)$

69. $\log_m \sqrt{\dfrac{5r^3}{z^5}}$ **70.** $\log_p \sqrt[3]{\dfrac{m^5 n^4}{t^2}}$ **71.** $\log_2 \dfrac{ab}{cd}$

72. $\log_2 \dfrac{xy}{tqr}$ **73.** $\log_3 \dfrac{\sqrt{x} \cdot \sqrt[3]{y}}{w^2 \sqrt{z}}$ **74.** $\log_4 \dfrac{\sqrt[3]{a} \cdot \sqrt[4]{b}}{\sqrt{c} \cdot \sqrt[3]{d^2}}$

Write each expression as a single logarithm with coefficient 1. Assume all variables represent positive real numbers. ***See Example 6.***

75. $\log_a x + \log_a y - \log_a m$ **76.** $\log_b k + \log_b m - \log_b a$

77. $\log_a m - \log_a n - \log_a t$ **78.** $\log_b p - \log_b q - \log_b r$

79. $\dfrac{1}{3} \log_b x^4 y^5 - \dfrac{3}{4} \log_b x^2 y$ **80.** $\dfrac{1}{2} \log_y p^3 q^4 - \dfrac{2}{3} \log_y p^4 q^3$

81. $2 \log_a(z + 1) + \log_a(3z + 2)$ **82.** $5 \log_a(z + 7) + \log_a(2z + 9)$

83. $-\dfrac{2}{3} \log_5 5m^2 + \dfrac{1}{2} \log_5 25m^2$ **84.** $-\dfrac{3}{4} \log_3 16p^4 - \dfrac{2}{3} \log_3 8p^3$

Given the approximations $\log_{10} 2 = 0.3010$ *and* $\log_{10} 3 = 0.4771$, *find each logarithm without using a calculator.* ***See Example 7.***

85. $\log_{10} 6$ **86.** $\log_{10} 12$ **87.** $\log_{10} \dfrac{3}{2}$ **88.** $\log_{10} \dfrac{2}{9}$

89. $\log_{10} \dfrac{9}{4}$ **90.** $\log_{10} \dfrac{20}{27}$ **91.** $\log_{10} \sqrt{30}$ **92.** $\log_{10} 36^{1/3}$

Solve each problem.

93. *(Modeling) Interest Rates of Treasury Securities* The table gives interest rates for various U.S. Treasury Securities on February 15, 2011.

(a) Make a scatter diagram of the data.

(b) Which type of function will model these data best: linear, exponential, or logarithmic?

Time	Yield
3-month	0.13%
6-month	0.17%
2-year	0.84%
5-year	2.35%
10-year	3.61%
30-year	4.66%

Source: www.federal
reserve.gov

94. *Concept Check* Use the graph to estimate each logarithm.

(a) $\log_3 0.3$ (b) $\log_3 0.8$

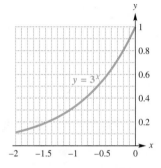

95. *Concept Check* Suppose $f(x) = \log_a x$ and $f(3) = 2$. Determine each function value.

(a) $f\left(\dfrac{1}{9}\right)$ (b) $f(27)$ (c) $f(9)$ (d) $f\left(\dfrac{\sqrt{3}}{3}\right)$

96. Use properties of logarithms to evaluate each expression.

(a) $100^{\log_{10} 3}$ (b) $\log_{10}(0.01)^3$ (c) $\log_{10}(0.0001)^5$ (d) $1000^{\log_{10} 5}$

97. Refer to the compound interest formula from **Section 4.2.** Show that the amount of time required for a deposit to double is

$$\frac{1}{\log_2\left(1 + \frac{r}{n}\right)^n}.$$

98. *Concept Check* If $(5, 4)$ is on the graph of the logarithmic function with base a, which of the following statements is true:

$$5 = \log_a 4 \quad \text{or} \quad 4 = \log_a 5?$$

Standard mathematical notation uses $\log x$ *as an abbreviation for* $\log_{10} x$. *Use the* $\log$ *key on your graphing calculator to graph each function.*

99. $f(x) = x \log_{10} x$ **100.** $f(x) = x^2 \log_{10} x$

Use a graphing calculator to find the solution set of each equation. Give solutions to the nearest hundredth.

101. $\log_{10} x = x - 2$ **102.** $2^{-x} = \log_{10} x$

103. Prove the quotient property of logarithms: $\log_a \dfrac{x}{y} = \log_a x - \log_a y$.

104. Prove the power property of logarithms: $\log_a x^r = r \log_a x$.

Summary Exercises on Inverse, Exponential, and Logarithmic Functions

The following exercises are designed to help solidify your understanding of inverse, exponential, and logarithmic functions from **Sections 4.1–4.3.**

Determine whether the functions in each pair are inverses of each other.

1. $f(x) = 3x - 4$, $\quad g(x) = \dfrac{1}{3}x + \dfrac{4}{3}$

2. $f(x) = 8 - 5x$, $\quad g(x) = 8 + \dfrac{1}{5}x$

3. $f(x) = 1 + \log_2 x$, $\quad g(x) = 2^{x-1}$

4. $f(x) = 3^{x/5} - 2$, $\quad g(x) = 5 \log_3(x + 2)$

Determine whether each function is one-to-one. If it is, then sketch the graph of its inverse function.

5.

6.

7.

8.

In Exercises 9–12, match the function with its graph from choices A–D.

9. $y = \log_3(x + 2)$

10. $y = 5 - 2^x$

11. $y = \log_2(5 - x)$

12. $y = 3^x - 2$

A.

B.

C.

D.

13. The functions in **Exercises 9–12** form two pairs of inverse functions. Determine which functions are inverses of each other.

14. Determine the inverse of the function $f(x) = \log_5 x$. (*Hint:* Replace $f(x)$ with y, and write in exponential form.)

For each function that is one-to-one, write an equation for its inverse function. Give the domain and range of both f and f⁻¹. If the function is not one-to-one, say so.

15. $f(x) = 3x - 6$

16. $f(x) = 2(x + 1)^3$

17. $f(x) = 3x^2$

18. $f(x) = \dfrac{2x - 1}{5 - 3x}$

19. $f(x) = \sqrt[3]{5 - x^4}$

20. $f(x) = \sqrt{x^2 - 9}, \ x \geq 3$

Write an equivalent statement in logarithmic form.

21. $\left(\dfrac{1}{10}\right)^{-3} = 1000$

22. $a^b = c$

23. $\left(\sqrt{3}\right)^4 = 9$

24. $4^{-3/2} = \dfrac{1}{8}$

25. $2^x = 32$

26. $27^{4/3} = 81$

Solve each equation.

27. $3x = 7^{\log_7 6}$

28. $x = \log_{10} 0.001$

29. $x = \log_6 \dfrac{1}{216}$

30. $\log_x 5 = \dfrac{1}{2}$

31. $\log_{10} 0.01 = x$

32. $\log_x 3 = -1$

33. $\log_x 1 = 0$

34. $x = \log_2 \sqrt{8}$

35. $\log_x \sqrt[3]{5} = \dfrac{1}{3}$

36. $\log_{1/3} x = -5$

37. $\log_{10}(\log_2 2^{10}) = x$

38. $x = \log_{4/5} \dfrac{25}{16}$

39. $2x - 1 = \log_6 6^x$

40. $x = \sqrt{\log_{1/2} \dfrac{1}{16}}$

41. $2^x = \log_2 16$

42. $\log_3 x = -2$

43. $\left(\dfrac{1}{3}\right)^{x+1} = 9^x$

44. $5^{2x-6} = 25^{x-3}$

4.4	**Evaluating Logarithms and the Change-of-Base Theorem**

■ Common Logarithms

■ Applications and Models with Common Logarithms

■ Natural Logarithms

■ Applications and Models with Natural Logarithms

■ Logarithms with Other Bases

Common Logarithms Two of the most important bases for logarithms are 10 and *e*. Base 10 logarithms are called **common logarithms.** The common logarithm of *x* is written log *x*, where the base is understood to be 10.

Common Logarithm

For all positive numbers *x*,

$$\log x = \log_{10} x.$$

A calculator with a log key can be used to find the base 10 logarithm of any positive number.

EXAMPLE 1 **Evaluating Common Logarithms with a Calculator**

Use a calculator to find the values of

$$\log 1000, \quad \log 142, \quad \text{and} \quad \log 0.005832.$$

SOLUTION **Figure 33** shows that the exact value of log 1000 is 3 (because $10^3 = 1000$), and that

$$\log 142 \approx 2.152288344$$

and $\qquad \log 0.005832 \approx -2.234182485.$

Most common logarithms that appear in calculations are approximations, as seen in the second and third displays.

Figure 33

✔ *Now Try Exercises 11, 15, and 17.*

Figure 34

Figure 34 reinforces the concept presented in the previous section: **log *x* is the exponent to which 10 must be raised to obtain *x*.**

NOTE *Base a logarithms of numbers between 0 and 1, where a > 1, are always negative,* as suggested by the graphs in **Section 4.3.**

Applications and Models with Common Logarithms In chemistry, the **pH** of a solution is defined as

$$\text{pH} = -\log[\text{H}_3\text{O}^+],$$

where $[\text{H}_3\text{O}^+]$ is the hydronium ion concentration in moles* per liter. The pH value is a measure of the acidity or alkalinity of a solution. Pure water has pH 7.0, substances with pH values greater than 7.0 are alkaline, and substances with pH values less than 7.0 are acidic. It is customary to round pH values to the nearest tenth.

1   7   14

Acidic   Neutral   Alkaline

EXAMPLE 2 **Finding pH**

(a) Find the pH of a solution with $[\text{H}_3\text{O}^+] = 2.5 \times 10^{-4}$.

(b) Find the hydronium ion concentration of a solution with pH = 7.1.

SOLUTION

(a) pH $= -\log[\text{H}_3\text{O}^+]$

$\qquad = -\log(2.5 \times 10^{-4})$ Substitute.

$\qquad = -(\log 2.5 + \log 10^{-4})$ Product property **(Section 4.3)**

$\qquad = -(0.3979 - 4)$ $\log 10^{-4} = -4$ **(Section 4.3)**

$\qquad = -0.3979 + 4$ Distributive property **(Section R.2)**

pH ≈ 3.6 Add.

---

*A *mole* is the amount of a substance that contains the same number of molecules as the number of atoms in exactly 12 grams of carbon-12.

(b)

$$\text{pH} = -\log\left[H_3O^+\right]$$

$$7.1 = -\log\left[H_3O^+\right] \qquad \text{Substitute.}$$

$$-7.1 = \log\left[H_3O^+\right] \qquad \text{Multiply by } -1. \textbf{ (Section 1.1)}$$

$$\left[H_3O^+\right] = 10^{-7.1} \qquad \text{Write in exponential form. \textbf{(Section 4.3)}}$$

$$\left[H_3O^+\right] \approx 7.9 \times 10^{-8} \qquad \text{Evaluate } 10^{-7.1} \text{ with a calculator.}$$

☑ *Now Try Exercises 29 and 33.*

NOTE   In the fourth line of the solution in **Example 2(a),** we use the equality symbol, $=$, rather than the approximate equality symbol, $\approx$, when replacing log 2.5 with 0.3979. This is often done for convenience, despite the fact that most logarithms used in applications are indeed approximations.

EXAMPLE 3   **Using pH in an Application**

Wetlands are classified as *bogs, fens, marshes,* and *swamps* based on pH values. A pH value between 6.0 and 7.5 indicates that the wetland is a "rich fen." When the pH is between 3.0 and 6.0, it is a "poor fen," and if the pH falls to 3.0 or less, the wetland is a "bog." (*Source:* R. Mohlenbrock, "Summerby Swamp, Michigan," *Natural History.*)

Suppose that the hydronium ion concentration of a sample of water from a wetland is 6.3×10^{-5}. How would this wetland be classified?

SOLUTION

$$\text{pH} = -\log\left[H_3O^+\right] \qquad \text{Definition of pH}$$

$$= -\log(6.3 \times 10^{-5}) \qquad \text{Substitute.}$$

$$= -(\log 6.3 + \log 10^{-5}) \qquad \text{Product property}$$

$$= -\log 6.3 - (-5) \qquad \text{Distributive property; } \log 10^n = n$$

$$= -\log 6.3 + 5 \qquad \text{Definition of subtraction}$$

$$\text{pH} \approx 4.2 \qquad \text{Use a calculator.}$$

Since the pH is between 3.0 and 6.0, the wetland is a poor fen.

☑ *Now Try Exercise 37.*

EXAMPLE 4   **Measuring the Loudness of Sound**

The loudness of sounds is measured in **decibels.** We first assign an intensity of I_0 to a very faint **threshold sound.** If a particular sound has intensity I, then the decibel rating d of this louder sound is given by the following formula.

$$d = 10 \log \frac{I}{I_0}$$

Find the decibel rating d of a sound with intensity $10{,}000I_0$.

SOLUTION

$$d = 10 \log \frac{10{,}000I_0}{I_0} \qquad \text{Let } I = 10{,}000I_0.$$

$$= 10 \log 10{,}000 \qquad \tfrac{I_0}{I_0} = 1$$

$$= 10(4) \qquad \log 10{,}000 = \log 10^4 = 4 \textbf{ (Section 4.3)}$$

$$= 40 \qquad \text{Multiply.}$$

The sound has a decibel rating of 40.

☑ *Now Try Exercise 63.*

LOOKING AHEAD TO CALCULUS

The natural logarithmic function $f(x) = \ln x$ and the reciprocal function $g(x) = \frac{1}{x}$ have an important relationship in calculus. The derivative of the natural logarithmic function is the reciprocal function. Using **Leibniz notation** (named after one of the co-inventors of calculus), we write this fact as $\frac{d}{dx}(\ln x) = \frac{1}{x}$.

Natural Logarithms In **Section 4.2**, we introduced the irrational number e. In most practical applications of logarithms, e is used as base. Logarithms with base e are called **natural logarithms,** since they occur in the life sciences and economics in natural situations that involve growth and decay. The base e logarithm of x is written **ln x** (read **"el-en x"**). *The expression **ln x** represents the exponent to which e must be raised to obtain x.*

Natural Logarithm

For all positive numbers x,

$$\ln x = \log_e x.$$

Figure 35

A graph of the natural logarithmic function $f(x) = \ln x$ is given in **Figure 35.**

EXAMPLE 5 **Evaluating Natural Logarithms with a Calculator**

Use a calculator to find the values of

$$\ln e^3, \quad \ln 142, \quad \text{and} \quad \ln 0.005832.$$

SOLUTION **Figure 36** shows that the exact value of $\ln e^3$ is 3, and that

$$\ln 142 \approx 4.955827058$$

and

$$\ln 0.005832 \approx -5.144395284.$$

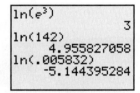

Figure 36

✔ *Now Try Exercises 45, 49, and 51.*

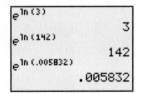

Figure 37

Figure 37 illustrates that **ln x is the exponent to which e must be raised to obtain x.**

Applications and Models with Natural Logarithms We now consider two applications of natural logarithms.

EXAMPLE 6 **Measuring the Age of Rocks**

Geologists sometimes measure the age of rocks by using "atomic clocks." By measuring the amounts of potassium-40 and argon-40 in a rock, it is possible to find the age t of the specimen in years with the formula

$$t = (1.26 \times 10^9) \frac{\ln\left(1 + 8.33\left(\frac{A}{K}\right)\right)}{\ln 2},$$

where A and K are the numbers of atoms of argon-40 and potassium-40, respectively, in the specimen.

(a) How old is a rock in which $A = 0$ and $K > 0$?

(b) The ratio $\frac{A}{K}$ for a sample of granite from New Hampshire is 0.212. How old is the sample?

SOLUTION

(a) If $A = 0$, then $\frac{A}{K} = 0$ and the equation is as follows.

$$t = (1.26 \times 10^9)\frac{\ln\left(1 + 8.33\left(\frac{A}{K}\right)\right)}{\ln 2} \qquad \text{Given formula}$$

$$= (1.26 \times 10^9)\frac{\ln 1}{\ln 2} \qquad \frac{A}{K} = 0, \text{ so } \ln(1 + 0) = \ln 1$$

$$= (1.26 \times 10^9)(0) \qquad \ln 1 = 0$$

$$t = 0$$

The rock is new (0 yr old).

(b) Since $\frac{A}{K} = 0.212$, we have the following.

$$t = (1.26 \times 10^9)\frac{\ln(1 + 8.33(0.212))}{\ln 2} \qquad \text{Substitute.}$$

$$t \approx 1.85 \times 10^9 \qquad \text{Use a calculator.}$$

The granite is about 1.85 billion yr old.

✔ *Now Try Exercise 77.*

EXAMPLE 7 **Modeling Global Temperature Increase**

Carbon dioxide in the atmosphere traps heat from the sun. The additional solar radiation trapped by carbon dioxide is called **radiative forcing.** It is measured in watts per square meter (w/m^2). In 1896 the Swedish scientist Svante Arrhenius modeled radiative forcing R caused by additional atmospheric carbon dioxide, using the logarithmic equation

$$R = k \ln \frac{C}{C_0},$$

where C_0 is the preindustrial amount of carbon dioxide, C is the current carbon dioxide level, and k is a constant. Arrhenius determined that $10 \le k \le 16$ when $C = 2C_0$. (*Source:* Clime, W., *The Economics of Global Warming,* Institute for International Economics, Washington, D.C.)

(a) Let $C = 2C_0$. Is the relationship between R and k linear or logarithmic?

(b) The average global temperature increase T (in °F) is given by $T(R) = 1.03R$. Write T as a function of k.

SOLUTION

(a) If $C = 2C_0$, then $\frac{C}{C_0} = 2$, so $R = k \ln 2$ is a linear relation, because $\ln 2$ is a constant.

(b) $$T(R) = 1.03R$$

$$T(k) = 1.03k \ln \frac{C}{C_0} \qquad \text{Use the given expression for } R.$$

✔ *Now Try Exercise 75.*

Logarithms with Other Bases We can use a calculator to find the values of either natural logarithms (base e) or common logarithms (base 10). However, sometimes we must use logarithms with other bases. The change-of-base theorem can be used to convert logarithms from one base to another.

LOOKING AHEAD TO CALCULUS

In calculus, natural logarithms are more convenient to work with than logarithms with other bases. The change-of-base theorem enables us to convert any logarithmic function to a *natural* logarithmic function.

Change-of-Base Theorem

For any positive real numbers x, a, and b, where $a \neq 1$ and $b \neq 1$, the following holds.

$$\log_a x = \frac{\log_b x}{\log_b a}$$

Proof   Let
$$y = \log_a x.$$

$a^y = x$	Change to exponential form.
$\log_b a^y = \log_b x$	Take the base b logarithm on each side.
$y \log_b a = \log_b x$	Power property **(Section 4.3)**
$y = \dfrac{\log_b x}{\log_b a}$	Divide each side by $\log_b a$.
$\log_a x = \dfrac{\log_b x}{\log_b a}$	Substitute $\log_a x$ for y.

Any positive number other than 1 can be used for base b in the change-of-base theorem, but usually the only practical bases are e and 10 since calculators give logarithms for these two bases.

For example, with the change-of-base theorem, we can now graph the equation $y = \log_2 x$ by directing the calculator to graph $y = \frac{\log x}{\log 2}$, or, equivalently, $y = \frac{\ln x}{\ln 2}$. ■

EXAMPLE 8   **Using the Change-of-Base Theorem**

Use the change-of-base theorem to find an approximation to four decimal places for each logarithm.

(a) $\log_5 17$ 　　　　　　　　　　　　　　　　　　　　**(b)** $\log_2 0.1$

SOLUTION

(a) We will arbitrarily use natural logarithms.

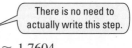
There is no need to actually write this step.

$$\log_5 17 = \frac{\ln 17}{\ln 5} \approx \frac{2.8332}{1.6094} \approx 1.7604$$

(b) Here, we use common logarithms.

$$\log_2 0.1 = \frac{\log 0.1}{\log 2} \approx -3.3219$$

✔ *Now Try Exercises 79 and 81.*

```
log(17)/log(5)
           1.7604
ln(.1)/ln(2)
          -3.3219
```

The screen shows how the result of **Example 8(a)** can be found using *common* logarithms, and how the result of **Example 8(b)** can be found using *natural* logarithms. The results are the same as those in **Example 8.**

NOTE   In **Example 8,** logarithms evaluated in the intermediate steps, such as ln 17 and ln 5, were shown to four decimal places. However, the final answers were obtained *without* rounding these intermediate values, using all the digits obtained with the calculator. *In general, it is best to wait until the final step to round off the answer; otherwise, a build-up of round-off errors may cause the final answer to have an incorrect digit in the final decimal place.*

EXAMPLE 9 **Modeling Diversity of Species**

One measure of the diversity of the species in an ecological community is modeled by the formula

$$H = -\left[P_1 \log_2 P_1 + P_2 \log_2 P_2 + \cdots + P_n \log_2 P_n\right],$$

where $P_1, P_2, \ldots, P_n$ are the proportions of a sample that belong to each of n species found in the sample. (*Source:* Ludwig, J., and J. Reynolds, *Statistical Ecology: A Primer on Methods and Computing,* New York, Wiley.)

Find the measure of diversity in a community with two species where there are 90 of one species and 10 of the other.

SOLUTION   Since there are 100 members in the community, $P_1 = \frac{90}{100} = 0.9$ and $P_2 = \frac{10}{100} = 0.1$, so

$$H = -\left[0.9 \log_2 0.9 + 0.1 \log_2 0.1\right]. \quad \text{Substitute for } P_1 \text{ and } P_2.$$

In **Example 8(b),** we found that $\log_2 0.1 \approx -3.32$. Now we find $\log_2 0.9$.

$$\log_2 0.9 = \frac{\log 0.9}{\log 2} \approx -0.152 \quad \text{Change-of-base theorem}$$

Therefore,

$$H = -\left[0.9 \log_2 0.9 + 0.1 \log_2 0.1\right]$$

$$H \approx -\left[0.9(-0.152) + 0.1(-3.32)\right] \quad \text{Substitute approximate values.}$$

$$H \approx 0.469. \quad \text{Simplify.}$$

Verify that $H \approx 0.971$ if there are 60 of one species and 40 of the other. As the proportions of n species get closer to $\frac{1}{n}$ each, the measure of diversity increases to a maximum of $\log_2 n$.

✔️ *Now Try Exercise 73.*

At the end of **Section 4.2,** we saw that graphing calculators are capable of fitting exponential curves to data that suggest such behavior. The same is true for logarithmic curves. For example, during the early 2000s on one particular day, interest rates for various U.S. Treasury Securities were as shown in the table.

Time	3-mo	6-mo	2-yr	5-yr	10-yr	30-yr
Yield	0.83%	0.91%	1.35%	2.46%	3.54%	4.58%

Source: U.S. Treasury.

Figure 38 shows how a calculator gives the best-fitting natural logarithmic curve for the data, as well as the data points and the graph of this curve.

Figure 38

4.4 Exercises

Concept Check *Answer each of the following.*

1. For the exponential function $f(x) = a^x$, where $a > 1$, is the function increasing or decreasing over its entire domain?

2. For the logarithmic function $g(x) = \log_a x$, where $a > 1$, is the function increasing or decreasing over its entire domain?

3. If $f(x) = 5^x$, what is the rule for $f^{-1}(x)$?

4. What is the name given to the exponent to which 4 must be raised to obtain 11?

5. A base e logarithm is called a(n) _____ logarithm, and a base 10 logarithm is called a(n) _____ logarithm.

6. How is $\log_3 12$ written in terms of natural logarithms?

7. Why is $\log_2 0$ undefined?

8. Between what two consecutive integers must $\log_2 12$ lie?

9. The graph of $y = \log x$ shows a point on the graph. Write the logarithmic equation associated with that point.

10. The graph of $y = \ln x$ shows a point on the graph. Write the logarithmic equation associated with that point.

Find each value. If applicable, give an approximation to four decimal places. ***See*** ***Example 1.***

11. $\log 10^{12}$	**12.** $\log 10^7$	**13.** $\log 0.1$	**14.** $\log 0.01$
15. $\log 63$	**16.** $\log 94$	**17.** $\log 0.0022$	**18.** $\log 0.0055$

19. $\log(387 \times 23)$ **20.** $\log(296 \times 12)$ **21.** $\log\left(\dfrac{518}{342}\right)$ **22.** $\log\left(\dfrac{643}{287}\right)$

23. $\log 387 + \log 23$ **24.** $\log 296 + \log 12$

25. $\log 518 - \log 342$ **26.** $\log 643 - \log 287$

27. Explain why the result in **Exercise 23** is the same as that in **Exercise 19**.

28. Explain why the result in **Exercise 25** is the same as that in **Exercise 21**.

For each substance, find the pH from the given hydronium ion concentration. **See Example 2(a).**

29. grapefruit, 6.3×10^{-4}

30. limes, 1.6×10^{-2}

31. crackers, 3.9×10^{-9}

32. sodium hydroxide (lye), 3.2×10^{-14}

Find the $\left[H_3O^+\right]$ *for each substance with the given pH.* **See Example 2(b).**

33. soda pop, 2.7

34. wine, 3.4

35. beer, 4.8

36. drinking water, 6.5

In Exercises 37–42, suppose that water from a wetland area is sampled and found to have the given hydronium ion concentration. Determine whether the wetland is a rich fen, a poor fen, or a bog. **See Example 3.**

37. 2.49×10^{-5}

38. 6.22×10^{-5}

39. 2.49×10^{-2}

40. 3.14×10^{-2}

41. 2.49×10^{-7}

42. 5.86×10^{-7}

43. Use your calculator to find an approximation for each logarithm.

  (a) log 398.4　　　　**(b)** log 39.84　　　　**(c)** log 3.984

  (d) From your answers to parts (a)–(c), make a conjecture concerning the decimal values in the approximations of common logarithms of numbers greater than 1 that have the same digits.

44. Given that $\log 25 \approx 1.3979$, $\log 250 \approx 2.3979$, and $\log 2500 \approx 3.3979$, make a conjecture for an approximation of log 25,000. Then explain why this pattern continues.

Find each value. If applicable, give an approximation to four decimal places. **See Example 5.**

45. $\ln e^{1.6}$

46. $\ln e^{5.8}$

47. $\ln\left(\dfrac{1}{e^2}\right)$

48. $\ln\left(\dfrac{1}{e^4}\right)$

49. $\ln 28$

50. $\ln 39$

51. $\ln 0.00013$

52. $\ln 0.0077$

53. $\ln(27 \times 943)$

54. $\ln(33 \times 568)$

55. $\ln\left(\dfrac{98}{13}\right)$

56. $\ln\left(\dfrac{84}{17}\right)$

57. $\ln 27 + \ln 943$

58. $\ln 33 + \ln 568$

59. $\ln 98 - \ln 13$

60. $\ln 84 - \ln 17$

61. Explain why the result in **Exercise 57** is the same as that in **Exercise 53**.

62. Explain why the result in **Exercise 59** is the same as that in **Exercise 55**.

Solve each application of logarithms. **See Examples 4, 6, 7, and 9.**

63. *Decibel Levels* Find the decibel ratings of sounds having the following intensities.

  (a) $100I_0$　　　**(b)** $1000I_0$　　　**(c)** $100,000I_0$　　　**(d)** $1,000,000I_0$

  (e) If the intensity of a sound is doubled, by how much is the decibel rating increased?

64. *Decibel Levels* Find the decibel ratings of the following sounds, having intensities as given. Round each answer to the nearest whole number.

  (a) whisper, $115I_0$　　　　　　　**(b)** busy street, $9,500,000I_0$

  (c) heavy truck, 20 m away, $1,200,000,000I_0$

  (d) rock music, $895,000,000,000I_0$

  (e) jetliner at takeoff, $109,000,000,000,000I_0$

65. *Earthquake Intensity* The magnitude of an earthquake, measured on the Richter scale, is $\log_{10} \frac{I}{I_0}$, where I is the amplitude registered on a seismograph 100 km from the epicenter of the earthquake, and I_0 is the amplitude of an earthquake of a certain (small) size. Find the Richter scale ratings for earthquakes having the following amplitudes.

    (a) $1000I_0$ **(b)** $1,000,000I_0$ **(c)** $100,000,000I_0$

66. *Earthquake Intensity* On December 26, 2004, an earthquake struck in the Indian Ocean with a magnitude of 9.1 on the Richter scale. The resulting tsunami killed an estimated 229,900 people in several countries. Express this reading in terms of I_0.

67. *Earthquake Intensity* On February 27, 2010, a massive earthquake struck Chile with a magnitude of 8.8 on the Richter scale. Express this reading in terms of I_0.

68. *Earthquake Intensity Comparison* Compare your answers to **Exercises 66 and 67.** How many times greater was the force of the 2004 earthquake than that of the 2010 earthquake?

69. *(Modeling) Bachelor's Degrees in Psychology* The table gives the number of bachelor's degrees in psychology (in thousands) earned at U.S. colleges and universities for selected years from 1980 through 2008. Suppose x represents the number of years since 1950. Thus, 1980 is represented by 30, 1990 is represented by 40, and so on.

Year	Degrees Earned (in thousands)
1980	42.1
1990	54.0
2000	74.2
2005	85.6
2007	90.0
2008	92.6

Source: U.S. National Center for Education Statistics.

The logarithmic function

$$f(x) = -228.1 + 78.19 \ln x$$

is the best-fitting logarithmic model for the data. Use this function to estimate the number of bachelor's degrees in psychology earned in the year 2012. What assumption must we make to estimate the number of degrees in years beyond 2012?

70. *(Modeling) Domestic Leisure Travel* The bar graph shows numbers of leisure trips within the United States (in millions of person-trips of 50 or more miles one-way) over the years 2003–2008. The function

$$f(t) = 1393 + 69.49 \ln t, \quad t \geq 1,$$

where t represents the number of years since 2002 and $f(t)$ is the number of person-trips, in millions, approximates the curve reasonably well.

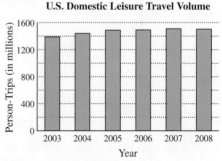

U.S. Domestic Leisure Travel Volume

Source: U.S. Travel Association.

(a) Use the function to approximate the number of person-trips in 2006. How does this approximation compare to the actual number of 1492 million?

(b) Explain why an exponential function would *not* provide a good model for these data.

71. *(Modeling) Diversity of Species* The number of species $S(n)$ in a sample is given by

$$S(n) = a \ln\left(1 + \frac{n}{a}\right),$$

where n is the number of individuals in the sample, and a is a constant that indicates the diversity of species in the community. If $a = 0.36$, find $S(n)$ for each value of n. (*Hint:* $S(n)$ must be a whole number.)

(a) 100   (b) 200   (c) 150   (d) 10

72. *(Modeling) Diversity of Species* In **Exercise 71,** find $S(n)$ if a changes to 0.88. Use the following values of n.

(a) 50   (b) 100   (c) 250

73. *(Modeling) Diversity of Species* Suppose a sample of a small community shows two species with 50 individuals each. Find the measure of diversity H.

74. *(Modeling) Diversity of Species* A virgin forest in northwestern Pennsylvania has 4 species of large trees with the following proportions of each: hemlock, 0.521; beech, 0.324; birch, 0.081; maple, 0.074. Find the measure of diversity H.

75. *(Modeling) Global Temperature Increase* In **Example 7,** we expressed the average global temperature increase T (in °F) as

$$T(k) = 1.03k \ln \frac{C}{C_0},$$

where C_0 is the preindustrial amount of carbon dioxide, C is the current carbon dioxide level, and k is a constant. Arrhenius determined that $10 \le k \le 16$ when C was double the value C_0. Use $T(k)$ to find the range of the rise in global temperature T (rounded to the nearest degree) that Arrhenius predicted. (*Source:* Clime, W., *The Economics of Global Warming,* Institute for International Economics, Washington, D.C.)

76. *(Modeling) Global Temperature Increase* (Refer to **Exercise 75.**) According to one study by the IPCC, future increases in average global temperatures (in °F) can be modeled by

$$T(C) = 6.489 \ln \frac{C}{280},$$

where C is the concentration of atmospheric carbon dioxide (in ppm). C can be modeled by the function

$$C(x) = 353(1.006)^{x-1990},$$

where x is the year. (*Source:* International Panel on Climate Change (IPCC).)

(a) Write T as a function of x.

(b) Using a graphing calculator, graph $C(x)$ and $T(x)$ on the interval [1990, 2275] using different coordinate axes. Describe the graph of each function. How are C and T related?

(c) Approximate the slope of the graph of T. What does this slope represent?

(d) Use graphing to estimate x and $C(x)$ when $T(x) = 10$°F.

77. *Age of Rocks* Use the formula of **Example 6** to estimate the age of a rock sample having $\frac{A}{K} = 0.103$.

78. *(Modeling) Planets' Distances from the Sun and Periods of Revolution* The table contains the planets' average distances D from the sun and their periods P of revolution around the sun in years. The distances have been normalized so that Earth is one unit away from the sun. For example, since Jupiter's distance is 5.2, its distance from the sun is 5.2 times farther than Earth's.

(a) Using a graphing calculator, make a scatter diagram by plotting the point $(\ln D, \ln P)$ for each planet on the xy-coordinate axes. Do the data points appear to be linear?

(b) Determine a linear equation that models the data points. Graph your line and the data on the same coordinate axes.

(c) Use this linear model to predict the period of Pluto if its distance is 39.5. Compare your answer to the actual value of 248.5 yr.

Planet	D	P
Mercury	0.39	0.24
Venus	0.72	0.62
Earth	1	1
Mars	1.52	1.89
Jupiter	5.2	11.9
Saturn	9.54	29.5
Uranus	19.2	84.0
Neptune	30.1	164.8

Source: Ronan, C., *The Natural History of the Universe,* MacMillan Publishing Co., New York.

Use the change-of-base theorem to find an approximation to four decimal places for each logarithm. ***See Example 8.***

79. $\log_2 5$

80. $\log_2 9$

81. $\log_8 0.59$

82. $\log_8 0.71$

83. $\log_{1/2} 3$

84. $\log_{1/3} 2$

85. $\log_\pi e$

86. $\log_\pi \sqrt{2}$

87. $\log_{\sqrt{13}} 12$

88. $\log_{\sqrt{19}} 5$

89. $\log_{0.32} 5$

90. $\log_{0.91} 8$

Let $u = \ln a$ and $v = \ln b$. Write each expression in terms of u and v without using the ln *function.*

91. $\ln\left(b^4\sqrt{a}\right)$

92. $\ln\dfrac{a^3}{b^2}$

93. $\ln\sqrt{\dfrac{a^3}{b^5}}$

94. $\ln\left(\sqrt[3]{a} \cdot b^4\right)$

Concept Check In Exercises 95–98, use the various properties of exponential and logarithmic functions to evaluate the expressions in parts (a)–(c).

95. Given $g(x) = e^x$, find　(a) $g(\ln 4)$　(b) $g(\ln(5^2))$　(c) $g\left(\ln\left(\frac{1}{e}\right)\right)$.

96. Given $f(x) = 3^x$, find　(a) $f(\log_3 2)$　(b) $f(\log_3(\ln 3))$　(c) $f(\log_3(2\ln 3))$.

97. Given $f(x) = \ln x$, find　(a) $f(e^6)$　(b) $f(e^{\ln 3})$　(c) $f(e^{2\ln 3})$.

98. Given $f(x) = \log_2 x$, find　(a) $f(2^7)$　(b) $f(2^{\log_2 2})$　(c) $f(2^{2\log_2 2})$.

Work each problem.

99. *Concept Check* Which of the following is equivalent to $2\ln(3x)$ for $x > 0$?

A. $\ln 9 + \ln x$　　**B.** $\ln(6x)$　　**C.** $\ln 6 + \ln x$　　**D.** $\ln(9x^2)$

100. *Concept Check* Which of the following is equivalent to $\ln(4x) - \ln(2x)$ for $x > 0$?

A. $2\ln x$　　**B.** $\ln(2x)$　　**C.** $\dfrac{\ln(4x)}{\ln(2x)}$　　**D.** $\ln 2$

101. The function $f(x) = \ln|x|$ plays a prominent role in calculus. Find its domain, its range, and the symmetries of its graph.

102. Consider the function $f(x) = \log_3|x|$.

(a) What is the domain of this function?

(b) Use a graphing calculator to graph $f(x) = \log_3|x|$ in the window $[-4, 4]$ by $[-4, 4]$.

(c) How might one easily misinterpret the domain of the function by merely observing the calculator graph?

Use the properties of logarithms and the terminology of **Chapter 2** *to describe how the graph of the given function compares to the graph of* $g(x) = \ln x$.

103. $f(x) = \ln(e^2 x)$ **104.** $f(x) = \ln \dfrac{x}{e}$ **105.** $f(x) = \ln \dfrac{x}{e^2}$

Chapter 4   Quiz (Sections 4.1–4.4)

1. For the one-to-one function $f(x) = \sqrt[3]{3x - 6}$, find $f^{-1}(x)$.

2. Solve $4^{2x+1} = 8^{3x-6}$.

3. Graph $f(x) = -3^x$. Give the domain and range.

4. Graph $f(x) = \log_4(x + 2)$. Give the domain and range.

5. *Future Value* Suppose that \$15,000 is deposited in a bank certificate of deposit at an annual rate of 3.5% for 8 yr. Find the future value if interest is compounded as follows.

    (a) annually   **(b)** quarterly   **(c)** monthly   **(d)** daily (365 days)

6. Use a calculator to evaluate each logarithm to four decimal places.

    (a) log 34.56   **(b)** ln 34.56

7. What is the meaning of the expression $\log_6 25$?

8. Solve.

    (a) $x = 3^{\log_3 4}$ **(b)** $\log_x 25 = 2$ **(c)** $\log_4 x = -2$

9. Assuming all variables represent positive real numbers, use properties of logarithms to rewrite

$$\log_3 \frac{\sqrt{x} \cdot y}{pq^4}.$$

10. Given $\log_b 9 = 3.1699$ and $\log_b 5 = 2.3219$, find the value of $\log_b 225$.

11. Find the value of $\log_3 40$ to four decimal places.

12. If $f(x) = 4^x$, what is the value of $f(\log_4 12)$?

4.5   Exponential and Logarithmic Equations

■ Exponential Equations

■ Logarithmic Equations

■ Applications and Models

Exponential Equations   We solved exponential equations in earlier sections. General methods for solving these equations depend on the property below, which follows from the fact that logarithmic functions are one-to-one.

Property of Logarithms

If $x > 0$, $y > 0$, $a > 0$, and $a \neq 1$, then the following holds.

$$x = y \quad \text{is equivalent to} \quad \log_a x = \log_a y.$$

EXAMPLE 1 Solving an Exponential Equation

Solve $7^x = 12$. Give the solution to the nearest thousandth.

SOLUTION The properties of exponents given in **Section 4.2** cannot be used to solve this equation, so we apply the preceding property of logarithms. While any appropriate base b can be used, the best practical base is base 10 or base e. We choose base e (natural) logarithms here.

$$7^x = 12$$

$$\ln 7^x = \ln 12 \qquad \text{Property of logarithms}$$

$$x \ln 7 = \ln 12 \qquad \text{Power property (Section 4.3)}$$

This is *exact.*
$$x = \frac{\ln 12}{\ln 7} \qquad \text{Divide by } \ln 7.$$

$$x \approx 1.277 \qquad \text{Use a calculator. (Section 4.4)}$$

The solution set is $\{1.277\}$. This is *approximate.*

Y = 7^X – 12

As seen in the display at the bottom of the screen, when rounded to three decimal places, the solution agrees with that found in **Example 1.**

☑ *Now Try Exercise 5.*

CAUTION When evaluating a quotient like $\frac{\ln 12}{\ln 7}$ in **Example 1,** do not confuse this quotient with $\ln \frac{12}{7}$, which can be written as $\ln 12 - \ln 7$. **We cannot change the quotient of two logarithms to a difference of logarithms.**

$$\frac{\ln 12}{\ln 7} \neq \ln \frac{12}{7}$$

EXAMPLE 2 Solving an Exponential Equation

Solve $3^{2x-1} = 0.4^{x+2}$. Give the solution to the nearest thousandth.

SOLUTION
$$3^{2x-1} = 0.4^{x+2}$$

$$\ln 3^{2x-1} = \ln 0.4^{x+2} \qquad \text{Take the natural logarithm on each side.}$$

$$(2x - 1) \ln 3 = (x + 2) \ln 0.4 \qquad \text{Power property}$$

$$2x \ln 3 - \ln 3 = x \ln 0.4 + 2 \ln 0.4 \qquad \text{Distributive property (Section R.2)}$$

$$2x \ln 3 - x \ln 0.4 = 2 \ln 0.4 + \ln 3 \qquad \text{Write the terms with } x \text{ on one side.}$$

$$x(2 \ln 3 - \ln 0.4) = 2 \ln 0.4 + \ln 3 \qquad \text{Factor out } x. \text{ (Section R.4)}$$

$$x = \frac{2 \ln 0.4 + \ln 3}{2 \ln 3 - \ln 0.4} \qquad \text{Divide by } 2 \ln 3 - \ln 0.4.$$

$$x = \frac{\ln 0.4^2 + \ln 3}{\ln 3^2 - \ln 0.4} \qquad \text{Power property}$$

$$x = \frac{\ln 0.16 + \ln 3}{\ln 9 - \ln 0.4} \qquad \text{Apply the exponents.}$$

This is *exact.*
$$x = \frac{\ln 0.48}{\ln 22.5} \qquad \text{Product and quotient properties (Section 4.3)}$$

$$x \approx -0.236 \qquad \text{Use a calculator.}$$

The solution set is $\{-0.236\}$. This is *approximate.*

Y = 3^{2X–1} – 0.4^{X+2}

This screen supports the solution found in **Example 2.**

☑ *Now Try Exercise 13.*

EXAMPLE 3 **Solving Base *e* Exponential Equations**

Solve each equation. Give solutions to the nearest thousandth.

(a) $e^{x^2} = 200$ **(b)** $e^{2x+1} \cdot e^{-4x} = 3e$

SOLUTION

(a)
$$e^{x^2} = 200$$
$$\ln e^{x^2} = \ln 200 \qquad \text{Take the natural logarithm on each side.}$$
$$x^2 = \ln 200 \qquad \ln e^{x^2} = x^2 \text{ (Section 4.3)}$$

Remember both roots.

$$x = \pm \sqrt{\ln 200} \qquad \text{Square root property (Section 1.4)}$$
$$x \approx \pm 2.302 \qquad \text{Use a calculator.}$$

The solution set is $\{\pm 2.302\}$.

(b)
$$e^{2x+1} \cdot e^{-4x} = 3e$$
$$e^{-2x+1} = 3e \qquad a^m \cdot a^n = a^{m+n} \text{ (Section R.3)}$$
$$e^{-2x} = 3 \qquad \text{Divide by } e; \frac{a^m}{a^n} = a^{m-n}. \text{ (Section R.3)}$$
$$\ln e^{-2x} = \ln 3 \qquad \text{Take the natural logarithm on each side.}$$
$$-2x \ln e = \ln 3 \qquad \text{Power property}$$
$$-2x = \ln 3 \qquad \ln e = 1$$
$$x = -\frac{1}{2} \ln 3 \qquad \text{Multiply by } -\frac{1}{2}.$$
$$x \approx -0.549 \qquad \text{Use a calculator.}$$

The solution set is $\{-0.549\}$.

Now Try Exercises 15 and 17.

EXAMPLE 4 **Solving an Exponential Equation Quadratic in Form**

Solve $e^{2x} - 4e^x + 3 = 0$. Give exact value(s) for *x*.

SOLUTION This is an equation that is quadratic in form, because it can be rewritten with $u = e^x$.

$$e^{2x} - 4e^x + 3 = 0$$
$$(e^x)^2 - 4e^x + 3 = 0 \qquad a^{mn} = (a^n)^m \text{ (Section R.3)}$$
$$u^2 - 4u + 3 = 0 \qquad \text{Let } u = e^x. \text{ (Section 1.6)}$$
$$(u-1)(u-3) = 0 \qquad \text{Factor. (Section R.4)}$$
$$u - 1 = 0 \quad \text{or} \quad u - 3 = 0 \qquad \text{Zero-factor property (Section 1.4)}$$
$$u = 1 \quad \text{or} \quad u = 3 \qquad \text{Solve for } u.$$
$$e^x = 1 \quad \text{or} \quad e^x = 3 \qquad \text{Substitute } e^x \text{ for } u.$$
$$\ln e^x = \ln 1 \quad \text{or} \quad \ln e^x = \ln 3 \qquad \text{Take the natural logarithm on each side.}$$
$$x = 0 \quad \text{or} \quad x = \ln 3 \qquad \ln e^x = x \text{ (Section 4.3)}$$

Both values check, so the solution set is $\{0, \ln 3\}$.

Now Try Exercise 29.

Logarithmic Equations The following equations involve logarithms of variable expressions.

EXAMPLE 5 **Solving Logarithmic Equations**

Solve each equation. Give exact values.

(a) $7 \ln x = 28$ **(b)** $\log_2(x^3 - 19) = 3$

SOLUTION

(a) $\qquad 7 \ln x = 28$

$\qquad\qquad \ln x = 4 \qquad$ Divide by 7.

$\qquad\qquad x = e^4 \qquad$ Write the natural logarithm in exponential form.

The solution set is $\{e^4\}$.

(b) $\qquad \log_2(x^3 - 19) = 3$

$\qquad\qquad x^3 - 19 = 2^3 \qquad$ Write in exponential form.

$\qquad\qquad x^3 - 19 = 8 \qquad$ Apply the exponent.

$\qquad\qquad\quad x^3 = 27 \qquad$ Add 19.

$\qquad\qquad\quad x = \sqrt[3]{27} \qquad$ Take cube roots. **(Section 1.6)**

$\qquad\qquad\quad x = 3 \qquad \sqrt[3]{27} = 3$

The solution set is $\{3\}$.

✔ *Now Try Exercises 35 and 43.*

EXAMPLE 6 **Solving a Logarithmic Equation**

Solve $\log(x + 6) - \log(x + 2) = \log x$. Give exact value(s).

SOLUTION Keep in mind that logarithms are defined only for nonnegative numbers.

$\qquad \log(x + 6) - \log(x + 2) = \log x$

$\qquad\qquad \log \dfrac{x + 6}{x + 2} = \log x \qquad$ Quotient property

$\qquad\qquad \dfrac{x + 6}{x + 2} = x \qquad$ Property of logarithms

$\qquad\qquad x + 6 = x(x + 2) \qquad$ Multiply by $x + 2$. **(Section 1.6)**

$\qquad\qquad x + 6 = x^2 + 2x \qquad$ Distributive property

$\qquad\qquad x^2 + x - 6 = 0 \qquad$ Standard form **(Section 1.4)**

$\qquad\qquad (x + 3)(x - 2) = 0 \qquad$ Factor.

$\qquad x + 3 = 0 \quad$ or $\quad x - 2 = 0 \qquad$ Zero-factor property

$\qquad\quad x = -3 \quad$ or $\qquad x = 2 \qquad$ Solve for x. **(Section 1.1)**

The proposed negative solution (-3) is not in the domain of $\log x$ in the original equation, so the only valid solution is the positive number 2. The solution set is $\{2\}$.

✔ *Now Try Exercise 63.*

CAUTION Recall that the domain of $y = \log_a x$ is $(0, \infty)$. For this reason, *it is always necessary to check that proposed solutions of a logarithmic equation result in logarithms of positive numbers in the original equation.*

EXAMPLE 7 **Solving a Logarithmic Equation**

Solve $\log_2\left[(3x - 7)(x - 4)\right] = 3$. Give exact value(s).

SOLUTION

$$\log_2\left[(3x - 7)(x - 4)\right] = 3$$

$$(3x - 7)(x - 4) = 2^3 \qquad \text{Write in exponential form.}$$

$$3x^2 - 19x + 28 = 8 \qquad \text{Multiply. (Section R.3)}$$

$$3x^2 - 19x + 20 = 0 \qquad \text{Standard form}$$

$$(3x - 4)(x - 5) = 0 \qquad \text{Factor.}$$

$$3x - 4 = 0 \quad \text{or} \quad x - 5 = 0 \qquad \text{Zero-factor property}$$

$$x = \frac{4}{3} \quad \text{or} \qquad x = 5 \qquad \text{Solve for } x.$$

A check is necessary to be sure that the argument of the logarithm in the given equation is positive. In both cases, the product $(3x - 7)(x - 4)$ leads to 8, and $\log_2 8 = 3$ is true. The solution set is $\left\{\frac{4}{3}, 5\right\}$.

✔ *Now Try Exercise 47.*

EXAMPLE 8 **Solving a Logarithmic Equation**

Solve $\log(3x + 2) + \log(x - 1) = 1$. Give exact value(s).

SOLUTION The notation $\log x$ is an abbreviation for $\log_{10} x$, and $1 = \log_{10} 10$.

$$\log(3x + 2) + \log(x - 1) = 1$$

$$\log(3x + 2) + \log(x - 1) = \log 10 \qquad \text{Substitute.}$$

$$\log\left[(3x + 2)(x - 1)\right] = \log 10 \qquad \text{Product property}$$

$$(3x + 2)(x - 1) = 10 \qquad \text{Property of logarithms}$$

$$3x^2 - x - 2 = 10 \qquad \text{Multiply.}$$

$$3x^2 - x - 12 = 0 \qquad \text{Subtract 10.}$$

$$x = \frac{1 \pm \sqrt{1 + 144}}{6} \qquad \text{Quadratic formula (Section 1.4)}$$

The two proposed solutions are

$$\frac{1 - \sqrt{145}}{6} \quad \text{and} \quad \frac{1 + \sqrt{145}}{6}.$$

The first of these proposed solutions, $\frac{1 - \sqrt{145}}{6}$, is negative and when substituted for x in $\log(x - 1)$ results in a negative argument, which is not allowed. Therefore, this solution must be rejected.

The second proposed solution, $\frac{1 + \sqrt{145}}{6}$, is positive. Substituting it for x in $\log(3x + 2)$ results in a positive argument, and substituting it for x in $\log(x + 1)$ also results in a positive argument, both of which are necessary conditions. Therefore, the solution set is $\left\{\frac{1 + \sqrt{145}}{6}\right\}$.

✔ *Now Try Exercise 71.*

NOTE We could have used the definition of logarithm in **Example 8** by first writing

$$\log(3x + 2) + \log(x - 1) = 1 \qquad \text{Equation from \textbf{Example 8}}$$

$$\log_{10}[(3x + 2)(x - 1)] = 1 \qquad \text{Product property}$$

$$(3x + 2)(x - 1) = 10^1, \qquad \text{Definition of logarithm (Section 4.3)}$$

and then continuing as shown on the preceding page.

EXAMPLE 9 Solving a Base *e* Logarithmic Equation

Solve $\ln e^{\ln x} - \ln(x - 3) = \ln 2$. Give exact value(s).

SOLUTION This logarithmic equation differs from those in **Examples 7 and 8** because the expression on the right side involves a logarithm.

$$\ln e^{\ln x} - \ln(x - 3) = \ln 2$$

$$\ln x - \ln(x - 3) = \ln 2 \qquad e^{\ln x} = x \text{ (Section 4.3)}$$

$$\ln \frac{x}{x - 3} = \ln 2 \qquad \text{Quotient property}$$

$$\frac{x}{x - 3} = 2 \qquad \text{Property of logarithms}$$

$$x = 2(x - 3) \qquad \text{Multiply by } x - 3.$$

$$x = 2x - 6 \qquad \text{Distributive property}$$

$$6 = x \qquad \text{Solve for } x.$$

Check that the solution set is $\{6\}$.

☑ *Now Try Exercise 73.*

Solving Exponential or Logarithmic Equations

To solve an exponential or logarithmic equation, change the given equation into one of the following forms, where a and b are real numbers, $a > 0$ and $a \neq 1$, and follow the guidelines.

1. $a^{f(x)} = b$

   Solve by taking logarithms on both sides.

2. $\log_a f(x) = b$

   Solve by changing to exponential form $a^b = f(x)$.

3. $\log_a f(x) = \log_a g(x)$

   The given equation is equivalent to the equation $f(x) = g(x)$. Solve algebraically.

4. In a more complicated equation, such as

   $$e^{2x+1} \cdot e^{-4x} = 3e$$

   in **Example 3(b)**, it may be necessary to first solve for $a^{f(x)}$ or $\log_a f(x)$ and then solve the resulting equation using one of the methods given above.

5. Check that each proposed solution is in the domain.

> **Applications and Models**

EXAMPLE 10 Applying an Exponential Equation to the Strength of a Habit

The strength of a habit is a function of the number of times the habit is repeated. If N is the number of repetitions and H is the strength of the habit, then, according to psychologist C.L. Hull,

$$H = 1000(1 - e^{-kN}),$$

where k is a constant. Solve this equation for k.

SOLUTION

$$H = 1000(1 - e^{-kN})$$ — *First solve for e^{-kN}.*

$$\frac{H}{1000} = 1 - e^{-kN}$$ Divide by 1000.

$$\frac{H}{1000} - 1 = -e^{-kN}$$ Subtract 1.

$$e^{-kN} = 1 - \frac{H}{1000}$$ Multiply by -1 and rewrite.

Now solve for k. $$\ln e^{-kN} = \ln\left(1 - \frac{H}{1000}\right)$$ Take the natural logarithm on each side.

$$-kN = \ln\left(1 - \frac{H}{1000}\right)$$ $\ln e^x = x$

$$k = -\frac{1}{N}\ln\left(1 - \frac{H}{1000}\right)$$ Multiply by $-\frac{1}{N}$.

With the final equation, if one pair of values for H and N is known, k can be found, and the equation can then be used to find either H or N for given values of the other variable.

✔ *Now Try Exercise 85.*

EXAMPLE 11 Modeling Coal Consumption in the U.S.

The table gives U.S. coal consumption (in quadrillions of British thermal units, or *quads*) for several years. The data can be modeled by the function

$$f(t) = 24.92 \ln t - 93.31, \quad t \geq 80,$$

where t is the number of years after 1900, and $f(t)$ is in quads.

(a) Approximately what amount of coal was consumed in the United States in 2003? How does this figure compare to the actual figure of 22.32 quads?

(b) If this trend continues, approximately when will annual consumption reach 25 quads?

SOLUTION

(a) The year 2003 is represented by $t = 2003 - 1900 = 103$.

$$f(103) = 24.92 \ln 103 - 93.31 \quad \text{Let } t = 103.$$

$$\approx 22.19 \quad\quad\quad\quad \text{Use a calculator.}$$

Based on this model, 22.19 quads were used in 2003. This figure is very close to the actual amount of 22.32 quads.

Year	Coal Consumption (in quads)
1980	15.42
1985	17.48
1990	19.17
1995	20.09
2000	22.58
2005	22.80
2008	22.39

Source: U.S. Energy Information Administration.

(b) Replace $f(t)$ with 25, and solve for t.

$$25 = 24.92 \ln t - 93.31 \qquad \text{$f(t) = 25$ in the given model.}$$

$$118.31 = 24.92 \ln t \qquad \text{Add 93.31.}$$

$$\ln t = \frac{118.31}{24.92} \qquad \text{Divide by 24.92. Rewrite.}$$

$$t = e^{118.31/24.92} \qquad \text{Write in exponential form.}$$

$$t \approx 115.3 \qquad \text{Use a calculator.}$$

Add 115 to 1900 to get 2015. Based on this model, annual consumption will reach 25 quads in 2015.

✔ *Now Try Exercise 105.*

4.5 Exercises

Concept Check An exponential equation such as

$$5^x = 9$$

can be solved for its exact solution using the meaning of logarithm and the change-of-base theorem. Since x is the exponent to which 5 must be raised in order to obtain 9, the exact solution is

$$\log_5 9, \quad or \quad \frac{\log 9}{\log 5}, \quad or \quad \frac{\ln 9}{\ln 5}.$$

For each equation, give the exact solution in three forms similar to the forms explained above.

1. $7^x = 19$ 　　　　**2.** $3^x = 10$ 　　　　**3.** $\left(\dfrac{1}{2}\right)^x = 12$ 　　　　**4.** $\left(\dfrac{1}{3}\right)^x = 4$

Solve each exponential equation. In Exercises 5–28, express irrational solutions as decimals correct to the nearest thousandth. In Exercises 29–34, express solutions in exact form. ***See Examples 1–4.***

5. $3^x = 7$ 　　　　　　　**6.** $5^x = 13$ 　　　　　　　**7.** $\left(\dfrac{1}{2}\right)^x = 5$

8. $\left(\dfrac{1}{3}\right)^x = 6$ 　　　　　**9.** $0.8^x = 4$ 　　　　　　**10.** $0.6^x = 3$

11. $4^{x-1} = 3^{2x}$ 　　　　　**12.** $2^{x+3} = 5^{2x}$ 　　　　**13.** $6^{x+1} = 4^{2x-1}$

14. $3^{x-4} = 7^{2x+5}$ 　　　　**15.** $e^{x^2} = 100$ 　　　　　**16.** $e^{x^4} = 1000$

17. $e^{3x-7} \cdot e^{-2x} = 4e$ 　　**18.** $e^{1-3x} \cdot e^{5x} = 2e$ 　　**19.** $\left(\dfrac{1}{3}\right)^x = -3$

20. $\left(\dfrac{1}{9}\right)^x = -9$ 　　　**21.** $0.05(1.15)^x = 5$ 　　**22.** $1.2(0.9)^x = 0.6$

23. $3(2)^{x-2} + 1 = 100$ 　　**24.** $5(1.2)^{3x-2} + 1 = 7$ 　　**25.** $2(1.05)^x + 3 = 10$

26. $3(1.4)^x - 4 = 60$ 　　　**27.** $5(1.015)^{x-1980} = 8$ 　　**28.** $6(1.024)^{x-1900} = 9$

29. $e^{2x} - 6e^x + 8 = 0$ 　　**30.** $e^{2x} - 8e^x + 15 = 0$ 　　**31.** $2e^{2x} + e^x = 6$

32. $3e^{2x} + 2e^x = 1$ 　　　**33.** $5^{2x} + 3(5^x) = 28$ 　　　**34.** $3^{2x} - 12(3^x) = -35$

Solve each logarithmic equation. Express all solutions in exact form. **See Examples 5–9.**

35. $5 \ln x = 10$

36. $3 \ln x = 9$

37. $\ln(4x) = 1.5$

38. $\ln(2x) = 5$

39. $\log(2 - x) = 0.5$

40. $\log(3 - x) = 0.75$

41. $\log_6(2x + 4) = 2$

42. $\log_5(8 - 3x) = 3$

43. $\log_4(x^3 + 37) = 3$

44. $\log_7(x^3 + 65) = 0$

45. $\ln x + \ln x^2 = 3$

46. $\log x + \log x^2 = 3$

47. $\log_3[(x + 5)(x - 3)] = 2$

48. $\log_4[(3x + 8)(x - 6)] = 3$

49. $\log_2[(2x + 8)(x + 4)] = 5$

50. $\log_5[(3x + 5)(x + 1)] = 1$

51. $\log x + \log(x + 15) = 2$

52. $\log x + \log(2x + 1) = 1$

53. $\log(x + 25) = \log(x + 10) + \log 4$

54. $\log(3x + 5) - \log(2x + 4) = 0$

55. $\log(x - 10) - \log(x - 6) = \log 2$

56. $\log(x^2 + 10x - 39) - \log(x - 3) = \log 10$

57. $\ln(5 - x) + \ln(-3 - x) = \ln(1 - 8x)$

58. $\ln(10 - x) + \ln(-6 - x) = \ln(-34 - 15x)$

59. $\log_8(x + 2) + \log_8(x + 4) = \log_8 8$

60. $\log_2(5x - 6) - \log_2(x + 1) = \log_2 3$

61. $\log_2(x^2 - 100) - \log_2(x + 10) = 1$

62. $\log_2(x - 2) + \log_2(x - 1) = 1$

63. $\log x + \log(x - 21) = \log 100$

64. $\log x + \log(3x - 13) = \log 10$

65. $\log(9x + 5) = 3 + \log(x + 2)$

66. $\log(11x + 9) = 3 + \log(x + 3)$

67. $\ln(4x - 2) - \ln 4 = -\ln(x - 2)$

68. $\ln(5 + 4x) - \ln(3 + x) = \ln 3$

69. $\log_5(x + 2) + \log_5(x - 2) = 1$

70. $\log_2(x - 7) + \log_2 x = 3$

71. $\log_2(2x - 3) + \log_2(x + 1) = 1$

72. $\log_5(3x + 2) + \log_5(x - 1) = 1$

73. $\ln e^x - 2 \ln e = \ln e^4$

74. $\ln e^x - \ln e^3 = \ln e^3$

75. $\log_2(\log_2 x) = 1$

76. $\log x = \sqrt{\log x}$

77. $\log x^2 = (\log x)^2$

78. $\log_2 \sqrt{2x^2} = \dfrac{3}{2}$

79. *Concept Check* Suppose you overhear the following statement: "I must reject any negative proposed solution when I solve an equation involving logarithms." Is this correct? Why or why not?

80. *Concept Check* What values of x could not possibly be solutions of the following equation?

$$\log_a(4x - 7) + \log_a(x^2 + 4) = 0$$

Solve each equation for the indicated variable. Use logarithms with the appropriate bases. **See Example 10.**

81. $p = a + \dfrac{k}{\ln x}$, for x

82. $r = p - k \ln t$, for t

83. $T = T_0 + (T_1 - T_0)10^{-kt}$, for t

84. $A = \dfrac{Pr}{1 - (1 + r)^{-n}}$, for n

85. $I = \dfrac{E}{R}(1 - e^{-Rt/2})$, for t

86. $y = \dfrac{K}{1 + ae^{-bx}}$, for b

87. $y = A + B(1 - e^{-Cx})$, for x

88. $m = 6 - 2.5 \log\left(\dfrac{M}{M_0}\right)$, for M

89. $\log A = \log B - C \log x$, for A

90. $d = 10 \log\left(\dfrac{I}{I_0}\right)$, for I

91. $A = P\left(1 + \dfrac{r}{n}\right)^{tn}$, for t

92. $D = 160 + 10 \log x$, for x

For Exercises 93–98, refer to the formulas for compound interest.

$$A = P\left(1 + \frac{r}{n}\right)^{tn} \quad \text{and} \quad A = Pe^{rt} \quad \text{(Section 4.2)}$$

93. *Compound Amount* If $10,000 is invested in an account at 3% annual interest compounded quarterly, how much will be in the account in 5 yr if no money is withdrawn?

94. *Compound Amount* If $5000 is invested in an account at 4% annual interest compounded continuously, how much will be in the account in 8 yr if no money is withdrawn?

95. *Investment Time* Kurt Daniels wants to buy a $30,000 car. He has saved $27,000. Find the number of years (to the nearest tenth) it will take for his $27,000 to grow to $30,000 at 4% interest compounded quarterly.

96. *Investment Time* Find t to the nearest hundredth of a year if $1786 becomes $2063 at 2.6%, with interest compounded monthly.

97. *Interest Rate* Find the interest rate to the nearest hundredth of a percent that will produce $2500, if $2000 is left at interest compounded semiannually for 4.5 yr.

98. *Interest Rate* At what interest rate, to the nearest hundredth of a percent, will $16,000 grow to $20,000 if invested for 5.25 yr and interest is compounded quarterly?

(Modeling) Solve each application. **See Example 11.**

99. In the central Sierra Nevada (a mountain range in California), the percent of moisture that falls as snow rather than rain is approximated reasonably well by

$$f(x) = 86.3 \ln x - 680,$$

where x is the altitude in feet and $f(x)$ is the percent of moisture that falls as snow. Find the percent of moisture that falls as snow at each altitude.

(a) 3000 ft   **(b)** 4000 ft   **(c)** 7000 ft

100. C. Horne Creations finds that its total sales in dollars, $T(x)$, from the distribution of x thousand catalogues is approximated by

$$T(x) = 5000 \log(x + 1).$$

Find the total sales resulting from the distribution of each number of catalogues.

(a) 5000   **(b)** 24,000   **(c)** 49,000

101. *Average Annual Public University Costs* The table shows the cost of a year's tuition, room and board, and fees at 4-year public colleges for the years 2002–2010. Letting *y* represent the cost and *x* the number of years since 2002, we find that the function

$$f(x) = 9318(1.06)^x$$

models the data quite well. According to this function, when will the cost in 2002 be doubled?

Year	Average Annual Cost
2002	$ 9,119
2003	$ 9,951
2004	$10,648
2005	$11,322
2006	$11,929
2007	$12,698
2008	$13,467
2009	$14,265
2010	$15,067

Source: The College Board, *Annual Survey of Colleges.*

102. *Race Speed* At the World Championship races held at Rome's Olympic Stadium in 1987, American sprinter Carl Lewis ran the 100-m race in 9.86 sec. His speed in meters per second after *t* seconds is closely modeled by the function

$$f(t) = 11.65(1 - e^{-t/1.27}).$$

(*Source:* Banks, Robert B., *Towing Icebergs, Falling Dominoes, and Other Adventures in Applied Mathematics,* Princeton University Press.)

(a) How fast was he running as he crossed the finish line?
(b) After how many seconds was he running at the rate of 10 m per sec?

103. *Women in Labor Force* The percent of women in the United States who were in the civilian labor force increased rapidly for several decades and then stabilized. If *x* represents the number of years since 1950, the function

$$f(x) = \frac{67.21}{1 + 1.081e^{-x/24.71}}$$

models the percent fairly well. (*Source: Monthly Labor Review,* U.S. Bureau of Labor Statistics.)

(a) What percent of U.S. women were in the civilian labor force in 2008?
(b) In what year were 55% of U.S. women in the civilian labor force?

104. *Height of the Eiffel Tower* One side of the Eiffel Tower in Paris has a shape that can be approximated by the graph of the function

$$f(x) = -301 \ln \frac{x}{207}, x > 0$$

See the figure. (*Source:* Banks, Robert B., *Towing Icebergs, Falling Dominoes, and Other Adventures in Applied Mathematics,* Princeton University Press.)

$f(x) = -301 \ln \frac{x}{207}, x > 0$

(a) Explain why the shape of the left side of the Eiffel Tower has the formula given by $f(-x)$.
(b) The short horizontal line at the top of the figure has length 7.8744 ft. Approximately how tall is the Eiffel Tower?
(c) Approximately how far from the center of the tower is the point on the right side that is 500 ft above the ground?

105. *CO₂ Emissions Tax* One action that government could take to reduce carbon emissions into the atmosphere is to levy a tax on fossil fuel. This tax would be based on the amount of carbon dioxide emitted into the air when the fuel is burned. The **cost-benefit equation**

$$\ln(1 - P) = -0.0034 - 0.0053T$$

models the approximate relationship between a tax of T dollars per ton of carbon and the corresponding percent reduction P (in decimal form) of emissions of carbon dioxide. (*Source:* Nordhause, W., "To Slow or Not to Slow: The Economics of the Greenhouse Effect," Yale University, New Haven, Connecticut.)

(a) Write P as a function of T.

(b) Graph P for $0 \le T \le 1000$. Discuss the benefit of continuing to raise taxes on carbon.

(c) Determine P when $T = \$60$, and interpret this result.

(d) What value of T will give a 50% reduction in carbon emissions?

106. *Radiative Forcing* (Refer to **Example 7** in **Section 4.4.**) Using computer models, the International Panel on Climate Change (IPCC) in 1990 estimated k to be 6.3 in the radiative forcing equation

$$R = k \ln \frac{C}{C_0},$$

where C_0 is the preindustrial amount of carbon dioxide and C is the current level. (*Source:* Clime, W., *The Economics of Global Warming,* Institute for International Economics, Washington, D.C.)

(a) Use the equation $R = 6.3 \ln \frac{C}{C_0}$ to determine the radiative forcing R (in watts per square meter) expected by the IPCC if the carbon dioxide level in the atmosphere doubles from its preindustrial level.

(b) Determine the global temperature increase T that the IPCC predicted would occur if atmospheric carbon dioxide levels were to double. (*Hint:* $T(R) = 1.03R$.)

Find $f^{-1}(x)$, and give the domain and range.

107. $f(x) = e^{x+1} - 4$

108. $f(x) = 2 \ln 3x$

Use a graphing calculator to solve each equation. Give irrational solutions correct to the nearest hundredth.

109. $e^x + \ln x = 5$

110. $e^x - \ln(x + 1) = 3$

111. $2e^x + 1 = 3e^{-x}$

112. $e^x + 6e^{-x} = 5$

113. $\log x = x^2 - 8x + 14$

114. $\ln x = -\sqrt[3]{x + 3}$

115. Explain the **error** in the following **"proof"** that $2 < 1$.

$$\frac{1}{9} < \frac{1}{3} \qquad \text{True statement}$$

$$\left(\frac{1}{3}\right)^2 < \frac{1}{3} \qquad \text{Rewrite the left side.}$$

$$\log\left(\frac{1}{3}\right)^2 < \log\frac{1}{3} \qquad \text{Take the logarithm on each side.}$$

$$2 \log\frac{1}{3} < 1 \log\frac{1}{3} \qquad \text{Property of logarithms; identity property}$$

$$2 < 1 \qquad \text{Divide each side by } \log\frac{1}{3}.$$

4.6 Applications and Models of Exponential Growth and Decay

- ■ The Exponential Growth or Decay Function
- ■ Growth Function Models
- ■ Decay Function Models

LOOKING AHEAD TO CALCULUS

The exponential growth and decay function formulas are studied in calculus in conjunction with the topic known as **differential equations.**

The Exponential Growth or Decay Function In many situations that occur in ecology, biology, economics, and the social sciences, a quantity changes at a rate proportional to the amount present. In such cases the amount present at time t is a special function of t called an **exponential growth or decay function.**

Exponential Growth or Decay Function

Let y_0 be the amount or number present at time $t = 0$. Then, under certain conditions, the amount present at any time t is modeled by

$$y = y_0 e^{kt}, \quad \text{where } k \text{ is a constant.}$$

When $k > 0$, the function describes growth. In **Section 4.2,** we saw examples of exponential growth: compound interest and atmospheric carbon dioxide, for example. When $k < 0$, the function describes decay; one example of exponential decay is radioactivity.

Growth Function Models The amount of time it takes for a quantity that grows exponentially to become twice its initial amount is its **doubling time.**

> **EXAMPLE 1** Determining an Exponential Function to Model the Increase of Carbon Dioxide

In **Example 11, Section 4.2,** we discussed the growth of atmospheric carbon dioxide over time. A function based on the data from the table was given in that example. Now we can see how to determine such a function from the data.

(a) Find an exponential function that gives the amount of carbon dioxide y in year x.

(b) Estimate the year when future levels of carbon dioxide will be double the preindustrial level of 280 ppm.

SOLUTION

(a) Recall that the graph of the data points showed exponential growth, so the equation will take the form

$$y = y_0 e^{kx}.$$

We must find the values of y_0 and k. The data begin with the year 1990, so to simplify our work we let 1990 correspond to $x = 0$, 1991 correspond to $x = 1$, and so on. Since y_0 is the initial amount, $y_0 = 353$ in 1990 when $x = 0$. Thus the equation is

$$y = 353 e^{kx}. \quad \text{Let } y_0 = 353.$$

From the last pair of values in the table, we know that in 2275 the carbon dioxide level is expected to be 2000 ppm. The year 2275 corresponds to $2275 - 1990 = 285$. Substitute 2000 for y and 285 for x, and solve for k.

Year	Carbon Dioxide (ppm)
1990	353
2000	375
2075	590
2175	1090
2275	2000

Source: International Panel on Climate Change (IPCC).

$$y = 353e^{kx}$$

$$2000 = 353e^{k(285)} \qquad \text{Substitute 2000 for } y \text{ and 285 for } x.$$

$$\frac{2000}{353} = e^{285k} \qquad \text{Divide by 353.}$$

$$\ln\left(\frac{2000}{353}\right) = \ln\left(e^{285k}\right) \qquad \text{Take the logarithm on each side. (Section 4.5)}$$

$$\ln\left(\frac{2000}{353}\right) = 285k \qquad \ln e^x = x, \text{ for all } x. \text{ (Section 4.3)}$$

$$k = \frac{1}{285} \cdot \ln\left(\frac{2000}{353}\right) \qquad \text{Multiply by } \tfrac{1}{285} \text{ and rewrite.}$$

$$k \approx 0.00609 \qquad \text{Use a calculator.}$$

A function that models the data is

$$y = 353e^{0.00609x}.$$

(b) Let $y = 2(280) = 560$ in $y = 353e^{0.00609x}$, and find x.

$$560 = 353e^{0.00609x} \qquad \text{Let } y = 560.$$

$$\frac{560}{353} = e^{0.00609x} \qquad \text{Divide by 353.}$$

$$\ln\left(\frac{560}{353}\right) = \ln e^{0.00609x} \qquad \text{Take the logarithm on each side.}$$

$$\ln\left(\frac{560}{353}\right) = 0.00609x \qquad \ln e^x = x, \text{ for all } x.$$

$$x = \frac{1}{0.00609} \cdot \ln\left(\frac{560}{353}\right) \qquad \text{Multiply by } \tfrac{1}{0.00609} \text{ and rewrite.}$$

$$x \approx 75.8 \qquad \text{Use a calculator.}$$

Since $x = 0$ corresponds to 1990, the preindustrial carbon dioxide level will double in the 75th year after 1990, or during 2065, according to this model.

✔ *Now Try Exercise 21.*

EXAMPLE 2 **Finding Doubling Time for Money**

How long will it take for the money in an account that accrues interest at a rate of 3%, compounded continuously, to double?

SOLUTION

$$A = Pe^{rt} \qquad \text{Continuous compounding formula (Section 4.2)}$$

$$2P = Pe^{0.03t} \qquad \text{Let } A = 2P \text{ and } r = 0.03.$$

$$2 = e^{0.03t} \qquad \text{Divide by } P.$$

$$\ln 2 = \ln e^{0.03t} \qquad \text{Take the logarithm on each side.}$$

$$\ln 2 = 0.03t \qquad \ln e^x = x$$

$$\frac{\ln 2}{0.03} = t \qquad \text{Divide by 0.03.}$$

$$23.10 \approx t \qquad \text{Use a calculator.}$$

It will take about 23 yr for the amount to double.

✔ *Now Try Exercise 27.*

EXAMPLE 3 Determining an Exponential Function to Model Population Growth

According to the U.S. Census Bureau, the world population reached 6 billion people during 1999 and was growing exponentially. By the end of 2010, the population had grown to 6.947 billion. The projected world population (in billions of people) t years after 2010 is given by the function

$$f(t) = 6.947e^{0.00745t}.$$

(a) Based on this model, what will the world population be in 2020?

(b) In what year will the world population reach 9 billion?

SOLUTION

(a) Since $t = 0$ represents the year 2010, in 2020, t would be $2020 - 2010 = 10$ yr. We must find $f(t)$ when t is 10.

$$f(t) = 6.947e^{0.00745t} \quad \text{Given formula}$$

$$f(10) = 6.947e^{0.00745(10)} \quad \text{Let } t = 10.$$

$$f(10) \approx 7.484 \quad \text{Use a calculator.}$$

The population will be 7.484 billion at the end of 2020.

(b)
$$f(t) = 6.947e^{0.00745t} \quad \text{Given formula}$$

$$9 = 6.947e^{0.00745t} \quad \text{Let } f(t) = 9.$$

$$\frac{9}{6.947} = e^{0.00745t} \quad \text{Divide by 6.947.}$$

$$\ln \frac{9}{6.947} = \ln e^{0.00745t} \quad \text{Take the logarithm on each side.}$$

$$\ln \frac{9}{6.947} = 0.00745t \quad \ln e^x = x, \text{ for all } x.$$

$$t = \frac{\ln \frac{9}{6.947}}{0.00745} \quad \text{Divide by 0.00745 and rewrite.}$$

$$t \approx 34.8 \quad \text{Use a calculator.}$$

Thus, 34.8 yr after 2010, during the year 2044, world population will reach 9 billion.

✔ *Now Try Exercise 35.*

Decay Function Models **Half-life** is the amount of time it takes for a quantity that decays exponentially to become half its initial amount.

EXAMPLE 4 Determining an Exponential Function to Model Radioactive Decay

Suppose 600 g of a radioactive substance are present initially and 3 yr later only 300 g remain.

(a) Determine the exponential equation that models this decay.

(b) How much of the substance will be present after 6 yr?

SOLUTION

(a) To express the situation as an exponential equation

$$y = y_0 e^{kt},$$

we use the given values first to find y_0 and then to find k.

$$y = y_0 e^{kt}$$

$$600 = y_0 e^{k(0)} \quad \text{Let } y = 600 \text{ and } t = 0.$$

$$600 = y_0 \quad\quad e^0 = 1 \text{ (Section R.3)}$$

Thus, $y_0 = 600$, which gives $y = 600e^{kt}$. Because the initial amount (600 g) decays to half that amount (300 g) in 3 yr, its half-life is 3 yr. Now we solve this exponential equation for k.

$$y = 600e^{kt} \quad\quad \text{Let } y_0 = 600.$$

$$300 = 600e^{3k} \quad\quad \text{Let } y = 300 \text{ and } t = 3.$$

$$0.5 = e^{3k} \quad\quad\quad \text{Divide by 600.}$$

$$\ln 0.5 = \ln e^{3k} \quad\quad \text{Take logarithms on each side.}$$

$$\ln 0.5 = 3k \quad\quad\quad \ln e^x = x, \text{ for all } x.$$

$$\frac{\ln 0.5}{3} = k \quad\quad\quad \text{Divide by 3.}$$

$$k \approx -0.231 \quad\quad \text{Use a calculator.}$$

Thus, the exponential decay equation is

$$y = 600e^{-0.231t}.$$

(b) To find the amount present after 6 yr, let $t = 6$.

$$y = 600e^{-0.231(6)} \quad \text{Let } t = 6.$$

$$y = 600e^{-1.386} \quad\quad \text{Multiply.}$$

$$y \approx 150 \quad\quad\quad\quad \text{Use a calculator.}$$

After 6 yr, 150 g of the substance will remain.

✔ *Now Try Exercise 15.*

NOTE In **Example 4** the initial amount of the substance was given as 600 g. Notice that it could have been *any* amount y_0, and substituting $\frac{1}{2}y_0$ for y on the left would allow the common factors of y to be divided out, leaving the equation $0.5 = e^{3k}$. The rest of the work would be the same.

EXAMPLE 5 **Solving a Carbon Dating Problem**

Carbon-14, also known as radiocarbon, is a radioactive form of carbon that is found in all living plants and animals. After a plant or animal dies, the radiocarbon disintegrates. Scientists can determine the age of the remains by comparing the amount of radiocarbon with the amount present in living plants and animals. This technique is called **carbon dating.** The amount of radiocarbon present after t years is given by

$$y = y_0 e^{-0.0001216t},$$

where y_0 is the amount present in living plants and animals.

(a) Find the half-life of carbon-14.

(b) Charcoal from an ancient fire pit on Java contained $\frac{1}{4}$ the carbon-14 of a living sample of the same size. Estimate the age of the charcoal.

SOLUTION

(a) If y_0 is the amount of radiocarbon present in a living thing, then $\frac{1}{2}y_0$ is half this initial amount. Thus, we substitute and solve the given equation for t.

$$y = y_0 e^{-0.0001216t} \qquad \text{Given equation}$$

$$\frac{1}{2}y_0 = y_0 e^{-0.0001216t} \qquad \text{Let } y = \tfrac{1}{2}y_0.$$

$$\frac{1}{2} = e^{-0.0001216t} \qquad \text{Divide by } y_0.$$

$$\ln \frac{1}{2} = \ln e^{-0.0001216t} \qquad \text{Take the logarithm on each side.}$$

$$\ln \frac{1}{2} = -0.0001216t \qquad \ln e^x = x, \text{ for all } x.$$

$$\frac{\ln \frac{1}{2}}{-0.0001216} = t \qquad \text{Divide by } -0.0001216.$$

$$5700 \approx t \qquad \text{Use a calculator.}$$

The half-life is about 5700 yr.

(b) Solve again for t, this time letting the amount $y = \frac{1}{4}y_0$.

$$y = y_0 e^{-0.0001216t} \qquad \text{Given equation}$$

$$\frac{1}{4}y_0 = y_0 e^{-0.0001216t} \qquad \text{Let } y = \tfrac{1}{4}y_0.$$

$$\frac{1}{4} = e^{-0.0001216t} \qquad \text{Divide by } y_0.$$

$$\ln \frac{1}{4} = \ln e^{-0.0001216t} \qquad \text{Take the logarithm on each side.}$$

$$\frac{\ln \frac{1}{4}}{-0.0001216} = t \qquad \ln e^x = x; \text{ Divide by } -0.0001216.$$

$$t \approx 11{,}400 \qquad \text{Use a calculator.}$$

The charcoal is about 11,400 yr old.

✔ *Now Try Exercise 17.*

EXAMPLE 6 **Modeling Newton's Law of Cooling**

Newton's law of cooling says that the rate at which a body cools is proportional to the difference in temperature between the body and the environment around it. The temperature $f(t)$ of the body at time t in appropriate units after being introduced into an environment having constant temperature T_0 is

$$f(t) = T_0 + Ce^{-kt}, \quad \text{where } C \text{ and } k \text{ are constants.}$$

A pot of coffee with a temperature of 100°C is set down in a room with a temperature of 20°C. The coffee cools to 60°C after 1 hr.

(a) Write an equation to model the data.

(b) Find the temperature after half an hour.

(c) How long will it take for the coffee to cool to 50°C?

SOLUTION

(a) We must find values for C and k in the given formula. From the given information, when $t = 0$, $T_0 = 20$, and the temperature of the coffee is $f(0) = 100$. Also, when $t = 1$, $f(1) = 60$. Substitute the first pair of values into the function along with $T_0 = 20$.

$$f(t) = T_0 + Ce^{-kt} \qquad \text{Given formula}$$

$$100 = 20 + Ce^{-0k} \qquad \text{Let } t = 0, \ f(0) = 100, \text{ and } T_0 = 20.$$

$$100 = 20 + C \qquad e^0 = 1$$

$$80 = C \qquad \text{Subtract 20.}$$

The function that models the data can now be given.

$$f(t) = 20 + 80e^{-kt} \qquad \text{Let } T_0 = 20 \text{ and } C = 80.$$

Now use the remaining pair of values in this function to find k.

$$f(t) = 20 + 80e^{-kt} \qquad \text{Formula with } T_0 = 20 \text{ and } C = 80$$

$$60 = 20 + 80e^{-1k} \qquad \text{Let } t = 1 \text{ and } f(1) = 60.$$

$$40 = 80e^{-k} \qquad \text{Subtract 20.}$$

$$\frac{1}{2} = e^{-k} \qquad \text{Divide by 80.}$$

$$\ln \frac{1}{2} = \ln e^{-k} \qquad \text{Take the logarithm on each side.}$$

$$\ln \frac{1}{2} = -k \qquad \ln e^x = x, \text{ for all } x.$$

$$k = -\ln \frac{1}{2} \qquad \text{Multiply by } -1 \text{ and rewrite.}$$

$$k \approx 0.693 \qquad \text{Use a calculator.}$$

Thus, the model is $f(t) = 20 + 80e^{-0.693t}$.

(b) To find the temperature after $\frac{1}{2}$ hr, let $t = \frac{1}{2}$ in the model from part (a).

$$f(t) = 20 + 80e^{-0.693t} \qquad \text{Model from part (a)}$$

$$f\left(\frac{1}{2}\right) = 20 + 80e^{(-0.693)(1/2)} \qquad \text{Let } t = \frac{1}{2}.$$

$$f\left(\frac{1}{2}\right) \approx 76.6°\text{C} \qquad \text{Use a calculator.}$$

(c) To find how long it will take for the coffee to cool to 50°C, let $f(t) = 50$.

$$50 = 20 + 80e^{-0.693t} \qquad \text{Let } f(t) = 50 \text{ in model from part (a).}$$

$$30 = 80e^{-0.693t} \qquad \text{Subtract 20.}$$

$$\frac{3}{8} = e^{-0.693t} \qquad \text{Divide by 80.}$$

$$\ln \frac{3}{8} = \ln e^{-0.693t} \qquad \text{Take the logarithm on each side.}$$

$$\ln \frac{3}{8} = -0.693t \qquad \ln e^x = x, \text{ for all } x.$$

$$t = \frac{\ln \frac{3}{8}}{-0.693} \qquad \text{Divide by } -0.693 \text{ and rewrite.}$$

$$t \approx 1.415 \text{ hr}, \quad \text{or} \quad \text{about 1 hr, 25 min}$$

☑ *Now Try Exercise 23.*

4.6 Exercises

Population Growth *A population is increasing according to the exponential function* $y = 2e^{0.02x}$, *where y is in millions and x is the number of years. Match each question in Column I with the correct procedure in Column II to answer the question.*

I

1. How long will it take for the population to triple?

2. When will the population reach 3 million?

3. How large will the population be in 3 yr?

4. How large will the population be in 4 months?

II

A. Evaluate $y = 2e^{0.02(1/3)}$.

B. Solve $2e^{0.02x} = 6$.

C. Evaluate $y = 2e^{0.02(3)}$.

D. Solve $2e^{0.02x} = 3$.

(Modeling) *The exercises in this set are grouped according to discipline. They involve exponential or logarithmic models.* **See Examples 1–6.**

Physical Sciences *(Exercises 5–24)*

In Exercises 5–10, an initial amount of a radioactive substance y_0 is given, along with information about the amount remaining after a given time t in appropriate units. For an equation of the form $y = y_0 e^{kt}$ that models the situation, give the exact value of k in terms of natural logarithms.

5. $y_0 = 60$ g; After 3 hr, 20 g remain.

6. $y_0 = 30$ g; After 6 hr, 10 g remain.

7. $y_0 = 10$ mg; The half-life is 100 days.

8. $y_0 = 20$ mg; The half-life is 200 days.

9. $y_0 = 2.56$ lb; After 2 yr, 0.64 lb remains.

10. $y_0 = 8.1$ kg; After 4 yr, 0.9 kg remains.

11. *Decay of Lead* A sample of 500 g of radioactive lead-210 decays to polonium-210 according to the function

$$A(t) = 500e^{-0.032t},$$

where t is time in years. Find the amount of radioactive lead remaining after

(a) 4 yr, (b) 8 yr, (c) 20 yr. (d) Find the half-life.

12. *Decay of Plutonium* Repeat **Exercise 11** for 500 g of plutonium-241, which decays according to the function $A(t) = A_0 e^{-0.053t}$, where t is time in years.

13. *Decay of Radium* Find the half-life of radium-226, which decays according to the function $A(t) = A_0 e^{-0.00043t}$, where t is time in years.

14. *Decay of Iodine* How long will it take any quantity of iodine-131 to decay to 25% of its initial amount, knowing that it decays according to the exponential function $A(t) = A_0 e^{-0.087t}$, where t is time in days?

15. *Radioactive Decay* If 12 g of a radioactive substance are present initially and 4 yr later only 6 g remain, how much of the substance will be present after 7 yr?

16. *Magnitude of a Star* The magnitude M of a star is modeled by

$$M = 6 - \frac{5}{2} \log \frac{I}{I_0},$$

where I_0 is the intensity of a just-visible star and I is the actual intensity of the star being measured. The dimmest stars are of magnitude 6, and the brightest are of magnitude 1. Determine the ratio of light intensities between a star of magnitude 1 and a star of magnitude 3.

17. *Carbon-14 Dating* Suppose an Egyptian mummy is discovered in which the amount of carbon-14 present is only about one-third the amount found in living human beings. About how long ago did the Egyptian die?

18. *Carbon-14 Dating* A sample from a refuse deposit near the Strait of Magellan had 60% of the carbon-14 of a contemporary sample. How old was the sample?

19. *Carbon-14 Dating* Paint from the Lascaux caves of France contains 15% of the normal amount of carbon-14. Estimate the age of the paintings.

20. *Dissolving a Chemical* The amount of a chemical that will dissolve in a solution increases exponentially as the (Celsius) temperature t is increased according to the model

$$A(t) = 10e^{0.0095t}.$$

At what temperature will 15 g dissolve?

21. *Sulfur Dioxide Levels* Sulfur dioxide (SO_2), a gas that is emitted by burning fossil fuels, is a source of air pollution and one of the major contributors to acid rain. Reductions in emissions of sulfur dioxide in the United States were mandated by the 1990 revision of the Clean Air Act. The graph displays approximate U.S. emissions of sulfur dioxide (in thousands of tons) over the period 1970–2008.

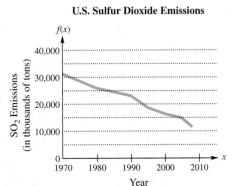

U.S. Sulfur Dioxide Emissions

Source: U.S. Environmental Protection Agency.

(a) Use the graph to find an exponential function

$$f(x) = A_0 a^{x-1970}$$

that models U.S. emissions of sulfur dioxide over the years 1970–2008.

(b) Approximate the average annual percent decrease in sulfur dioxide emissions over this time period.

22. *Rock Sample Age* Estimate the age of a rock sample using the function

$$t = T \frac{\ln\left(1 + 8.33\left(\frac{A}{K}\right)\right)}{\ln 2}.$$

Tests show that $\frac{A}{K}$ is 0.175 for the sample. Let $T = 1.26 \times 10^9$.

23. *Newton's Law of Cooling* Boiling water, at 100°C, is placed in a freezer at 0°C. The temperature of the water is 50°C after 24 min. Find the temperature of the water after 96 min. (*Hint:* Change minutes to hours.)

24. *Newton's Law of Cooling* A piece of metal is heated to 300°C and then placed in a cooling liquid at 50°C. After 4 min, the metal has cooled to 175°C. Find its temperature after 12 min. (*Hint:* Change minutes to hours.)

Finance (Exercises 25–30)

25. *Comparing Investments* Russ McClelland, who is self-employed, wants to invest $60,000 in a pension plan. One investment offers 5% compounded quarterly. Another offers 4.75% compounded continuously.

(a) Which investment will earn more interest in 5 yr?

(b) How much more will the better plan earn?

26. *Growth of an Account* If Russ (see **Exercise 25**) chooses the plan with continuous compounding, how long will it take for his $60,000 to grow to $80,000?

27. *Doubling Time* Find the doubling time of an investment earning 2.5% interest if interest is compounded continuously.

28. *Doubling Time* If interest is compounded continuously and the interest rate is tripled, what effect will this have on the time required for an investment to double?

29. *Growth of an Account* How long will it take an investment to triple if interest is compounded continuously at 5%?

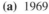 **30.** *Growth of an Account* Use the Table feature of a graphing calculator to find how long it will take $1500 invested at 5.75% compounded daily to triple in value. Zoom in on the solution by systematically decreasing the increment for *x*. Find the answer to the nearest day. (Find the answer to the nearest day by eventually letting the increment of *x* equal $\frac{1}{365}$. The decimal part of the solution can be multiplied by 365 to determine the number of days greater than the nearest year. For example, if the solution is determined to be 16.2027 yr, then multiply 0.2027 by 365 to get 73.9855. The solution is then, to the nearest day, 16 yr, 74 days.) Confirm the answer algebraically.

Social Sciences (Exercises 31–38)

31. *Legislative Turnover* The turnover of legislators is a problem of interest to political scientists. In a study during the 1970s, it was found that one model of legislative turnover in the U.S. House of Representatives was

$$M(t) = 434e^{-0.08t},$$

where $M(t)$ represents the number of continuously serving members at time *t*. Here, $t = 0$ represents 1965, $t = 1$ represents 1966, and so on. Use this model to approximate the number of continuously serving members in each year.

(a) 1969 **(b)** 1973 **(c)** 1979

32. *Legislative Turnover* Use the model in **Exercise 31** to determine the year in which the number of continuously serving members was 338.

33. *Population Growth* In 2000 India's population reached 1 billion, and it is projected to be 1.4 billion in 2025. (*Source:* U.S. Census Bureau.)

(a) Find values for P_0 and *a* so that $f(x) = P_0 a^{x-2000}$ models the population of India in year *x*.

(b) Predict India's population in 2020 to the nearest tenth of a billion.

(c) Use *f* to determine the year when India's population might reach 1.5 billion.

34. *Population Decline* A midwestern city finds its residents moving to the suburbs. Its population is declining according to the function defined by

$$P(t) = P_0 e^{-0.04t},$$

where *t* is time measured in years and P_0 is the population at time $t = 0$. Assume that $P_0 = 1{,}000{,}000$.

(a) Find the population at time $t = 1$.

(b) Estimate the time it will take for the population to decline to 750,000.

(c) How long will it take for the population to decline to half the initial number?

35. *Health Care Spending* Out-of-pocket spending in the United States for health care increased between 2004 and 2008. The function

$$f(x) = 2572e^{0.0359x}$$

models average annual expenditures per household, in dollars. In this model, *x* represents the year, where $x = 0$ corresponds to 2004. (*Source*: U.S. Bureau of Labor Statistics.)

(a) Estimate out-of-pocket household spending on health care in 2008.

(b) Determine the year when spending reached $2775 per household.

36. *Recreational Expenditures* Personal consumption expenditures for recreation in billions of dollars in the United States during the years 2004–2008 can be approximated by the function

$$A(t) = 769.5e^{0.0503t},$$

where $t = 0$ corresponds to the year 2004. Based on this model, how much were personal consumption expenditures in 2008? (*Source:* U.S. Bureau of Economic Analysis.)

37. *Housing Costs* Average annual per-household spending on housing over the years 1990–2008 is approximated by

$$H = 8790e^{0.0382t},$$

where t is the number of years since 1990. Find H to the nearest dollar for each year. (*Source:* U.S. Bureau of Labor Statistics.)

(a) 2000   **(b)** 2005   **(c)** 2008

38. *Evolution of Language* The number of years, n, since two independently evolving languages split off from a common ancestral language is approximated by

$$n \approx -7600 \log r,$$

where r is the proportion of words from the ancestral language common to both languages.

(a) Find n if $r = 0.9$.   **(b)** Find n if $r = 0.3$.

(c) How many years have elapsed since the split if half of the words of the ancestral language are common to both languages?

Life Sciences *(Exercises 39–44)*

39. *Spread of Disease* During an epidemic, the number of people who have never had the disease and who are not immune (they are *susceptible*) decreases exponentially according to the function

$$f(t) = 15{,}000e^{-0.05t},$$

where t is time in days. Find the number of susceptible people at each time.

(a) at the beginning of the epidemic   **(b)** after 10 days   **(c)** after 3 weeks

40. *Spread of Disease* Refer to **Exercise 39** and determine how long it will take for the initial number of people susceptible to decrease to half its amount.

41. *Growth of Bacteria* The growth of bacteria makes it necessary to time-date some food products so that they will be sold and consumed before the bacteria count is too high. Suppose for a certain product the number of bacteria present is given by

$$f(t) = 500e^{0.1t},$$

where t is time in days and the value of $f(t)$ is in millions. Find the number of bacteria present at each time.

(a) 2 days   **(b)** 4 days   **(c)** 1 week

42. *Growth of Bacteria* How long will it take the bacteria population in **Exercise 41** to double?

43. *Medication Effectiveness* Drug effectiveness decreases over time. If, each hour, a drug is only 90% as effective as the previous hour, at some point the patient will not be receiving enough medication and must receive another dose. If the initial dose was 200 mg and the drug was administered 3 hr ago, the expression $200(0.90)^3$, which equals 145.8, represents the amount of effective medication still in the system. (The exponent is equal to the number of hours since the drug was administered.)

The amount of medication still available in the system is given by the function

$$f(t) = 200(0.90)^t.$$

In this model, t is in hours and $f(t)$ is in milligrams. How long will it take for this initial dose to reach the dangerously low level of 50 mg?

Population Size *Many environmental situations place effective limits on the growth of the number of an organism in an area. Many such limited-growth situations are described by the* **logistic function**

$$G(x) = \frac{MG_0}{G_0 + (M - G_0)e^{-kMx}},$$

where G_0 is the initial number present, M is the maximum possible size of the population, and k is a positive constant. The screens illustrate a typical logistic function calculation and graph.

44. Assume that $G_0 = 100$, $M = 2500$, $k = 0.0004$, and $x =$ time in decades (10-yr periods).

  (a) Use a calculator to graph the function, using $0 \le x \le 8$, $0 \le y \le 2500$.

  (b) Estimate the value of $G(2)$ from the graph. Then evaluate $G(2)$ algebraically to find the population after 20 yr.

  (c) Find the x-coordinate of the intersection of the curve with the horizontal line $y = 1000$ to estimate the number of decades required for the population to reach 1000. Then solve $G(x) = 1000$ algebraically to obtain the exact value of x.

Economics (Exercises 45–50)

45. *Consumer Price Index* The U.S. Consumer Price Index for the years 1990–2009 is approximated by

$$A(t) = 100e^{0.026t},$$

where t represents the number of years after 1990. (Since $A(16)$ is about 152, the amount of goods that could be purchased for \$100 in 1990 cost about \$152 in 2006.) Use the function to determine the year in which costs will be 100% higher than in 1990. (*Source:* U.S. Bureau of Labor Statistics.)

46. *Product Sales* Sales of a product, under relatively stable market conditions but in the absence of promotional activities such as advertising, tend to decline at a constant yearly rate. This rate of sales decline varies considerably from product to product, but it seems to remain the same for any particular product. The sales decline can be expressed by the function

$$S(t) = S_0 e^{-at},$$

where $S(t)$ is the rate of sales at time t measured in years, S_0 is the rate of sales at time $t = 0$, and a is the sales decay constant.

  (a) Suppose the sales decay constant for a particular product is $a = 0.10$. Let $S_0 = 50,000$ and find $S(1)$ and $S(3)$.

  (b) Find $S(2)$ and $S(10)$ if $S_0 = 80,000$ and $a = 0.05$.

47. *Product Sales* Use the sales decline function given in **Exercise 46.** If $a = 0.10$, $S_0 = 50,000$, and t is time measured in years, find the number of years it will take for sales to fall to half the initial sales.

48. *Cost of Bread* Assume the cost of a loaf of bread is $4. With continuous compounding, find the time it would take for the cost to triple at an annual inflation rate of 4%.

49. *Electricity Consumption* Suppose that in a certain area the consumption of electricity has increased at a continuous rate of 6% per year. If it continued to increase at this rate, find the number of years before twice as much electricity would be needed.

50. *Electricity Consumption* Suppose a conservation campaign, together with higher rates, caused demand for electricity to increase at only 2% per year. (See **Exercise 49.**) Find the number of years before twice as much electricity would be needed.

(Modeling) Exercises 51 and 52 use another type of logistic function.

51. *Heart Disease* As age increases, so does the likelihood of coronary heart disease (CHD). The fraction of people x years old with some CHD is modeled by

$$f(x) = \frac{0.9}{1 + 271e^{-0.122x}}.$$

(*Source:* Hosmer, D., and S. Lemeshow, *Applied Logistic Regression,* John Wiley and Sons.)

(a) Evaluate $f(25)$ and $f(65)$. Interpret the results.

(b) At what age does this likelihood equal 50%?

52. *Tree Growth* The height of a certain tree in feet after x years is modeled by

$$f(x) = \frac{50}{1 + 47.5e^{-0.22x}}.$$

(a) Make a table for f starting at $x = 10$, and incrementing by 10. What appears to be the maximum height of the tree?

(b) Graph f and identify the horizontal asymptote. Explain its significance.

(c) After how long was the tree 30 ft tall?

Summary Exercises on Functions: Domains and Defining Equations

Finding the Domain of a Function: A Summary To find the domain of a function, given the equation that defines the function, remember that the value of x input into the equation must yield a real number for y when the function is evaluated. For the functions studied so far in this book, there are three cases to consider when determining domains.

Guidelines for Domain Restrictions

1. No input value can lead to 0 in a denominator, because division by 0 is undefined.

2. No input value can lead to an even root of a negative number, because this situation does not yield a real number.

3. No input value can lead to the logarithm of a negative number or 0, because this situation does not yield a real number.

Unless domains are otherwise specified, we can determine domains as follows.

- The domain of a **polynomial function** is the set of all real numbers.

- The domain of an **absolute value function** is the set of all real numbers for which the expression inside the absolute value bars (the argument) is defined.

- If a **function is defined by a rational expression,** the domain is the set of all real numbers for which the denominator is not zero.

- The domain of a **function defined by a radical with *even* root index** is the set of all real numbers that make the radicand greater than or equal to zero. **If the root index is** *odd,* the domain is the set of all real numbers for which the radicand is itself a real number.

- For an **exponential function** with constant base, the domain is the set of all real numbers for which the exponent is a real number.

- For a **logarithmic function,** the domain is the set of all real numbers that make the argument of the logarithm greater than zero.

Determining Whether an Equation Defines *y* as a Function of *x* *For y to be a function of x, it is necessary that every input value of x in the domain leads to one and only one value of y.* To determine whether an equation such as

$$x - y^3 = 0 \quad \text{or} \quad x - y^2 = 0$$

represents a function, solve the equation for *y*. In the first equation above, doing so leads to

$$y = \sqrt[3]{x}.$$

Notice that every value of *x* in the domain (that is, all real numbers) leads to one and only one value of *y*. So in the first equation, we can write *y* as a function of *x*. However, in the second equation above, solving for *y* leads to

$$y = \pm \sqrt{x}.$$

If we let *x* = 4, for example, we get two values of *y*: −2 and 2. Thus, in the second equation, we cannot write *y* as a function of *x*.

Find the domain of each function. Write answers using interval notation.

1. $f(x) = 3x - 6$
2. $f(x) = \sqrt{2x - 7}$
3. $f(x) = |x + 4|$

4. $f(x) = \dfrac{x + 2}{x - 6}$
5. $f(x) = \dfrac{-2}{x^2 + 7}$
6. $f(x) = \sqrt{x^2 - 9}$

7. $f(x) = \dfrac{x^2 + 7}{x^2 - 9}$
8. $f(x) = \sqrt[3]{x^3 + 7x - 4}$
9. $f(x) = \log_5(16 - x^2)$

10. $f(x) = \log \dfrac{x + 7}{x - 3}$
11. $f(x) = \sqrt{x^2 - 7x - 8}$
12. $f(x) = 2^{1/x}$

13. $f(x) = \dfrac{1}{2x^2 - x + 7}$
14. $f(x) = \dfrac{x^2 - 25}{x + 5}$
15. $f(x) = \sqrt{x^3 - 1}$

16. $f(x) = \ln |x^2 - 5|$
17. $f(x) = e^{x^2 + x + 4}$
18. $f(x) = \dfrac{x^3 - 1}{x^2 - 1}$

19. $f(x) = \sqrt{\dfrac{-1}{x^3 - 1}}$

20. $f(x) = \sqrt[3]{\dfrac{1}{x^3 - 8}}$

21. $f(x) = \ln(x^2 + 1)$

22. $f(x) = \sqrt{(x - 3)(x + 2)(x - 4)}$

23. $f(x) = \log\left(\dfrac{x + 2}{x - 3}\right)^2$

24. $f(x) = \sqrt[12]{(4 - x)^2(x + 3)}$

25. $f(x) = e^{|1/x|}$

26. $f(x) = \dfrac{1}{|x^2 - 7|}$

27. $f(x) = x^{100} - x^{50} + x^2 + 5$

28. $f(x) = \sqrt{-x^2 - 9}$

29. $f(x) = \sqrt[4]{16 - x^4}$

30. $f(x) = \sqrt[3]{16 - x^4}$

31. $f(x) = \sqrt{\dfrac{x^2 - 2x - 63}{x^2 + x - 12}}$

32. $f(x) = \sqrt[5]{5 - x}$

33. $f(x) = |\sqrt{5 - x}|$

34. $f(x) = \sqrt{\dfrac{-1}{x - 3}}$

35. $f(x) = \log\left|\dfrac{1}{4 - x}\right|$

36. $f(x) = 6^{x^2 - 9}$

37. $f(x) = 6^{\sqrt{x^2 - 25}}$

38. $f(x) = 6^{\sqrt[3]{x^2 - 25}}$

39. $f(x) = \ln\left(\dfrac{-3}{(x + 2)(x - 6)}\right)$

40. $f(x) = \dfrac{-2}{\log x}$

Determine which one of the choices (A, B, C, or D) is an equation in which y can be written as a function of x.

41. A. $3x + 2y = 6$ **B.** $x = \sqrt{|y|}$ **C.** $x = |y + 3|$ **D.** $x^2 + y^2 = 9$

42. A. $3x^2 + 2y^2 = 36$ **B.** $x^2 + y - 2 = 0$ **C.** $x - |y| = 0$ **D.** $x = y^2 - 4$

43. A. $x = \sqrt{y^2}$ **B.** $x = \log y^2$ **C.** $x^3 + y^3 = 5$ **D.** $x = \dfrac{1}{y^2 + 3}$

44. A. $\dfrac{x^2}{4} + \dfrac{y^2}{4} = 1$ **B.** $x = 5y^2 - 3$ **C.** $\dfrac{x^2}{4} - \dfrac{y^2}{9} = 1$ **D.** $x = 10^y$

45. A. $x = \dfrac{2 - y}{y + 3}$ **B.** $x = \ln(y + 1)^2$ **C.** $\sqrt{x} = |y + 1|$ **D.** $\sqrt[4]{x} = y^2$

46. A. $e^{y^2} = x$ **B.** $e^{y + 2} = x$ **C.** $e^{|y|} = x$ **D.** $10^{|y + 2|} = x$

47. A. $x^2 = \dfrac{1}{y^2}$ **B.** $x + 2 = \dfrac{1}{y^2}$ **C.** $3x = \dfrac{1}{y^4}$ **D.** $2x = \dfrac{1}{y^3}$

48. A. $|x| = |y|$ **B.** $x = |y^2|$ **C.** $x = \dfrac{1}{y}$ **D.** $x^4 + y^4 = 81$

49. A. $\dfrac{x^2}{4} - \dfrac{y^2}{9} = 1$ **B.** $\dfrac{y^2}{4} - \dfrac{x^2}{9} = 1$ **C.** $\dfrac{x}{4} - \dfrac{y}{9} = 0$ **D.** $\dfrac{x^2}{4} - \dfrac{y^2}{9} = 0$

50. A. $y^2 - \sqrt{(x + 2)^2} = 0$ **B.** $y - \sqrt{(x + 2)^2} = 0$

    C. $y^6 - \sqrt{(x + 1)^2} = 0$ **D.** $y^4 - \sqrt{x^2} = 0$

Chapter 4 Test Prep

Key Terms

4.1 one-to-one function inverse function **4.2** exponential function exponential equation compound interest	future value present value compound amount continuous compounding	**4.3** logarithm base argument logarithmic equation logarithmic function	**4.4** common logarithm pH natural logarithm **4.6** doubling time half-life

New Symbols

$f^{-1}(x)$	the inverse of $f(x)$
e	a constant, approximately 2.718281828459045
$\log_a x$	the logarithm of x with the base a

$\log x$	common (base 10) logarithm of x
$\ln x$	natural (base e) logarithm of x

Quick Review

Concepts	Examples

4.1 Inverse Functions

One-to-One Function

In a one-to-one function, each x-value corresponds to only one y-value, and each y-value corresponds to only one x-value.

A function f is one-to-one if, for elements a and b in the domain of f, $a \neq b$ implies $f(a) \neq f(b)$.

The function $y = x^2$ is not one-to-one, because $y = 16$, for example, corresponds to both $x = 4$ and $x = -4$.

Horizontal Line Test

A function is one-to-one if every horizontal line intersects the graph of the function at most once.

The graph of $f(x) = 2x - 1$ is a straight line with slope 2. f is a one-to-one function by the horizontal line test.

Inverse Functions

Let f be a one-to-one function. Then g is the inverse function of f if

$$(f \circ g)(x) = x \quad \text{for every } x \text{ in the domain of } g$$

and

$$(g \circ f)(x) = x \quad \text{for every } x \text{ in the domain of } f.$$

To find $g(x)$, interchange x and y in $y = f(x)$, solve for y, and replace y with $g(x)$, which is $f^{-1}(x)$.

Find the inverse of $y = f(x) = 2x - 1$.

$$x = 2y - 1 \qquad \text{Interchange } x \text{ and } y.$$

$$y = \frac{x + 1}{2} \qquad \text{Solve for } y.$$

$$f^{-1}(x) = \frac{x + 1}{2} \qquad \text{Replace } y \text{ with } f^{-1}(x).$$

$$f^{-1}(x) = \frac{1}{2}x + \frac{1}{2} \qquad \frac{x+1}{2} = \frac{x}{2} + \frac{1}{2} = \frac{1}{2}x + \frac{1}{2}$$

(continued)

Concepts	Examples

4.2 Exponential Functions

Additional Properties of Exponents

For any real number $a > 0$, $a \neq 1$, the following are true.

(a) a^x is a unique real number for all real numbers x.
(b) $a^b = a^c$ if and only if $b = c$.
(c) If $a > 1$ and $m < n$, then $a^m < a^n$.
(d) If $0 < a < 1$ and $m < n$, then $a^m > a^n$.

(a) 2^x is a unique real number for all real numbers x.
(b) $2^x = 2^3$ if and only if $x = 3$.
(c) $2^5 < 2^{10}$, because $2 > 1$ and $5 < 10$.
(d) $\left(\frac{1}{2}\right)^5 > \left(\frac{1}{2}\right)^{10}$ because $0 < \frac{1}{2} < 1$ and $5 < 10$.

Exponential Function

If $a > 0$ and $a \neq 1$, then $\boldsymbol{f(x) = a^x}$ defines the exponential function with base a.

$f(x) = 3^x$ is the exponential function with base 3.

Graph of $\boldsymbol{f(x) = a^x}$

1. The graph contains the points $\left(-1, \frac{1}{a}\right)$, $(0, 1)$, and $(1, a)$.
2. If $a > 1$, then f is an increasing function.
   If $0 < a < 1$, then f is a decreasing function.
3. The x-axis is a horizontal asymptote.
4. The domain is $(-\infty, \infty)$, and the range is $(0, \infty)$.

x	y
-1	$\frac{1}{3}$
0	1
1	3

4.3 Logarithmic Functions

Logarithmic Function

If $a > 0$, $a \neq 1$, and $x > 0$, then $\boldsymbol{f(x) = \log_a x}$ defines the logarithmic function with base a.

$f(x) = \log_3 x$ is the logarithmic function with base 3.

Graph of $\boldsymbol{f(x) = \log_a x}$

1. The graph contains the points $\left(\frac{1}{a}, -1\right)$, $(1, 0)$, and $(a, 1)$.
2. If $a > 1$, then f is an increasing function.
   If $0 < a < 1$, then f is a decreasing function.
3. The y-axis is a vertical asymptote.
4. The domain is $(0, \infty)$, and the range is $(-\infty, \infty)$.

x	y
$\frac{1}{3}$	-1
1	0
3	1

Properties of Logarithms

For $x > 0$, $y > 0$, $a > 0$, $a \neq 1$, and any real number r, the following properties hold.

$\log_a xy = \log_a x + \log_a y$ Product property

$\log_a \dfrac{x}{y} = \log_a x - \log_a y$ Quotient property

$\log_a x^r = r \log_a x$ Power property

$\log_a 1 = 0$ Logarithm of 1

$\log_a a = 1$ Base a logarithm of a

$\log_2(3 \cdot 5) = \log_2 3 + \log_2 5$

$\log_2 \dfrac{3}{5} = \log_2 3 - \log_2 5$

$\log_6 3^5 = 5 \log_6 3$

$\log_{10} 1 = 0$

$\log_{10} 10 = 1$

Theorem on Inverses

For $a > 0$ and $a \neq 1$, the following properties hold.

$$a^{\log_a x} = x \ (x > 0) \quad \text{and} \quad \log_a a^x = x$$

$2^{\log_2 5} = 5 \quad \text{and} \quad \log_2 2^5 = 5$

Concepts	Examples

4.4 Evaluating Logarithms and the Change-of-Base Theorem

Common and Natural Logarithms

For all positive numbers x, base 10 logarithms and base e logarithms are written as follows.

$$\log x = \log_{10} x \quad \text{Common logarithm}$$

$$\ln x = \log_e x \quad \text{Natural logarithm}$$

Approximate $\log 0.045$ and $\ln 247.1$.

$$\log 0.045 \approx -1.3468$$
$$\ln 247.1 \approx 5.5098$$
Use a calculator.

Change-of-Base Theorem

For any positive real numbers x, a, and b, where $a \neq 1$ and $b \neq 1$, the following holds.

$$\log_a x = \frac{\log_b x}{\log_b a}$$

Approximate $\log_8 7$.

$$\log_8 7 = \frac{\log 7}{\log 8} = \frac{\ln 7}{\ln 8} \approx 0.9358 \quad \text{Use a calculator.}$$

4.5 Exponential and Logarithmic Equations

Property of Logarithms

If $x > 0$, $y > 0$, $a > 0$, and $a \neq 1$, then the following holds.

$$x = y \quad \text{is equivalent to} \quad \log_a x = \log_a y.$$

Solve $e^{5x} = 10$.

$$\ln e^{5x} = \ln 10 \qquad \text{Take logarithms.}$$
$$5x = \ln 10 \qquad \ln e^x = x, \text{ for all } x.$$
$$x = \frac{\ln 10}{5} \approx 0.461 \qquad \text{Use a calculator.}$$

The solution set can be written with the exact value, $\left\{\frac{\ln 10}{5}\right\}$, or with the approximate value, $\{0.461\}$.

Solve $\log_2(x^2 - 3) = \log_2 6$.

$$x^2 - 3 = 6 \qquad \text{Property of logarithms}$$
$$x^2 = 9 \qquad \text{Add 3.}$$
$$x = \pm 3 \qquad \text{Take square roots.}$$

Both values check, so the solution set is $\{\pm 3\}$.

4.6 Applications and Models of Exponential Growth and Decay

Exponential Growth or Decay Function

The exponential growth or decay function is

$$y = y_0 e^{kt},$$

for constant k, where y_0 is the amount or number present at time $t = 0$.

The formula for continuous compounding,

$$A = Pe^{rt},$$

is an example of exponential growth. Here, A is the compound amount if P is invested at an annual interest rate r for t years.

If $P = \$200$, $r = 3\%$, and $t = 5$ yr, find A.

$$A = 200e^{0.03(5)} \qquad \text{Substitute.}$$
$$A \approx \$232.37 \qquad \text{Use a calculator.}$$

| Chapter 4 | Review Exercises |

Decide whether each function as defined or graphed is one-to-one.

1.

2.

3. $y = 5x - 4$

4. $y = x^3 + 1$

5. $y = (x + 3)^2$

6. $y = \sqrt{3x^2 + 2}$

Write an equation, if possible, for each inverse function in the form $y = f^{-1}(x)$.

7. $f(x) = x^3 - 3$

8. $f(x) = \sqrt{25 - x^2}$

9. *Concept Check* Suppose $f(t)$ is the amount an investment will grow to t years after 2004. What does $f^{-1}(\$50,000)$ represent?

10. *Concept Check* The graphs of two functions are shown. Based on their graphs, are these functions inverses?

11. *Concept Check* To have an inverse, a function must be a(n) _____ function.

12. *Concept Check* Is it true that the x-coordinate of the x-intercept of the graph of $y = f(x)$ is the y-coordinate of the y-intercept of the graph of $y = f^{-1}(x)$?

Match each equation with the figure that most closely resembles its graph.

13. $y = \log_{0.3} x$

14. $y = e^x$

15. $y = \ln x$

16. $y = 0.3^x$

A.

B.

C.

D.

Write each equation in logarithmic form.

17. $2^5 = 32$

18. $100^{1/2} = 10$

19. $\left(\dfrac{3}{4}\right)^{-1} = \dfrac{4}{3}$

20. Graph $y = (1.5)^{x+2}$.

Write each equation in exponential form.

21. $\log 1000 = 3$

22. $\log_9 27 = \dfrac{3}{2}$

23. $\ln \sqrt{e} = \dfrac{1}{2}$

24. *Concept Check* What is the base of the logarithmic function whose graph contains the point $(81, 4)$?

25. *Concept Check* What is the base of the exponential function whose graph contains the point $\left(-4, \frac{1}{16}\right)$?

Use properties of logarithms to rewrite each expression. Simplify the result, if possible. Assume all variables represent positive numbers.

26. $\log_5\left(x^2 y^4 \sqrt[5]{m^3 p}\right)$　　**27.** $\log_3 \dfrac{mn}{5r}$　　**28.** $\log_7(7k + 5r^2)$

Find each logarithm. Round to four decimal places.

29. $\log 0.0411$　　　　**30.** $\log 45.6$　　　　**31.** $\ln 144{,}000$

32. $\ln 470$　　　　**33.** $\log_{2/3} \dfrac{5}{8}$　　　　**34.** $\log_3 769$

Solve each exponential equation. Unless otherwise specified, express irrational solutions as decimals correct to the nearest thousandth.

35. $16^{x+4} = 8^{3x-2}$　　**36.** $4^x = 12$　　　**37.** $3^{2x-5} = 13$

38. $2^{x+3} = 5^x$　　　**39.** $6^{x+3} = 4^x$　　**40.** $e^{x-1} = 4$

41. $e^{2-x} = 12$　　　**42.** $2e^{5x+2} = 8$　　**43.** $10e^{3x-7} = 5$

44. $5^{x+2} = 2^{2x-1}$　　**45.** $6^{x-3} = 3^{4x+1}$　　**46.** $e^{8x} \cdot e^{2x} = e^{20}$

47. $e^{6x} \cdot e^x = e^{21}$　　**48.** $100(1.02)^{x/4} = 200$　　**49.** $2e^{2x} - 5e^x - 3 = 0$
　　　　　　　　　　　　　　　　　　　　　　　(Give exact form.)

50. $\left(\dfrac{1}{2}\right)^x + 2 = 0$　　　　　　　**51.** $4(1.06)^x + 2 = 8$

52. *Concept Check* Which one or more of the following choices is the solution set of $5^x = 9$?

　　A. $\{\log_5 9\}$　　　**B.** $\{\log_9 5\}$　　　**C.** $\left\{\dfrac{\log 9}{\log 5}\right\}$　　　**D.** $\left\{\dfrac{\ln 9}{\ln 5}\right\}$

Solve each logarithmic equation. Express all solutions in exact form.

53. $3 \ln x = 13$　　　　　　　　**54.** $\ln(5x) = 16$

55. $\log(2x + 7) = 0.25$　　　　　**56.** $\ln x + \ln x^3 = 12$

57. $\log_2(x^3 + 5) = 5$　　　　　**58.** $\log_3(x^2 - 9) = 3$

59. $\log_4\left[(3x + 1)(x - 4)\right] = 2$　　**60.** $\ln e^{\ln x} - \ln(x - 4) = \ln 3$

61. $\log x + \log(13 - 3x) = 1$　　　**62.** $\log_7(3x + 2) - \log_7(x - 2) = 1$

63. $\ln(6x) - \ln(x + 1) = \ln 4$　　　**64.** $\log_{16} \sqrt{x + 1} = \dfrac{1}{4}$

65. $\ln\left[\ln(e^{-x})\right] = \ln 3$　　　　**66.** $S = a \ln\left(1 + \dfrac{n}{a}\right)$,　　for n

67. $d = 10 \log\left(\dfrac{I}{I_0}\right)$,　　for I_0　　**68.** $D = 200 + 100 \log x$,　　for x

69. Use a graphing calculator to solve the equation $e^x = 4 - \ln x$. Give solution(s) to the nearest thousandth.

Solve each problem.

70. *Earthquake Intensity* On September 30, 2009, the region of Indonesia known as Sumatra was shaken by an earthquake that measured 7.5 on the Richter scale.

　　(a) Express this reading in terms of I_0. (See **Section 4.4** Exercises.)

　　(b) On May 12, 2008, a quake measuring 7.9 on the Richter scale killed about 88,000 people in East Sichuan Province, China. Express the magnitude of a 7.9 reading in terms of I_0.

　　(c) How much greater was the force of the earthquake that measured 7.9?

71. *Earthquake Intensity*

    (a) The San Francisco earthquake of 1906 had a Richter scale rating of 8.3. Express the magnitude of this earthquake as a multiple of I_0.

    (b) In 1989, the San Francisco region experienced an earthquake with a Richter scale rating of 7.1. Express the magnitude of this earthquake as a multiple of I_0.

    (c) Compare the magnitudes of the two San Francisco earthquakes discussed in parts (a) and (b).

72. *(Modeling) Decibel Levels* In **Section 4.4** we found that the model for the decibel rating of the loudness of a sound is

$$d = 10 \log \frac{I}{I_0}.$$

A few years ago, there was a controversy about a proposed government limit on factory noise. One group wanted a maximum of 89 decibels, while another group wanted 86. This difference seemed very small to many people. Find the percent by which the 89-decibel intensity exceeds that for 86 decibels.

73. *Interest Rate* What annual interest rate, to the nearest tenth, will produce $5760 if $3500 is left at interest compounded annually for 10 yr?

74. *Growth of an Account* Find the number of years (to the nearest tenth) needed for $48,000 to become $58,344 at 5% interest compounded semiannually.

75. *Growth of an Account* Manuel deposits $10,000 for 12 yr in an account paying 4% compounded annually. He then puts this total amount on deposit in another account paying 5% compounded semiannually for another 9 yr. Find the total amount on deposit after the entire 21-yr period.

76. *Growth of an Account* Anne Kelly deposits $12,000 for 8 yr in an account paying 5% compounded annually. She then leaves the money alone with no further deposits at 6% compounded annually for an additional 6 yr. Find the total amount on deposit after the entire 14-yr period.

77. *Cost from Inflation* Suppose the inflation rate is 4%. Use the formula for continuous compounding to find the number of years, to the nearest tenth, for a $1 item to cost $2.

78. *(Modeling) Drug Level in the Bloodstream* After a medical drug is injected directly into the bloodstream, it is gradually eliminated from the body. Graph the following functions on the interval $[0, 10]$. Use $[0, 500]$ for the range of $A(t)$. Determine the function that best models the amount $A(t)$ (in milligrams) of a drug remaining in the body after t hours if 350 mg were initially injected.

    (a) $A(t) = t^2 - t + 350$ **(b)** $A(t) = 350 \log(t + 1)$

    (c) $A(t) = 350(0.75)^t$ **(d)** $A(t) = 100(0.95)^t$

79. *(Modeling) Chicago Cubs' Payroll* The table shows the total payroll (in millions of dollars) of the Chicago Cubs baseball team for the years 2007–2010.

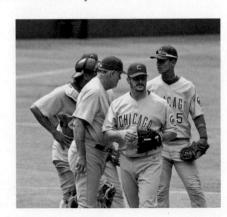

Year	Total Payroll (millions of dollars)
2007	115.9
2008	118.3
2009	134.8
2010	146.6

Source: www.baseball.about.com;
www.espn.go.com

Letting y represent the total payroll and x represent the number of years since 2007, we find that the function

$$f(x) = 113.2e^{0.0836x}$$

models the data quite well. According to this function, when would the total payroll double its 2010 value?

80. *(Modeling) Transistors on Computer Chips* Computing power of personal computers has increased dramatically as a result of the ability to place an increasing number of transistors on a single processor chip. The table lists the number of transistors on some popular computer chips made by Intel.

Year	Chip	Transistors
1985	386DX	275,000
1989	486DX	1,200,000
1994	Pentium	3,300,000
1999	Pentium 3	9,500,000
2000	Pentium 4	42,000,000
2006	Core 2 Duo	291,000,000
2008	Core 2 Quad	820,000,000

Source: Intel.

(a) Make a scatter diagram of the data. Let the x-axis represent the year, where $x = 0$ corresponds to 1985, and let the y-axis represent the number of transistors.

(b) Decide whether a linear, a logarithmic, or an exponential function describes the data best.

(c) Determine a function f that approximates these data. Plot f and the data on the same coordinate axes.

(d) Assuming that this trend continues, use f to predict the number of transistors on a chip in the year 2012.

81. *Financial Planning* The traditional IRA (individual retirement account) is a common tax-deferred saving plan in the United States. Earned income deposited into an IRA is not taxed in the current year, and no taxes are incurred on the interest paid in subsequent years. However, when you withdraw the money from the account after age $59\frac{1}{2}$, you pay taxes on the entire amount you withdraw.

Suppose you deposited $5000 of earned income into an IRA, you can earn an annual interest rate of 4%, and you are in a 25% tax bracket. (*Note:* Interest rates and tax brackets are subject to change over time, but some assumptions must be made to evaluate the investment.) Also, suppose that you deposit the $5000 at age 25 and withdraw it at age 60, and that interest is compounded continuously.

(a) How much money will remain after you pay the taxes at age 60?

(b) Suppose that instead of depositing the money into an IRA, you pay taxes on the money and the annual interest. How much money will you have at age 60? (*Note:* You effectively start with $3750 (75% of $5000), and the money earns 3% (75% of 4%) interest after taxes.)

(c) To the nearest dollar, how much additional money will you earn with the IRA?

(d) Suppose you pay taxes on the original $5000 but are then able to earn 4% in a tax-free investment. Compare your balance at age 60 with the IRA balance.

82. Consider $f(x) = \log_4(2x^2 - x)$.

(a) Use the change-of-base theorem with base e to write $\log_4(2x^2 - x)$ in a suitable form to graph with a calculator.

(b) Graph the function using a graphing calculator. Use the window $[-2.5, 2.5]$ by $[-5, 2.5]$.

(c) What are the x-intercepts?

(d) Give the equations of the vertical asymptotes.

(e) Explain why there is no y-intercept.

Chapter 4 Test

1. Consider the function $f(x) = \sqrt[3]{2x - 7}$.
   (a) What are the domain and range of f?
   (b) Explain why f^{-1} exists.
   (c) Find the rule for $f^{-1}(x)$.
   (d) What are the domain and range of f^{-1}?
   (e) Graph both f and f^{-1}. How are the two graphs related to the line $y = x$?

2. Match each equation with its graph.

   (a) $y = \log_{1/3} x$ (b) $y = e^x$ (c) $y = \ln x$ (d) $y = \left(\dfrac{1}{3}\right)^x$

   A.   **B.**   **C.**   **D.**

3. Solve $\left(\dfrac{1}{8}\right)^{2x-3} = 16^{x+1}$.

4. (a) Write $4^{3/2} = 8$ in logarithmic form.

   (b) Write $\log_8 4 = \dfrac{2}{3}$ in exponential form.

5. Graph $f(x) = \left(\dfrac{1}{2}\right)^x$ and $g(x) = \log_{1/2} x$ on the same axes. What is their relationship?

6. Use properties of logarithms to write

$$\log_7 \frac{x^2 \sqrt[4]{y}}{z^3}$$

   as a sum, difference, or product of logarithms. Assume all variables represent positive numbers.

Use a calculator to find an approximation for each logarithm. Express answers to four decimal places.

7. $\log 2388$ 8. $\ln 2388$ 9. $\log_9 13$

10. Solve the following.

$$x^{2/3} = 25$$

Solve each exponential equation. If applicable, express solutions to the nearest thousandth.

11. $12^x = 1$ 12. $9^x = 4$ 13. $16^{2x+1} = 8^{3x}$

14. $2^{x+1} = 3^{x-4}$ 15. $e^{0.4x} = 4^{x-2}$

16. $2e^{2x} - 5e^x + 3 = 0$ (Give both exact and approximate values.)

Solve each logarithmic equation. Express all solutions in exact form.

17. $\log_x \dfrac{9}{16} = 2$ 18. $\log_2(x - 4)(x - 2) = 3$

19. $\log_2 x + \log_2(x + 2) = 3$ 20. $\ln x - 4 \ln 3 = \ln\left(\dfrac{1}{5}x\right)$

21. $\log_3(x + 1) - \log_3(x - 3) = 2$

22. One of your friends is taking another mathematics course and tells you, "I have no idea what an expression like $\log_5 27$ really means." Write an explanation of what it means, and tell how you can find an approximation for it with a calculator.

Solve each problem.

23. *(Modeling) Skydiver Fall Speed* A skydiver in free fall travels at a speed modeled by

$$v(t) = 176(1 - e^{-0.18t})$$

feet per second after t seconds. How long will it take for the skydiver to attain a speed of 147 ft per sec (100 mph)?

24. *Growth of an Account* How many years, to the nearest tenth, will be needed for $5000 to increase to $18,000 at 3.8% annual interest compounded **(a)** monthly **(b)** continuously?

25. *Tripling Time* For *any* amount of money invested at 3.8% annual interest compounded continuously, how long will it take to triple?

26. *(Modeling) Radioactive Decay* The amount of radioactive material, in grams, present after t days is modeled by

$$A(t) = 600e^{-0.05t}.$$

(a) Find the amount present after 12 days.

(b) Find the half-life of the material.

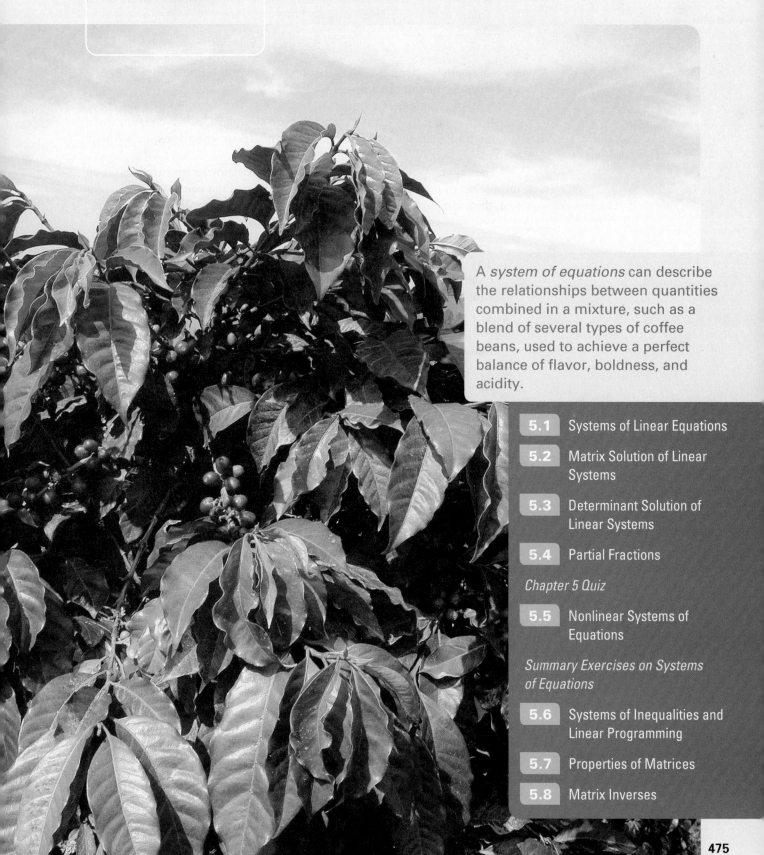

5 Systems and Matrices

A *system of equations* can describe the relationships between quantities combined in a mixture, such as a blend of several types of coffee beans, used to achieve a perfect balance of flavor, boldness, and acidity.

5.1 Systems of Linear Equations

5.2 Matrix Solution of Linear Systems

5.3 Determinant Solution of Linear Systems

5.4 Partial Fractions

Chapter 5 Quiz

5.5 Nonlinear Systems of Equations

Summary Exercises on Systems of Equations

5.6 Systems of Inequalities and Linear Programming

5.7 Properties of Matrices

5.8 Matrix Inverses

5.1 Systems of Linear Equations

■ Linear Systems
■ Substitution Method
■ Elimination Method
■ Special Systems
■ Applying Systems of Equations
■ Solving Linear Systems with Three Unknowns (Variables)
■ Using Systems of Equations to Model Data

Linear Systems   The definition of a linear equation given in **Chapter 1** can be extended to more variables. Any equation of the form

$$a_1 x_1 + a_2 x_2 + \cdots + a_n x_n = b,$$

for real numbers $a_1, a_2, \ldots, a_n$ (all nonzero) and b, is a **linear equation,** or a **first-degree equation, in** n **unknowns.**

A set of equations is a **system of equations.** The **solutions** of a system of equations must satisfy every equation in the system. If all the equations in a system are linear, the system is a **system of linear equations,** or a **linear system.**

The solution set of a linear equation in two unknowns (or variables) is an infinite set of ordered pairs. Since the graph of such an equation is a straight line, there are three possibilities for the number of elements in the solution set of a system of two linear equations in two unknowns, as shown in **Figure 1.** The possible graphs of a linear system in two unknowns are as follows.

1. **The graphs intersect at exactly one point,** which gives the (single) ordered-pair solution of the system. The **system is consistent** and the **equations are independent.** See **Figure 1(a).**

2. **The graphs are parallel lines,** so there is no solution and the solution set is $\emptyset$. The **system is inconsistent** and the **equations are independent.** See **Figure 1(b).**

3. **The graphs are the same line,** and there are an infinite number of solutions. The **system is consistent** and the **equations are dependent.** See **Figure 1(c).**

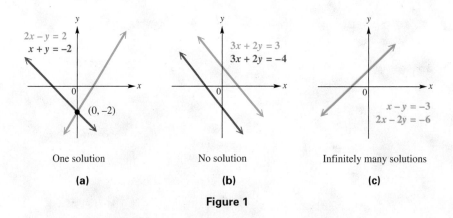

One solution
(a)

No solution
(b)

Infinitely many solutions
(c)

Figure 1

Using graphs to find the solution set of a linear system in two unknowns provides a good visual perspective, but may be inefficient when the solution set contains non-integer values. Thus, we introduce two algebraic methods for solving systems: *substitution* and *elimination*.

Substitution Method   In a system of two equations with two variables, the **substitution method** involves using one equation to find an expression for one variable in terms of the other, and then substituting into the other equation of the system.

EXAMPLE 1 Solving a System by Substitution

Solve the system.

$$3x + 2y = 11 \quad (1)$$
$$-x + y = 3 \quad (2)$$

SOLUTION Begin by solving one of the equations for one of the variables. We solve equation (2) for y.

$$-x + y = 3 \qquad (2)$$
$$y = x + 3 \quad \text{Add } x. \text{ (Section 1.1)} \quad (3)$$

Now replace y with $x + 3$ in equation (1), and solve for x.

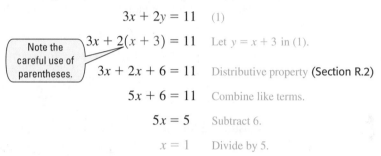

$$3x + 2y = 11 \quad (1)$$
$$3x + 2(x + 3) = 11 \quad \text{Let } y = x + 3 \text{ in (1).}$$

Note the careful use of parentheses.

$$3x + 2x + 6 = 11 \quad \text{Distributive property (Section R.2)}$$
$$5x + 6 = 11 \quad \text{Combine like terms.}$$
$$5x = 5 \quad \text{Subtract 6.}$$
$$x = 1 \quad \text{Divide by 5.}$$

Replace x with 1 in equation (3) to obtain

$$y = x + 3 = 1 + 3 = 4.$$

The solution of the system is the ordered pair $(1, 4)$. ***Check this solution in both equations (1) and (2).***

CHECK

$$3x + 2y = 11 \quad (1) \qquad\qquad -x + y = 3 \quad (2)$$
$$3(1) + 2(4) \stackrel{?}{=} 11 \qquad\qquad -1 + 4 \stackrel{?}{=} 3$$
$$11 = 11 \checkmark \text{ True} \qquad\qquad 3 = 3 \checkmark \text{ True}$$

True statements result when the solution is substituted in both equations, confirming that the solution set is $\{(1, 4)\}$.

☑ *Now Try Exercise 7.*

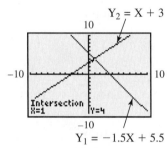

$Y_2 = X + 3$

$Y_1 = -1.5X + 5.5$

To solve the system in **Example 1** graphically, solve both equations for y:

$3x + 2y = 11$ leads to

$$Y_1 = -1.5X + 5.5.$$

$-x + y = 3$ leads to

$$Y_2 = X + 3.$$

Graph both Y_1 and Y_2 in the standard window to find that their point of intersection is $(1, 4)$.

Elimination Method Another way to solve a system of two equations, called the **elimination method,** uses multiplication and addition to eliminate a variable from one equation. To eliminate a variable, the coefficients of that variable in the two equations must be additive inverses. To achieve this, we use properties of algebra to change the system to an **equivalent system,** one with the same solution set.

The three transformations that produce an equivalent system are listed here.

Transformations of a Linear System

1. Interchange any two equations of the system.

2. Multiply or divide any equation of the system by a nonzero real number.

3. Replace any equation of the system by the sum of that equation and a multiple of another equation in the system.

EXAMPLE 2 Solving a System by Elimination

Solve the system.

$$3x - 4y = 1 \quad (1)$$
$$2x + 3y = 12 \quad (2)$$

SOLUTION One way to eliminate a variable is to use the second transformation and multiply both sides of equation (2) by -3, to get an equivalent system.

$$3x - 4y = 1 \quad (1)$$
$$-6x - 9y = -36 \quad \text{Multiply (2) by } -3. \quad (3)$$

Now multiply both sides of equation (1) by 2, and use the third transformation to add the result to equation (3), eliminating x. Solve the result for y.

$$6x - 8y = 2 \qquad \text{Multiply (1) by 2.}$$
$$\underline{-6x - 9y = -36} \quad (3)$$
$$-17y = -34 \qquad \text{Add.}$$
$$y = 2 \qquad \text{Solve for } y.$$

Substitute 2 for y in either of the original equations and solve for x.

$$3x - 4y = 1 \quad (1)$$
$$3x - 4(2) = 1 \quad \text{Let } y = 2 \text{ in (1).}$$
$$3x - 8 = 1 \quad \text{Multiply.}$$
$$3x = 9 \quad \text{Add 8.}$$
$$x = 3 \quad \text{Divide by 3.}$$

> Write the *x*-value first.

A check shows that $(3, 2)$ satisfies both equations (1) and (2). Therefore, the solution set is $\{(3, 2)\}$. The graph in **Figure 2** confirms this.

✔ *Now Try Exercise 21.*

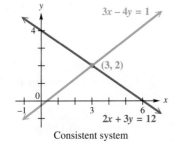

Consistent system

Figure 2

Special Systems The systems in **Examples 1 and 2** were both consistent, having a single solution. This is not always the case.

EXAMPLE 3 Solving an Inconsistent System

Solve the system.

$$3x - 2y = 4 \quad (1)$$
$$-6x + 4y = 7 \quad (2)$$

SOLUTION To eliminate the variable x, multiply both sides of equation (1) by 2, and add the result to equation (2).

$$6x - 4y = 8 \quad \text{Multiply (1) by 2.}$$
$$\underline{-6x + 4y = 7} \quad (2)$$
$$0 = 15 \quad \text{False}$$

Since $0 = 15$ is false, the system is inconsistent and has no solution. As suggested by **Figure 3,** this means that the graphs of the equations of the system never intersect. (The lines are parallel.) The solution set is $\varnothing$.

Inconsistent system

Figure 3

✔ *Now Try Exercise 31.*

EXAMPLE 4 Solving a System with Infinitely Many Solutions

Solve the system.

$$8x - 2y = -4 \quad (1)$$
$$-4x + y = 2 \quad (2)$$

ALGEBRAIC SOLUTION

Divide both sides of equation (1) by 2, and add the result to equation (2).

$$4x - y = -2 \quad \text{Divide (1) by 2.}$$
$$\underline{-4x + y = 2} \quad (2)$$
$$0 = 0 \quad \text{True}$$

The result, $0 = 0$, is a true statement, which indicates that the equations are equivalent. Any ordered pair (x, y) that satisfies either equation will satisfy the system. Solve for y in equation (2).

$$-4x + y = 2 \quad (2)$$
$$y = 4x + 2$$

The solutions of the system can be written in the form of a set of ordered pairs $(x, 4x + 2)$, for any real number x. Some ordered pairs in the solution set are $(0, 4 \cdot 0 + 2)$, or $(0, 2)$, and $(1, 4 \cdot 1 + 2)$, or $(1, 6)$, as well as $(3, 14)$, and $(-2, -6)$.

As shown in **Figure 4,** the equations of the original system are dependent and lead to the same straight-line graph. Using this method, the solution set can be written $\{(x, 4x + 2)\}$.

Infinitely many solutions

Figure 4

GRAPHING CALCULATOR SOLUTION

Solving the equations for y gives

$$Y_1 = 4X + 2$$

and

$$Y_2 = 4X + 2.$$

When written in this form, we can immediately determine that the equations are identical. Each has slope 4 and y-intercept 2.

As expected, the graphs coincide. See the top screen in **Figure 5.** The table indicates that

$$Y_1 = Y_2 \text{ for selected values of X,}$$

providing another way to show that the two equations lead to the same graph.

Figure 5

Refer to the algebraic solution to see how the solution set can be written using an arbitrary variable.

✔ *Now Try Exercise 33.*

NOTE In the algebraic solution for **Example 4,** we wrote the solution set with the variable x arbitrary. We could write the solution set with y arbitrary.

$$\left\{ \left(\frac{y - 2}{4}, y \right) \right\}$$

By selecting values for y and solving for x in this ordered pair, we can find individual solutions. Verify again that $(0, 2)$ is a solution by letting $y = 2$ and solving for x to obtain $\frac{2 - 2}{4} = 0$.

Applying Systems of Equations   Many applied problems involve more than one unknown quantity. Although some problems with two unknowns can be solved using just one variable, it is often easier to use two variables.

To solve a problem with two unknowns, we must write two equations that relate the unknown quantities. The system formed by the pair of equations can then be solved using the methods of this chapter. The following steps, based on the six-step problem-solving method first introduced in **Section 1.2,** give a strategy for solving such applied problems.

Solving an Applied Problem by Writing a System of Equations

Step 1   **Read** the problem carefully until you understand what is given and what is to be found.

Step 2   **Assign variables** to represent the unknown values, using diagrams or tables as needed. *Write down* what each variable represents.

Step 3   **Write a system of equations** that relates the unknowns.

Step 4   **Solve** the system of equations.

Step 5   **State the answer** to the problem. Does it seem reasonable?

Step 6   **Check** the answer in the words of the original problem.

EXAMPLE 5   **Using a Linear System to Solve an Application**

Salaries for the same position can vary depending on the location. In 2010, the average of the salaries for the position of Accountant I in San Diego, California, and Salt Lake City, Utah, was $45,091.50. The salary in San Diego, however, exceeded the salary in Salt Lake City by $5231. Determine the salary for the Accountant I position in San Diego and in Salt Lake City. (*Source:* www.salary.com)

SOLUTION

Step 1   **Read** the problem. We must find the salary of the Accountant I position in San Diego and in Salt Lake City.

Step 2   **Assign variables.** Let x represent the salary of the Accountant I position in San Diego and y represent the salary for the same position in Salt Lake City.

Step 3   **Write a system of equations.** Since the average of the salaries for the Accountant I position in San Diego and Salt Lake City was $45,091.50, one equation is as follows.

$$\frac{x + y}{2} = 45{,}091.50$$

Multiply both sides of this equation by 2 to clear the fraction and get an equivalent equation.

$$x + y = 90{,}183 \quad (1)$$

The salary in San Diego exceeded the salary in Salt Lake City by $5231. Thus, $x - y = 5231$, which gives the following system of equations.

$$x + y = 90{,}183 \quad (1)$$
$$x - y = 5231 \quad (2)$$

Step 4 **Solve** the system. To eliminate y, add the two equations.

$$
\begin{array}{ll}
x + y = 90{,}183 & (1) \\
\underline{x - y = 5231} & (2) \\
2x = 95{,}414 & \text{Add.} \\
x = 47{,}707 & \text{Solve for } x.
\end{array}
$$

To find y, substitute 47,707 for x in equation (2).

$$
\begin{array}{ll}
x - y = 5231 & (2) \\
47{,}707 - y = 5231 & \text{Let } x = 47{,}707. \\
-y = -42{,}476 & \text{Subtract } 47{,}707. \\
y = 42{,}476 & \text{Multiply by } -1.
\end{array}
$$

Step 5 **State the answer.** The salary for the position of Accountant I was $47,707 in San Diego and $42,476 in Salt Lake City.

Step 6 **Check.** The average of $47,707 and $42,476 is

$$
\frac{\$47{,}707 + \$42{,}476}{2} = \$45{,}091.50.
$$

Also,

$$
\$47{,}707 - \$42{,}476 = \$5231, \quad \text{as required.}
$$

 Now Try Exercise 95.

Solving Linear Systems with Three Unknowns (Variables) Earlier, we saw that the graph of a linear equation in two unknowns is a straight line. The graph of a linear equation in three unknowns requires a three-dimensional coordinate system. The three number lines are placed at right angles. The graph of a linear equation in three unknowns is a plane. Some possible intersections of planes representing three equations in three variables are shown in **Figure 6.**

A single solution

Points of a line in common

All points in common

No points in common

No points in common

No points in common

No points in common

Figure 6

Solve a linear system with three unknowns as follows.

Step 1 Eliminate a variable from any two of the equations.

Step 2 Eliminate the *same variable* from a different pair of equations.

Step 3 Eliminate a second variable using the resulting two equations in two variables to get an equation with just one variable whose value we can now determine.

Step 4 Find the values of the remaining variables by substitution. Write the solution of the system as an **ordered triple.**

EXAMPLE 6 **Solving a System of Three Equations with Three Variables**

Solve the system.

$$3x + 9y + 6z = 3 \quad (1)$$
$$2x + y - z = 2 \quad (2)$$
$$x + y + z = 2 \quad (3)$$

SOLUTION Eliminate z by adding equations (2) and (3). (Step 1)

$$3x + 2y = 4 \quad (4)$$

To eliminate z from another pair of equations, multiply both sides of equation (2) by 6 and add the result to equation (1). (Step 2)

Make sure equation (5) has the *same* two variables as equation (4).

$$
\begin{aligned}
12x + 6y - 6z &= 12 \quad \text{Multiply (2) by 6.} \\
\underline{3x + 9y + 6z} &= \underline{3} \quad (1) \\
15x + 15y &= 15 \quad (5)
\end{aligned}
$$

To eliminate x from equations (4) and (5), multiply both sides of equation (4) by -5 and add the result to equation (5). Solve the resulting equation for y. (Step 3)

$$
\begin{aligned}
-15x - 10y &= -20 \quad \text{Multiply (4) by } -5. \\
\underline{15x + 15y} &= \underline{15} \quad (5) \\
5y &= -5 \quad \text{Add.} \\
y &= -1 \quad \text{Divide by 5.}
\end{aligned}
$$

Using $y = -1$, find x from equation (4) by substitution. (Step 4)

$$3x + 2(-1) = 4 \quad \text{(4) with } y = -1$$
$$x = 2 \quad \text{Solve for } x.$$

Substitute 2 for x and -1 for y in equation (3) to find z.

Write the values of x, y, and z in the correct order.

$$2 + (-1) + z = 2 \quad \text{(3) with } x = 2, y = -1$$
$$z = 1 \quad \text{Solve for } z.$$

Verify that the ordered triple $(2, -1, 1)$ satisfies all three equations in the *original* system. The solution set is $\{(2, -1, 1)\}$.

☑ *Now Try Exercise 47.*

CAUTION *When eliminating a variable from any two equations to create equations like (4) and (5) in Example 6, be sure to eliminate the same variable from both equations.* Otherwise, the result will include two equations that still have three variables, and no progress will have been made.

EXAMPLE 7 Solving a System of Two Equations with Three Variables

Solve the system.

$$x + 2y + \ z = 4 \qquad (1)$$

$$3x - \ y - 4z = -9 \qquad (2)$$

SOLUTION Geometrically, the solution is the intersection of the two planes given by equations (1) and (2). The intersection of two different nonparallel planes is a line. Thus there will be an infinite number of ordered triples in the solution set, representing the points on the line of intersection.

To eliminate x, multiply both sides of equation (1) by -3 and add the result to equation (2). (Either y or z could have been eliminated instead.)

$$
\begin{array}{ll}
-3x - 6y - 3z = -12 & \text{Multiply (1) by } -3. \\
\underline{3x - \ y - 4z = -9} & (2) \\
-7y - 7z = -21 & (3) \\
-7z = 7y - 21 & \text{Add } 7y. \\
z = -y + 3 & \text{Divide } each \text{ term by } -7.
\end{array}
$$

Solve this equation for z.

This gives z in terms of y. Express x also in terms of y by solving equation (1) for x and substituting $-y + 3$ for z in the result.

$$
\begin{array}{ll}
x + 2y + z = 4 & (1) \\
x = -2y - z + 4 & \text{Solve for } x. \\
x = -2y - (-y + 3) + 4 & \text{Substitute } (-y + 3) \text{ for } z. \\
x = -y + 1 & \text{Simplify.}
\end{array}
$$

Use parentheses around $-y + 3$.

The system has an infinite number of solutions. For any value of y, the value of z is $-y + 3$ and the value of x is $-y + 1$. For example, if $y = 1$, then $x = -1 + 1 = 0$ and $z = -1 + 3 = 2$, giving the solution $(0, 1, 2)$. Verify that another solution is $(-1, 2, 1)$.

With y arbitrary, the solution set is of the form $\{(-y + 1, y, -y + 3)\}$.

✔ *Now Try Exercise 59.*

NOTE Had we solved equation (3) in **Example 7** for y instead of z, the solution would have had a different form but would have led to the same set of solutions. In that case we would have z arbitrary, and the solution set would be of the form

$$\{(z - 2, -z + 3, z)\}.$$

By choosing $z = 2$, one solution would be $(0, 1, 2)$, which was found above.

Using Systems of Equations to Model Data Applications with three unknowns usually require solving a system of three equations. If we know three points on the graph, we can find the equation of a parabola in the form

$$y = ax^2 + bx + c \qquad \text{(Section 3.1)}$$

by solving a system of three equations with three variables.

EXAMPLE 8 **Using Modeling to Find an Equation through Three Points**

Find the equation of the parabola $y = ax^2 + bx + c$ that passes through the points $(2, 4)$, $(-1, 1)$, and $(-2, 5)$.

SOLUTION Since the three points lie on the graph of the given equation $y = ax^2 + bx + c$, they must satisfy the equation. Substituting each ordered pair into the equation gives three equations with three unknowns.

$$4 = a(2)^2 + b(2) + c, \quad \text{or} \quad 4 = 4a + 2b + c \quad (1)$$
$$1 = a(-1)^2 + b(-1) + c, \quad \text{or} \quad 1 = a - b + c \quad (2)$$
$$5 = a(-2)^2 + b(-2) + c, \quad \text{or} \quad 5 = 4a - 2b + c \quad (3)$$

To solve this system, first eliminate c using equations (1) and (2).

$$4 = 4a + 2b + c \quad (1)$$
$$\underline{-1 = -a + b - c} \quad \text{Multiply (2) by } -1.$$
$$3 = 3a + 3b \quad (4)$$

Now, use equations (2) and (3) to eliminate the same unknown, c.

> Equation (5) must have the *same* two unknowns as equation (4).

$$1 = a - b + c \quad (2)$$
$$\underline{-5 = -4a + 2b - c} \quad \text{Multiply (3) by } -1.$$
$$-4 = -3a + b \quad (5)$$

Solve the system of equations (4) and (5) in two unknowns by eliminating a.

$$3 = 3a + 3b \quad (4)$$
$$\underline{-4 = -3a + b} \quad (5)$$
$$-1 = 4b \quad \text{Add.}$$
$$-\frac{1}{4} = b \quad \text{Divide by 4.}$$

Find a by substituting $-\frac{1}{4}$ for b in equation (4).

$$1 = a + b \quad \text{Equation (4) divided by 3}$$
$$1 = a - \frac{1}{4} \quad \text{Let } b = -\tfrac{1}{4}.$$
$$\frac{5}{4} = a \quad \text{Add } \tfrac{1}{4}.$$

Finally, find c by substituting $a = \frac{5}{4}$ and $b = -\frac{1}{4}$ in equation (2).

$$1 = a - b + c \quad (2)$$
$$1 = \frac{5}{4} - \left(-\frac{1}{4}\right) + c \quad \text{Let } a = \tfrac{5}{4},\ b = -\tfrac{1}{4}.$$
$$1 = \frac{6}{4} + c \quad \text{Add.}$$
$$-\frac{1}{2} = c \quad \text{Subtract } \tfrac{6}{4}.$$

This graph/table screen shows that the points $(2, 4)$, $(-1, 1)$, and $(-2, 5)$ lie on the graph of $Y_1 = 1.25X^2 - 0.25X - 0.5$. This supports the result of **Example 8.**

The required equation is $y = \frac{5}{4}x^2 - \frac{1}{4}x - \frac{1}{2}$, or $y = 1.25x^2 - 0.25x - 0.5$.

☑ *Now Try Exercise 79.*

> **EXAMPLE 9** **Solving an Application Using a System of Three Equations**

An animal feed is made from three ingredients: corn, soybeans, and cottonseed. One unit of each ingredient provides units of protein, fat, and fiber as shown in the table. How many units of each ingredient should be used to make a feed that contains 22 units of protein, 28 units of fat, and 18 units of fiber?

	Corn	Soybeans	Cottonseed	Total
Protein	0.25	0.4	0.2	22
Fat	0.4	0.2	0.3	28
Fiber	0.3	0.2	0.1	18

SOLUTION

Step 1 **Read** the problem. We must determine the number of units of corn, soybeans, and cottonseed.

Step 2 **Assign variables.** Let x represent the number of units of corn, y the number of units of soybeans, and z the number of units of cottonseed.

Step 3 **Write a system of equations.** The total amount of protein is to be 22 units, so we use the first row of the table to write equation (1).

$$0.25x + 0.4y + 0.2z = 22 \quad (1)$$

We use the second row of the table to obtain 28 units of fat.

$$0.4x + 0.2y + 0.3z = 28 \quad (2)$$

Finally, we use the third row of the table to obtain 18 units of fiber.

$$0.3x + 0.2y + 0.1z = 18 \quad (3)$$

Multiply equation (1) on both sides by 100, and equations (2) and (3) by 10, to get an equivalent system.

$$25x + 40y + 20z = 2200 \quad (4)$$
$$4x + 2y + 3z = 280 \quad (5)$$
$$3x + 2y + z = 180 \quad (6)$$

Eliminate the decimal points in equations (1), (2), and (3) by multiplying each equation by an appropriate power of 10.

Step 4 **Solve** the system. Using the methods described earlier in this section, we find that

$$x = 40, \quad y = 15, \quad \text{and} \quad z = 30.$$

Step 5 **State the answer.** The feed should contain 40 units of corn, 15 units of soybeans, and 30 units of cottonseed.

Step 6 **Check.** Show that the ordered triple (40, 15, 30) satisfies the system formed by equations (1), (2), and (3).

> ✔ *Now Try Exercise 101.*

NOTE Notice how the table in **Example 9** is used to set up the equations of the system. The coefficients in each equation are read from left to right. This idea is extended in the next section, where we introduce the solution of systems by matrices.

5.1 Exercises

Changes in Population *Many factors may contribute to population changes in metropolitan areas. The graph shows the populations of the New Orleans, Louisiana, and the Jacksonville, Florida, metropolitan areas over the years 2004–2009.*

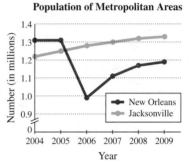

Population of Metropolitan Areas

Source: U.S. Census Bureau.

1. In what years was the population of the Jacksonville metropolitan area greater than that of the New Orleans metropolitan area?

2. At the time when the populations of the two metropolitan areas were equal, what was the approximate population of each area?

3. Express the solution of the system as an ordered pair.

4. Use the terms *increasing, decreasing,* and *constant* to describe the trends for the population of the New Orleans metropolitan area.

5. If equations of the form $y = f(t)$ were determined that modeled either of the two graphs, then the variable t would represent _____ and the variable y would represent _____ .

6. Explain why each graph is that of a function.

Solve each system by substitution. **See Example 1.**

7. $4x + 3y = -13$
   $-x + y = 5$

8. $3x + 4y = 4$
   $x - y = 13$

9. $x - 5y = 8$
   $x = 6y$

10. $6x - y = 5$
    $y = 11x$

11. $8x - 10y = -22$
    $3x + y = 6$

12. $4x - 5y = -11$
    $2x + y = 5$

13. $7x - y = -10$
    $3y - x = 10$

14. $4x + 5y = 7$
    $9y = 31 + 2x$

15. $-2x = 6y + 18$
    $-29 = 5y - 3x$

16. $3x - 7y = 15$
    $3x + 7y = 15$

17. $3y = 5x + 6$
    $x + y = 2$

18. $4y = 2x - 4$
    $x - y = 4$

Solve each system by elimination. In Exercises 27–30, first clear denominators. **See Example 2.**

19. $3x - y = -4$
    $x + 3y = 12$

20. $4x + y = -23$
    $x - 2y = -17$

21. $2x - 3y = -7$
    $5x + 4y = 17$

22. $4x + 3y = -1$
    $2x + 5y = 3$

23. $5x + 7y = 6$
    $10x - 3y = 46$

24. $12x - 5y = 9$
    $3x - 8y = -18$

25. $6x + 7y + 2 = 0$
    $7x - 6y - 26 = 0$

26. $5x + 4y + 2 = 0$
    $4x - 5y - 23 = 0$

27. $\dfrac{x}{2} + \dfrac{y}{3} = 4$
    $\dfrac{3x}{2} + \dfrac{3y}{2} = 15$

28. $\dfrac{3x}{2} + \dfrac{y}{2} = -2$
    $\dfrac{x}{2} + \dfrac{y}{2} = 0$

29. $\dfrac{2x - 1}{3} + \dfrac{y + 2}{4} = 4$
    $\dfrac{x + 3}{2} - \dfrac{x - y}{3} = 3$

30. $\dfrac{x + 6}{5} + \dfrac{2y - x}{10} = 1$
    $\dfrac{x + 2}{4} + \dfrac{3y + 2}{5} = -3$

Solve each system. State whether it is inconsistent or has infinitely many solutions. If the system has infinitely many solutions, write the solution set with y arbitrary. **See Examples 3 and 4.**

31. $9x - 5y = 1$
$-18x + 10y = 1$

32. $3x + 2y = 5$
$6x + 4y = 8$

33. $4x - y = 9$
$-8x + 2y = -18$

34. $3x + 5y = -2$
$9x + 15y = -6$

35. $5x - 5y - 3 = 0$
$x - y - 12 = 0$

36. $2x - 3y - 7 = 0$
$-4x + 6y - 14 = 0$

37. $7x + 2y = 6$
$14x + 4y = 12$

38. $2x - 8y = 4$
$x - 4y = 2$

39. $2x - 6y = 0$
$-7x + 21y = 10$

40. *Concept Check* Which screen gives the correct graphical solution of the system? (*Hint:* Solve for y first in each equation and use the slope-intercept forms to help you answer the question.)

$$4x - 5y = -11$$
$$2x + y = 5$$

A.

B.

C. **D.**

Connecting Graphs with Equations Determine the system of equations illustrated in each graph. Write equations in standard form.

41.

42.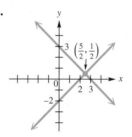

Use a graphing calculator to solve each system. Express solutions with approximations to the nearest thousandth.

43. $\dfrac{11}{3}x + y = 0.5$
$0.6x - y = 3$

44. $\sqrt{3}x - y = 5$
$100x + y = 9$

45. $\sqrt{7}x + \sqrt{2}y - 3 = 0$
$\sqrt{6}x - y - \sqrt{3} = 0$

46. $0.2x + \sqrt{2}y = 1$
$\sqrt{5}x + 0.7y = 1$

Solve each system. **See Example 6.**

47. $x + y + z = 2$
$2x + y - z = 5$
$x - y + z = -2$

48. $2x + y + z = 9$
$-x - y + z = 1$
$3x - y + z = 9$

49. $x + 3y + 4z = 14$
$2x - 3y + 2z = 10$
$3x - y + z = 9$

50. $4x - y + 3z = -2$
$3x + 5y - z = 15$
$-2x + y + 4z = 14$

51. $x + 4y - z = 6$
$2x - y + z = 3$
$3x + 2y + 3z = 16$

52. $4x - 3y + z = 9$
$3x + 2y - 2z = 4$
$x - y + 3z = 5$

53. $x - 3y - 2z = -3$
$3x + 2y - z = 12$
$-x - y + 4z = 3$

54. $x + y + z = 3$
$3x - 3y - 4z = -1$
$x + y + 3z = 11$

55. $2x + 6y - z = 6$
$4x - 3y + 5z = -5$
$6x + 9y - 2z = 11$

56. $8x - 3y + 6z = -2$
$4x + 9y + 4z = 18$
$12x - 3y + 8z = -2$

57. $2x - 3y + 2z - 3 = 0$
$4x + 8y + z - 2 = 0$
$-x - 7y + 3z - 14 = 0$

58. $-x + 2y - z - 1 = 0$
$-x - y - z + 2 = 0$
$x - y + 2z - 2 = 0$

Solve each system in terms of the arbitrary variable x. See Example 7.

59. $x - 2y + 3z = 6$
$2x - y + 2z = 5$

60. $3x - 2y + z = 15$
$x + 4y - z = 11$

61. $5x - 4y + z = 9$
$y + z = 15$

62. $3x - 5y - 4z = -7$
$y - z = -13$

63. $3x + 4y - z = 13$
$x + y + 2z = 15$

64. $x - y + z = -6$
$4x + y + z = 7$

Solve each system. State whether it is inconsistent or has infinitely many solutions. If the system has infinitely many solutions, write the solution set with z arbitrary. See Examples 3, 4, 6, and 7.

65. $3x + 5y - z = -2$
$4x - y + 2z = 1$
$-6x - 10y + 2z = 0$

66. $3x + y + 3z = 1$
$x + 2y - z = 2$
$2x - y + 4z = 4$

67. $5x - 4y + z = 0$
$x + y = 0$
$-10x + 8y - 2z = 0$

68. $2x + y - 3z = 0$
$4x + 2y - 6z = 0$
$x - y + z = 0$

Solve each system. (Hint: In Exercises 69–72, let $\frac{1}{x} = t$ and $\frac{1}{y} = u$.)

69. $\dfrac{2}{x} + \dfrac{1}{y} = \dfrac{3}{2}$
$\dfrac{3}{x} - \dfrac{1}{y} = 1$

70. $\dfrac{1}{x} + \dfrac{3}{y} = \dfrac{16}{5}$
$\dfrac{5}{x} + \dfrac{4}{y} = 5$

71. $\dfrac{2}{x} + \dfrac{1}{y} = 11$
$\dfrac{3}{x} - \dfrac{5}{y} = 10$

72. $\dfrac{2}{x} + \dfrac{3}{y} = 18$
$\dfrac{4}{x} - \dfrac{5}{y} = -8$

73. $\dfrac{2}{x} + \dfrac{3}{y} - \dfrac{2}{z} = -1$
$\dfrac{8}{x} - \dfrac{12}{y} + \dfrac{5}{z} = 5$
$\dfrac{6}{x} + \dfrac{3}{y} - \dfrac{1}{z} = 1$

74. $-\dfrac{5}{x} + \dfrac{4}{y} + \dfrac{3}{z} = 2$
$\dfrac{10}{x} + \dfrac{3}{y} - \dfrac{6}{z} = 7$
$\dfrac{5}{x} + \dfrac{2}{y} - \dfrac{9}{z} = 6$

75. *Concept Check* For what value(s) of k will the following system of linear equations have no solution? infinitely many solutions?

$$x - 2y = 3$$
$$-2x + 4y = k$$

76. *Concept Check* Consider the linear equation in three variables $x + y + z = 4$. Find a pair of linear equations in three variables that, when considered together with the given equation, form a system having **(a)** exactly one solution, **(b)** no solution, **(c)** infinitely many solutions.

(Modeling) *Use a system of equations to solve each problem. See Example 8.*

77. Find the equation of the line $y = ax + b$ that passes through the points $(-2, 1)$ and $(-1, -2)$.

78. Find the equation of the line $y = ax + b$ that passes through the points $(3, -4)$ and $(-1, 4)$.

79. Find the equation of the parabola $y = ax^2 + bx + c$ that passes through the points $(2, 3)$, $(-1, 0)$, and $(-2, 2)$.

80. Find the equation of the parabola $y = ax^2 + bx + c$ that passes through the points $(-2, 4)$, $(2, 2)$, and $(4, 9)$.

81. *Connecting Graphs with Equations* Use a system to find the equation of the line through the given points.

82. *Connecting Graphs with Equations* Use a system to find the equation of the parabola through the given points.

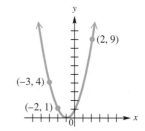

83. *Connecting Graphs with Equations* Find the equation of the parabola. Three views of the same curve are given.

84. *(Modeling)* The table was generated using a function defined by $Y_1 = aX^2 + bX + c$. Use any three points from the table to find the equation that defines the function.

(Modeling) *Given three noncollinear points, there is one and only one circle that passes through them. Knowing that the equation of a circle may be written in the form*

$$x^2 + y^2 + ax + by + c = 0, \quad \text{(Section 2.2)}$$

find the equation of the circle passing through the given points.

85. $(-1, 3)$, $(6, 2)$, and $(-2, -4)$

86. $(-1, 5)$, $(6, 6)$, and $(7, -1)$

87. $(2, 1)$, $(-1, 0)$, and $(3, 3)$

88. $(-5, 0)$, $(2, -1)$, and $(4, 3)$

89. *Connecting Graphs with Equations*

90. *Connecting Graphs with Equations*

 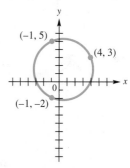

(Modeling) Use the method of **Example 8** to work Exercises 91 and 92.

91. *Atmospheric Carbon Dioxide* Carbon dioxide concentrations (in parts per million) have been measured directly from the atmosphere since 1960. This concentration has increased quadratically. The table lists readings for three years.

Year	CO_2
1960	317
1980	339
2009	385

Source: U.S. Department of Energy; Carbon Dioxide Information Analysis Center.

(a) If the quadratic relationship between the carbon dioxide concentration C and the year t is expressed as $C = at^2 + bt + c$, where $t = 0$ corresponds to 1960, use a system of linear equations to determine the constants a, b, and c, and give the equation.

(b) Predict when the amount of carbon dioxide in the atmosphere will be double its 1960 level.

92. *Aircraft Speed and Altitude* For certain aircraft there exists a quadratic relationship between an airplane's maximum speed S (in knots) and its ceiling C, or highest altitude possible (in thousands of feet). The table lists three airplanes that conform to this relationship.

Airplane	Max Speed (S)	Ceiling (C)
Hawkeye	320	33
Corsair	600	40
Tomcat	1283	50

Source: Sanders, D., *Statistics: A First Course,* Sixth Edition, McGraw Hill.

(a) If the quadratic relationship between C and S is written as $C = aS^2 + bS + c$, use a system of linear equations to determine the constants a, b and c, and give the equation.

(b) A new aircraft of this type has a ceiling of 45,000 ft. Predict its top speed.

Solve each problem. *See Examples 5 and 9.*

93. *Unknown Numbers* The sum of two numbers is 47, and the difference between the numbers is 1. Find the numbers.

94. *Costs of Goats and Sheep* At the Brendan Berger ranch, 6 goats and 5 sheep sell for $305, while 2 goats and 9 sheep sell for $285. Find the cost of a single goat and of a single sheep.

95. *Fan Cost Index* The Fan Cost Index (FCI) is a measure of how much it will cost a family of four to attend a professional sports event. In 2010, the FCI prices for Major League Baseball and the National Football League averaged $307.76. The FCI for baseball was $225.56 less than that for football. What were the FCIs for these sports? (*Source:* Team Marketing Report.)

96. *Money Denominations* A cashier has a total of 30 bills, made up of ones, fives, and twenties. The number of twenties is 9 more than the number of ones. The total value of the money is $351. How many of each denomination of bill are there?

97. *Mixing Water* A sparkling-water distributor wants to make up 300 gal of sparkling water to sell for $6.00 per gallon. She wishes to mix three grades of water selling for $9.00, $3.00, and $4.50 per gallon, respectively. She must use twice as much of the $4.50 water as of the $3.00 water. How many gallons of each should she use?

98. *Mixing Glue* A glue company needs to make some glue that it can sell for $120 per barrel. It wants to use 150 barrels of glue worth $100 per barrel, along with some glue worth $150 per barrel and some glue worth $190 per barrel. It must use the same number of barrels of $150 and $190 glue. How much of the $150 and $190 glue will be needed? How many barrels of $120 glue will be produced?

99. *Triangle Dimensions* The perimeter of a triangle is 59 in. The longest side is 11 in. longer than the medium side, and the medium side is 3 in. longer than the shortest side. Find the length of each side of the triangle.

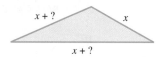

100. *Triangle Dimensions* The sum of the measures of the angles of any triangle is 180°. In a certain triangle, the largest angle measures 55° less than twice the medium angle, and the smallest angle measures 25° less than the medium angle. Find the measures of all three angles.

101. *Investment Decisions* Patrick Summers wins $200,000 in the Louisiana state lottery. He invests part of the money in real estate with an annual return of 3% and another part in a money market account at 2.5% interest. He invests the rest, which amounts to $80,000 less than the sum of the other two parts, in certificates of deposit that pay 1.5%. If the total annual interest on the money is $4900, how much was invested at each rate?

	Amount Invested	Rate (as a decimal)	Annual Interest
Real Estate		0.03	
Money Market		0.025	
CDs		0.015	

102. *Investment Decisions* Jane Hooker invests $40,000 received as an inheritance in three parts. With one part she buys mutual funds that offer a return of 2% per year. The second part, which amounts to twice the first, is used to buy government bonds paying 2.5% per year. She puts the rest of the money into a savings account that pays 1.25% annual interest. During the first year, the total interest is $825. How much did she invest at each rate?

Relating Concepts

For individual or collaborative investigation *(Exercises 103–108)*

Supply and Demand *In many applications of economics, as the price of an item goes up, demand for the item goes down and supply of the item goes up. The price where supply and demand are equal is the **equilibrium price**, and the resulting supply or demand is the **equilibrium supply** or **equilibrium demand**.*

Suppose the supply of a product is related to its price by the equation

$$p = \frac{2}{3}q,$$

where p is in dollars and q is supply in appropriate units. (Here, q stands for quantity.)

(continued)

Furthermore, suppose demand and price for the same product are related by

$$p = -\frac{1}{3}q + 18,$$

where p is price and q is demand. The system formed by these two equations has solution (18, 12), as seen in the graph. Use this information to **work Exercises 103–108 in order.**

103. Suppose the demand and price for a certain model of electric can opener are related by $p = 16 - \frac{5}{4}q$, where p is price, in dollars, and q is demand, in appropriate units. Find the price when the demand is at each level.

    (a) 0 units   **(b)** 4 units   **(c)** 8 units

104. Find the demand for the electric can opener at each price.

    (a) $6   **(b)** $11   **(c)** $16

105. Graph $p = 16 - \frac{5}{4}q$.

106. Suppose the price and supply of the can opener are related by $p = \frac{3}{4}q$, where q represents the supply and p the price. Find the supply at each price.

    (a) $0   **(b)** $10   **(c)** $20

107. Graph $p = \frac{3}{4}q$ on the same axes used for **Exercise 105.**

108. Use the result of **Exercise 107** to find the equilibrium price and the equilibrium demand.

109. Solve the system of equations (4), (5), and (6) from **Example 9.**

$$25x + 40y + 20z = 2200 \quad (4)$$
$$4x + 2y + 3z = 280 \quad (5)$$
$$3x + 2y + z = 180 \quad (6)$$

110. Check your solution in **Exercise 109,** showing that it satisfies all three equations of the system.

111. *Blending Coffee Beans* Three varieties of coffee—Arabian Mocha Sanani, Organic Shade Grown Mexico, and Guatemala Antigua—are combined and roasted, yielding a 50-lb batch of coffee beans. Twice as many pounds of Guatemala Antigua, which retails for $10.19 per lb, are needed as of Arabian Mocha Sanani, which retails for $15.99 per lb. Organic Shade Grown Mexico retails for $12.99 per lb. How many pounds of each coffee should be used in a blend that sells for $12.37 per lb?

112. *Blending Coffee Beans* Rework **Exercise 111** if Guatemala Antigua retails for $12.49 per lb instead of $10.19 per lb. Does your answer seem reasonable? Explain.

5.2 Matrix Solution of Linear Systems

- The Gauss-Jordan Method
- Special Systems

$$\begin{bmatrix} 2 & 3 & 7 \\ 5 & -1 & 10 \end{bmatrix} \quad \text{Matrix}$$

Since systems of linear equations occur in so many practical situations, computer methods have been developed for efficiently solving linear systems. Computer solutions of linear systems depend on the idea of a **matrix** (plural **matrices**), a rectangular array of numbers enclosed in brackets. Each number is an **element** of the matrix.

The Gauss-Jordan Method In this section, we develop a method for solving linear systems using matrices. We start with a system and write the coefficients of the variables and the constants as an **augmented matrix** of the system.

Linear system of equations

$$\begin{aligned} x + 3y + 2z &= 1 \\ 2x + y - z &= 2 \quad \text{can be written as} \\ x + y + z &= 2 \end{aligned}$$

Augmented matrix

$$\begin{bmatrix} 1 & 3 & 2 & | & 1 \\ 2 & 1 & -1 & | & 2 \\ 1 & 1 & 1 & | & 2 \end{bmatrix} \leftarrow \text{Rows}$$

Columns

The vertical line, which is optional, separates the coefficients from the constants. Because this matrix has 3 rows (horizontal) and 4 columns (vertical), we say its **dimension*** is 3×4 (read "three by four"). *The number of rows is always given first.* To refer to a number in the matrix, use its row and column numbers. For example, the number 3 is in the first row, second column.

We can treat the rows of this matrix just like the equations of the corresponding system of linear equations. Since the augmented matrix is nothing more than a shorthand form of the system, any transformation of the matrix that results in an equivalent system of equations can be performed.

Matrix Row Transformations

For any augmented matrix of a system of linear equations, the following row transformations will result in the matrix of an equivalent system.

1. Interchange any two rows.

2. Multiply or divide the elements of any row by a nonzero real number.

3. Replace any row of the matrix by the sum of the elements of that row and a multiple of the elements of another row.

These transformations are restatements in matrix form of the transformations of systems discussed in the previous section. From now on, when referring to the third transformation, we will abbreviate "a multiple of the elements of a row" as "a multiple of a row."

Before matrices can be used to solve a linear system, the system must be arranged in the proper form, with variable terms on the left side of the equation and constant terms on the right. The variable terms must be in the same order in each of the equations.

---

*Other terms used to describe the dimension of a matrix are *order* and *size*.

The **Gauss-Jordan method** is a systematic technique for applying matrix row transformations in an attempt to reduce a matrix to **diagonal form,** with 1s along the diagonal, from which the solutions are easily obtained.

$$\begin{bmatrix} 1 & 0 & | & a \\ 0 & 1 & | & b \end{bmatrix} \quad \text{or} \quad \begin{bmatrix} 1 & 0 & 0 & | & a \\ 0 & 1 & 0 & | & b \\ 0 & 0 & 1 & | & c \end{bmatrix} \qquad \text{Diagonal form, or} \atop \text{reduced-row echelon form}$$

This form is also called **reduced-row echelon form.**

Using the Gauss-Jordan Method to Put a Matrix into Diagonal Form

Step 1 Obtain 1 as the first element of the first column.

Step 2 Use the first row to transform the remaining entries in the first column to 0.

Step 3 Obtain 1 as the second entry in the second column.

Step 4 Use the second row to transform the remaining entries in the second column to 0.

Step 5 Continue in this manner as far as possible.

NOTE *The Gauss-Jordan method proceeds column by column, from left to right.* In each column, we work to obtain 1 in the appropriate diagonal location, and then use it to transform the remaining elements in that column to 0s. When we are working with a particular column, no row operation should undo the form of a preceding column.

EXAMPLE 1 Using the Gauss-Jordan Method

Solve the system.

$$3x - 4y = 1$$

$$5x + 2y = 19$$

SOLUTION Both equations are in the same form, with variable terms in the same order on the left, and constant terms on the right.

$$\begin{bmatrix} 3 & -4 & | & 1 \\ 5 & 2 & | & 19 \end{bmatrix} \qquad \text{Write the augmented matrix.}$$

The goal is to transform the augmented matrix into one in which the value of the variables will be easy to see. That is, since each of the first two columns in the matrix represents the coefficients of one variable, the augmented matrix should be transformed so that it is of the following form.

This form is our goal. $\longrightarrow$ $\begin{bmatrix} \mathbf{1} & \mathbf{0} & | & k \\ \mathbf{0} & \mathbf{1} & | & j \end{bmatrix}$ Here k and j are real numbers.

Once the augmented matrix is in this form, the matrix can be rewritten as a linear system

$$x = k$$

$$y = j.$$

It is best to work in columns, beginning in each column with the element that is to become **1.** In the augmented matrix

$$\begin{bmatrix} 3 & -4 & | & 1 \\ 5 & 2 & | & 19 \end{bmatrix},$$

3 is in the first row, first column position. Use transformation 2, multiplying each entry in the first row by $\frac{1}{3}$ to get 1 in this position. (This step is abbreviated as $\frac{1}{3}$ R1.)

$$\begin{bmatrix} 1 & -\frac{4}{3} & | & \frac{1}{3} \\ 5 & 2 & | & 19 \end{bmatrix} \quad \frac{1}{3} R1$$

Introduce 0 in the second row, first column by multiplying each element of the first row by -5 and adding the result to the corresponding element in the second row, using transformation 3.

$$\begin{bmatrix} 1 & -\frac{4}{3} & | & \frac{1}{3} \\ 0 & \frac{26}{3} & | & \frac{52}{3} \end{bmatrix} \quad -5R1 + R2$$

Obtain 1 in the second row, second column by multiplying each element of the second row by $\frac{3}{26}$, using transformation 2.

$$\begin{bmatrix} 1 & -\frac{4}{3} & | & \frac{1}{3} \\ 0 & 1 & | & 2 \end{bmatrix} \quad \frac{3}{26} R2$$

Finally, get 0 in the first row, second column by multiplying each element of the second row by $\frac{4}{3}$ and adding the result to the corresponding element in the first row.

$$\begin{bmatrix} 1 & 0 & | & 3 \\ 0 & 1 & | & 2 \end{bmatrix} \quad \frac{4}{3} R2 + R1$$

This last matrix corresponds to the system

$$x = 3$$
$$y = 2,$$

which indicates the solution $(3, 2)$. We can read this solution directly from the third column of the final matrix.

CHECK   Substitute the solution in both equations of the *original* system.

$$3x - 4y = 1 \qquad\qquad 5x + 2y = 19$$
$$3(3) - 4(2) \overset{?}{=} 1 \qquad\qquad 5(3) + 2(2) \overset{?}{=} 19$$
$$9 - 8 \overset{?}{=} 1 \qquad\qquad 15 + 4 \overset{?}{=} 19$$
$$1 = 1 \checkmark \text{ True} \qquad\qquad 19 = 19 \checkmark \text{ True}$$

Since true statements result, the solution set is $\{(3, 2)\}$.

✔️ *Now Try Exercise 17.*

NOTE Using row operations to write a matrix in diagonal form requires effective use of the inverse properties of addition and multiplication. See **Section R.2.**

A linear system with three equations is solved in a similar way. Row transformations are used to get 1s down the diagonal from left to right and 0s above and below each 1.

EXAMPLE 2 **Using the Gauss-Jordan Method**

Solve the system.

$$x - y + 5z = -6$$

$$3x + 3y - z = 10$$

$$x + 3y + 2z = 5$$

SOLUTION
$$\begin{bmatrix} 1 & -1 & 5 & | & -6 \\ 3 & 3 & -1 & | & 10 \\ 1 & 3 & 2 & | & 5 \end{bmatrix}$$ Write the augmented matrix.

There is already a 1 in the first row, first column. Introduce 0 in the second row of the first column by multiplying each element in the first row by -3 and adding the result to the corresponding element in the second row.

$$\begin{bmatrix} 1 & -1 & 5 & | & -6 \\ 0 & 6 & -16 & | & 28 \\ 1 & 3 & 2 & | & 5 \end{bmatrix} \quad -3R1 + R2$$

To change the third element in the first column to 0, multiply each element of the first row by -1. Add the result to the corresponding element of the third row.

$$\begin{bmatrix} 1 & -1 & 5 & | & -6 \\ 0 & 6 & -16 & | & 28 \\ 0 & 4 & -3 & | & 11 \end{bmatrix} \quad -1R1 + R3$$

Use the same procedure to transform the second and third columns. For both of these columns, perform the additional step of getting 1 in the appropriate position of each column. Do this by multiplying the elements of the row by the reciprocal of the number in that position.

$$\begin{bmatrix} 1 & -1 & 5 & | & -6 \\ 0 & 1 & -\frac{8}{3} & | & \frac{14}{3} \\ 0 & 4 & -3 & | & 11 \end{bmatrix} \quad \frac{1}{6}R2$$

$$\begin{bmatrix} 1 & 0 & \frac{7}{3} & | & -\frac{4}{3} \\ 0 & 1 & -\frac{8}{3} & | & \frac{14}{3} \\ 0 & 4 & -3 & | & 11 \end{bmatrix} \quad R2 + R1$$

$$\begin{bmatrix} 1 & 0 & \frac{7}{3} & | & -\frac{4}{3} \\ 0 & 1 & -\frac{8}{3} & | & \frac{14}{3} \\ 0 & 0 & \frac{23}{3} & | & -\frac{23}{3} \end{bmatrix} \quad -4R2 + R3$$

$$\begin{bmatrix} 1 & 0 & \frac{7}{3} & | & -\frac{4}{3} \\ 0 & 1 & -\frac{8}{3} & | & \frac{14}{3} \\ 0 & 0 & 1 & | & -1 \end{bmatrix} \quad \frac{3}{23}R3$$

$$\begin{bmatrix} 1 & 0 & 0 & | & 1 \\ 0 & 1 & -\frac{8}{3} & | & \frac{14}{3} \\ 0 & 0 & 1 & | & -1 \end{bmatrix} \quad -\frac{7}{3}R3 + R1$$

$$\begin{bmatrix} 1 & 0 & 0 & | & 1 \\ 0 & 1 & 0 & | & 2 \\ 0 & 0 & 1 & | & -1 \end{bmatrix} \quad \frac{8}{3}R3 + R2$$

The linear system associated with this final matrix is

$$x = 1$$
$$y = 2$$
$$z = -1.$$

The solution set is $\{(1, 2, -1)\}$. Check the solution in the original system.

✔ Now Try Exercise 25.

A TI-83/84 Plus graphing calculator with matrix capability is able to perform row operations. See **Figure 7(a).** The screen in **Figure 7(b)** shows typical entries for the matrix in the second step of the solution in **Example 2.** The entire Gauss-Jordan method can be carried out in one step with the rref (reduced-row echelon form) command, as shown in **Figure 7(c).**

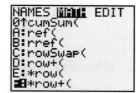

This typical menu shows various options for matrix row transformations in choices C–F.

(a)

This is the matrix after column 1 has been transformed.

(b)

This screen shows the final matrix in the algebraic solution, found by using the rref command.

(c)

Figure 7

Special Systems   The next two examples show how to recognize inconsistent systems or systems with infinitely many solutions when solving such systems using row transformations.

EXAMPLE 3   Solving an Inconsistent System

Use the Gauss-Jordan method to solve the system.

$$x + \ \ y = 2$$
$$2x + 2y = 5$$

SOLUTION
$$\begin{bmatrix} 1 & 1 & | & 2 \\ 2 & 2 & | & 5 \end{bmatrix} \quad \text{Write the augmented matrix.}$$

$$\begin{bmatrix} 1 & 1 & | & 2 \\ 0 & 0 & | & 1 \end{bmatrix} \quad -2R1 + R2$$

The next step would be to get 1 in the second row, second column. Because of the 0 there, it is impossible to go further. Since the second row corresponds to the equation

$$0x + 0y = 1,$$

which has no solution, the system is inconsistent and the solution set is $\emptyset$.

✔ Now Try Exercise 21.

EXAMPLE 4 **Solving a System with Infinitely Many Solutions**

Use the Gauss-Jordan method to solve the system.

$$2x - 5y + 3z = 1$$
$$x - 2y - 2z = 8$$

SOLUTION Recall from the previous section that a system with two equations in three variables usually has an infinite number of solutions. We can use the Gauss-Jordan method to give the solution with z arbitrary.

$$\begin{bmatrix} 2 & -5 & 3 & \big| & 1 \\ 1 & -2 & -2 & \big| & 8 \end{bmatrix} \quad \text{Write the augmented matrix.}$$

$$\begin{bmatrix} 1 & -2 & -2 & \big| & 8 \\ 2 & -5 & 3 & \big| & 1 \end{bmatrix} \quad \begin{array}{l} \text{Interchange rows to get 1 in the} \\ \text{first row, first column position.} \end{array}$$

$$\begin{bmatrix} 1 & -2 & -2 & \big| & 8 \\ 0 & -1 & 7 & \big| & -15 \end{bmatrix} \quad -2R1 + R2$$

$$\begin{bmatrix} 1 & -2 & -2 & \big| & 8 \\ 0 & 1 & -7 & \big| & 15 \end{bmatrix} \quad -1R2$$

$$\begin{bmatrix} 1 & 0 & -16 & \big| & 38 \\ 0 & 1 & -7 & \big| & 15 \end{bmatrix} \quad 2R2 + R1$$

It is not possible to go further with the Gauss-Jordan method. The equations that correspond to the final matrix are

$$x - 16z = 38 \quad \text{and} \quad y - 7z = 15.$$

Solve these equations for x and y, respectively.

$$\begin{array}{l|l} x - 16z = 38 & y - 7z = 15 \\ \quad\quad x = 16z + 38 & \quad\quad y = 7z + 15 \quad \text{(Section 1.1)} \end{array}$$

The solution set, written with z arbitrary, is $\{(16z + 38, 7z + 15, z)\}$.

✔ *Now Try Exercise 37.*

Summary of Possible Cases

When matrix methods are used to solve a system of linear equations and the resulting matrix is written in diagonal form:

1. If the number of rows with nonzero elements to the left of the vertical line is equal to the number of variables in the system, then the system has a single solution. **See Examples 1 and 2.**

2. If one of the rows has the form $\begin{bmatrix} 0 & 0 & \cdots & 0 & \big| & a \end{bmatrix}$ with $a \neq 0$, then the system has no solution. **See Example 3.**

3. If there are fewer rows in the matrix containing nonzero elements than the number of variables, then the system has either no solution or infinitely many solutions. If there are infinitely many solutions, give the solutions in terms of one or more arbitrary variables. **See Example 4.**

5.2 Exercises

Use the given row transformation to change each matrix as indicated. **See Example 1.**

1. $\begin{bmatrix} 1 & 4 \\ 4 & 7 \end{bmatrix}$; $\begin{array}{l} -4 \text{ times row } 1 \\ \text{added to row } 2 \end{array}$

2. $\begin{bmatrix} 1 & -4 \\ 7 & 0 \end{bmatrix}$; $\begin{array}{l} -7 \text{ times row } 1 \\ \text{added to row } 2 \end{array}$

3. $\begin{bmatrix} 1 & 5 & 6 \\ -2 & 3 & -1 \\ 4 & 7 & 0 \end{bmatrix}$; $\begin{array}{l} 2 \text{ times row } 1 \\ \text{added to row } 2 \end{array}$

4. $\begin{bmatrix} 1 & 5 & 6 \\ -4 & -1 & 2 \\ 3 & 7 & 1 \end{bmatrix}$; $\begin{array}{l} 4 \text{ times row } 1 \\ \text{added to row } 2 \end{array}$

Concept Check *Write the augmented matrix for each system and give its dimension. Do not solve.*

5. $2x + 3y = 11$
$x + 2y = 8$

6. $3x + 5y = -13$
$2x + 3y = -9$

7. $2x + y + z - 3 = 0$
$3x - 4y + 2z + 7 = 0$
$x + y + z - 2 = 0$

8. $4x - 2y + 3z - 4 = 0$
$3x + 5y + z - 7 = 0$
$5x - y + 4z - 7 = 0$

Concept Check *Write the system of equations associated with each augmented matrix. Do not solve.*

9. $\begin{bmatrix} 3 & 2 & 1 & | & 1 \\ 0 & 2 & 4 & | & 22 \\ -1 & -2 & 3 & | & 15 \end{bmatrix}$

10. $\begin{bmatrix} 2 & 1 & 3 & | & 12 \\ 4 & -3 & 0 & | & 10 \\ 5 & 0 & -4 & | & -11 \end{bmatrix}$

11. $\begin{bmatrix} 1 & 0 & 0 & | & 2 \\ 0 & 1 & 0 & | & 3 \\ 0 & 0 & 1 & | & -2 \end{bmatrix}$

12. $\begin{bmatrix} 1 & 0 & 0 & | & 4 \\ 0 & 1 & 0 & | & 2 \\ 0 & 0 & 1 & | & 3 \end{bmatrix}$

13. [A]
$\begin{bmatrix} 1 & 1 & 0 & 3 \\ 0 & 2 & 1 & -4 \\ 1 & 0 & -1 & 5 \end{bmatrix}$

14. [B]
$\begin{bmatrix} 2 & 0 & 1 & 9 \\ 0 & -1 & -1 & 5 \\ 3 & 1 & 0 & 8 \end{bmatrix}$

Use the Gauss-Jordan method to solve each system of equations. For systems in two variables with infinitely many solutions, give the solution with y arbitrary. For systems in three variables with infinitely many solutions, give the solution with z arbitrary. **See Examples 1–4.**

15. $x + y = 5$
$x - y = -1$

16. $x + 2y = 5$
$2x + y = -2$

17. $3x + 2y = -9$
$2x - 5y = -6$

18. $2x - 5y = 10$
$3x + y = 15$

19. $6x - 3y - 4 = 0$
$3x + 6y - 7 = 0$

20. $2x - 3y - 10 = 0$
$2x + 2y - 5 = 0$

21. $2x - y = 6$
$4x - 2y = 0$

22. $3x - 2y = 1$
$6x - 4y = -1$

23. $\dfrac{3}{8}x - \dfrac{1}{2}y = \dfrac{7}{8}$
$-6x + 8y = -14$

24. $\dfrac{1}{2}x + \dfrac{3}{5}y = \dfrac{1}{4}$
$10x + 12y = 5$

25. $x + y - 5z = -18$
$3x - 3y + z = 6$
$x + 3y - 2z = -13$

26. $-x + 2y + 6z = 2$
$3x + 2y + 6z = 6$
$x + 4y - 3z = 1$

27. $x + y - z = 6$
$2x - y + z = -9$
$x - 2y + 3z = 1$

28. $x + 3y - 6z = 7$
$2x - y + z = 1$
$x + 2y + 2z = -1$

29. $x - z = -3$
$y + z = 9$
$x + z = 7$

30. $-x + y = -1$
$\quad\quad y - z = 6$
$\quad\quad x + z = -1$

31. $y = -2x - 2z + 1$
$\quad\quad x = -2y - z + 2$
$\quad\quad z = x - y$

32. $x = -y + 1$
$\quad\quad z = 2x$
$\quad\quad y = -2z - 2$

33. $2x - y + 3z = 0$
$\quad\quad x + 2y - z = 5$
$\quad\quad 2y + z = 1$

34. $4x + 2y - 3z = 6$
$\quad\quad x - 4y + z = -4$
$\quad\quad -x + 2z = 2$

35. $\quad 3x + 5y - z + 2 = 0$
$\quad\quad 4x - y + 2z - 1 = 0$
$\quad\quad -6x - 10y + 2z = 0$

36. $3x + y + 3z - 1 = 0$
$\quad\quad x + 2y - z - 2 = 0$
$\quad\quad 2x - y + 4z - 4 = 0$

37. $x - 8y + z = 4$
$\quad\quad 3x - y + 2z = -1$

38. $5x - 3y + z = 1$
$\quad\quad 2x + y - z = 4$

39. $x - y + 2z + w = 4$
$\quad\quad y + z = 3$
$\quad\quad z - w = 2$
$\quad\quad x - y = 0$

40. $x + 2y + z - 3w = 7$
$\quad\quad y + z = 0$
$\quad\quad x - w = 4$
$\quad\quad -x + y = -3$

41. $\quad x + 3y - 2z - w = 9$
$\quad\quad 4x + y + z + 2w = 2$
$\quad\quad -3x - y + z - w = -5$
$\quad\quad x - y - 3z - 2w = 2$

42. $2x + y - z + 3w = 0$
$\quad\quad 3x - 2y + z - 4w = -24$
$\quad\quad x + y - z + w = 2$
$\quad\quad x - y + 2z - 5w = -16$

Solve each system using a graphing calculator capable of performing row operations. Give solutions with values correct to the nearest thousandth.

43. $0.3x + 2.7y - \sqrt{2}z = 3$
$\quad\quad \sqrt{7}x - 20y + 12z = -2$
$\quad\quad 4x + \sqrt{3}y - 1.2z = \dfrac{3}{4}$

44. $\sqrt{5}x - 1.2y + z = -3$
$\quad\quad \dfrac{1}{2}x - 3y + 4z = \dfrac{4}{3}$
$\quad\quad 4x + 7y - 9z = \sqrt{2}$

Graph each system of three equations together on the same axes, and determine the number of solutions (exactly one, none, or infinitely many). If there is exactly one solution, estimate the solution. Then confirm your answer by solving the system with the Gauss-Jordan method.

45. $\quad 2x + 3y = 5$
$\quad\quad -3x + 5y = 22$
$\quad\quad 2x + y = -1$

46. $\quad 3x - 2y = 3$
$\quad\quad -2x + 4y = 14$
$\quad\quad x + y = 11$

For each equation, determine the constants A and B that make the equation an identity. (Hint: Combine terms on the right, and set coefficients of corresponding terms in the numerators equal.)

47. $\dfrac{1}{(x-1)(x+1)} = \dfrac{A}{x-1} + \dfrac{B}{x+1}$

48. $\dfrac{x+4}{x^2} = \dfrac{A}{x} + \dfrac{B}{x^2}$

49. $\dfrac{x}{(x-a)(x+a)} = \dfrac{A}{x-a} + \dfrac{B}{x+a}$

50. $\dfrac{2x}{(x+2)(x-1)} = \dfrac{A}{x+2} + \dfrac{B}{x-1}$

Solve each problem using matrices.

51. *Daily Wages* Dan Abbey is a building contractor. If he hires 7 day laborers and 2 concrete finishers, his payroll for the day is $1384. If he hires 1 day laborer and 5 concrete finishers, his daily cost is $952. Find the daily wage for each type of worker.

52. *Mixing Nuts* At the Everglades Nut Company, 5 lb of peanuts and 6 lb of cashews cost $33.60, while 3 lb of peanuts and 7 lb of cashews cost $32.40. Find the cost of a single pound of peanuts and a single pound of cashews.

53. *Unknown Numbers* Find three numbers whose sum is 20, if the first number is three times the difference between the second and the third, and the second number is two more than twice the third.

54. *Car Sales Quota* To meet a sales quota, a car salesperson must sell 24 new cars, consisting of small, medium, and large cars. She must sell 3 more small cars than medium cars, and the same number of medium cars as large cars. How many of each size must she sell?

55. *Mixing Acid Solutions* A chemist has two prepared acid solutions, one of which is 2% acid by volume, the other 7% acid. How many cubic centimeters of each should the chemist mix together to obtain 40 cm^3 of a 3.2% acid solution?

56. *Financing an Expansion* To get the necessary funds for a planned expansion, a small company took out three loans totaling $25,000. The company was able to borrow some of the money at 4% interest. It borrowed $2000 more than one-half the amount of the 4% loan at 6%, and the rest at 5%. The total annual interest was $1220. How much did the company borrow at each rate?

57. *Financing an Expansion* In **Exercise 56,** suppose we drop the condition that the amount borrowed at 6% is $2000 more than one-half the amount borrowed at 4%. How is the solution changed?

58. *Financing an Expansion* Suppose the company in **Exercise 56** can borrow only $6000 at 5%. Is a solution possible that still meets the given conditions? Explain.

59. *Planning a Diet* In a special diet for a hospital patient, the total amount per meal of food groups A, B, and C must equal 400 g. The diet should include one-third as much of group A as of group B. The sum of the amounts of group A and group C should equal twice the amount of group B. How many grams of each food group should be included? (Give answers to the nearest tenth.)

60. *Planning a Diet* In **Exercise 59,** suppose that, in addition to the conditions given there, foods A and B cost $0.02 per gram, food C costs $0.03 per gram, and a meal must cost $8. Is a solution possible? Explain.

(Modeling) Age Distribution in the United States As people live longer, a larger percentage of the U.S. population is 65 or over and a smaller percentage is in younger age brackets. Use matrices to solve the problems in Exercises 61 and 62. Let $x = 0$ represent 2010 and $x = 40$ represent 2050. Express percents in decimal form.

61. In 2010, 13.0% of the population was 65 or older. By 2050, this percentage is expected to be 20.2%. The percentage of the population aged 25–34 in 2010 was 13.5%. That age group is expected to include 12.8% of the population in 2050. (*Source:* U.S. Census Bureau.)

   (a) Assuming these population changes are linear, use the data for the 65-or-older age group to write a linear equation. Then do the same for the 25–34 age group.

   (b) Solve the system of linear equations from part (a). In what year will the two age groups include the same percentage of the population? What is that percentage?

   (c) Does your answer to part (b) suggest that the *number* of people in the U.S. population aged 25–34 is decreasing? Explain.

62. In 2010, 13.3% of the U.S. population was aged 35–44. This percentage is expected to decrease to 12.5% in 2050. (*Source:* U.S. Census Bureau.)

   (a) Write a linear equation representing this population change.

   (b) Solve the system containing the equation from part (a) and the equation from **Exercise 61** for the 65-or-older age group. Give the year and percentage when these two age groups will include the same portion of the population.

63. *(Modeling) Athlete's Weight and Height*
The relationship between a professional basketball player's height H (in inches) and weight W (in pounds) was modeled using two different samples of players. The resulting equations that modeled the two samples were

$$W = 7.46H - 374$$

and $\quad W = 7.93H - 405.$

(a) Use each equation to predict the weight of a 6 ft 11 in. professional basketball player.

(b) According to each model, what change in weight is associated with a 1-in. increase in height?

(c) Determine the weight and height where the two models agree.

64. *(Modeling) Traffic Congestion*
At rush hours, substantial traffic congestion is encountered at the traffic intersections shown in the figure. (All streets are one-way.) The city wishes to improve the signals at these corners to speed the flow of traffic. The traffic engineers first gather data. As the figure shows, 700 cars per hour come down M Street to intersection A, and 300 cars per hour come to intersection A on 10th Street. A total of x_1 of these cars leave A on M Street, while x_4 cars leave A on 10th Street. The number of cars entering A must equal the number leaving, which suggests the following equation.

$$x_1 + x_4 = 700 + 300$$
$$x_1 + x_4 = 1000$$

For intersection B, x_1 cars enter B on M Street, and x_2 cars enter B on 11th Street. As the figure shows, 900 cars leave B on 11th, while 200 leave on M, which leads to the following equation.

$$x_1 + x_2 = 900 + 200$$
$$x_1 + x_2 = 1100$$

At intersection C, 400 cars enter on N Street and 300 on 11th Street, while x_2 leave on 11th Street and x_3 leave on N Street.

$$x_2 + x_3 = 400 + 300$$
$$x_2 + x_3 = 700$$

Finally, intersection D has x_3 cars entering on N and x_4 entering on 10th. There are 400 cars leaving D on 10th and 200 leaving on N.

(a) Set up an equation for intersection D.

(b) Use the four equations to write an augmented matrix, and then transform it so that 1s are on the diagonal and 0s are below. This is **triangular form.**

(c) Since you got a row of all 0s, the system of equations does not have a unique solution. Write three equations, corresponding to the three nonzero rows of the matrix. Solve each of the equations for x_4.

(d) One of your equations should have been $x_4 = 1000 - x_1$. What is the greatest possible value of x_1 so that x_4 is not negative?

(e) Another equation should have been $x_4 = x_2 - 100$. Find the least possible value of x_2 so that x_4 is not negative.

(f) Find the greatest possible values of x_3 and x_4 so that neither variable is negative.

(g) Use the results of parts (a)–(f) to give a solution for the problem in which all the equations are satisfied and all variables are nonnegative. Is the solution unique?

Relating Concepts

For individual or collaborative investigation *(Exercises 65–68)*

(Modeling) Number of Fawns To model the spring fawn count F from the adult pronghorn population A, the precipitation P, and the severity of the winter W, environmentalists have used the equation

$$F = a + bA + cP + dW,$$

where the coefficients a, b, c, and d are constants that must be determined before using the equation. *(Winter severity is scaled between 1 and 5, with 1 being mild and 5 being severe.)* **Work Exercises 65–68 in order.** (*Source:* Brase, C. and C. Brase, *Understandable Statistics*, D.C. Heath and Company; Bureau of Land Management.)

65. Substitute the values for F, A, P, and W from the table for Years 1–4 into the equation

$$F = a + bA + cP + dW$$

and obtain four linear equations involving a, b, c, and d.

Year	Fawns	Adults	Precip. (in inches)	Winter Severity
1	239	871	11.5	3
2	234	847	12.2	2
3	192	685	10.6	5
4	343	969	14.2	1
5	?	960	12.6	3

66. Write an augmented matrix representing the system in **Exercise 65,** and solve for a, b, c, and d.

67. Write the equation for F using the values found in **Exercise 66** for the coefficients.

68. Use the information in the table to predict the spring fawn count in Year 5. (Compare this with the actual count of 320.)

(Modeling) Number of Calculations When computers are programmed to solve large linear systems involved in applications like designing aircraft or electrical circuits, they frequently use an algorithm that is similar to the Gauss-Jordan method presented in this section. Solving a linear system with n equations and n variables requires the computer to perform a total of

$$T(n) = \frac{2}{3}n^3 + \frac{3}{2}n^2 - \frac{7}{6}n$$

arithmetic calculations, including additions, subtractions, multiplications, and divisions. Use this model to solve the problems in Exercises 69–72. (*Source:* Burden, R. and J. Faires, *Numerical Analysis*, Sixth Edition, Brooks/Cole Publishing Company.)

69. Compute T for the following values of n.

$n = 3, 6, 10, 29, 100, 200, 400, 1000, 5000, 10,000, 100,000$

Write the results in a table.

70. In 1940, John Atanasoff, a physicist from Iowa State University, wanted to solve a 29×29 linear system of equations. How many arithmetic operations would this have required? Is this too many to do by hand? (Atanasoff's work led to the invention of the first fully electronic digital computer.) (*Source: The Gazette.*)

Atanasoff-Berry Computer

71. If the number of equations and variables is doubled, does the number of arithmetic operations double?

72. A Cray-T90 supercomputer can execute up to 60 billion arithmetic operations per second. How many hours would be required to solve a linear system with 100,000 variables?

5.3 Determinant Solution of Linear Systems

- Determinants
- Cofactors
- Evaluating $n \times n$ Determinants
- Determinant Theorems
- Cramer's Rule

Determinants Every $n \times n$ matrix A is associated with a real number called the **determinant** of A, written $|A|$. The determinant of a 2×2 matrix is defined as follows.

Determinant of a 2 × 2 Matrix

If $\quad A = \begin{bmatrix} a_{11} & a_{12} \\ a_{21} & a_{22} \end{bmatrix}, \quad$ then $\quad |A| = \begin{vmatrix} a_{11} & a_{12} \\ a_{21} & a_{22} \end{vmatrix} = a_{11}a_{22} - a_{21}a_{12}.$

NOTE *Matrices are enclosed with square brackets, while determinants are denoted with vertical bars.* A matrix is an *array* of numbers, but its determinant is a *single* number.

The arrows in the diagram in the margin will remind you which products to find when evaluating a 2×2 determinant.

EXAMPLE 1 Evaluating a 2 × 2 Determinant

Let $A = \begin{bmatrix} -3 & 4 \\ 6 & 8 \end{bmatrix}$. Find $|A|$.

ALGEBRAIC SOLUTION

Use the definition with

$$a_{11} = -3, \quad a_{12} = 4, \quad a_{21} = 6, \quad a_{22} = 8.$$

$$|A| = \underset{\underset{a_{11} \ a_{22}}{\uparrow \quad \uparrow}}{-3 \cdot 8} \ - \ \underset{\underset{a_{21} \ a_{12}}{\uparrow \quad \uparrow}}{6 \cdot 4}$$

$$= -24 - 24$$

$$= -48$$

GRAPHING CALCULATOR SOLUTION

We can define a matrix and then use the capability of a graphing calculator to find the determinant of the matrix. In the screen in **Figure 8,** the symbol $\det([A])$ represents the determinant of $[A]$.

Figure 8

✔ *Now Try Exercise 1.*

LOOKING AHEAD TO CALCULUS

Determinants are used in calculus to find **vector cross products,** which are used to study the effect of forces in the plane or in space.

Determinant of a 3 × 3 Matrix

If $A = \begin{bmatrix} a_{11} & a_{12} & a_{13} \\ a_{21} & a_{22} & a_{23} \\ a_{31} & a_{32} & a_{33} \end{bmatrix}$, then the determinant of A, symbolized $|A|$, is defined as follows.

$$|A| = \begin{vmatrix} a_{11} & a_{12} & a_{13} \\ a_{21} & a_{22} & a_{23} \\ a_{31} & a_{32} & a_{33} \end{vmatrix} = (a_{11}a_{22}a_{33} + a_{12}a_{23}a_{31} + a_{13}a_{21}a_{32}) - (a_{31}a_{22}a_{13} + a_{32}a_{23}a_{11} + a_{33}a_{21}a_{12})$$

The terms on the right side of the equation in the definition of $|A|$ can be rearranged to get the following.

$$\begin{vmatrix} a_{11} & a_{12} & a_{13} \\ a_{21} & a_{22} & a_{23} \\ a_{31} & a_{32} & a_{33} \end{vmatrix} = a_{11}(a_{22}a_{33} - a_{32}a_{23}) - a_{21}(a_{12}a_{33} - a_{32}a_{13}) + a_{31}(a_{12}a_{23} - a_{22}a_{13})$$

Each quantity in parentheses represents the determinant of a 2 × 2 matrix that is the part of the 3 × 3 matrix remaining when the row and column of the multiplier are eliminated, as shown below.

$$a_{11}(a_{22}a_{33} - a_{32}a_{23}) \qquad \begin{bmatrix} a_{11} & a_{12} & a_{13} \\ a_{21} & a_{22} & a_{23} \\ a_{31} & a_{32} & a_{33} \end{bmatrix}$$

$$a_{21}(a_{12}a_{33} - a_{32}a_{13}) \qquad \begin{bmatrix} a_{11} & a_{12} & a_{13} \\ a_{21} & a_{22} & a_{23} \\ a_{31} & a_{32} & a_{33} \end{bmatrix}$$

$$a_{31}(a_{12}a_{23} - a_{22}a_{13}) \qquad \begin{bmatrix} a_{11} & a_{12} & a_{13} \\ a_{21} & a_{22} & a_{23} \\ a_{31} & a_{32} & a_{33} \end{bmatrix}$$

Cofactors   The determinant of each 2 × 2 matrix above is the **minor** of the associated element in the 3 × 3 matrix. The symbol M_{ij} represents the minor that results when row i and column j are eliminated. The following list gives some of the minors from the matrix above.

Element	Minor	Element	Minor
a_{11}	$M_{11} = \begin{vmatrix} a_{22} & a_{23} \\ a_{32} & a_{33} \end{vmatrix}$	a_{22}	$M_{22} = \begin{vmatrix} a_{11} & a_{13} \\ a_{31} & a_{33} \end{vmatrix}$
a_{21}	$M_{21} = \begin{vmatrix} a_{12} & a_{13} \\ a_{32} & a_{33} \end{vmatrix}$	a_{23}	$M_{23} = \begin{vmatrix} a_{11} & a_{12} \\ a_{31} & a_{32} \end{vmatrix}$
a_{31}	$M_{31} = \begin{vmatrix} a_{12} & a_{13} \\ a_{22} & a_{23} \end{vmatrix}$	a_{33}	$M_{33} = \begin{vmatrix} a_{11} & a_{12} \\ a_{21} & a_{22} \end{vmatrix}$

In a 4×4 matrix, the minors are determinants of 3×3 matrices. Similarly, an $n \times n$ matrix has minors that are determinants of $(n - 1) \times (n - 1)$ matrices.

To find the determinant of a 3×3 or larger matrix, first choose any row or column. Then the minor of each element in that row or column must be multiplied by $+1$ or -1, depending on whether the sum of the row number and column number is even or odd. The product of a minor and the number $+1$ or -1 is called a *cofactor*.

Cofactor

Let M_{ij} be the minor for element a_{ij} in an $n \times n$ matrix. The **cofactor** of a_{ij}, written A_{ij}, is defined as follows.

$$A_{ij} = (-1)^{i+j} \cdot M_{ij}$$

EXAMPLE 2 **Finding Cofactors of Elements**

Find the cofactor of each of the following elements of the given matrix.

$$\begin{bmatrix} 6 & 2 & 4 \\ 8 & 9 & 3 \\ 1 & 2 & 0 \end{bmatrix}$$

(a) 6 **(b)** 3 **(c)** 8

SOLUTION

(a) Since 6 is in the first row, first column of the matrix, $i = 1$ and $j = 1$, so

$$M_{11} = \begin{vmatrix} 9 & 3 \\ 2 & 0 \end{vmatrix} = -6. \text{ The cofactor is}$$

$$(-1)^{1+1}(-6) = 1(-6) = -6.$$

(b) Here $i = 2$ and $j = 3$, so $M_{23} = \begin{vmatrix} 6 & 2 \\ 1 & 2 \end{vmatrix} = 10.$ The cofactor is

$$(-1)^{2+3}(10) = -1(10) = -10.$$

(c) We have $i = 2$ and $j = 1$, and $M_{21} = \begin{vmatrix} 2 & 4 \\ 2 & 0 \end{vmatrix} = -8.$ The cofactor is

$$(-1)^{2+1}(-8) = -1(-8) = 8.$$

✔ *Now Try Exercise 11.*

Evaluating $n \times n$ Determinants The determinant of a 3×3 or larger matrix is found as follows.

Finding the Determinant of a Matrix

Multiply each element in any row or column of the matrix by its cofactor. The sum of these products gives the value of the determinant.

The process of forming this sum of products is called **expansion by a given row or column.**

EXAMPLE 3 Evaluating a 3 × 3 Determinant

Evaluate $\begin{vmatrix} 2 & -3 & -2 \\ -1 & -4 & -3 \\ -1 & 0 & 2 \end{vmatrix}$, expanding by the second column.

SOLUTION To find this determinant, first find the minor of each element in the second column.

$$M_{12} = \begin{vmatrix} -1 & -3 \\ -1 & 2 \end{vmatrix} = -1(2) - (-1)(-3) = -5$$

$$M_{22} = \begin{vmatrix} 2 & -2 \\ -1 & 2 \end{vmatrix} = 2(2) - (-1)(-2) = 2$$

$$M_{32} = \begin{vmatrix} 2 & -2 \\ -1 & -3 \end{vmatrix} = 2(-3) - (-1)(-2) = -8$$

> Use parentheses, and keep track of all negative signs to avoid errors.

Now find the cofactor of each element of these minors.

$$A_{12} = (-1)^{1+2} \cdot M_{12} = (-1)^3 \cdot (-5) = -1(-5) = 5$$
$$A_{22} = (-1)^{2+2} \cdot M_{22} = (-1)^4 \cdot 2 = 1 \cdot 2 = 2$$
$$A_{32} = (-1)^{3+2} \cdot M_{32} = (-1)^5 \cdot (-8) = -1(-8) = 8$$

Find the determinant by multiplying each cofactor by its corresponding element in the matrix and finding the sum of these products.

$$\begin{vmatrix} 2 & -3 & -2 \\ -1 & -4 & -3 \\ -1 & 0 & 2 \end{vmatrix} = a_{12} \cdot A_{12} + a_{22} \cdot A_{22} + a_{32} \cdot A_{32}$$

$$= -3(5) + (-4)2 + 0(8)$$
$$= -15 + (-8) + 0$$
$$= -23$$

✔ *Now Try Exercise 15.*

```
[B]
   [2  -3  -2]
   [-1 -4  -3]
   [-1  0   2]
det([B])
              -23
```

The matrix in **Example 3** can be entered as [B] and its determinant found using a graphing calculator, as shown in the screen above.

In **Example 3,** we would have found the same answer using any row or column of the matrix. One reason we used column 2 is that it contains a 0 element, so it was not really necessary to calculate M_{32} and A_{32}.

Instead of calculating $(-1)^{i+j}$ for a given element, we can use the sign checkerboard shown below. The signs alternate for each row and column, beginning with + in the first row, first column position. If we expand a 3 × 3 matrix about row 3, for example, the first minor would have a + sign associated with it, the second minor a − sign, and the third minor a + sign.

Sign array
for 3 × 3 matrices

$$\begin{array}{ccc} + & - & + \\ - & + & - \\ + & - & + \end{array}$$

This array of signs can be extended for determinants of 4 × 4 and larger matrices.

Determinant Theorems   The following theorems are true for square matrices of any dimension and can be used to simplify the process of finding determinants.

Determinant Theorems

1. If every element in a row (or column) of matrix A is 0, then $|A| = 0$.
2. If the rows of matrix A are the corresponding columns of matrix B, then $|B| = |A|$.
3. If any two rows (or columns) of matrix A are interchanged to form matrix B, then $|B| = -|A|$.
4. Suppose matrix B is formed by multiplying every element of a row (or column) of matrix A by the real number k. Then $|B| = k \cdot |A|$.
5. If two rows (or columns) of matrix A are identical, then $|A| = 0$.
6. Changing a row (or column) of a matrix by adding to it a constant times another row (or column) does not change the determinant of the matrix.
7. If matrix A is in triangular form, having zeros either above or below the main diagonal, then $|A|$ is the product of the elements on the main diagonal of A.

EXAMPLE 4   **Using the Determinant Theorems**

Use the determinant theorems to find the value of each determinant.

(a) $\begin{vmatrix} -2 & 4 & 2 \\ 6 & 7 & 3 \\ 0 & 16 & 8 \end{vmatrix}$ (b) $\begin{vmatrix} 3 & -7 & 4 & 10 \\ 0 & 1 & 8 & 3 \\ 0 & 0 & -5 & 2 \\ 0 & 0 & 0 & 6 \end{vmatrix}$

SOLUTION

(a) Use determinant theorem 6 to obtain a 0 in the second row of the first column. Multiply each element in the first row by 3, and add the result to the corresponding element in the second row.

$$\begin{vmatrix} -2 & 4 & 2 \\ 0 & 19 & 9 \\ 0 & 16 & 8 \end{vmatrix} \quad 3R1 + R2$$

Now, find the determinant by expanding by column 1.

$$-2(-1)^{1+1}\begin{vmatrix} 19 & 9 \\ 16 & 8 \end{vmatrix} = -2(1)(8) = -16 \quad 19(8) - 16(9) = 152 - 144 = 8$$

(b) Use determinant theorem 7 to find the determinant of this triangular matrix by multiplying the elements on the main diagonal.

$$\begin{vmatrix} 3 & -7 & 4 & 10 \\ 0 & 1 & 8 & 3 \\ 0 & 0 & -5 & 2 \\ 0 & 0 & 0 & 6 \end{vmatrix} = 3(1)(-5)(6) = -90$$

☑ *Now Try Exercises 65 and 67.*

Cramer's Rule   The elimination method can be used to develop a process for solving a linear system in two unknowns using determinants. Consider the following system.

$$a_1x + b_1y = c_1 \quad \text{(1)}$$

$$a_2x + b_2y = c_2 \quad \text{(2)}$$

The variable y in this system of equations can be eliminated by using multiplication to create coefficients that are additive inverses and by adding the two equations.

$$a_1b_2x + b_1b_2y = c_1b_2 \qquad \text{Multiply (1) by } b_2.$$

$$\underline{-a_2b_1x - b_1b_2y = -c_2b_1} \qquad \text{Multiply (2) by } -b_1.$$

$$(a_1b_2 - a_2b_1)x \qquad\qquad = c_1b_2 - c_2b_1 \quad \text{Add.}$$

$$x = \frac{c_1b_2 - c_2b_1}{a_1b_2 - a_2b_1}, \quad \text{if } a_1b_2 - a_2b_1 \neq 0$$

Similarly, the variable x can be eliminated.

$$-a_1a_2x - a_2b_1y = -a_2c_1 \qquad \text{Multiply (1) by } -a_2.$$

$$\underline{a_1a_2x + a_1b_2y = a_1c_2} \qquad \text{Multiply (2) by } a_1.$$

$$(a_1b_2 - a_2b_1)y = a_1c_2 - a_2c_1 \quad \text{Add.}$$

$$y = \frac{a_1c_2 - a_2c_1}{a_1b_2 - a_2b_1}, \quad \text{if } a_1b_2 - a_2b_1 \neq 0$$

Both numerators and the common denominator of these values for x and y can be written as determinants.

$$c_1b_2 - c_2b_1 = \begin{vmatrix} c_1 & b_1 \\ c_2 & b_2 \end{vmatrix}, \quad a_1c_2 - a_2c_1 = \begin{vmatrix} a_1 & c_1 \\ a_2 & c_2 \end{vmatrix}, \quad \text{and } a_1b_2 - a_2b_1 = \begin{vmatrix} a_1 & b_1 \\ a_2 & b_2 \end{vmatrix}$$

The solutions for x and y can be written using these determinants.

$$x = \frac{\begin{vmatrix} c_1 & b_1 \\ c_2 & b_2 \end{vmatrix}}{\begin{vmatrix} a_1 & b_1 \\ a_2 & b_2 \end{vmatrix}} \quad \text{and} \quad y = \frac{\begin{vmatrix} a_1 & c_1 \\ a_2 & c_2 \end{vmatrix}}{\begin{vmatrix} a_1 & b_1 \\ a_2 & b_2 \end{vmatrix}}, \quad \text{if } \begin{vmatrix} a_1 & b_1 \\ a_2 & b_2 \end{vmatrix} \neq 0.$$

We denote the three determinants in the solution as follows.

$$\begin{vmatrix} a_1 & b_1 \\ a_2 & b_2 \end{vmatrix} = D, \quad \begin{vmatrix} c_1 & b_1 \\ c_2 & b_2 \end{vmatrix} = D_x, \quad \text{and} \quad \begin{vmatrix} a_1 & c_1 \\ a_2 & c_2 \end{vmatrix} = D_y$$

NOTE   The elements of D are the four coefficients of the variables in the given system. The elements of D_x are obtained by replacing the coefficients of x in D by the respective constants, and the elements of D_y are obtained by replacing the coefficients of y in D by the respective constants.

These results are summarized as **Cramer's rule.**

Cramer's Rule for Two Equations in Two Variables

Given the system

$$a_1x + b_1y = c_1$$
$$a_2x + b_2y = c_2,$$

if $D \neq 0$, then the system has the unique solution

$$x = \frac{D_x}{D} \quad \text{and} \quad y = \frac{D_y}{D},$$

where $D = \begin{vmatrix} a_1 & b_1 \\ a_2 & b_2 \end{vmatrix}$, $D_x = \begin{vmatrix} c_1 & b_1 \\ c_2 & b_2 \end{vmatrix}$, and $D_y = \begin{vmatrix} a_1 & c_1 \\ a_2 & c_2 \end{vmatrix}$.

CAUTION *Cramer's rule does not apply if $D = 0$.* When $D = 0$, the system is inconsistent or has infinitely many solutions. Evaluate D first.

EXAMPLE 5 **Applying Cramer's Rule to a 2 × 2 System**

Use Cramer's rule to solve the system.

$$5x + 7y = -1$$
$$6x + 8y = 1$$

SOLUTION By Cramer's rule, $x = \frac{D_x}{D}$ and $y = \frac{D_y}{D}$. Find D first, since if $D = 0$, Cramer's rule does not apply. If $D \neq 0$, then find D_x and D_y.

$$D = \begin{vmatrix} 5 & 7 \\ 6 & 8 \end{vmatrix} = 5(8) - 6(7) = -2$$

$$D_x = \begin{vmatrix} -1 & 7 \\ 1 & 8 \end{vmatrix} = -1(8) - 1(7) = -15$$

$$D_y = \begin{vmatrix} 5 & -1 \\ 6 & 1 \end{vmatrix} = 5(1) - 6(-1) = 11$$

$$x = \frac{D_x}{D} = \frac{-15}{-2} = \frac{15}{2} \quad \text{and} \quad y = \frac{D_y}{D} = \frac{11}{-2} = -\frac{11}{2} \quad \text{Cramer's rule}$$

The solution set is $\left\{ \left(\frac{15}{2}, -\frac{11}{2} \right) \right\}$. Verify by substituting in the given system.

☑ *Now Try Exercise 81.*

Because graphing calculators can evaluate determinants, they can also be used to apply Cramer's rule to solve a system of linear equations. The screens above support the result in **Example 5.**

Cramer's rule can be generalized.

General Form of Cramer's Rule

Let an $n \times n$ system have linear equations of the form

$$a_1x_1 + a_2x_2 + a_3x_3 + \cdots + a_nx_n = b.$$

Define D as the determinant of the $n \times n$ matrix of all coefficients of the variables. Define D_{x_1} as the determinant obtained from D by replacing the entries in column 1 of D with the constants of the system. Define D_{x_i} as the determinant obtained from D by replacing the entries in column i with the constants of the system. If $D \neq 0$, the unique solution of the system is

$$x_1 = \frac{D_{x_1}}{D}, \quad x_2 = \frac{D_{x_2}}{D}, \quad x_3 = \frac{D_{x_3}}{D}, \quad \ldots, \quad x_n = \frac{D_{x_n}}{D}.$$

EXAMPLE 6 Applying Cramer's Rule to a 3 × 3 System

Use Cramer's rule to solve the system.

$$x + y - z + 2 = 0$$

$$2x - y + z + 5 = 0$$

$$x - 2y + 3z - 4 = 0$$

SOLUTION

$$x + y - z = -2$$

$$2x - y + z = -5 \qquad \text{Rewrite each equation in the form } ax + by + cz + \cdots = k.$$

$$x - 2y + 3z = 4$$

Verify the required determinants.

$$D = \begin{vmatrix} 1 & 1 & -1 \\ 2 & -1 & 1 \\ 1 & -2 & 3 \end{vmatrix} = -3, \qquad D_x = \begin{vmatrix} -2 & 1 & -1 \\ -5 & -1 & 1 \\ 4 & -2 & 3 \end{vmatrix} = 7,$$

$$D_y = \begin{vmatrix} 1 & -2 & -1 \\ 2 & -5 & 1 \\ 1 & 4 & 3 \end{vmatrix} = -22, \qquad D_z = \begin{vmatrix} 1 & 1 & -2 \\ 2 & -1 & -5 \\ 1 & -2 & 4 \end{vmatrix} = -21$$

$$x = \frac{D_x}{D} = \frac{7}{-3} = -\frac{7}{3}, \quad y = \frac{D_y}{D} = \frac{-22}{-3} = \frac{22}{3}, \quad z = \frac{D_z}{D} = \frac{-21}{-3} = 7$$

The solution set is $\left\{ \left(-\frac{7}{3}, \frac{22}{3}, 7 \right) \right\}$.

✔ *Now Try Exercise 93.*

CAUTION As shown in **Example 6,** each equation in the system must be written in the form $ax + by + cz + \cdots = k$ before Cramer's rule is used.

EXAMPLE 7 Showing That Cramer's Rule Does Not Apply

Show that Cramer's rule does not apply to the following system.

$$2x - 3y + 4z = 10$$

$$6x - 9y + 12z = 24$$

$$x + 2y - 3z = 5$$

SOLUTION We need to show that $D = 0$. Expand about column 1.

$$D = \begin{vmatrix} 2 & -3 & 4 \\ 6 & -9 & 12 \\ 1 & 2 & -3 \end{vmatrix} = 2 \begin{vmatrix} -9 & 12 \\ 2 & -3 \end{vmatrix} - 6 \begin{vmatrix} -3 & 4 \\ 2 & -3 \end{vmatrix} + 1 \begin{vmatrix} -3 & 4 \\ -9 & 12 \end{vmatrix}$$

$$= 2(3) - 6(1) + 1(0)$$

$$= 0$$

Since $D = 0$, Cramer's rule does not apply.

✔ *Now Try Exercise 95.*

NOTE *When $D = 0$, the system is either inconsistent or has infinitely many solutions.* Use the elimination method to tell which is the case. Verify that the system in **Example 7** is inconsistent, and thus the solution set is $\varnothing$.

5.3 Exercises

*Find the value of each determinant. **See Example 1.***

1. $\begin{vmatrix} -5 & 9 \\ 4 & -1 \end{vmatrix}$

2. $\begin{vmatrix} -1 & 3 \\ -2 & 9 \end{vmatrix}$

3. $\begin{vmatrix} -1 & -2 \\ 5 & 3 \end{vmatrix}$

4. $\begin{vmatrix} 6 & -4 \\ 0 & -1 \end{vmatrix}$

5. $\begin{vmatrix} 9 & 3 \\ -3 & -1 \end{vmatrix}$

6. $\begin{vmatrix} 0 & 2 \\ 1 & 5 \end{vmatrix}$

7. $\begin{vmatrix} 3 & 4 \\ 5 & -2 \end{vmatrix}$

8. $\begin{vmatrix} -9 & 7 \\ 2 & 6 \end{vmatrix}$

9. $\begin{vmatrix} -7 & 0 \\ 3 & 0 \end{vmatrix}$

10. *Concept Check* Refer to **Exercise 5.** Make a conjecture about the value of the determinant of a matrix in which one row is a multiple of another row.

*Find the cofactor of each element in the second row for each determinant. **See Example 2.***

11. $\begin{vmatrix} -2 & 0 & 1 \\ 1 & 2 & 0 \\ 4 & 2 & 1 \end{vmatrix}$

12. $\begin{vmatrix} 1 & -1 & 2 \\ 1 & 0 & 2 \\ 0 & -3 & 1 \end{vmatrix}$

13. $\begin{vmatrix} 1 & 2 & -1 \\ 2 & 3 & -2 \\ -1 & 4 & 1 \end{vmatrix}$

14. $\begin{vmatrix} 2 & -1 & 4 \\ 3 & 0 & 1 \\ -2 & 1 & 4 \end{vmatrix}$

*Find the value of each determinant. **See Example 3.***

15. $\begin{vmatrix} 4 & -7 & 8 \\ 2 & 1 & 3 \\ -6 & 3 & 0 \end{vmatrix}$

16. $\begin{vmatrix} 8 & -2 & -4 \\ 7 & 0 & 3 \\ 5 & -1 & 2 \end{vmatrix}$

17. $\begin{vmatrix} 1 & 2 & 0 \\ -1 & 2 & -1 \\ 0 & 1 & 4 \end{vmatrix}$

18. $\begin{vmatrix} 2 & 1 & -1 \\ 4 & 7 & -2 \\ 2 & 4 & 0 \end{vmatrix}$

19. $\begin{vmatrix} 10 & 2 & 1 \\ -1 & 4 & 3 \\ -3 & 8 & 10 \end{vmatrix}$

20. $\begin{vmatrix} 7 & -1 & 1 \\ 1 & -7 & 2 \\ -2 & 1 & 1 \end{vmatrix}$

21. $\begin{vmatrix} 1 & -2 & 3 \\ 0 & 0 & 0 \\ 1 & 10 & -12 \end{vmatrix}$

22. $\begin{vmatrix} 2 & 3 & 0 \\ 1 & 9 & 0 \\ -1 & -2 & 0 \end{vmatrix}$

23. $\begin{vmatrix} 3 & 3 & -1 \\ 2 & 6 & 0 \\ -6 & -6 & 2 \end{vmatrix}$

24. $\begin{vmatrix} 5 & -3 & 2 \\ -5 & 3 & -2 \\ 1 & 0 & 1 \end{vmatrix}$

25. $\begin{vmatrix} 1 & 0 & 0 \\ 0 & 1 & 0 \\ 0 & 0 & 1 \end{vmatrix}$

26. $\begin{vmatrix} 1 & 0 & 0 \\ 0 & -1 & 0 \\ 1 & 0 & 1 \end{vmatrix}$

27. $\begin{vmatrix} -2 & 0 & 1 \\ 0 & 1 & 0 \\ 0 & 0 & -1 \end{vmatrix}$

28. $\begin{vmatrix} 0 & 0 & -1 \\ -1 & 0 & 1 \\ 0 & -1 & 0 \end{vmatrix}$

29. $\begin{vmatrix} \sqrt{2} & 4 & 0 \\ 1 & -\sqrt{5} & 7 \\ -5 & \sqrt{5} & 1 \end{vmatrix}$

30. $\begin{vmatrix} \sqrt{3} & 1 & 0 \\ \sqrt{7} & 4 & -1 \\ 5 & 0 & -\sqrt{7} \end{vmatrix}$

🔲 31. $\begin{vmatrix} 0.4 & -0.8 & 0.6 \\ 0.3 & 0.9 & 0.7 \\ 3.1 & 4.1 & -2.8 \end{vmatrix}$

🔲 32. $\begin{vmatrix} -0.3 & -0.1 & 0.9 \\ 2.5 & 4.9 & -3.2 \\ -0.1 & 0.4 & 0.8 \end{vmatrix}$

Relating Concepts

For individual or collaborative investigation (Exercises 33–36)

The determinant of a 3×3 matrix A is defined as follows.

If $A = \begin{bmatrix} a_{11} & a_{12} & a_{13} \\ a_{21} & a_{22} & a_{23} \\ a_{31} & a_{32} & a_{33} \end{bmatrix}$, then $|A| = \begin{vmatrix} a_{11} & a_{12} & a_{13} \\ a_{21} & a_{22} & a_{23} \\ a_{31} & a_{32} & a_{33} \end{vmatrix}$

$$= (a_{11}a_{22}a_{33} + a_{12}a_{23}a_{31} + a_{13}a_{21}a_{32})$$
$$- (a_{31}a_{22}a_{13} + a_{32}a_{23}a_{11} + a_{33}a_{21}a_{12}).$$

Work these exercises in order.

33. The determinant of a 3×3 matrix can also be found using the method of "diagonals." To use this method, first rewrite columns 1 and 2 of matrix A to the right of matrix A. Next, identify the diagonals d_1 through d_6 and multiply their elements. Find the sum of the products from d_1, d_2, and d_3. Then subtract the sum of the products from d_4, d_5, and d_6 from that sum: $(d_1 + d_2 + d_3) - (d_4 + d_5 + d_6)$. Verify that this method produces the same results as the previous method given.

34. Find the determinant $\begin{vmatrix} 1 & 3 & 2 \\ 0 & 2 & 6 \\ 7 & 1 & 5 \end{vmatrix}$ using the method of "diagonals."

35. Refer to **Exercise 34.** Find the determinant by expanding about column 1 and using the method of cofactors. Do these methods give the same determinant for 3×3 matrices?

36. *Concept Check* Does the method of finding a determinant using "diagonals" extend to 4×4 matrices?

Solve each equation for x.

37. $\begin{vmatrix} 5 & x \\ -3 & 2 \end{vmatrix} = 6$

38. $\begin{vmatrix} -0.5 & 2 \\ x & x \end{vmatrix} = 0$

39. $\begin{vmatrix} x & 3 \\ x & x \end{vmatrix} = 4$

40. $\begin{vmatrix} 2x & x \\ 11 & x \end{vmatrix} = 6$

41. $\begin{vmatrix} -2 & 0 & 1 \\ -1 & 3 & x \\ 5 & -2 & 0 \end{vmatrix} = 3$

42. $\begin{vmatrix} 4 & 3 & 0 \\ 2 & 0 & 1 \\ -3 & x & -1 \end{vmatrix} = 5$

43. $\begin{vmatrix} 5 & 3x & -3 \\ 0 & 2 & -1 \\ 4 & -1 & x \end{vmatrix} = -7$

44. $\begin{vmatrix} 2x & 1 & -1 \\ 0 & 4 & x \\ 3 & 0 & 2 \end{vmatrix} = x$

45. $\begin{vmatrix} x & 0 & -1 \\ 2 & -3 & x \\ x & 0 & 7 \end{vmatrix} = 12$

46. *Concept Check* Write the sign array representing $(-1)^{i+j}$ for each element of a 4×4 matrix.

Area of a Triangle *A triangle with vertices at (x_1, y_1), (x_2, y_2), and (x_3, y_3), as shown in the figure, has area equal to the absolute value of D, where*

$$D = \frac{1}{2} \begin{vmatrix} x_1 & y_1 & 1 \\ x_2 & y_2 & 1 \\ x_3 & y_3 & 1 \end{vmatrix}.$$

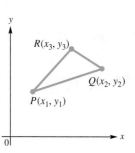

Find the area of each triangle having vertices at P, Q, and R.

47. $P(0, 0), Q(0, 2), R(1, 4)$

48. $P(0, 1), Q(2, 0), R(1, 5)$

49. $P(2, 5), Q(-1, 3), R(4, 0)$

50. $P(2, -2), Q(0, 0), R(-3, -4)$

51. *Area of a Triangle* Find the area of a triangular lot whose vertices have coordinates in feet of

$$(101.3, 52.7), \quad (117.2, 253.9), \quad \text{and} \quad (313.1, 301.6).$$

(*Source:* Al-Khafaji, A. and J. Tooley, *Numerical Methods in Engineering Practice*, Holt, Rinehart, and Winston.)

52. Let $A = \begin{bmatrix} a_{11} & a_{12} & a_{13} \\ a_{21} & a_{22} & a_{23} \\ a_{31} & a_{32} & a_{33} \end{bmatrix}$. Find $|A|$ by expansion about row 3 of the matrix. Show that your result is really equal to $|A|$ as given in the definition of the determinant of a 3×3 matrix at the beginning of this section.

Use the determinant theorems and the fact that $\begin{vmatrix} 1 & 2 & 3 \\ 4 & 5 & 6 \\ 7 & 9 & 10 \end{vmatrix} = 3$ *to find the value of each determinant.*

53. $\begin{vmatrix} 4 & 5 & 6 \\ 1 & 2 & 3 \\ 7 & 9 & 10 \end{vmatrix}$
54. $\begin{vmatrix} 3 & 2 & 1 \\ 6 & 5 & 4 \\ 10 & 9 & 7 \end{vmatrix}$
55. $\begin{vmatrix} 5 & 10 & 15 \\ 4 & 5 & 6 \\ 7 & 9 & 10 \end{vmatrix}$

56. $\begin{vmatrix} 1 & 20 & 3 \\ 4 & 50 & 6 \\ 7 & 90 & 10 \end{vmatrix}$
57. $\begin{vmatrix} 1 & 2 & 3 \\ 4 & 5 & 6 \\ 8 & 11 & 13 \end{vmatrix}$
58. $\begin{vmatrix} 1 & 2 & 0 \\ 4 & 5 & -6 \\ 7 & 9 & -11 \end{vmatrix}$

Use the determinant theorems to find the value of each determinant. ***See Example 4.***

59. $\begin{vmatrix} 1 & 0 & 0 \\ 1 & 0 & 1 \\ 3 & 0 & 0 \end{vmatrix}$
60. $\begin{vmatrix} -1 & 2 & 4 \\ 4 & -8 & -16 \\ 0 & 0 & 0 \end{vmatrix}$
61. $\begin{vmatrix} 6 & 8 & -12 \\ -1 & 0 & 2 \\ 4 & 0 & -8 \end{vmatrix}$

62. $\begin{vmatrix} 4 & 8 & 0 \\ -1 & -2 & 1 \\ 2 & 4 & 3 \end{vmatrix}$
63. $\begin{vmatrix} -4 & 1 & 4 \\ 2 & 0 & 1 \\ 0 & 2 & 4 \end{vmatrix}$
64. $\begin{vmatrix} 6 & 3 & 2 \\ 1 & 0 & 2 \\ 5 & 7 & 3 \end{vmatrix}$

65. $\begin{vmatrix} 0 & 1 & -3 \\ 7 & 5 & 2 \\ 1 & -2 & 6 \end{vmatrix}$
66. $\begin{vmatrix} 7 & 9 & -3 \\ 7 & -6 & 2 \\ 8 & 1 & 0 \end{vmatrix}$
67. $\begin{vmatrix} 1 & 6 & 7 \\ 0 & 6 & 7 \\ 0 & 0 & 9 \end{vmatrix}$

68. $\begin{vmatrix} 7 & 0 & 0 \\ 1 & 6 & 0 \\ 4 & 2 & 4 \end{vmatrix}$
69. $\begin{vmatrix} 2 & -1 & 3 \\ 6 & 4 & 10 \\ 4 & 5 & 7 \end{vmatrix}$
70. $\begin{vmatrix} 9 & 1 & 7 \\ 12 & 5 & 2 \\ 11 & 4 & 3 \end{vmatrix}$

71. $\begin{vmatrix} -1 & 0 & 2 & 3 \\ 5 & 4 & -3 & 7 \\ 8 & 2 & 9 & -5 \\ 4 & 4 & -1 & 10 \end{vmatrix}$
72. $\begin{vmatrix} 5 & 1 & 4 & 2 \\ 4 & -3 & 7 & -4 \\ 5 & 8 & -3 & 6 \\ 9 & 9 & 0 & 8 \end{vmatrix}$
73. $\begin{vmatrix} 4 & 0 & 0 & 2 \\ -1 & 0 & 3 & 0 \\ 2 & 4 & 0 & 1 \\ 0 & 0 & 1 & 2 \end{vmatrix}$

74. $\begin{vmatrix} -2 & 0 & 4 & 2 \\ 3 & 6 & 0 & 4 \\ 0 & 0 & 0 & 3 \\ 9 & 0 & 2 & -1 \end{vmatrix}$
75. $\begin{vmatrix} 3 & -6 & 5 & -1 \\ 0 & 2 & -1 & 3 \\ -6 & 4 & 2 & 0 \\ -7 & 3 & 1 & 1 \end{vmatrix}$
76. $\begin{vmatrix} 4 & 5 & -1 & -1 \\ 2 & -3 & 1 & 0 \\ -5 & 1 & 3 & 9 \\ 0 & -2 & 1 & 5 \end{vmatrix}$

77. *Concept Check* For the system below, match each determinant in (a)–(d) with its equivalent from choices A–D.

$$4x + 3y - 2z = 1$$
$$7x - 4y + 3z = 2$$
$$-2x + y - 8z = 0$$

(a) D **(b)** D_x **(c)** D_y **(d)** D_z

A. $\begin{vmatrix} 1 & 3 & -2 \\ 2 & -4 & 3 \\ 0 & 1 & -8 \end{vmatrix}$ **B.** $\begin{vmatrix} 4 & 3 & 1 \\ 7 & -4 & 2 \\ -2 & 1 & 0 \end{vmatrix}$

C. $\begin{vmatrix} 4 & 1 & -2 \\ 7 & 2 & 3 \\ -2 & 0 & -8 \end{vmatrix}$ **D.** $\begin{vmatrix} 4 & 3 & -2 \\ 7 & -4 & 3 \\ -2 & 1 & -8 \end{vmatrix}$

78. *Concept Check* For the following system, $D = -43$, $D_x = -43$, $D_y = 0$, and $D_z = 43$. What is the solution set of the system?

$$x + 3y - 6z = 7$$
$$2x - y + z = 1$$
$$x + 2y + 2z = -1$$

Use Cramer's rule to solve each system of equations. If $D = 0$, use another method to determine the solution set. ***See Examples 5–7.***

79. $x + y = 4$
$2x - y = 2$

80. $3x + 2y = -4$
$2x - y = -5$

81. $4x + 3y = -7$
$2x + 3y = -11$

82. $4x - y = 0$
$2x + 3y = 14$

83. $5x + 4y = 10$
$3x - 7y = 6$

84. $3x + 2y = -4$
$5x - y = 2$

85. $1.5x + 3y = 5$
$2x + 4y = 3$

86. $12x + 8y = 3$
$1.5x + y = 0.9$

87. $3x + 2y = 4$
$6x + 4y = 8$

88. $4x + 3y = 9$
$12x + 9y = 27$

89. $\dfrac{1}{2}x + \dfrac{1}{3}y = 2$
$\dfrac{3}{2}x - \dfrac{1}{2}y = -12$

90. $-\dfrac{3}{4}x + \dfrac{2}{3}y = 16$
$\dfrac{5}{2}x + \dfrac{1}{2}y = -37$

91. $2x - y + 4z = -2$
$3x + 2y - z = -3$
$x + 4y + 2z = 17$

92. $x + y + z = 4$
$2x - y + 3z = 4$
$4x + 2y - z = -15$

93. $4x - 3y + z + 1 = 0$
$5x + 7y + 2z + 2 = 0$
$3x - 5y - z - 1 = 0$

94. $2x - 3y + z - 8 = 0$
$-x - 5y + z + 4 = 0$
$3x - 5y + 2z - 12 = 0$

95. $x + 2y + 3z = 4$
$4x + 3y + 2z = 1$
$-x - 2y - 3z = 0$

96. $2x - y + 3z = 1$
$-2x + y - 3z = 2$
$5x - y + z = 2$

97. $-2x - 2y + 3z = 4$
$5x + 7y - z = 2$
$2x + 2y - 3z = -4$

98. $3x - 2y + 4z = 1$
$4x + y - 5z = 2$
$-6x + 4y - 8z = -2$

99. $5x - y = -4$
$3x + 2z = 4$
$4y + 3z = 22$

100. $3x + 5y = -7$
$2x + 7z = 2$
$4y + 3z = -8$

101. $x + 2y = 10$
$3x + 4z = 7$
$-y - z = 1$

102. $5x - 2y = 3$
$4y + z = 8$
$x + 2z = 4$

(Modeling) *Solve the systems in Exercises 103 and 104.*

103. *Roof Trusses* The simplest type of roof truss is a triangle. The truss shown in the figure is used to frame roofs of small buildings. If a 100-pound force is applied at the peak of the truss, then the forces or weights W_1 and W_2 exerted parallel to each rafter of the truss are determined by the following linear system of equations.

$$\frac{\sqrt{3}}{2}(W_1 + W_2) = 100$$

$$W_1 - W_2 = 0$$

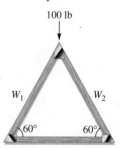

Solve the system to find W_1 and W_2. (*Source:* Hibbeler, R., *Structural Analysis,* Fourth Edition. Reprinted by permission of Pearson Education, Inc., Upper Saddle River, NJ.)

104. *Roof Trusses* **(Refer to Exercise 103.)** Use the following system of equations to determine the forces or weights W_1 and W_2 exerted on each rafter for the truss shown in the figure.

$$W_1 + \sqrt{2}W_2 = 300$$
$$\sqrt{3}W_1 - \sqrt{2}W_2 = 0$$

Solve each system for x and y using Cramer's rule. Assume a and b are nonzero constants.

105. $bx + y = a^2$
　　$ax + y = b^2$

106. $ax + by = \dfrac{b}{a}$

　　$x + \ y = \dfrac{1}{b}$

107. $b^2x + a^2y = b^2$
　　$ax + \ by = a$

108. $\ x + by = b$
　　$ax + \ y = a$

109. Use Cramer's Rule to find the solution set if $a, b, c, d, e,$ and f are consecutive integers.

$$ax + by = c$$
$$dx + ey = f$$

110. In the following system, $a, b, c, \dots, l$ are consecutive integers. Express the solution set in terms of z.

$$ax + by + cz = d$$
$$ex + fy + gz = h$$
$$ix + jy + kz = l$$

5.4 Partial Fractions

- **Decomposition of Rational Expressions**
- **Distinct Linear Factors**
- **Repeated Linear Factors**
- **Distinct Linear and Quadratic Factors**
- **Repeated Quadratic Factors**

Decomposition of Rational Expressions The sums of rational expressions are found by combining two or more rational expressions into one rational expression. Here, the reverse process is considered:

> *Given one rational expression, express it as the sum of two or more rational expressions.*

A special type of sum involving rational expressions is a **partial fraction decomposition**—each term in the sum is a **partial fraction.** The technique of decomposing a rational expression into partial fractions is useful in calculus and other areas of mathematics.

Add rational expressions ➜

$$\frac{2}{x+1} + \frac{3}{x} = \frac{5x+3}{x(x+1)}$$

⬅ Partial fraction decomposition

LOOKING AHEAD TO CALCULUS

In calculus, partial fraction decomposition provides a powerful technique for determining integrals of rational functions.

Partial Fraction Decomposition of $\frac{f(x)}{g(x)}$

To form a partial fraction decomposition of a rational expression, follow these steps.

Step 1 If $\frac{f(x)}{g(x)}$ is not a proper fraction (a fraction with the numerator of lesser degree than the denominator), divide $f(x)$ by $g(x)$. For example,

$$\frac{x^4 - 3x^3 + x^2 + 5x}{x^2 + 3} = x^2 - 3x - 2 + \frac{14x + 6}{x^2 + 3}.$$

Then apply the following steps to the remainder, which is a proper fraction.

Step 2 Factor the denominator $g(x)$ completely into factors of the form $(ax + b)^m$ or $(cx^2 + dx + e)^n$, where $cx^2 + dx + e$ is irreducible and m and n are positive integers.

Step 3 **(a)** For each distinct linear factor $(ax + b)$, the decomposition must include the term $\frac{A}{ax + b}$.

(b) For each repeated linear factor $(ax + b)^m$, the decomposition must include the terms

$$\frac{A_1}{ax + b} + \frac{A_2}{(ax + b)^2} + \cdots + \frac{A_m}{(ax + b)^m}.$$

Step 4 **(a)** For each distinct quadratic factor $(cx^2 + dx + e)$, the decomposition must include the term $\frac{Bx + C}{cx^2 + dx + e}$.

(b) For each repeated quadratic factor $(cx^2 + dx + e)^n$, the decomposition must include the terms

$$\frac{B_1x + C_1}{cx^2 + dx + e} + \frac{B_2x + C_2}{(cx^2 + dx + e)^2} + \cdots + \frac{B_nx + C_n}{(cx^2 + dx + e)^n}.$$

Step 5 Use algebraic techniques to solve for the constants in the numerators of the decomposition.

To find the constants in Step 5, the goal is to get a system of equations with as many equations as there are unknowns in the numerators.

Distinct Linear Factors

EXAMPLE 1 Finding a Partial Fraction Decomposition

Find the partial fraction decomposition of $\dfrac{2x^4 - 8x^2 + 5x - 2}{x^3 - 4x}$.

SOLUTION The given fraction is not a proper fraction—the numerator has greater degree than the denominator. Perform the division.

$$\begin{array}{r} 2x \\ x^3 - 4x \overline{)2x^4 - 8x^2 + 5x - 2} \\ \underline{2x^4 - 8x^2 } \\ 5x - 2 \end{array}$$ (Section R.3)

The quotient is $\dfrac{2x^4 - 8x^2 + 5x - 2}{x^3 - 4x} = 2x + \dfrac{5x - 2}{x^3 - 4x}.$

Now, work with the remainder fraction. Factor the denominator as

$$x^3 - 4x = x(x + 2)(x - 2).$$

The factors are distinct linear factors. Use Step 3(a) to write the decomposition as

$$\frac{5x - 2}{x^3 - 4x} = \frac{A}{x} + \frac{B}{x + 2} + \frac{C}{x - 2}, \quad (1)$$

where A, B, and C are constants that need to be found. Multiply each side of equation (1) by $x(x + 2)(x - 2)$ to obtain

$$5x - 2 = A(x + 2)(x - 2) + Bx(x - 2) + Cx(x + 2). \quad (2)$$

Equation (1) is an identity since each side represents the same rational expression. Thus, equation (2) is also an identity. Equation (1) holds for all values of x except 0, -2, and 2. However, equation (2) holds for all values of x. In particular, substituting 0 for x in equation (2) gives $-2 = -4A$, so $A = \frac{1}{2}$. Similarly, choosing $x = -2$ gives $-12 = 8B$, so $B = -\frac{3}{2}$. Finally, choosing $x = 2$ gives $8 = 8C$, so $C = 1$. The remainder rational expression can be written as the following sum of partial fractions.

$$\frac{5x - 2}{x^3 - 4x} = \frac{1}{2x} + \frac{-3}{2(x + 2)} + \frac{1}{x - 2}$$

The given rational expression can be written as follows.

$$\frac{2x^4 - 8x^2 + 5x - 2}{x^3 - 4x} = 2x + \frac{1}{2x} + \frac{-3}{2(x + 2)} + \frac{1}{x - 2}$$

Check the work by combining the terms on the right.

✔ *Now Try Exercise 15.*

Repeated Linear Factors

EXAMPLE 2 **Finding a Partial Fraction Decomposition**

Find the partial fraction decomposition of $\dfrac{2x}{(x - 1)^3}$.

SOLUTION This is a proper fraction. The denominator is already factored with repeated linear factors. Write the decomposition as shown, by using Step 3(b).

$$\frac{2x}{(x - 1)^3} = \frac{A}{x - 1} + \frac{B}{(x - 1)^2} + \frac{C}{(x - 1)^3}$$

Clear the denominators by multiplying each side of this equation by $(x - 1)^3$.

$$2x = A(x - 1)^2 + B(x - 1) + C \quad \text{(Section R.5)}$$

Substituting 1 for x leads to $C = 2$.

$$2x = A(x - 1)^2 + B(x - 1) + 2 \quad (1)$$

The only root has been substituted, and values for A and B still need to be found. However, *any* number can be substituted for x. For example, when we choose $x = -1$ (because it is easy to substitute), equation (1) becomes the following.

$$-2 = 4A - 2B + 2 \quad \text{Let } x = -1 \text{ in (1).}$$

$$-4 = 4A - 2B \quad \text{Subtract 2.}$$

$$-2 = 2A - B \quad \text{Divide by 2.} \quad (2)$$

Substituting 0 for x in equation (1) gives another equation in A and B.

$$0 = A - B + 2 \quad \text{Let } x = 0 \text{ in (1).}$$

$$2 = -A + B \quad \text{Subtract 2. Multiply by } -1. \quad (3)$$

Now, solve the system of equations (2) and (3) as shown in the margin to get $A = 0$ and $B = 2$. Substitute these values for A and B and 2 for C to obtain the partial fraction decomposition.

$$\frac{2x}{(x-1)^3} = \frac{2}{(x-1)^2} + \frac{2}{(x-1)^3}$$

We needed three substitutions because there were three constants to evaluate: A, B, and C. To check this result, we could combine the terms on the right.

☑ *Now Try Exercise 11.*

$$-2 = 2A - B \quad (2)$$

$$\underline{2 = -A + B} \quad (3)$$

$$0 = A \quad \text{Add.}$$

If $A = 0$, then using equation (3),

$$2 = 0 + B$$

$$2 = B.$$

Distinct Linear and Quadratic Factors

EXAMPLE 3 **Finding a Partial Fraction Decomposition**

Find the partial fraction decomposition of $\dfrac{x^2 + 3x - 1}{(x+1)(x^2+2)}$.

SOLUTION   The denominator $(x+1)(x^2+2)$ has distinct linear and quadratic factors, where neither is repeated. Since $x^2 + 2$ cannot be factored, it is irreducible. The partial fraction decomposition is of the following form.

$$\frac{x^2 + 3x - 1}{(x+1)(x^2+2)} = \frac{A}{x+1} + \frac{Bx + C}{x^2 + 2}$$

Multiply each side by $(x+1)(x^2+2)$.

$$x^2 + 3x - 1 = A(x^2 + 2) + (Bx + C)(x + 1) \quad (1)$$

First, substitute -1 for x.

Use parentheses around substituted values to avoid errors.

$$(-1)^2 + 3(-1) - 1 = A[(-1)^2 + 2] + 0$$

$$-3 = 3A$$

$$A = -1$$

Replace A with -1 in equation (1) and substitute any value for x. Let $x = 0$.

$$0^2 + 3(0) - 1 = -1(0^2 + 2) + (B \cdot 0 + C)(0 + 1)$$

$$-1 = -2 + C$$

$$C = 1$$

Now, letting $A = -1$ and $C = 1$, substitute again in equation (1), using another value for x. Let $x = 1$.

$$3 = -3 + (B + 1)(2)$$

$$6 = 2B + 2$$

$$B = 2$$

Use $A = -1$, $B = 2$, and $C = 1$ to find the partial fraction decomposition.

$$\frac{x^2 + 3x - 1}{(x+1)(x^2+2)} = \frac{-1}{x+1} + \frac{2x + 1}{x^2 + 2} \quad \text{Check by combining the terms on the right.}$$

☑ *Now Try Exercise 21.*

For fractions with denominators that have quadratic factors, another method is often more convenient. The system of equations is formed by equating coefficients of like terms on each side of the partial fraction decomposition. For instance, in **Example 3,** equation (1) was

$$x^2 + 3x - 1 = A(x^2 + 2) + (Bx + C)(x + 1). \quad (1)$$

Multiply on the right and collect like terms.

$$x^2 + 3x - 1 = Ax^2 + 2A + Bx^2 + Bx + Cx + C$$

$$1x^2 + 3x - 1 = (A + B)x^2 + (B + C)x + (C + 2A)$$

Now, equate the coefficients of like powers of x to obtain three equations.

$$1 = A + B$$

$$3 = B + C$$

$$-1 = C + 2A$$

Solving this system for A, B, and C gives the partial fraction decomposition.

Repeated Quadratic Factors

EXAMPLE 4 **Finding a Partial Fraction Decomposition**

Find the partial fraction decomposition of $\dfrac{2x}{(x^2 + 1)^2(x - 1)}$.

SOLUTION This expression has both a linear factor and a repeated quadratic factor. Use Steps 3(a) and 4(b) from the beginning of this section.

$$\frac{2x}{(x^2 + 1)^2(x - 1)} = \frac{Ax + B}{x^2 + 1} + \frac{Cx + D}{(x^2 + 1)^2} + \frac{E}{x - 1}$$

Multiply each side by $(x^2 + 1)^2(x - 1)$.

$$2x = (Ax + B)(x^2 + 1)(x - 1) + (Cx + D)(x - 1) + E(x^2 + 1)^2 \quad (1)$$

If $x = 1$, then equation (1) reduces to $2 = 4E$, or $E = \frac{1}{2}$. Substitute $\frac{1}{2}$ for E in equation (1), and expand and combine like terms on the right.

$$2x = \left(A + \frac{1}{2}\right)x^4 + (-A + B)x^3 + (A - B + C + 1)x^2$$

$$+ (-A + B + D - C)x + \left(-B - D + \frac{1}{2}\right) \quad (2)$$

To get additional equations involving the unknowns, equate the coefficients of like powers of x on the two sides of equation (2). Setting corresponding coefficients of x^4 equal, $0 = A + \frac{1}{2}$, or $A = -\frac{1}{2}$. From the corresponding coefficients of x^3, $0 = -A + B$. Since $A = -\frac{1}{2}$, $B = -\frac{1}{2}$.

Using the coefficients of x^2, $0 = A - B + C + 1$. Since $A = -\frac{1}{2}$ and $B = -\frac{1}{2}$, $C = -1$. Finally, from the coefficients of x, $2 = -A + B + D - C$. Substituting for A, B, and C gives $D = 1$. With $A = -\frac{1}{2}$, $B = -\frac{1}{2}$, $C = -1$, $D = 1$, and $E = \frac{1}{2}$, the given fraction has the following partial fraction decomposition.

$$\frac{2x}{(x^2 + 1)^2(x - 1)} = \frac{-\frac{1}{2}x - \frac{1}{2}}{x^2 + 1} + \frac{-x + 1}{(x^2 + 1)^2} + \frac{\frac{1}{2}}{x - 1},$$

or $\dfrac{2x}{(x^2 + 1)^2(x - 1)} = \dfrac{-(x + 1)}{2(x^2 + 1)} + \dfrac{-x + 1}{(x^2 + 1)^2} + \dfrac{1}{2(x - 1)}$

Simplify complex fractions. **(Section R.5)**

✔ *Now Try Exercise 25.*

In summary, to solve for the constants in the numerators of a partial fraction decomposition, use either of the following methods or a combination of the two.

Techniques for Decomposition into Partial Fractions

Method 1   For Linear Factors

Step 1 Multiply each side of the resulting rational equation by the common denominator.

Step 2 Substitute the zero of each factor in the resulting equation. For repeated linear factors, substitute as many other numbers as necessary to find all the constants in the numerators. The number of substitutions required will equal the number of constants $A, B, \ldots$.

Method 2   For Quadratic Factors

Step 1 Multiply each side of the resulting rational equation by the common denominator.

Step 2 Collect like terms on the right side of the equation.

Step 3 Equate the coefficients of like terms to get a system of equations.

Step 4 Solve the system to find the constants in the numerators.

5.4   Exercises

Find the partial fraction decomposition for each rational expression. **See Examples 1–4.**

1. $\dfrac{5}{3x(2x+1)}$

2. $\dfrac{3x-1}{x(x+1)}$

3. $\dfrac{4x+2}{(x+2)(2x-1)}$

4. $\dfrac{x+2}{(x+1)(x-1)}$

5. $\dfrac{x}{x^2+4x-5}$

6. $\dfrac{5x-3}{x^2-2x-3}$

7. $\dfrac{4}{x(1-x)}$

8. $\dfrac{9}{x(x-3)}$

9. $\dfrac{4x^2-x-15}{x(x+1)(x-1)}$

10. $\dfrac{3}{(x+1)(x+3)}$

11. $\dfrac{2x+1}{(x+2)^3}$

12. $\dfrac{2x}{(x+1)(x+2)^2}$

13. $\dfrac{x^2}{x^2+2x+1}$

14. $\dfrac{2}{x^2(x+3)}$

15. $\dfrac{2x^5+3x^4-3x^3-2x^2+x}{2x^2+5x+2}$

16. $\dfrac{6x^5+7x^4-x^2+2x}{3x^2+2x-1}$

17. $\dfrac{x^3+4}{9x^3-4x}$

18. $\dfrac{x^3+2}{x^3-3x^2+2x}$

19. $\dfrac{-3}{x^2(x^2+5)}$

20. $\dfrac{1}{x^2(x^2-2)}$

21. $\dfrac{3x-2}{(x+4)(3x^2+1)}$

22. $\dfrac{2x+1}{(x+1)(x^2+2)}$

23. $\dfrac{1}{x(2x+1)(3x^2+4)}$

24. $\dfrac{3}{x(x+1)(x^2+1)}$

25. $\dfrac{3x - 1}{x(2x^2 + 1)^2}$

26. $\dfrac{x^4 + 1}{x(x^2 + 1)^2}$

27. $\dfrac{-x^4 - 8x^2 + 3x - 10}{(x + 2)(x^2 + 4)^2}$

28. $\dfrac{3x^4 + x^3 + 5x^2 - x + 4}{(x - 1)(x^2 + 1)^2}$

29. $\dfrac{5x^5 + 10x^4 - 15x^3 + 4x^2 + 13x - 9}{x^3 + 2x^2 - 3x}$

30. $\dfrac{3x^6 + 3x^4 + 3x}{x^4 + x^2}$

31. $\dfrac{x^2}{x^4 - 1}$

32. $\dfrac{-2x^2 - 24}{x^4 - 16}$

33. $\dfrac{4x^2 - 3x - 4}{x^3 + x^2 - 2x}$

34. $\dfrac{2x + 4}{x^3 - 2x^2}$

Chapter 5   Quiz (Sections 5.1–5.4)

Solve each system, using the method indicated, if possible.

1. (Substitution)
$$2x + y = -4$$
$$-x + 2y = 2$$

2. (Substitution)
$$5x + 10y = 10$$
$$x + 2y = 2$$

3. (Elimination)
$$x - y = 6$$
$$x - y = 4$$

4. (Elimination)
$$2x - 3y = 18$$
$$5x + 2y = 7$$

5. (Gauss-Jordan)
$$3x + 5y = -5$$
$$-2x + 3y = 16$$

6. (Cramer's rule)
$$5x + 2y = -3$$
$$4x - 3y = -30$$

7. (Elimination)
$$x + y + z = 1$$
$$-x + y + z = 5$$
$$y + 2z = 5$$

8. (Gauss-Jordan)
$$2x + 4y + 4z = 4$$
$$x + 3y + z = 4$$
$$-x + 3y + 2z = -1$$

9. (Cramer's rule)
$$7x + y - z = 4$$
$$2x - 3y + z = 2$$
$$-6x + 9y - 3z = -6$$

Solve each problem.

10. *Spending on Food* In 2008, the amount spent by a typical American household on food was about $6450. For every $10 spent on food away from home, about $14 was spent on food at home. Find the amount of household spending on food in each category. (*Source:* U.S. Bureau of Labor Statistics.)

11. *Investments* A sum of $5000 is invested in three accounts that pay 3%, 4%, and 6% interest rates. The amount of money invested in the account paying 6% equals the total amount of money invested in the other two accounts, and the total annual interest from all three investments is $240. Find the amount invested at each rate.

12. Let $A = \begin{bmatrix} -5 & 4 \\ 2 & -1 \end{bmatrix}$. Find $|A|$.

13. Evaluate $\begin{vmatrix} -1 & 2 & 4 \\ -3 & -2 & -3 \\ 2 & -1 & 5 \end{vmatrix}$. Use determinant theorems if desired.

Find the partial fraction decomposition for each rational expression.

14. $\dfrac{10x + 13}{x^2 - x - 20}$

15. $\dfrac{2x^2 - 15x - 32}{(x - 1)(x^2 + 6x + 8)}$

5.5 Nonlinear Systems of Equations

- Solving Nonlinear Systems with Real Solutions
- Solving Nonlinear Systems with Nonreal Complex Solutions
- Applying Nonlinear Systems

Solving Nonlinear Systems with Real Solutions A system of equations in which at least one equation is *not* linear is a **nonlinear system.**

$$x^2 - y = 4 \qquad x^2 + y^2 = 16$$
$$x + y = -2 \qquad |x| + y = 4$$

Nonlinear systems

The substitution method works well for solving many such systems, particularly when one of the equations is linear, as in the next example.

EXAMPLE 1 **Solving a Nonlinear System by Substitution**

Solve the system.

$$x^2 - y = 4 \qquad (1)$$
$$x + y = -2 \qquad (2)$$

ALGEBRAIC SOLUTION

When one of the equations in a nonlinear system is linear, it is usually best to begin by solving the linear equation for one of the variables.

$$y = -2 - x \quad \text{Solve equation (2) for } y.$$

Substitute this result for y in equation (1).

$$x^2 - y = 4 \quad (1)$$
$$x^2 - (-2 - x) = 4 \quad \text{Let } y = -2 - x.\ \textbf{(Section 5.1)}$$
$$x^2 + 2 + x = 4 \quad \text{Distributive property } \textbf{(Section R.2)}$$
$$x^2 + x - 2 = 0 \quad \text{Standard form } \textbf{(Section 1.4)}$$
$$(x + 2)(x - 1) = 0 \quad \text{Factor. } \textbf{(Section R.4)}$$
$$x + 2 = 0 \quad \text{or} \quad x - 1 = 0$$
$$\qquad\qquad \text{Zero-factor property } \textbf{(Section 1.4)}$$
$$x = -2 \quad \text{or} \quad x = 1$$

Substituting -2 for x in equation (2) gives $y = 0$. If $x = 1$, then $y = -3$. The solution set of the given system is $\{(-2, 0), (1, -3)\}$. A graph of the system is shown in **Figure 9.**

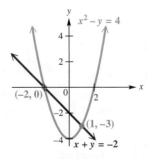

Figure 9

GRAPHING CALCULATOR SOLUTION

Solve each equation for y and graph them in the same viewing window. We obtain

$$Y_1 = X^2 - 4$$

and

$$Y_2 = -X - 2.$$

The screens in **Figure 10,** which indicate that the points of intersection are $(-2, 0)$ and $(1, -3)$, support the solution found algebraically.

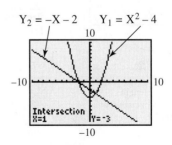

Figure 10

✔ *Now Try Exercise 9.*

LOOKING AHEAD TO CALCULUS

In calculus, finding the maximum and minimum points for a function of several variables usually requires solving a nonlinear system of equations.

CAUTION If we had solved for x in equation (2) to begin the algebraic solution in **Example 1,** we would have found $y = 0$ or $y = -3$. Substituting $y = 0$ into equation (1) gives $x^2 = 4$, so $x = 2$ or $x = -2$, leading to the ordered pairs $(2, 0)$ and $(-2, 0)$. The ordered pair $(2, 0)$ does not satisfy equation (2), however. ***This illustrates the necessity of checking by substituting all proposed solutions into each equation of the system.***

Visualizing the types of graphs involved in a nonlinear system helps predict the possible numbers of ordered pairs of real numbers that may be in the solution set of the system. For example, a line and a parabola may have 0, 1, or 2 points of intersection, as shown in **Figure 11.**

No points of intersection   One point of intersection   Two points of intersection

Figure 11

Nonlinear systems where both variables are squared in both equations are best solved by elimination, as shown in the next example.

To solve the system in **Example 2** graphically, solve equation (1) for y to get

$$y = \pm\sqrt{4 - x^2}.$$

Similarly, equation (2) yields

$$y = \pm\sqrt{-8 + 2x^2}.$$

Graph these *four* functions to find the two solutions.

| EXAMPLE 2 | Solving a Nonlinear System by Elimination |

Solve the system.

$$x^2 + y^2 = 4 \quad (1)$$
$$2x^2 - y^2 = 8 \quad (2)$$

SOLUTION The graph of equation (1) is a circle **(Section 2.2)**, and the graph of equation (2) is a *hyperbola.* These graphs may intersect in 0, 1, 2, 3, or 4 points. We add to eliminate y^2.

$$
\begin{array}{ll}
x^2 + y^2 = 4 & (1) \\
2x^2 - y^2 = 8 & (2) \\
\hline
3x^2 \qquad = 12 & \text{Add. (Section 5.1)} \\
x^2 = 4 & \text{Divide by 3.} \\
x = \pm 2 & \text{Square root property (Section 1.4)}
\end{array}
$$

Remember to find *both* square roots.

Find y by substituting the values of x in either equation (1) or equation (2).

Let $x = 2$.

$$2^2 + y^2 = 4 \quad (1)$$
$$y^2 = 0$$
$$y = 0$$

Let $x = -2$.

$$(-2)^2 + y^2 = 4 \quad (1)$$
$$y^2 = 0$$
$$y = 0$$

The proposed solutions are $(2, 0)$ and $(-2, 0)$. These satisfy both equations, confirming that the solution set is $\{(2, 0), (-2, 0)\}$.

✔ *Now Try Exercise 17.*

NOTE The elimination method works with the system in **Example 2** since the system can be thought of as a system of linear equations where the variables are x^2 and y^2. To see this, substitute u for x^2 and v for y^2. The resulting system is linear in u and v.

Sometimes a combination of the elimination method and the substitution method is effective in solving a system, as illustrated in **Example 3.**

EXAMPLE 3   **Solving a Nonlinear System by a Combination of Methods**

Solve the system.

$$x^2 + 3xy + y^2 = 22 \quad (1)$$
$$x^2 - xy + y^2 = 6 \quad (2)$$

SOLUTION   Begin as with the elimination method.

$$x^2 + 3xy + y^2 = 22 \quad (1)$$
$$\underline{-x^2 + xy - y^2 = -6} \quad \text{Multiply (2) by } -1.$$
$$4xy \qquad = 16 \quad \text{Add.} \quad (3)$$

$$y = \frac{4}{x} \quad \text{Solve for } y \ (x \neq 0). \quad (4)$$

Now substitute $\frac{4}{x}$ for y in either equation (1) or equation (2). We use equation (2).

$$x^2 - x\left(\frac{4}{x}\right) + \left(\frac{4}{x}\right)^2 = 6 \qquad \text{Let } y = \frac{4}{x} \text{ in (2).}$$

$$x^2 - 4 + \frac{16}{x^2} = 6 \qquad \text{Multiply and square.}$$

$$x^4 - 4x^2 + 16 = 6x^2 \qquad \text{Multiply by } x^2 \text{ to clear fractions.} \\ \text{(Section 1.6)}$$

> This equation is quadratic in form.

$$x^4 - 10x^2 + 16 = 0 \qquad \text{Subtract } 6x^2.$$

$$(x^2 - 2)(x^2 - 8) = 0 \qquad \text{Factor.}$$

$$x^2 - 2 = 0 \quad \text{or} \quad x^2 - 8 = 0 \qquad \text{Zero-factor property}$$

$$x^2 = 2 \quad \text{or} \quad x^2 = 8 \qquad \text{Solve each equation.}$$

> For each equation, include *both* square roots.

$$x = \pm\sqrt{2} \quad \text{or} \quad x = \pm 2\sqrt{2} \qquad \text{Square root property;} \\ \pm\sqrt{8} = \pm\sqrt{4} \cdot \sqrt{2} = \pm 2\sqrt{2} \\ \text{(Section R.7)}$$

Substitute these x-values into equation (4) to find corresponding values of y.

Let $x = \sqrt{2}$.	Let $x = -\sqrt{2}$.	Let $x = 2\sqrt{2}$.	Let $x = -2\sqrt{2}$.
$y = \dfrac{4}{\sqrt{2}} = 2\sqrt{2}$	$y = \dfrac{4}{-\sqrt{2}} = -2\sqrt{2}$	$y = \dfrac{4}{2\sqrt{2}} = \sqrt{2}$	$y = \dfrac{4}{-2\sqrt{2}} = -\sqrt{2}$

The solution set of the system is

$$\left\{\left(\sqrt{2}, 2\sqrt{2}\right), \left(-\sqrt{2}, -2\sqrt{2}\right), \left(2\sqrt{2}, \sqrt{2}\right), \left(-2\sqrt{2}, -\sqrt{2}\right)\right\}.$$

Verify these solutions by substitution in the original system.

✔ *Now Try Exercise 35.*

EXAMPLE 4 **Solving a Nonlinear System with an Absolute Value Equation**

Solve the system.

$$x^2 + y^2 = 16 \quad \text{(1)}$$
$$|x| + y = 4 \quad \text{(2)}$$

SOLUTION Use the substitution method. Begin by solving equation (2) for $|x|$.

$$|x| = 4 - y \quad \text{(3)}$$

In equation (1), the first term is x^2, which is the same as $|x|^2$. Therefore, we substitute $4 - y$ for x in equation (1).

$$x^2 + y^2 = 16 \quad \text{(1)}$$
$$(4 - y)^2 + y^2 = 16 \qquad \text{Let } x = 4 - y.$$
$$(16 - 8y + y^2) + y^2 = 16 \qquad \text{Square the binomial. (Section R.3)}$$

> Remember the middle term.

$$2y^2 - 8y = 0 \qquad \text{Combine like terms.}$$
$$y^2 - 4y = 0 \qquad \text{Divide by 2.}$$
$$y(y - 4) = 0 \qquad \text{Factor.}$$
$$y = 0 \quad \text{or} \quad y - 4 = 0 \qquad \text{Zero-factor property}$$
$$y = 4 \qquad \text{Add 4.}$$

To solve for the corresponding values of x, use either equation (1) or equation (2).

Let $y = 0$.

$$x^2 + 0^2 = 16 \quad \text{(1)}$$
$$x^2 = 16$$
$$x = \pm 4$$

Let $y = 4$.

$$x^2 + 4^2 = 16 \quad \text{(1)}$$
$$x^2 = 0$$
$$x = 0$$

The solution set, $\{(4, 0), (-4, 0), (0, 4)\}$, includes the points of intersection shown in **Figure 12.** Check the solutions in the original system.

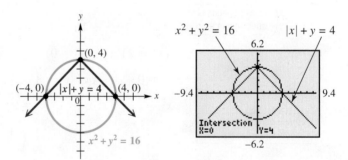

Figure 12

✔️ *Now Try Exercise 39.*

NOTE After solving for y in **Example 4,** the corresponding values of x can be found by using equation (2) instead of equation (1). As shown below, the same values for x result.

Let $y = 0$.

$$|x| + 0 = 4 \quad \text{(2)}$$
$$|x| = 4$$
$$x = \pm 4$$

Let $y = 4$.

$$|x| + 4 = 4 \quad \text{(2)}$$
$$|x| = 0$$
$$x = 0$$

EXAMPLE 5 Solving a Nonlinear System with Nonreal Complex Numbers as Solutions

Solve the system.

$$x^2 + y^2 = 5 \quad (1)$$

$$4x^2 + 3y^2 = 11 \quad (2)$$

SOLUTION Begin by eliminating a variable.

$$-3x^2 - 3y^2 = -15 \qquad \text{Multiply (1) by } -3.$$

$$\underline{4x^2 + 3y^2 = \quad 11} \qquad (2)$$

$$x^2 \qquad = -4 \qquad \text{Add.}$$

$$x = \pm\sqrt{-4} \qquad \text{Square root property}$$

$$x = \pm 2i \qquad \sqrt{-4} = i\sqrt{4} = 2i \text{ (Section 1.3)}$$

To find the corresponding values of y, substitute into equation (1).

Let $x = 2i$.

$$(2i)^2 + y^2 = 5 \quad (1)$$

$i^2 = -1$ → $-4 + y^2 = 5$

$$y^2 = 9$$

$$y = \pm 3$$

Let $x = -2i$.

$$(-2i)^2 + y^2 = 5 \quad (1)$$

$$-4 + y^2 = 5$$

$$y^2 = 9$$

$$y = \pm 3$$

Checking the proposed solutions confirms the following solution set.

$$\{(2i, 3), (2i, -3), (-2i, 3), (-2i, -3)\}$$

Note that these solutions with nonreal complex number components do not appear as intersection points on the graph of the system.

☑ *Now Try Exercise 37.*

3.1

−4.7 ———— 4.7

−3.1

The graphs of the two equations in **Example 5** do not intersect, as seen here. The graphs are obtained by graphing

$$y = \pm\sqrt{5 - x^2}$$

and

$$y = \pm\sqrt{\frac{11 - 4x^2}{3}}.$$

EXAMPLE 6 Using a Nonlinear System to Find the Dimensions of a Box

A box with an open top has a square base and four sides of equal height. The volume of the box is 75 in.^{3}, and the surface area is 85 in.^{2}. Find the dimensions of the box.

SOLUTION

Step 1 **Read** the problem. We must find the box width, length, and height.

Step 2 **Assign variables.** Let x represent the length and width of the square base, and let y represent the height. See **Figure 13.**

Step 3 **Write a system of equations.** Use the formula for the volume of a box, $V = LWH$, to write one equation using the given volume, 75 in.^{3}.

$$x^2y = 75 \qquad \text{Volume formula}$$

The surface consists of the base, whose area is x^2, and four sides, each having area xy. The total surface area of 85 in.^{2} is used to write a second equation.

$$x^2 + 4xy = 85 \qquad \text{Sum of areas of base and sides}$$

Figure 13

The two equations form this system.

$$x^2y = 75 \quad (1)$$

$$x^2 + 4xy = 85 \quad (2)$$

Step 4 **Solve** the system. Solve equation (1) for y to get $y = \frac{75}{x^2}$, and substitute into equation (2).

$$x^2 + 4x\left(\frac{75}{x^2}\right) = 85 \qquad \text{Let } y = \frac{75}{x^2} \text{ in (2)}.$$

$$x^2 + \frac{300}{x} = 85 \qquad \text{Multiply.}$$

$$x^3 + 300 = 85x \qquad \text{Multiply by } x, \text{ where } x \neq 0.$$

$$x^3 - 85x + 300 = 0 \qquad \text{Subtract } 85x.$$

$$
\begin{array}{r}
5)\overline{1 \quad 0 \quad -85 \quad 300} \\
5 \quad 25 \quad -300 \\
\hline
1 \quad 5 \quad -60 \quad 0
\end{array}
$$

Coefficients of a
quadratic polynomial factor

(Section 3.2)

We are restricted to positive values for x, and considering the nature of the problem, any solution should be relatively small. By the rational zeros theorem, factors of 300 are the only possible rational solutions. Using synthetic division, as shown in the margin, we see that 5 is a solution. Therefore, one value of x is 5, and $y = \frac{75}{5^2} = 3$. We must now solve

$$x^2 + 5x - 60 = 0$$

for any other possible positive solutions. Use the quadratic formula to find the positive solution.

$$x = \frac{-5 + \sqrt{5^2 - 4(1)(-60)}}{2(1)} \approx 5.639 \qquad \begin{array}{l} \text{Quadratic formula with} \\ a = 1, b = 5, c = -60 \\ \text{(Section 1.4)} \end{array}$$

This value of x leads to $y \approx 2.359$.

Step 5 **State the answer.** There are two possible answers.

First answer: length = width = 5 in.; height = 3 in.

Second answer: length = width $\approx$ 5.639 in.; height $\approx$ 2.359 in.

Step 6 **Check.** The check is left for **Exercise 63.**

✔ *Now Try Exercises 61 and 63.*

5.5 Exercises

Concept Check In Exercises 1–6 a nonlinear system is given, along with the graphs of both equations in the system. Verify that the points of intersection specified on the graph are solutions of the system by substituting directly into both equations.

1. $x^2 = y - 1$
$\ y = 3x + 5$

2. $2x^2 = 3y + 23$
$\ y = 2x - 5$

3. $x^2 + y^2 = 5$
$-3x + 4y = 2$

$\left(\frac{38}{25}, \frac{41}{25}\right)$

$(-2, -1)$

$x^2 + y^2 = 5$
$-3x + 4y = 2$

4. $x + y = -3$
$x^2 + y^2 = 45$

$(-6, 3)$

$(3, -6)$

$x + y = -3$
$x^2 + y^2 = 45$

5. $y = \log x$
$x^2 - y^2 = 4$

Intersection
X=2.0232821 Y=.30605644

6. $y = \ln(2x + 3)$
$y = \sqrt{2 - 0.5x^2}$

Intersection
X=.47504984 Y=1.3737408

7. *Concept Check* In **Example 1,** we solved the following system. How can we tell, before doing any work, that this system cannot have more than two solutions?

$$x^2 - y = 4$$
$$x + y = -2$$

8. *Concept Check* In **Example 5,** there were four solutions to the system, but there were no points of intersection of the graphs. If a nonlinear system has nonreal complex numbers as components of its solutions, will they appear as intersection points of the graphs?

Give all solutions of each nonlinear system of equations, including those with nonreal complex components. **See Examples 1–5.**

9. $x^2 - y = 0$
$x + y = 2$

10. $x^2 + y = 2$
$x - y = 0$

11. $y = x^2 - 2x + 1$
$x - 3y = -1$

12. $y = x^2 + 6x + 9$
$x + 2y = -2$

13. $y = x^2 + 4x$
$2x - y = -8$

14. $y = 6x + x^2$
$4x - y = -3$

15. $3x^2 + 2y^2 = 5$
$x - y = -2$

16. $x^2 + y^2 = 5$
$-3x + 4y = 2$

17. $x^2 + y^2 = 8$
$x^2 - y^2 = 0$

18. $x^2 + y^2 = 10$
$2x^2 - y^2 = 17$

19. $5x^2 - y^2 = 0$
$3x^2 + 4y^2 = 0$

20. $x^2 + y^2 = 0$
$2x^2 - 3y^2 = 0$

21. $3x^2 + y^2 = 3$
$4x^2 + 5y^2 = 26$

22. $x^2 + 2y^2 = 9$
$x^2 + y^2 = 25$

23. $2x^2 + 3y^2 = 5$
$3x^2 - 4y^2 = -1$

24. $3x^2 + 5y^2 = 17$
$2x^2 - 3y^2 = 5$

25. $2x^2 + 2y^2 = 20$
$4x^2 + 4y^2 = 30$

26. $x^2 + y^2 = 4$
$5x^2 + 5y^2 = 28$

27. $2x^2 - 3y^2 = 12$
$6x^2 + 5y^2 = 36$

28. $5x^2 - 2y^2 = 25$
$10x^2 + y^2 = 50$

29. $xy = -15$
$4x + 3y = 3$

30. $xy = 8$
$3x + 2y = -16$

31. $2xy + 1 = 0$
$x + 16y = 2$

32. $-5xy + 2 = 0$
$x - 15y = 5$

33. $3x^2 - y^2 = 11$
$xy = 12$

34. $5x^2 - 2y^2 = 6$
$xy = 2$

35. $x^2 - xy + y^2 = 5$
$2x^2 + xy - y^2 = 10$

36. $3x^2 + 2xy - y^2 = 9$
$x^2 - xy + y^2 = 9$

37. $x^2 + 2xy - y^2 = 14$
$x^2 - y^2 = -16$

38. $x^2 + 3xy - y^2 = 12$
$x^2 - y^2 = -12$

39. $x^2 + y^2 = 25$
$|x| - y = 5$

40. $x^2 + y^2 = 9$
$|x| + y = 3$

41. $x = |y|$
$x^2 + y^2 = 18$

42. $2x + |y| = 4$
$x^2 + y^2 = 5$

43. $2x^2 - y^2 = 4$
$|x| = |y|$

44. $x^2 + y^2 = 9$
$|x| = |y|$

Many nonlinear systems cannot be solved algebraically, so graphical analysis is the only way to determine the solutions of such systems. Use a graphing calculator to solve each nonlinear system. Give x- and y-coordinates to the nearest hundredth.

45. $y = \log(x + 5)$
$y = x^2$

46. $y = 5^x$
$xy = 1$

47. $y = e^{x+1}$
$2x + y = 3$

48. $y = \sqrt[3]{x - 4}$
$x^2 + y^2 = 6$

Solve each problem using a system of equations in two variables. **See Example 6.**

49. *Unknown Numbers* Find two numbers whose sum is 17 and whose product is 42.

50. *Unknown Numbers* Find two numbers whose sum is 10 and whose squares differ by 20.

51. *Unknown Numbers* Find two numbers whose squares have a sum of 100 and a difference of 28.

52. *Unknown Numbers* Find two numbers whose squares have a sum of 194 and a difference of 144.

53. *Unknown Numbers* Find two numbers whose ratio is 9 to 2 and whose product is 162.

54. *Unknown Numbers* Find two numbers whose ratio is 4 to 3 and are such that the sum of their squares is 100.

55. *Triangle Dimensions* The longest side of a right triangle is 13 m in length. One of the other sides is 7 m longer than the shortest side. Find the lengths of the two shorter sides of the triangle.

56. *Triangle Dimensions* The longest side of a right triangle is 29 ft in length. One of the other two sides is 1 ft longer than the shortest side. Find the lengths of the two shorter sides of the triangle.

57. Does the straight line $3x - 2y = 9$ intersect the circle $x^2 + y^2 = 25$? (*Hint:* To find out, solve the system formed by these two equations.)

58. For what value(s) of b will the line $x + 2y = b$ touch the circle $x^2 + y^2 = 9$ in only one point?

59. Find the equation of the line passing through the points of intersection of the graphs of $y = x^2$ and $x^2 + y^2 = 90$.

60. Suppose you are given the equations of two circles that are known to intersect in exactly two points. Explain how you would find the equation of the only chord common to these circles.

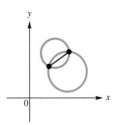

61. *Dimensions of a Box* A box with an open top has a square base and four sides of equal height. The volume of the box is 360 ft^3. If the surface area is 276 ft^2, find the dimensions of the box. (Round answers to the nearest thousandth, if necessary.)

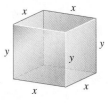

62. *Dimensions of a Cylinder* Find the radius and height (to the nearest thousandth) of an open-ended cylinder with volume 50 in.^{3} and lateral surface area 65 in.^{2}.

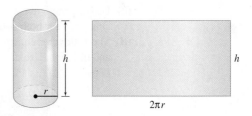

63. *Checking Answers* Check the two answers in **Example 6.**

64. *(Modeling) Equilibrium Demand and Price* The supply and demand equations for a certain commodity are given.

$$\text{supply: } p = \frac{2000}{2000 - q} \quad \text{and} \quad \text{demand: } p = \frac{7000 - 3q}{2q}$$

(a) Find the equilibrium demand.
(b) Find the equilibrium price (in dollars).

65. *(Modeling) Equilibrium Demand and Price* The supply and demand equations for a certain commodity are given.

$$\text{supply: } p = \sqrt{0.1q + 9} - 2 \quad \text{and} \quad \text{demand: } p = \sqrt{25 - 0.1q}$$

(a) Find the equilibrium demand.
(b) Find the equilibrium price (in dollars).

66. *(Modeling) Circuit Gain* In electronics, circuit gain is modeled by

$$G = \frac{Bt}{R + R_t},$$

where R is the value of a resistor, t is temperature, R_t is the value of R at temperature t, and B is a constant. The sensitivity of the circuit to temperature is modeled by

$$S = \frac{BR}{(R + R_t)^2}.$$

If $B = 3.7$ and t is 90 K (kelvins), find the values of R and R_t that will result in the values $G = 0.4$ and $S = 0.001$.

67. *(Modeling) Carbon Dioxide Emissions* The emissions of carbon dioxide from consumption of fossil fuels from 1990 through 2007 are modeled in the graph for both the United States and China.

Carbon Dioxide Emissions from Consumption of Fossil Fuels

Source: U.S. Energy Information Administration.

(a) Interpret this graph. How are emissions changing with time?

(b) During what years were the carbon dioxide emissions in China decreasing?

(c) Use the graph to estimate the year and the amount of emissions when the carbon emissions for both countries were equal.

(d) The following equations model carbon dioxide emissions in the United States and in China. Use the equations to determine the year and emissions levels when $U = C$.

$$U = 5092(1.0105)^{(x-1990)} \quad \text{United States}$$

$$C = 2082(1.0638)^{(x-1990)} \quad \text{China}$$

Relating Concepts

For individual or collaborative investigation *(Exercises 68–74)*

Consider the following nonlinear system.

$$y = |x - 1|$$
$$y = x^2 - 4$$

Work Exercises 68–74 in order, *to see how concepts from previous chapters are related to the graphs and the solutions of this system.*

68. How is the graph of $y = |x - 1|$ obtained by transforming the graph of $y = |x|$?

69. How is the graph of $y = x^2 - 4$ obtained by transforming the graph of $y = x^2$?

70. Use the definition of absolute value to write $y = |x - 1|$ as a piecewise-defined function.

71. Write two quadratic equations that will be used to solve the system. (*Hint:* Set both parts of the piecewise-defined function in **Exercise 70** equal to $x^2 - 4$.)

72. Use the quadratic formula to solve both equations from **Exercise 71.** Pay close attention to the restriction on x.

73. Use the values of x found in **Exercise 72** to find the solution set of the system.

74. Check the two answers in **Exercise 73.**

Summary Exercises on Systems of Equations

This chapter has introduced methods for solving systems of equations, including substitution and elimination, and matrix methods such as the Gauss-Jordan method and Cramer's rule. Use each method at least once when solving the systems below. Include solutions with nonreal complex number components. For systems with infinitely many solutions, write the solution set using an arbitrary variable.

1. $2x + 5y = 4$
$3x - 2y = -13$

2. $x - 3y = 7$
$-3x + 4y = -1$

3. $2x^2 + y^2 = 5$
$3x^2 + 2y^2 = 10$

4. $2x - 3y = -2$
$x + y = -16$
$3x - 2y + z = 7$

5. $6x - y = 5$
$xy = 4$

6. $4x + 2z = -12$
$x + y - 3z = 13$
$-3x + y - 2z = 13$

7. $x + 2y + z = 5$
$y + 3z = 9$

8. $x - 4 = 3y$
$x^2 + y^2 = 8$

9. $3x + 6y - 9z = 1$
$2x + 4y - 6z = 1$
$3x + 4y + 5z = 0$

10. $x + 2y + z = 0$
$x + 2y - z = 6$
$2x - y = -9$

11. $x^2 + y^2 = 4$
$y = x + 6$

12. $x + 5y = -23$
$4y - 3z = -29$
$2x + 5z = 19$

13. $y + 1 = x^2 + 2x$
$y + 2x = 4$

14. $3x + 6z = -3$
$y - z = 3$
$2x + 4z = -1$

15. $2x + 3y + 4z = 3$
$-4x + 2y - 6z = 2$
$4x + 3z = 0$

16. $\dfrac{3}{x} + \dfrac{4}{y} = 4$

$\dfrac{1}{x} + \dfrac{2}{y} = \dfrac{2}{3}$

17. $-5x + 2y + z = 5$
$-3x - 2y - z = 3$
$-x + 6y = 1$

18. $x + 5y + 3z = 9$
$2x + 9y + 7z = 5$

19. $2x^2 + y^2 = 9$
$3x - 2y = -6$

20. $2x - 4y - 6 = 0$
$-x + 2y + 3 = 0$

21. $x + y - z = 0$
$2y - z = 1$
$2x + 3y - 4z = -4$

22. $2y = 3x - x^2$
$x + 2y = 12$

23. $4x - z = -6$
$\dfrac{3}{5}y + \dfrac{1}{2}z = 0$
$\dfrac{1}{3}x + \dfrac{2}{3}z = -5$

24. $x - y + 3z = 3$
$-2x + 3y - 11z = -4$
$x - 2y + 8z = 6$

25. $x^2 + 3y^2 = 28$
$y - x = -2$

26. $5x - 2z = 8$
$4y + 3z = -9$
$\dfrac{1}{2}x + \dfrac{2}{3}y = -1$

27. $2x^2 + 3y^2 = 20$
$x^2 + 4y^2 = 5$

28. $x + y + z = -1$
$2x + 3y + 2z = 3$
$2x + y + 2z = -7$

29. $x + 2z = 9$
$y + z = 1$
$3x - 2y = 9$

30. $x^2 - y^2 = 15$
$x - 2y = 2$

31. $-x + y = -1$
$x + z = 4$
$6x - 3y + 2z = 10$

32. $2x - y - 2z = -1$
$-x + 2y + 13z = 12$
$3x + 9z = 6$

33. $xy = -3$
$x + y = -2$

34. $-3x + 2z = 1$
$4x + y - 2z = -6$
$x + y + 4z = 3$

35. $y = x^2 + 6x + 9$
$x + y = 3$

36. $x^2 + y^2 = 9$
$2x - y = 3$

37. $2x - 3y = -2$
$x + y - 4z = -16$
$3x - 2y + z = 7$

38. $3x - y = -2$
$y + 5z = -4$
$-2x + 3y - z = -8$

39. $y = (x - 1)^2 + 2$
$y = 2x - 1$

■ Solving Linear Inequalities

■ Solving Systems of Inequalities

■ Linear Programming

Solving Linear Inequalities   A line divides a plane into three sets of points: the points of the line itself and the points belonging to the two regions determined by the line. Each of these two regions is a **half-plane**. In **Figure 14,** line *r* divides the plane into three different sets of points: line *r*, half-plane *P*, and half-plane *Q*. The points on *r* belong neither to *P* nor to *Q*. Line *r* is the **boundary** of each half-plane.

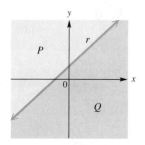

Figure 14

Linear Inequality in Two Variables

A **linear inequality in two variables** is an inequality of the form

$$Ax + By \leq C,$$

where *A*, *B*, and *C* are real numbers, with *A* and *B* not both equal to 0. (The symbol $\leq$ could be replaced with $\geq$, $<$, or $>$.)

The graph of a linear inequality is a half-plane, perhaps with its boundary.

EXAMPLE 1   **Graphing a Linear Inequality**

Graph $3x - 2y \leq 6$.

SOLUTION   First graph the boundary, $3x - 2y = 6$, as shown in **Figure 15.** Since the points of the line $3x - 2y = 6$ satisfy $3x - 2y \leq 6$, this line is part of the solution set. To decide which half-plane (the one above the line $3x - 2y = 6$ or the one below the line) is part of the solution set, solve the original inequality for *y*.

$$3x - 2y \leq 6$$

$$-2y \leq -3x + 6 \qquad \text{Subtract } 3x.$$

> Reverse the inequality symbol when dividing by a negative number.

$$y \geq \frac{3}{2}x - 3 \qquad \text{Divide by } -2. \text{ (Section 1.7)}$$

For a particular value of *x*, the inequality will be satisfied by all values of *y* that are *greater than* or equal to $\frac{3}{2}x - 3$. Thus, the solution set contains the half-plane *above* the line, as shown in **Figure 16.**

Figure 15   **Figure 16**

Coordinates for *x* and *y* from the solution set (the shaded region) satisfy the original inequality, while coordinates outside the solution set do not.

✔ *Now Try Exercise 1.*

CAUTION *A linear inequality must be in slope-intercept form (solved for y) to determine, from the presence of a $<$ symbol or a $>$ symbol, whether to shade the lower or the upper half-plane.* In **Figure 16,** the upper half-plane is shaded, even though the inequality is $3x - 2y \leq 6$ (with a $<$ symbol) in standard form. Only when we write the inequality as

$$y \geq \frac{3}{2}x - 3 \quad \text{Slope-intercept form}$$

does the $>$ symbol indicate to shade the upper half-plane.

EXAMPLE 2 Graphing a Linear Inequality

Graph $x + 4y > 4$.

SOLUTION The boundary of the graph is the straight line $x + 4y = 4$. Since points on this line do *not* satisfy $x + 4y > 4$, it is customary to make the line dashed, as in **Figure 17.**

Figure 17

To decide which half-plane represents the solution set, solve for y.

$$x + 4y > 4$$
$$4y > -x + 4 \quad \text{Subtract } x.$$
$$y > -\frac{1}{4}x + 1 \quad \text{Divide by 4.}$$

Since y is *greater than* $-\frac{1}{4}x + 1$, the graph of the solution set is the half-plane *above* the boundary, as shown in **Figure 17.**

Alternatively, or as a check, choose a test point not on the boundary line and substitute into the inequality. The point $(0, 0)$ is a good choice if it does not lie on the boundary, since the substitution is easily done.

$$x + 4y > 4 \quad \text{Original inequality}$$
$$0 + 4(0) \overset{?}{>} 4 \quad \text{Use } (0, 0) \text{ as a test point.}$$
$$0 > 4 \quad \text{False}$$

Since the point $(0, 0)$ is below the boundary, the points that satisfy the inequality must be above the boundary, which agrees with the result above.

✔ *Now Try Exercise 5.*

To graph the inequality in **Example 2** using a graphing calculator, solve for y, and then direct the calculator to shade *above* the boundary line, $y = -\frac{1}{4}x + 1$. The calculator graph does not distinguish between solid boundary lines and dashed boundary lines. We must understand the mathematics to interpret a calculator graph correctly.

An inequality containing $<$ or $>$ is a **strict inequality** and does not include the boundary in its solution set. This is indicated with a dashed boundary, as shown in **Example 2.** A **nonstrict inequality** contains $\leq$ or $\geq$ and does include its boundary in the solution set. This is indicated with a solid boundary, as shown in **Example 1.**

This discussion is summarized in the box on the next page.

Graphing an Inequality in Two Variables

Method 1 If the inequality is or can be solved for y, then the following hold.

- The graph of $y < f(x)$ consists of all the points that are *below* the graph of $y = f(x)$.

- The graph of $y > f(x)$ consists of all the points that are *above* the graph of $y = f(x)$.

Method 2 If the inequality is not or cannot be solved for y, choose a test point not on the boundary.

- If the test point satisfies the inequality, the graph includes all points on the *same* side of the boundary as the test point.

- If the test point does not satisfy the inequality, the graph includes all points on the *other* side of the boundary.

Solving Systems of Inequalities The solution set of a **system of inequalities**, such as

$$x > 6 - 2y$$
$$x^2 < 2y,$$

is the intersection of the solution sets of its members. We find this intersection by graphing the solution sets of all inequalities on the same coordinate axes and identifying, by shading, the region common to all graphs.

EXAMPLE 3 **Graphing Systems of Inequalities**

Graph the solution set of each system.

(a) $x > 6 - 2y$
$\quad\ x^2 < 2y$

(b) $|x| \leq 3$
$\quad\ y \leq 0$
$\quad\ y \geq |x| + 1$

SOLUTION

(a) **Figures 18(a) and (b)** show the graphs of $x > 6 - 2y$ and $x^2 < 2y$. The methods presented earlier in this chapter can be used to show that the boundaries intersect at the points $(2, 2)$ and $\left(-3, \frac{9}{2}\right)$.

The solution set of the system is shown in **Figure 18(c).** Since the points on the boundaries of $x > 6 - 2y$ and $x^2 < 2y$ do not belong to the graph of the solution set, the boundaries are dashed.

The cross-hatched region is the solution set of the system in **Example 3(a).**

(a)

(b)

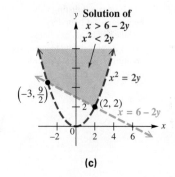

(c)

Figure 18

(b) Writing $|x| \leq 3$ as $-3 \leq x \leq 3$ shows that this inequality is satisfied by points in the region between and including $x = -3$ and $x = 3$. See **Figure 19(a).** The set of points that satisfies $y \leq 0$ includes the points below or on the x-axis. See **Figure 19(b).**

Graph $y = |x| + 1$ and use a test point to verify that the solutions of $y \geq |x| + 1$ are on or above the boundary. See **Figure 19(c).** Since the solution sets of $y \leq 0$ and $y \geq |x| + 1$ shown in **Figures 19(b) and (c)** have no points in common, *the solution set of the system is $\emptyset$.*

The solution set of the system is $\emptyset$, because there are no points common to all three regions.

(a)   (b)   (c)

Figure 19

✔ *Now Try Exercises 35, 41, and 43.*

NOTE While we gave three graphs in the solutions of **Example 3,** in practice we usually give only a final graph showing the solution set of the system.

Linear Programming   An important application of mathematics is *linear programming.* We use **linear programming** to find an optimum value such as minimum cost or maximum profit. It was first developed to solve problems in allocating supplies for the U.S. Air Force during World War II.

EXAMPLE 4   **Finding a Maximum Profit**

The Charlson Company makes two products: MP3 players and DVD players. Each MP3 player gives a profit of \$30, while each DVD player produces \$70 profit. The company must manufacture at least 10 MP3 players per day to satisfy one of its customers, but it can manufacture no more than 50 per day because of production restrictions. The number of DVD players produced cannot exceed 60 per day, and the number of MP3 players cannot exceed the number of DVD players. How many of each should the company manufacture to obtain maximum profit?

SOLUTION   First we translate the statement of the problem into symbols.

Let x = the number of MP3 players to be produced daily,

and y = the number of DVD players to be produced daily.

The company must produce at least 10 MP3 players (10 or more), so

$$x \geq 10.$$

Since no more than 50 MP3 players may be produced,

$$x \leq 50.$$

No more than 60 DVD players may be made in one day, so

$$y \le 60.$$

The number of MP3 players may not exceed the number of DVD players, so

$$x \le y.$$

The numbers of MP3 players and of DVD players cannot be negative, so

$$x \ge 0 \quad \text{and} \quad y \ge 0.$$

These restrictions, or **constraints,** form the following system of inequalities.

$$x \ge 10, \quad x \le 50, \quad y \le 60, \quad x \le y, \quad x \ge 0, \quad y \ge 0$$

Each MP3 player gives a profit of \$30, so the daily profit from production of x MP3 players is $30x$ dollars. Also, the profit from production of y DVD players will be $70y$ dollars per day. Therefore, the total daily profit is

$$\text{profit} = 30x + 70y.$$

This equation defines the **objective function** to be maximized.

To find the maximum possible profit, subject to these constraints, we sketch the graph of each constraint. The only feasible values of x and y are those that satisfy all constraints—that is, the values that lie in the intersection of the graphs.

The intersection is shown in **Figure 20.** Any point lying inside the shaded region or on the boundary in the figure satisfies the restrictions as to the number of MP3 and DVD players that can be produced. (For practical purposes, however, only points with integer coefficients are useful.) This region is called the **region of feasible solutions.**

The **vertices** (singular **vertex**), or **corner points,** of the region of feasible solutions have coordinates

$$(10, 10), \quad (10, 60), \quad (50, 50), \quad \text{and} \quad (50, 60).$$

We must find the value of the objective function $30x + 70y$ for each vertex. We want the vertex that produces the maximum possible value of

$$30x + 70y. \quad \text{Objective function}$$

$$(10, 10): \quad 30(10) + 70(10) = 1000$$

$$(10, 60): \quad 30(10) + 70(60) = 4500$$

$$(50, 50): \quad 30(50) + 70(50) = 5000$$

$$(50, 60): \quad 30(50) + 70(60) = 5700 \leftarrow \text{Maximum}$$

The maximum profit, obtained when 50 MP3 players and 60 DVD players are produced each day, will be

$$30(50) + 70(60) = \$5700 \text{ per day.}$$

✔ *Now Try Exercise 77.*

Figure 20

To justify the procedure used in **Example 4,** consider the following. The Charlson Company needed to find values of x and y in the shaded region of **Figure 20** that produce the maximum profit—that is, the maximum value of $30x + 70y$. To locate the point (x, y) that gives the maximum profit, add to the graph of **Figure 20** lines corresponding to arbitrarily chosen profits of \$0, \$1000, \$3000, and \$7000:

$$30x + 70y = 0, \quad 30x + 70y = 1000,$$

$$30x + 70y = 3000, \quad \text{and} \quad 30x + 70y = 7000.$$

For instance, each point on the line $30x + 70y = 3000$ corresponds to production values that yield a profit of \$3000.

Figure 21 shows the region of feasible solutions, together with these lines. The lines are parallel, and the higher the line, the greater the profit. The line $30x + 70y = 7000$ yields the greatest profit but does not contain any points of the region of feasible solutions. To find the feasible solution of greatest profit, lower the line $30x + 70y = 7000$ until it contains a feasible solution—that is, until it just touches the region of feasible solutions. This occurs at point A, a vertex of the region. See **Figure 22.**

| Figure 21 | Figure 22 |

This discussion can be generalized.

Fundamental Theorem of Linear Programming

If an optimal value for a linear programming problem exists, it occurs at a vertex of the region of feasible solutions.

Use the following method to solve a linear programming problem.

Solving a Linear Programming Problem

Step 1 Write the objective function and all necessary constraints.

Step 2 Graph the region of feasible solutions.

Step 3 Identify all vertices (corner points).

Step 4 Find the value of the objective function at each vertex.

Step 5 The solution is given by the vertex producing the optimal value of the objective function.

EXAMPLE 5 **Finding a Minimum Cost**

Robin takes multivitamins each day. She wants at least 16 units of vitamin A, at least 5 units of vitamin B_1, and at least 20 units of vitamin C. Capsules, costing $0.10 each, contain 8 units of A, 1 of B_1, and 2 of C. Chewable tablets, costing $0.20 each, contain 2 units of A, 1 of B_1, and 7 of C. How many of each should she take each day to minimize her cost and yet fulfill her daily requirements?

SOLUTION

Step 1 Let x represent the number of capsules to take each day, and let y represent the number of chewable tablets to take. Then the cost in *pennies* per day is

$$\text{cost} = 10x + 20y. \quad \text{Objective function}$$

Robin takes x of the \$0.10 capsules and y of the \$0.20 chewable tablets, and she gets 8 units of vitamin A from each capsule and 2 units of vitamin A from each tablet. Altogether she gets $8x + 2y$ units of A per day. Since she wants at least 16 units,

$$8x + 2y \geq 16.$$

Each capsule and each tablet supplies 1 unit of vitamin B_1. Robin wants at least 5 units per day, so

$$x + y \geq 5.$$

For vitamin C, the inequality is

$$2x + 7y \geq 20.$$

Since Robin cannot take negative numbers of multivitamins,

$$x \geq 0 \quad \text{and} \quad y \geq 0.$$

Figure 23

Step 2 **Figure 23** shows the intersection of the graphs of

$$8x + 2y \geq 16, \quad x + y \geq 5, \quad 2x + 7y \geq 20, \quad x \geq 0, \quad \text{and} \quad y \geq 0.$$

Step 3 The vertices are $(0, 8)$, $(1, 4)$, $(3, 2)$, and $(10, 0)$.

Steps 4 and 5 See the table. The minimum cost occurs at $(3, 2)$.

Point	Cost = 10x + 20y	
$(0, 8)$	$10(0) + 20(8) = 160$	
$(1, 4)$	$10(1) + 20(4) = 90$	
$(3, 2)$	$10(3) + 20(2) = 70$	← Minimum
$(10, 0)$	$10(10) + 20(0) = 100$	

Robin's best choice is to take 3 capsules and 2 chewable tablets each day, for a total cost of \$0.70 per day. She receives just the minimum amounts of vitamins B_1 and C, and an excess of vitamin A.

☑ *Now Try Exercises 71 and 81.*

5.6 Exercises

*Graph each inequality. **See Examples 1–3.***

1. $x + 2y \leq 6$ **2.** $x - y \geq 2$

3. $2x + 3y \geq 4$ **4.** $4y - 3x \leq 5$

5. $3x - 5y > 6$ **6.** $x < 3 + 2y$

7. $5x < 4y - 2$ **8.** $2x > 3 - 4y$

9. $x \leq 3$ **10.** $y \leq -2$

11. $y < 3x^2 + 2$

12. $y \leq x^2 - 4$

13. $y > (x - 1)^2 + 2$

14. $y > 2(x + 3)^2 - 1$

15. $x^2 + (y + 3)^2 \leq 16$

16. $(x - 4)^2 + y^2 \geq 9$

17. $y > 2^x + 1$

18. $y \leq \log(x - 1) - 2$

19. Explain how to determine whether the boundary of the graph of an inequality is solid or dashed.

20. When graphing $y \leq 3x - 6$, would you shade above or below the line $y = 3x - 6$? Explain your answer.

Concept Check *Work each problem.*

21. For $Ax + By \geq C$, if $B > 0$, would you shade above or below the line?

22. For $Ax + By \geq C$, if $B < 0$, would you shade above or below the line?

23. Which one of the following is a description of the graph of the inequality

$$(x - 5)^2 + (y - 2)^2 < 4?$$

   A. the region inside a circle with center $(-5, -2)$ and radius 2

   B. the region inside a circle with center $(5, 2)$ and radius 2

   C. the region inside a circle with center $(-5, -2)$ and radius 4

   D. the region outside a circle with center $(5, 2)$ and radius 4

24. Find an example of a linear inequality in two variables whose graph does not intersect the graph of $y \geq 3x + 5$.

Concept Check *In Exercises 25–28, match each inequality with the appropriate calculator graph. Do not use your calculator. Instead, use your knowledge of the concepts involved in graphing inequalities.*

25. $y \leq 3x - 6$

A.

26. $y \geq 3x - 6$

B.

27. $y \leq -3x - 6$

C.

28. $y \geq -3x - 6$

D.

Graph the solution set of each system of inequalities. ***See Example 3.***

29. $x + y \geq 0$
$2x - y \geq 3$

30. $x + y \leq 4$
$x - 2y \geq 6$

31. $2x + y > 2$
$x - 3y < 6$

32. $4x + 3y < 12$
$y + 4x > -4$

33. $3x + 5y \leq 15$
$x - 3y \geq 9$

34. $y \leq x$
$x^2 + y^2 < 1$

35. $4x - 3y \leq 12$
$y \leq x^2$

36. $y \leq -x^2$
$y \geq x^2 - 6$

37. $x + 2y \leq 4$
$y \geq x^2 - 1$

38. $x + y \leq 9$
$x \leq -y^2$

39. $y \leq (x + 2)^2$
$y \geq -2x^2$

40. $x - y < 1$
$-1 < y < 1$

41. $x + y \leq 36$
$-4 \leq x \leq 4$

42. $y \geq (x - 2)^2 + 3$
$y \leq -(x - 1)^2 + 6$

43. $y \geq x^2 + 4x + 4$
$y < -x^2$

44. $x \geq 0$
$x + y \leq 4$
$2x + y \leq 5$

45. $3x - 2y \geq 6$
$x + y \leq -5$
$y \leq 4$

46. $-2 < x < 3$
$-1 \leq y \leq 5$
$2x + y < 6$

47. $-2 < x < 2$
$y > 1$
$x - y > 0$

48. $x + y \leq 4$
$x - y \leq 5$
$4x + y \leq -4$

49. $x \leq 4$
$x \geq 0$
$y \geq 0$
$x + 2y \geq 2$

50. $2y + x \geq -5$
$y \leq 3 + x$
$x \leq 0$
$y \leq 0$

51. $2x + 3y \leq 12$
$2x + 3y > -6$
$3x + y < 4$
$x \geq 0$
$y \geq 0$

52. $y \geq 3^x$
$y \geq 2$

53. $y \leq \left(\dfrac{1}{2}\right)^x$
$y \geq 4$

54. $\ln x - y \geq 1$
$x^2 - 2x - y \leq 1$

55. $y \leq \log x$
$y \geq |x - 2|$

56. $e^{-x} - y \leq 1$
$x - 2y \geq 4$

57. $y > x^3 + 1$
$y \geq -1$

58. $y \leq x^3 - x$
$y > -3$

Use the shading capabilities of your graphing calculator to graph each inequality or system of inequalities.

59. $3x + 2y \geq 6$

60. $y \leq x^2 + 5$

61. $x + y \geq 2$
$x + y \leq 6$

62. $y \geq |x + 2|$
$y \leq 6$

Connecting Graphs with Equations In Exercises 63–66, determine the system of inequalities illustrated in each graph. Write inequalities in standard form.

63.

64.

65.

66.

67. *Concept Check* Write a system of inequalities for which the graph is the region in the first quadrant inside and including the circle with radius 2 centered at the origin, and above (not including) the line that passes through the points $(0, -1)$ and $(2, 2)$.

68. *Cost of Vitamins* The figure shows the region of feasible solutions for the vitamin problem of **Example 5** and the straight-line graph of all combinations of capsules and chewable tablets for which the cost is $0.40.

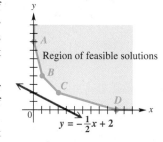

Region of feasible solutions

$y = -\frac{1}{2}x + 2$

(a) The cost function is $10x + 20y$. Give the linear equation (in slope-intercept form) of the line of constant cost c.

(b) As c increases, does the line of constant cost move up or down?

(c) By inspection, find the vertex of the region of feasible solutions that gives the optimal solution.

The graphs show regions of feasible solutions. Find the maximum and minimum values of each objective function. **See Examples 4 and 5.**

69. objective function $= 3x + 5y$

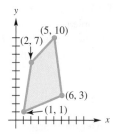

(5, 10)
(2, 7)
(6, 3)
(1, 1)

70. objective function $= 6x + y$

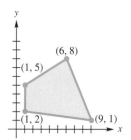

(6, 8)
(1, 5)
(1, 2)
(9, 1)

Find the maximum and minimum values of each objective function over the region of feasible solutions shown at the right. **See Examples 4 and 5.**

71. objective function $= 3x + 5y$

72. objective function $= 5x + 5y$

73. objective function $= 10y$

74. objective function $= 3x - y$

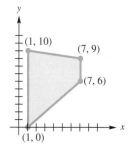

(1, 10)
(7, 9)
(7, 6)
(1, 0)

Write a system of inequalities for each problem, and then graph the region of feasible solutions of the system. **See Examples 4 and 5.**

75. *Vitamin Requirements* Jane Lukas was given the following advice. She should supplement her daily diet with at least 6000 USP units of vitamin A, at least 195 mg of vitamin C, and at least 600 USP units of vitamin D. She finds that Mason's Pharmacy carries Brand X and Brand Y vitamins. Each Brand X pill contains 3000 USP units of A, 45 mg of C, and 75 USP units of D, while the Brand Y pills contain 1000 USP units of A, 50 mg of C, and 200 USP units of D.

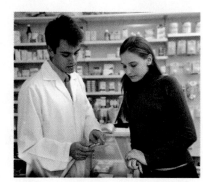

76. *Shipping Requirements* The California Almond Growers have 2400 boxes of almonds to be shipped from their plant in Sacramento to Des Moines and San Antonio. The Des Moines market needs at least 1000 boxes, while the San Antonio market must have at least 800 boxes.

*Solve each linear programming problem. **See Examples 4 and 5.***

77. *Aid to Disaster Victims* An agency wants to ship food and clothing to tsunami victims in Japan. Commercial carriers have volunteered to transport the packages, provided they fit in the available cargo space. Each 20-ft^3 box of food weighs 40 lb and each 30-ft^3 box of clothing weighs 10 lb. The total weight cannot exceed 16,000 lb, and the total volume must be at most 18,000 ft^3. Each carton of food will feed 10 people, and each carton of clothing will help 8 people.

(a) How many cartons of food and clothing should be sent to maximize the number of people assisted?

(b) What is the maximum number assisted?

78. *Aid to Disaster Victims* Earthquake victims in Haiti need medical supplies and bottled water. Each medical kit measures 1 ft^3 and weighs 10 lb. Each container of water is also 1 ft^3 but weighs 20 lb. The plane can carry only 80,000 lb with a total volume of 6000 ft^3. Each medical kit will aid 4 people, and each container of water will serve 10 people.

(a) How many of each should be sent to maximize the number of people helped?

(b) If each medical kit could aid 6 people instead of 4, how would the results from part (a) change?

79. *Storage Capacity* An office manager wants to buy some filing cabinets. He knows that cabinet A costs $10 each, requires 6 ft^2 of floor space, and holds 8 ft^3 of files. Cabinet B costs $20 each, requires 8 ft^2 of floor space, and holds 12 ft^3 of files. He can spend no more than $140 due to budget limitations, and his office has room for no more than 72 ft^2 of cabinets. He wants to maximize storage capacity within the limits imposed by funds and space. How many of each type of cabinet should he buy?

80. *Gasoline Revenues* The manufacturing process requires that oil refineries manufacture at least 2 gal of gasoline for each gallon of fuel oil. To meet the winter demand for fuel oil, at least 3 million gal per day must be produced. The demand for gasoline is no more than 6.4 million gal per day. If the price of gasoline is $2.90 per gal and the price of fuel oil is $2.50 per gal, how much of each should be produced to maximize revenue?

81. *Diet Requirements* Theo, who is dieting, requires two food supplements, I and II. He can get these supplements from two different products, *A* and *B*, as shown in the table. Theo's physician has recommended that he include at least 15 g of each supplement in his daily diet. If product *A* costs $0.25 per serving and product *B* costs $0.40 per serving, how can he satisfy his requirements most economically?

Supplement (g/serving)	I	II
Product *A*	3	2
Product *B*	2	4

82. *Profit from Televisions* Seall Manufacturing Company makes television monitors. It produces a bargain monitor that sells for $100 profit and a deluxe monitor that sells for $150 profit. On the assembly line the bargain monitor requires 3 hr, and the deluxe monitor takes 5 hr. The cabinet shop spends 1 hr on the cabinet for the bargain monitor and 3 hr on the cabinet for the deluxe monitor. Both models require 2 hr of time for testing and packing. On a particular production run, the Seall Company has available 3900 work hours on the assembly line, 2100 work hours in the cabinet shop, and 2200 work hours in the testing and packing department. How many of each model should it produce to make the maximum profit? What is the maximum profit?

5.7 **Properties of Matrices**

■ Basic Definitions
■ Adding Matrices
■ Special Matrices
■ Subtracting Matrices
■ Multiplying Matrices
■ Applying Matrix Algebra

We used matrix notation to solve a system of linear equations earlier in this chapter. In this section and the next, we discuss algebraic properties of matrices.

Basic Definitions It is customary to use capital letters to name matrices and to use subscript notation to name elements of a matrix, as in the following matrix A.

$$A = \begin{bmatrix} a_{11} & a_{12} & a_{13} & \cdots & a_{1n} \\ a_{21} & a_{22} & a_{23} & \cdots & a_{2n} \\ a_{31} & a_{32} & a_{33} & \cdots & a_{3n} \\ \vdots & \vdots & \vdots & & \vdots \\ a_{m1} & a_{m2} & a_{m3} & \cdots & a_{mn} \end{bmatrix}$$

The first row, first column element is a_{11} (read "a-sub-one-one"); the second row, third column element is a_{23}; and in general, the ith row, jth column element is a_{ij}.

Certain matrices have special names. An $n \times n$ matrix is a **square matrix** because the number of rows is equal to the number of columns. A matrix with just one row is a **row matrix,** and a matrix with just one column is a **column matrix.**

Two matrices are equal if they have the same dimension and if corresponding elements, position by position, are equal. Using this definition, the matrices

$$\begin{bmatrix} 2 & 1 \\ 3 & -5 \end{bmatrix} \quad \text{and} \quad \begin{bmatrix} 1 & 2 \\ -5 & 3 \end{bmatrix} \quad \text{are } not \text{ equal}$$

(even though they contain the same elements and have the same dimension), since the corresponding elements differ.

EXAMPLE 1 **Finding Values to Make Two Matrices Equal**

Find the values of the variables for which each statement is true, if possible.

(a) $\begin{bmatrix} 2 & 1 \\ p & q \end{bmatrix} = \begin{bmatrix} x & y \\ -1 & 0 \end{bmatrix}$ **(b)** $\begin{bmatrix} x \\ y \end{bmatrix} = \begin{bmatrix} 1 \\ 4 \\ 0 \end{bmatrix}$

SOLUTION

(a) From the definition of equality given above, the only way that the statement can be true is if $2 = x$, $1 = y$, $p = -1$, and $q = 0$.

(b) This statement can never be true since the two matrices have different dimensions. (One is 2×1 and the other is 3×1.)

✔ *Now Try Exercises 1 and 7.*

Adding Matrices Addition of matrices is defined as follows.

Addition of Matrices

To add two matrices of the same dimension, add corresponding elements. *Only matrices of the same dimension can be added.*

It can be shown that matrix addition satisfies the commutative, associative, closure, identity, and inverse properties. **(See Exercises 87 and 88.)**

EXAMPLE 2 **Adding Matrices**

Find each sum, if possible.

(a) $\begin{bmatrix} 5 & -6 \\ 8 & 9 \end{bmatrix} + \begin{bmatrix} -4 & 6 \\ 8 & -3 \end{bmatrix}$ (b) $\begin{bmatrix} 2 \\ 5 \\ 8 \end{bmatrix} + \begin{bmatrix} -6 \\ 3 \\ 12 \end{bmatrix}$

(c) $A + B$, if $A = \begin{bmatrix} 5 & 8 \\ 6 & 2 \end{bmatrix}$ and $B = \begin{bmatrix} 3 & 9 & 1 \\ 4 & 2 & 5 \end{bmatrix}$

ALGEBRAIC SOLUTION

(a) $\begin{bmatrix} 5 & -6 \\ 8 & 9 \end{bmatrix} + \begin{bmatrix} -4 & 6 \\ 8 & -3 \end{bmatrix}$

$= \begin{bmatrix} 5 + (-4) & -6 + 6 \\ 8 + 8 & 9 + (-3) \end{bmatrix}$

$= \begin{bmatrix} 1 & 0 \\ 16 & 6 \end{bmatrix}$

(b) $\begin{bmatrix} 2 \\ 5 \\ 8 \end{bmatrix} + \begin{bmatrix} -6 \\ 3 \\ 12 \end{bmatrix} = \begin{bmatrix} -4 \\ 8 \\ 20 \end{bmatrix}$

(c) The matrices

$$A = \begin{bmatrix} 5 & 8 \\ 6 & 2 \end{bmatrix}$$

and $$B = \begin{bmatrix} 3 & 9 & 1 \\ 4 & 2 & 5 \end{bmatrix}$$

have different dimensions, so A and B cannot be added. The sum $A + B$ does not exist.

GRAPHING CALCULATOR SOLUTION

(a) **Figure 24(b)** shows the sum of matrices A and B as defined in **Figure 24(a)**.

(a)   (b)

Figure 24

(b) The screen in **Figure 25** shows how the sum of two column matrices entered directly on the home screen is displayed.

Figure 25   **Figure 26**

(c) A graphing calculator will return an ERROR message if it is directed to perform an operation on matrices that is not possible because of dimension mismatch. See **Figure 26**.

✔ *Now Try Exercises 21, 23, and 25.*

Special Matrices   A matrix containing only zero elements is a **zero matrix.** A zero matrix can be written with any dimension.

1×3 zero matrix 2×3 zero matrix

$$O = \begin{bmatrix} 0 & 0 & 0 \end{bmatrix} \qquad O = \begin{bmatrix} 0 & 0 & 0 \\ 0 & 0 & 0 \end{bmatrix}$$

By the additive inverse property, each real number has an additive inverse: If a is a real number, then there is a real number $-a$ such that

$$a + (-a) = 0 \quad \text{and} \quad -a + a = 0. \quad \text{(Section R.2)}$$

Given matrix A, there is a matrix $-A$ such that $A + (-A) = O$. The matrix $-A$ has as elements the additive inverses of the elements of A. (Remember, each element of A is a real number and therefore has an additive inverse.) For example, if

$$A = \begin{bmatrix} -5 & 2 & -1 \\ 3 & 4 & -6 \end{bmatrix}, \quad \text{then} \quad -A = \begin{bmatrix} 5 & -2 & 1 \\ -3 & -4 & 6 \end{bmatrix}.$$

CHECK Test that $A + (-A)$ equals the zero matrix, O.

$$A + (-A) = \begin{bmatrix} -5 & 2 & -1 \\ 3 & 4 & -6 \end{bmatrix} + \begin{bmatrix} 5 & -2 & 1 \\ -3 & -4 & 6 \end{bmatrix} = \begin{bmatrix} 0 & 0 & 0 \\ 0 & 0 & 0 \end{bmatrix} = O \checkmark$$

Matrix $-A$ is the **additive inverse,** or **negative,** of matrix A. Every matrix has an additive inverse.

Subtracting Matrices The real number b is subtracted from the real number a, written $a - b$, by adding a and the additive inverse of b.

$$a - b = a + (-b) \quad \text{Real number subtraction}$$

The same definition applies to subtraction of matrices.

Subtraction of Matrices

If A and B are two matrices of the same dimension, then

$$A - B = A + (-B).$$

In practice, the difference of two matrices of the same dimension is found by subtracting corresponding elements.

EXAMPLE 3 **Subtracting Matrices**

Find each difference, if possible.

(a) $\begin{bmatrix} -5 & 6 \\ 2 & 4 \end{bmatrix} - \begin{bmatrix} -3 & 2 \\ 5 & -8 \end{bmatrix}$ **(b)** $\begin{bmatrix} 8 & 6 & -4 \end{bmatrix} - \begin{bmatrix} 3 & 5 & -8 \end{bmatrix}$

(c) $A - B$, if $A = \begin{bmatrix} -2 & 5 \\ 0 & 1 \end{bmatrix}$ and $B = \begin{bmatrix} 3 \\ 5 \end{bmatrix}$

SOLUTION

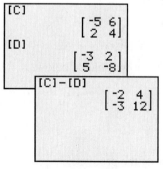

These screens support the result in **Example 3(a).**

(a) $\begin{bmatrix} -5 & 6 \\ 2 & 4 \end{bmatrix} - \begin{bmatrix} -3 & 2 \\ 5 & -8 \end{bmatrix} = \begin{bmatrix} -5 - (-3) & 6 - 2 \\ 2 - 5 & 4 - (-8) \end{bmatrix}$

$$= \begin{bmatrix} -2 & 4 \\ -3 & 12 \end{bmatrix}$$

(b) $\begin{bmatrix} 8 & 6 & -4 \end{bmatrix} - \begin{bmatrix} 3 & 5 & -8 \end{bmatrix} = \begin{bmatrix} 5 & 1 & 4 \end{bmatrix}$

(c) The matrices

$$A = \begin{bmatrix} -2 & 5 \\ 0 & 1 \end{bmatrix} \quad \text{and} \quad B = \begin{bmatrix} 3 \\ 5 \end{bmatrix}$$

have different dimensions and cannot be subtracted, so the difference $A - B$ does not exist.

✔ *Now Try Exercises 27, 29, and 31.*

Multiplying Matrices In work with matrices, a real number is called a **scalar** to distinguish it from a matrix. The product of a scalar k and a matrix X is the matrix kX, each of whose elements is k times the corresponding element of X.

These screens support the results in **Example 4**.

EXAMPLE 4 **Multiplying Matrices by Scalars**

Find each product.

(a) $5\begin{bmatrix} 2 & -3 \\ 0 & 4 \end{bmatrix}$

(b) $\dfrac{3}{4}\begin{bmatrix} 20 & 36 \\ 12 & -16 \end{bmatrix}$

SOLUTION

(a) $5\begin{bmatrix} 2 & -3 \\ 0 & 4 \end{bmatrix} = \begin{bmatrix} 5(2) & 5(-3) \\ 5(0) & 5(4) \end{bmatrix}$ Multiply each element of the matrix by the scalar 5.

$= \begin{bmatrix} 10 & -15 \\ 0 & 20 \end{bmatrix}$

(b) $\dfrac{3}{4}\begin{bmatrix} 20 & 36 \\ 12 & -16 \end{bmatrix} = \begin{bmatrix} \frac{3}{4}(20) & \frac{3}{4}(36) \\ \frac{3}{4}(12) & \frac{3}{4}(-16) \end{bmatrix}$ Multiply each element of the matrix by the scalar $\frac{3}{4}$.

$= \begin{bmatrix} 15 & 27 \\ 9 & -12 \end{bmatrix}$

✔ *Now Try Exercises 37 and 39.*

The proofs of the following properties of scalar multiplication are left for **Exercises 91–94.**

Properties of Scalar Multiplication

Let A and B be matrices of the same dimension, and let c and d be scalars. Then these properties hold.

$$(c + d)A = cA + dA \qquad (cA)d = (cd)A$$

$$c(A + B) = cA + cB \qquad (cd)A = c(dA)$$

We have seen how to multiply a real number (scalar) and a matrix. To find the product of two matrices, such as

$$A = \begin{bmatrix} -3 & 4 & 2 \\ 5 & 0 & 4 \end{bmatrix} \quad \text{and} \quad B = \begin{bmatrix} -6 & 4 \\ 2 & 3 \\ 3 & -2 \end{bmatrix},$$

first locate *row* 1 of A and *column* 1 of B, which are shown shaded below.

$$A = \begin{bmatrix} -3 & 4 & 2 \\ 5 & 0 & 4 \end{bmatrix} \qquad B = \begin{bmatrix} -6 & 4 \\ 2 & 3 \\ 3 & -2 \end{bmatrix}$$

Multiply corresponding elements, and find the sum of the products.

$$-3(-6) + 4(2) + 2(3) = 32$$

This result is the element for row 1, column 1 of the product matrix.

Now use *row* 1 of A and *column* 2 of B to determine the element in row 1, column 2 of the product matrix.

$$\begin{bmatrix} -3 & 4 & 2 \\ 5 & 0 & 4 \end{bmatrix}\begin{bmatrix} -6 & 4 \\ 2 & 3 \\ 3 & -2 \end{bmatrix} \qquad -3(4) + 4(3) + 2(-2) = -4$$

Next, use *row* 2 of *A* and *column* 1 of *B*. This will give the row 2, column 1 entry of the product matrix.

$$\begin{bmatrix} -3 & 4 & 2 \\ 5 & 0 & 4 \end{bmatrix} \begin{bmatrix} -6 & 4 \\ 2 & 3 \\ 3 & -2 \end{bmatrix} \qquad 5(-6) + 0(2) + 4(3) = -18$$

Finally, use *row* 2 of *A* and *column* 2 of *B* to find the entry for row 2, column 2 of the product matrix.

$$\begin{bmatrix} -3 & 4 & 2 \\ 5 & 0 & 4 \end{bmatrix} \begin{bmatrix} -6 & 4 \\ 2 & 3 \\ 3 & -2 \end{bmatrix} \qquad 5(4) + 0(3) + 4(-2) = 12$$

The product matrix can now be written.

$$\begin{bmatrix} -3 & 4 & 2 \\ 5 & 0 & 4 \end{bmatrix} \begin{bmatrix} -6 & 4 \\ 2 & 3 \\ 3 & -2 \end{bmatrix} = \begin{bmatrix} 32 & -4 \\ -18 & 12 \end{bmatrix} \qquad \text{Product of } A \text{ and } B$$

NOTE As seen here, the product of a 2×3 matrix and a 3×2 matrix is a 2×2 matrix. The dimension of a product matrix AB is given by the number of rows of A and the number of columns of B, respectively.

By definition, the product AB of an $m \times n$ matrix A and an $n \times p$ matrix B is found as follows. Multiply each element of the first row of A by the corresponding element of the first column of B. The sum of these n products is the first row, first column element of AB. Also, the sum of the products found by multiplying the elements of the first row of A times the corresponding elements of the second column of B gives the first row, second column element of AB, and so on.

To find the ith row, jth column element of AB, multiply each element in the ith row of A by the corresponding element in the jth column of B. (Note the shaded areas in the matrices below.) The sum of these products will give the row i, column j element of AB.

$$A = \begin{bmatrix} a_{11} & a_{12} & a_{13} & \cdots & a_{1n} \\ a_{21} & a_{22} & a_{23} & \cdots & a_{2n} \\ \vdots & & & & \\ a_{i1} & a_{i2} & a_{i3} & \cdots & a_{in} \\ \vdots & & & & \\ a_{m1} & a_{m2} & a_{m3} & \cdots & a_{mn} \end{bmatrix} \qquad B = \begin{bmatrix} b_{11} & b_{12} & \cdots & b_{1j} & \cdots & b_{1p} \\ b_{21} & b_{22} & \cdots & b_{2j} & \cdots & b_{2p} \\ \vdots & & & & & \\ b_{n1} & b_{n2} & \cdots & b_{nj} & \cdots & b_{np} \end{bmatrix}$$

Matrix Multiplication

The number of columns of an $m \times n$ matrix A is the same as the number of rows of an $n \times p$ matrix B (i.e., both n). The element c_{ij} of the product matrix $C = AB$ is found as follows.

$$c_{ij} = a_{i1}b_{1j} + a_{i2}b_{2j} + \cdots + a_{in}b_{nj}$$

Matrix AB will be an $m \times p$ matrix.

EXAMPLE 5 Deciding Whether Two Matrices Can Be Multiplied

Suppose A is a 3×2 matrix, while B is a 2×4 matrix.

(a) Can the product AB be calculated?

(b) If AB can be calculated, what is its dimension?

(c) Can BA be calculated?

(d) If BA can be calculated, what is its dimension?

SOLUTION

(a) The following diagram shows that AB can be calculated, because the number of columns of A is equal to the number of rows of B. (Both are 2.)

Matrix A Matrix B

3×2 2×4

must match

dimension of AB

3×4

(b) As indicated in the diagram above, the product AB is a 3×4 matrix.

(c) The diagram below shows that BA cannot be calculated.

Matrix B Matrix A

2×4 3×2

different

(d) The product BA cannot be calculated, because B has 4 columns and A has only 3 rows.

✔ *Now Try Exercises 45 and 47.*

EXAMPLE 6 Multiplying Matrices

Let $A = \begin{bmatrix} 1 & -3 \\ 7 & 2 \end{bmatrix}$ and $B = \begin{bmatrix} 1 & 0 & -1 & 2 \\ 3 & 1 & 4 & -1 \end{bmatrix}$. Find each product, if possible.

(a) AB **(b)** BA

SOLUTION

(a) First decide whether AB can be found. Since A is 2×2 and B is 2×4, the product can be found and will be a 2×4 matrix.

$$AB = \begin{bmatrix} 1 & -3 \\ 7 & 2 \end{bmatrix}\begin{bmatrix} 1 & 0 & -1 & 2 \\ 3 & 1 & 4 & -1 \end{bmatrix}$$

$$= \begin{bmatrix} 1(1) + (-3)3 & 1(0) + (-3)1 & 1(-1) + (-3)4 & 1(2) + (-3)(-1) \\ 7(1) + 2(3) & 7(0) + 2(1) & 7(-1) + 2(4) & 7(2) + 2(-1) \end{bmatrix}$$

Use the definition of matrix multiplication.

$$= \begin{bmatrix} -8 & -3 & -13 & 5 \\ 13 & 2 & 1 & 12 \end{bmatrix}$$ Perform the operations.

(b) Since B is a 2×4 matrix, and A is a 2×2 matrix, the number of columns of B (here 4) does not equal the number of rows of A (here 2). Therefore, the product BA cannot be found.

✔ *Now Try Exercises 65 and 69.*

The three screens here support the results of the matrix multiplication in **Example 6.** The final screen indicates that the product BA cannot be found.

EXAMPLE 7 Multiplying Square Matrices in Different Orders

Let $A = \begin{bmatrix} 1 & 3 \\ -2 & 5 \end{bmatrix}$ and $B = \begin{bmatrix} -2 & 7 \\ 0 & 2 \end{bmatrix}$. Find each product.

(a) AB **(b)** BA

SOLUTION

(a) $AB = \begin{bmatrix} 1 & 3 \\ -2 & 5 \end{bmatrix}\begin{bmatrix} -2 & 7 \\ 0 & 2 \end{bmatrix}$

$= \begin{bmatrix} 1(-2) + 3(0) & 1(7) + 3(2) \\ -2(-2) + 5(0) & -2(7) + 5(2) \end{bmatrix}$ Multiply elements of each row of A by elements of each column of B.

$= \begin{bmatrix} -2 & 13 \\ 4 & -4 \end{bmatrix}$

(b) $BA = \begin{bmatrix} -2 & 7 \\ 0 & 2 \end{bmatrix}\begin{bmatrix} 1 & 3 \\ -2 & 5 \end{bmatrix}$

$= \begin{bmatrix} -2(1) + 7(-2) & -2(3) + 7(5) \\ 0(1) + 2(-2) & 0(3) + 2(5) \end{bmatrix}$ Multiply elements of each row of B by elements of each column of A.

$= \begin{bmatrix} -16 & 29 \\ -4 & 10 \end{bmatrix}$

Note that $AB \neq BA$.

✔ *Now Try Exercise 75.*

When multiplying matrices, it is important to pay special attention to the dimensions of the matrices as well as the order in which they are to be multiplied. **Examples 5 and 6** showed that the order in which two matrices are to be multiplied may determine whether their product can be found. **Example 7** showed that even when both products AB and BA can be found, they may not be equal.

In general, if A and B are matrices, then $AB \neq BA$. ***Matrix multiplication is not commutative.***

Matrix multiplication does, however, satisfy the associative and distributive properties.

Properties of Matrix Multiplication

If A, B, and C are matrices such that all the following products and sums exist, then these properties hold.

$$(AB)C = A(BC), \quad A(B + C) = AB + AC, \quad (B + C)A = BA + CA$$

For proofs of the first two results for the special cases when A, B, and C are square matrices, see **Exercises 89 and 90.** The identity and inverse properties for matrix multiplication are discussed in the next section.

Applying Matrix Algebra

EXAMPLE 8 | Using Matrix Multiplication to Model Plans for a Subdivision

A contractor builds three kinds of houses, models A, B, and C, with a choice of two styles, colonial or ranch. Matrix P below shows the number of each kind of house the contractor is planning to build for a new 100-home subdivision. The amounts for each of the main materials used depend on the style of the house. These amounts are shown in matrix Q, while matrix R gives the cost in dollars for each kind of material. Concrete is measured here in cubic yards, lumber in 1000 board feet, brick in 1000s, and shingles in 100 square feet.

$$\begin{array}{c} \\ \text{Model A} \\ \text{Model B} \\ \text{Model C} \end{array} \begin{array}{c} \text{Colonial Ranch} \\ \begin{bmatrix} 0 & 30 \\ 10 & 20 \\ 20 & 20 \end{bmatrix} \end{array} = P$$

$$\begin{array}{c} \\ \text{Colonial} \\ \text{Ranch} \end{array} \begin{array}{c} \text{Concrete Lumber Brick Shingles} \\ \begin{bmatrix} 10 & 2 & 0 & 2 \\ 50 & 1 & 20 & 2 \end{bmatrix} \end{array} = Q$$

$$\begin{array}{c} \\ \text{Concrete} \\ \text{Lumber} \\ \text{Brick} \\ \text{Shingles} \end{array} \begin{array}{c} \text{Cost per Unit} \\ \begin{bmatrix} 20 \\ 180 \\ 60 \\ 25 \end{bmatrix} \end{array} = R$$

(a) What is the total cost of materials for all houses of each model?

(b) How much of each of the four kinds of material must be ordered?

(c) What is the total cost of the materials?

SOLUTION

(a) To find the materials cost for each model, first find matrix PQ, which will show the total amount of each material needed for all houses of each model.

$$PQ = \begin{bmatrix} 0 & 30 \\ 10 & 20 \\ 20 & 20 \end{bmatrix} \begin{bmatrix} 10 & 2 & 0 & 2 \\ 50 & 1 & 20 & 2 \end{bmatrix} = \begin{array}{c} \text{Concrete Lumber Brick Shingles} \\ \begin{bmatrix} 1500 & 30 & 600 & 60 \\ 1100 & 40 & 400 & 60 \\ 1200 & 60 & 400 & 80 \end{bmatrix} \end{array} \begin{array}{c} \text{Model A} \\ \text{Model B} \\ \text{Model C} \end{array}$$

Multiplying PQ and the cost matrix R gives the total cost of materials for each model.

$$(PQ)R = \begin{bmatrix} 1500 & 30 & 600 & 60 \\ 1100 & 40 & 400 & 60 \\ 1200 & 60 & 400 & 80 \end{bmatrix} \begin{bmatrix} 20 \\ 180 \\ 60 \\ 25 \end{bmatrix} = \begin{array}{c} \text{Cost} \\ \begin{bmatrix} 72{,}900 \\ 54{,}700 \\ 60{,}800 \end{bmatrix} \end{array} \begin{array}{c} \text{Model A} \\ \text{Model B} \\ \text{Model C} \end{array}$$

(b) To find how much of each kind of material to order, refer to the columns of matrix PQ. The sums of the elements of the columns will give a matrix whose elements represent the total amounts of all materials needed for the subdivision. Call this matrix T, and write it as a row matrix.

$$T = \begin{bmatrix} 3800 & 130 & 1400 & 200 \end{bmatrix}$$

(c) The total cost of all the materials is given by the product of matrix T, the total amounts matrix, and matrix R, the cost matrix. To multiply these matrices and get a 1×1 matrix, representing the total cost, requires multiplying a 1×4 matrix and a 4×1 matrix. This is why in part (b) a row matrix was written rather than a column matrix.

The total materials cost is given by *TR*, so

$$TR = \begin{bmatrix} 3800 & 130 & 1400 & 200 \end{bmatrix} \begin{bmatrix} 20 \\ 180 \\ 60 \\ 25 \end{bmatrix} = \begin{bmatrix} 188{,}400 \end{bmatrix}.$$

The total cost of materials is $188,400. This total may also be found by summing the elements of the column matrix *(PQ)R*.

✔ *Now Try Exercise 81.*

5.7 Exercises

Find the values of the variables for which each statement is true, if possible. **See Examples 1 and 2.**

1. $\begin{bmatrix} -3 & a \\ b & 5 \end{bmatrix} = \begin{bmatrix} c & 0 \\ 4 & d \end{bmatrix}$

2. $\begin{bmatrix} w & x \\ 8 & -12 \end{bmatrix} = \begin{bmatrix} 9 & 17 \\ y & z \end{bmatrix}$

3. $\begin{bmatrix} x+2 & y-6 \\ z-3 & w+5 \end{bmatrix} = \begin{bmatrix} -2 & 8 \\ 0 & 3 \end{bmatrix}$

4. $\begin{bmatrix} 6 & a+3 \\ b+2 & 9 \end{bmatrix} = \begin{bmatrix} c-3 & 4 \\ -2 & d-4 \end{bmatrix}$

5. $\begin{bmatrix} x & y & z \end{bmatrix} = \begin{bmatrix} 21 & 6 \end{bmatrix}$

6. $\begin{bmatrix} p \\ q \\ r \end{bmatrix} = \begin{bmatrix} 3 \\ -9 \end{bmatrix}$

7. $\begin{bmatrix} 0 & 5 & x \\ -1 & 3 & y+2 \\ 4 & 1 & z \end{bmatrix} = \begin{bmatrix} 0 & w & 6 \\ -1 & 3 & 0 \\ 4 & 1 & 8 \end{bmatrix}$

8. $\begin{bmatrix} 5 & x-4 & 9 \\ 2 & -3 & 8 \\ 6 & 0 & 5 \end{bmatrix} = \begin{bmatrix} y+3 & 2 & 9 \\ z+4 & -3 & 8 \\ 6 & 0 & w \end{bmatrix}$

9. $\begin{bmatrix} -7+z & 4r & 8s \\ 6p & 2 & 5 \end{bmatrix} + \begin{bmatrix} -9 & 8r & 3 \\ 2 & 5 & 4 \end{bmatrix} = \begin{bmatrix} 2 & 36 & 27 \\ 20 & 7 & 12a \end{bmatrix}$

10. $\begin{bmatrix} a+2 & 3z+1 & 5m \\ 8k & 0 & 3 \end{bmatrix} + \begin{bmatrix} 3a & 2z & 5m \\ 2k & 5 & 6 \end{bmatrix} = \begin{bmatrix} 10 & -14 & 80 \\ 10 & 5 & 9 \end{bmatrix}$

11. *Concept Check* Two matrices are equal if they have the same _____ and if corresponding elements are _____.

12. *Concept Check* In order to add two matrices, they must have the same _____.

Concept Check *Find the dimension of each matrix. Identify any square, column, or row matrices.*

13. $\begin{bmatrix} -4 & 8 \\ 2 & 3 \end{bmatrix}$

14. $\begin{bmatrix} -9 & 6 & 2 \\ 4 & 1 & 8 \end{bmatrix}$

15. $\begin{bmatrix} -6 & 8 & 0 & 0 \\ 4 & 1 & 9 & 2 \\ 3 & -5 & 7 & 1 \end{bmatrix}$

16. $\begin{bmatrix} 8 & -2 & 4 & 6 & 3 \end{bmatrix}$

17. $\begin{bmatrix} 2 \\ 4 \end{bmatrix}$

18. $\begin{bmatrix} -9 \end{bmatrix}$

📄 **19.** Your friend missed the lecture on adding matrices. Explain to him how to add two matrices.

📄 **20.** Explain how to subtract two matrices.

Perform each operation when possible. ***See Examples 2 and 3.***

21. $\begin{bmatrix} -4 & 3 \\ 12 & -6 \end{bmatrix} + \begin{bmatrix} 2 & -8 \\ 5 & 10 \end{bmatrix}$

22. $\begin{bmatrix} 9 & 4 \\ -8 & 2 \end{bmatrix} + \begin{bmatrix} -3 & 2 \\ -4 & 7 \end{bmatrix}$

23. $\begin{bmatrix} 6 & -9 & 2 \\ 4 & 1 & 3 \end{bmatrix} + \begin{bmatrix} -8 & 2 & 5 \\ 6 & -3 & 4 \end{bmatrix}$

24. $\begin{bmatrix} 4 & -3 \\ 7 & 2 \\ -6 & 8 \end{bmatrix} + \begin{bmatrix} 9 & -10 \\ 0 & 5 \\ -1 & 6 \end{bmatrix}$

25. $\begin{bmatrix} 2 & 4 & 6 \end{bmatrix} + \begin{bmatrix} -2 \\ -4 \\ -6 \end{bmatrix}$

26. $\begin{bmatrix} 3 \\ 1 \\ 0 \end{bmatrix} + \begin{bmatrix} 2 \\ -6 \end{bmatrix}$

27. $\begin{bmatrix} -6 & 8 \\ 0 & 0 \end{bmatrix} - \begin{bmatrix} 0 & 0 \\ -4 & -2 \end{bmatrix}$

28. $\begin{bmatrix} 11 & 0 \\ -4 & 0 \end{bmatrix} - \begin{bmatrix} 0 & 12 \\ 0 & -14 \end{bmatrix}$

29. $\begin{bmatrix} 12 \\ -1 \\ 3 \end{bmatrix} - \begin{bmatrix} 8 \\ 4 \\ -1 \end{bmatrix}$

30. $\begin{bmatrix} 10 & -4 & 6 \end{bmatrix} - \begin{bmatrix} -2 & 5 & 3 \end{bmatrix}$

31. $\begin{bmatrix} -4 & 3 \end{bmatrix} - \begin{bmatrix} 5 & 8 & 2 \end{bmatrix}$

32. $\begin{bmatrix} 4 & 6 \end{bmatrix} - \begin{bmatrix} 2 \\ 3 \end{bmatrix}$

33. $\begin{bmatrix} \sqrt{3} & -4 \\ 2 & -\sqrt{5} \\ -8 & \sqrt{8} \end{bmatrix} - \begin{bmatrix} 2\sqrt{3} & 9 \\ -2 & \sqrt{5} \\ -7 & 3\sqrt{2} \end{bmatrix}$

34. $\begin{bmatrix} 2 & \sqrt{7} \\ 3\sqrt{28} & -6 \end{bmatrix} - \begin{bmatrix} -1 & 5\sqrt{7} \\ 2\sqrt{7} & 2 \end{bmatrix}$

35. $\begin{bmatrix} 3x + y & -2y \\ x + 2y & 3y \end{bmatrix} + \begin{bmatrix} 2x & 3y \\ 5x & x \end{bmatrix}$

36. $\begin{bmatrix} 4k - 8y \\ 6z - 3x \\ 2k + 5a \\ -4m + 2n \end{bmatrix} - \begin{bmatrix} 5k + 6y \\ 2z + 5x \\ 4k + 6a \\ 4m - 2n \end{bmatrix}$

Let $A = \begin{bmatrix} -2 & 4 \\ 0 & 3 \end{bmatrix}$ *and* $B = \begin{bmatrix} -6 & 2 \\ 4 & 0 \end{bmatrix}$. *Find each of the following.* ***See Example 4.***

37. $2A$ **38.** $-3B$ **39.** $\dfrac{3}{2}B$ **40.** $-\dfrac{3}{2}A$

41. $2A - 3B$ **42.** $-2A + 4B$ **43.** $-A + \dfrac{1}{2}B$ **44.** $\dfrac{3}{4}A - B$

Suppose that matrix A has dimension 2×3, *B has dimension* 3×5, *and C has dimension* 5×2. *Decide whether the given product can be calculated. If it can, determine its dimension.* ***See Example 5.***

45. AB **46.** CA **47.** BA **48.** AC **49.** BC **50.** CB

Find each matrix product when possible. ***See Examples 5–7.***

51. $\begin{bmatrix} 1 & 2 \\ 3 & 4 \end{bmatrix} \begin{bmatrix} -1 \\ 7 \end{bmatrix}$

52. $\begin{bmatrix} -1 & 5 \\ 7 & 0 \end{bmatrix} \begin{bmatrix} 6 \\ 2 \end{bmatrix}$

53. $\begin{bmatrix} 3 & -4 & 1 \\ 5 & 0 & 2 \end{bmatrix} \begin{bmatrix} -1 \\ 4 \\ 2 \end{bmatrix}$

54. $\begin{bmatrix} -6 & 3 & 5 \\ 2 & 9 & 1 \end{bmatrix} \begin{bmatrix} -2 \\ 0 \\ 3 \end{bmatrix}$

55. $\begin{bmatrix} \sqrt{2} & \sqrt{2} & -\sqrt{18} \\ \sqrt{3} & \sqrt{27} & 0 \end{bmatrix} \begin{bmatrix} 8 & -10 \\ 9 & 12 \\ 0 & 2 \end{bmatrix}$

56. $\begin{bmatrix} -9 & 2 & 1 \\ 3 & 0 & 0 \end{bmatrix} \begin{bmatrix} \sqrt{5} \\ \sqrt{20} \\ -2\sqrt{5} \end{bmatrix}$

57. $\begin{bmatrix} \sqrt{3} & 1 \\ 2\sqrt{5} & 3\sqrt{2} \end{bmatrix} \begin{bmatrix} \sqrt{3} & -\sqrt{6} \\ 4\sqrt{3} & 0 \end{bmatrix}$

58. $\begin{bmatrix} \sqrt{7} & 0 \\ 2 & \sqrt{28} \end{bmatrix} \begin{bmatrix} 2\sqrt{3} & -\sqrt{7} \\ 0 & -6 \end{bmatrix}$

59. $\begin{bmatrix} -3 & 0 & 2 & 1 \\ 4 & 0 & 2 & 6 \end{bmatrix} \begin{bmatrix} -4 & 2 \\ 0 & 1 \end{bmatrix}$

60. $\begin{bmatrix} -1 & 2 & 4 & 1 \\ 0 & 2 & -3 & 5 \end{bmatrix} \begin{bmatrix} 1 & 2 & 4 \\ -2 & 5 & 1 \end{bmatrix}$

61. $\begin{bmatrix} -2 & 4 & 1 \end{bmatrix} \begin{bmatrix} 3 & -2 & 4 \\ 2 & 1 & 0 \\ 0 & -1 & 4 \end{bmatrix}$

62. $\begin{bmatrix} 0 & 3 & -4 \end{bmatrix} \begin{bmatrix} -2 & 6 & 3 \\ 0 & 4 & 2 \\ -1 & 1 & 4 \end{bmatrix}$

63. $\begin{bmatrix} -2 & -3 & -4 \\ 2 & -1 & 0 \\ 4 & -2 & 3 \end{bmatrix} \begin{bmatrix} 0 & 1 & 4 \\ 1 & 2 & -1 \\ 3 & 2 & -2 \end{bmatrix}$

64. $\begin{bmatrix} -1 & 2 & 0 \\ 0 & 3 & 2 \\ 0 & 1 & 4 \end{bmatrix} \begin{bmatrix} 2 & -1 & 2 \\ 0 & 2 & 1 \\ 3 & 0 & -1 \end{bmatrix}$

Given $A = \begin{bmatrix} 4 & -2 \\ 3 & 1 \end{bmatrix}$, $B = \begin{bmatrix} 5 & 1 \\ 0 & -2 \\ 3 & 7 \end{bmatrix}$, *and* $C = \begin{bmatrix} -5 & 4 & 1 \\ 0 & 3 & 6 \end{bmatrix}$, *find each product when possible.* ***See Examples 5–7.***

65. *BA*

66. *AC*

67. *BC*

68. *CB*

69. *AB*

70. *CA*

71. A^2

72. A^3
(*Hint:* $A^3 = A^2 \cdot A$)

73. *Concept Check* Compare the answers to **Exercises 65 and 69, 67 and 68,** and **66 and 70.** How do they show that matrix multiplication is not commutative?

74. Your friend missed the lecture on multiplying matrices. Explain to her the process of matrix multiplication.

For each pair of matrices A and B, find (**a**) *AB and* (**b**) *BA.* ***See Example 7.***

75. $A = \begin{bmatrix} 3 & 4 \\ -2 & 1 \end{bmatrix}$, $B = \begin{bmatrix} 6 & 0 \\ 5 & -2 \end{bmatrix}$

76. $A = \begin{bmatrix} 0 & -5 \\ -4 & 2 \end{bmatrix}$, $B = \begin{bmatrix} 3 & -1 \\ -5 & 4 \end{bmatrix}$

77. $A = \begin{bmatrix} 0 & 1 & -1 \\ 0 & 1 & 0 \\ 0 & 0 & 1 \end{bmatrix}$, $B = \begin{bmatrix} 1 & 0 & 0 \\ 0 & 1 & 0 \\ 0 & 0 & 1 \end{bmatrix}$

78. $A = \begin{bmatrix} -1 & 0 & 1 \\ 0 & 1 & 1 \\ -1 & -1 & 0 \end{bmatrix}$, $B = \begin{bmatrix} 0 & 0 & 1 \\ 0 & 1 & 0 \\ 1 & 0 & 0 \end{bmatrix}$

79. *Concept Check* In **Exercise 77,** $AB = A$ and $BA = A$. For this pair of matrices, *B* acts the same way for matrix multiplication as the number _____ acts for multiplication of real numbers.

80. *Concept Check* Find *AB* and *BA* for

$$A = \begin{bmatrix} a & b \\ c & d \end{bmatrix} \quad \text{and} \quad B = \begin{bmatrix} 1 & 0 \\ 0 & 1 \end{bmatrix}.$$

What do you notice? Matrix *B* acts as the multiplicative _____ element for 2×2 square matrices.

Solve each problem. ***See Example 8.***

81. *Income from Yogurt* Yagel's Yogurt sells three types of yogurt: nonfat, regular, and super creamy, at three locations. Location I sells 50 gal of nonfat, 100 gal of regular, and 30 gal of super creamy each day. Location II sells 10 gal of nonfat, and Location III sells 60 gal of nonfat each day. Daily sales of regular yogurt are 90 gal at Location II and 120 gal at Location III. At Location II, 50 gal of super creamy are sold each day, and 40 gal of super creamy are sold each day at Location III.

(**a**) Write a 3×3 matrix that shows the sales figures for the three locations, with the rows representing the three locations.

(**b**) The incomes per gallon for nonfat, regular, and super creamy are $12, $10, and $15, respectively. Write a 1×3 or 3×1 matrix displaying the incomes.

(**c**) Find a matrix product that gives the daily income at each of the three locations.

(**d**) What is Yagel's Yogurt's total daily income from the three locations?

82. *Purchasing Costs* The Bread Box, a small neighborhood bakery, sells four main items: sweet rolls, bread, cakes, and pies. The amount of each ingredient (in cups, except for eggs) required for these items is given by matrix A.

$$
\begin{array}{c}
\begin{array}{ccccc}
\text{Eggs} & \text{Flour} & \text{Sugar} & \text{Shortening} & \text{Milk}
\end{array} \\
\begin{array}{c}
\text{Rolls (doz)} \\
\text{Bread (loaf)} \\
\text{Cake} \\
\text{Pie (crust)}
\end{array}
\begin{bmatrix}
1 & 4 & \frac{1}{4} & \frac{1}{4} & 1 \\
0 & 3 & 0 & \frac{1}{4} & 0 \\
4 & 3 & 2 & 1 & 1 \\
0 & 1 & 0 & \frac{1}{3} & 0
\end{bmatrix} = A
\end{array}
$$

The cost (in cents) for each ingredient when purchased in large lots or small lots is given by matrix B.

$$
\begin{array}{c}
\text{Cost} \\
\begin{array}{cc}
\text{Large Lot} & \text{Small Lot}
\end{array} \\
\begin{array}{c}
\text{Eggs} \\
\text{Flour} \\
\text{Sugar} \\
\text{Shortening} \\
\text{Milk}
\end{array}
\begin{bmatrix}
5 & 5 \\
8 & 10 \\
10 & 12 \\
12 & 15 \\
5 & 6
\end{bmatrix} = B
\end{array}
$$

(a) Use matrix multiplication to find a matrix giving the comparative cost per bakery item for the two purchase options.

(b) Suppose a day's orders consist of 20 dozen sweet rolls, 200 loaves of bread, 50 cakes, and 60 pies. Write the orders as a 1×4 matrix, and, using matrix multiplication, write as a matrix the amount of each ingredient needed to fill the day's orders.

(c) Use matrix multiplication to find a matrix giving the costs under the two purchase options to fill the day's orders.

83. *(Modeling) Northern Spotted Owl Population* Several years ago, mathematical ecologists created a model to analyze population dynamics of the endangered northern spotted owl in the Pacific Northwest. The ecologists divided the female owl population into three categories: juvenile (up to 1 yr old), subadult (1 to 2 yr old), and adult (over 2 yr old). They concluded that the change in the makeup of the northern spotted owl population in successive years could be described by the following matrix equation.

$$
\begin{bmatrix} j_{n+1} \\ s_{n+1} \\ a_{n+1} \end{bmatrix} = \begin{bmatrix} 0 & 0 & 0.33 \\ 0.18 & 0 & 0 \\ 0 & 0.71 & 0.94 \end{bmatrix} \begin{bmatrix} j_n \\ s_n \\ a_n \end{bmatrix}
$$

The numbers in the column matrices give the numbers of females in the three age groups after n years and $n + 1$ years. Multiplying the matrices yields the following.

$j_{n+1} = 0.33a_n$ Each year 33 juvenile females are born for each 100 adult females.

$s_{n+1} = 0.18j_n$ Each year 18% of the juvenile females survive to become subadults.

$a_{n+1} = 0.71s_n + 0.94a_n$ Each year 71% of the subadults survive to become adults, and 94% of the adults survive.

(*Source:* Lamberson, R. H., R. McKelvey, B. R. Noon, and C. Voss, "A Dynamic Analysis of Northern Spotted Owl Viability in a Fragmented Forest Landscape," *Conservation Biology,* Vol. 6, No. 4.)

(a) Suppose there are currently 3000 female northern spotted owls made up of 690 juveniles, 210 subadults, and 2100 adults. Use the matrix equation on the preceding page to determine the total number of female owls for each of the next 5 yr.

(b) Using advanced techniques from linear algebra, we can show that in the long run,

$$\begin{bmatrix} j_{n+1} \\ s_{n+1} \\ a_{n+1} \end{bmatrix} \approx 0.98359 \begin{bmatrix} j_n \\ s_n \\ a_n \end{bmatrix}.$$

What can we conclude about the long-term fate of the northern spotted owl?

(c) In the model, the main impediment to the survival of the northern spotted owl is the number 0.18 in the second row of the 3×3 matrix. This number is low for two reasons.

- The first year of life is precarious for most animals living in the wild.

- In addition, juvenile owls must eventually leave the nest and establish their own territory. If much of the forest near their original home has been cleared, then they are vulnerable to predators while searching for a new home.

Suppose that, thanks to better forest management, the number 0.18 can be increased to 0.3. Rework part (a) under this new assumption.

84. *(Modeling) Predator-Prey Relationship* In certain parts of the Rocky Mountains, deer provide the main food source for mountain lions. When the deer population is large, the mountain lions thrive. However, a large mountain lion population reduces the size of the deer population. Suppose the fluctuations of the two populations from year to year can be modeled with the matrix equation

$$\begin{bmatrix} m_{n+1} \\ d_{n+1} \end{bmatrix} = \begin{bmatrix} 0.51 & 0.4 \\ -0.05 & 1.05 \end{bmatrix} \begin{bmatrix} m_n \\ d_n \end{bmatrix}.$$

The numbers in the column matrices give the numbers of animals in the two populations after n years and $n + 1$ years, where the number of deer is measured in hundreds.

(a) Give the equation for d_{n+1} obtained from the second row of the square matrix. Use this equation to determine the rate at which the deer population will grow from year to year if there are no mountain lions.

(b) Suppose we start with a mountain lion population of 2000 and a deer population of 500,000 (that is, 5000 hundred deer). How large would each population be after 1 yr? 2 yr?

(c) Consider part (b) but change the initial mountain lion population to 4000. Show that the populations would both grow at a steady annual rate of 1.01.

85. *Northern Spotted Owl Population* Refer to **Exercise 83(b).** Show that the number 0.98359 is an approximate zero of the polynomial represented by

$$\begin{vmatrix} -x & 0 & 0.33 \\ 0.18 & -x & 0 \\ 0 & 0.71 & 0.94 - x \end{vmatrix}.$$

86. *Predator-Prey Relationship* Refer to **Exercise 84(c).** Show that the number 1.01 is a zero of the polynomial represented by

$$\begin{vmatrix} 0.51 - x & 0.4 \\ -0.05 & 1.05 - x \end{vmatrix}.$$

For Exercises 87–94, let

$$A = \begin{bmatrix} a_{11} & a_{12} \\ a_{21} & a_{22} \end{bmatrix}, \quad B = \begin{bmatrix} b_{11} & b_{12} \\ b_{21} & b_{22} \end{bmatrix}, \quad and \quad C = \begin{bmatrix} c_{11} & c_{12} \\ c_{21} & c_{22} \end{bmatrix},$$

where all the elements are real numbers. Use these matrices to show that each statement is true for 2×2 matrices.

87. $A + B = B + A$
(commutative property)

88. $A + (B + C) = (A + B) + C$
(associative property)

89. $(AB)C = A(BC)$
(associative property)

90. $A(B + C) = AB + AC$
(distributive property)

91. $c(A + B) = cA + cB$,
for any real number c.

92. $(c + d)A = cA + dA$,
for any real numbers c and d.

93. $(cA)d = (cd)A$, for any real numbers
c and d.

94. $(cd)A = c(dA)$, for any real numbers
c and d.

5.8 Matrix Inverses

- Identity Matrices
- Multiplicative Inverses
- Solving Systems Using Inverse Matrices

In the previous section, we saw several parallels between the set of real numbers and the set of matrices. Another similarity is that both sets have identity and inverse elements for multiplication.

Identity Matrices   By the identity property for real numbers,

$$a \cdot 1 = a \quad and \quad 1 \cdot a = a \quad \text{(Section R.2)}$$

for any real number a. If there is to be a multiplicative **identity matrix** I, such that

$$AI = A \quad and \quad IA = A,$$

for any matrix A, then A and I must be square matrices of the same dimension.

2×2 Identity Matrix

If I_2 represents the 2×2 identity matrix, then

$$I_2 = \begin{bmatrix} 1 & 0 \\ 0 & 1 \end{bmatrix}.$$

To verify that I_2 is the 2×2 identity matrix, we must show that $AI = A$ and $IA = A$ for any 2×2 matrix A. Let

$$A = \begin{bmatrix} x & y \\ z & w \end{bmatrix}.$$

Then

$$AI = \begin{bmatrix} x & y \\ z & w \end{bmatrix}\begin{bmatrix} 1 & 0 \\ 0 & 1 \end{bmatrix} = \begin{bmatrix} x \cdot 1 + y \cdot 0 & x \cdot 0 + y \cdot 1 \\ z \cdot 1 + w \cdot 0 & z \cdot 0 + w \cdot 1 \end{bmatrix} = \begin{bmatrix} x & y \\ z & w \end{bmatrix} = A,$$

and

$$IA = \begin{bmatrix} 1 & 0 \\ 0 & 1 \end{bmatrix}\begin{bmatrix} x & y \\ z & w \end{bmatrix} = \begin{bmatrix} 1 \cdot x + 0 \cdot z & 1 \cdot y + 0 \cdot w \\ 0 \cdot x + 1 \cdot z & 0 \cdot y + 1 \cdot w \end{bmatrix} = \begin{bmatrix} x & y \\ z & w \end{bmatrix} = A.$$

Generalizing, there is an $n \times n$ identity matrix for every $n \times n$ square matrix. The $n \times n$ identity matrix has 1s on the main diagonal and 0s elsewhere.

$n \times n$ Identity Matrix

The $n \times n$ identity matrix is I_n, where

$$I_n = \begin{bmatrix} 1 & 0 & \cdots & 0 \\ 0 & 1 & \cdots & 0 \\ \vdots & \vdots & a_{ij} & \vdots \\ 0 & 0 & \cdots & 1 \end{bmatrix}.$$

The element $a_{ij} = 1$ when $i = j$ (the diagonal elements), and $a_{ij} = 0$ otherwise.

EXAMPLE 1 Verifying the Identity Property of I_3

Let $A = \begin{bmatrix} -2 & 4 & 0 \\ 3 & 5 & 9 \\ 0 & 8 & -6 \end{bmatrix}$. Give the 3×3 identity matrix I_3 and show that $AI_3 = A$.

ALGEBRAIC SOLUTION

The 3×3 identity matrix is

$$I_3 = \begin{bmatrix} 1 & 0 & 0 \\ 0 & 1 & 0 \\ 0 & 0 & 1 \end{bmatrix}.$$

By the definition of matrix multiplication,

$$AI_3 = \begin{bmatrix} -2 & 4 & 0 \\ 3 & 5 & 9 \\ 0 & 8 & -6 \end{bmatrix}\begin{bmatrix} 1 & 0 & 0 \\ 0 & 1 & 0 \\ 0 & 0 & 1 \end{bmatrix}$$

$$= \begin{bmatrix} -2 & 4 & 0 \\ 3 & 5 & 9 \\ 0 & 8 & -6 \end{bmatrix} = A.$$

GRAPHING CALCULATOR SOLUTION

The calculator screen in **Figure 27(a)** shows the identity matrix for $n = 3$. The screens in **Figures 27(b) and (c)** support the algebraic result.

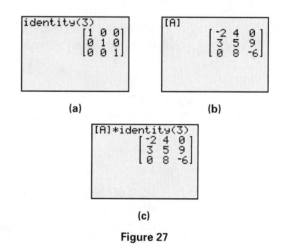

(a)

(b)

(c)

Figure 27

☑ *Now Try Exercise 1.*

Multiplicative Inverses For every *nonzero* real number a, there is a multiplicative inverse $\frac{1}{a}$ that satisfies both of the following.

$$a \cdot \frac{1}{a} = 1 \quad \text{and} \quad \frac{1}{a} \cdot a = 1 \quad \text{(Section R.2)}$$

(Recall: $\frac{1}{a}$ is also written a^{-1}.) In a similar way, if A is an $n \times n$ matrix, then its **multiplicative inverse,** written A^{-1}, must satisfy both of the following.

$$AA^{-1} = I_n \quad \text{and} \quad A^{-1}A = I_n$$

This means that only a square matrix can have a multiplicative inverse.

CAUTION Although $a^{-1} = \frac{1}{a}$ for any nonzero real number a, if A is a matrix, then $A^{-1} \neq \frac{1}{A}$. *We do not use the symbol $\frac{1}{A}$, since 1 is a number and A is a matrix.*

To find the matrix A^{-1}, we use row transformations, introduced earlier in this chapter. As an example, we find the inverse of

$$A = \begin{bmatrix} 2 & 4 \\ 1 & -1 \end{bmatrix}.$$

Let the unknown inverse matrix be symbolized as follows.

$$A^{-1} = \begin{bmatrix} x & y \\ z & w \end{bmatrix}$$

By the definition of matrix inverse, $AA^{-1} = I_2$.

$$AA^{-1} = \begin{bmatrix} 2 & 4 \\ 1 & -1 \end{bmatrix} \begin{bmatrix} x & y \\ z & w \end{bmatrix} = \begin{bmatrix} 1 & 0 \\ 0 & 1 \end{bmatrix}$$

Use matrix multiplication for the product above.

$$\begin{bmatrix} 2x + 4z & 2y + 4w \\ x - z & y - w \end{bmatrix} = \begin{bmatrix} 1 & 0 \\ 0 & 1 \end{bmatrix} \quad \text{(Section 5.7)}$$

Set the corresponding elements equal to obtain a system of equations.

$$2x + 4z = 1 \quad (1)$$
$$2y + 4w = 0 \quad (2)$$
$$x - z = 0 \quad (3)$$
$$y - w = 1 \quad (4)$$

Since equations (1) and (3) involve only x and z, while equations (2) and (4) involve only y and w, these four equations lead to two systems of equations.

$$\begin{array}{cc} 2x + 4z = 1 & 2y + 4w = 0 \\ & \text{and} \\ x - z = 0 & y - w = 1 \end{array}$$

Write the two systems as augmented matrices.

$$\begin{bmatrix} 2 & 4 & | & 1 \\ 1 & -1 & | & 0 \end{bmatrix} \quad \text{and} \quad \begin{bmatrix} 2 & 4 & | & 0 \\ 1 & -1 & | & 1 \end{bmatrix} \quad \text{(Section 5.2)}$$

Each of these systems can be solved by the Gauss-Jordan method. However, since the elements to the left of the vertical bar are identical, the two systems can be combined into one matrix.

$$\begin{bmatrix} 2 & 4 & | & 1 \\ 1 & -1 & | & 0 \end{bmatrix} \quad \text{and} \quad \begin{bmatrix} 2 & 4 & | & 0 \\ 1 & -1 & | & 1 \end{bmatrix} \quad \text{yields} \quad \begin{bmatrix} 2 & 4 & | & 1 & 0 \\ 1 & -1 & | & 0 & 1 \end{bmatrix}$$

We can solve simultaneously using matrix row transformations. We need to change the numbers on the left of the vertical bar to the 2×2 identity matrix.

$$\begin{bmatrix} 1 & -1 & | & 0 & 1 \\ 2 & 4 & | & 1 & 0 \end{bmatrix} \quad \text{Interchange R1 and R2 to get 1 in the upper left-hand corner. (Section 5.2)}$$

$$\begin{bmatrix} 1 & -1 & | & 0 & 1 \\ 0 & 6 & | & 1 & -2 \end{bmatrix} \quad -2\text{R1} + \text{R2}$$

$$\begin{bmatrix} 1 & -1 & | & 0 & 1 \\ 0 & 1 & | & \frac{1}{6} & -\frac{1}{3} \end{bmatrix} \quad \frac{1}{6}\text{R2}$$

$$\begin{bmatrix} 1 & 0 & | & \frac{1}{6} & \frac{2}{3} \\ 0 & 1 & | & \frac{1}{6} & -\frac{1}{3} \end{bmatrix} \quad \text{R2} + \text{R1}$$

The numbers in the first column to the right of the vertical bar in the final matrix give the values of x and z. The second column gives the values of y and w. That is,

$$\begin{bmatrix} 1 & 0 & | & x & y \\ 0 & 1 & | & z & w \end{bmatrix} = \begin{bmatrix} 1 & 0 & | & \frac{1}{6} & \frac{2}{3} \\ 0 & 1 & | & \frac{1}{6} & -\frac{1}{3} \end{bmatrix}$$

so that

$$A^{-1} = \begin{bmatrix} x & y \\ z & w \end{bmatrix} = \begin{bmatrix} \frac{1}{6} & \frac{2}{3} \\ \frac{1}{6} & -\frac{1}{3} \end{bmatrix}.$$

To check, multiply A by A^{-1}. The result should be I_2.

$$AA^{-1} = \begin{bmatrix} 2 & 4 \\ 1 & -1 \end{bmatrix} \begin{bmatrix} \frac{1}{6} & \frac{2}{3} \\ \frac{1}{6} & -\frac{1}{3} \end{bmatrix}$$

$$= \begin{bmatrix} \frac{1}{3} + \frac{2}{3} & \frac{4}{3} - \frac{4}{3} \\ \frac{1}{6} - \frac{1}{6} & \frac{2}{3} + \frac{1}{3} \end{bmatrix}$$

$$= \begin{bmatrix} 1 & 0 \\ 0 & 1 \end{bmatrix} = I_2$$

Thus,

$$A^{-1} = \begin{bmatrix} \frac{1}{6} & \frac{2}{3} \\ \frac{1}{6} & -\frac{1}{3} \end{bmatrix}.$$

This process is summarized below.

Finding an Inverse Matrix

To obtain A^{-1} for any $n \times n$ matrix A for which A^{-1} exists, follow these steps.

Step 1 Form the augmented matrix $[A \mid I_n]$, where I_n is the $n \times n$ identity matrix.

Step 2 Perform row transformations on $[A \mid I_n]$ to obtain a matrix of the form $[I_n \mid B]$.

Step 3 Matrix B is A^{-1}.

NOTE To confirm that two $n \times n$ matrices A and B are inverses of each other, it is sufficient to show that $AB = I_n$. It is not necessary to show also that $BA = I_n$.

EXAMPLE 2 **Finding the Inverse of a 3 × 3 Matrix**

Find A^{-1} if $A = \begin{bmatrix} 1 & 0 & 1 \\ 2 & -2 & -1 \\ 3 & 0 & 0 \end{bmatrix}$.

SOLUTION Use row transformations as follows.

Step 1 Write the augmented matrix $[A \mid I_3]$.

$$\begin{bmatrix} 1 & 0 & 1 & | & 1 & 0 & 0 \\ 2 & -2 & -1 & | & 0 & 1 & 0 \\ 3 & 0 & 0 & | & 0 & 0 & 1 \end{bmatrix}$$

Step 2 Since 1 is already in the upper left-hand corner as desired, begin by using the row transformation that will result in 0 for the first element in the second row. Multiply the elements of the first row by -2, and add the result to the second row.

$$\begin{bmatrix} 1 & 0 & 1 & \vline & 1 & 0 & 0 \\ 0 & -2 & -3 & \vline & -2 & 1 & 0 \\ 3 & 0 & 0 & \vline & 0 & 0 & 1 \end{bmatrix} \quad -2R1 + R2$$

To get 0 for the first element in the third row, multiply the elements of the first row by -3 and add to the third row.

$$\begin{bmatrix} 1 & 0 & 1 & \vline & 1 & 0 & 0 \\ 0 & -2 & -3 & \vline & -2 & 1 & 0 \\ 0 & 0 & -3 & \vline & -3 & 0 & 1 \end{bmatrix} \quad -3R1 + R3$$

To get 1 for the second element in the second row, multiply the elements of the second row by $-\frac{1}{2}$.

$$\begin{bmatrix} 1 & 0 & 1 & \vline & 1 & 0 & 0 \\ 0 & 1 & \frac{3}{2} & \vline & 1 & -\frac{1}{2} & 0 \\ 0 & 0 & -3 & \vline & -3 & 0 & 1 \end{bmatrix} \quad -\frac{1}{2}R2$$

To get 1 for the third element in the third row, multiply the elements of the third row by $-\frac{1}{3}$.

$$\begin{bmatrix} 1 & 0 & 1 & \vline & 1 & 0 & 0 \\ 0 & 1 & \frac{3}{2} & \vline & 1 & -\frac{1}{2} & 0 \\ 0 & 0 & 1 & \vline & 1 & 0 & -\frac{1}{3} \end{bmatrix} \quad -\frac{1}{3}R3$$

To get 0 for the third element in the first row, multiply the elements of the third row by -1 and add to the first row.

$$\begin{bmatrix} 1 & 0 & 0 & \vline & 0 & 0 & \frac{1}{3} \\ 0 & 1 & \frac{3}{2} & \vline & 1 & -\frac{1}{2} & 0 \\ 0 & 0 & 1 & \vline & 1 & 0 & -\frac{1}{3} \end{bmatrix} \quad -1R3 + R1$$

To get 0 for the third element in the second row, multiply the elements of the third row by $-\frac{3}{2}$ and add to the second row.

$$\begin{bmatrix} 1 & 0 & 0 & \vline & 0 & 0 & \frac{1}{3} \\ 0 & 1 & 0 & \vline & -\frac{1}{2} & -\frac{1}{2} & \frac{1}{2} \\ 0 & 0 & 1 & \vline & 1 & 0 & -\frac{1}{3} \end{bmatrix} \quad -\frac{3}{2}R3 + R2$$

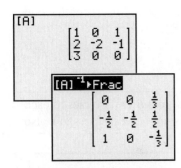

A graphing calculator can be used to find the inverse of a matrix. The screens support the result in **Example 2.** The elements of the inverse are expressed as fractions, so it is easier to compare with the inverse matrix found in the example.

Step 3 The last transformation shows that the inverse is

$$A^{-1} = \begin{bmatrix} 0 & 0 & \frac{1}{3} \\ -\frac{1}{2} & -\frac{1}{2} & \frac{1}{2} \\ 1 & 0 & -\frac{1}{3} \end{bmatrix}.$$

Confirm this by forming the product $A^{-1}A$ or AA^{-1}, each of which should equal the matrix I_3.

✔ *Now Try Exercises 11 and 19.*

As illustrated by the examples, the most efficient order for the transformations in Step 2 is to make the changes column by column from left to right, so that for each column the required 1 is the result of the first change. Next, perform the steps that obtain the 0s in that column. Then proceed to the next column.

EXAMPLE 3   **Identifying a Matrix with No Inverse**

Find A^{-1}, if possible, given that $A = \begin{bmatrix} 2 & -4 \\ 1 & -2 \end{bmatrix}$.

ALGEBRAIC SOLUTION

Using row transformations to change the first column of the augmented matrix

$$\begin{bmatrix} 2 & -4 & | & 1 & 0 \\ 1 & -2 & | & 0 & 1 \end{bmatrix}$$

results in the following matrices.

$$\begin{bmatrix} 1 & -2 & | & \frac{1}{2} & 0 \\ 1 & -2 & | & 0 & 1 \end{bmatrix} \quad \text{and} \quad \begin{bmatrix} 1 & -2 & | & \frac{1}{2} & 0 \\ 0 & 0 & | & -\frac{1}{2} & 1 \end{bmatrix}$$

(We multiplied the elements in row one by $\frac{1}{2}$ in the first step, and in the second step we added the negative of row one to row two.) At this point, the matrix should be changed so that the second row, second element will be 1. Since that element is now 0, there is no way to complete the desired transformation, so A^{-1} does not exist for this matrix A. Just as there is no multiplicative inverse for the real number 0, not every matrix has a multiplicative inverse. Matrix A is an example of such a matrix.

GRAPHING CALCULATOR SOLUTION

If the inverse of a matrix does not exist, the matrix is called **singular,** as shown in **Figure 28** for matrix [A]. This occurs when the determinant of the matrix is 0. **(See Exercises 29–34.)**

Figure 28

☑ *Now Try Exercise 15.*

If the inverse of a matrix exists, it is unique. That is, any given square matrix has no more than one inverse. The proof of this is left as **Exercise 70.**

> **Solving Systems Using Inverse Matrices**   Matrix inverses can be used to solve square linear systems of equations. (A square system has the same number of equations as variables.) For example, consider the following linear system of three equations with three variables.

$$a_{11}x + a_{12}y + a_{13}z = b_1$$

$$a_{21}x + a_{22}y + a_{23}z = b_2$$

$$a_{31}x + a_{32}y + a_{33}z = b_3$$

The definition of matrix multiplication can be used to rewrite the system using matrices.

$$\begin{bmatrix} a_{11} & a_{12} & a_{13} \\ a_{21} & a_{22} & a_{23} \\ a_{31} & a_{32} & a_{33} \end{bmatrix} \cdot \begin{bmatrix} x \\ y \\ z \end{bmatrix} = \begin{bmatrix} b_1 \\ b_2 \\ b_3 \end{bmatrix} \quad (1)$$

(To see this, multiply the matrices on the left.)

$$\text{If} \quad A = \begin{bmatrix} a_{11} & a_{12} & a_{13} \\ a_{21} & a_{22} & a_{23} \\ a_{31} & a_{32} & a_{33} \end{bmatrix}, \quad X = \begin{bmatrix} x \\ y \\ z \end{bmatrix}, \quad \text{and} \quad B = \begin{bmatrix} b_1 \\ b_2 \\ b_3 \end{bmatrix},$$

then the system given in (1) becomes $AX = B$. If A^{-1} exists, then both sides of $AX = B$ can be multiplied on the left as shown on the next page.

$$A^{-1}(AX) = A^{-1}B$$

$$(A^{-1}A)X = A^{-1}B \quad \text{Associative property (Section 5.7)}$$

$$I_3X = A^{-1}B \quad \text{Inverse property}$$

$$X = A^{-1}B \quad \text{Identity property}$$

Matrix $A^{-1}B$ gives the solution of the system.

Solution of the Matrix Equation $AX = B$

If A is an $n \times n$ matrix with inverse A^{-1}, X is an $n \times 1$ matrix of variables, and B is an $n \times 1$ matrix, then the matrix equation

$$AX = B$$

has the solution $\qquad X = A^{-1}B.$

This method of using matrix inverses to solve systems of equations is useful when the inverse is already known or when many systems of the form $AX = B$ must be solved and only B changes.

EXAMPLE 4 **Solving Systems of Equations Using Matrix Inverses**

Use the inverse of the coefficient matrix to solve each system.

(a) $2x - 3y = 4$

$\qquad x + 5y = 2$

(b) $\qquad x + z = -1$

$\qquad 2x - 2y - z = 5$

$\qquad 3x = 6$

ALGEBRAIC SOLUTION

(a) The system can be written in matrix form as

$$\begin{bmatrix} 2 & -3 \\ 1 & 5 \end{bmatrix} \begin{bmatrix} x \\ y \end{bmatrix} = \begin{bmatrix} 4 \\ 2 \end{bmatrix},$$

where

$$A = \begin{bmatrix} 2 & -3 \\ 1 & 5 \end{bmatrix}, \quad X = \begin{bmatrix} x \\ y \end{bmatrix}, \quad \text{and} \quad B = \begin{bmatrix} 4 \\ 2 \end{bmatrix}.$$

An equivalent matrix equation is $AX = B$ with solution $X = A^{-1}B$. Use the methods described in this section to determine that

$$A^{-1} = \begin{bmatrix} \frac{5}{13} & \frac{3}{13} \\ -\frac{1}{13} & \frac{2}{13} \end{bmatrix}.$$

Now, find $A^{-1}B$.

$$A^{-1}B = \begin{bmatrix} \frac{5}{13} & \frac{3}{13} \\ -\frac{1}{13} & \frac{2}{13} \end{bmatrix} \begin{bmatrix} 4 \\ 2 \end{bmatrix} = \begin{bmatrix} 2 \\ 0 \end{bmatrix}$$

Since $X = A^{-1}B$, $\qquad X = \begin{bmatrix} x \\ y \end{bmatrix} = \begin{bmatrix} 2 \\ 0 \end{bmatrix}.$

The final matrix shows that the solution set of the system is $\{(2, 0)\}$.

GRAPHING CALCULATOR SOLUTION

(a) Enter $[A]$ and $[B]$, and then find the product $[A]^{-1}[B]$ as shown in **Figure 29.**

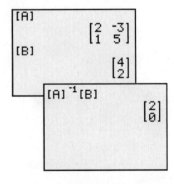

Figure 29

The column matrix indicates the solution set, $\{(2, 0)\}$.

(b) The coefficient matrix A for this system is

$$A = \begin{bmatrix} 1 & 0 & 1 \\ 2 & -2 & -1 \\ 3 & 0 & 0 \end{bmatrix},$$

and its inverse A^{-1} was found in **Example 2.** Let

$$X = \begin{bmatrix} x \\ y \\ z \end{bmatrix} \quad \text{and} \quad B = \begin{bmatrix} -1 \\ 5 \\ 6 \end{bmatrix}.$$

Since $X = A^{-1}B$, we have

$$\begin{bmatrix} x \\ y \\ z \end{bmatrix} = \underbrace{\begin{bmatrix} 0 & 0 & \frac{1}{3} \\ -\frac{1}{2} & -\frac{1}{2} & \frac{1}{2} \\ 1 & 0 & -\frac{1}{3} \end{bmatrix}}_{A^{-1} \text{ from } \textbf{Example 2}} \begin{bmatrix} -1 \\ 5 \\ 6 \end{bmatrix}$$

$$= \begin{bmatrix} 2 \\ 1 \\ -3 \end{bmatrix}.$$

The solution set is $\{(2, 1, -3)\}$.

(b) Figure 30 shows the coefficient matrix $[A]$ and the column matrix of constants $[B]$. Note that when $[A]^{-1}[B]$ is determined, it is not necessary to display $[A]^{-1}$. As long as the inverse exists, the product will be computed.

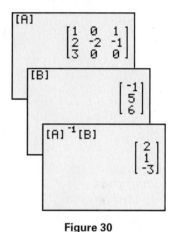

Figure 30

✔️ *Now Try Exercises 37 and 51.*

5.8 Exercises

1. Show that for

$$A = \begin{bmatrix} -2 & 4 & 0 \\ 3 & 5 & 9 \\ 0 & 8 & -6 \end{bmatrix} \quad \text{and} \quad I_3 = \begin{bmatrix} 1 & 0 & 0 \\ 0 & 1 & 0 \\ 0 & 0 & 1 \end{bmatrix},$$

$I_3A = A$. (This result, along with that of **Example 1,** illustrates that the commutative property holds when one of the matrices is an identity matrix.)

2. Let $A = \begin{bmatrix} a & b \\ c & d \end{bmatrix}$ and $I_2 = \begin{bmatrix} 1 & 0 \\ 0 & 1 \end{bmatrix}$. Show that $AI_2 = I_2A = A$, thus proving that I_2 is the identity element for matrix multiplication for 2×2 square matrices.

Decide whether or not the given matrices are inverses of each other. (Hint: Check to see whether their products are the identity matrix I_n.)

3. $\begin{bmatrix} 5 & 7 \\ 2 & 3 \end{bmatrix}$ and $\begin{bmatrix} 3 & -7 \\ -2 & 5 \end{bmatrix}$

4. $\begin{bmatrix} 2 & 3 \\ 1 & 1 \end{bmatrix}$ and $\begin{bmatrix} -1 & 3 \\ 1 & -2 \end{bmatrix}$

5. $\begin{bmatrix} -1 & 2 \\ 3 & -5 \end{bmatrix}$ and $\begin{bmatrix} -5 & -2 \\ -3 & -1 \end{bmatrix}$

6. $\begin{bmatrix} 2 & 1 \\ 3 & 2 \end{bmatrix}$ and $\begin{bmatrix} 2 & 1 \\ -3 & 2 \end{bmatrix}$

7. $\begin{bmatrix} 0 & 1 & 0 \\ 0 & 0 & -2 \\ 1 & -1 & 0 \end{bmatrix}$ and $\begin{bmatrix} 1 & 0 & 1 \\ 1 & 0 & 0 \\ 0 & -1 & 0 \end{bmatrix}$

8. $\begin{bmatrix} 1 & 2 & 0 \\ 0 & 1 & 0 \\ 0 & 1 & 0 \end{bmatrix}$ and $\begin{bmatrix} 1 & -2 & 0 \\ 0 & 1 & 0 \\ 0 & -1 & 1 \end{bmatrix}$

9. $\begin{bmatrix} 1 & 0 & 0 \\ 0 & -1 & 0 \\ 1 & 0 & 1 \end{bmatrix}$ and $\begin{bmatrix} 1 & 0 & 0 \\ 0 & -1 & 0 \\ -1 & 0 & 1 \end{bmatrix}$

10. $\begin{bmatrix} 1 & 3 & 3 \\ 1 & 4 & 3 \\ 1 & 3 & 4 \end{bmatrix}$ and $\begin{bmatrix} 7 & -3 & -3 \\ -1 & 1 & 0 \\ -1 & 0 & 1 \end{bmatrix}$

Find the inverse, if it exists, for each matrix. **See Examples 2 and 3.**

11. $\begin{bmatrix} -1 & 2 \\ -2 & -1 \end{bmatrix}$
12. $\begin{bmatrix} 1 & -1 \\ 2 & 0 \end{bmatrix}$
13. $\begin{bmatrix} -1 & -2 \\ 3 & 4 \end{bmatrix}$

14. $\begin{bmatrix} 3 & -1 \\ -5 & 2 \end{bmatrix}$
15. $\begin{bmatrix} 5 & 10 \\ -3 & -6 \end{bmatrix}$
16. $\begin{bmatrix} -6 & 4 \\ -3 & 2 \end{bmatrix}$

17. $\begin{bmatrix} 1 & 0 & 1 \\ 0 & -1 & 0 \\ 2 & 1 & 1 \end{bmatrix}$
18. $\begin{bmatrix} 1 & 0 & 0 \\ 0 & -1 & 0 \\ 1 & 0 & 1 \end{bmatrix}$
19. $\begin{bmatrix} 2 & 3 & 3 \\ 1 & 4 & 3 \\ 1 & 3 & 4 \end{bmatrix}$

20. $\begin{bmatrix} -2 & 2 & 4 \\ -3 & 4 & 5 \\ 1 & 0 & 2 \end{bmatrix}$
21. $\begin{bmatrix} 2 & 2 & -4 \\ 2 & 6 & 0 \\ -3 & -3 & 5 \end{bmatrix}$
22. $\begin{bmatrix} 2 & 4 & 6 \\ -1 & -4 & -3 \\ 0 & 1 & -1 \end{bmatrix}$

23. $\begin{bmatrix} 1 & 1 & 0 & 2 \\ 2 & -1 & 1 & -1 \\ 3 & 3 & 2 & -2 \\ 1 & 2 & 1 & 0 \end{bmatrix}$
24. $\begin{bmatrix} 1 & -2 & 3 & 0 \\ 0 & 1 & -1 & 1 \\ -2 & 2 & -2 & 4 \\ 0 & 2 & -3 & 1 \end{bmatrix}$
25. $\begin{bmatrix} 3 & 2 & 0 & -1 \\ 2 & 0 & 1 & 2 \\ 1 & 2 & -1 & 0 \\ 2 & -1 & 1 & 1 \end{bmatrix}$

26. Explain the process for finding the inverse of a matrix.

Each graphing calculator screen shows A^{-1} for some matrix A. Find each matrix A. (*Hint:* $(A^{-1})^{-1} = A$.)

27.
```
[A]⁻¹▶Frac
      ⎡ 1     7  ⎤
      ⎢ ──   -── ⎥
      ⎢ 11    22 ⎥
      ⎢ 1     2  ⎥
      ⎢ ──    ── ⎥
      ⎣ 11    11 ⎦
```

28.
```
[A]⁻¹▶Frac
   ⎡ 2    1    ⎤
   ⎢ ──  -──  0⎥
   ⎢ 3    3    ⎥
   ⎢ 1    5    ⎥
   ⎢ ──  -──  1⎥
   ⎢ 3    3    ⎥
   ⎢ 1    1    ⎥
   ⎢ ──   ──  0⎥
   ⎣ 3    3    ⎦
```

Relating Concepts

For individual or collaborative investigation *(Exercises 29–34)*

It can be shown that the inverse of matrix $A = \begin{bmatrix} a & b \\ c & d \end{bmatrix}$ *is*

$$A^{-1} = \begin{bmatrix} \dfrac{d}{ad - bc} & \dfrac{-b}{ad - bc} \\[2ex] \dfrac{-c}{ad - bc} & \dfrac{a}{ad - bc} \end{bmatrix}.$$

Work Exercises 29–34 in order, *to discover a connection between the inverse and the determinant of a* 2×2 *matrix.*

29. With respect to the matrix $A = \begin{bmatrix} a & b \\ c & d \end{bmatrix}$, what do we call $ad - bc$?

30. Refer to A^{-1} as given above, and write it using determinant notation.

31. Write A^{-1} using scalar multiplication, where the scalar is $\frac{1}{|A|}$.

32. Explain how the inverse of matrix A can be found using a determinant.

33. Use the method described here to find the inverse of $A = \begin{bmatrix} 4 & 2 \\ 7 & 3 \end{bmatrix}$.

34. Complete the following statement: The inverse of a 2×2 matrix A does not exist if the determinant of A has value _____. (*Hint:* Look at the denominators in A^{-1} as given earlier.)

Solve each system by using the inverse of the coefficient matrix. ***See Example 4.***

35. $-x + y = 1$
$2x - y = 1$

36. $x + y = 5$
$x - y = -1$

37. $2x - y = -8$
$3x + y = -2$

38. $x + 3y = -12$
$2x - y = 11$

39. $3x + 4y = -3$
$-5x + 8y = 16$

40. $2x - 3y = 10$
$2x + 2y = 5$

41. $6x + 9y = 3$
$-8x + 3y = 6$

42. $5x - 3y = 0$
$10x + 6y = -4$

43. $0.2x + 0.3y = -1.9$
$0.7x - 0.2y = 4.6$

44. $0.5x + 0.2y = 0.8$
$0.3x - 0.1y = 0.7$

45. $\dfrac{1}{2}x + \dfrac{1}{3}y = \dfrac{49}{18}$
$\dfrac{1}{2}x + 2y = \dfrac{4}{3}$

46. $\dfrac{1}{5}x + \dfrac{1}{7}y = \dfrac{12}{5}$
$\dfrac{1}{10}x + \dfrac{1}{3}y = \dfrac{5}{6}$

Concept Check *Show that the matrix inverse method cannot be used to solve each system.*

47. $7x - 2y = 3$
$14x - 4y = 1$

48. $x - 2y + 3z = 4$
$2x - 4y + 6z = 8$
$3x - 6y + 9z = 14$

Solve each system by using the inverse of the coefficient matrix. For Exercises 51–56, the inverses were found in ***Exercises 19–24.*** ***See Example 4.***

49. $x + y + z = 6$
$2x + 3y - z = 7$
$3x - y - z = 6$

50. $2x + 5y + 2z = 9$
$4x - 7y - 3z = 7$
$3x - 8y - 2z = 9$

51. $2x + 3y + 3z = 1$
$x + 4y + 3z = 0$
$x + 3y + 4z = -1$

52. $-2x + 2y + 4z = 3$
$-3x + 4y + 5z = 1$
$x + 2z = 2$

53. $2x + 2y - 4z = 12$
$2x + 6y = 16$
$-3x - 3y + 5z = -20$

54. $2x + 4y + 6z = 4$
$-x - 4y - 3z = 8$
$y - z = -4$

55. $x + y + 2w = 3$
$2x - y + z - w = 3$
$3x + 3y + 2z - 2w = 5$
$x + 2y + z = 3$

56. $x - 2y + 3z = 1$
$y - z + w = -1$
$-2x + 2y - 2z + 4w = 2$
$2y - 3z + w = -3$

(Modeling) *Solve each problem.*

57. *Plate-Glass Sales* The amount of plate-glass sales S (in millions of dollars) can be affected by the number of new building contracts B issued (in millions) and automobiles A produced (in millions). A plate-glass company in California wants to forecast future sales by using the past three years of sales. The totals for the three years are given in the table. To describe the relationship among these variables, we can use the equation

S	A	B
602.7	5.543	37.14
656.7	6.933	41.30
778.5	7.638	45.62

$$S = a + bA + cB,$$

where the coefficients a, b, and c are constants that must be determined before the equation can be used. (*Source:* Makridakis, S., and S. Wheelwright, *Forecasting Methods for Management,* John Wiley and Sons.)

(a) Substitute the values for S, A, and B for each year from the table into the equation $S = a + bA + cB$, and obtain three linear equations involving a, b, and c.

(b) Use a graphing calculator to solve this linear system for a, b, and c. Use matrix inverse methods.

(c) Write the equation for S using these values for the coefficients.

(d) For the next year it is estimated that $A = 7.752$ and $B = 47.38$. Predict S. (The actual value for S was 877.6.)

(e) It is predicted that in 6 yr, $A = 8.9$ and $B = 66.25$. Find the value of S in this situation and discuss its validity.

58. *Tire Sales* The number of automobile tire sales is dependent on several variables. In one study the relationship among annual tire sales S (in thousands of dollars), automobile registrations R (in millions), and personal disposable income I (in millions of dollars) was investigated. The results for three years are given in the table. To describe the relationship among these variables, we can use the equation

S	R	I
10,170	112.9	307.5
15,305	132.9	621.63
21,289	155.2	1937.13

$$S = a + bR + cI,$$

where the coefficients a, b, and c are constants that must be determined before the equation can be used. (*Source:* Jarrett, J., *Business Forecasting Methods*, Basil Blackwell, Ltd.)

(a) Substitute the values for S, R, and I for each year from the table into the equation $S = a + bR + cI$, and obtain three linear equations involving a, b, and c.

(b) Use a graphing calculator to solve this linear system for a, b, and c. Use matrix inverse methods.

(c) Write the equation for S using these values for the coefficients.

(d) If $R = 117.6$ and $I = 310.73$, predict S. (The actual value for S was 11,314.)

(e) If $R = 143.8$ and $I = 829.06$, predict S. (The actual value for S was 18,481.)

59. *Social Security Numbers* In **Exercise 89 of Section 3.4**, construction of your own personal Social Security polynomial was discussed. It is also possible to find a polynomial that goes through a given set of points in the plane by using a process called **polynomial interpolation.**

Recall that three points define a second-degree polynomial, four points define a third-degree polynomial, and so on. The only restriction on the points, since polynomials define functions, is that no two distinct points can have the same x-coordinate. Using the same SSN (539-58-0954) as in that exercise, we can find an eighth-degree polynomial that lies on the nine points with x-coordinates 1 through 9 and y-coordinates that are digits of the SSN: $(1, 5)$, $(2, 3)$, $(3, 9)$, ..., $(9, 4)$. This is done by writing a system of nine equations with nine variables, which is then solved by the inverse matrix method. The graph of this polynomial is shown. Find such a polynomial using your own SSN.

Use a graphing calculator to find the inverse of each matrix. Give as many decimal places as the calculator shows. See Example 2.

60. $\begin{bmatrix} \sqrt{2} & 0.5 \\ -17 & \frac{1}{2} \end{bmatrix}$

61. $\begin{bmatrix} \frac{2}{3} & 0.7 \\ 22 & \sqrt{3} \end{bmatrix}$

62. $\begin{bmatrix} 1.4 & 0.5 & 0.59 \\ 0.84 & 1.36 & 0.62 \\ 0.56 & 0.47 & 1.3 \end{bmatrix}$

63. $\begin{bmatrix} \frac{1}{2} & \frac{1}{4} & \frac{1}{3} \\ 0 & \frac{1}{4} & \frac{1}{3} \\ \frac{1}{2} & \frac{1}{2} & \frac{1}{3} \end{bmatrix}$

Use a graphing calculator and the method of matrix inverses to solve each system. Give as many decimal places as the calculator shows. See Example 4.

64. $\quad x - \sqrt{2}y = 2.6$
$\quad 0.75x + \quad y = -7$

65. $\quad 2.1x + \quad y = \sqrt{5}$
$\quad \sqrt{2}x - 2y = 5$

66. $\quad \pi x + ey + \sqrt{2}z = 1$
$\quad ex + \pi y + \sqrt{2}z = 2$
$\quad \sqrt{2}x + ey + \quad \pi z = 3$

67. $\quad (\log 2)x + (\ln 3)y + (\ln 4)z = 1$
$\quad (\ln 3)x + (\log 2)y + (\ln 8)z = 5$
$\quad (\log 12)x + (\ln 4)y + (\ln 8)z = 9$

Let $A = \begin{bmatrix} a & b \\ c & d \end{bmatrix}$, and let O be the 2×2 zero matrix. Show that the statements in Exercises 68 and 69 are true.

68. $A \cdot O = O \cdot A = O$

69. For square matrices A and B of the same dimension, if $AB = O$ and if A^{-1} exists, then $B = O$.

70. Prove that any square matrix has no more than one inverse.

71. Give an example of two matrices A and B, where $(AB)^{-1} \neq A^{-1}B^{-1}$.

72. Suppose A and B are matrices, where A^{-1}, B^{-1}, and AB all exist. Show that $(AB)^{-1} = B^{-1}A^{-1}$.

73. Let $A = \begin{bmatrix} a & 0 & 0 \\ 0 & b & 0 \\ 0 & 0 & c \end{bmatrix}$, where a, b, and c are nonzero real numbers. Find A^{-1}.

74. Let $A = \begin{bmatrix} 1 & 0 & 0 \\ 0 & 0 & -1 \\ 0 & 1 & -1 \end{bmatrix}$. Show that $A^3 = I_3$, and use this result to find the inverse of A.

75. What are the inverses of I_n, $-A$ (in terms of A), and kA (k a scalar)?

76. Discuss the similarities and differences between solving the linear equation $ax = b$ and solving the matrix equation $AX = B$.

Chapter 5 Test Prep

Key Terms

5.1 linear equation (first-degree equation) in *n* unknowns system of equations solutions of a system of equations system of linear equations (linear system) consistent system independent equations inconsistent system dependent equations equivalent systems ordered triple	**5.2** matrix (matrices) element (of a matrix) augmented matrix dimension (of a matrix) **5.3** determinant minor cofactor expansion by a row or column Cramer's rule **5.4** partial fraction decomposition partial fraction	**5.5** nonlinear system **5.6** half-plane boundary linear inequality in two variables system of inequalities linear programming constraints objective function region of feasible solutions vertex (corner point)	**5.7** square matrix row matrix column matrix zero matrix additive inverse (negative) of a matrix scalar **5.8** identity matrix multiplicative inverse (of a matrix)

New Symbols

(a, b, c)	ordered triple
$[A]$	matrix A (graphing calculator symbolism)
$\lvert A \rvert$	determinant of matrix A
a_{ij}	the element in row i, column j, of a matrix

D, D_x, D_y	determinants used in Cramer's rule
I_2, I_3	identity matrices
A^{-1}	multiplicative inverse of matrix A

(continued)

Quick Review

Concepts	Examples

5.1 **Systems of Linear Equations**

Transformations of a Linear System

1. Interchange any two equations of the system.
2. Multiply or divide any equation of the system by a non-zero real number.
3. Replace any equation of the system by the sum of that equation and a multiple of another equation in the system.

Systems may be solved by substitution, elimination, or a combination of the two methods.

Substitution Method
Use one equation to find an expression for one variable in terms of the other, and then substitute into the other equation of the system.

Elimination Method
Use multiplication and addition to eliminate a variable from one equation. To eliminate a variable, the coefficients of that variable in the equations must be additive inverses.

Solve by substitution.

$$4x - y = 7 \quad (1)$$
$$3x + 2y = 30 \quad (2)$$

Solve for y in equation (1).

$$y = 4x - 7$$

Substitute $4x - 7$ for y in equation (2), and solve for x.

$$3x + 2(4x - 7) = 30$$
$$3x + 8x - 14 = 30 \quad \text{Distribute.}$$
$$11x - 14 = 30 \quad \text{Combine like terms.}$$
$$11x = 44 \quad \text{Add 14.}$$
$$x = 4 \quad \text{Divide by 11.}$$

Substitute 4 for x in the equation $y = 4x - 7$ to find that $y = 9$. The solution set is $\{(4, 9)\}$.

Solve the system.

$$x + 2y - z = 6 \quad (1)$$
$$x + y + z = 6 \quad (2)$$
$$2x + y - z = 7 \quad (3)$$

Add equations (1) and (2); z is eliminated and the result is $2x + 3y = 12$.

Eliminate z again by adding equations (2) and (3) to get $3x + 2y = 13$. Now solve the system

$$2x + 3y = 12 \quad (4)$$
$$3x + 2y = 13. \quad (5)$$

$$-6x - 9y = -36 \quad \text{Multiply (4) by } -3.$$
$$\underline{6x + 4y = 26} \quad \text{Multiply (5) by 2.}$$
$$-5y = -10 \quad \text{Add.}$$
$$y = 2 \quad \text{Divide by } -5.$$

Let $y = 2$ in equation (4).

$$2x + 3(2) = 12 \quad \text{(4) with } y = 2$$
$$2x + 6 = 12 \quad \text{Multiply.}$$
$$2x = 6 \quad \text{Subtract 6.}$$
$$x = 3 \quad \text{Divide by 2.}$$

Let $y = 2$ and $x = 3$ in any of the original equations to find $z = 1$. The solution set is $\{(3, 2, 1)\}$.

Concepts	Examples

5.2 Matrix Solution of Linear Systems

Matrix Row Transformations

For any augmented matrix of a system of linear equations, the following row transformations will result in the matrix of an equivalent system.

1. Interchange any two rows.
2. Multiply or divide the elements of any row by a nonzero real number.
3. Replace any row of the matrix by the sum of the elements of that row and a multiple of the elements of another row.

Gauss-Jordan Method

The Gauss-Jordan method is a systematic technique for applying matrix row transformations in an attempt to reduce a matrix to diagonal form, with 1s along the diagonal.

Solve the system.

$$x + 3y = 7$$
$$2x + \ y = 4$$

$$\begin{bmatrix} 1 & 3 & | & 7 \\ 2 & 1 & | & 4 \end{bmatrix} \qquad \text{Augmented matrix}$$

$$\begin{bmatrix} 1 & 3 & | & 7 \\ 0 & -5 & | & -10 \end{bmatrix} \qquad -2R1 + R2$$

$$\begin{bmatrix} 1 & 3 & | & 7 \\ 0 & 1 & | & 2 \end{bmatrix} \qquad -\tfrac{1}{5}R2$$

$$\begin{bmatrix} 1 & 0 & | & 1 \\ 0 & 1 & | & 2 \end{bmatrix} \qquad -3R2 + R1$$

This leads to

$$x = 1$$
$$y = 2,$$

and the solution set is $\{(1, 2)\}$.

5.3 Determinant Solution of Linear Systems

Determinant of a 2×2 Matrix

If $A = \begin{bmatrix} a_{11} & a_{12} \\ a_{21} & a_{22} \end{bmatrix}$, then

$$|A| = \begin{vmatrix} a_{11} & a_{12} \\ a_{21} & a_{22} \end{vmatrix} = a_{11}a_{22} - a_{21}a_{12}.$$

Determinant of a 3×3 Matrix

If $A = \begin{bmatrix} a_{11} & a_{12} & a_{13} \\ a_{21} & a_{22} & a_{23} \\ a_{31} & a_{32} & a_{33} \end{bmatrix}$, then

$$|A| = \begin{vmatrix} a_{11} & a_{12} & a_{13} \\ a_{21} & a_{22} & a_{23} \\ a_{31} & a_{32} & a_{33} \end{vmatrix} = (a_{11}a_{22}a_{33} + a_{12}a_{23}a_{31} + a_{13}a_{21}a_{32}) - (a_{31}a_{22}a_{13} + a_{32}a_{23}a_{11} + a_{33}a_{21}a_{12}).$$

In practice, we usually evaluate determinants by expansion by minors.

Properties of Determinants

See the discussion in **Section 5.3.**

Evaluate.

$$\begin{vmatrix} 3 & 5 \\ -2 & 6 \end{vmatrix} = 3(6) - (-2)5 = 28$$

Evaluate by expanding about the second column.

$$\begin{vmatrix} 2 & -3 & -2 \\ -1 & -4 & -3 \\ -1 & 0 & 2 \end{vmatrix} = -(-3)\begin{vmatrix} -1 & -3 \\ -1 & 2 \end{vmatrix} + (-4)\begin{vmatrix} 2 & -2 \\ -1 & 2 \end{vmatrix}$$
$$- 0\begin{vmatrix} 2 & -2 \\ -1 & -3 \end{vmatrix}$$
$$= 3(-5) - 4(2) - 0(-8)$$
$$= -15 - 8 + 0$$
$$= -23$$

(continued)

Concepts	Examples

Cramer's Rule for Two Equations in Two Variables

Given the system

$$a_1x + b_1y = c_1$$
$$a_2x + b_2y = c_2,$$

if $D \neq 0$, then the system has the unique solution

$$x = \frac{D_x}{D} \quad \text{and} \quad y = \frac{D_y}{D},$$

where $D = \begin{vmatrix} a_1 & b_1 \\ a_2 & b_2 \end{vmatrix}$, $D_x = \begin{vmatrix} c_1 & b_1 \\ c_2 & b_2 \end{vmatrix}$, and $D_y = \begin{vmatrix} a_1 & c_1 \\ a_2 & c_2 \end{vmatrix}$.

Solve using Cramer's rule.

$$x - 2y = -1$$
$$2x + 5y = 16$$

$$x = \frac{\begin{vmatrix} -1 & -2 \\ 16 & 5 \end{vmatrix}}{\begin{vmatrix} 1 & -2 \\ 2 & 5 \end{vmatrix}} = \frac{-5 + 32}{5 + 4} = \frac{27}{9} = 3$$

$$y = \frac{\begin{vmatrix} 1 & -1 \\ 2 & 16 \end{vmatrix}}{\begin{vmatrix} 1 & -2 \\ 2 & 5 \end{vmatrix}} = \frac{16 + 2}{5 + 4} = \frac{18}{9} = 2$$

The solution set is $\{(3, 2)\}$.

General Form of Cramer's Rule

Let an $n \times n$ system have linear equations of the form

$$a_1x_1 + a_2x_2 + a_3x_3 + \cdots + a_nx_n = b.$$

Define D as the determinant of the $n \times n$ matrix of coefficients of the variables. Define D_{x_1} as the determinant obtained from D by replacing the entries in column 1 of D with the constants of the system. Define D_{x_i} as the determinant obtained from D by replacing the entries in column i with the constants of the system. If $D \neq 0$, the unique solution of the system is

$$x_1 = \frac{D_{x_1}}{D}, \quad x_2 = \frac{D_{x_2}}{D}, \quad x_3 = \frac{D_{x_3}}{D}, \quad \ldots, \quad x_n = \frac{D_{x_n}}{D}.$$

Solve using Cramer's rule.

$$3x + 2y + z = -5$$
$$x - y + 3z = -5$$
$$2x + 3y + z = 0$$

Using the method of expansion by minors, it can be shown that $D_x = 45$, $D_y = -30$, $D_z = 0$, and $D = -15$. Thus,

$$x = \frac{D_x}{D} = \frac{45}{-15} = -3, \quad y = \frac{D_y}{D} = \frac{-30}{-15} = 2,$$

$$z = \frac{D_z}{D} = \frac{0}{-15} = 0.$$

The solution set is $\{(-3, 2, 0)\}$.

5.4 Partial Fractions

To solve for the constants in the numerators of a partial fraction decomposition, use either of the following methods or a combination of the two. (See **Section 5.4** for more details.)

Method 1 **For Linear Factors**

Step 1 Multiply each side by the common denominator.

Step 2 Substitute the zero of each factor in the resulting equation. For repeated linear factors, substitute as many other numbers as necessary to find all the constants in the numerators. The number of substitutions required will equal the number of constants $A, B, \ldots$.

Method 2 **For Quadratic Factors**

Step 1 Multiply each side by the common denominator.

Step 2 Collect like terms on the right side of the resulting equation.

Step 3 Equate the coefficients of like terms to get a system of equations.

Step 4 Solve the system to find the constants in the numerators.

Decompose $\dfrac{9}{2x^2 + 9x + 9}$ into partial fractions.

$$\frac{9}{2x^2 + 9x + 9} = \frac{9}{(2x + 3)(x + 3)}$$

$$\frac{9}{(2x + 3)(x + 3)} = \frac{A}{2x + 3} + \frac{B}{x + 3} \qquad *$$

Multiply by $(2x + 3)(x + 3)$.

$$9 = A(x + 3) + B(2x + 3)$$
$$9 = Ax + 3A + 2Bx + 3B$$
$$9 = (A + 2B)x + (3A + 3B)$$

Now solve the system

$$A + 2B = 0$$
$$3A + 3B = 9$$

to obtain $A = 6$ and $B = -3$.

$$\frac{9}{2x^2 + 9x + 9} = \frac{6}{2x + 3} + \frac{-3}{x + 3} \qquad \text{Substitute into *.}$$

Concepts	Examples

5.5 Nonlinear Systems of Equations

Solving a Nonlinear System

A nonlinear system can be solved by the substitution method, the elimination method, or a combination of the two.

Solve the system.

$$x^2 + 2xy - y^2 = 14 \quad (1)$$
$$x^2 - y^2 = -16 \quad (2)$$

$$
\begin{array}{ll}
x^2 + 2xy - y^2 = 14 & \\
\underline{-x^2 \qquad\; + y^2 = 16} & \text{Multiply (2) by } -1. \\
\qquad\quad 2xy \qquad\;\; = 30 & \text{Add to eliminate } x^2 \text{ and } y^2. \\
\qquad\qquad\;\; xy = 15 & \text{Divide by 2.}
\end{array}
$$

Solve for y to obtain $y = \frac{15}{x}$. Substitute into equation (2).

$$
\begin{array}{ll}
x^2 - \left(\dfrac{15}{x}\right)^2 = -16 & \quad (2) \\[2mm]
x^2 - \dfrac{225}{x^2} = -16 & \quad \text{Square.} \\[2mm]
x^4 + 16x^2 - 225 = 0 & \quad \text{Multiply by } x^2 \text{ and} \\
& \quad \text{add } 16x^2. \\
(x^2 - 9)(x^2 + 25) = 0 & \quad \text{Factor.} \\
x = \pm 3 \quad \text{or} \quad x = \pm 5i & \quad \text{Zero-factor property}
\end{array}
$$

Find corresponding y-values to obtain the solution set.

$$\{(3, 5), (-3, -5), (5i, -3i), (-5i, 3i)\}$$

5.6 Systems of Inequalities and Linear Programming

Graphing an Inequality

Method 1

If the inequality is or can be solved for y, then the following hold.

- The graph of $y < f(x)$ consists of all the points that are *below* the graph of $y = f(x)$.

- The graph of $y > f(x)$ consists of all the points that are *above* the graph of $y = f(x)$.

Method 2

If the inequality is not or cannot be solved for y, choose a test point not on the boundary.

- If the test point satisfies the inequality, the graph includes all points on the *same* side of the boundary as the test point.

- If the test point does not satisfy the inequality, the graph includes all points on the *other* side of the boundary.

Solving Systems of Inequalities

To solve a system of inequalities, graph all inequalities on the same axes, and find the intersection of their solution sets.

Graph $y \geq x^2 - 2x + 3$.

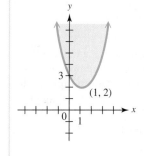

Graph the solution set of the system.

$$3x - 5y > -15$$
$$x^2 + y^2 \leq 25$$

(continued)

Concepts	Examples

Fundamental Theorem of Linear Programming
The optimal value for a linear programming problem occurs at a vertex of the region of feasible solutions.

The region of feasible solutions for

$$x + 2y \le 14$$
$$3x + 4y \le 36$$
$$x \ge 0$$
$$y \ge 0$$

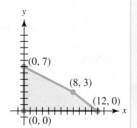

is given in the figure. Maximize the objective function $8x + 12y$.

Solving a Linear Programming Problem

Step 1 Write the objective function and all necessary constraints.

Step 2 Graph the region of feasible solutions.

Step 3 Identify all vertices (corner points).

Step 4 Find the value of the objective function at each vertex.

Step 5 The solution is given by the vertex producing the optimal value of the objective function.

Vertex Point	Value of $8x + 12y$
$(0, 0)$	0
$(0, 7)$	84
$(8, 3)$	100 ← Maximum
$(12, 0)$	96

The objective function is maximized for $x = 8$ and $y = 3$.

5.7 Properties of Matrices

Addition and Subtraction of Matrices
To add (subtract) matrices of the same dimension, add (subtract) the corresponding elements.

Find the sum or difference.

$$\begin{bmatrix} 2 & 3 & -1 \\ 0 & 4 & 9 \end{bmatrix} + \begin{bmatrix} -8 & 12 & 1 \\ 5 & 3 & -3 \end{bmatrix} = \begin{bmatrix} -6 & 15 & 0 \\ 5 & 7 & 6 \end{bmatrix}$$

$$\begin{bmatrix} 5 & -1 \\ -8 & 8 \end{bmatrix} - \begin{bmatrix} -2 & 4 \\ 3 & -6 \end{bmatrix} = \begin{bmatrix} 7 & -5 \\ -11 & 14 \end{bmatrix}$$

Scalar Multiplication
To multiply a matrix by a scalar, multiply each element of the matrix by the scalar.

Find the scalar product.

$$3 \begin{bmatrix} 6 & 2 \\ 1 & -2 \\ 0 & 8 \end{bmatrix} = \begin{bmatrix} 18 & 6 \\ 3 & -6 \\ 0 & 24 \end{bmatrix}$$

Multiplication of Matrices
The product AB of an $m \times n$ matrix A and an $n \times p$ matrix B is found as follows. To get the ith row, jth column element of the $m \times p$ matrix AB, multiply each element in the ith row of A by the corresponding element in the jth column of B. The sum of these products will give the element of row i, column j of AB.

Find the matrix product.

$$\underbrace{\begin{bmatrix} 1 & -2 & 3 \\ 5 & 0 & 4 \\ -8 & 7 & -7 \end{bmatrix}}_{3 \times 3} \underbrace{\begin{bmatrix} 1 \\ -2 \\ 3 \end{bmatrix}}_{3 \times 1} = \begin{bmatrix} 1(1) + (-2)(-2) + 3(3) \\ 5(1) + 0(-2) + 4(3) \\ -8(1) + 7(-2) + (-7)(3) \end{bmatrix}$$

$$= \underbrace{\begin{bmatrix} 14 \\ 17 \\ -43 \end{bmatrix}}_{3 \times 1}$$

Concepts	Examples

5.8 Matrix Inverses

Finding an Inverse Matrix

To obtain A^{-1} for any $n \times n$ matrix A for which A^{-1} exists, follow these steps.

Step 1 Form the augmented matrix $[A \,|\, I_n]$, where I_n is the $n \times n$ identity matrix.

Step 2 Perform row transformations on $[A \,|\, I_n]$ to obtain a matrix of the form $[I_n \,|\, B]$.

Step 3 Matrix B is A^{-1}.

Find A^{-1} if $A = \begin{bmatrix} 5 & 2 \\ 2 & 1 \end{bmatrix}$.

$$\left[\begin{array}{cc|cc} 5 & 2 & 1 & 0 \\ 2 & 1 & 0 & 1 \end{array} \right]$$

$$\left[\begin{array}{cc|cc} 1 & 0 & 1 & -2 \\ 2 & 1 & 0 & 1 \end{array} \right] \quad -2R2 + R1$$

$$\left[\begin{array}{cc|cc} 1 & 0 & 1 & -2 \\ 0 & 1 & -2 & 5 \end{array} \right] \quad -2R1 + R2$$

$\underbrace{}_{I_2} \quad \underbrace{}_{A^{-1}}$

Therefore, $A^{-1} = \begin{bmatrix} 1 & -2 \\ -2 & 5 \end{bmatrix}$.

Chapter 5 Review Exercises

Use the substitution or elimination method to solve each linear system. Identify any inconsistent systems or systems with infinitely many solutions. If the system has infinitely many solutions, write the solution set with y arbitrary.

1. $2x + 6y = 6$
$5x + 9y = 9$

2. $3x - 5y = 7$
$2x + 3y = 30$

3. $x + 5y = 9$
$2x + 10y = 18$

4. $\dfrac{1}{6}x + \dfrac{1}{3}y = 8$
$\dfrac{1}{4}x + \dfrac{1}{2}y = 12$

5. $y = -x + 3$
$2x + 2y = 1$

6. $0.2x + 0.5y = 6$
$0.4x + y = 9$

7. $3x - 2y = 0$
$9x + 8y = 7$

8. $6x + 10y = -11$
$9x + 6y = -3$

9. $2x - 5y + 3z = -1$
$x + 4y - 2z = 9$
$-x + 2y + 4z = 5$

10. $4x + 3y + z = -8$
$3x + y - z = -6$
$x + y + 2z = -2$

11. $5x - y = 26$
$4y + 3z = -4$
$3x + 3z = 15$

12. $x + z = 2$
$2y - z = 2$
$-x + 2y = -4$

13. *Concept Check* Create your own inconsistent system of two equations.

14. *Connecting Graphs with Equations* Determine the system of equations illustrated in the graph. Write equations in standard form.

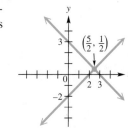

Write a system of linear equations, and then use the system to solve the problem.

15. *Meal Planning* A cup of uncooked rice contains 15 g of protein and 810 calories. A cup of uncooked soybeans contains 22.5 g of protein and 270 calories. How many cups of each should be used for a meal containing 9.5 g of protein and 324 calories?

16. *Order Quantities* A company sells recordable CDs for $0.80 each and play-only CDs for $0.60 each. The company receives $76.00 for an order of 100 CDs. However, the customer neglected to specify how many of each type to send. Determine the number of each type of CD that should be sent.

17. *Indian Weavers* The Waputi Indians make woven blankets, rugs, and skirts. Each blanket requires 24 hr for spinning the yarn, 4 hr for dyeing the yarn, and 15 hr for weaving. Rugs require 30, 5, and 18 hr and skirts 12, 3, and 9 hr, respectively. If there are 306, 59, and 201 hr available for spinning, dyeing, and weaving, respectively, how many of each item can be made? (*Hint:* Simplify the equations you write, if possible, before solving the system.)

18. *(Modeling) Populations of Minorities in the United States* The current and estimated resident populations (in percent) of blacks and Hispanics in the United States for the years 1990–2050 are modeled by the following linear functions.

$$y = 0.0515x + 12.3 \quad \text{Blacks}$$
$$y = 0.255x + 9.01 \quad \text{Hispanics}$$

In each case, x represents the number of years since 1990. (*Source:* U.S. Census Bureau.)

(a) Solve the system to find the year when these population percents will be equal.

(b) What percent of the U.S. resident population will be black or Hispanic in the year found in part (a)?

(c) Use a calculator graph of the system to support your algebraic solution.

(d) Which population is increasing more rapidly? (*Hint:* Consider the slopes of the lines.)

19. *(Modeling) Heart Rate* In a study, a group of athletes was exercised to exhaustion. Let x represent an athlete's heart rate 5 sec after stopping exercise and y this rate 10 sec after stopping. It was found that the maximum heart rate H for these athletes satisfied the two equations

$$H = 0.491x + 0.468y + 11.2$$
$$H = -0.981x + 1.872y + 26.4.$$

If an athlete had maximum heart rate $H = 180$, determine x and y graphically. Interpret your answer. (*Source:* Thomas, V., *Science and Sport,* Faber and Faber.)

20. *(Modeling) Equilibrium Supply and Demand* Let the supply and demand equations for units of backpacks be

$$\text{supply: } p = \frac{3}{2}q \quad \text{and} \quad \text{demand: } p = 81 - \frac{3}{4}q.$$

(a) Graph these equations on the same axes.

(b) Find the equilibrium demand.

(c) Find the equilibrium price.

21. *(Modeling)* Find the equation of the vertical parabola that passes through the points shown in the table.

X	Y₁
1	-2.3
2	-1.3
3	4.5
4	15.1
5	30.5
6	50.7
7	75.7

X=1

22. *(Modeling)* Knowing that the equation of a circle may be written in the form

$$x^2 + y^2 + ax + by + c = 0,$$

find the equation of the circle passing through the points $(-3, -7)$, $(4, -8)$, and $(1, 1)$.

Find solutions for each system with the specified arbitrary variable.

23. $3x - 4y + z = 2$
$2x + y = 1$; x

24. $2x - 6y + 4z = 5$
$5x + y - 3z = 1$; z

Use the Gauss-Jordan method to solve each system.

25. $5x + 2y = -10$
$3x - 5y = -6$

26. $2x + 3y = 10$
$-3x + y = 18$

27. $3x + y = -7$
$x - y = -5$

28. $2x - y + 4z = -1$
$-3x + 5y - z = 5$
$2x + 3y + 2z = 3$

29. $x - z = -3$
$y + z = 6$
$2x - 3z = -9$

30. $2x - y + z = 4$
$x + 2y - z = 0$
$3x + y - 2z = 1$

Solve each problem by using the Gauss-Jordan method to solve a system of equations.

31. *Mixing Teas* Three kinds of tea worth $4.60, $5.75, and $6.50 per lb are to be mixed to get 20 lb of tea worth $5.25 per lb. The amount of $4.60 tea used is to be equal to the total amount of the other two kinds together. How many pounds of each tea should be used?

32. *Mixing Solutions* A 5% solution of a drug is to be mixed with some 15% solution and some 10% solution to get 20 ml of 8% solution. The amount of 5% solution used must be 2 ml more than the sum of the other two solutions. How many milliliters of each solution should be used?

33. *(Modeling) Master's Degrees* During the period 1970–2008, the numbers of master's degrees awarded to both males and females grew, but degrees earned by females grew at a greater rate. If $x = 0$ represents 1970 and $x = 38$ represents 2008, the number of master's degrees earned (in thousands) are closely modeled by the following system.

$$y = 2.81x + 121 \quad \text{Males}$$
$$y = 7.25x + 65.5 \quad \text{Females}$$

Solve the system to find the year in which males and females earned the same number of master's degrees. What was the total number of master's degrees earned in that year? (*Source:* U.S. Census Bureau.)

34. *(Modeling) Comparing Prices* One refrigerator sells for $700 and uses $85 worth of electricity per year. A second refrigerator is $100 more expensive but costs only $25 per year to run. Assuming that there are no repair costs, the costs to run the refrigerators over a 10-yr period are given by the following system of equations. Here, y represents the total cost in dollars, and x is time in years.

$$y = 700 + 85x$$
$$y = 800 + 25x$$

In how many years will the costs for the two refrigerators be equal? What are the equivalent costs at that time?

Evaluate each determinant.

35. $\begin{vmatrix} -1 & 8 \\ 2 & 9 \end{vmatrix}$

36. $\begin{vmatrix} -2 & 4 \\ 0 & 3 \end{vmatrix}$

37. $\begin{vmatrix} x & 4x \\ 2x & 8x \end{vmatrix}$

38. $\begin{vmatrix} -2 & 4 & 1 \\ 3 & 0 & 2 \\ -1 & 0 & 3 \end{vmatrix}$

39. $\begin{vmatrix} -1 & 2 & 3 \\ 4 & 0 & 3 \\ 5 & -1 & 2 \end{vmatrix}$

40. $\begin{vmatrix} -3 & 2 & 7 \\ 6 & -4 & -14 \\ 7 & 1 & 4 \end{vmatrix}$

Solve each determinant equation.

41. $\begin{vmatrix} 3x & 7 \\ -x & 4 \end{vmatrix} = 8$

42. $\begin{vmatrix} 6x & 2 & 0 \\ 1 & 5 & 3 \\ x & 2 & -1 \end{vmatrix} = 2x$

Solve each system by Cramer's rule if possible. Identify any inconsistent systems or systems with infinitely many solutions. (Use another method if Cramer's rule cannot be used.)

43. $3x + 7y = 2$
$\quad 5x - y = -22$

44. $3x + y = -1$
$\quad 5x + 4y = 10$

45. $6x + y = -3$
$\quad 12x + 2y = 1$

46. $3x + 2y + z = 2$
$\quad 4x - y + 3z = -16$
$\quad x + 3y - z = 12$

47. $x + y = -1$
$\quad 2y + z = 5$
$\quad 3x - 2z = -28$

48. $5x - 2y - z = 8$
$\quad -5x + 2y + z = -8$
$\quad x - 4y - 2z = 0$

Find the partial fraction decomposition for each rational expression.

49. $\dfrac{2}{3x^2 - 5x + 2}$

50. $\dfrac{11 - 2x}{x^2 - 8x + 16}$

51. $\dfrac{5 - 2x}{(x^2 + 2)(x - 1)}$

52. $\dfrac{x^3 + 2x^2 - 3}{x^4 - 4x^2 + 4}$

Solve each nonlinear system.

53. $y = 2x + 10$
$\quad x^2 + y = 13$

54. $x^2 = 2y - 3$
$\quad x + y = 3$

55. $x^2 + y^2 = 17$
$\quad 2x^2 - y^2 = 31$

56. $2x^2 + 3y^2 = 30$
$\quad x^2 + y^2 = 13$

57. $xy = -10$
$\quad x + 2y = 1$

58. $xy + 2 = 0$
$\quad y - x = 3$

59. $x^2 + 2xy + y^2 = 4$
$\quad x - 3y = -2$

60. $x^2 + 2xy = 15 + 2x$
$\quad xy - 3x + 3 = 0$

61. $2x^2 - 3y^2 = 18$
$\quad 2x^2 - 2y^2 = 14$

62. Find all values of b such that the straight line $3x - y = b$ touches the circle $x^2 + y^2 = 25$ at only one point.

63. Do the circle $x^2 + y^2 = 144$ and the line $x + 2y = 8$ have any points in common? If so, what are they?

64. Find the equation of the line passing through the points of intersection of the graphs of $x^2 + y^2 = 20$ and $x^2 - y = 0$.

Graph the solution set of each system of inequalities.

65. $x + y \leq 6$
$2x - y \geq 3$

66. $y \leq \dfrac{1}{3}x - 2$
$y^2 \leq 16 - x^2$

67. Find $x \geq 0$ and $y \geq 0$ such that

$$3x + 2y \leq 12$$
$$5x + y \geq 5$$

and $2x + 4y$ is maximized.

68. *Connecting Graphs with Equations* Determine the system of inequalities illustrated in the graph. Write them in standard form.

Solve each linear programming problem.

69. *Cost of Nutrients* Certain laboratory animals must have at least 30 g of protein and at least 20 g of fat per feeding period. These nutrients come from food A, which costs $0.18 per unit and supplies 2 g of protein and 4 g of fat; and from food B, which costs $0.12 per unit and has 6 g of protein and 2 g of fat. Food B is purchased under a long-term contract requiring that at least 2 units of B be used per serving. How much of each food must be purchased to produce the minimum cost per serving? What is the minimum cost?

70. *Profit from Farm Animals* A 4-H member raises only geese and pigs. She wants to raise no more than 16 animals, including no more than 10 geese. She spends $5 to raise a goose and $15 to raise a pig, and she has $180 available for this project. Each goose produces $6 in profit, and each pig produces $20 in profit. How many of each animal should she raise to maximize her profit? What is her maximum profit?

Find the value of each variable.

71. $\begin{bmatrix} 5 & x+2 \\ -6y & z \end{bmatrix} = \begin{bmatrix} a & 3x-1 \\ 5y & 9 \end{bmatrix}$

72. $\begin{bmatrix} -6+k & 2 & a+3 \\ -2+m & 3p & 2r \end{bmatrix} + \begin{bmatrix} 3-2k & 5 & 7 \\ 5 & 8p & 5r \end{bmatrix} = \begin{bmatrix} 5 & y & 6a \\ 2m & 11 & -35 \end{bmatrix}$

Perform each operation when possible.

73. $\begin{bmatrix} 3 \\ 2 \\ 5 \end{bmatrix} - \begin{bmatrix} 8 \\ -4 \\ 6 \end{bmatrix} + \begin{bmatrix} 1 \\ 0 \\ 2 \end{bmatrix}$

74. $4\begin{bmatrix} 3 & -4 & 2 \\ 5 & -1 & 6 \end{bmatrix} + \begin{bmatrix} -3 & 2 & 5 \\ 1 & 0 & 4 \end{bmatrix}$

75. $\begin{bmatrix} 2 & 5 & 8 \\ 1 & 9 & 2 \end{bmatrix} - \begin{bmatrix} 3 & 4 \\ 7 & 1 \end{bmatrix}$

76. $\begin{bmatrix} -3 & 4 \\ 2 & 8 \end{bmatrix}\begin{bmatrix} -1 & 0 \\ 2 & 5 \end{bmatrix}$

77. $\begin{bmatrix} -1 & 0 \\ 2 & 5 \end{bmatrix} \begin{bmatrix} -3 & 4 \\ 2 & 8 \end{bmatrix}$

78. $\begin{bmatrix} 1 & 2 \\ 3 & 0 \\ -6 & 5 \end{bmatrix} \begin{bmatrix} 4 & 8 \\ -1 & 2 \end{bmatrix}$

79. $\begin{bmatrix} 3 & 2 & -1 \\ 4 & 0 & 6 \end{bmatrix} \begin{bmatrix} -2 & 0 \\ 0 & 2 \\ 3 & 1 \end{bmatrix}$

80. $\begin{bmatrix} 1 & -2 & 4 & 2 \\ 0 & 1 & -1 & 8 \end{bmatrix} \begin{bmatrix} -1 \\ 2 \\ 0 \\ 1 \end{bmatrix}$

81. $\begin{bmatrix} -2 & 5 & 5 \\ 0 & 1 & 4 \\ 3 & -4 & -1 \end{bmatrix} \begin{bmatrix} 1 & 0 & -1 \\ -1 & 0 & 0 \\ 1 & 1 & -1 \end{bmatrix}$

82. $\begin{bmatrix} 0 & 1 & 4 \\ 7 & -2 & 9 \\ 10 & 0 & 1 \end{bmatrix} \begin{bmatrix} 3 & 2 & 1 \\ -4 & 7 & 6 \end{bmatrix}$

Find the inverse of each matrix that has an inverse.

83. $\begin{bmatrix} 2 & 1 \\ 5 & 3 \end{bmatrix}$

84. $\begin{bmatrix} -4 & 2 \\ 0 & 3 \end{bmatrix}$

85. $\begin{bmatrix} 2 & -1 & 0 \\ 1 & 0 & 1 \\ 1 & -2 & 0 \end{bmatrix}$

86. $\begin{bmatrix} 2 & 3 & 5 \\ -2 & -3 & -5 \\ 1 & 4 & 2 \end{bmatrix}$

Use the method of matrix inverses to solve each system.

87. $5x - 4y = 1$
$\quad\;\; x + 4y = 3$

88. $2x + y = 5$
$\quad\; 3x - 2y = 4$

89. $3x + 2y + z = -5$
$\quad\;\; x - y + 3z = -5$
$\quad\; 2x + 3y + z = 0$

90. $x + y + z = 1$
$\quad 2x - y = -2$
$\quad 3y + z = 2$

Chapter 5 Test

Use substitution or elimination to solve each system. Identify any system that is inconsistent or has infinitely many solutions. If a system has infinitely many solutions, express the solution set with y arbitrary.

1. $3x - y = 9$
$\quad x + 2y = 10$

2. $6x + 9y = -21$
$\quad 4x + 6y = -14$

3. $\dfrac{1}{4}x - \dfrac{1}{3}y = -\dfrac{5}{12}$
$\quad \dfrac{1}{10}x + \dfrac{1}{5}y = \dfrac{1}{2}$

4. $\quad x - 2y = 4$
$\quad -2x + 4y = 6$

5. $2x + y + z = 3$
$\quad x + 2y - z = 3$
$\quad 3x - y + z = 5$

Use the Gauss-Jordan method to solve each system.

6. $3x - 2y = 13$
$\quad 4x - y = 19$

7. $3x - 4y + 2z = 15$
$\quad 2x - y + z = 13$
$\quad x + 2y - z = 5$

8. *Connecting Graphs with Equations* Find the equation that defines the parabola shown.

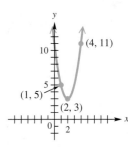

9. *Ordering Supplies* A knitting shop orders yarn from three suppliers in Toronto, Montreal, and Ottawa. One month the shop ordered a total of 100 units of yarn from these suppliers. The delivery costs were $80, $50, and $65 per unit for the orders from Toronto, Montreal, and Ottawa, respectively, with total delivery costs of $5990. The shop ordered the same amount from Toronto and Ottawa. How many units were ordered from each supplier?

Evaluate each determinant.

10. $\begin{vmatrix} 6 & 8 \\ 2 & -7 \end{vmatrix}$

11. $\begin{vmatrix} 2 & 0 & 8 \\ -1 & 7 & 9 \\ 12 & 5 & -3 \end{vmatrix}$

Solve each system by Cramer's rule.

12. $2x - 3y = -33$
$4x + 5y = 11$

13. $x + y - z = -4$
$2x - 3y - z = 5$
$x + 2y + 2z = 3$

14. Find the partial fraction decomposition of $\dfrac{x + 2}{x^3 + 2x^2 + x}$.

15. *Connecting Graphs with Equations* Determine the system of equations illustrated in the graph. Write equations in standard form.

Solve each nonlinear system of equations.

16. $2x^2 + y^2 = 6$
$x^2 - 4y^2 = -15$

17. $x^2 + y^2 = 25$
$x + y = 7$

18. *Unknown Numbers* Find two numbers such that their sum is -1 and the sum of their squares is 61.

19. Graph the solution set.

$x - 3y \geq 6$
$y^2 \leq 16 - x^2$

20. Find $x \geq 0$ and $y \geq 0$ such that

$x + 2y \leq 24$
$3x + 4y \leq 60$

and $2x + 3y$ is maximized.

21. *Jewelry Profits* The J. J. Gravois Company designs and sells two types of rings: the VIP and the SST. The company can produce up to 24 rings each day using up to 60 total hours of labor. It takes 3 hr to make one VIP ring and 2 hr to make one SST ring. How many of each type of ring should be made daily in order to maximize the company's profit, if the profit on one VIP ring is $30 and the profit on one SST ring is $40? What is the maximum profit?

22. Find the value of each variable in the equation $\begin{bmatrix} 5 & x+6 \\ 0 & 4 \end{bmatrix} = \begin{bmatrix} y-2 & 4-x \\ 0 & w+7 \end{bmatrix}$.

Perform each operation when possible.

23. $3\begin{bmatrix} 2 & 3 \\ 1 & -4 \\ 5 & 9 \end{bmatrix} - \begin{bmatrix} -2 & 6 \\ 3 & -1 \\ 0 & 8 \end{bmatrix}$

24. $\begin{bmatrix} 1 \\ 2 \end{bmatrix} + \begin{bmatrix} 4 \\ -6 \end{bmatrix} + \begin{bmatrix} 2 & 8 \\ -7 & 5 \end{bmatrix}$

25. $\begin{bmatrix} 2 & 1 & -3 \\ 4 & 0 & 5 \end{bmatrix} \begin{bmatrix} 1 & 3 \\ 2 & 4 \\ 3 & -2 \end{bmatrix}$

26. $\begin{bmatrix} 2 & -4 \\ 3 & 5 \end{bmatrix} \begin{bmatrix} 4 \\ 2 \\ 7 \end{bmatrix}$

27. *Concept Check* Which of the following properties does not apply to multiplication of matrices?

    A. commutative   **B.** associative   **C.** distributive   **D.** identity

Find the inverse, if it exists, of each matrix.

28. $\begin{bmatrix} -8 & 5 \\ 3 & -2 \end{bmatrix}$

29. $\begin{bmatrix} 4 & 12 \\ 2 & 6 \end{bmatrix}$

30. $\begin{bmatrix} 1 & 3 & 4 \\ 2 & 7 & 8 \\ -2 & -5 & -7 \end{bmatrix}$

Use matrix inverses to solve each system.

31. $2x + y = -6$
    $3x - y = -29$

32. $x + y = 5$
    $y - 2z = 23$
    $x + 3z = -27$

Glossary

For a more complete discussion, see the section(s) in parentheses.

A

absolute value   The absolute value of a real number is the distance between 0 and the number on the number line. (Sections R.2, 1.8)

additive inverse (negative) of a matrix   If the sum of two matrices is the zero matrix, then those two matrices are additive inverses (negatives) of each other. (Section 5.7)

algebraic expression   Any collection of numbers or variables joined by the basic operations of addition, subtraction, multiplication, or division (except by 0), or by the operations of raising to powers or taking roots, formed according to the rules of algebra, is an algebraic expression. (Section R.3)

argument   In the expression $\log_a x$, x is the argument. (Section 4.3)

augmented matrix   An augmented matrix is a matrix whose elements are the coefficients of the variables and the constants of a system of equations. An augmented matrix is written with a vertical bar that separates the coefficients of the variables from the constants. (Section 5.2)

average rate of change   The slope of a line gives the average rate of change in y per unit of change in x, where the value of y depends on the value of x. (Section 2.4)

axis (of symmetry) of a parabola   The line of symmetry for a parabola is the axis of the parabola. (Section 3.1)

B

base of an exponential   The base is the number that is a repeated factor in exponential notation. In the expression a^n, a is the base. (Section R.2)

base of a logarithm   In the expression $\log_a x$, a is the base. (Section 4.3)

binomial   A binomial is a polynomial containing exactly two terms. (Section R.3)

boundary   A line that separates a plane into two half-planes is the boundary of each half-plane. (Section 5.6)

break-even point   The break-even point is the point where the revenue from selling a product is equal to the cost of producing it. (Sections 1.7, 2.4)

C

center of a circle   The center of a circle is the given point that is a given distance from all points on the circle. (Section 2.2)

change in x   The change in x is the horizontal difference (the difference in x-coordinates) between two points on a line. (Section 2.4)

change in y   The change in y is the vertical difference (the difference in y-coordinates) between two points on a line. (Section 2.4)

circle   A circle is the set of all points in a plane that lie a given distance from a given point. (Section 2.2)

closed interval   A closed interval is an interval that includes both of its endpoints. (Section 1.7)

coefficient (numerical coefficient)   The real number factor in a term of an algebraic expression is the coefficient of the other factors. (Section R.3)

cofactor   The product of a minor of an element of a square matrix and the number $+1$ (if the sum of the row number and column number is even), or -1 (if the sum of the row number and column number is odd), is a cofactor. (Section 5.3)

collinear   Two points are collinear if they lie on the same line. (Section 2.1)

column matrix   A matrix with exactly one column is a column matrix. (Section 5.7)

combined variation   Variation in which one variable depends on more than one other variable is combined variation. (Section 3.6)

common logarithm   A base 10 logarithm is a common logarithm. (Section 4.4)

complement of a set   The set of all elements in the universal set U that do not belong to set A is the complement of A, written A'. (Section R.1)

completing the square   The process of adding to a binomial the expression that makes it a perfect square trinomial is called completing the square. (Sections 1.4, 3.1)

complex conjugates   The complex numbers $a + bi$ and $a - bi$ are complex conjugates. (Section 1.3)

complex fraction   A complex fraction is a quotient of two rational expressions. (Section R.5)

complex number   A complex number is a number of the form $a + bi$, where a and b are real numbers and $i = \sqrt{-1}$. (Section 1.3)

composite function (composition)   If f and g are functions, then the composite function, or composition, of g and f is symbolized by $(g \circ f)(x)$, and defined to be $g(f(x))$. (Section 2.8)

compound amount   In an investment paying compound interest, the compound amount is the balance *after* interest has been earned. (The compound amount is sometimes called the *future value*.) (Section 4.2)

compound interest   For compound interest, interest is paid both on the principal and on any previously earned interest. (Section 4.2)

conditional equation   An equation that is satisfied by some numbers but not by others is a conditional equation. (Section 1.1)

conjugates   The expressions $a - b$ and $a + b$ are conjugates. (Section R.7)

consistent system   A consistent system is a system of equations with at least one solution. (Section 5.1)

constant function   A function f is constant on an open interval I if, for every x_1 and x_2 in I, $f(x_1) = f(x_2)$. (Section 2.3)

constant of variation   In a variation equation, such as $y = kx$, $y = \frac{k}{x}$, or $y = kx^n z^m$, the real number k is the constant of variation. (Section 3.6)

constraints   In linear programming, the inequalities that represent restrictions on a particular situation are the constraints. (Section 5.6)

continuous compounding   Continuous compounding of money involves the computation of interest as the frequency of compounding approaches infinity, leading to the formula $A = Pe^{rt}$. (Section 4.2)

continuous function   (informal definition) A function is continuous over an interval of its domain if its hand-drawn graph over that interval can be sketched without lifting the pencil from the paper. (Section 2.6)

contradiction An equation that has no solution is a contradiction. (Section 1.1)

coordinate (on a number line) A number that corresponds to a particular point on a number line is the coordinate of the point. (Section R.2)

coordinate plane (*xy*-plane) The plane into which the rectangular coordinate system is introduced is the coordinate plane, or *xy*-plane. (Section 2.1)

coordinates (in the *xy*-plane) The coordinates of a point in the *xy*-plane are the numbers in the ordered pair that correspond to that point. (Section 2.1)

coordinate system (on a number line) The correspondence between points on a number line and the real numbers is a coordinate system. (Section R.2)

Cramer's rule Cramer's rule is a method of using determinants to solve a system of linear equations. (Section 5.3)

D

decreasing function A function f is decreasing on an open interval I if, whenever $x_1 < x_2$ in I, $f(x_1) > f(x_2)$. (Section 2.3)

degree of a polynomial The greatest degree of any term in a polynomial is the degree of the polynomial. (Section R.3)

degree of a term The degree of a term is the sum of the exponents on the variable factors in the term. (Section R.3)

dependent equations Two equations are dependent if any solution of one equation is also a solution of the other. (Section 5.1)

dependent variable If the value of the variable y depends on the value of the variable x, then y is the dependent variable. (Section 2.3)

determinant Every $n \times n$ matrix A is associated with a single real number called the determinant of A, written $|A|$. (Section 5.3)

difference quotient If f is a function and h is a positive number, then the expression $\frac{f(x + h) - f(x)}{h}$ is the difference quotient. (Section 2.8)

dimension of a matrix The dimension of a matrix indicates the number of rows and columns that the matrix has, with the number of rows given first. (Section 5.2)

discriminant The quantity under the radical in the quadratic formula, $b^2 - 4ac$, is the discriminant of the expression $ax^2 + bx + c$. (Section 1.4)

disjoint sets Two sets that have no elements in common are disjoint sets. (Section R.1)

domain of a rational expression The domain of a rational expression is the set of real numbers for which the expression is defined. (Section R.5)

domain of a relation In a relation, the set of all values of the independent variable (x) is the domain. (Section 2.3)

dominating term In a polynomial function, the dominating term is the term of greatest degree. (Section 3.4)

doubling time The amount of time that it takes for a quantity that grows exponentially to become twice its initial amount is its doubling time. (Section 4.6)

E

element of a matrix Each entry in a matrix is an element of the matrix. (Section 5.2)

elements (members) The objects that belong to a set are the elements (members) of the set. (Section R.1)

empty set (null set) The empty set or null set, written $\emptyset$ or { }, is the set containing no elements. (Sections R.1, 1.1)

end behavior The end behavior of a graph of a polynomial function describes how the values of y increase or decrease as $|x|$ increases without bound. (Section 3.4)

equation An equation is a statement that two expressions are equal. (Section 1.1)

equation with a rational exponent An equation that contains a variable, or variable expression, raised to an exponent that is a rational number is an equation with a rational exponent. (Section 1.6)

equivalent equations Equations with the same solution set are equivalent equations. (Section 1.1)

equivalent systems Equivalent systems are systems of equations that have the same solution set. (Section 5.1)

even function A function f is an even function if $f(-x) = f(x)$ for all x in the domain of f. (Section 2.7)

expansion by a row or column Expansion by a row or column is a method for evaluating a determinant of a 3×3 or larger matrix. This process involves multiplying each element of any row or column of the matrix by its cofactor and then adding these products. (Section 5.3)

exponent In the expression a^n, the exponent n indicates the number of times that the base a is used as a factor. (Section R.2)

exponential equation An exponential equation is an equation with a variable in an exponent. (Section 4.2)

exponential function If $a > 0$ and $a \neq 1$, then $f(x) = a^x$ defines the exponential function with base a. (Section 4.2)

exponential growth or decay function An exponential growth or decay function models a situation in which a quantity changes at a rate proportional to the amount present. Such a function is defined by an equation of the form $y = y_0 e^{kt}$, where y_0 is the amount or number present at time $t = 0$ and k is a constant. (Section 4.6)

F

factored completely A polynomial is factored completely when it is written as a product of prime polynomials. (Section R.4)

factored form A polynomial is in factored form when it is written as a product of polynomials. (Section R.4)

factoring The process of finding polynomials whose product equals a given polynomial is called factoring. (Section R.4)

factoring by grouping Factoring by grouping is a method of grouping the terms of a polynomial in such a way that the polynomial can be factored even though the greatest common factor of its terms is 1. (Section R.4)

finite set A finite set is a set that has a limited number of elements. (Section R.1)

function A function is a relation (set of ordered pairs) in which, for each value of the first component of the ordered pairs, there is *exactly one* value of the second component. (Section 2.3)

function notation Function notation $f(x)$ (read "f of x") represents the y-value of the function f for the indicated x-value. (Section 2.3)

future value In an investment paying compound interest, the future value is the balance *after* interest has been earned. (The future value is sometimes called the *compound amount*.) (Section 4.2)

G

graph of an equation The graph of an equation is the set of all points that correspond to the ordered pairs that satisfy the equation. (Section 2.1)

H

half-life The period of time that it takes for a quantity that decays exponentially to become half its initial amount is its half-life. (Section 4.6)

half-plane A line separates a plane into two regions, each of which is a half-plane. (Section 5.6)

horizontal asymptote A horizontal line that a graph approaches as $|x|$ increases without bound is a horizontal asymptote. The line $y = b$ is a horizontal asymptote if $y \to b$ as $|x| \to \infty$. (Section 3.5)

I

identity An equation satisfied by every number that is a meaningful replacement for the variable is an identity. (Section 1.1)

identity matrix (multiplicative identity matrix) An identity matrix is an $n \times n$ matrix with 1s on the main diagonal and 0s everywhere else. (Section 5.8)

imaginary part In the complex number $a + bi$, b is the imaginary part. (Section 1.3)

imaginary unit The number i, defined by $i = \sqrt{-1}$ (so $i^2 = -1$), is the imaginary unit. (Section 1.3)

inconsistent system An inconsistent system is a system of equations with no solution. (Section 5.1)

increasing function A function f is increasing on an open interval I if, whenever $x_1 < x_2$ in I, $f(x_1) < f(x_2)$. (Section 2.3)

independent variable If the value of the variable y depends on the value of the variable x, then x is the independent variable. (Section 2.3)

index of a radical In a radical of the form $\sqrt[n]{a}$, n is the index. (Section R.7)

inequality An inequality says that one expression is greater than, greater than or equal to, less than, or less than or equal to, another. (Sections R.2, 1.7)

infinite set (This is an informal definition.) An infinite set is a set that has an unending list of distinct elements. (Section R.1)

integers The set of integers is $\{\ldots, -3, -2, -1, 0, 1, 2, 3, \ldots\}$. (Section R.2)

intersection The intersection of sets A and B, written $A \cap B$, is the set of elements that belong to both A and B. (Section R.1)

interval An interval is a portion of the real number line, which may or may not include its endpoint(s). (Section 1.7)

interval notation Interval notation is a simplified notation for writing intervals. It uses parentheses and brackets to show whether the endpoints are included. (Section 1.7)

inverse function Let f be a one-to-one function. Then g is the inverse function of f if $(f \circ g)(x) = x$ for every x in the domain of g, and $(g \circ f)(x) = x$ for every x in the domain of f. (Section 4.1)

irrational numbers Real numbers that cannot be represented as quotients of integers are irrational numbers. (Section R.2)

L

leading coefficient In a polynomial function of degree n, the leading coefficient is a_n. It is the coefficient of the term of greatest degree. (Section 3.1)

like radicals Radicals with the same radicand and the same index are like radicals. (Section R.7)

like terms Terms with the same variables each raised to the same powers are like terms. (Section R.3)

linear cost function A linear cost function is a linear function that has the form $C(x) = mx + b$, where x represents the number of items produced, m represents the cost per item, and b represents the fixed cost. (Section 2.4)

linear equation (first-degree equation) in n unknowns Any equation of the form $a_1x_1 + a_2x_2 + \cdots + a_nx_n = b$, for real numbers $a_1, a_2, \ldots, a_n$ (not all of which are 0) and b, is a linear equation, or a first-degree equation, in n unknowns. (Section 5.1)

linear equation (first-degree equation) in one variable A linear equation in one variable is an equation that can be written in the form $ax + b = 0$, where a and b are real numbers with $a \neq 0$. (Section 1.1)

linear function A function f of the form $f(x) = ax + b$, for real numbers a and b, is a linear function. (Section 2.4)

linear inequality in one variable A linear inequality in one variable is an inequality that can be written in the form $ax + b > 0$, where a and b are real numbers with $a \neq 0$. (Any of the symbols $<$, $\geq$, and $\leq$ may also be used.) (Section 1.7)

linear inequality in two variables A linear inequality in two variables is an inequality of the form $Ax + By \leq C$, where A, B, and C are real numbers with A and B not both equal to 0. (Any of the symbols $\geq$, $<$, and $>$ may also be used.) (Section 5.6)

linear programming Linear programming, an application of a system of linear inequalities, is a method for finding an optimum value, such as minimum cost or maximum profit. (Section 5.6)

literal equation A literal equation is an equation that relates two or more variables (letters). (Section 1.1)

logarithm A logarithm is an exponent. The expression $\log_a x$ is the power to which the base a must be raised to obtain x. (Section 4.3)

logarithmic equation A logarithmic equation is an equation with a logarithm of a variable expression in at least one term. (Section 4.5)

logarithmic function If $a > 0$, $a \neq 1$, and $x > 0$, then $f(x) = \log_a x$ defines the logarithmic function with base a. (Section 4.3)

lowest terms A rational expression is in lowest terms when the greatest common factor of its numerator and denominator is 1. (Section R.5)

M

mathematical model A mathematical model is an equation (or inequality) that describes the relationship between two or more quantities. (Section 1.2)

matrix (plural, matrices) A matrix is a rectangular array of numbers enclosed in brackets. (Section 5.2)

minor In an $n \times n$ matrix with $n \geq 3$, the minor of a particular element is the determinant of the $(n - 1) \times (n - 1)$ matrix that results when the row and column that contain the chosen element are eliminated. (Section 5.3)

monomial A monomial is a polynomial containing exactly one term. (Section R.3)

multiplicative inverse of a matrix If A is an $n \times n$ matrix, then its multiplicative inverse, written A^{-1}, is the matrix that satisfies both $AA^{-1} = I_n$ and $A^{-1}A = I_n$. It is possible that A^{-1} does not exist. (Section 5.8)

N

natural logarithm A base e logarithm is a natural logarithm. (Section 4.4)

natural numbers (counting numbers) The natural numbers, or counting numbers, form the set of numbers $\{1, 2, 3, 4, \ldots\}$. (Sections R.1, R.2)

nonlinear system A system of equations in which at least one equation is *not* linear is a nonlinear system. (Section 5.5)

nonreal complex number A complex number $a + bi$ in which $b \neq 0$ is a nonreal complex number. (Section 1.3)

nonstrict inequality An inequality in which the symbol is either $\leq$ or $\geq$ is a nonstrict inequality. (Sections 1.7, 5.6)

O

objective function In linear programming, the function to be maximized or minimized is the objective function. (Section 5.6)

oblique asymptote A nonvertical, nonhorizontal line that a graph approaches as $|x|$ increases without bound is an oblique asymptote. (Section 3.5)

odd function   A function f is an odd function if $f(-x) = -f(x)$ for all x in the domain of f. (Section 2.7)

one-to-one function   In a one-to-one function, each x-value corresponds to only one y-value, and each y-value corresponds to only one x-value. (Section 4.1)

open interval   An open interval is an interval that does not include its endpoint(s). (Section 1.7)

ordered pair   An ordered pair consists of two components, written inside parentheses, called the first and second components. Ordered pairs are used to identify points in the coordinate plane. (Section 2.1)

ordered triple   An ordered triple consists of three components, written inside parentheses, called the first, second, and third components. Ordered triples are used to identify points in space. (Section 5.1)

origin (of a rectangular coordinate system)   The point of intersection of the x-axis and the y-axis of a rectangular coordinate system is the origin. (Section 2.1)

P

parabola   A parabola is a curve that is the graph of a quadratic function defined over the set of all real numbers. (Sections 2.5, 3.1)

partial fraction   Each term in the decomposition of a rational expression is a partial fraction. (Section 5.4)

partial fraction decomposition   When one rational expression is expressed as the sum of two or more rational expressions, the sum is the partial fraction decomposition. (Section 5.4)

piecewise-defined function   A piecewise-defined function is a function that is defined by different rules over different intervals of its domain. (Section 2.6)

point-slope form   The point-slope form of the equation of the line with slope m through the point (x_1, y_1) is $y - y_1 = m(x - x_1)$. (Section 2.5)

polynomial   A polynomial is a term or a finite sum of terms, with only positive or zero integer exponents present on the variables. (Section R.3)

polynomial function of degree n   A polynomial function of degree n, where n is a nonnegative integer, is a function defined by an expression of the form $f(x) = a_n x^n + a_{n-1} x^{n-1} + \cdots + a_1 x + a_0$, where $a_n, a_{n-1}, \ldots, a_1$, and a_0 are real numbers, with $a_n \neq 0$. (Section 3.1)

power (exponential expression, exponential)   An expression of the form a^n is called a power, an exponential expression, or an exponential. (Section R.2)

present value   In an investment paying compound interest, the principal is also called the present value. (Section 4.2)

prime polynomial   A polynomial with variable terms that cannot be written as a product of two polynomials of lesser degree is a prime polynomial. (Section R.4)

principal nth root   For even values of n (square roots, fourth roots, and so on), when a is positive, there are two real nth roots, one positive and one negative.

In such cases, the notation $\sqrt[n]{a}$ represents the positive root, or principal nth root. (Section R.7)

pure imaginary number   A complex number $a + bi$ in which $a = 0$ and $b \neq 0$ is a pure imaginary number. (Section 1.3)

Pythagorean theorem   The Pythagorean theorem states that in a right triangle, the sum of the squares of the lengths of the legs is equal to the square of the length of the hypotenuse. (Section 1.5)

Q

quadrants   The quadrants of a rectangular coordinate system are the four regions into which the x-axis and y-axis divide the plane. (Section 2.1)

quadratic equation (second-degree equation)   An equation that can be written in the form $ax^2 + bx + c = 0$, where a, b, and c are real numbers with $a \neq 0$, is a quadratic equation. (Section 1.4)

quadratic in form   An equation is quadratic in form if it can be written as $au^2 + bu + c = 0$, where $a \neq 0$ and u is some algebraic expression. (Section 1.6)

quadratic formula   The quadratic formula $x = \dfrac{-b \pm \sqrt{b^2 - 4ac}}{2a}$ is a general formula that can be used to solve a quadratic equation of the form $ax^2 + bx + c = 0$. (Section 1.4)

quadratic function   A function f of the form $f(x) = ax^2 + bx + c$, where a, b, and c are real numbers, with $a \neq 0$, is a quadratic function. (Section 3.1)

quadratic inequality   A quadratic inequality is an inequality that can be written in the form $ax^2 + bx + c < 0$, for real numbers a, b, and c, with $a \neq 0$. (The symbol $<$ can be replaced with $>$, $\leq$, and $\geq$.) (Section 1.7)

R

radicand   The number or expression under a radical symbol is the radicand. (Section R.7)

radius   The radius of a circle is the given distance between the center and any point on the circle. (Section 2.2)

range   In a relation, the set of all values of the dependent variable (y) is the range. (Section 2.3)

rational equation   A rational equation is an equation that has a rational expression for one or more terms. (Section 1.6)

rational expression   The quotient of two polynomials P and Q, with $Q \neq 0$, is a rational expression. (Section R.5)

rational function   A function f of the form $f(x) = \dfrac{p(x)}{q(x)}$, where $p(x)$ and $q(x)$ are polynomials, with $q(x) \neq 0$, is a rational function. (Section 3.5)

rational inequality   A rational inequality is an inequality in which one or both sides contain rational expressions. (Section 1.7)

rationalizing the denominator   Rationalizing a denominator is the process of writing a radical expression so that there are no radicals in the denominator. (Section R.7)

rational numbers   The rational numbers are the set of numbers $\frac{p}{q}$, where p and q are integers and $q \neq 0$. (Section R.2)

real numbers   The set of all numbers that correspond to points on a number line is the real numbers. (Section R.2)

real part   In the complex number $a + bi$, a is the real part. (Section 1.3)

reciprocal function   The function $f(x) = \frac{1}{x}$ is the reciprocal function. (Section 3.5)

rectangular (Cartesian) coordinate system   Two perpendicular number lines intersecting at their origins form a rectangular coordinate system. (Section 2.1)

region of feasible solutions   In linear programming, the region of feasible solutions is the region of the graph that satisfies all of the constraints. (Section 5.6)

relation   A relation is a set of ordered pairs. (Section 2.3)

root (solution) of an equation   A root (solution) k of $f(x)$ is a number such that $f(k) = 0$ is true. (Section 3.2)

row matrix   A matrix with exactly one row is a row matrix. (Section 5.7)

S

scalar In work with matrices, a real number is called a scalar to distinguish it from a matrix. (Section 5.7)

scatter diagram (of ordered pairs) A scatter diagram is a graph of specific ordered pairs of data. (Section 2.5)

set A set is a collection of objects. (Section R.1)

set-builder notation Set-builder notation uses the form $\{x \mid x$ has a certain property$\}$ to describe a set without having to list all of its elements. (Section R.1)

set operations The processes of finding the complement of a set, the intersection of two sets, and the union of two sets are set operations. (Section R.1)

simple interest In computing simple interest, interest is paid only on the principal, and not on previously earned interest. (Section 1.1)

slope The slope of a nonvertical line is the ratio of the change in y to the change in x. (Section 2.4)

slope-intercept form The slope-intercept form of the equation of the line with slope m and y-intercept $(0, b)$ is $y = mx + b$. (Section 2.5)

solution (root) A solution or root of an equation is a number that makes the equation a true statement. (Section 1.1)

solution set The solution set of an equation is the set of all numbers that satisfy the equation. (Section 1.1)

solution of a system of equations A solution of a system of equations must satisfy every equation in the system. (Section 5.1)

square matrix A matrix with the same number of columns as rows is a square matrix. (Section 5.7)

standard form of a complex number A complex number written in the form $a + bi$ (or $a + ib$) is in standard form. (Section 1.3)

standard form of a linear equation The form $Ax + By = C$ is the standard form of a linear equation. (Section 2.4)

step function A step function is a function that is defined using the greatest integer function, and whose graph resembles a series of steps. (Section 2.6)

strict inequality An inequality in which the symbol is either $<$ or $>$ is a strict inequality. (Sections 1.7, 5.6)

subset If every element of set A is also an element of set B, then A is a subset of B, written $A \subseteq B$. (Section R.1)

synthetic division Synthetic division is a shortcut method of dividing a polynomial by a binomial of the form $x - k$. (Section 3.2)

system of equations A group of equations that are considered at the same time is a system of equations. (Section 5.1)

system of inequalities A group of inequalities that are considered at the same time is a system of inequalities. (Section 5.6)

system of linear equations (linear system) If all the equations in a system are linear, then the system is a system of linear equations, or a linear system. (Section 5.1)

T

term The product of a real number and one or more variables raised to powers is a term. (Section R.3)

trinomial A trinomial is a polynomial containing exactly three terms. (Section R.3)

turning points The points on the graph of a function where the function changes from increasing to decreasing or from decreasing to increasing are turning points of the graph. (Section 3.4)

U

union The union of sets A and B, written $A \cup B$, is the set of all elements that belong to set A or set B (or both). (Sections R.1, 1.7)

universal set The universal set, written U, contains all the elements under discussion in a particular situation. (Section R.1)

V

varies directly (is directly proportional to) y varies directly as x, or y is directly proportional to x, if there exists a nonzero real number k such that $y = kx$. (Section 3.6)

varies inversely (is inversely proportional to) y varies inversely as x, or y is inversely proportional to x, if there exists a nonzero real number k such that $y = \frac{k}{x}$. (Section 3.6)

varies jointly In joint variation, one variable depends on the product of two or more other variables. If m and n are real numbers, then y varies jointly as the nth power of x and the mth power of z if there exists a nonzero real number k such that $y = kx^n z^m$. (Section 3.6)

Venn diagram A Venn diagram is a diagram used to illustrate relationships among sets or probability concepts. (Section R.1)

V (vertex)

vertex (corner point) In linear programming, a vertex, or corner point, is one of the vertices of the region of feasible solutions. (Section 5.6)

vertex of a parabola The vertex of a parabola is the point where the axis of symmetry intersects the parabola. For the graph of a quadratic function, this is the turning point of the parabola. (Sections 2.6, 3.1)

vertical asymptote A vertical line that a graph approaches, but never touches or intersects, is a vertical asymptote. The line $x = a$ is a vertical asymptote for a function f if $|f(x)| \to \infty$ as $x \to a$. (Section 3.5)

W

whole numbers The set of whole numbers $\{0, 1, 2, 3, 4, \dots\}$ is the union of the set of natural numbers and $\{0\}$. (Section R.2)

X

x-axis The horizontal number line in a rectangular coordinate system is the x-axis. (Section 2.1)

x-intercept An x-intercept is a point where the graph of an equation intersects the x-axis. (Section 2.1)

Y

y-axis The vertical number line in a rectangular coordinate system is the y-axis. (Section 2.1)

y-intercept A y-intercept is a point where the graph of an equation intersects the y-axis. (Section 2.1)

Z

zero-factor property The zero-factor property states that if the product of two (or more) complex numbers is 0, then at least one of the numbers must be 0. (Sections R.5, 1.4)

zero matrix A matrix containing only zero elements is a zero matrix. (Section 5.7)

zero of multiplicity n A polynomial function has a zero k of multiplicity n if the zero k occurs exactly n times. The polynomial has exactly n factors of $x - k$. (Section 3.3)

zero polynomial The function f defined by $f(x) = 0$ is the zero polynomial. (Section 3.1)

zero of a polynomial function A zero of a polynomial function f is a number k such that $f(k) = 0$ is true. (Section 3.2)

Chapter R Review of Basic Concepts

R.2 Exercises (pages 17–20)

61. No; in general $a - b \neq b - a$.

Examples:

$a = 15, b = 0$: $a - b = 15 - 0 = 15$, but
$b - a = 0 - 15 = -15$.

$a = 12, b = 7$: $a - b = 12 - 7 = 5$, but
$b - a = 7 - 12 = -5$.

$a = -6, b = 4$: $a - b = -6 - 4 = -10$, but
$b - a = 4 - (-6) = 10$.

$a = -18, b = -3$: $a - b = -18 - (-3) = -15$, but
$b - a = -3 - (-18) = 15$.

73. The process in your head should be like the following.

$$72 \cdot 17 + 28 \cdot 17 = 17(72 + 28)$$
$$= 17(100)$$
$$= 1700$$

113. $\dfrac{x^3}{y} > 0$

The quotient of two numbers is positive if they have the same sign (both positive or both negative). The sign of x^3 is the same as the sign of x. Therefore, $\dfrac{x^3}{y} > 0$ if x and y have the same sign.

R.3 Exercises (pages 28–31)

15. $-(4m^3n^0)^2 = -\left[4^2(m^3)^2(n^0)^2\right]$ Power rule 2

$\qquad = -(4^2)m^6n^0$ Power rule 1

$\qquad = -(4^2)m^6 \cdot 1$ Zero exponent

$\qquad = -4^2m^6,$ or $-16m^6$

43. $(6m^4 - 3m^2 + m) - (2m^3 + 5m^2 + 4m) + (m^2 - m)$

$= (6m^4 - 3m^2 + m) + (-2m^3 - 5m^2 - 4m)$
$\quad + (m^2 - m)$

$= 6m^4 - 2m^3 + (-3 - 5 + 1)m^2 + (1 - 4 - 1)m$

$= 6m^4 - 2m^3 - 7m^2 - 4m$

67. $\left[(2p - 3) + q\right]^2$

$= (2p - 3)^2 + 2(2p - 3)(q) + q^2$
   Square of a binomial, treating $(2p - 3)$ as one term

$= (2p)^2 - 2(2p)(3) + 3^2 + 2(2p - 3)(q) + q^2$
   Square the binomial $(2p - 3)$.

$= 4p^2 - 12p + 9 + 4pq - 6q + q^2$

87. $\dfrac{-4x^7 - 14x^6 + 10x^4 - 14x^2}{-2x^2}$

$= \dfrac{-4x^7}{-2x^2} + \dfrac{-14x^6}{-2x^2} + \dfrac{10x^4}{-2x^2} + \dfrac{-14x^2}{-2x^2}$

$= 2x^5 + 7x^4 - 5x^2 + 7$

R.4 Exercises (pages 38–40)

39. $24a^4 + 10a^3b - 4a^2b^2$

$= 2a^2(12a^2 + 5ab - 2b^2)$ Factor out the GCF, $2a^2$.

$= 2a^2(4a - b)(3a + 2b)$ Factor the trinomial.

47. $(a - 3b)^2 - 6(a - 3b) + 9$

$= \left[(a - 3b) - 3\right]^2$ Factor the perfect square trinomial.

$= (a - 3b - 3)^2$

69. $27 - (m + 2n)^3$

$= 3^3 - (m + 2n)^3$ Write as a difference of cubes.

$= \left[3 - (m + 2n)\right]\left[3^2 + 3(m + 2n) + (m + 2n)^2\right]$
   Factor the difference of cubes.

$= (3 - m - 2n)(9 + 3m + 6n + m^2 + 4mn + 4n^2)$
   Distributive property; square the binomial $(m + 2n)$.

81. $9(a - 4)^2 + 30(a - 4) + 25$

$= 9u^2 + 30u + 25$ Replace $a - 4$ with u.

$= (3u)^2 + 2(3u)(5) + 5^2$

$= (3u + 5)^2$ Factor the perfect square trinomial.

$= \left[3(a - 4) + 5\right]^2$ Replace u with $a - 4$.

$= (3a - 12 + 5)^2$ Distributive property

$= (3a - 7)^2$ Add.

R.5 Exercises (pages 47–49)

19. $\dfrac{8m^2 + 6m - 9}{16m^2 - 9} = \dfrac{(2m + 3)(4m - 3)}{(4m + 3)(4m - 3)}$ Factor.

$= \dfrac{2m + 3}{4m + 3}$ Lowest terms

35. $\dfrac{x^3 + y^3}{x^3 - y^3} \cdot \dfrac{x^2 - y^2}{x^2 + 2xy + y^2}$

$= \dfrac{(x + y)(x^2 - xy + y^2)}{(x - y)(x^2 + xy + y^2)} \cdot \dfrac{(x + y)(x - y)}{(x + y)(x + y)}$ Factor.

$= \dfrac{x^2 - xy + y^2}{x^2 + xy + y^2}$ Lowest terms

57. $\dfrac{4}{x+1} + \dfrac{1}{x^2-x+1} - \dfrac{12}{x^3+1}$

$= \dfrac{4}{x+1} + \dfrac{1}{x^2-x+1} - \dfrac{12}{(x+1)(x^2-x+1)}$

 Factor the sum of cubes.

$= \dfrac{4(x^2-x+1)}{(x+1)(x^2-x+1)} + \dfrac{1(x+1)}{(x+1)(x^2-x+1)}$

$\quad - \dfrac{12}{(x+1)(x^2-x+1)}$

 Write each fraction with the common denominator.

$= \dfrac{4(x^2-x+1) + 1(x+1) - 12}{(x+1)(x^2-x+1)}$

 Add and subtract numerators.

$= \dfrac{4x^2-4x+4+x+1-12}{(x+1)(x^2-x+1)}$

 Distributive property

$= \dfrac{4x^2-3x-7}{(x+1)(x^2-x+1)}$ Combine like terms.

$= \dfrac{(4x-7)(x+1)}{(x+1)(x^2-x+1)}$ Factor the numerator.

$= \dfrac{4x-7}{x^2-x+1}$ Lowest terms

75. $\dfrac{\dfrac{1}{x+h} - \dfrac{1}{x}}{h}$

Multiply both numerator and denominator by the LCD of all the fractions, $x(x+h)$.

$\dfrac{\dfrac{1}{x+h} - \dfrac{1}{x}}{h} = \dfrac{x(x+h)\left(\dfrac{1}{x+h} - \dfrac{1}{x}\right)}{x(x+h)(h)}$

$= \dfrac{x(x+h)\left(\dfrac{1}{x+h}\right) - x(x+h)\left(\dfrac{1}{x}\right)}{x(x+h)(h)}$

 Distributive property

$= \dfrac{x-(x+h)}{x(x+h)(h)}$ Divide common factors in the numerator.

$= \dfrac{-h}{x(x+h)(h)}$ Subtract.

$= \dfrac{-1}{x(x+h)}$ Lowest terms

R.6 Exercises (pages 55–58)

43. $\left(-\dfrac{64}{27}\right)^{1/3} = -\dfrac{4}{3}$ because $\left(-\dfrac{4}{3}\right)^3 = -\dfrac{64}{27}$.

71. $\dfrac{p^{1/5}p^{7/10}p^{1/2}}{(p^3)^{-1/5}} = \dfrac{p^{1/5+7/10+1/2}}{p^{-3/5}}$ Product rule; power rule 1

$= \dfrac{p^{2/10+7/10+5/10}}{p^{-6/10}}$ Write all fractions with the LCD, 10.

$= \dfrac{p^{14/10}}{p^{-6/10}}$ Add exponents.

$= p^{(14/10)-(-6/10)}$ Quotient rule

$= p^{20/10}$ Subtract and write exponent in lowest terms.

$= p^2$

81. $(r^{1/2} - r^{-1/2})^2 = (r^{1/2})^2 - 2(r^{1/2})(r^{-1/2}) + (r^{-1/2})^2$

 Square of a binomial

$= r - 2r^0 + r^{-1}$ Power rule 1; product rule

$= r - 2 \cdot 1 + r^{-1}$ Zero exponent

$= r - 2 + r^{-1}$, or $r - 2 + \dfrac{1}{r}$

 Negative exponent

103. $\dfrac{x - 9y^{-1}}{(x-3y^{-1})(x+3y^{-1})}$

$= \dfrac{x - \dfrac{9}{y}}{\left(x - \dfrac{3}{y}\right)\left(x + \dfrac{3}{y}\right)}$ Definition of negative exponent

$= \dfrac{x - \dfrac{9}{y}}{x^2 - \dfrac{9}{y^2}}$ Multiply in the denominator.

$= \dfrac{y^2\left(x - \dfrac{9}{y}\right)}{y^2\left(x^2 - \dfrac{9}{y^2}\right)}$ Multiply numerator and denominator by the LCD, y^2.

$= \dfrac{y^2x - 9y}{y^2x^2 - 9}$ Distributive property

$= \dfrac{y(xy - 9)}{x^2y^2 - 9}$ Factor the numerator.

R.7 Exercises (pages 66–69)

61. $\sqrt[4]{\dfrac{g^3h^5}{9r^6}} = \dfrac{\sqrt[4]{g^3h^5}}{\sqrt[4]{9r^6}}$ Quotient rule

$= \dfrac{\sqrt[4]{h^4(g^3h)}}{\sqrt[4]{r^4(9r^2)}}$ Factor out perfect fourth powers.

$= \dfrac{h\sqrt[4]{g^3h}}{r\sqrt[4]{9r^2}}$ Remove all perfect fourth powers from the radicals.

$= \dfrac{h\sqrt[4]{g^3h}}{r\sqrt[4]{9r^2}} \cdot \dfrac{\sqrt[4]{9r^2}}{\sqrt[4]{9r^2}}$ Rationalize the denominator.

$= \dfrac{h\sqrt[4]{9g^3hr^2}}{r\sqrt[4]{81r^4}}$ Product rule

$= \dfrac{h\sqrt[4]{9g^3hr^2}}{3r^2}$ Simplify.

79. $\left(\sqrt[3]{11} - 1\right)\left(\sqrt[3]{11^2} + \sqrt[3]{11} + 1\right)$

This product is a difference of cubes.

$$(x - y)(x^2 + xy + y^2) = x^3 - y^3$$

Thus,

$$\left(\sqrt[3]{11} - 1\right)\left(\sqrt[3]{11^2} + \sqrt[3]{11} + 1\right) = \left(\sqrt[3]{11}\right)^3 - 1^3$$

$$= 11 - 1$$

$$= 10.$$

81. $\left(\sqrt{3} + \sqrt{8}\right)^2 = \left(\sqrt{3}\right)^2 + 2\left(\sqrt{3}\right)\left(\sqrt{8}\right) + \left(\sqrt{8}\right)^2$

Square of a binomial

$$= 3 + 2\sqrt{24} + 8$$

$$= 3 + 2\sqrt{4 \cdot 6} + 8$$

$$= 3 + 2\left(2\sqrt{6}\right) + 8 \quad \text{Product rule}$$

$$= 11 + 4\sqrt{6}$$

91. $\dfrac{-4}{\sqrt[3]{3}} + \dfrac{1}{\sqrt[3]{24}} - \dfrac{2}{\sqrt[3]{81}} = \dfrac{-4}{\sqrt[3]{3}} + \dfrac{1}{\sqrt[3]{8 \cdot 3}} - \dfrac{2}{\sqrt[3]{27 \cdot 3}}$

$$= \dfrac{-4}{\sqrt[3]{3}} + \dfrac{1}{2\sqrt[3]{3}} - \dfrac{2}{3\sqrt[3]{3}}$$

Simplify radicals.

$$= \dfrac{-4 \cdot 6}{\sqrt[3]{3} \cdot 6} + \dfrac{1 \cdot 3}{2\sqrt[3]{3} \cdot 3} - \dfrac{2 \cdot 2}{3\sqrt[3]{3} \cdot 2}$$

Write fractions with common denominator, $6\sqrt[3]{3}$.

$$= \dfrac{-24}{6\sqrt[3]{3}} + \dfrac{3}{6\sqrt[3]{3}} - \dfrac{4}{6\sqrt[3]{3}}$$

$$= \dfrac{-25}{6\sqrt[3]{3}}$$

$$= \dfrac{-25}{6\sqrt[3]{3}} \cdot \dfrac{\sqrt[3]{3^2}}{\sqrt[3]{3^2}}$$

Rationalize the denominator.

$$= \dfrac{-25\sqrt[3]{9}}{6\sqrt[3]{27}}$$

$$= \dfrac{-25\sqrt[3]{9}}{6 \cdot 3}$$

$$= \dfrac{-25\sqrt[3]{9}}{18}$$

95. $\dfrac{\sqrt{7} - 1}{2\sqrt{7} + 4\sqrt{2}}$

$$= \dfrac{\sqrt{7} - 1}{2\sqrt{7} + 4\sqrt{2}} \cdot \dfrac{2\sqrt{7} - 4\sqrt{2}}{2\sqrt{7} - 4\sqrt{2}}$$

Multiply numerator and denominator by the conjugate of the denominator.

$$= \dfrac{\left(\sqrt{7} - 1\right)\left(2\sqrt{7} - 4\sqrt{2}\right)}{\left(2\sqrt{7} + 4\sqrt{2}\right)\left(2\sqrt{7} - 4\sqrt{2}\right)}$$

Multiply numerators. Multiply denominators.

$$= \dfrac{\sqrt{7} \cdot 2\sqrt{7} - \sqrt{7} \cdot 4\sqrt{2} - 1 \cdot 2\sqrt{7} + 1 \cdot 4\sqrt{2}}{\left(2\sqrt{7}\right)^2 - \left(4\sqrt{2}\right)^2}$$

Use FOIL in the numerator. Find the product of the sum and difference of two terms in the denominator.

$$= \dfrac{2 \cdot 7 - 4\sqrt{14} - 2\sqrt{7} + 4\sqrt{2}}{4 \cdot 7 - 16 \cdot 2}$$

$$= \dfrac{14 - 4\sqrt{14} - 2\sqrt{7} + 4\sqrt{2}}{-4}$$

$$= \dfrac{-2\left(-7 + 2\sqrt{14} + \sqrt{7} - 2\sqrt{2}\right)}{-2 \cdot 2}$$

Factor the numerator and denominator.

$$= \dfrac{-7 + 2\sqrt{14} + \sqrt{7} - 2\sqrt{2}}{2}$$

Chapter 1 Equations and Inequalities

1.1 Exercises *(pages 84–86)*

25. $0.5x + \dfrac{4}{3}x = x + 10$

$$\dfrac{1}{2}x + \dfrac{4}{3}x = x + 10 \quad \text{Change decimal to fraction.}$$

$$6\left(\dfrac{1}{2}x + \dfrac{4}{3}x\right) = 6(x + 10) \quad \text{Multiply by the LCD, 6.}$$

$$3x + 8x = 6x + 60 \quad \text{Distributive property}$$

$$11x = 6x + 60 \quad \text{Combine like terms.}$$

$$5x = 60 \quad \text{Subtract } 6x.$$

$$x = 12 \quad \text{Divide by 5.}$$

Solution set: $\{12\}$

33. $0.3(x + 2) - 0.5(x + 2) = -0.2x - 0.4$

$$10\left[0.3(x + 2) - 0.5(x + 2)\right] = 10(-0.2x - 0.4)$$

Multiply by 10 to clear decimals.

$$3(x + 2) - 5(x + 2) = -2x - 4$$

Distributive property

$$3x + 6 - 5x - 10 = -2x - 4$$

Distributive property

$$-2x - 4 = -2x - 4$$

Combine like terms.

$$0 = 0 \quad \text{Add } 2x. \text{ Add } 4.$$

$0 = 0$ is a true statement, so the equation is an identity.

Solution set: $\{\text{all real numbers}\}$

57. $3x = (2x - 1)(m + 4), \quad \text{for } x$

$$3x = 2xm + 8x - m - 4 \quad \text{FOIL}$$

$$m + 4 = 2xm + 5x \quad \text{Add } m + 4. \text{ Subtract } 3x.$$

$$m + 4 = x(2m + 5) \quad \text{Factor out } x.$$

$$\dfrac{m + 4}{2m + 5} = x, \quad \text{or} \quad x = \dfrac{m + 4}{2m + 5} \quad \text{Divide by } 2m + 5.$$

1.2 Exercises (pages 92–97)

17. Let h = the height of the box. Use the formula for the surface area of a rectangular box.

$$S = 2lw + 2wh + 2hl$$
$$496 = 2 \cdot 18 \cdot 8 + 2 \cdot 8 \cdot h + 2 \cdot h \cdot 18$$

Let $S = 496$, $l = 18$, $w = 8$.

$496 = 288 + 16h + 36h$	Multiply.
$496 = 288 + 52h$	Combine like terms.
$208 = 52h$	Subtract 288.
$4 = h$	Divide by 52.

The height of the box is 4 ft.

21. Let x = David's biking speed.

Then $x + 4.5$ = David's driving speed.

Summarize the given information in a table, using the equation $d = rt$. Because the speeds are given in miles per hour, the times must be changed from minutes to hours.

	r	t	d
Driving	$x + 4.5$	$\frac{1}{3}$	$\frac{1}{3}(x + 4.5)$
Biking	x	$\frac{3}{4}$	$\frac{3}{4}x$

Distance driving = Distance biking

$$\frac{1}{3}(x + 4.5) = \frac{3}{4}x$$
$$12\left(\frac{1}{3}(x + 4.5)\right) = 12\left(\frac{3}{4}x\right)$$

Multiply by the LCD, 12.

$4(x + 4.5) = 9x$	Multiply.
$4x + 18 = 9x$	Distributive property
$18 = 5x$	Subtract $4x$.
$\dfrac{18}{5} = x$	Divide by 5.

Now find the distance.

$$d = rt = \frac{3}{4}x = \frac{3}{4}\left(\frac{18}{5}\right) = \frac{27}{10} = 2.7$$

David travels 2.7 mi to work.

33. Let x = the number of milliliters of water to be added.

Strength	Milliliters of Solution	Milliliters of Salt
6%	8	0.06(8)
0%	x	$0(x)$
4%	$8 + x$	$0.04(8 + x)$

The number of milliliters of salt in the 6% solution plus the number of milliliters of salt in the water (0% solution) must equal the number of milliliters of salt in the 4% solution.

$$0.06(8) + 0(x) = 0.04(8 + x)$$
$$0.48 = 0.32 + 0.04x$$
$$0.16 = 0.04x$$
$$4 = x$$

To reduce the saline concentration to 4%, 4 mL of water should be added.

43. (a) Since 1 ft^2 of plywood flooring can emit 140 μg/hr of formaldehyde, 100 ft^2 of flooring can emit

$$100 \text{ ft}^2 \times \frac{140 \; \mu\text{g/hr}}{1 \text{ ft}^2} = 14{,}000 \; \mu\text{g/hr}.$$

Thus, the linear equation that computes the amount of formaldehyde, in micrograms, emitted into the room in x hours is

$$F = 14{,}000x.$$

(b) The amount of formaldehyde necessary to produce a concentration of 33 μg/ft^3 in a room with 800 ft^3 of air is

$$800 \text{ ft}^3 \times 33 \; \mu\text{g/ft}^3 = 26{,}400 \; \mu\text{g}.$$

In the linear equation from part (a), substitute this value for F and solve for x.

$$F = 14{,}000x$$
$$26{,}400 = 14{,}000x$$
$$x \approx 1.9 \text{ hr}$$

1.3 Exercises (pages 102–104)

49. $-i\sqrt{2} - 2 - \left(6 - 4i\sqrt{2}\right) - \left(5 - i\sqrt{2}\right)$

$$= (-2 - 6 - 5) + \left[-\sqrt{2} - \left(-4\sqrt{2}\right) - \left(-\sqrt{2}\right)\right]i$$
$$= -13 + 4i\sqrt{2}$$

67. $(2 + i)(2 - i)(4 + 3i)$

$= \left[(2 + i)(2 - i)\right](4 + 3i)$	Associative property
$= (2^2 - i^2)(4 + 3i)$	Product of the sum and difference of two terms
$= \left[4 - (-1)\right](4 + 3i)$	$i^2 = -1$
$= 5(4 + 3i)$	
$= 20 + 15i$	Distributive property

95.

$\dfrac{1}{i^{-11}} = i^{11}$	Negative exponent rule
$= i^8 \cdot i^3$	$a^{m+n} = a^m \cdot a^n$
$= (i^4)^2 \cdot i^3$	$a^{mn} = (a^m)^n$
$= 1(-i)$	$(i^4)^2 = 1^2 = 1$
$= -i$	Multiply.

99.

$$\left(\frac{\sqrt{2}}{2} + \frac{\sqrt{2}}{2}i\right)^2 = \left(\frac{\sqrt{2}}{2}\right)^2 + 2 \cdot \frac{\sqrt{2}}{2} \cdot \frac{\sqrt{2}}{2}i + \left(\frac{\sqrt{2}}{2}i\right)^2$$

Square of a binomial;

$$= \frac{2}{4} + 2 \cdot \frac{2}{4}i + \frac{2}{4}i^2$$ Apply exponents and multiply.

$$= \frac{1}{2} + i + \frac{1}{2}i^2$$ Simplify.

$$= \frac{1}{2} + i + \frac{1}{2}(-1)$$ $i^2 = -1$

$$= \frac{1}{2} + i - \frac{1}{2}$$ Multiply and combine real parts.

$$= i$$

Thus, $\frac{\sqrt{2}}{2} + \frac{\sqrt{2}}{2}i$ is a square root of i.

1.4 Exercises *(pages 111–113)*

17. $-4x^2 + x = -3$

$$0 = 4x^2 - x - 3$$ Standard form

$$0 = (4x + 3)(x - 1)$$ Factor.

$$4x + 3 = 0 \quad \text{or} \quad x - 1 = 0$$

Zero-factor property

$$x = -\frac{3}{4} \quad \text{or} \quad x = 1$$

Solution set: $\left\{-\frac{3}{4}, 1\right\}$

59. $\frac{1}{2}x^2 + \frac{1}{4}x - 3 = 0$

$$4\left(\frac{1}{2}x^2 + \frac{1}{4}x - 3\right) = 4 \cdot 0$$ Multiply by the LCD, 4.

$$2x^2 + x - 12 = 0$$ Distributive property; standard form

$$x = \frac{-b \pm \sqrt{b^2 - 4ac}}{2a}$$

Quadratic formula

$$x = \frac{-1 \pm \sqrt{1^2 - 4(2)(-12)}}{2(2)}$$

$a = 2, \ b = 1, \ c = -12$

$$x = \frac{-1 \pm \sqrt{97}}{4}$$

Solution set: $\left\{\dfrac{-1 \pm \sqrt{97}}{4}\right\}$

79. $4x^2 - 2xy + 3y^2 = 2$

(a) Solve for x in terms of y.

$$4x^2 - 2yx + 3y^2 - 2 = 0$$ Standard form

$$4x^2 - (2y)x + (3y^2 - 2) = 0$$

$a = 4, \ b = -2y, \ c = 3y^2 - 2$

$$x = \frac{-b \pm \sqrt{b^2 - 4ac}}{2a}$$

$$= \frac{-(-2y) \pm \sqrt{(-2y)^2 - 4(4)(3y^2 - 2)}}{2(4)}$$

$$= \frac{2y \pm \sqrt{4y^2 - 16(3y^2 - 2)}}{8}$$

$$= \frac{2y \pm \sqrt{4y^2 - 48y^2 + 32}}{8}$$

$$= \frac{2y \pm \sqrt{32 - 44y^2}}{8}$$

$$= \frac{2y \pm \sqrt{4(8 - 11y^2)}}{8}$$

$$= \frac{2y \pm 2\sqrt{8 - 11y^2}}{8}$$

$$= \frac{2\left(y \pm \sqrt{8 - 11y^2}\right)}{2(4)}$$

$$x = \frac{y \pm \sqrt{8 - 11y^2}}{4}$$

(b) Solve for y in terms of x.

$$3y^2 - 2xy + 4x^2 - 2 = 0$$ Standard form

$$3y^2 - (2x)y + (4x^2 - 2) = 0$$

$a = 3, \ b = -2x, \ c = 4x^2 - 2$

$$y = \frac{-b \pm \sqrt{b^2 - 4ac}}{2a}$$

$$= \frac{-(-2x) \pm \sqrt{(-2x)^2 - 4(3)(4x^2 - 2)}}{2(3)}$$

$$= \frac{2x \pm \sqrt{4x^2 - 12(4x^2 - 2)}}{6}$$

$$= \frac{2x \pm \sqrt{4x^2 - 48x^2 + 24}}{6}$$

$$= \frac{2x \pm \sqrt{24 - 44x^2}}{6}$$

$$= \frac{2x \pm \sqrt{4(6 - 11x^2)}}{6}$$

$$= \frac{2x \pm 2\sqrt{6 - 11x^2}}{6}$$

$$= \frac{2\left(x \pm \sqrt{6 - 11x^2}\right)}{2(3)}$$

$$y = \frac{x \pm \sqrt{6 - 11x^2}}{3}$$

95. $x = 4 \quad \text{or} \quad x = 5$

$$x - 4 = 0 \quad \text{or} \quad x - 5 = 0$$ Zero-factor property in reverse

$$(x - 4)(x - 5) = 0$$

$$x^2 - 9x + 20 = 0$$ Multiply.

$$a = 1, \ b = -9, \ c = 20$$

(Any nonzero constant multiple of these numbers will also work.)

1.5 Exercises *(pages 118–124)*

25.

$$S = 2\pi rh + 2\pi r^2 \qquad \text{Surface area of a cylinder}$$
$$8\pi = 2\pi r \cdot 3 + 2\pi r^2 \qquad \text{Let } S = 8\pi, h = 3.$$
$$8\pi = 6\pi r + 2\pi r^2 \qquad \text{Multiply.}$$
$$0 = 2\pi r^2 + 6\pi r - 8\pi \qquad \text{Standard form}$$
$$0 = r^2 + 3r - 4 \qquad \text{Divide by } 2\pi.$$
$$0 = (r + 4)(r - 1) \qquad \text{Factor.}$$
$$r + 4 = 0 \quad \text{or} \quad r - 1 = 0 \qquad \text{Zero-factor property}$$
$$r = -4 \quad \text{or} \qquad r = 1 \qquad \text{Solve each equation.}$$

Because r represents the radius of a cylinder, -4 is not reasonable. The radius is 1 ft.

33. Let $r =$ the radius of the circle.

$$\mathcal{A} = s^2 \qquad \text{Area of square}$$
$$800 = s^2 \qquad \text{Let } \mathcal{A} = 800.$$
$$800 = (2r)^2 \qquad \text{Let } s = 2r.$$
$$800 = 4r^2 \qquad (ab)^2 = a^2b^2$$
$$200 = r^2 \qquad \text{Divide by 4.}$$
$$r = \pm\sqrt{200} = \pm 10\sqrt{2}$$

Because r represents the radius of a circle, $-10\sqrt{2}$ is not reasonable. The radius is $10\sqrt{2}$ ft.

55. Let $x =$ number of passengers in excess of 75.
Then $225 - 5x =$ the cost per passenger (in dollars)
and $75 + x =$ the number of passengers.
(Number of passengers)(Cost per passenger)
$$= \text{Revenue}$$
$$(75 + x)(225 - 5x) = 16{,}000$$
$$16{,}875 - 150x - 5x^2 = 16{,}000$$
$$0 = 5x^2 + 150x - 875$$
$$0 = x^2 + 30x - 175$$
$$0 = (x + 35)(x - 5)$$
$$x + 35 = 0 \quad \text{or} \quad x - 5 = 0$$
$$x = -35 \quad \text{or} \qquad x = 5$$

The negative solution is not meaningful. Since there are 5 passengers in excess of 75, the total number of passengers is 80.

1.6 Exercises *(pages 133–136)*

13.
$$\frac{4}{x^2 + x - 6} - \frac{1}{x^2 - 4} = \frac{2}{x^2 + 5x + 6}$$

$$\frac{4}{(x+3)(x-2)} - \frac{1}{(x+2)(x-2)} = \frac{2}{(x+3)(x+2)}$$
$$\qquad\qquad\qquad\qquad \text{Factor denominators.}$$

$$(x+3)(x-2)(x+2)\left(\frac{4}{(x+3)(x-2)} - \frac{1}{(x+2)(x-2)}\right)$$

$$= (x+3)(x-2)(x+2)\left(\frac{2}{(x+3)(x+2)}\right)$$

Multiply by the LCD, $(x+3)(x-2)(x+2)$, where $x \neq -3$, $x \neq 2$, $x \neq -2$.

$$4(x+2) - 1(x+3) = 2(x-2)$$
$$\qquad\qquad\qquad\qquad \text{Simplify each side.}$$

$$4x + 8 - x - 3 = 2x - 4$$
$$\qquad\qquad\qquad \text{Distributive property}$$
$$3x + 5 = 2x - 4$$
$$\qquad\qquad \text{Combine like terms.}$$
$$x = -9$$
$$\qquad\qquad \text{Subtract } 2x \text{ and subtract 5.}$$

The restrictions $x \neq -3$, $x \neq 2$, $x \neq -2$ do not affect the proposed solution.

Solution set: $\{-9\}$

31. Let $x =$ the number of hours to fill the pool with both pipes open.

	Rate	Time	Part of the Job Accomplished
Inlet Pipe	$\frac{1}{5}$	x	$\frac{1}{5}x$
Outlet Pipe	$\frac{1}{8}$	x	$\frac{1}{8}x$

Filling the pool is 1 whole job, but because the outlet pipe empties the pool, its contribution should be subtracted from the contribution of the inlet pipe.

$$\frac{1}{5}x - \frac{1}{8}x = 1$$
$$40\left(\frac{1}{5}x - \frac{1}{8}x\right) = 40 \cdot 1 \qquad \text{Multiply by the LCD, 40.}$$
$$8x - 5x = 40 \qquad \text{Distributive property}$$
$$3x = 40 \qquad \text{Combine like terms.}$$
$$x = \frac{40}{3} = 13\frac{1}{3} \qquad \text{Divide by 3.}$$

It took $13\frac{1}{3}$ hr to fill the pool.

53.
$$\sqrt{2\sqrt{7x+2}} = \sqrt{3x+2}$$
$$\left(\sqrt{2\sqrt{7x+2}}\right)^2 = \left(\sqrt{3x+2}\right)^2 \qquad \text{Square each side.}$$
$$2\sqrt{7x+2} = 3x+2$$
$$\left(2\sqrt{7x+2}\right)^2 = (3x+2)^2 \qquad \text{Square each side again.}$$
$$4(7x+2) = 9x^2 + 12x + 4$$
$$\qquad\qquad\qquad \text{Square the binomial on the right.}$$
$$28x + 8 = 9x^2 + 12x + 4$$
$$0 = 9x^2 - 16x - 4$$
$$0 = (9x + 2)(x - 2)$$
$$9x + 2 = 0 \quad \text{or} \quad x - 2 = 0$$
$$x = -\frac{2}{9} \quad \text{or} \qquad x = 2$$

Check both proposed solutions by substituting first $-\frac{2}{9}$ and then 2 in the *original* equation. These checks will verify that both of these numbers are solutions.

Solution set: $\left\{-\frac{2}{9}, 2\right\}$

89. $x^{-2/3} + x^{-1/3} - 6 = 0$

Since $(x^{-1/3})^2 = x^{-2/3}$, let $u = x^{-1/3}$.

$u^2 + u - 6 = 0$ Substitute.

$(u + 3)(u - 2) = 0$ Factor.

$u + 3 = 0$ or $u - 2 = 0$

$u = -3$ or $u = 2$

Now replace u with $x^{-1/3}$.

$x^{-1/3} = -3$ or $x^{-1/3} = 2$

$(x^{-1/3})^{-3} = (-3)^{-3}$ or $(x^{-1/3})^{-3} = 2^{-3}$

Raise each side of each equation to the -3 power.

$x = \dfrac{1}{(-3)^3}$ or $x = \dfrac{1}{2^3}$

$x = -\dfrac{1}{27}$ or $x = \dfrac{1}{8}$

A check will show that both $-\frac{1}{27}$ and $\frac{1}{8}$ satisfy the original equation.

Solution set: $\left\{ -\frac{1}{27}, \frac{1}{8} \right\}$

99. $m^{3/4} + n^{3/4} = 1$,   for m

$m^{3/4} = 1 - n^{3/4}$

$(m^{3/4})^{4/3} = (1 - n^{3/4})^{4/3}$ Raise each side to the $\frac{4}{3}$ power.

$m = (1 - n^{3/4})^{4/3}$

1.7 Exercises *(pages 145–149)*

51. $x^2 - 2x \le 1$

Step 1 Solve the corresponding quadratic equation.

$x^2 - 2x = 1$

$x^2 - 2x - 1 = 0$

The trinomial $x^2 - 2x - 1$ cannot be factored, so solve this equation with the quadratic formula.

$x = \dfrac{-b \pm \sqrt{b^2 - 4ac}}{2a}$

$x = \dfrac{-(-2) \pm \sqrt{(-2)^2 - 4(1)(-1)}}{2(1)}$

$a = 1,\ b = -2,\ c = -1$

Simplify to obtain $x = 1 \pm \sqrt{2}$—that is, $x = 1 - \sqrt{2}$ or $x = 1 + \sqrt{2}$.

Step 2 Identify the intervals determined by the solutions of the equation. The values $1 - \sqrt{2}$ and $1 + \sqrt{2}$ divide a number line into three intervals: $(-\infty, 1 - \sqrt{2})$, $(1 - \sqrt{2}, 1 + \sqrt{2})$, and $(1 + \sqrt{2}, \infty)$. Use solid circles on $1 - \sqrt{2}$ and $1 + \sqrt{2}$ because the inequality symbol includes equality. (Note that $1 - \sqrt{2} \approx -0.4$ and $1 + \sqrt{2} \approx 2.4$.)

Step 3 Use a test value from each interval to determine which intervals form the solution set.

Interval	Test Value	Is $x^2 - 2x \le 1$ True or False?
A: $(-\infty, 1 - \sqrt{2})$	-1	$(-1)^2 - 2(-1) \overset{?}{\le} 1$ $\quad 3 \le 1$ **False**
B: $(1 - \sqrt{2}, 1 + \sqrt{2})$	0	$0^2 - 2(0) \overset{?}{\le} 1$ $\quad 0 \le 1$ **True**
C: $(1 + \sqrt{2}, \infty)$	3	$3^2 - 2(3) \overset{?}{\le} 1$ $\quad 3 \le 1$ **False**

Only Interval B makes the inequality true. Both endpoints are included because the given inequality is a nonstrict inequality.

Solution set: $[1 - \sqrt{2}, 1 + \sqrt{2}]$

61. $4x - x^3 \ge 0$

Step 1 $4x - x^3 = 0$ Corresponding equation

$x(4 - x^2) = 0$ Factor out the GCF, x.

$x(2 - x)(2 + x) = 0$

Factor the difference of squares.

$x = 0$ or $2 - x = 0$ or $2 + x = 0$

Zero-factor property

$x = 0$ or $x = 2$ or $x = -2$

Step 2 The values -2, 0, and 2 divide the number line into four intervals.

Interval A: $(-\infty, -2)$; Interval B: $(-2, 0)$;
Interval C: $(0, 2)$; Interval D: $(2, \infty)$

Step 3 Use a test value from each interval to determine which intervals form the solution set.

Interval	Test Value	Is $4x - x^3 \ge 0$ True or False?
A: $(-\infty, -2)$	-3	$4(-3) - (-3)^3 \overset{?}{\ge} 0$ $\quad 15 \ge 0$ **True**
B: $(-2, 0)$	-1	$4(-1) - (-1)^3 \overset{?}{\ge} 0$ $\quad -3 \ge 0$ **False**
C: $(0, 2)$	1	$4(1) - 1^3 \overset{?}{\ge} 0$ $\quad 3 \ge 0$ **True**
D: $(2, \infty)$	3	$4(3) - 3^3 \overset{?}{\ge} 0$ $\quad -15 \ge 0$ **False**

Intervals A and C make the inequality true. The endpoints -2, 0, and 2 are all included because the given inequality is a nonstrict inequality.

Solution set: $(-\infty, -2] \cup [0, 2]$

79. $\dfrac{7}{x+2} \geq \dfrac{1}{x+2}$

Step 1 Rewrite the inequality so that 0 is on one side and there is a single fraction on the other side.

$$\frac{7}{x+2} - \frac{1}{x+2} \geq 0$$

$$\frac{6}{x+2} \geq 0$$

Step 2 Determine the values that will cause either the numerator or the denominator to equal 0.

Since $6 \neq 0$, the numerator is never equal to 0. The denominator is equal to 0 when $x+2=0$, or $x=-2$.

-2 divides a number line into two intervals, $(-\infty, -2)$ and $(-2, \infty)$. Use an open circle on -2 because it makes the denominator 0.

Interval A | Interval B
$(-\infty, -2)$ | $(-2, \infty)$

$$\begin{array}{cccc} + & + & \circ & + + + \rightarrow \\ & & -2 & \;\; 0 \end{array}$$

Step 3 Use a test value from each interval to determine which intervals form the solution set.

Interval	Test Value	Is $\dfrac{7}{x+2} \geq \dfrac{1}{x+2}$ True or False?
A: $(-\infty, -2)$	-3	$\dfrac{7}{-3+2} \overset{?}{\geq} \dfrac{1}{-3+2}$ $-7 \geq -1$ False
B: $(-2, \infty)$	0	$\dfrac{7}{0+2} \overset{?}{\geq} \dfrac{1}{0+2}$ $\dfrac{7}{2} \geq \dfrac{1}{2}$ True

Interval B satisfies the original inequality. The endpoint -2 is not included because it makes the denominator 0.

Solution set: $(-2, \infty)$

87. $\dfrac{2x-3}{x^2+1} \geq 0$

The inequality already has 0 on one side, so set the numerator and denominator equal to 0. If $2x-3=0$, then $x=\frac{3}{2}$. $x^2+1=0$ has no real solutions. $\frac{3}{2}$ divides a number line into two intervals, $\left(-\infty, \frac{3}{2}\right)$ and $\left(\frac{3}{2}, \infty\right)$.

Interval	Test Value	Is $\dfrac{2x-3}{x^2+1} \geq 0$ True or False?
A: $\left(-\infty, \frac{3}{2}\right)$	0	$\dfrac{2(0)-3}{0^2+1} \overset{?}{\geq} 0$ $-3 \geq 0$ False
B: $\left(\frac{3}{2}, \infty\right)$	2	$\dfrac{2(2)-3}{2^2+1} \overset{?}{\geq} 0$ $\dfrac{1}{5} \geq 0$ True

Interval B satisfies the original inequality, along with the endpoint $\frac{3}{2}$.

Solution set: $\left[\frac{3}{2}, \infty\right)$

1.8 Exercises *(pages 153–155)*

33. $\qquad 4|x-3| > 12$

$\qquad |x-3| > 3$ $\qquad\qquad$ Divide by 4.

$x-3 < -3 \quad$ or $\quad x-3 > 3 \qquad$ Case 3

$\quad x < 0 \qquad$ or $\qquad x > 6 \qquad$ Add 3.

Solution set: $(-\infty, 0) \cup (6, \infty)$

49. $\left|5x + \dfrac{1}{2}\right| - 2 < 5$

$\qquad \left|5x + \dfrac{1}{2}\right| < 7 \qquad\qquad$ Add 2.

$\qquad -7 < \quad 5x + \dfrac{1}{2} \quad < 7 \qquad$ Case 2

$\qquad 2(-7) < 2\left(5x + \dfrac{1}{2}\right) < 2(7)$

$\qquad\qquad\qquad\qquad\qquad$ Multiply each part by 2.

$\qquad -14 < \quad 10x + 1 \quad < 14$

$\qquad\qquad\qquad\qquad$ Distributive property

$\qquad -15 < 10x < 13 \qquad$ Subtract 1 from each part.

$\qquad \dfrac{-15}{10} < \ x \ < \dfrac{13}{10} \qquad$ Divide each part by 10.

$\qquad -\dfrac{3}{2} < \ x \ < \dfrac{13}{10} \qquad$ Lowest terms

Solution set: $\left(-\frac{3}{2}, \frac{13}{10}\right)$

63. $|2x+1| \leq 0$

Since the absolute value of a number is always non-negative, $|2x+1| < 0$ is never true, so $|2x+1| \leq 0$ is true only when $|2x+1| = 0$. Solve this equation.

$\qquad |2x+1| = 0$

$\qquad 2x+1 = 0 \qquad$ If $|a|=0$, then $a=0$.

$\qquad\qquad x = -\dfrac{1}{2} \quad$ Solve.

Solution set: $\left\{-\frac{1}{2}\right\}$

75. $|x^2+1| - |2x| = 0$

$\qquad |x^2+1| = |2x|$

If $|a|=|b|$, then $a=b$ or $a=-b$.

$\qquad x^2+1 = 2x \quad$ or $\qquad x^2+1 = -2x$

$\qquad x^2-2x+1 = 0 \quad$ or $\quad x^2+2x+1 = 0$

$\qquad (x-1)^2 = 0 \quad$ or $\qquad (x+1)^2 = 0$

$\qquad x-1 = 0 \quad$ or $\qquad x+1 = 0$

$\qquad\qquad x = 1 \quad$ or $\qquad\qquad x = -1$

Solution set: $\{-1, 1\}$

Chapter 2 Graphs and Functions

2.1 Exercises *(pages 178–180)*

17. $P(3\sqrt{2}, 4\sqrt{5}), Q(\sqrt{2}, -\sqrt{5})$

(a) $d(P, Q) = \sqrt{(\sqrt{2} - 3\sqrt{2})^2 + (-\sqrt{5} - 4\sqrt{5})^2}$

Let $x_1 = 3\sqrt{2}, y_1 = 4\sqrt{5}, x_2 = \sqrt{2}, y_2 = -\sqrt{5}$.

$= \sqrt{(-2\sqrt{2})^2 + (-5\sqrt{5})^2}$

$= \sqrt{8 + 125} \quad (ab)^2 = a^2 b^2; \left(\sqrt{a}\right)^2 = a$

$= \sqrt{133}$

(b) The midpoint M of segment PQ has these coordinates.

$M = \left(\dfrac{3\sqrt{2} + \sqrt{2}}{2}, \dfrac{4\sqrt{5} + (-\sqrt{5})}{2}\right) = \left(2\sqrt{2}, \dfrac{3\sqrt{5}}{2}\right)$

2.2 Exercises *(pages 185–187)*

49. Let $P(x, y)$ be a point whose distance from $A(1, 0)$ is $\sqrt{10}$ and whose distance from $B(5, 4)$ is also $\sqrt{10}$.

$d(P, A) = \sqrt{10}$, so

$\sqrt{(1 - x)^2 + (0 - y)^2} = \sqrt{10}$ Distance formula

$(1 - x)^2 + y^2 = 10.$ Square each side.

$d(P, B) = \sqrt{10}$, so

$\sqrt{(5 - x)^2 + (4 - y)^2} = \sqrt{10}$

$(5 - x)^2 + (4 - y)^2 = 10.$

Set these distances equal to each other. Remember the middle term when squaring each binomial.

$(1 - x)^2 + y^2 = (5 - x)^2 + (4 - y)^2$

$1 - 2x + x^2 + y^2 = 25 - 10x + x^2 + 16 - 8y + y^2$

$1 - 2x = 41 - 10x - 8y$ Subtract x^2 and y^2. Add.

$8y = 40 - 8x$ Solve for y.

$y = 5 - x$ Divide by 8.

Substitute $5 - x$ for y in the equation $(1 - x)^2 + y^2 = 10$ and solve for x.

$(1 - x)^2 + (5 - x)^2 = 10$

$1 - 2x + x^2 + 25 - 10x + x^2 = 10$

$2x^2 - 12x + 26 = 10$

$2x^2 - 12x + 16 = 0$

$x^2 - 6x + 8 = 0$

$(x - 2)(x - 4) = 0$

$x - 2 = 0 \quad$ or $\quad x - 4 = 0$

$x = 2 \quad$ or $\quad x = 4$

To find the corresponding values of y, substitute in the equation $y = 5 - x$.

If $x = 2$, then $y = 5 - 2 = 3$.

If $x = 4$, then $y = 5 - 4 = 1$.

The points satisfying the given conditions are $(2, 3)$ and $(4, 1)$.

2.4 Exercises *(pages 211–216)*

49. $5x - 2y = 10$

(a) Write the equation in slope-intercept form.

$5x - 2y = 10$

$-2y = -5x + 10$ Subtract $5x$.

$y = \dfrac{5}{2}x - 5$ Divide by -2.

The slope is $\frac{5}{2}$, the coefficient of x.

(b) Plot the y-intercept $(0, -5)$. The slope m is $\frac{5}{2}$, so move 5 units up and 2 units to the right to locate a second point, $(2, 0)$. Draw a line through those two points.

89. fixed cost = \$1650; cost = \$400; price of item = \$305

(a) $C(x) = 400x + 1650$ $m = 400, b = 1650$

(b) $R(x) = 305x$

(c) $P(x) = R(x) - C(x)$

$P(x) = 305x - (400x + 1650)$

$P(x) = -95x - 1650$

(d) $C(x) = R(x)$

$400x + 1650 = 305x$

$95x = -1650$

$x \approx -17.4$

This result indicates a negative "break-even point," but the number of units produced must be a positive number. A calculator graph of the lines $y = 400x + 1650$ and $y = 305x$ on the same screen or solving the inequality $305x < 400x + 1650$ will show that $R(x) < C(x)$ for all positive values of x (in fact, whenever x is greater than about -17.4). Do not produce the product since it is impossible to make a profit.

2.5 Exercises *(pages 226–230)*

15. The line passes through the points $(3, 0)$ and $(0, -2)$. Use these points to find the slope.

$$m = \dfrac{-2 - 0}{0 - 3} = \dfrac{-2}{-3} = \dfrac{2}{3}$$

The slope is $\frac{2}{3}$ and the y-intercept is $(0, -2)$, so the equation of the line in slope-intercept form is as follows.

$$y = \dfrac{2}{3}x - 2$$

53. (a) Find the slope of the line $3y + 2x = 6$.

$$3y + 2x = 6$$
$$3y = -2x + 6$$
$$y = -\frac{2}{3}x + 2$$

Thus, $m = -\frac{2}{3}$. A line parallel to $3y + 2x = 6$ will also have slope $-\frac{2}{3}$. Use the points $(4, -1)$ and $(k, 2)$ and the definition of slope.

$$\frac{2 - (-1)}{k - 4} = -\frac{2}{3}$$

Solve this equation for k.

$$\frac{3}{k - 4} = -\frac{2}{3}$$
$$3(k - 4)\left(\frac{3}{k - 4}\right) = 3(k - 4)\left(-\frac{2}{3}\right)$$
$$9 = -2(k - 4)$$
$$9 = -2k + 8$$
$$2k = -1$$
$$k = -\frac{1}{2}$$

(b) Find the slope of the line $2y - 5x = 1$.

$$2y - 5x = 1$$
$$2y = 5x + 1$$
$$y = \frac{5}{2}x + \frac{1}{2}$$

Thus, $m = \frac{5}{2}$. A line perpendicular to $2y - 5x = 1$ will have slope $-\frac{2}{5}$ since $\frac{5}{2}\left(-\frac{2}{5}\right) = -1$. Use the points $(4, -1)$ and $(k, 2)$ and the definition of slope.

$$\frac{2 - (-1)}{k - 4} = -\frac{2}{5}$$

Solve this equation for k.

$$\frac{3}{k - 4} = -\frac{2}{5}$$
$$5(k - 4)\left(\frac{3}{k - 4}\right) = 5(k - 4)\left(-\frac{2}{5}\right)$$
$$15 = -2(k - 4)$$
$$15 = -2k + 8$$
$$2k = -7$$
$$k = -\frac{7}{2}$$

77. $A(-1, 4), B(-2, -1), C(1, 14)$

For A and B, $m = \dfrac{-1 - 4}{-2 - (-1)} = \dfrac{-5}{-1} = 5$.

For B and C, $m = \dfrac{14 - (-1)}{1 - (-2)} = \dfrac{15}{3} = 5$.

For A and C, $m = \dfrac{14 - 4}{1 - (-1)} = \dfrac{10}{2} = 5$.

All three slopes are the same, so the three points are collinear.

2.6 Exercises *(pages 238–242)*

35. First, notice that for all x-values less than or equal to 0, the function value is -1. Next, notice that for all x-values greater than 0, the function value is 1. Thus, we have a function f that is defined in two pieces.

$$f(x) = \begin{cases} -1 & \text{if } x \leq 0 \\ 1 & \text{if } x > 0 \end{cases}$$

The domain is $(-\infty, \infty)$ and the range is $\{-1, 1\}$.

2.7 Exercises *(pages 253–257)*

91. $f(x) = 2x + 5$

Translate the graph of $f(x)$ up 2 units to obtain the graph of

$$t(x) = (2x + 5) + 2$$
$$= 2x + 7.$$

Now translate the graph of $t(x)$ left 3 units to obtain the graph of

$$g(x) = 2(x + 3) + 7$$
$$= 2x + 6 + 7$$
$$= 2x + 13.$$

(Note that if the original graph is first translated left 3 units and then up 2 units, the final result is the same.)

2.8 Exercises *(pages 266–271)*

25. (a) From the graph, $f(-1) = 0$ and $g(-1) = 3$.

$$(f + g)(-1) = f(-1) + g(-1)$$
$$= 0 + 3$$
$$= 3$$

(b) From the graph, $f(-2) = -1$ and $g(-2) = 4$.

$$(f - g)(-2) = f(-2) - g(-2)$$
$$= -1 - 4$$
$$= -5$$

(c) From the graph, $f(0) = 1$ and $g(0) = 2$.

$$(fg)(0) = f(0) \cdot g(0)$$
$$= 1 \cdot 2$$
$$= 2$$

(d) From the graph, $f(2) = 3$ and $g(2) = 0$.

$$\left(\frac{f}{g}\right)(2) = \frac{f(2)}{g(2)} = \frac{3}{0}, \quad \text{which is undefined.}$$

39. $f(x) = \dfrac{1}{x}$

(a) $f(x + h) = \dfrac{1}{x + h}$

(b) $f(x + h) - f(x) = \dfrac{1}{x + h} - \dfrac{1}{x}$ Substitute.

$$= \frac{x - (x + h)}{x(x + h)}$$

Use a common denominator.

$$= \frac{-h}{x(x + h)}$$ Simplify.

(c) $\dfrac{f(x+h)-f(x)}{h} = \dfrac{-h}{x(x+h)} \cdot \dfrac{1}{h}$

Substitute in the numerator.
Multiply by the reciprocal of the denominator.

$= \dfrac{-1}{x(x+h)}$

Multiply. Write in lowest terms.

79. $g(f(2)) = g(1) = 2;\ g(f(3)) = g(2) = 5$
Since $g(f(1)) = 7$ and $f(1) = 3,\ g(3) = 7$.
Completed table:

x	$f(x)$	$g(x)$	$g(f(x))$
1	3	2	7
2	1	5	2
3	2	7	5

Chapter 3 Polynomial and Rational Functions

3.1 Exercises *(pages 292–300)*

3. $f(x) = -2(x+3)^2 + 2$

(a) domain: $(-\infty, \infty)$
  range: $(-\infty, 2]$

(b) vertex: $(h, k) = (-3, 2)$

(c) axis: $x = -3$

(d) $y = -2(0+3)^2 + 2$ Let $x = 0$.
  $y = -16$
  y-intercept: $(0, -16)$

(e) $0 = -2(x+3)^2 + 2$ Set $f(x) = 0$.
  $2(x+3)^2 = 2$ Add $2(x+3)^2$.
  $(x+3)^2 = 1$ Divide by 2.
  $x+3 = \pm\sqrt{1}$ Take square roots.
  $x+3 = 1$ or $x+3 = -1$
  $x = -2$ or $x = -4$
  x-intercepts: $(-4, 0),\ (-2, 0)$

59. $y = \dfrac{-16x^2}{0.434v^2} + 1.15x + 8$

(a) $10 = \dfrac{-16(15)^2}{0.434v^2} + 1.15(15) + 8$
  Let $y = 10,\ x = 15$.

  $10 = \dfrac{-3600}{0.434v^2} + 17.25 + 8$

  $\dfrac{3600}{0.434v^2} = 15.25$ Isolate the variable term.

  $3600 = 6.6185v^2$ Multiply by $0.434v^2$.

  $v^2 = \dfrac{3600}{6.6185}$ Divide by 6.6185.

  $v = \pm\sqrt{\dfrac{3600}{6.6185}} \approx \pm 23.32$
  Take square roots.

Because v represents velocity, only the positive square root is meaningful. The basketball should have an initial velocity of approximately 23.32 ft per sec.

(b) $y = \dfrac{-16x^2}{0.434(23.32)^2} + 1.15x + 8$

Since the coefficient of x^2 is negative, the graph of this function is a parabola that opens downward. Find the coordinates of the vertex to answer the question.

$$x = -\dfrac{b}{2a} = -\dfrac{1.15}{2\left(\dfrac{-16}{0.434(23.32)^2}\right)} \approx 8.48$$

The y-value of the vertex is found by evaluating the function at $x = 8.48$.

$$f(8.48) = \dfrac{-16(8.48)^2}{0.434(23.32)^2} + 1.15(8.48) + 8$$

$$\approx 12.88$$

The maximum height of the basketball is approximately 12.88 ft. This is verified by the calculator screen.

69. $y = x^2 - 10x + c$
An x-intercept occurs where $y = 0$, or

$$0 = x^2 - 10x + c.$$

There will be exactly one x-intercept if this equation has exactly one solution, meaning that the discriminant is 0.

$$b^2 - 4ac = 0$$
$$(-10)^2 - 4(1)c = 0$$
$$100 = 4c$$
$$c = 25$$

3.2 Exercises *(pages 305–306)*

7. $\dfrac{x^5 + 3x^4 + 2x^3 + 2x^2 + 3x + 1}{x + 2}$

$x + 2 = x - (-2)$

$-2)$	1	3	2	2	3	1
		-2	-2	0	-4	2
	1	1	0	2	-1	3

In the last row of the synthetic division, all numbers except the last one give the coefficients of the quotient, and the last number gives the remainder. The quotient is

$$1x^4 + 1x^3 + 0x^2 + 2x - 1, \quad \text{or} \quad x^4 + x^3 + 2x - 1,$$

and the remainder is 3. Thus,

$$\frac{x^5 + 3x^4 + 2x^3 + 2x^2 + 3x + 1}{x + 2}$$

$$= x^4 + x^3 + 2x - 1 + \frac{3}{x + 2}.$$

19. $f(x) = 2x^3 + x^2 + x - 8; \quad k = -1$

$$
\begin{array}{r|rrrr}
-1) & 2 & 1 & 1 & -8 \\
    &   & -2 & 1 & -2 \\
\hline
    & 2 & -1 & 2 & -10
\end{array}
$$

$$f(x) = (x + 1)(2x^2 - x + 2) + (-10)$$
$$f(x) = (x + 1)(2x^2 - x + 2) - 10$$

27. $f(x) = x^2 + 5x + 6; \quad k = -2$

$$
\begin{array}{r|rrr}
-2) & 1 & 5 & 6 \\
    &   & -2 & -6 \\
\hline
    & 1 & 3 & 0
\end{array}
$$

The last number in the bottom row of the synthetic division gives the remainder, 0. Therefore, by the remainder theorem, $f(-2) = 0$.

3.3 Exercises (pages 315–317)

23. $f(x) = x^3 + (7 - 3i)x^2 + (12 - 21i)x - 36i; \quad k = 3i$

$$
\begin{array}{r|rrrr}
3i) & 1 & 7 - 3i & 12 - 21i & -36i \\
    &   & 3i & 21i & 36i \\
\hline
    & 1 & 7 & 12 & 0
\end{array}
$$

The quotient is $x^2 + 7x + 12$, so

$$f(x) = (x - 3i)(x^2 + 7x + 12)$$
$$f(x) = (x - 3i)(x + 4)(x + 3).$$

45. $f(x) = 3x(x - 2)(x + 3)(x^2 - 1)$

$0 = 3x(x - 2)(x + 3)(x^2 - 1)$ Set $f(x) = 0$.

$0 = 3x(x - 2)(x + 3)(x - 1)(x + 1)$
       Factor the difference of squares.

$x = 0$ or $x - 2 = 0$ or $x + 3 = 0$ or $x - 1 = 0$ or $x + 1 = 0$
       Zero-factor property

$x = 0$ or $x = 2$ or $x = -3$ or $x = 1$ or $x = -1$

The zeros are 0, 2, −3, 1, and −1, all of multiplicity 1.

51. Zeros of −2, 1, and 0; $f(-1) = -1$

The factors of $f(x)$ are $x - (-2) = x + 2$, $x - 1$, and $x - 0 = x$.

$$f(x) = a(x + 2)(x - 1)(x)$$
$$f(-1) = a(-1 + 2)(-1 - 1)(-1) = -1$$
$$a(1)(-2)(-1) = -1$$
$$2a = -1$$
$$a = -\frac{1}{2}$$

Therefore,

$$f(x) = -\frac{1}{2}(x + 2)(x - 1)(x)$$

$$f(x) = -\frac{1}{2}(x^2 + x - 2)(x)$$

$$f(x) = -\frac{1}{2}(x^3 + x^2 - 2x)$$

$$f(x) = -\frac{1}{2}x^3 - \frac{1}{2}x^2 + x.$$

3.4 Exercises (pages 330–337)

37. $f(x) = x^3 + 5x^2 - x - 5$
   $f(x) = x^2(x + 5) - 1(x + 5)$
   $f(x) = (x + 5)(x^2 - 1)$ Factor by grouping.
   $f(x) = (x + 5)(x + 1)(x - 1)$ Factor the difference
                              of squares.

Find the real zeros of f. (All three zeros of this function are real.)

$x + 5 = 0$ or $x + 1 = 0$ or $x - 1 = 0$
                              Zero-factor property

$x = -5$ or $x = -1$ or $x = 1$

The zeros of f are −5, −1, and 1, so the x-intercepts of the graph are $(-5, 0)$, $(-1, 0)$, and $(1, 0)$. Plot these points.

Now find the y-intercept.

$$f(0) = 0^3 + 5 \cdot 0^2 - 0 - 5 = -5$$

Plot the y-intercept $(0, -5)$.

The dominating term of $f(x)$ is x^3, so the end behavior of the graph is ↗. Each zero is of multiplicity 1, indicating that the graph will cross the x-axis at each zero. Begin at either end of the graph with the appropriate end behavior and draw a smooth curve that crosses the x-axis at each zero, has a turning point between successive zeros, and passes through the y-intercept.

Additional points may be used to verify whether the graph is above or below the x-axis between the zeros and to add detail to the sketch of the graph. A typical selection of test points and the results are shown in the table.

Interval	Test Point x	Value of $f(x)$	Sign of $f(x)$	Graph Above or Below x-axis
$(-\infty, -5)$	-6	-35	Negative	Below
$(-5, -1)$	-2	9	Positive	Above
$(-1, 1)$	0	-5	Negative	Below
$(1, \infty)$	2	21	Positive	Above

$f(x) = x^3 + 5x^2 - x - 5$

83. $f(x) = x^3 + 4x^2 - 8x - 8$; $[-3.8, -3]$

Graph this function in a window that will produce a comprehensive graph, such as $[-10, 10]$ by $[-50, 50]$, with X-scale = 1, Y-scale = 10.

From this graph, we can see that the turning point in the interval $[-3.8, -3]$ is a maximum. Use "Maximum" from the CALC menu to approximate the coordinates of this turning point.

To the nearest hundredth, the turning point in the interval $[-3.8, -3]$ is $(-3.44, 26.15)$.

101. Use the following volume formulas:

$$V_{\text{cylinder}} = \pi r^2 h$$

$$V_{\text{hemisphere}} = \frac{1}{2} V_{\text{sphere}} = \frac{1}{2}\left(\frac{4}{3}\pi r^3\right) = \frac{2}{3}\pi r^3$$

$$\pi r^2 h + 2\left(\frac{2}{3}\pi r^3\right) = \text{Total volume of tank}$$

$$\pi x^2 (12) + \frac{4}{3}\pi x^3 = 144\pi \qquad \text{Let } V = 144\pi, h = 12, \text{ and } r = x.$$

$$\frac{4}{3}\pi x^3 + 12\pi x^2 - 144\pi = 0 \qquad \text{Subtract } 144\pi \text{ and rewrite.}$$

$$\frac{4}{3}x^3 + 12x^2 - 144 = 0 \qquad \text{Divide by } \pi.$$

$$4x^3 + 36x^2 - 432 = 0 \qquad \text{Multiply by 3.}$$

$$x^3 + 9x^2 - 108 = 0 \qquad \text{Divide by 4.}$$

Synthetic division or graphing $f(x) = x^3 + 9x^2 - 108$ with a graphing calculator will show that this function has zeros of -6 (multiplicity 2) and 3. Because x represents the radius of the hemispheres, a negative solution for the equation $x^3 + 9x^2 - 108 = 0$ is not meaningful. In order to get a volume of 144π ft^3, a radius of 3 ft should be used.

3.5 Exercises (pages 351–359)

81. $f(x) = \dfrac{1}{x^2 + 1}$

Because $x^2 + 1 = 0$ has no real solutions, the denominator is never 0, so there are no vertical asymptotes. As $|x| \to \infty$, $y \to 0$, and thus the line $y = 0$ (the x-axis) is the horizontal asymptote.

$f(0) = \dfrac{1}{0^2 + 1} = 1$, so the y-intercept is $(0, 1)$.

The equation $f(x) = 0$ has no solutions (notice that the numerator has no zeros), so there are no x-intercepts.

$f(-x) = \dfrac{1}{(-x)^2 + 1} = \dfrac{1}{x^2 + 1} = f(x)$, so the graph is symmetric with respect to the y-axis.

Use a table of values to obtain several additional points on the graph. Use the asymptote, the y-intercept, and these additional points (which reflect the symmetry of the graph) to sketch the graph of the function.

x	y
± 0.5	0.8
± 1	0.5
± 2	0.2

$f(x) = \dfrac{1}{x^2 + 1}$

89. $f(x) = \dfrac{x^2 + 2x}{2x - 1}$

Step 1 Find any vertical asymptotes.

$$2x - 1 = 0$$
$$x = \frac{1}{2}$$

Step 2 Find any horizontal or oblique asymptotes. Because the numerator has degree exactly one more than the denominator, there is an oblique asymptote. Because $2x - 1$ is not of the form $x - a$, use polynomial long division rather than synthetic division to find the equation of this asymptote.

$$
\begin{array}{r}
\frac{1}{2}x + \frac{5}{4} \\
2x - 1{\overline{\smash{\big)}\,x^2 + 2x + 0}} \\
\underline{x^2 - \frac{1}{2}x} \\
\frac{5}{2}x + 0 \\
\underline{\frac{5}{2}x - \frac{5}{4}} \\
\frac{5}{4}
\end{array}
$$

Disregard the remainder. The equation of the oblique asymptote is $y = \frac{1}{2}x + \frac{5}{4}$.

Step 3 Find the y-intercept.

$$f(0) = \frac{0^2 + 2(0)}{2(0) - 1} = \frac{0}{-1} = 0,$$

so the y-intercept is $(0, 0)$.

Step 4 Find the x-intercepts, if any, by finding the zeros of the numerator.

$$x^2 + 2x = 0$$
$$x(x + 2) = 0$$
$$x = 0 \quad \text{or} \quad x = -2$$

There are two x-intercepts, $(0, 0)$ and $(-2, 0)$.

Step 5 The graph does not intersect its oblique asymptote because the following equation has no solution.

$$\frac{x^2 + 2x}{2x - 1} = \frac{1}{2}x + \frac{5}{4}$$
$$x^2 + 2x = x^2 + 2x - \frac{5}{4} \quad \text{Multiply by } (2x - 1).$$
$$0 = -\frac{5}{4} \quad \text{False}$$

Step 6

x	y
-3	$-\frac{3}{7} \approx -0.43$
-1	$\frac{1}{3} \approx 0.33$
$\frac{1}{4} = 0.25$	$-\frac{9}{8} = -1.125$
1	3
3	3

Use the asymptotes, the intercepts, and these additional points to sketch the graph.

101. The graph has one vertical asymptote, $x = 2$, so $x - 2$ is a factor of the denominator of the rational expression. There is a point of discontinuity ("hole") in the graph at $x = -2$, so there is a factor of $x + 2$ in both numerator and denominator. There is one x-intercept, $(3, 0)$, so 3 is a zero of the numerator, which means that $x - 3$ is a factor of the numerator. Putting all of this information together, we have a possible function for the graph:

$$f(x) = \frac{(x - 3)(x + 2)}{(x - 2)(x + 2)}, \quad \text{or} \quad f(x) = \frac{x^2 - x - 6}{x^2 - 4}.$$

Note: From the second form of the function, we can see that the graph of this function has a horizontal asymptote of $y = 1$, which is consistent with the given graph.

3.6 Exercises *(pages 365–369)*

19. ***Step 1*** Write the general relationship among the variables as an equation. Use the constant k.

$$a = \frac{kmn^2}{y^3}$$

Step 2 Substitute $a = 9$, $m = 4$, $n = 9$, and $y = 3$ to find k.

$$9 = \frac{k \cdot 4 \cdot 9^2}{3^3}$$
$$9 = 12k$$
$$k = \frac{3}{4}$$

Step 3 Substitute this value of k into the equation from Step 1, obtaining a specific formula.

$$a = \frac{3}{4} \cdot \frac{mn^2}{y^3}$$

$$a = \frac{3mn^2}{4y^3}$$

Step 4 Substitute $m = 6$, $n = 2$, and $y = 5$ and solve for a.

$$a = \frac{3mn^2}{4y^3} = \frac{3 \cdot 6 \cdot 2^2}{4 \cdot 5^3} = \frac{18}{125}$$

35. *Step 1* Let F represent the force of the wind, $\mathcal{A}$ represent the area of the surface, and v represent the velocity of the wind.

$$F = k\mathcal{A}v^2$$

Step 2 $50 = k\left(\frac{1}{2}\right)(40)^2$ Let $F = 50$, $\mathcal{A} = \frac{1}{2}$, $v = 40$.

$$50 = 800k$$

$$k = \frac{50}{800}, \quad \text{or} \quad \frac{1}{16}$$

Step 3 $F = \frac{1}{16}\mathcal{A}v^2$

Step 4 $F = \frac{1}{16} \cdot 2 \cdot 80^2 = 800$

The force of the wind would be 800 lb.

43. Let t represent the Kelvin temperature.

$$R = kt^4$$

$$213.73 = k \cdot 293^4 \quad \text{Let } R = 213.73, \ t = 293.$$

$$k = \frac{213.73}{293^4} \approx 2.9 \times 10^{-8}$$

Thus, $R = (2.9 \times 10^{-8})t^4$.

If $t = 335$, then

$$R = (2.9 \times 10^{-8})(335^4) \approx 365.24.$$

Chapter 4 Inverse, Exponential, and Logarithmic Functions

4.1 Exercises *(pages 393–397)*

47. $f(x) = \dfrac{2}{x + 6}$, $g(x) = \dfrac{6x + 2}{x}$

$$(f \circ g)(x) = f(g(x)) \qquad \text{Definition}$$

$$= f\left(\frac{6x + 2}{x}\right) \qquad \text{Replace } g(x) \text{ with } \tfrac{6x + 2}{x}.$$

$$= \frac{2}{\dfrac{6x + 2}{x} + 6} \qquad \text{Use } f(x) = \tfrac{2}{x + 6}.$$

$$= \frac{2}{\dfrac{6x + 2 + 6x}{x}} \qquad \begin{array}{l}\text{Add terms in the}\\\text{denominator.}\end{array}$$

$$= \frac{2}{1} \cdot \frac{x}{12x + 2} \qquad \text{Definition of division}$$

$$= \frac{2x}{12x + 2} \qquad \text{Multiply.}$$

$$= \frac{2x}{2(6x + 1)} \qquad \text{Factor.}$$

$$= \frac{x}{6x + 1} \qquad \text{Lowest terms}$$

$$(f \circ g)(x) \neq x$$

Since $(f \circ g)(x) \neq x$, the functions are not inverses. It is not necessary to check $(g \circ f)(x)$.

69. $f(x) = \dfrac{1}{x - 3}$, $x \neq 3$

(a)

$$y = \frac{1}{x - 3}, \ x \neq 3 \qquad y = f(x)$$

$$x = \frac{1}{y - 3}, \ y \neq 3 \qquad \text{Interchange } x \text{ and } y.$$

$$x(y - 3) = 1 \qquad \text{Solve for } y.$$

$$xy - 3x = 1$$

$$xy = 1 + 3x$$

$$y = \frac{1 + 3x}{x}$$

$$f^{-1}(x) = \frac{1 + 3x}{x}, \ x \neq 0 \quad \text{Replace } y \text{ with } f^{-1}(x).$$

(b)

(c) For both f and f^{-1}, the domain contains all real numbers except those for which the denominator is equal to 0.

Domain of f = range of $f^{-1} = (-\infty, 3) \cup (3, \infty)$

Domain of f^{-1} = range of $f = (-\infty, 0) \cup (0, \infty)$

4.2 Exercises *(pages 409–413)*

7. $g(x) = \left(\dfrac{1}{4}\right)^x$

$$g(-2) = \left(\frac{1}{4}\right)^{-2} = 4^2 = 16$$

73. $\left(\dfrac{1}{e}\right)^{-x} = \left(\dfrac{1}{e^2}\right)^{x+1}$

$$(e^{-1})^{-x} = (e^{-2})^{x+1} \quad \text{Definition of negative exponent}$$

$$e^x = e^{-2(x+1)} \qquad (a^m)^n = a^{mn}$$

$$e^x = e^{-2x-2} \qquad \text{Distributive property}$$

$$x = -2x - 2 \qquad \text{Property (b)}$$

$$3x = -2 \qquad \text{Add } 2x.$$

$$x = -\frac{2}{3} \qquad \text{Divide by 3.}$$

Solution set: $\left\{-\frac{2}{3}\right\}$

75. $\left(\sqrt{2}\right)^{x+4} = 4^x$

$\left(2^{1/2}\right)^{x+4} = \left(2^2\right)^x$ Definition of $a^{1/n}$; Write each side as a power of a common base.

$2^{(1/2)(x+4)} = 2^{2x}$ $(a^m)^n = a^{mn}$

$2^{(1/2)x+2} = 2^{2x}$ Distributive property

$\dfrac{1}{2}x + 2 = 2x$ Property (b)

$2 = \dfrac{3}{2}x$ Subtract $\frac{1}{2}x$.

$\dfrac{4}{3} = x$ Multiply by $\frac{2}{3}$.

Solution set: $\left\{\frac{4}{3}\right\}$

77. $\dfrac{1}{27} = x^{-3}$

$3^{-3} = x^{-3}$ Write $\frac{1}{27}$ as a power of 3.

$x = 3$ Property of logarithms

Solution set: $\{3\}$

Alternative solution:

$\dfrac{1}{27} = x^{-3}$

$\dfrac{1}{27} = \dfrac{1}{x^3}$ Definition of negative exponent

$27 = x^3$ Take reciprocals.

$x = \sqrt[3]{27}$ Find cube roots.

$x = 3$ Evaluate.

Solution set: $\{3\}$

4.3 Exercises (pages 422–425)

23. $\log_x 25 = -2$

We must consider only positive values for x, $x \neq 1$, because it is the base of the logarithm.

$x^{-2} = 25$ Write in exponential form.

$\left(x^{-2}\right)^{-1/2} = 25^{-1/2}$ Raise each side to the same power.

$x = \dfrac{1}{25^{1/2}}$ Definition of negative exponent

$x = \dfrac{1}{5}$ $25^{1/2} = \sqrt{25} = 5$

Solution set: $\left\{\frac{1}{5}\right\}$

83. $-\dfrac{2}{3}\log_5 5m^2 + \dfrac{1}{2}\log_5 25m^2$

$= \log_5(5m^2)^{-2/3} + \log_5(25m^2)^{1/2}$ Power property

$= \log_5\left[(5m^2)^{-2/3} \cdot (25m^2)^{1/2}\right]$ Product property

$= \log_5(5^{-2/3}m^{-4/3} \cdot 5m)$ Properties of exponents

$= \log_5(5^{-2/3} \cdot 5^1 \cdot m^{-4/3} \cdot m^1)$

$= \log_5(5^{1/3} \cdot m^{-1/3})$ $a^m \cdot a^n = a^{m+n}$

$= \log_5\dfrac{5^{1/3}}{m^{1/3}}$, or $\log_5\sqrt[3]{\dfrac{5}{m}}$

91. $\log_{10}\sqrt{30} = \log_{10} 30^{1/2}$ Definition of $a^{1/n}$

$= \dfrac{1}{2}\log_{10} 30$ Power property

$= \dfrac{1}{2}\log_{10}(10 \cdot 3)$

$= \dfrac{1}{2}(\log_{10} 10 + \log_{10} 3)$ Product property

$= \dfrac{1}{2}(1 + 0.4771)$ $\log_a a = 1$; Substitute.

$= \dfrac{1}{2}(1.4771)$

≈ 0.7386

4.4 Exercises (pages 434–439)

87. $\log_{\sqrt{13}} 12 = \dfrac{\ln 12}{\ln \sqrt{13}}$ Change-of-base theorem

$= \dfrac{\ln 12}{\ln 13^{1/2}}$ Definition of $a^{1/n}$

$= \dfrac{\ln 12}{0.5 \ln 13}$ Power property

≈ 1.9376 Use a calculator and round answer to four decimal places.

The required logarithm can also be found by entering $\ln \sqrt{13}$ into the calculator directly.

$\log_{\sqrt{13}} 12 = \dfrac{\ln 12}{\ln \sqrt{13}}$ Change-of-base theorem

≈ 1.9376 Use a calculator and round answer to four decimal places.

91. $\ln\left(b^4\sqrt{a}\right) = \ln(b^4 a^{1/2})$ Definition of $a^{1/n}$

$= \ln b^4 + \ln a^{1/2}$ Product property

$= 4\ln b + \dfrac{1}{2}\ln a$ Power property

$= 4v + \dfrac{1}{2}u$ Substitute v for $\ln b$ and u for $\ln a$.

4.5 Exercises (pages 446–450)

23. $3(2)^{x-2} + 1 = 100$

$3(2)^{x-2} = 99$ Subtract 1.

$2^{x-2} = 33$ Divide by 3.

$\log 2^{x-2} = \log 33$ Take the logarithm on each side.

$(x - 2)\log 2 = \log 33$ Power property

$x\log 2 - 2\log 2 = \log 33$ Distributive property

$x\log 2 = \log 33 + 2\log 2$ Add $2\log 2$.

$x = \dfrac{\log 33 + 2\log 2}{\log 2}$ Divide by $\log 2$.

$x \approx 7.044$

Solution set: $\{7.044\}$

75. $\log_2(\log_2 x) = 1$

$\quad\quad \log_2 x = 2^1$ Write in exponential form.

$\quad\quad\quad\quad x = 2^2$ Write in exponential form.

$\quad\quad\quad\quad x = 4$ Evaluate.

Solution set: $\{4\}$

77.

$\quad\quad\quad\quad \log x^2 = (\log x)^2$

$\quad\quad\quad 2 \log x = (\log x)^2$ Power property

$\quad (\log x)^2 - 2 \log x = 0$ Subtract 2 log x and rewrite.

$\quad (\log x)(\log x - 2) = 0$ Factor.

$\log x = 0 \quad$ or $\quad \log x - 2 = 0$

$\quad\quad\quad\quad\quad\quad\quad\quad\quad$ Zero-factor property

$\quad\quad\quad\quad\quad\quad \log x = 2$

$\quad x = 10^0 \quad$ or $\quad\quad x = 10^2$

$\quad\quad\quad\quad\quad\quad\quad\quad\quad$ Write in exponential form.

$\quad x = 1 \quad$ or $\quad\quad x = 100$

Solution set: $\{1, 100\}$

81.

$\quad\quad\quad\quad p = a + \dfrac{k}{\ln x}, \quad$ for x

$\quad\quad\quad p - a = \dfrac{k}{\ln x}$ Subtract a.

$\quad (\ln x)(p - a) = k$ Multiply by ln x.

$\quad\quad\quad \ln x = \dfrac{k}{p - a}$ Divide by $p - a$.

$\quad\quad\quad\quad x = e^{k/(p-a)}$ Change to exponential form.

4.6 Exercises *(pages 457–462)*

29. $\quad A = Pe^{rt}$ Continuous compounding formula

$\quad 3P = Pe^{0.05t}$ Let $A = 3P$ and $r = 0.05$.

$\quad\quad 3 = e^{0.05t}$ Divide by P.

$\quad \ln 3 = \ln e^{0.05t}$ Take the logarithm on each side.

$\quad \ln 3 = 0.05t$ $\ln e^x = x$, for all x.

$\quad \dfrac{\ln 3}{0.05} = t$ Divide by 0.05.

$\quad\quad t \approx 21.97$ Use a calculator.

It will take about 21.97 yr for the investment to triple.

43. $f(t) = 200(0.90)^t$

Find t when $f(t) = 50$.

$\quad\quad 50 = 200(0.90)^t$ Let $f(t) = 50$.

$\quad\quad 0.25 = (0.90)^t$ Divide by 200.

$\quad \ln 0.25 = \ln(0.90)^t$ Take the logarithm on each side.

$\quad \ln 0.25 = t \ln 0.90$ Power property

$\quad \dfrac{\ln 0.25}{\ln 0.90} = t$ Divide by ln 0.90.

$\quad\quad t \approx 13.2$ Use a calculator.

The initial dose will reach a level of 50 mg in about 13.2 hr.

Chapter 5 Systems and Matrices

5.1 Exercises *(pages 486–492)*

27. $\dfrac{x}{2} + \dfrac{y}{3} = 4$ (1)

$\quad \dfrac{3x}{2} + \dfrac{3y}{2} = 15$ (2)

It is easier to work with equations that do not contain fractions. To clear fractions, multiply each equation by the LCD of all the fractions in the equation.

$\quad 6\left(\dfrac{x}{2} + \dfrac{y}{3}\right) = 6(4)$ Multiply by the LCD, 6.

$\quad 2\left(\dfrac{3x}{2} + \dfrac{3y}{2}\right) = 2(15)$ Multiply by the LCD, 2.

This gives the system

$\quad\quad 3x + 2y = 24$ (3)

$\quad\quad 3x + 3y = 30.$ (4)

To solve this system by elimination, multiply equation (3) by -1 and add the result to equation (4), eliminating x.

$\quad\quad -3x - 2y = -24$

$\quad\quad \underline{3x + 3y = 30}$

$\quad\quad\quad\quad\quad\quad y = 6$

Substitute 6 for y in equation (1).

$\quad\quad \dfrac{x}{2} + \dfrac{6}{3} = 4$ Let $y = 6$.

$\quad\quad \dfrac{x}{2} + 2 = 4$ $\frac{6}{3} = 2$

$\quad\quad\quad \dfrac{x}{2} = 2$ Subtract 2.

$\quad\quad\quad\quad x = 4$ Multiply by 2.

Solution set: $\{(4, 6)\}$

69. $\dfrac{2}{x} + \dfrac{1}{y} = \dfrac{3}{2}$ (1)

$\quad\quad \dfrac{3}{x} - \dfrac{1}{y} = 1$ (2)

Let $\frac{1}{x} = t$ and $\frac{1}{y} = u$. With these substitutions, the system becomes

$\quad\quad 2t + u = \dfrac{3}{2}$ (3)

$\quad\quad 3t - u = 1.$ (4)

Add these equations, eliminating u, and solve for t.

$\quad\quad 5t = \dfrac{5}{2}$

$\quad\quad\quad t = \dfrac{1}{2}$ Multiply by $\frac{1}{5}$.

Substitute $\frac{1}{2}$ for t in equation (3) and solve for u.

$$2\left(\frac{1}{2}\right) + u = \frac{3}{2} \quad \text{Let } t = \frac{1}{2}.$$

$$1 + u = \frac{3}{2} \quad \text{Multiply.}$$

$$u = \frac{1}{2} \quad \text{Subtract 1.}$$

Now find the values of x and y, the variables in the original system. Since $t = \frac{1}{x}$, $tx = 1$, and $x = \frac{1}{t}$. Likewise, $y = \frac{1}{u}$.

$$x = \frac{1}{t} = \frac{1}{\frac{1}{2}} = 2 \quad \text{Let } t = \frac{1}{2}.$$

$$y = \frac{1}{u} = \frac{1}{\frac{1}{2}} = 2 \quad \text{Let } u = \frac{1}{2}.$$

Solution set: $\{(2, 2)\}$

5.2 Exercises (pages 499–504)

47. $\dfrac{1}{(x-1)(x+1)} = \dfrac{A}{x-1} + \dfrac{B}{x+1}$

$$\frac{1}{(x-1)(x+1)}$$

$$= \frac{A(x+1)}{(x-1)(x+1)} + \frac{B(x-1)}{(x-1)(x+1)}$$

Write terms on the right with the LCD, $(x-1)(x+1)$.

$$= \frac{A(x+1) + B(x-1)}{(x-1)(x+1)}$$

Combine terms on the right.

$1 = A(x+1) + B(x-1)$ Denominators are equal, so numerators must be equal.

$1 = Ax + A + Bx - B$ Distributive property

$1 = (Ax + Bx) + (A - B)$ Group like terms.

$1 = (A + B)x + (A - B)$ Factor $Ax + Bx$.

Since $1 = 0x + 1$, we can equate the coefficients of like powers of x to obtain the following system.

$$A + B = 0$$
$$A - B = 1$$

Solve this system by the Gauss-Jordan method.

$$\begin{bmatrix} 1 & 1 & | & 0 \\ 1 & -1 & | & 1 \end{bmatrix}$$

$$\begin{bmatrix} 1 & 1 & | & 0 \\ 0 & -2 & | & 1 \end{bmatrix} \quad -R1 + R2$$

$$\begin{bmatrix} 1 & 1 & | & 0 \\ 0 & 1 & | & -\frac{1}{2} \end{bmatrix} \quad -\frac{1}{2}R2$$

$$\begin{bmatrix} 1 & 0 & | & \frac{1}{2} \\ 0 & 1 & | & -\frac{1}{2} \end{bmatrix} \quad -R2 + R1$$

From the final matrix, $A = \frac{1}{2}$ and $B = -\frac{1}{2}$.

55. Let x = number of cubic centimeters of 2% solution and y = number of cubic centimeters of 7% solution. From the given information, we can write the system

$$x + y = 40$$
$$0.02x + 0.07y = 0.032(40),$$

or

$$x + y = 40$$
$$0.02x + 0.07y = 1.28.$$

Solve this system by the Gauss-Jordan method.

$$\begin{bmatrix} 1 & 1 & | & 40 \\ 0.02 & 0.07 & | & 1.28 \end{bmatrix}$$

$$\begin{bmatrix} 1 & 1 & | & 40 \\ 0 & 0.05 & | & 0.48 \end{bmatrix} \quad -0.02R1 + R2$$

$$\begin{bmatrix} 1 & 1 & | & 40 \\ 0 & 1 & | & 9.6 \end{bmatrix} \quad \frac{1}{0.05}R2 \text{ (or 20R2)}$$

$$\begin{bmatrix} 1 & 0 & | & 30.4 \\ 0 & 1 & | & 9.6 \end{bmatrix} \quad -R2 + R1$$

From the final matrix, $x = 30.4$ and $y = 9.6$, so the chemist should mix 30.4 cm^3 of the 2% solution with 9.6 cm^3 of the 7% solution.

5.3 Exercises (pages 512–516)

63. $\begin{vmatrix} -4 & 1 & 4 \\ 2 & 0 & 1 \\ 0 & 2 & 4 \end{vmatrix} = \begin{vmatrix} 0 & 1 & 6 \\ 2 & 0 & 1 \\ 0 & 2 & 4 \end{vmatrix}$ Use determinant theorem 6; 2R2 + R1.

$$= 0\begin{vmatrix} 0 & 1 \\ 2 & 4 \end{vmatrix} - 2\begin{vmatrix} 1 & 6 \\ 2 & 4 \end{vmatrix} + 0\begin{vmatrix} 1 & 6 \\ 0 & 1 \end{vmatrix}$$

Expand about column 1.

$$= 0 - 2(4 - 12) + 0$$
$$= -2(-8)$$
$$= 16$$

85. $1.5x + 3y = 5$ (1)
$2x + 4y = 3$ (2)

$$D = \begin{vmatrix} 1.5 & 3 \\ 2 & 4 \end{vmatrix} = 1.5(4) - 2(3) = 6 - 6 = 0$$

Because $D = 0$, Cramer's rule does not apply. To determine whether the system is inconsistent or has infinitely many solutions, use the elimination method.

$6x + 12y = 20$ Multiply equation (1) by 4.
$\underline{-6x - 12y = -9}$ Multiply equation (2) by -3.
$0 = 11$ False

The system is inconsistent.
Solution set: $\emptyset$

5.4 Exercises *(pages 521–522)*

13. $\dfrac{x^2}{x^2 + 2x + 1}$

This is not a proper fraction; the numerator has degree greater than or equal to that of the denominator. Divide the numerator by the denominator.

$$
\begin{array}{r}
1 \\
x^2 + 2x + 1\overline{\smash{)}x^2} \\
\underline{x^2 + 2x + 1} \\
-2x - 1
\end{array}
$$

Find the partial fraction decomposition for

$$\frac{-2x - 1}{x^2 + 2x + 1} = \frac{-2x - 1}{(x + 1)^2}.$$

$\dfrac{-2x - 1}{(x + 1)^2} = \dfrac{A}{x + 1} + \dfrac{B}{(x + 1)^2}$ Factor the denominator of the given fraction.

$-2x - 1 = A(x + 1) + B$ Multiply by the LCD, $(x + 1)^2$.

Use this equation to find the value of B.

$-2(-1) - 1 = A(-1 + 1) + B$ Let $x = -1$.

$1 = B$

Now use the same equation and the value of B to find the value of A.

$-2(2) - 1 = A(2 + 1) + 1$ Let $x = 2$ and $B = 1$.

$-5 = 3A + 1$

$-6 = 3A$

$A = -2$

Thus,

$$\frac{x^2}{x^2 + 2x + 1} = 1 + \frac{-2}{x + 1} + \frac{1}{(x + 1)^2}.$$

23. $\dfrac{1}{x(2x + 1)(3x^2 + 4)}$

The denominator contains two linear factors and one quadratic factor. All factors are distinct, and $3x^2 + 4$ cannot be factored, so it is irreducible. The partial fraction decomposition is of the form

$$\frac{1}{x(2x + 1)(3x^2 + 4)} = \frac{A}{x} + \frac{B}{2x + 1} + \frac{Cx + D}{3x^2 + 4}.$$

We need to find the values of A, B, C, and D.

$1 = A(2x + 1)(3x^2 + 4) + Bx(3x^2 + 4)$
$\quad + (Cx + D)(x)(2x + 1)$ Multiply by the LCD, $x(2x + 1)(3x^2 + 4)$.

$1 = A(6x^3 + 3x^2 + 8x + 4) + B(3x^3 + 4x)$
$\quad + C(2x^3 + x^2) + D(2x^2 + x)$ Multiply on the right.

$1 = 6Ax^3 + 3Ax^2 + 8Ax + 4A + 3Bx^3 + 4Bx$
$\quad + 2Cx^3 + Cx^2 + 2Dx^2 + Dx$ Distributive property

$1 = (6A + 3B + 2C)x^3 + (3A + C + 2D)x^2$
$\quad + (8A + 4B + D)x + 4A$ Combine like terms.

Since $1 = 0x^3 + 0x^2 + 0x + 1$, equating the coefficients of like powers of x produces the following system of equations:

$$
\begin{aligned}
6A + 3B + 2C &= 0 \\
3A + C + 2D &= 0 \\
8A + 4B + D &= 0 \\
4A &= 1.
\end{aligned}
$$

Any of the methods from this chapter can be used to solve this system. If the Gauss-Jordan method is used, begin by representing the system with the augmented matrix

$$
\left[\begin{array}{cccc|c}
6 & 3 & 2 & 0 & 0 \\
3 & 0 & 1 & 2 & 0 \\
8 & 4 & 0 & 1 & 0 \\
4 & 0 & 0 & 0 & 1
\end{array}\right].
$$

After performing a series of row operations, we obtain the following final matrix:

$$
\left[\begin{array}{cccc|c}
1 & 0 & 0 & 0 & \frac{1}{4} \\
0 & 1 & 0 & 0 & -\frac{8}{19} \\
0 & 0 & 1 & 0 & -\frac{9}{76} \\
0 & 0 & 0 & 1 & -\frac{6}{19}
\end{array}\right],
$$

from which we read the values of the four variables: $A = \frac{1}{4}$, $B = -\frac{8}{19}$, $C = -\frac{9}{76}$, $D = -\frac{6}{19}$. Substitute these values into the form given at the beginning of this solution for the partial fraction decomposition.

$$\frac{1}{x(2x + 1)(3x^2 + 4)} = \frac{A}{x} + \frac{B}{2x + 1} + \frac{Cx + D}{3x^2 + 4}$$

$$\frac{1}{x(2x + 1)(3x^2 + 4)} = \frac{\frac{1}{4}}{x} + \frac{-\frac{8}{19}}{2x + 1} + \frac{-\frac{9}{76}x - \frac{6}{19}}{3x^2 + 4}$$

$$= \frac{\frac{1}{4}}{x} + \frac{-\frac{8}{19}}{2x + 1} + \frac{-\frac{9}{76}x - \frac{24}{76}}{3x^2 + 4}$$
Get a common denominator for the numerator of the last term.

$$= \frac{1}{4x} + \frac{-8}{19(2x + 1)} + \frac{-9x - 24}{76(3x^2 + 4)}$$
Simplify the complex fractions.

5.5 Exercises *(pages 528–532)*

53. Let x and y represent the two numbers.

$$\frac{x}{y} = \frac{9}{2} \quad (1)$$

$$xy = 162 \quad (2)$$

Solve equation (1) for x.

$$y\left(\frac{x}{y}\right) = y\left(\frac{9}{2}\right) \quad \text{Multiply by } y.$$

$$x = \frac{9}{2}y$$

Substitute $\frac{9}{2}y$ for x in equation (2).

$$\left(\frac{9}{2}y\right)y = 162$$

$$\frac{9}{2}y^2 = 162 \qquad \text{Multiply.}$$

$$\frac{2}{9}\left(\frac{9}{2}y^2\right) = \frac{2}{9}(162) \qquad \text{Multiply by } \frac{2}{9}.$$

$$y^2 = 36$$

$$y = \pm 6 \qquad \text{Take square roots.}$$

If $y = 6$, then $x = \frac{9}{2}(6) = 27$.

If $y = -6$, then $x = \frac{9}{2}(-6) = -27$. The two numbers are either 27 and 6, or -27 and -6.

65. supply: $p = \sqrt{0.1q + 9} - 2$

demand: $p = \sqrt{25 - 0.1q}$

(a) Equilibrium occurs when supply = demand, so solve the system formed by the supply and demand equations. This system can be solved by substitution. Substitute $\sqrt{0.1q + 9} - 2$ for p in the demand equation, and solve the resulting equation for q.

$$\sqrt{0.1q + 9} - 2 = \sqrt{25 - 0.1q}$$

$$\left(\sqrt{0.1q + 9} - 2\right)^2 = \left(\sqrt{25 - 0.1q}\right)^2$$

Square each side.

$$0.1q + 9 - 4\sqrt{0.1q + 9} + 4 = 25 - 0.1q$$

$(x - y)^2 = x^2 - 2xy + y^2$

$$0.2q - 12 = 4\sqrt{0.1q + 9}$$

Combine like terms, and isolate the radical.

$$(0.2q - 12)^2 = \left(4\sqrt{0.1q + 9}\right)^2$$

Square each side again.

$$0.04q^2 - 4.8q + 144 = 16(0.1q + 9)$$

$$0.04q^2 - 4.8q + 144 = 1.6q + 144$$

Distributive property

$$0.04q^2 - 6.4q = 0 \qquad \text{Combine like terms.}$$

$$0.04q(q - 160) = 0 \qquad \text{Factor.}$$

$$q = 0 \quad \text{or} \quad q = 160 \qquad \text{Zero-factor property}$$

Disregard an equilibrium demand of 0. The equilibrium demand is 160 units.

(b) Substitute 160 for q in either equation and solve for p. Using the supply equation, we obtain the following.

$$p = \sqrt{0.1q + 9} - 2$$

$$= \sqrt{0.1(160) + 9} - 2 \qquad \text{Let } q = 160.$$

$$= \sqrt{16 + 9} - 2$$

$$= \sqrt{25} - 2$$

$$= 5 - 2$$

$$= 3$$

The equilibrium price is $3.

5.6 Exercises *(pages 540–544)*

55. $y \leq \log x$

$y \geq |x - 2|$

Graph $y = \log x$ as a solid curve because $y \leq \log x$ is a nonstrict inequality.

(Recall that "$\log x$" means $\log_{10} x$.) This graph contains the points $(0.1, -1)$, $(1, 0)$, and $(10, 1)$ because $10^{-1} = 0.1$, $10^0 = 1$, and $10^1 = 10$. Because the symbol is $\leq$, shade the region *below* the curve.

Now graph $y = |x - 2|$. Make this boundary solid because $y \geq |x - 2|$ is also a nonstrict inequality. This graph can be obtained by translating the graph of $y = |x|$ to the right 2 units. It contains the points $(0, 2)$, $(2, 0)$, and $(4, 2)$. Because the symbol is $\geq$, shade the region *above* the absolute value graph.

The solution set of the system is the intersection (or overlap) of the two shaded regions, which is shown in the final graph.

79. Let $x =$ the number of cabinet A and $y =$ the number of cabinet B.

The cost constraint is

$$10x + 20y \leq 140.$$

The space constraint is

$$6x + 8y \leq 72.$$

Since the numbers of cabinets cannot be negative, we also have the constraints $x \geq 0$ and $y \geq 0$. We want to maximize the objective function,

$$\text{storage capacity} = 8x + 12y.$$

Find the region of feasible solutions by graphing the system of inequalities that is made up of the constraints.

To graph $10x + 20y \leq 140$, draw the line with x-intercept $(14, 0)$ and y-intercept $(0, 7)$ as a solid line. Because the symbol is $\leq$, shade the region *below* the line.

To graph $6x + 8y \leq 72$, draw the line with x-intercept $(12, 0)$ and y-intercept $(0, 9)$ as a solid line. Because the symbol is $\leq$, shade the region *below* the line.

The constraints $x \geq 0$ and $y \geq 0$ restrict the graph to the first quadrant. The graph of the feasible region is the intersection of the regions that are the graphs of the individual constraints.

From the graph, observe that three of the vertices are $(0, 0)$, $(0, 7)$, and $(12, 0)$. The fourth vertex is the intersection point of the lines $10x + 20y = 140$ and $6x + 8y = 72$. To find this point, solve the system

$$10x + 20y = 140$$
$$6x + 8y = 72$$

to obtain the ordered pair $(8, 3)$. Evaluate the objective function at each vertex.

Point	Storage Capacity $= 8x + 12y$
$(0, 0)$	$8(0) + 12(0) = 0$
$(0, 7)$	$8(0) + 12(7) = 84$
$(8, 3)$	$8(8) + 12(3) = 100$
$(12, 0)$	$8(12) + 12(0) = 96$

The maximum value of $8x + 12y$ occurs at the vertex $(8, 3)$, so the office manager should buy 8 of cabinet A and 3 of cabinet B for a maximum storage capacity of 100 ft^3.

5.8 Exercises *(pages 565–569)*

23. $A = \begin{bmatrix} 1 & 1 & 0 & 2 \\ 2 & -1 & 1 & -1 \\ 3 & 3 & 2 & -2 \\ 1 & 2 & 1 & 0 \end{bmatrix}$

$\left[\begin{array}{cccc|cccc} 1 & 1 & 0 & 2 & 1 & 0 & 0 & 0 \\ 2 & -1 & 1 & -1 & 0 & 1 & 0 & 0 \\ 3 & 3 & 2 & -2 & 0 & 0 & 1 & 0 \\ 1 & 2 & 1 & 0 & 0 & 0 & 0 & 1 \end{array}\right]$ Write the augmented matrix $[A \mid I_4]$.

$\left[\begin{array}{cccc|cccc} 1 & 1 & 0 & 2 & 1 & 0 & 0 & 0 \\ 0 & -3 & 1 & -5 & -2 & 1 & 0 & 0 \\ 0 & 0 & 2 & -8 & -3 & 0 & 1 & 0 \\ 0 & 1 & 1 & -2 & -1 & 0 & 0 & 1 \end{array}\right]$ $\begin{array}{l} -2R1 + R2 \\ -3R1 + R3 \\ -1R1 + R4 \end{array}$

$\left[\begin{array}{cccc|cccc} 1 & 1 & 0 & 2 & 1 & 0 & 0 & 0 \\ 0 & 1 & -\frac{1}{3} & \frac{5}{3} & \frac{2}{3} & -\frac{1}{3} & 0 & 0 \\ 0 & 0 & 2 & -8 & -3 & 0 & 1 & 0 \\ 0 & 1 & 1 & -2 & -1 & 0 & 0 & 1 \end{array}\right]$ $-\frac{1}{3}R2$

$\left[\begin{array}{cccc|cccc} 1 & 0 & \frac{1}{3} & \frac{1}{3} & \frac{1}{3} & \frac{1}{3} & 0 & 0 \\ 0 & 1 & -\frac{1}{3} & \frac{5}{3} & \frac{2}{3} & -\frac{1}{3} & 0 & 0 \\ 0 & 0 & 2 & -8 & -3 & 0 & 1 & 0 \\ 0 & 0 & \frac{4}{3} & -\frac{11}{3} & -\frac{5}{3} & \frac{1}{3} & 0 & 1 \end{array}\right]$ $\begin{array}{l} -1R2 + R1 \\ \\ \\ -1R2 + R4 \end{array}$

$\left[\begin{array}{cccc|cccc} 1 & 0 & \frac{1}{3} & \frac{1}{3} & \frac{1}{3} & \frac{1}{3} & 0 & 0 \\ 0 & 1 & -\frac{1}{3} & \frac{5}{3} & \frac{2}{3} & -\frac{1}{3} & 0 & 0 \\ 0 & 0 & 1 & -4 & -\frac{3}{2} & 0 & \frac{1}{2} & 0 \\ 0 & 0 & \frac{4}{3} & -\frac{11}{3} & -\frac{5}{3} & \frac{1}{3} & 0 & 1 \end{array}\right]$ $\frac{1}{2}R3$

$\left[\begin{array}{cccc|cccc} 1 & 0 & 0 & \frac{5}{3} & \frac{5}{6} & \frac{1}{3} & -\frac{1}{6} & 0 \\ 0 & 1 & 0 & \frac{1}{3} & \frac{1}{6} & -\frac{1}{3} & \frac{1}{6} & 0 \\ 0 & 0 & 1 & -4 & -\frac{3}{2} & 0 & \frac{1}{2} & 0 \\ 0 & 0 & 0 & \frac{5}{3} & \frac{1}{3} & \frac{1}{3} & -\frac{2}{3} & 1 \end{array}\right]$ $\begin{array}{l} -\frac{1}{3}R3 + R1 \\ \frac{1}{3}R3 + R2 \\ \\ -\frac{4}{3}R3 + R4 \end{array}$

$\left[\begin{array}{cccc|cccc} 1 & 0 & 0 & \frac{5}{3} & \frac{5}{6} & \frac{1}{3} & -\frac{1}{6} & 0 \\ 0 & 1 & 0 & \frac{1}{3} & \frac{1}{6} & -\frac{1}{3} & \frac{1}{6} & 0 \\ 0 & 0 & 1 & -4 & -\frac{3}{2} & 0 & \frac{1}{2} & 0 \\ 0 & 0 & 0 & 1 & \frac{1}{5} & \frac{1}{5} & -\frac{2}{5} & \frac{3}{5} \end{array}\right]$ $\frac{3}{5}R4$

$\left[\begin{array}{cccc|cccc} 1 & 0 & 0 & 0 & \frac{1}{2} & 0 & \frac{1}{2} & -1 \\ 0 & 1 & 0 & 0 & \frac{1}{10} & -\frac{2}{5} & \frac{3}{10} & -\frac{1}{5} \\ 0 & 0 & 1 & 0 & -\frac{7}{10} & \frac{4}{5} & -\frac{11}{10} & \frac{12}{5} \\ 0 & 0 & 0 & 1 & \frac{1}{5} & \frac{1}{5} & -\frac{2}{5} & \frac{3}{5} \end{array}\right]$ $\begin{array}{l} -\frac{5}{3}R4 + R1 \\ -\frac{1}{3}R4 + R2 \\ 4R4 + R3 \end{array}$

$$A^{-1} = \begin{bmatrix} \frac{1}{2} & 0 & \frac{1}{2} & -1 \\ \frac{1}{10} & -\frac{2}{5} & \frac{3}{10} & -\frac{1}{5} \\ -\frac{7}{10} & \frac{4}{5} & -\frac{11}{10} & \frac{12}{5} \\ \frac{1}{5} & \frac{1}{5} & -\frac{2}{5} & \frac{3}{5} \end{bmatrix}$$

43. $0.2x + 0.3y = -1.9$
$0.7x - 0.2y = 4.6$

$A = \begin{bmatrix} 0.2 & 0.3 \\ 0.7 & -0.2 \end{bmatrix}$, $X = \begin{bmatrix} x \\ y \end{bmatrix}$, $B = \begin{bmatrix} -1.9 \\ 4.6 \end{bmatrix}$

Find A^{-1}.

$\left[\begin{array}{cc|cc} 0.2 & 0.3 & 1 & 0 \\ 0.7 & -0.2 & 0 & 1 \end{array}\right]$ Write the augmented matrix $[A \mid I_2]$.

$\left[\begin{array}{cc|cc} 1 & 1.5 & 5 & 0 \\ 0.7 & -0.2 & 0 & 1 \end{array}\right]$ $5R1$

$\left[\begin{array}{cc|cc} 1 & 1.5 & 5 & 0 \\ 0 & -1.25 & -3.5 & 1 \end{array}\right]$ $-0.7R1 + R2$

$\left[\begin{array}{cc|cc} 1 & 1.5 & 5 & 0 \\ 0 & 1 & 2.8 & -0.8 \end{array}\right]$ $\frac{1}{-1.25}R2$, or $-0.8R2$

$\left[\begin{array}{cc|cc} 1 & 0 & 0.8 & 1.2 \\ 0 & 1 & 2.8 & -0.8 \end{array}\right]$ $-1.5R2 + R1$

$A^{-1} = \begin{bmatrix} 0.8 & 1.2 \\ 2.8 & -0.8 \end{bmatrix}$

$X = A^{-1}B = \begin{bmatrix} 0.8 & 1.2 \\ 2.8 & -0.8 \end{bmatrix}\begin{bmatrix} -1.9 \\ 4.6 \end{bmatrix} = \begin{bmatrix} 4 \\ -9 \end{bmatrix}$

Solution set: $\{(4, -9)\}$

To The Student

In this section we provide the answers that we think most students will obtain when they work the exercises using the methods explained in the text. If your answer does not look exactly like the one given here, it is not necessarily wrong. In many cases there are equivalent forms of the answer. For example, if the answer section shows $\frac{3}{4}$ and your answer is 0.75, you have obtained the correct answer but written it in a different (yet equivalent) form. Unless the directions specify otherwise, 0.75 is just as valid an answer as $\frac{3}{4}$. In general, if your answer does not agree with the one given in the text, see whether it can be transformed into the other form. If it can, then it is the correct answer. If you still have doubts, talk with your instructor. For further assistance, you may want to refer to the *Student's Solution Manual* that goes with this book.

Chapter R Review of Basic Concepts

R.1 Exercises (pages 6–7)

1. finite; yes **3.** infinite; no **5.** infinite; no
7. infinite; no **9.** $\{12, 13, 14, 15, 16, 17, 18, 19, 20\}$
11. $\{1, \frac{1}{2}, \frac{1}{4}, \frac{1}{8}, \frac{1}{16}, \frac{1}{32}\}$ **13.** $\{17, 22, 27, 32, 37, 42, 47\}$
15. $\{8, 9, 10, 11, 12, 13, 14\}$ **17.** $\in$ **19.** $\notin$ **21.** $\in$
23. $\notin$ **25.** $\notin$ **27.** $\notin$ **29.** false **31.** true **33.** true
35. true **37.** false **39.** true **41.** true **43.** true
45. false **47.** true **49.** true **51.** false **53.** $\subseteq$
55. $\not\subseteq$ **57.** $\subseteq$ **59.** true **61.** false **63.** true
65. false **67.** true **69.** true **71.** $\{0, 2, 4\}$
73. $\{0, 1, 2, 3, 4, 5, 6, 7, 8, 9, 11, 13\}$ **75.** $\emptyset$; M and N are disjoint sets. **77.** $\{0, 1, 2, 3, 4, 5, 7, 9, 11, 13\}$
79. Q, or $\{0, 2, 4, 6, 8, 10, 12\}$ **81.** $\{10, 12\}$ **83.** $\emptyset$; $\emptyset$ and R are disjoint. **85.** N, or $\{1, 3, 5, 7, 9, 11, 13\}$; N and $\emptyset$ are disjoint. **87.** R, or $\{0, 1, 2, 3, 4\}$
89. $\{0, 1, 2, 3, 4, 6, 8\}$ **91.** R, or $\{0, 1, 2, 3, 4\}$
93. $\emptyset$; Q' and $(N' \cap U)$ are disjoint.

R.2 Exercises (pages 17–20)

1. (a) B, C, D, F **(b)** A, B, C, D, F **(c)** D, F
(d) A, B, C, D, F **(e)** E, F **(f)** D, F **3.** false; Some are whole numbers, but negative integers are not. **5.** false; No irrational number is an integer. **7.** true **9.** true **11.** 1, 3
13. $-6, -\frac{12}{4}$ (or -3), 0, 1, 3 **15.** $-\sqrt{3}, 2\pi, \sqrt{12}$
17. -16 **19.** 16 **21.** -243 **23.** -162 **25.** -6
27. -60 **29.** -12 **31.** $-\frac{25}{36}$ **33.** $-\frac{6}{7}$ **35.** 28

37. $-\frac{1}{2}$ **39.** $-\frac{23}{20}$ **41.** $-\frac{25}{11}$ **43.** $-\frac{11}{3}$ **45.** $-\frac{13}{3}$
47. 6 **49.** distributive **51.** inverse **53.** identity
55. commutative **57.** associative **59.** closure
63. $20z$ **65.** $m + 11$ **67.** $\frac{2}{3}y + \frac{4}{9}z - \frac{5}{3}$
69. $(8 - 14)p = -6p$ **71.** $-4z + 4y$ **73.** 1700
75. 150 **77.** false; $|6 - 8| = |8| - |6|$ **79.** true
81. false; $|a - b| = |b| - |a|$ **83.** 10 **85.** $-\frac{4}{7}$ **87.** -8
89. 9 **91.** 16 **93.** 20 **95.** -1 **97.** -5 **99.** Property 2
101. Property 3 **103.** Property 1 **105.** 3 **107.** 9
109. x and y have the same sign. **111.** x and y have different signs. **113.** x and y have the same sign.
115. 17; This represents the number of strokes between their scores. **117.** 0.031 **119.** 0.024; 0.023; Increased weight results in lower BACs. **121.** 111.0 **123.** 101.2
125. 97.0 **127.** 93.6 **129.** 92.0 **131.** 91.0 **133.** 121.1

R.3 Exercises (pages 28–31)

1. $-16x^7$ **3.** n^{11} **5.** 9^8 **7.** $72m^{11}$ **9.** $-15x^5y^5$
11. 2^{10} **13.** $(-6)^3x^6$, or $-216x^6$ **15.** -4^2m^6, or $-16m^6$
17. $\frac{r^{24}}{s^6}$ **19.** $\frac{(-4)^4m^8}{t^4p^8}$, or $\frac{256m^8}{t^4p^8}$ **21.** -1 **23. (a)** B **(b)** C
(c) B **(d)** C **27.** polynomial; degree 11; monomial
29. polynomial; degree 4; binomial **31.** polynomial; degree 5; trinomial **33.** polynomial; degree 11; none of these **35.** not a polynomial **37.** polynomial; degree 0; monomial **39.** $x^2 - x + 2$ **41.** $12y^2 + 4$
43. $6m^4 - 2m^3 - 7m^2 - 4m$ **45.** $28r^2 + r - 2$
47. $15x^4 - \frac{7}{3}x^3 - \frac{2}{9}x^2$ **49.** $12x^5 + 8x^4 - 20x^3 + 4x^2$
51. $-2z^3 + 7z^2 - 11z + 4$
53. $m^2 + mn - 2n^2 - 2km + 5kn - 3k^2$
55. $16x^4 - 72x^2 + 81$ **57.** $x^4 - 2x^2 + 1$
59. $4m^2 - 9$ **61.** $16x^4 - 25y^2$ **63.** $16m^2 + 16mn + 4n^2$
65. $25r^2 - 30rt^2 + 9t^4$
67. $4p^2 - 12p + 9 + 4pq - 6q + q^2$
69. $9q^2 + 30q + 25 - p^2$
71. $9a^2 + 6ab + b^2 - 6a - 2b + 1$
73. $y^3 + 6y^2 + 12y + 8$ **75.** $q^4 - 8q^3 + 24q^2 - 32q + 16$
77. $p^3 - 7p^2 - p - 7$ **79.** $49m^2 - 4n^2$
81. $-14q^2 + 11q - 14$ **83.** $4p^2 - 16$
85. $11y^3 - 18y^2 + 4y$ **87.** $2x^5 + 7x^4 - 5x^2 + 7$
89. $4x^2 + 5x + 10 + \frac{21}{x - 2}$ **91.** $2m^2 + m - 2 + \frac{6}{3m + 2}$
93. $x^2 + 2 + \frac{5x + 21}{x^2 + 3}$ **95.** 9999 **96.** 3591 **97.** 10,404
98. 5041 **99. (a)** $(x + y)^2$ **(b)** $x^2 + 2xy + y^2$
(d) the special product for squaring a binomial

101. (a) approximately 60,501,000 ft^3 **(b)** The shape becomes a rectangular box with a square base, with volume $V = b^2h$. **(c)** If we let $a = b$, then $V = \frac{1}{3}h(a^2 + ab + b^2)$ becomes $V = \frac{1}{3}h(b^2 + bb + b^2)$, which simplifies to $V = hb^2$. Yes, the Egyptian formula gives the same result. **103.** 6.3; exact **105.** 2.1; 0.3 low **107.** 1,000,000 **109.** 32

R.4 Exercises (pages 38–40)

1. $12(m + 5)$ **3.** $8k(k^2 + 3)$ **5.** $xy(1 - 5y)$ **7.** $-2p^2q^4(2p + q)$ **9.** $4k^2m^3(1 + 2k^2 - 3m)$ **11.** $2(a + b)(1 + 2m)$ **13.** $(r + 3)(3r - 5)$ **15.** $(m - 1)(2m^2 - 7m + 7)$ **17.** The *completely* factored form is $4xy^3(xy^2 - 2)$. **19.** $(2s + 3)(3t - 5)$ **21.** $(m^4 + 3)(2 - a)$ **23.** $(p^2 - 2)(q^2 + 5)$ **25.** $(2a - 1)(3a - 4)$ **27.** $(3m + 2)(m + 4)$ **29.** prime **31.** $2a(3a + 7)(2a - 3)$ **33.** $(3k - 2p)(2k + 3p)$ **35.** $(5a + 3b)(a - 2b)$ **37.** $(4x + y)(3x - y)$ **39.** $2a^2(4a - b)(3a + 2b)$ **41.** $(3m - 2)^2$ **43.** $2(4a + 3b)^2$ **45.** $(2xy + 7)^2$ **47.** $(a - 3b - 3)^2$ **49. (a)** B **(b)** C **(c)** A **(d)** D **51.** $(3a + 4)(3a - 4)$ **53.** $(x^2 + 4)(x + 2)(x - 2)$ **55.** $(5s^2 + 3t)(5s^2 - 3t)$ **57.** $(a + b + 4)(a + b - 4)$ **59.** $(p^2 + 25)(p + 5)(p - 5)$ **61.** $(2 - a)(4 + 2a + a^2)$ **63.** $(5x - 3)(25x^2 + 15x + 9)$ **65.** $(3y^3 + 5z^2)(9y^6 - 15y^3z^2 + 25z^4)$ **67.** $r(r^2 + 18r + 108)$ **69.** $(3 - m - 2n)(9 + 3m + 6n + m^2 + 4mn + 4n^2)$ **71.** B **73.** $(x - 1)(x^2 + x + 1)(x + 1)(x^2 - x + 1)$ **74.** $(x - 1)(x + 1)(x^4 + x^2 + 1)$ **75.** $(x^2 - x + 1)(x^2 + x + 1)$ **76.** additive inverse property (0 in the form $x^2 - x^2$ was added on the right.); associative property of addition; factoring a perfect square trinomial; factoring a difference of squares; commutative property of addition **77.** They are the same. **78.** $(x^4 - x^2 + 1)(x^2 + x + 1)(x^2 - x + 1)$ **79.** $9(7k - 3)(k + 1)$ **81.** $(3a - 7)^2$ **83.** $(m^2 - 5)(m^2 + 2)$ **85.** $(2b + c + 4)(2b + c - 4)$ **87.** $(x + y)(x - 5)$ **89.** $(m - 2n)(p^4 + q)$ **91.** $(2z + 7)^2$ **93.** $(10x + 7y)(100x^2 - 70xy + 49y^2)$ **95.** $(5m^2 - 6)(25m^4 + 30m^2 + 36)$ **97.** $9(x + 2)(3x^2 + 4)$ **99.** $2y(3x^2 + y^2)$ **101.** prime **103.** $4xy$ **107.** $\left(7x + \frac{1}{5}\right)\left(7x - \frac{1}{5}\right)$ **109.** $\left(\frac{5}{3}x^2 + 3y\right)\left(\frac{5}{3}x^2 - 3y\right)$ **111.** ± 36 **113.** 9

R.5 Exercises (pages 47–49)

1. $\{x \mid x \neq 6\}$ **3.** $\{x \mid x \neq -\frac{1}{2}, 1\}$ **5.** $\{x \mid x \neq -2, -3\}$ **7.** $\{x \mid x \neq -1\}$ **9.** $\{x \mid x \neq 1\}$ **11.** $\frac{2x + 4}{x}$ **13.** $\frac{-3}{t + 5}$ **15.** $\frac{8}{9}$ **17.** $\frac{m - 2}{m + 3}$ **19.** $\frac{2m + 3}{4m + 3}$ **21.** $x^2 - 4x + 16$ **23.** $\frac{25p^2}{9}$ **25.** $\frac{2}{9}$ **27.** $\frac{5x}{y}$ **29.** $\frac{2a + 8}{a - 3}$, or $\frac{2(a + 4)}{a - 3}$ **31.** 1 **33.** $\frac{m + 6}{m + 3}$ **35.** $\frac{x^2 - xy + y^2}{x^2 + xy + y^2}$ **37.** $\frac{x + 2y}{4 - x}$

39. B, C **41.** $\frac{19}{6k}$ **43.** $\frac{137}{30m}$ **45.** $\frac{a - b}{a^2}$ **47.** $\frac{5 - 22x}{12x^2y}$ **49.** 3 **51.** $\frac{2x}{(x + z)(x - z)}$ **53.** $\frac{4}{a - 2}$, or $\frac{-4}{2 - a}$ **55.** $\frac{3x + y}{2x - y}$, or $\frac{-3x - y}{y - 2x}$ **57.** $\frac{4x - 7}{x^2 - x + 1}$ **59.** $\frac{2x^2 - 9x}{(x - 3)(x + 4)(x - 4)}$ **61.** $\frac{x + 1}{x - 1}$ **63.** $\frac{-1}{x + 1}$ **65.** $\frac{(2 - b)(1 + b)}{b(1 - b)}$ **67.** $\frac{a + b}{a^2 - ab + b^2}$ **69.** $\frac{m^3 - 4m - 1}{m - 2}$ **71.** $\frac{p^3 - 16p + 3}{p + 4}$ **73.** $\frac{y^2 - 2y - 3}{y^2 + y - 1}$ **75.** $\frac{-1}{x(x + h)}$ **77.** $\frac{-2x - h}{(x^2 + 9)[(x + h)^2 + 9]}$ **79.** 0 mi **81.** 20.1 (thousand dollars)

R.6 Exercises (pages 55–58)

1. (a) B **(b)** D **(c)** B **(d)** D **3.** $\frac{1}{(-4)^3}$, or $-\frac{1}{64}$ **5.** $-\frac{1}{5^4}$, or $-\frac{1}{625}$ **7.** 3^2, or 9 **9.** $\frac{1}{16x^2}$ **11.** $\frac{4}{x^2}$ **13.** $-\frac{1}{a^3}$ **15.** 4^2, or 16 **17.** x^4 **19.** $\frac{1}{r^3}$ **21.** 6^6 **23.** $\frac{2r^3}{3}$ **25.** $\frac{4n^7}{3m^7}$ **27.** $-4r^6$ **29.** $\frac{5^4}{a^{10}}$ **31.** $\frac{p^4}{5}$ **33.** $\frac{1}{2pq}$ **35.** $\frac{4}{a^2}$ **37.** $\frac{5}{x^2}$ **39.** 13 **41.** 2 **43.** $-\frac{4}{3}$ **45.** This expression is not a real number. **47. (a)** E **(b)** G **(c)** F **(d)** F **49.** 4 **51.** 1000 **53.** -27 **55.** $\frac{256}{81}$ **57.** 9 **59.** 4 **61.** y **63.** $k^{2/3}$ **65.** x^3y^8 **67.** $\frac{1}{x^{10/3}}$ **69.** $\frac{6}{m^{1/4}n^{3/4}}$ **71.** p^2 **73. (a)** approximately 250 sec **(b)** $2^{-1.5} \approx 0.3536$ **75.** $y - 10y^2$ **77.** $-4k^{10/3} + 24k^{4/3}$ **79.** $x^2 - x$ **81.** $r - 2 + r^{-1}$, or $r - 2 + \frac{1}{r}$ **83.** $k^{-2}(4k + 1)$ **85.** $4t^{-4}(t^2 + 2)$ **87.** $z^{-1/2}(9 + 2z)$ **89.** $p^{-7/4}(p - 2)$ **91.** $4a^{-7/5}(-a + 4)$ **93.** $(p + 4)^{-3/2}(p^2 + 9p + 21)$ **95.** $6(3x + 1)^{-3/2}(9x^2 + 8x + 2)$ **97.** $2x(2x + 3)^{-5/9}(-16x^4 - 48x^3 - 30x^2 + 9x + 2)$ **99.** $b + a$ **101.** -1 **103.** $\frac{y(xy - 9)}{x^2y^2 - 9}$ **105.** $\frac{2x(1 - 3x^2)}{(x^2 + 1)^5}$ **107.** $\frac{1 + 2x^3 - 2x}{4}$ **109.** $\frac{3x - 5}{(2x - 3)^{4/3}}$ **111.** 27 **113.** 4 **115.** $\frac{1}{100}$

R.7 Exercises (pages 66–69)

1. 5 **3.** 3 **5.** -5 **7.** This expression is not a real number. **9.** 2 **11.** 2 **13. (a)** F **(b)** H **(c)** G **(d)** C **15.** $\sqrt[3]{m^2}$, or $(\sqrt[3]{m})^2$ **17.** $\sqrt[3]{(2m + p)^2}$, or $(\sqrt[3]{2m + p})^2$ **19.** $k^{2/5}$ **21.** $-3 \cdot 5^{1/2}p^{3/2}$ **23.** A **25.** $x \geq 0$ **27.** $|x|$ **29.** $5k^2|m|$ **31.** $|4x - y|$ **33.** $3\sqrt[3]{3}$ **35.** $-2\sqrt[4]{2}$ **37.** $\sqrt{42pqr}$ **39.** $\sqrt[3]{14xy}$ **41.** $-\frac{3}{5}$ **43.** $-\frac{\sqrt[3]{5}}{2}$ **45.** $\frac{\sqrt[4]{m}}{n}$ **47.** -15 **49.** $32\sqrt[3]{2}$ **51.** $2x^2z^4\sqrt{2x}$ **53.** This expression cannot be simplified further. **55.** $\frac{\sqrt{6x}}{3x}$ **57.** $\frac{x^2y\sqrt{xy}}{z}$ **59.** $\frac{2\sqrt[3]{x^2}}{x^2}$ **61.** $\frac{h\sqrt[4]{9g^3hr^2}}{3r^2}$ **63.** $\sqrt{3}$ **65.** $\sqrt[6]{2}$ **67.** $\sqrt[12]{2}$ **69.** $12\sqrt{2x}$ **71.** $7\sqrt[3]{3}$ **73.** $3x\sqrt[4]{x^2y^3} - 2x^2\sqrt[4]{x^2y^3}$ **75.** This expression cannot be simplified further. **77.** -7 **79.** 10 **81.** $11 + 4\sqrt{6}$ **83.** $5\sqrt{6}$ **85.** $\frac{m\sqrt[3]{n^2}}{n}$ **87.** $\frac{x\sqrt[3]{2} - \sqrt[3]{5}}{x^3}$ **89.** $\frac{11\sqrt{2}}{8}$

91. $-\frac{25\sqrt[3]{9}}{18}$ **93.** $\frac{\sqrt{15}-3}{2}$ **95.** $\frac{-7+2\sqrt{14}+\sqrt{7}-2\sqrt{2}}{2}$

97. $\sqrt{p}-2$ **99.** $\frac{3m(2-\sqrt{m+n})}{4-m-n}$ **101.** $\frac{5\sqrt{x}(2\sqrt{x}-\sqrt{y})}{4x-y}$

103. 17.7 ft per sec **105.** $-12°F$ **107.** 2 **109.** 2
111. 3 **113.** It gives six decimal places of accuracy.
115. It first differs in the fourth decimal place.

Chapter R Review Exercises *(pages 73–77)*

1. $\{6, 8, 10, 12, 14, 16, 18, 20\}$ **3.** true **5.** true **7.** false
9. true **11.** true **13.** $\{2, 6, 9, 10\}$ **15.** $\emptyset$ **17.** $\emptyset$
19. $\{1, 2, 3, 4, 6, 8\}$ **21.** $\{1, 2, 3, 4, 5, 6, 7, 8, 9, 10\}$, or U
23. $-12, -6, -\sqrt{4}$ (or -2), $0, 6$ **25.** irrational number,
real number **27.** whole number, integer, rational number,
real number **35.** commutative **37.** associative
39. identity **41.** 3550 **43.** 32 **45.** $-\frac{37}{18}$ **47.** $-\frac{12}{5}$
49. -32 **51.** -13 **53.** $7q^3 - 9q^2 - 8q + 9$
55. $16y^2 + 42y - 49$ **57.** $9k^2 - 30km + 25m^2$
59. $6m^2 - 3m + 5$ **61.** $3b - 8 + \frac{2}{b^2 + 4}$
63. $3(z-4)^2(3z-11)$ **65.** $(z-8k)(z+2k)$
67. $6a^6(4a+5b)(2a-3b)$ **69.** $(7m^4+3n)(7m^4-3n)$
71. $3(9r-10)(2r+1)$ **73.** $(x-1)(y+2)$
75. $(3x-4)(9x-34)$ **77.** $\frac{1}{2k^2(k-1)}$ **79.** $\frac{x+1}{x+4}$
81. $\frac{p+6q}{p+q}$ **83.** $\frac{2m}{m-4}$, or $\frac{-2m}{4-m}$ **85.** $\frac{q+p}{pq-1}$ **87.** $\frac{16}{25}$
89. $-10z^8$ **91.** 1 **93.** $-8y^{11}p$ **95.** $\frac{1}{(p+q)^5}$
97. $-14r^{17/12}$ **99.** $y^{1/2}$ **101.** $\frac{1}{p^2 q^4}$ **103.** $10\sqrt{2}$
105. $5\sqrt[4]{2}$ **107.** $-\frac{\sqrt[3]{50p}}{5p}$ **109.** $\sqrt[12]{m}$ **111.** 66
113. $-9m\sqrt{2m} + 5m\sqrt{m}$, or $m(-9\sqrt{2m} + 5\sqrt{m})$
115. $\frac{6(3+\sqrt{2})}{7}$

In Exercises 117–125, we give only the corrected right-
hand sides of the equations.
117. $x^3 + 5x$ **119.** m^6 **121.** $\frac{a}{2b}$ **123.** $(-2)^{-3}$
125. $\frac{7b}{8b + 7a}$

Chapter R Test *(pages 77–78)*

[R.1] 1. false **2.** true **3.** false **4.** true
[R.2] 5. (a) $-13, -\frac{12}{4}$ (or -3), $0, \sqrt{49}$ (or 7)
(b) $-13, -\frac{12}{4}$ (or -3), $0, \frac{3}{5}, 5.9, \sqrt{49}$ (or 7)
(c) All are real numbers. **6.** 4 **7. (a)** associative
(b) commutative **(c)** distributive **(d)** inverse **8.** 109.6
[R.3] 9. $11x^2 - x + 2$ **10.** $36r^2 - 60r + 25$
11. $3t^3 + 5t^2 + 2t + 8$ **12.** $2x^2 - x - 5 + \frac{3}{x-5}$
13. 8760 **14.** $10,823$ **[R.4] 15.** $(3x-7)(2x-1)$
16. $(x^2+4)(x+2)(x-2)$ **17.** $2m(4m+3)(3m-4)$
18. $(x-2)(x^2+2x+4)(y+3)(y-3)$
19. $(a-b)(a-b+2)$ **20.** $(1-3x^2)(1+3x^2+9x^4)$
[R.5] 21. $\frac{x^4(x+1)}{3(x^2+1)}$ **22.** $\frac{x(4x+1)}{(x+2)(x+1)(2x-3)}$
23. $\frac{2a}{2a-3}$, or $\frac{-2a}{3-2a}$ **24.** $\frac{y}{y+2}$
[R.6, R.7] 25. $3x^2 y^4 \sqrt{2x}$ **26.** $2\sqrt{2x}$ **27.** $x - y$

28. $\frac{7(\sqrt{11}+\sqrt{7})}{2}$ **29.** $\frac{y}{x}$ **30.** $\frac{9}{16}$ **31.** false
32. approximately 2.1 sec

Chapter 1 Equations and Inequalities

1.1 Exercises *(pages 84–86)*

1. true **3.** false **7.** B **9.** $\{-4\}$ **11.** $\{1\}$
13. $\{-\frac{2}{7}\}$ **15.** $\{-\frac{7}{8}\}$ **17.** $\{-1\}$ **19.** $\{10\}$
21. $\{75\}$ **23.** $\{0\}$ **25.** $\{12\}$ **27.** $\{50\}$ **29.** identity;
$\{$all real numbers$\}$ **31.** conditional equation; $\{0\}$
33. identity; $\{$all real numbers$\}$ **35.** contradiction; $\emptyset$
39. $l = \frac{V}{wh}$ **41.** $c = P - a - b$ **43.** $B = \frac{2\mathcal{A} - hb}{h}$, or
$B = \frac{2\mathcal{A}}{h} - b$ **45.** $h = \frac{S - 2\pi r^2}{2\pi r}$, or $h = \frac{S}{2\pi r} - r$
47. $h = \frac{S - 2lw}{2w + 2l}$

Answers in Exercises 49–57 exist in equivalent forms.
49. $x = -3a + b$ **51.** $x = \frac{3a+b}{3-a}$ **53.** $x = \frac{3-3a}{a^2-a-1}$
55. $x = \frac{2a^2}{a^2+3}$ **57.** $x = \frac{m+4}{2m+5}$ **59. (a)** $63 **(b)** $3213
61. $104°F$ **63.** $10°C$ **65.** $37.8°C$ **67.** $463.9°C$
69. $-13.9°C$

1.2 Exercises *(pages 92–97)*

1. 20 mi **3.** $40 **5.** A **7.** D **9.** 90 cm **11.** 6 cm
13. 600 ft, 800 ft, 1000 ft **15.** width: 1.28 cm; length:
1.7 cm **17.** 4 ft **19.** 50 mi **21.** 2.7 mi **23.** 45 min
25. 1 hr, 7 min, 34 sec; It is about $\frac{1}{2}$ the world record time.
27. 35 km per hr **29.** $7\frac{1}{2}$ gal **31.** 2 L **33.** 4 mL
35. short-term note: $100,000; long-term note: $140,000
37. $10,000 at 2.5%; $20,000 at 3% **39.** $50,000 at 1.5%;
$90,000 at 4% **41. (a)** $52 **(b)** $2500 **(c)** $5000
43. (a) $F = 14,000x$ **(b)** about 1.9 hr
45. (a) 19.5 million **(b)** 2016 **(c)** They are quite close.
(d) 15.6 million

1.3 Exercises *(pages 102–104)*

1. true **3.** true **5.** false; *Every* real number is a complex
number. **7.** real, complex **9.** pure imaginary, nonreal
complex, complex **11.** nonreal complex, complex
13. real, complex **15.** pure imaginary, nonreal complex,
complex **17.** $5i$ **19.** $i\sqrt{10}$ **21.** $12i\sqrt{2}$ **23.** $-3i\sqrt{2}$
25. -13 **27.** $-2\sqrt{6}$ **29.** $\sqrt{3}$ **31.** $i\sqrt{3}$ **33.** $\frac{1}{2}$
35. -2 **37.** $-3 - i\sqrt{6}$ **39.** $2 + 2i\sqrt{2}$
41. $-\frac{1}{8} + \frac{\sqrt{2}}{8}i$ **43.** $12 - i$ **45.** 2 **47.** $1 - 10i$
49. $-13 + 4i\sqrt{2}$ **51.** $8 - i$ **53.** $-14 + 2i$
55. $5 - 12i$ **57.** 10 **59.** 13 **61.** 7 **63.** $25i$
65. $12 + 9i$ **67.** $20 + 15i$ **69.** $2 - 2i$ **71.** $\frac{3}{5} - \frac{4}{5}i$
73. $-1 - 2i$ **75.** $5i$ **77.** $8i$ **79.** $-\frac{2}{3}i$
81. $E = 2 + 62i$ **83.** $Z = 12 + 8i$ **85.** i **87.** -1
89. $-i$ **91.** 1 **93.** $-i$ **95.** $-i$

1.4 Exercises *(pages 111–113)*

1. G **3.** C **5.** H **7.** D **9.** D; $\left\{\frac{1}{3}, 7\right\}$

11. C; $\{-4, 3\}$ **13.** $\{2, 3\}$ **15.** $\left\{-\frac{2}{5}, 1\right\}$

17. $\left\{-\frac{3}{4}, 1\right\}$ **19.** $\{\pm 10\}$ **21.** $\left\{\frac{1}{2}\right\}$ **23.** $\left\{-\frac{3}{5}\right\}$

25. $\{\pm 4\}$ **27.** $\left\{\pm 3\sqrt{3}\right\}$ **29.** $\{\pm 9i\}$ **31.** $\left\{\frac{1 \pm 2\sqrt{3}}{3}\right\}$

33. $\left\{-5 \pm i\sqrt{3}\right\}$ **35.** $\left\{\frac{3}{5} \pm \frac{\sqrt{3}}{5}i\right\}$ **37.** $\{1, 3\}$

39. $\left\{-\frac{7}{2}, 4\right\}$ **41.** $\left\{1 \pm \sqrt{3}\right\}$ **43.** $\left\{-\frac{5}{2}, 2\right\}$

45. $\left\{\frac{2 \pm \sqrt{10}}{2}\right\}$ **47.** $\left\{1 \pm \frac{\sqrt{3}}{2}i\right\}$ **49.** He is incorrect

because $c = 0$. **51.** $\left\{\frac{1 \pm \sqrt{5}}{2}\right\}$ **53.** $\left\{3 \pm \sqrt{2}\right\}$

55. $\{1 \pm 2i\}$ **57.** $\left\{\frac{3}{2} \pm \frac{\sqrt{2}}{2}i\right\}$ **59.** $\left\{\frac{-1 \pm \sqrt{97}}{4}\right\}$

61. $\left\{\frac{-2 \pm \sqrt{10}}{2}\right\}$ **63.** $\left\{\frac{-3 \pm \sqrt{41}}{8}\right\}$ **65.** $\{5\}$

67. $\left\{2, -1 \pm i\sqrt{3}\right\}$ **69.** $\left\{-3, \frac{3}{2} \pm \frac{3\sqrt{3}}{2}i\right\}$

71. $t = \frac{\pm \sqrt{2sg}}{g}$ **73.** $v = \frac{\pm \sqrt{FrkM}}{kM}$ **75.** $t = \frac{\pm \sqrt{2a(r - r_0)}}{a}$

77. $t = \frac{v_0 \pm \sqrt{v_0^2 - 64h + 64s_0}}{32}$

79. (a) $x = \frac{y \pm \sqrt{8 - 11y^2}}{4}$ **(b)** $y = \frac{x \pm \sqrt{6 - 11x^2}}{3}$

81. (a) $x = \frac{-2y \pm \sqrt{10y^2 + 4}}{2}$ **(b)** $y = \frac{2x \pm \sqrt{10x^2 - 6}}{3}$

83. 0; one rational solution (a double solution)

85. 1; two distinct rational solutions

87. 84; two distinct irrational solutions

89. -23; two distinct nonreal complex solutions

91. 2304; two distinct rational solutions

In Exercises 95 and 97, there are other possible answers.
95. $a = 1$, $b = -9$, $c = 20$ **97.** $a = 1$, $b = -2$, $c = -1$

Chapter 1 Quiz *(pages 113–114)*

[1.1] 1. $\{2\}$ **2. (a)** contradiction; $\emptyset$ **(b)** identity;
{all real numbers} **(c)** conditional equation; $\left\{\frac{11}{4}\right\}$

3. $y = \frac{3x}{a - 1}$ **[1.2] 4.** $10,000 at 2.5\%$; $20,000 at 3\%$

5. 5.62; The model predicts a wage that is $0.47 greater
than the actual wage. **[1.3] 6.** $-\frac{1}{2} + \frac{\sqrt{6}}{4}i$

7. $\frac{3}{10} - \frac{8}{5}i$ **[1.4] 8.** $\left\{\frac{1}{6} \pm \frac{\sqrt{11}}{6}i\right\}$ **9.** $\left\{\pm \sqrt{29}\right\}$

10. $r = \frac{\pm \sqrt{2s\mathcal{A}\theta}}{\theta}$

1.5 Exercises *(pages 118–124)*

1. A **3.** D **5.** 7, 8 or -8, -7 **7.** 12, 14 or -14, -12
9. 7, 9 or -9, -7 **11.** 9, 11 or -11, -9 **13.** 20, 22
15. 6, 8, 10 **17.** 7 in., 10 in. **19.** 100 yd by 400 yd
21. 9 ft by 12 ft **23.** 20 in. by 30 in. **25.** 1 ft **27.** 4
29. 3.75 cm **31.** 5 ft **33.** $10\sqrt{2}$ ft **35.** 16.4 ft
37. 3000 yd **39. (a)** 1 sec, 5 sec **(b)** 6 sec
41. (a) It will not reach 80 ft. **(b)** 2 sec
43. (a) 0.19 sec, 10.92 sec **(b)** 11.32 sec
45. (a) approximately 19.2 hr **(b)** 84.3 ppm (109.8 is
not in the interval $[50, 100]$.)
47. (a) 781.8 million metric tons **(b)** 2005

49. $13,094 million **51.** $80 - x$ **52.** $300 + 20x$
53. $R = (80 - x)(300 + 20x) = 24,000 + 1300x - 20x^2$
54. 10, 55; Because of the restriction, only $x = 10$ is valid.
The number of apartments rented is 70. **55.** 80 **57.** 4

1.6 Exercises *(pages 133–136)*

1. $-\frac{3}{2}, 6$ **3.** $2, -1$ **5.** 0 **7.** $\{-10\}$ **9.** $\emptyset$ **11.** $\emptyset$

13. $\{-9\}$ **15.** $\{-2\}$ **17.** $\emptyset$ **19.** $\left\{-\frac{5}{2}, \frac{1}{9}\right\}$

21. $\left\{\frac{3}{4}, 1\right\}$ **23.** $\{3, 5\}$ **25.** $\left\{-2, \frac{5}{4}\right\}$ **27.** $1\frac{7}{8}$ hr

29. 78 hr **31.** $13\frac{1}{3}$ hr **33.** 10 min **35.** $\{3\}$ **37.** $\{-1\}$

39. $\{5\}$ **41.** $\{9\}$ **43.** $\{9\}$ **45.** $\emptyset$ **47.** $\{\pm 2\}$

49. $\{0, 3\}$ **51.** $\{-2\}$ **53.** $\left\{-\frac{2}{9}, 2\right\}$ **55.** $\{4\}$

57. $\{-2\}$ **59.** $\left\{\frac{2}{5}, 1\right\}$ **61.** $\{31\}$ **63.** $\{-3, 1\}$

65. $\{25\}$ **67.** $\{-27, 3\}$ **69.** $\{-29, 35\}$ **71.** $\left\{\frac{3}{2}\right\}$

73. $\left\{\frac{1}{4}, 1\right\}$ **75.** $\{0, 8\}$ **77.** $\left\{\pm 1, \pm \frac{\sqrt{10}}{2}\right\}$

79. $\left\{\pm \sqrt{3}, \pm i\sqrt{5}\right\}$ **81.** $\{-63, 28\}$ **83.** $\{0, 31\}$

85. $\left\{\frac{-6 \pm 2\sqrt{3}}{3}, \frac{-4 \pm \sqrt{2}}{2}\right\}$ **87.** $\left\{-\frac{2}{7}, 5\right\}$

89. $\left\{-\frac{1}{27}, \frac{1}{8}\right\}$ **91.** $\left\{\pm \frac{1}{2}, \pm 4\right\}$ **93.** $\{16\}$; $u = -3$
does not lead to a solution of the equation. **94.** $\{16\}$;
9 does not satisfy the equation. **96.** $\{4\}$ **97.** $h = \frac{d^2}{k^2}$

99. $m = (1 - n^{3/4})^{4/3}$ **101.** $e = \frac{Er}{R + r}$

Summary Exercises on Solving Equations *(page 136)*

1. $\{3\}$ **2.** $\{-1\}$ **3.** $\left\{-3 \pm 3\sqrt{2}\right\}$ **4.** $\{2, 6\}$ **5.** $\emptyset$

6. $\{-31\}$ **7.** $\{-6\}$ **8.** $\{6\}$ **9.** $\left\{\frac{1}{5} \pm \frac{2}{5}i\right\}$

10. $\{-2, 1\}$ **11.** $\left\{-\frac{1}{243}, \frac{1}{3125}\right\}$ **12.** $\{-1\}$

13. $\{\pm i, \pm 2\}$ **14.** $\{-2.4\}$ **15.** $\{4\}$ **16.** $\left\{\frac{1}{3} \pm \frac{\sqrt{2}}{3}i\right\}$

17. $\left\{\frac{15}{7}\right\}$ **18.** $\{4\}$ **19.** $\{3, 11\}$ **20.** $\{1\}$

21. $\{x \mid x \neq 3\}$ **22.** $a = \pm \sqrt{c^2 - b^2}$

1.7 Exercises *(pages 145–149)*

1. F **3.** A **5.** I **7.** B **9.** E **13.** $[-4, \infty)$

15. $[-1, \infty)$ **17.** $(-\infty, \infty)$ **19.** $(-\infty, 4)$

21. $\left[-\frac{11}{5}, \infty\right)$ **23.** $\left(-\infty, \frac{48}{7}\right]$ **25.** $[500, \infty)$

27. The product will never break even. **29.** $(-5, 3)$

31. $[3, 6]$ **33.** $(4, 6)$ **35.** $[-9, 9]$ **37.** $(-16, 19]$

39. $(-\infty, -2) \cup (3, \infty)$ **41.** $\left[-\frac{3}{2}, 6\right]$

43. $(-\infty, -3] \cup [-1, \infty)$ **45.** $[-2, 3]$ **47.** $[-3, 3]$

49. $\emptyset$ **51.** $\left[1 - \sqrt{2}, 1 + \sqrt{2}\right]$ **53.** A

55. $\left\{\frac{4}{3}, -2, -6\right\}$ **56.**

57. In the interval $(-\infty, -6)$, choose $x = -10$, for
example. It satisfies the original inequality. In the interval
$(-6, -2)$, choose $x = -4$, for example. It does not satisfy
the inequality. In the interval $\left(-2, \frac{4}{3}\right)$, choose $x = 0$, for
example. It satisfies the original inequality. In the interval
$\left(\frac{4}{3}, \infty\right)$, choose $x = 4$, for example. It does not satisfy the
original inequality.

58. $(-\infty, -6] \cup [-2, \frac{4}{3}]$

59. $[-2, \frac{3}{2}] \cup [3, \infty)$ **61.** $(-\infty, -2] \cup [0, 2]$

63. $(-\infty, -1) \cup (-1, 3)$ **65.** $[-4, -3] \cup [3, \infty)$

67. $(-\infty, \infty)$ **69.** $(-5, 3]$ **71.** $(-\infty, -2)$

73. $(-\infty, 6) \cup [\frac{15}{2}, \infty)$ **75.** $(-\infty, 1) \cup (\frac{9}{5}, \infty)$

77. $(-\infty, -\frac{3}{2}) \cup [-\frac{1}{2}, \infty)$ **79.** $(-2, \infty)$

81. $(0, \frac{4}{11}) \cup (\frac{1}{2}, \infty)$ **83.** $(-\infty, -2] \cup (1, 2)$

85. $(-\infty, 5)$ **87.** $[\frac{3}{2}, \infty)$ **89.** $(\frac{5}{2}, \infty)$

91. $[-\frac{8}{3}, \frac{3}{2}] \cup (6, \infty)$ **93. (a)** 1995 **(b)** 2007

95. between (and inclusive of) 4 sec and 9.75 sec

97. between -1.5 sec and 4 sec

1.8 Exercises *(pages 153–155)*

1. F **3.** D **5.** G **7.** C **9.** $\{-\frac{1}{3}, 1\}$ **11.** $\{\frac{2}{3}, \frac{8}{3}\}$

13. $\{-6, 14\}$ **15.** $\{\frac{5}{2}, \frac{7}{2}\}$ **17.** $\{-\frac{4}{3}, \frac{2}{9}\}$

19. $\{-\frac{7}{3}, -\frac{1}{7}\}$ **21.** $\{1\}$ **23.** $(-\infty, \infty)$ **27.** $(-4, -1)$

29. $(-\infty, -4] \cup [-1, \infty)$ **31.** $(-\frac{3}{2}, \frac{5}{2})$

33. $(-\infty, 0) \cup (6, \infty)$ **35.** $(-\infty, -\frac{2}{3}) \cup (4, \infty)$

37. $[-\frac{2}{3}, 4]$ **39.** $[-1, -\frac{1}{2}]$ **41.** $(-101, -99)$

43. $\{-1, -\frac{1}{2}\}$ **45.** $\{2, 4\}$ **47.** $(-\frac{4}{3}, \frac{2}{3})$

49. $(-\frac{3}{2}, \frac{13}{10})$ **51.** $(-\infty, \frac{3}{2}] \cup [\frac{7}{2}, \infty)$ **53.** $\emptyset$

55. $(-\infty, \infty)$ **57.** $\emptyset$ **59.** $\{-\frac{5}{8}\}$ **61.** $\emptyset$ **63.** $\{-\frac{1}{2}\}$

65. $(-\infty, -\frac{2}{3}) \cup (-\frac{2}{3}, \infty)$ **67.** -6 or 6

68. $x^2 - x = 6; \{-2, 3\}$ **69.** $x^2 - x = -6; \{\frac{1}{2} \pm \frac{\sqrt{23}}{2}i\}$

70. $\{-2, 3, \frac{1}{2} \pm \frac{\sqrt{23}}{2}i\}$ **71.** $\{-\frac{7}{3}, 2, -\frac{1}{6} \pm \frac{\sqrt{167}}{6}i\}$

73. $\{-\frac{1}{4}, 6\}$ **75.** $\{-1, 1\}$ **77.** $\emptyset$

79. $(-\infty, -\frac{1}{3}) \cup (-\frac{1}{3}, \infty)$

In Exercises 81–87, the expression in absolute value bars may be replaced by its additive inverse. For example, in Exercise 81, $p - q$ may be written $q - p$.

81. $|p - q| = 2$ **83.** $|m - 7| \le 2$ **85.** $|p - 9| < 0.0001$

87. $|r - 29| \ge 1$ **89.** $(0.9996, 1.0004)$ **91.** $[6.7, 9.7]$

93. $|F - 730| \le 50$ **95.** $25.33 \le R_L \le 28.17$; $36.58 \le R_E \le 40.92$

Chapter 1 Review Exercises *(pages 160–166)*

1. $\{6\}$ **3.** $\{-\frac{11}{3}\}$ **5.** $f = \frac{AB(p + 1)}{24}$ **7.** A, B

9. 13 in. on each side **11.** $3\frac{3}{7}$ L **13.** 15 mph

15. (a) $A = 36.525x$ **(b)** 2629.8 mg

17. (a) $4.31; The model gives a figure that is $0.51 more than the actual figure of $3.80. **(b)** 47.6 yr after 1956, which is mid-2003. This is close to the minimum wage changing to $5.85 in 2007. **19.** $13 - 3i$ **21.** $-14 + 13i$

23. $19 + 17i$ **25.** 146 **27.** $-30 - 40i$ **29.** $1 - 2i$

31. $-i$ **33.** i **35.** i **37.** $\{-7 \pm \sqrt{5}\}$ **39.** $\{-3, \frac{5}{2}\}$

41. $\{-\frac{3}{2}, 7\}$ **43.** $\{2 \pm \sqrt{6}\}$ **45.** $\{\frac{\sqrt{5} \pm 3}{2}\}$ **47.** D

49. A **51.** 76; two distinct irrational solutions

53. -124; two distinct nonreal complex solutions

55. 0; one rational solution (a double solution)

57. 6.25 sec and 7.5 sec **59.** $\frac{1}{2}$ ft **61.** $565.9 billion

63. $\{\pm i, \pm \frac{1}{2}\}$ **65.** $\{-\frac{7}{24}\}$ **67.** $\emptyset$ **69.** $\{-\frac{7}{4}\}$

71. $\{-15, \frac{5}{2}\}$ **73.** $\{3\}$ **75.** $\{-2, -1\}$ **77.** $\emptyset$

79. $\{-4, 1\}$ **81.** $\{-1\}$ **83.** $(-\frac{7}{13}, \infty)$ **85.** $(-\infty, 1]$

87. $[4, 5]$ **89.** $[-4, 1]$ **91.** $(-\frac{2}{3}, \frac{5}{2})$

93. $(-\infty, -4] \cup [0, 4]$ **95.** $(-\infty, -2) \cup (5, \infty)$

97. $(-2, 0)$ **99.** $(-3, 1) \cup [7, \infty)$ **101. (a)** 79.8 ppb

(b) 87.7 ppb **103. (a)** 20 sec **(b)** between 2 sec and 18 sec **107.** $c \ge 30$ **109.** $x > 14,000$ **111.** $\{-11, 3\}$

113. $\{\frac{11}{27}, \frac{25}{27}\}$ **115.** $\{-\frac{2}{7}, \frac{4}{3}\}$ **117.** $[-6, -3]$

119. $(-\infty, -\frac{1}{7}) \cup (1, \infty)$ **121.** $\{-4, -\frac{2}{3}\}$

123. $(-\infty, \infty)$ **125.** $\{0, -4\}$ **127.** $|k - 6| = 12$ (or $|6 - k| = 12$) **129.** $|t - 5| \ge 0.01$ (or $|5 - t| \ge 0.01$)

Chapter 1 Test *(pages 166–168)*

[1.1] **1.** $\{0\}$ **2.** $\{-12\}$ [1.4] **3.** $\{-\frac{1}{2}, \frac{7}{3}\}$

4. $\{\frac{-1 \pm 2\sqrt{2}}{3}\}$ **5.** $\{-\frac{1}{3} \pm \frac{\sqrt{5}}{3}i\}$ [1.6] **6.** $\emptyset$

7. $\{-\frac{3}{4}\}$ **8.** $\{4\}$ **9.** $\{-3, 1\}$ **10.** $\{-2\}$

11. $\{\pm 1, \pm 4\}$ **12.** $\{-30, 5\}$ [1.8] **13.** $\{-\frac{5}{2}, 1\}$

14. $\{-6, \frac{4}{3}\}$ [1.1] **15.** $W = \frac{S - 2LH}{2H + 2L}$

[1.3] **16. (a)** $5 - 8i$ **(b)** $-29 - 3i$ **(c)** $55 + 48i$

(d) $6 + i$ **17. (a)** -1 **(b)** i **(c)** i

[1.2] **18. (a)** $A = 806,400x$ **(b)** 24,192,000 gal

(c) $P = 40.32x$; approximately 40 pools

(d) approximately 24.8 days **19.** length: 200 m; width: 110 m **20.** cashews: $23\frac{1}{3}$ lb; walnuts: $11\frac{2}{3}$ lb

21. 560 km per hr [1.5] **22. (a)** 1 sec and 5 sec

(b) 6 sec [1.2, 1.5] **23.** B [1.7] **24.** $(-3, \infty)$

25. $[-10, 2]$ **26.** $(-\infty, -1] \cup [\frac{3}{2}, \infty)$

27. $(-\infty, 3) \cup (4, \infty)$ [1.8] **28.** $(-2, 7)$

29. $(-\infty, -6] \cup [5, \infty)$ **30.** $\{-\frac{7}{3}\}$

Chapter 2 Graphs and Functions

2.1 Exercises *(pages 178–180)*

1. false; $(-1, 3)$ lies in quadrant II. **3.** true **5.** true

7. any three of the following: $(2, -5)$, $(-1, 7)$, $(3, -9)$, $(5, -17)$, $(6, -21)$ **9.** any three of the following: $(1997, 36)$, $(1999, 35)$, $(2001, 29)$, $(2003, 22)$ $(2005, 23)$, $(2007, 20)$ **11. (a)** $8\sqrt{2}$ **(b)** $(-9, -3)$

13. (a) $\sqrt{34}$ **(b)** $(\frac{11}{2}, \frac{7}{2})$ **15. (a)** $3\sqrt{41}$ **(b)** $(0, \frac{5}{2})$

17. (a) $\sqrt{133}$ **(b)** $(2\sqrt{2}, \frac{3\sqrt{5}}{2})$ **19.** yes **21.** no

23. yes **25.** yes **27.** no **29.** no **31.** $(-3, 6)$

33. $(5, -4)$ **35.** $(2a - p, 2b - q)$ **37.** 25.35%; This is very close to the actual figure of 25.2%. **39.** \$20,666

Other ordered pairs are possible in Exercises 43–53.

43. (a)

x	y
0	−2
4	0
2	−1

(b) $y = \frac{1}{2}x - 2$

45. (a)

x	y
0	$\frac{5}{3}$
$\frac{5}{2}$	0
4	−1

(b) $2x + 3y = 5$

47. (a)

x	y
0	0
1	1
−2	4

(b) $y = x^2$

49. (a)

x	y
3	0
4	1
7	2

(b) $y = \sqrt{x - 3}$

51. (a)

x	y
4	2
−2	4
0	2

(b) $y = |x - 2|$

53. (a)

x	y
0	0
−1	−1
2	8

(b) $y = x^3$

55. $(4, 0)$ **57.** III; I; IV; IV **59.** yes; no

2.2 Exercises (pages 185–187)

1. (a) $x^2 + y^2 = 36$

(b) $x^2 + y^2 = 36$

3. (a) $(x - 2)^2 + y^2 = 36$

(b) $(x - 2)^2 + y^2 = 36$

5. (a) $x^2 + (y - 4)^2 = 16$ **(b)** $x^2 + (y - 4)^2 = 16$

7. (a) $(x + 2)^2 + (y - 5)^2 = 16$ **(b)** $(x + 2)^2 + (y - 5)^2 = 16$

9. (a) $(x - 5)^2 + (y + 4)^2 = 49$ **(b)** $(x - 5)^2 + (y + 4)^2 = 49$

11. (a) $(x - \sqrt{2})^2 + (y - \sqrt{2})^2 = 2$ **(b)** $(x - \sqrt{2})^2 + (y - \sqrt{2})^2 = 2$

13. (a) $(x - 3)^2 + (y - 1)^2 = 4$
(b) $x^2 + y^2 - 6x - 2y + 6 = 0$
15. (a) $(x + 2)^2 + (y - 2)^2 = 4$
(b) $x^2 + y^2 + 4x - 4y + 4 = 0$
17. B **19.** yes; center: $(-3, -4)$; radius: 4 **21.** yes; center: $(2, -6)$; radius: 6 **23.** yes; center: $\left(-\frac{1}{2}, 2\right)$; radius: 3 **25.** no; The graph is nonexistent.
27. no; The graph is the point $(3, 3)$. **29.** yes; center: $\left(\frac{1}{3}, -\frac{1}{3}\right)$; radius: $\frac{5}{3}$ **31.** $(2, -3)$ **32.** $3\sqrt{5}$
33. $3\sqrt{5}$ **34.** $3\sqrt{5}$ **35.** $(x - 2)^2 + (y + 3)^2 = 45$
36. $(x + 2)^2 + (y + 1)^2 = 41$
37. $(x - 5)^2 + \left(y - \frac{9}{2}\right)^2 = \frac{169}{4}$
39. $(x - 3)^2 + \left(y - \frac{5}{2}\right)^2 = \frac{25}{4}$
41. at $(3, 1)$ **43.** at $(-2, -2)$
45. $(x - 3)^2 + (y - 2)^2 = 4$
47. $\left(2 + \sqrt{7}, 2 + \sqrt{7}\right), \left(2 - \sqrt{7}, 2 - \sqrt{7}\right)$
49. $(2, 3)$ and $(4, 1)$ **51.** $9 + \sqrt{119}, 9 - \sqrt{119}$
53. $\sqrt{113} - 5$

2.3 Exercises (pages 199–203)

1. function **3.** not a function **5.** function **7.** function
9. not a function; domain: $\{0, 1, 2\}$;
range: $\{-4, -1, 0, 1, 4\}$ **11.** function; domain: $\{2, 3, 5, 11, 17\}$; range: $\{1, 7, 20\}$ **13.** function; domain: $\{0, -1, -2\}$; range: $\{0, 1, 2\}$ **15.** function; domain: $\{2005, 2006, 2007, 2008\}$; range: $\{63.5, 60.4, 62.3, 61.2\}$

17. function; domain: $(-\infty, \infty)$; range: $(-\infty, \infty)$
19. not a function; domain: $[3, \infty)$; range: $(-\infty, \infty)$
21. function; domain: $(-\infty, \infty)$; range: $(-\infty, \infty)$
23. function; domain: $(-\infty, \infty)$; range: $[0, \infty)$
25. not a function; domain: $[0, \infty)$; range: $(-\infty, \infty)$
27. function; domain: $(-\infty, \infty)$; range: $(-\infty, \infty)$
29. not a function; domain: $(-\infty, \infty)$; range: $(-\infty, \infty)$
31. function; domain: $[0, \infty)$; range: $[0, \infty)$
33. function; domain: $(-\infty, 0) \cup (0, \infty)$; range:
$(-\infty, 0) \cup (0, \infty)$ **35.** function; domain: $\left[-\frac{1}{4}, \infty\right)$;
range: $[0, \infty)$ **37.** function; domain: $(-\infty, 3) \cup (3, \infty)$;
range: $(-\infty, 0) \cup (0, \infty)$ **39.** B **41.** 4 **43.** -11
45. 3 **47.** $\frac{11}{4}$ **49.** $-3p + 4$ **51.** $3x + 4$ **53.** $-3x - 2$
55. $-6m + 13$ **57. (a)** 2 **(b)** 3 **59. (a)** 15 **(b)** 10
61. (a) 3 **(b)** -3 **63. (a)** $f(x) = -\frac{1}{3}x + 4$ **(b)** 3
65. (a) $f(x) = -2x^2 - x + 3$ **(b)** -18
67. (a) $f(x) = \frac{4}{3}x - \frac{8}{3}$ **(b)** $\frac{4}{3}$ **69.** $f(3) = 4$ **71.** -4
73. (a) 0 **(b)** 4 **(c)** 2 **(d)** 4 **75. (a)** -3 **(b)** -2
(c) 0 **(d)** 2 **77. (a)** $(4, \infty)$ **(b)** $(-\infty, -1)$
(c) $(-1, 4)$ **79. (a)** $(-\infty, 4)$ **(b)** $(4, \infty)$ **(c)** none
81. (a) none **(b)** $(-\infty, -2); (3, \infty)$ **(c)** $(-2, 3)$
83. (a) yes **(b)** $[0, 24]$ **(c)** 1200 megawatts
(d) at 17 hr or 5 P.M.; at 4 A.M. **(e)** $f(12) = 1900$; At
12 noon, electricity use is 1900 megawatts. **(f)** increasing
from 4 A.M. to 5 P.M.; decreasing from midnight to 4 A.M.
and from 5 P.M. to midnight **85. (a)** about 12 noon to
about 8 P.M. **(b)** from midnight until about 6 A.M.
and after 10 P.M. **(c)** about 10 A.M. and 8:30 P.M. **(d)** The
temperature is just below 40° from midnight to 6 A.M.,
when it begins to rise until it reaches a maximum of just
below 65° at 4 P.M. It then begins to fall until it reaches
just under 40° again at midnight.

2.4 Exercises *(pages 211–216)*
1. B **3.** C **5.** A

In Exercises 7–23, we give the domain first and then the range.

7. $(-\infty, \infty); (-\infty, \infty)$ **9.** $(-\infty, \infty); (-\infty, \infty)$

11. $(-\infty, \infty); (-\infty, \infty)$ **13.** $(-\infty, \infty); \{-4\};$ constant function

15. $(-\infty, \infty); (-\infty, \infty)$ **17.** $(-\infty, \infty); (-\infty, \infty)$

19. $\{3\}; (-\infty, \infty)$ **21.** $\{-2\}; (-\infty, \infty)$

23. $\{5\}; (-\infty, \infty)$ **25.** A **27.** D

29. **31.**

33. A, C, D, E **35.** $\frac{2}{5}$ **37.** -2 **39.** 0 **41.** 0 **43.** undefined
45. (a) $m = 3$ **47. (a)** $m = -\frac{3}{2}$
(b) **(b)**

49. (a) $m = \frac{5}{2}$ **51.**
(b)

53. **55.**

57. **59.** D **61.** A **63.** E

65. −$4000 per yr; The value of the machine is decreasing $4000 each year during these years. **67.** 0% per yr (or no change); The percent of pay raise is not changing—it is 3% each year during these years. **69.** zero **71. (b)** 196.6; This means that the average rate of change in the number of radio stations per year is an increase of 196.6.
73. (a) −28.58 thousand mobile homes per yr **(b)** The negative slope means that the number of mobile homes *decreased* by an average of 28.58 thousand each year from 1999 to 2009. **75.** −74.8 thousand per yr; The number of high school dropouts decreased by an average of 74.8 thousand per yr from 1980 to 2008. **77.** 3 **78.** 3
79. the same **80.** $\sqrt{10}$ **81.** $2\sqrt{10}$ **82.** $3\sqrt{10}$
83. The sum is $3\sqrt{10}$, which is equal to the answer in **Exercise 82.** **84.** B; C; A; C **85.** The midpoint is $(3, 3)$, which is the same as the middle entry in the table. **86.** 7.5
87. (a) $C(x) = 10x + 500$ **(b)** $R(x) = 35x$
(c) $P(x) = 25x - 500$ **(d)** 20 units; do not produce
89. (a) $C(x) = 400x + 1650$ **(b)** $R(x) = 305x$
(c) $P(x) = -95x - 1650$ **(d)** $R(x) < C(x)$ for all positive x; don't produce, impossible to make a profit
91. 25 units; $6000

Chapter 2 Quiz *(pages 216–217)*

[2.1] **1.** $\sqrt{41}$ **2.** 2002: 6.25 million; 2004: 6.76 million
3. [2.2] **4.**

5. radius: $\sqrt{17}$; center: $(2, -4)$ [2.3] **6.** 2
7. domain: $(-\infty, \infty)$; range: $[0, \infty)$ **8. (a)** $(-\infty, -3)$
(b) $(-3, \infty)$ **(c)** none [2.4] **9. (a)** $\frac{3}{2}$ **(b)** 0
(c) undefined **10.** −1711 thousand per yr; The number of new motor vehicles sold in the United States decreased an average of 1711 thousand per yr from 2005 to 2009.

2.5 Exercises *(pages 226–230)*

1. D **3.** C **5.** $2x + y = 5$ **7.** $3x + 2y = -7$
9. $x = -8$ **11.** $y = -8$ **13.** $x - 4y = -13$
15. $y = \frac{2}{3}x - 2$ **17.** $x = -6$ (cannot be written in slope-intercept form) **19.** $y = 4$ **21.** $y = 5x + 15$
23. $y = -4x - 3$ **25.** $y = \frac{3}{2}$ **27.** $(-2, 0)$; does not; undefined; $\left(0, \frac{1}{2}\right)$; does not; 0

29. slope: 3;
y-intercept: $(0, -1)$

31. slope: 4;
y-intercept: $(0, -7)$

33. slope: $-\frac{3}{4}$;
y-intercept: $(0, 0)$

35. slope: $-\frac{1}{2}$;
y-intercept: $(0, -2)$

37. slope: $\frac{3}{2}$;
y-intercept: $(0, 1)$

39. (a) -2; $(0, 1)$; $\left(\frac{1}{2}, 0\right)$
(b) $f(x) = -2x + 1$
41. (a) $-\frac{1}{3}$; $(0, 2)$; $(6, 0)$
(b) $f(x) = -\frac{1}{3}x + 2$
43. (a) -200; $(0, 300)$; $\left(\frac{3}{2}, 0\right)$
(b) $f(x) = -200x + 300$
45. (a) $x + 3y = 11$
(b) $y = -\frac{1}{3}x + \frac{11}{3}$
47. (a) $5x - 3y = -13$ **(b)** $y = \frac{5}{3}x + \frac{13}{3}$
49. (a) $y = 1$ **(b)** $y = 1$ **51. (a)** $y = 6$ **(b)** $y = 6$
53. (a) $-\frac{1}{2}$ **(b)** $-\frac{7}{2}$ **55.** $y = 389.5x + 5492$; $7440;
The result is $165 less than the actual figure.
57. (a) $f(x) = 797.3x + 12,881$

The average tuition increase is about $797 per year for the period, because this is the slope of the line.

(b) $f(11) = \$21,651$; This is a fairly good approximation.
(c) $f(x) = 802.3x + 12,432$ **59. (a)** $F = \frac{9}{5}C + 32$
(b) $C = \frac{5}{9}(F - 32)$ **(c)** $-40°$
61. (a) $C = 0.8636I - 296.3$ **(b)** 0.8636 **63.** $\{1\}$
65. $\{4\}$ **67. (a)** $\{12\}$ **69.** $\sqrt{x_1^2 + m_1^2 x_1^2}$
70. $\sqrt{x_2^2 + m_2^2 x_2^2}$ **71.** $\sqrt{(x_2 - x_1)^2 + (m_2 x_2 - m_1 x_1)^2}$
73. $-2x_1 x_2 (m_1 m_2 + 1) = 0$ **74.** Since $x_1 \neq 0$, $x_2 \neq 0$, we have $m_1 m_2 + 1 = 0$, implying that $m_1 m_2 = -1$.
75. If two nonvertical lines are perpendicular, then the product of the slopes of these lines is -1. **77.** yes **79.** no

Summary Exercises on Graphs, Circles, Functions, and Equations *(pages 230–231)*

1. (a) $\sqrt{65}$ **(b)** $\left(\frac{5}{2}, 1\right)$ **(c)** $y = 8x - 19$ **2. (a)** $\sqrt{29}$

(b) $\left(\frac{3}{2}, -1\right)$ **(c)** $y = -\frac{2}{5}x - \frac{2}{5}$ **3. (a)** 5 **(b)** $\left(\frac{1}{2}, 2\right)$

(c) $y = 2$ **4. (a)** $\sqrt{10}$ **(b)** $\left(\frac{3\sqrt{2}}{2}, 2\sqrt{2}\right)$

(c) $y = -2x + 5\sqrt{2}$ **5. (a)** 2 **(b)** $(5, 0)$ **(c)** $x = 5$

6. (a) $4\sqrt{2}$ **(b)** $(-1, -1)$ **(c)** $y = x$

7. (a) $4\sqrt{3}$ **(b)** $\left(4\sqrt{3}, 3\sqrt{5}\right)$ **(c)** $y = 3\sqrt{5}$

8. (a) $\sqrt{34}$ **(b)** $\left(\frac{3}{2}, -\frac{3}{2}\right)$ **(c)** $y = \frac{5}{3}x - 4$

9. $y = -\frac{1}{3}x + \frac{1}{3}$　　　**10.** $y = 3$

11. $(x - 2)^2 + (y + 1)^2 = 9$

12. $x^2 + (y - 2)^2 = 4$

13. $y = -\frac{5}{6}x - \frac{5}{2}$　　　**14.** $y = -\frac{4}{3}x$

15. $y = -\frac{2}{3}x$　　　**16.** $x = -4$

17. yes; center: $(2, -1)$; radius: 3 **18.** no **19.** yes; center: $(6, 0)$; radius: 4 **20.** yes; center: $(-1, -8)$; radius: 2 **21.** no **22.** yes; center: $(0, 4)$; radius: 5 **23.** $\left(4 - \sqrt{7}, 2\right), \left(4 + \sqrt{7}, 2\right)$ **24.** 8

25. (a) domain: $(-\infty, \infty)$; range: $(-\infty, \infty)$

(b) $f(x) = \frac{1}{4}x + \frac{3}{2}$; 1 **26. (a)** domain: $[-5, \infty)$; range: $(-\infty, \infty)$ **(b)** y is not a function of x.

27. (a) domain: $[-7, 3]$; range: $[-5, 5]$

(b) y is not a function of x. **28. (a)** domain: $(-\infty, \infty)$; range: $\left[-\frac{3}{2}, \infty\right)$ **(b)** $f(x) = \frac{1}{2}x^2 - \frac{3}{2}$; $\frac{1}{2}$

2.6 Exercises *(pages 238–242)*

1. E; $(-\infty, \infty)$ **3.** A; $(-\infty, \infty)$ **5.** F; $y = x$

7. H; no **9.** B; $\{\ldots, -3, -2, -1, 0, 1, 2, 3, \ldots\}$

11. $(-\infty, \infty)$ **13.** $[0, \infty)$ **15.** $(-\infty, 3)$; $(3, \infty)$

17. (a) -10 **(b)** -2 **(c)** -1 **(d)** 2

19. (a) -3 **(b)** 1 **(c)** 0 **(d)** 9

21.
$$f(x) = \begin{cases} x - 1 & \text{if } x \le 3 \\ 2 & \text{if } x > 3 \end{cases}$$

23.
$$f(x) = \begin{cases} 4 - x & \text{if } x < 2 \\ 1 + 2x & \text{if } x \ge 2 \end{cases}$$

25.
$$f(x) = \begin{cases} -3 & \text{if } x \le 1 \\ -1 & \text{if } x > 1 \end{cases}$$

27.
$$f(x) = \begin{cases} 2 + x & \text{if } x < -4 \\ -x & \text{if } -4 \le x \le 5 \\ 3x & \text{if } x > 5 \end{cases}$$

29.
$$f(x) = \begin{cases} -\frac{1}{2}x^2 + 2 & \text{if } x \le 2 \\ \frac{1}{2}x & \text{if } x > 2 \end{cases}$$

31.
$$f(x) = \begin{cases} 2x & \text{if } -5 \le x < -1 \\ -2 & \text{if } -1 \le x < 0 \\ x^2 - 2 & \text{if } 0 \le x \le 2 \end{cases}$$

33.
$$f(x) = \begin{cases} x^3 + 3 & \text{if } -2 \le x \le 0 \\ x + 3 & \text{if } 0 < x < 1 \\ 4 + x - x^2 & \text{if } 1 \le x \le 3 \end{cases}$$

In Exercises 35–41, we give one of the possible rules, then the domain, and then the range.

35. $f(x) = \begin{cases} -1 & \text{if } x \le 0 \\ 1 & \text{if } x > 0 \end{cases}$; $(-\infty, \infty)$; $\{-1, 1\}$

37. $f(x) = \begin{cases} 2 & \text{if } x \le 0 \\ -1 & \text{if } x > 1 \end{cases}$; $(-\infty, 0] \cup (1, \infty)$; $\{-1, 2\}$

39. $f(x) = \begin{cases} x & \text{if } x \le 0 \\ 2 & \text{if } x > 0 \end{cases}$; $(-\infty, \infty)$; $(-\infty, 0] \cup \{2\}$

41. $f(x) = \begin{cases} \sqrt[3]{x} & \text{if } x < 1 \\ x + 1 & \text{if } x \ge 1 \end{cases}$; $(-\infty, \infty)$; $(-\infty, 1) \cup [2, \infty)$

43. $(-\infty, \infty)$; $\{\ldots, -2, -1, 0, 1, 2, \ldots\}$

45. $(-\infty, \infty)$; $\{\ldots, -2, -1, 0, 1, 2, \ldots\}$

47.

49.

51. (a) for $[0, 4]$: $y = -0.9x + 42.8$; for $[4, 8]$:
$y = -1.625x + 45.7$

(b) $f(x) = \begin{cases} -0.9x + 42.8 & \text{if } 0 \leq x \leq 4 \\ -1.625x + 45.7 & \text{if } 4 < x \leq 8 \end{cases}$

53. (a) 50,000 gal; 30,000 gal   **(b)** during the first and
fourth days   **(c)** 45,000; 40,000   **(d)** 5000 gal per day
55. (a) $f(x) = 0.80[\![\frac{x}{2}]\!]$ if $6 \leq x \leq 18$ **(b)** \$3.20; \$5.60

2.7 Exercises *(pages 253–257)*

1. (a) B **(b)** D **(c)** E **(d)** A **(e)** C **3. (a)** B **(b)** A
(c) G **(d)** C **(e)** F **(f)** D **(g)** H **(h)** E **(i)** I
5. (a) F **(b)** C **(c)** H **(d)** D **(e)** G **(f)** A
(g) E **(h)** I **(i)** B

7.

9.

11.

13.

15.

17.

19.

21.

23.

25. (a) $(4, 12)$ **(b)** $(8, 16)$
27. (a) $(2, 12)$ **(b)** $(32, 12)$

29.

31.

33. $x = 2$ **35.** y-axis   **37.** x-axis, y-axis, origin   **39.** origin
41. none of these   **43.** odd   **45.** even   **47.** neither

49.

51.

53.

55.

57.

59.

61.

63.

65.

67.

69.

71.

73.

75.

77. (a)

The graph of $g(x)$ is reflected
across the y-axis.

(b)

The graph of $g(x)$ is translated 2 units to the right.

(c)

The graph of $g(x)$ is reflected across the x-axis and translated 2 units up.

79. It is the graph of $f(x) = |x|$ translated 1 unit to the left, reflected across the x-axis, and translated 3 units up. The equation is $y = -|x+1| + 3$. **81.** It is the graph of $g(x) = \sqrt{x}$ translated 1 unit to the right and translated 3 units down. The equation is $y = \sqrt{x-1} - 3$.
83. It is the graph of $g(x) = \sqrt{x}$ translated 4 units to the left, stretched vertically by a factor of 2, and translated 4 units down. The equation is $y = 2\sqrt{x+4} - 4$.
85. $f(-3) = -6$ **87.** $f(9) = 6$ **89.** $f(-3) = -6$
91. $g(x) = 2x + 13$

93. (a)

(b)

Chapter 2 Quiz *(pages 257–258)*

[2.5] 1. (a) $y = 2x + 11$ **(b)** $\left(-\frac{11}{2}, 0\right)$ **2.** $y = -\frac{2}{3}x$
3. (a) $x = -8$ **(b)** $y = 5$ **[2.6] 4. (a)** cubing function; domain: $(-\infty, \infty)$; range: $(-\infty, \infty)$; increasing over $(-\infty, \infty)$ **(b)** absolute value function; domain: $(-\infty, \infty)$; range: $[0, \infty)$; decreasing over $(-\infty, 0)$; increasing over $(0, \infty)$ **(c)** cube root function; domain: $(-\infty, \infty)$; range: $(-\infty, \infty)$; increasing over $(-\infty, \infty)$ **5.** \$2.75

6.

$f(x) = \begin{cases} \sqrt{x} & \text{if } x \geq 0 \\ 2x + 3 & \text{if } x < 0 \end{cases}$

[2.7] 7.

$f(x) = -x^3 + 1$

8.

$f(x) = 2|x-1| + 3$

9. $y = -\sqrt{x+4} - 2$
10. (a) even **(b)** neither **(c)** odd

2.8 Exercises *(pages 266–271)*
1. 12 **3.** -4 **5.** -38 **7.** $\frac{1}{2}$ **9.** $5x - 1$; $x + 9$; $6x^2 - 7x - 20$; $\frac{3x+4}{2x-5}$; All domains are $(-\infty, \infty)$ except for that of $\frac{f}{g}$, which is $\left(-\infty, \frac{5}{2}\right) \cup \left(\frac{5}{2}, \infty\right)$.
11. $3x^2 - 4x + 3$; $x^2 - 2x - 3$; $2x^4 - 5x^3 + 9x^2 - 9x$; $\frac{2x^2 - 3x}{x^2 - x + 3}$; All domains are $(-\infty, \infty)$. **13.** $\sqrt{4x - 1} + \frac{1}{x}$; $\sqrt{4x - 1} - \frac{1}{x}$; $\frac{\sqrt{4x-1}}{x}$; $x\sqrt{4x - 1}$; All domains are $\left[\frac{1}{4}, \infty\right)$. **15.** 260; 400; 660 (all in thousands)
17. 2000–2004 **19.** 6; It represents the dollars (in billions) spent for general science in 2000. **21.** space and other technologies; 1995–2000 **23. (a)** 2 **(b)** 4 **(c)** 0 **(d)** $-\frac{1}{3}$
25. (a) 3 **(b)** -5 **(c)** 2 **(d)** undefined
27. (a) 5 **(b)** 5 **(c)** 0 **(d)** undefined
29.

x	$(f+g)(x)$	$(f-g)(x)$	$(fg)(x)$	$\left(\frac{f}{g}\right)(x)$
-2	6	-6	0	0
0	5	5	0	undefined
2	5	9	-14	-3.5
4	15	5	50	2

33. (a) $2 - x - h$ **(b)** $-h$ **(c)** -1 **35. (a)** $6x + 6h + 2$ **(b)** $6h$ **(c)** 6 **37. (a)** $-2x - 2h + 5$ **(b)** $-2h$ **(c)** -2
39. (a) $\frac{1}{x+h}$ **(b)** $\frac{-h}{x(x+h)}$ **(c)** $\frac{-1}{x(x+h)}$
41. (a) $x^2 + 2xh + h^2$ **(b)** $2xh + h^2$ **(c)** $2x + h$
43. (a) $1 - x^2 - 2xh - h^2$ **(b)** $-2xh - h^2$ **(c)** $-2x - h$
45. (a) $x^2 + 2xh + h^2 + 3x + 3h + 1$
(b) $2xh + h^2 + 3h$ **(c)** $2x + h + 3$ **47.** -5 **49.** 7
51. 6 **53.** -1 **55.** 1 **57.** 9 **59.** 1 **61.** $g(1) = 9$, and $f(9)$ cannot be determined from the table given.
63. (a) $-30x - 33$; $(-\infty, \infty)$ **(b)** $-30x + 52$; $(-\infty, \infty)$
65. (a) $\sqrt{x + 3}$; $[-3, \infty)$ **(b)** $\sqrt{x} + 3$; $[0, \infty)$
67. (a) $(x^2 + 3x - 1)^3$; $(-\infty, \infty)$
(b) $x^6 + 3x^3 - 1$; $(-\infty, \infty)$
69. (a) $\sqrt{3x - 1}$; $\left[\frac{1}{3}, \infty\right)$ **(b)** $3\sqrt{x - 1}$; $[1, \infty)$
71. (a) $\frac{2}{x+1}$; $(-\infty, -1) \cup (-1, \infty)$
(b) $\frac{2}{x} + 1$; $(-\infty, 0) \cup (0, \infty)$
73. (a) $\sqrt{-\frac{1}{x} + 2}$; $(-\infty, 0) \cup \left[\frac{1}{2}, \infty\right)$
(b) $-\frac{1}{\sqrt{x+2}}$; $(-2, \infty)$
75. (a) $\sqrt{\frac{1}{x+5}}$; $(-5, \infty)$ **(b)** $\frac{1}{\sqrt{x+5}}$; $[0, \infty)$
77. (a) $\frac{x}{1 - 2x}$; $(-\infty, 0) \cup \left(0, \frac{1}{2}\right) \cup \left(\frac{1}{2}, \infty\right)$
(b) $x - 2$; $(-\infty, 2) \cup (2, \infty)$
79.

x	$f(x)$	$g(x)$	$g(f(x))$
1	3	2	7
2	1	5	2
3	2	7	5

81. $(f \circ g)(x) = 6x - 11$ and $(g \circ f)(x) = 6x - 7$, so they are not equivalent.

In Exercises 87–91, we give only one of the many possible ways.

87. $g(x) = 6x - 2$, $f(x) = x^2$

89. $g(x) = x^2 - 1$, $f(x) = \sqrt{x}$

91. $g(x) = 6x$, $f(x) = \sqrt{x} + 12$

93. $(f \circ g)(x) = 63{,}360x$ computes the number of inches in x miles.   **95. (a)** $\mathcal{A}(2x) = \sqrt{3}x^2$ **(b)** $64\sqrt{3}$ square units

97. (a) $(\mathcal{A} \circ r)(t) = 16\pi t^2$ **(b)** It defines the area of the leak in terms of the time t, in minutes.   **(c)** 144π ft^2

99. (a) $N(x) = 100 - x$ **(b)** $G(x) = 20 + 5x$

(c) $C(x) = (100 - x)(20 + 5x)$ **(d)** \$9600

Chapter 2 Review Exercises *(pages 275–280)*

1. $\sqrt{85}$; $\left(-\frac{1}{2}, 2\right)$ **3.** 5; $\left(-6, \frac{11}{2}\right)$ **5.** $(22, -6)$

7. $(x + 2)^2 + (y - 3)^2 = 225$

9. $(x + 8)^2 + (y - 1)^2 = 289$ **11.** $x^2 + y^2 = 34$

13. $x^2 + (y - 3)^2 = 13$ **15.** $(2, -3); 1$

17. $\left(-\frac{7}{2}, -\frac{3}{2}\right); \frac{3\sqrt{6}}{2}$ **19.** $3 + 2\sqrt{5}; 3 - 2\sqrt{5}$

21. no; $[-6, 6]; [-6, 6]$ **23.** no; $(-\infty, \infty)$;

$(-\infty, -1] \cup [1, \infty)$ **25.** no; $[0, \infty); (-\infty, \infty)$

27. function of x **29.** not a function of x

31. $(-\infty, 8) \cup (8, \infty)$ **33. (a)** $(2, \infty)$ **(b)** $(-\infty, -2)$

(c) $(-2, 2)$ **35.** -15 **37.** $-2k^2 + 3k - 6$

39.   **41.**

43.   **45.**

47. $y = 0$ **49.**

51. -2 **53.** 0 **55.** $-\frac{11}{2}$ **57.** undefined   **59.** Initially, the car is at home. After traveling 30 mph for 1 hr, the car is 30 mi away from home. During the second hour, the car travels 20 mph until it is 50 mi away. During the third hour, the car travels toward home at 30 mph until it is 20 mi away. During the fourth hour, the car travels away from home at 40 mph until it is 60 mi away from home. During the last hour, the car travels 60 mi at 60 mph until it arrives home.

61. (a) $y = 4.56x + 30.7$; The slope 4.56 indicates that the percent of returns filed electronically rose an average of 4.56% per year during this period.   **(b)** 58.1%

63. (a) $y = -2x + 1$ **(b)** $2x + y = 1$

65. (a) $y = 3x - 7$ **(b)** $3x - y = 7$ **67. (a)** $y = -10$

(b) $y = -10$ **69. (a)** not possible   **(b)** $x = -7$

71.   **73.**

$f(x) = |x| - 3$

$f(x) = -(x + 1)^2 + 3$

75.   **77.**

$f(x) = [\![x - 3]\!]$

$f(x) = \begin{cases} -4x + 2 \text{ if } x \le 1 \\ 3x - 5 \text{ if } x > 1 \end{cases}$

79. $2x$ **81.** true   **83.** false; For example, $f(x) = x^2$ is even, and $(2, 4)$ is on the graph but $(2, -4)$ is not.

85. true   **87.** x-axis   **89.** y-axis   **91.** none of these

93. y-axis   **95.** Reflect the graph of $f(x) = |x|$ across the x-axis.   **97.** Translate the graph of $f(x) = |x|$ to the right 4 units and stretch it vertically by a factor of 2.

99. $y = -3x - 4$

101. (a)   **(b)**

(c)   **(d)**

103. $3x^4 - 9x^3 - 16x^2 + 12x + 16$ **105.** 68   **107.** $-\frac{23}{4}$

109. $(-\infty, \infty)$ **111.** C and D   **113.** $2x + h - 5$

115. $x - 2$ **117.** 1   **119.** $\left(-\infty, -\sqrt{2}\right] \cup \left[\sqrt{2}, \infty\right)$

121. 0   **123.** undefined   **125.** 2   **127.** 1

129. $f(x) = 36x$; $g(x) = 1760x$; $(f \circ g)(x) = f(g(x)) = f(1760x) = 36(1760x) = 63{,}360x$

131. $V(r) = \frac{4}{3}\pi(r + 3)^3 - \frac{4}{3}\pi r^3$

Chapter 2 Test *(pages 280–282)*

[2.3] 1. (a) D   **(b)** D   **(c)** C   **(d)** B   **(e)** C

(f) C   **(g)** C   **(h)** D   **(i)** D   **(j)** C   **[2.4] 2.** $\frac{3}{5}$

[2.1] 3. $\sqrt{34}$ **4.** $\left(\frac{1}{2}, \frac{5}{2}\right)$ **[2.5] 5.** $3x - 5y = -11$

6. $f(x) = \frac{3}{5}x + \frac{11}{5}$ **[2.2] 7. (a)** $x^2 + y^2 = 4$

(b) $(x - 1)^2 + (y - 4)^2 = 1$

8.

$x^2 + y^2 + 4x - 10y + 13 = 0$

[2.3] 9. (a) not a function; domain: $[0, 4]$; range: $[-4, 4]$

(b) function; domain: $(-\infty, -1) \cup (-1, \infty)$; range: $(-\infty, 0) \cup (0, \infty)$; decreasing on $(-\infty, -1)$ and on $(-1, \infty)$

[2.5] 10. (a) $x = 5$ **(b)** $y = -3$

11. (a) $y = -3x + 9$ **(b)** $y = \frac{1}{3}x + \frac{7}{3}$

[2.3] 12. (a) $(-\infty, -3)$ **(b)** $(4, \infty)$ **(c)** $(-3, 4)$

(d) $(-\infty, -3); [-3, 4]; (4, \infty)$ **(e)** $(-\infty, \infty)$

(f) $(-\infty, 2)$

[2.6, 2.7]

13.

14.

15.

$f(x) = \begin{cases} 3 & \text{if } x < -2 \\ 2 - \frac{1}{2}x & \text{if } x \geq -2 \end{cases}$

16. (a)

(b)

(c)

(d)

(e)

17. It is translated 2 units to the left, stretched vertically by a factor of 2, reflected across the x-axis, and translated 3 units down. **18. (a)** yes **(b)** yes **(c)** yes

[2.8] 19. (a) $2x^2 - x + 1$ **(b)** $\frac{2x^2 - 3x + 2}{-2x + 1}$

(c) $\left(-\infty, \frac{1}{2}\right) \cup \left(\frac{1}{2}, \infty\right)$ **(d)** $4x + 2h - 3$

20. (a) 0 **(b)** -12 **(c)** 1 **21.** $\sqrt{2x - 6}; [3, \infty)$

22. $2\sqrt{x + 1} - 7; [-1, \infty)$ **[2.6] 23.** \$6.75

[2.4] 24. (a) $C(x) = 3300 + 4.50x$ **(b)** $R(x) = 10.50x$

(c) $R(x) - C(x) = 6.00x - 3300$ **(d)** 551

Chapter 3 Polynomial and Rational Functions

3.1 Exercises *(pages 292–300)*

1. (a) domain: $(-\infty, \infty)$; range: $[-4, \infty)$ **(b)** $(-3, -4)$

(c) $x = -3$ **(d)** $(0, 5)$ **(e)** $(-5, 0), (-1, 0)$

3. (a) domain: $(-\infty, \infty)$; range: $(-\infty, 2]$ **(b)** $(-3, 2)$

(c) $x = -3$ **(d)** $(0, -16)$

(e) $(-4, 0), (-2, 0)$ **5.** B **7.** D **9.**

(d) The greater $|a|$ is, the narrower the parabola will be. The smaller $|a|$ is, the wider the parabola will be.

11.

(d) The graph of $y = (x - h)^2$ is translated h units to the right if h is positive and $|h|$ units to the left if h is negative.

13. (a) $(2, 0)$

(b) $x = 2$

(c) $(-\infty, \infty)$

(d) $[0, \infty)$

(e) $(2, \infty)$

(f) $(-\infty, 2)$

15. (a) $(-3, -4)$

(b) $x = -3$

(c) $(-\infty, \infty)$

(d) $[-4, \infty)$

(e) $(-3, \infty)$

(f) $(-\infty, -3)$

$f(x) = (x - 2)^2$

$f(x) = (x + 3)^2 - 4$

17. (a) $(-1, -3)$

(b) $x = -1$

(c) $(-\infty, \infty)$

(d) $(-\infty, -3]$

(e) $(-\infty, -1)$

(f) $(-1, \infty)$

19. (a) $(1, 2)$

(b) $x = 1$

(c) $(-\infty, \infty)$

(d) $[2, \infty)$

(e) $(1, \infty)$

(f) $(-\infty, 1)$

$f(x) = -\frac{1}{2}(x + 1)^2 - 3$

$f(x) = x^2 - 2x + 3$

21. (a) $(5, -4)$

(b) $x = 5$

(c) $(-\infty, \infty)$

(d) $[-4, \infty)$

(e) $(5, \infty)$

(f) $(-\infty, 5)$

23. (a) $(-3, 2)$

(b) $x = -3$

(c) $(-\infty, \infty)$

(d) $(-\infty, 2]$

(e) $(-\infty, -3)$

(f) $(-3, \infty)$

$f(x) = x^2 - 10x + 21$

$f(x) = -2x^2 - 12x - 16$

25. (a) $(-3, 4)$

(b) $x = -3$

(c) $(-\infty, \infty)$

(d) $(-\infty, 4]$

(e) $(-\infty, -3)$

(f) $(-3, \infty)$

$f(x) = -\frac{1}{2}x^2 - 3x - \frac{1}{2}$

27. 3 **29.** none **31.** E **33.** D **35.** C

37. $f(x) = \frac{1}{4}(x - 2)^2 - 1$, or $f(x) = \frac{1}{4}x^2 - x$

39. $f(x) = -2(x - 1)^2 + 4$, or $f(x) = -2x^2 + 4x + 2$

41. linear; positive **43.** quadratic; positive **45.** quadratic; negative **47.** 10 and 10 **49.** 20 units; $210 **51.** 5 in.; 25 million mosquitos **53. (a)** $f(t) = -16t^2 + 200t + 50$ **(b)** 6.25 sec; 675 ft **(c)** between approximately 1.41 and 11.09 sec **(d)** approximately 12.75 sec

55. (a) $640 - 2x$ **(b)** $0 < x < 320$ **(c)** $\mathscr{A}(x) = -2x^2 + 640x$ **(d)** between approximately 57.04 ft and 85.17 ft or 234.83 ft and 262.96 ft **(e)** 160 ft by 320 ft; The maximum area is 51,200 ft^2.

57. (a) $2x$ **(b)** length: $2x - 4$; width: $x - 4$; $x > 4$ **(c)** $V(x) = 4x^2 - 24x + 32$ **(d)** 8 in. by 20 in. **(e)** 13.0 in. to 14.2 in. **59. (a)** approximately 23.32 ft per sec **(b)** approximately 12.88 ft **61.** 49.0 (percent)

63. 2007

65. (a) 1,100,000

(b) quadratic; The rate at which the number of cases is increasing appears to be slowing down.

(c) $f(x) = -1328x^2 + 71,882x + 181,195$

(d) 1,100,000

The graph of f models the data very well.

(e) 2009: 1,067,545; 2010: 1,087,635 **(f)** 20,090

67. (a) 15

(b) $f(x) = 0.0055(x - 40)^2 + 4.7$

(c) There is a good fit.

(d) $g(x) = 0.0045x^2 - 0.3357x + 11.62$

(e) $f(85) = 15.8\%$; $g(85) = 15.6\%$

69. $c = 25$ **71.** $f(x) = \frac{1}{2}x^2 - \frac{7}{2}x + 5$ **73.** $(3, 6)$

75.

$f(x) = x^2 + 2x - 8$

76. $(-4, 0)$ and $(2, 0)$; $\{-4, 2\}$

77. $(-4, 2)$

78. $(-\infty, -4) \cup (2, \infty)$

79. $(-2, 3)$

81. $\left(-\infty, -\frac{3}{2}\right] \cup [6, \infty)$

83. $\left[2 - \sqrt{5}, 2 + \sqrt{5}\right]$

3.2 Exercises (pages 305–306)

1. $x^2 + 2x + 9$ **3.** $5x^3 + 2x - 3$ **5.** $x^3 + 2x + 1$

7. $x^4 + x^3 + 2x - 1 + \frac{3}{x + 2}$

9. $-9x^2 - 10x - 27 + \frac{-52}{x - 2}$ **11.** $\frac{1}{3}x^2 - \frac{1}{9}x + \frac{1}{27}$

13. $x^3 - x^2 - 6x$ **15.** $x^2 + x + 1$

17. $x^4 - x^3 + x^2 - x + 1$

19. $f(x) = (x + 1)(2x^2 - x + 2) - 10$

21. $f(x) = (x + 2)(x^2 + 2x + 1) + 0$

23. $f(x) = (x - 3)(4x^3 + 9x^2 + 7x + 20) + 60$

25. $f(x) = (x + 1)(3x^3 + x^2 - 11x + 11) + 4$

27. 0 **29.** -1 **31.** -6 **33.** -5 **35.** 7 **37.** $-6 - i$

39. 0 **41.** yes **43.** yes **45.** no; -9 **47.** yes **49.** yes

51. no; $\frac{357}{125}$ **53.** yes **55.** no; $13 + 7i$ **57.** no; $-2 + 7i$

59. -12; $(-2, -12)$ **60.** 0; $(-1, 0)$ **61.** $\frac{15}{8}$; $\left(-\frac{1}{2}, \frac{15}{8}\right)$

62. 2; $(0, 2)$ **63.** 0; $(1, 0)$ **64.** $-\frac{5}{8}$; $\left(\frac{3}{2}, -\frac{5}{8}\right)$

65. 0; $(2, 0)$ **66.** 8; $(3, 8)$

67.

$f(x) = x^3 - 2x^2 - x + 2$

68.

$f(x) = -x^3 - x^2 + 2x$

3.3 Exercises (pages 315–317)

1. true **3.** false; -2 is a zero of multiplicity 4. **5.** yes

7. no **9.** yes **11.** no **13.** no **15.** yes

17. $f(x) = (x - 2)(2x - 5)(x + 3)$

19. $f(x) = (x + 3)(3x - 1)(2x - 1)$

21. $f(x) = (x + 4)(3x - 1)(2x + 1)$

23. $f(x) = (x - 3i)(x + 4)(x + 3)$

25. $f(x) = [x - (1 + i)](2x - 1)(x + 3)$

27. $f(x) = (x + 2)^2(x + 1)(x - 3)$

29. $-1 \pm i$ **31.** $3, 2 + i$ **33.** $i, \pm 2i$

35. (a) $\pm 1, \pm 2, \pm 5, \pm 10$ **(b)** $-1, -2, 5$

(c) $f(x) = (x + 1)(x + 2)(x - 5)$ **37. (a)** $\pm 1, \pm 2, \pm 3, \pm 5, \pm 6, \pm 10, \pm 15, \pm 30$ **(b)** $-5, -3, 2$

(c) $f(x) = (x + 5)(x + 3)(x - 2)$ **39. (a)** $\pm 1, \pm 2, \pm 3, \pm 4, \pm 6, \pm 12, \pm \frac{1}{2}, \pm \frac{3}{2}, \pm \frac{1}{3}, \pm \frac{2}{3}, \pm \frac{4}{3}, \pm \frac{1}{6}$

(b) $-4, -\frac{1}{3}, \frac{3}{2}$ **(c)** $f(x) = (x + 4)(3x + 1)(2x - 3)$

41. (a) $\pm 1, \pm 2, \pm 3, \pm 4, \pm 6, \pm 12, \pm\frac{1}{2}, \pm\frac{3}{2}, \pm\frac{1}{3},$
$\pm\frac{2}{3}, \pm\frac{4}{3}, \pm\frac{1}{4}, \pm\frac{3}{4}, \pm\frac{1}{6}, \pm\frac{1}{8}, \pm\frac{3}{8}, \pm\frac{1}{12}, \pm\frac{1}{24}$
(b) $-\frac{3}{2}, -\frac{2}{3}, \frac{1}{2}$ **(c)** $f(x) = 2(2x + 3)(3x + 2)(2x - 1)$
43. 2 (multiplicity 3), $\pm\sqrt{7}$ **45.** $0, 2, -3, 1, -1$
47. -2 (multiplicity 5), 1 (multiplicity 5), $1 - \sqrt{3}$
(multiplicity 2) **49.** $f(x) = -3x^3 + 6x^2 + 33x - 36$
51. $f(x) = -\frac{1}{2}x^3 - \frac{1}{2}x^2 + x$
53. $f(x) = \frac{1}{6}x^3 + \frac{3}{2}x^2 + \frac{9}{2}x + \frac{9}{2}$
55. $f(x) = 5x^3 - 10x^2 + 5x$

In Exercises 57–73, we give only one possible answer.
57. $f(x) = x^2 - 10x + 26$
59. $f(x) = x^5 - 2x^4 + 3x^3 - 2x^2 + 2x$
61. $f(x) = x^3 - 3x^2 + x + 1$
63. $f(x) = x^4 - 6x^3 + 10x^2 + 2x - 15$
65. $f(x) = x^3 - 8x^2 + 22x - 20$
67. $f(x) = x^4 - 4x^3 + 5x^2 - 2x - 2$
69. $f(x) = x^4 - 16x^3 + 98x^2 - 240x + 225$
71. $f(x) = x^5 - 12x^4 + 74x^3 - 248x^2 + 445x - 500$
73. $f(x) = x^4 - 6x^3 + 17x^2 - 28x + 20$

75.

Positive	Negative	Nonreal Complex
2	1	0
0	1	2

77.

Positive	Negative	Nonreal Complex
3	0	0
1	0	2

79.

Positive	Negative	Nonreal Complex
1	1	2

81.

Positive	Negative	Nonreal Complex
4	0	0
2	0	2
0	0	4

83.

Positive	Negative	Nonreal Complex
2	3	0
2	1	2
0	3	2
0	1	4

85.

Positive	Negative	Nonreal Complex
0	5	0
0	3	2
0	1	4

87.

Positive	Negative	Nonreal Complex
2	3	0
2	1	2
0	3	2
0	1	4

89.

Positive	Negative	Nonreal Complex
4	2	0
4	0	2
2	2	2
2	0	4
0	2	4
0	0	6

91. $-5, 3, \pm 2i\sqrt{3}$ **93.** $1, 1, 1, -4$ **95.** $-3, -3, 0,$
$\frac{1 \pm i\sqrt{31}}{4}$ **97.** $2, 2, 2, \pm i\sqrt{2}$ **99.** $-\frac{1}{2}, 1, \pm 2i$
101. $-\frac{1}{5}, 1 \pm i\sqrt{5}$ **103.** $\pm 2i, \pm 5i$ **105.** $\pm i, \pm i$
107. $0, 0, 3 \pm \sqrt{2}$ **109.** $3, 3, 1 \pm i\sqrt{7}$ **111.** $\pm 2,$
$\pm 3, \pm 2i$ **117.** 2

3.4 Exercises *(pages 330–337)*
1. A **3.** one **5.** B and D **7.** $f(x) = x(x + 5)^2(x - 3)$

9.

(a) $(0, \infty)$ **(b)** $(-\infty, 0)$

11.

(a) none **(b)** $(-\infty, \infty)$

13.

(a) $(-\infty, \infty)$ **(b)** none

15.

(a) none **(b)** $(-\infty, \infty)$

17. $f(x) = (x - 1)^4 + 2$

(a) $(1, \infty)$
(b) $(-\infty, 1)$

19. $f(x) = \frac{1}{2}(x - 2)^2 + 4$

(a) $(2, \infty)$
(b) $(-\infty, 2)$

21. **23.** **25.** **27.**

29.

$f(x) = x^3 + 5x^2 + 2x - 8$

31.

$f(x) = 2x(x - 3)(x + 2)$

33.

$f(x) = x^2(x - 2)(x + 3)^2$

35.

$f(x) = (3x - 1)(x + 2)^2$

37.

$f(x) = x^3 + 5x^2 - x - 5$

39.

$f(x) = x^3 - x^2 - 2x$

41.

$f(x) = 2x^3(x^2 - 4)(x - 1)$

43.

$f(x) = 2x^3 - 5x^2 - x + 6$

45.

$f(x) = 3x^4 - 7x^3 - 6x^2 + 12x + 8$

47. $f(2) = -2 < 0;\ f(3) = 1 > 0$

49. $f(0) = 7 > 0;\ f(1) = -1 < 0$

51. $f(1) = -6 < 0;\ f(2) = 16 > 0$

53. $f(3.2) = -3.8144 < 0;\ f(3.3) = 7.1891 > 0$

55. $f(-1) = -35 < 0;\ f(0) = 12 > 0$

65. $f(x) = \frac{1}{2}(x + 6)(x - 2)(x - 5)$, or
$f(x) = \frac{1}{2}x^3 - \frac{1}{2}x^2 - 16x + 30$

67. $f(x) = (x - 1)^3(x + 1)^3$, or
$f(x) = x^6 - 3x^4 + 3x^2 - 1$

69. $f(x) = (x - 3)^2(x + 3)^2$, or $f(x) = x^4 - 18x^2 + 81$

71. $f(1.25) = -14.21875$ **73.** $f(1.25) = 29.046875$

$f(x) = 2x(x - 3)(x + 2)$

$f(x) = (3x - 1)(x + 2)^2$

75. 2.7807764   **77.** 1.543689   **79.** −3.0, −1.4, 1.4

81. −1.1, 1.2   **83.** (−3.44, 26.15)   **85.** (−0.09, 1.05)

87. (−0.20, −28.62)   **89.** Answers will vary.

91.

$f(x) = x^3 - 3x^2 - 6x + 8$
$= (x - 4)(x - 1)(x + 2)$

(a) $\{-2, 1, 4\}$

(b) $(-\infty, -2) \cup (1, 4)$

(c) $(-2, 1) \cup (4, \infty)$

92.

$f(x) = x^3 + 4x^2 - 11x - 30$
$= (x - 3)(x + 2)(x + 5)$

(a) $\{-5, -2, 3\}$

(b) $(-\infty, -5) \cup (-2, 3)$

(c) $(-5, -2) \cup (3, \infty)$

93.

$f(x) = 2x^4 - 9x^3 - 5x^2 + 57x - 45$
$= (x - 3)^2(2x + 5)(x - 1)$

(a) $\{-2.5, 1, 3 \text{ (multiplicity 2)}\}$

(b) $(-2.5, 1)$

(c) $(-\infty, -2.5) \cup (1, 3) \cup (3, \infty)$

94.

$f(x) = 4x^4 + 27x^3 - 42x^2 - 445x - 300$
$= (x + 5)^2(4x + 3)(x - 4)$

(a) $\{-5 \text{ (multiplicity 2)}, -0.75, 4\}$

(b) $(-0.75, 4)$

(c) $(-\infty, -5) \cup (-5, -0.75) \cup (4, \infty)$

95.

$f(x) = -x^4 - 4x^3 + 3x^2 + 18x$
$= x(2 - x)(x + 3)^2$

(a) $\{-3 \text{ (multiplicity 2)}, 0, 2\}$

(b) $\{-3\} \cup [0, 2]$

(c) $(-\infty, 0] \cup [2, \infty)$

96.

$f(x) = -x^4 + 2x^3 + 8x^2$
$= x^2(4 - x)(x + 2)$

(a) $\{-2, 0 \text{ (multiplicity 2)}, 4\}$

(b) $[-2, 4]$

(c) $(-\infty, -2] \cup \{0\} \cup [4, \infty)$

97. (a) $0 < x < 6$ (b) $V(x) = x(18 - 2x)(12 - 2x)$, or
$V(x) = 4x^3 - 60x^2 + 216x$ (c) $x \approx 2.35$; about
228.16 in.³   (d) $0.42 < x < 5$ **99.** (a) $x - 1$; $(1, \infty)$

(b) $\sqrt{x^2 - (x - 1)^2}$ (c) $2x^3 - 5x^2 + 4x - 28{,}225 = 0$

(d) hypotenuse: 25 in.; legs: 24 in. and 7 in.   **101.** 3 ft

103. (a) about 7.13 cm; The ball floats partly above the
surface.   (b) The sphere is more dense than water and
sinks below the surface.   (c) 10 cm; The balloon is
submerged with its top even with the surface.

105. (a)

(b)

$y = 33.93x + 113.4$

(c) $y = -0.0032x^3 + 0.4245x^2 + 16.64x + 323.1$

(d) linear: 1572 ft; cubic: 1569 ft   **(e)** The cubic function appears slightly better because only one data point is not on the curve.   **107.** C

Summary Exercises on Polynomial Functions, Zeros, and Graphs (pages 338–339)

1. (a)

Positive	Negative	Nonreal Complex
2	1	0
0	1	2

(b) $\pm 1,\ \pm 2,\ \pm 3,\ \pm 4,\ \pm 6,\ \pm 8,\ \pm 12,\ \pm 24,$ $\pm\frac{1}{2},\ \pm\frac{3}{2},\ \pm\frac{1}{3},\ \pm\frac{2}{3},\ \pm\frac{4}{3},\ \pm\frac{8}{3},\ \pm\frac{1}{6}$
(c) $-\frac{1}{2}, \frac{4}{3}, 6$ **(d)** no other complex zeros

2. (a)

Positive	Negative	Nonreal Complex
2	1	0
0	1	2

(b) $\pm 1,\ \pm 3,\ \pm\frac{1}{2},\ \pm\frac{3}{2}$ **(c)** $-1, \frac{1}{2}, 3$
(d) no other complex zeros

3. (a)

Positive	Negative	Nonreal Complex
4	0	0
2	0	2
0	0	4

(b) $\pm 1,\ \pm 2,\ \pm 4,\ \pm 8,\ \pm\frac{1}{3},\ \pm\frac{2}{3},\ \pm\frac{4}{3},\ \pm\frac{8}{3}$
(c) $\frac{2}{3}, 1$ **(d)** $-2i, 2i$

4. (a)

Positive	Negative	Nonreal Complex
3	1	0
1	1	2

(b) $\pm 1,\ \pm 2,\ \pm 3,\ \pm 6,\ \pm 9,\ \pm 18,\ \pm\frac{1}{2},\ \pm\frac{3}{2},\ \pm\frac{9}{2}$
(c) $-\frac{1}{2}, 2$ **(d)** $-3i, 3i$

5. (a)

Positive	Negative	Nonreal Complex
3	1	0
1	1	2

(b) $\pm 1,\ \pm 2,\ \pm\frac{1}{2},\ \pm\frac{1}{3},\ \pm\frac{2}{3},\ \pm\frac{1}{6}$ **(c)** $\frac{1}{3}, \frac{1}{2}$
(d) $-\sqrt{2}, \sqrt{2}$

6. (a)

Positive	Negative	Nonreal Complex
2	2	0
2	0	2
0	2	2
0	0	4

(b) $\pm 1,\ \pm 2,\ \pm 3,\ \pm 4,\ \pm 6,\ \pm 12,\ \pm\frac{1}{5},\ \pm\frac{2}{5},\ \pm\frac{3}{5},\ \pm\frac{4}{5},$ $\pm\frac{6}{5},\ \pm\frac{12}{5}$ **(c)** $-2, \frac{2}{5}$ **(d)** $-\sqrt{3}, \sqrt{3}$

7. (a)

Positive	Negative	Nonreal Complex
4	0	0
2	0	2
0	0	4

(b) $0,\ \pm 1,\ \pm 2,\ \pm 4,\ \pm 8,\ \pm 16$
(c) $0, 2$ (multiplicity 2)   **(d)** $1 - i\sqrt{3}, 1 + i\sqrt{3}$

8. (a)

Positive	Negative	Nonreal Complex
1	3	0
1	1	2

(b) $\pm 1,\ \pm 3,\ \pm 9,\ \pm\frac{1}{2},\ \pm\frac{3}{2},\ \pm\frac{9}{2}$ **(c)** -3 (multiplicity 2)
(d) $\frac{2 - \sqrt{6}}{2}, \frac{2 + \sqrt{6}}{2}$

9. (a)

Positive	Negative	Nonreal Complex
1	3	0
1	1	2

(b) $\pm 1,\ \pm\frac{1}{2},\ \pm\frac{1}{4},\ \pm\frac{1}{8}$ **(c)** $-1, \frac{1}{2}$
(d) $-\frac{1}{4} - \frac{\sqrt{3}}{4}i,\ -\frac{1}{4} + \frac{\sqrt{3}}{4}i$

10. (a)

Positive	Negative	Nonreal Complex
3	2	0
3	0	2
1	2	2
1	0	4

(b) $\pm 1,\ \pm 2,\ \pm 3,\ \pm 6,\ \pm\frac{1}{2},\ \pm\frac{3}{2}$ **(c)** $-3, -2, \frac{1}{2}, 1$
(multiplicity 2)   **(d)** no other complex zeros

11. (a)

Positive	Negative	Nonreal Complex
1	3	0
1	1	2

(b) $\pm 1,\ \pm 2,\ \pm 3,\ \pm 6$ **(c)** $-3, -1$ (multiplicity 2), 2
(d) no other real zeros   **(e)** no other complex zeros
(f) $(-3, 0), (-1, 0), (2, 0)$ **(g)** $(0, -6)$
(h) $f(4) = 350; (4, 350)$
(i) ↺↗   **(j)**

$f(x) = x^4 + 3x^3 - 3x^2 - 11x - 6$

12. (a)

Positive	Negative	Nonreal Complex
3	2	0
3	0	2
1	2	2
1	0	4

(b) $\pm 1,\ \pm 3,\ \pm 5,\ \pm 9,\ \pm 15,\ \pm 45,\ \pm\frac{1}{2},\ \pm\frac{3}{2},\ \pm\frac{5}{2},$ $\pm\frac{9}{2},\ \pm\frac{15}{2},\ \pm\frac{45}{2}$ **(c)** $-3, \frac{1}{2}, 5$ **(d)** $-\sqrt{3}, \sqrt{3}$
(e) no other complex zeros   **(f)** $(-3, 0), \left(\frac{1}{2}, 0\right), (5, 0),$ $\left(-\sqrt{3}, 0\right), \left(\sqrt{3}, 0\right)$

12. (g) $(0, 45)$ **(h)** $f(4) = 637; (4, 637)$ **(i)**
(j)

$f(x) = -2x^5 + 5x^4 + 34x^3 - 30x^2 - 84x + 45$

13. (a)

Positive	Negative	Nonreal Complex
4	1	0
2	1	2
0	1	4

(b) $\pm 1, \pm 5, \pm \frac{1}{2}, \pm \frac{5}{2}$ **(c)** 5 **(d)** $-\frac{\sqrt{2}}{2}, \frac{\sqrt{2}}{2}$
(e) $-i, i$ **(f)** $\left(-\frac{\sqrt{2}}{2}, 0\right), \left(\frac{\sqrt{2}}{2}, 0\right), (5, 0)$ **(g)** $(0, 5)$
(h) $f(4) = -527; (4, -527)$ **(i)**

(j)

$f(x) = 2x^5 - 10x^4 + x^3 - 5x^2 - x + 5$

14. (a)

Positive	Negative	Nonreal Complex
2	2	0
2	0	2
0	2	2
0	0	4

(b) $\pm 1, \pm 2, \pm 3, \pm 6, \pm 9, \pm 18, \pm \frac{1}{3}, \pm \frac{2}{3}$
(c) $-\frac{2}{3}, 3$ **(d)** $\frac{-1 + \sqrt{13}}{2}, \frac{-1 - \sqrt{13}}{2}$ **(e)** no other complex
zeros **(f)** $\left(-\frac{2}{3}, 0\right), (3, 0), \left(\frac{-1 \pm \sqrt{13}}{2}, 0\right)$ **(g)** $(0, 18)$
(h) $f(4) = 238; (4, 238)$ **(i)** **(j)**

$f(x) = 3x^4 - 4x^3 - 22x^2 + 15x + 18$

15. (a)

Positive	Negative	Nonreal Complex
1	3	0
1	1	2

(b) $\pm 1, \pm 2, \pm \frac{1}{2}$ **(c)** $-1, 1$ **(d)** no other real zeros
(e) $-\frac{1}{4} + \frac{\sqrt{15}}{4}i, -\frac{1}{4} - \frac{\sqrt{15}}{4}i$ **(f)** $(-1, 0), (1, 0)$
(g) $(0, 2)$ **(h)** $f(4) = -570; (4, -570)$ **(i)**
(j)

$f(x) = -2x^4 - x^3 + x + 2$

16. (a)

Positive	Negative	Nonreal Complex
0	4	0
0	2	2
0	0	4

(b) $0, \pm 1, \pm 3, \pm 9, \pm 27, \pm \frac{1}{2}, \pm \frac{3}{2}, \pm \frac{9}{2}, \pm \frac{27}{2},$
$\pm \frac{1}{4}, \pm \frac{3}{4}, \pm \frac{9}{4}, \pm \frac{27}{4}$ **(c)** $0, -\frac{3}{2}$ (multiplicity 2)
(d) no other real zeros **(e)** $\frac{1}{2} + \frac{\sqrt{11}}{2}i, \frac{1}{2} - \frac{\sqrt{11}}{2}i$
(f) $(0, 0), \left(-\frac{3}{2}, 0\right)$ **(g)** $(0, 0)$
(h) $f(4) = 7260; (4, 7260)$ **(i)** **(j)**

$f(x) = 4x^5 + 8x^4 + 9x^3 + 27x^2 + 27x$

17. (a)

Positive	Negative	Nonreal Complex
1	1	2

(b) $\pm 1, \pm 5, \pm \frac{1}{3}, \pm \frac{5}{3}$ **(c)** no rational zeros
(d) $-\sqrt{5}, \sqrt{5}$ **(e)** $-\frac{\sqrt{3}}{3}i, \frac{\sqrt{3}}{3}i$ **(f)** $\left(-\sqrt{5}, 0\right),$
$\left(\sqrt{5}, 0\right)$ **(g)** $(0, -5)$ **(h)** $f(4) = 539; (4, 539)$
(i) **(j)**

$f(x) = 3x^4 - 14x^2 - 5$

18. (a)

Positive	Negative	Nonreal Complex
2	3	0
2	1	2
0	3	2
0	1	4

(b) $\pm 1, \pm 3, \pm 9$ **(c)** $-3, -1$ (multiplicity 2), 1, 3
(d) no other real zeros **(e)** no other complex zeros
(f) $(-3, 0), (-1, 0), (1, 0), (3, 0)$ **(g)** $(0, -9)$
(h) $f(4) = -525; (4, -525)$
(i) **(j)**

$f(x) = -x^5 - x^4 + 10x^3 + 10x^2 - 9x - 9$

19. (a)

Positive	Negative	Nonreal Complex
4	0	0
2	0	2
0	0	4

(b) $\pm 1, \pm 2, \pm 3, \pm 4, \pm 6, \pm 12, \pm \frac{1}{3}, \pm \frac{2}{3}, \pm \frac{4}{3}$
(c) $\frac{1}{3}, 2$ (multiplicity 2), 3 **(d)** no other real zeros
(e) no other complex zeros **(f)** $\left(\frac{1}{3}, 0\right), (2, 0), (3, 0)$
(g) $(0, -12)$

(h) $f(4) = -44$; $(4, -44)$ **(i)** ⌢⌣

(j)

$f(x) = -3x^4 + 22x^3 - 55x^2$
$+ 52x - 12$

20. For the function in **Exercise 12:** ± 1.732; for the function in **Exercise 13:** ± 0.707; for the function in **Exercise 14:** -2.303, 1.303; for the function in **Exercise 17:** ± 2.236

3.5 Exercises *(pages 351–359)*

1. $(-\infty, 0) \cup (0, \infty)$; $(-\infty, 0) \cup (0, \infty)$ **3.** none; $(-\infty, 0)$ and $(0, \infty)$; none   **5.** $x = 3$; $y = 2$ **7.** even; symmetry with respect to the y-axis   **9.** A, B, C   **11.** A
13. A   **15.** A, C, D
17. To obtain the graph of f, stretch the graph of $y = \frac{1}{x}$ vertically by a factor of 2.
(a) $(-\infty, 0) \cup (0, \infty)$
(b) $(-\infty, 0) \cup (0, \infty)$ **(c)** none
(d) $(-\infty, 0)$ and $(0, \infty)$

$f(x) = \frac{2}{x}$

19. To obtain the graph of f, shift the graph of $y = \frac{1}{x}$ to the left 2 units.
(a) $(-\infty, -2) \cup (-2, \infty)$
(b) $(-\infty, 0) \cup (0, \infty)$ **(c)** none
(d) $(-\infty, -2)$ and $(-2, \infty)$

$f(x) = \frac{1}{x+2}$ $x = -2$

21. To obtain the graph of f, shift the graph of $y = \frac{1}{x}$ up 1 unit.
(a) $(-\infty, 0) \cup (0, \infty)$
(b) $(-\infty, 1) \cup (1, \infty)$ **(c)** none
(d) $(-\infty, 0)$ and $(0, \infty)$

$y = 1$ $f(x) = \frac{1}{x} + 1$

23. To obtain the graph of f, stretch the graph of $y = \frac{1}{x^2}$ vertically by a factor of 2 and reflect across the x-axis.
(a) $(-\infty, 0) \cup (0, \infty)$
(b) $(-\infty, 0)$ **(c)** $(0, \infty)$ **(d)** $(-\infty, 0)$

$f(x) = -\frac{2}{x^2}$

25. To obtain the graph of f, shift the graph of $y = \frac{1}{x^2}$ to the right 3 units.
(a) $(-\infty, 3) \cup (3, \infty)$
(b) $(0, \infty)$ **(c)** $(-\infty, 3)$ **(d)** $(3, \infty)$

$f(x) = \frac{1}{(x-3)^2}$ $x = 3$

27. To obtain the graph of f, shift the graph of $y = \frac{1}{x^2}$ to the left 2 units, reflect across the x-axis, and shift 3 units down.
(a) $(-\infty, -2) \cup (-2, \infty)$
(b) $(-\infty, -3)$ **(c)** $(-2, \infty)$
(d) $(-\infty, -2)$

$x = -2$ $y = -3$ $f(x) = \frac{-1}{(x+2)^2} - 3$

29. D   **31.** G   **33.** E   **35.** F

In selected Exercises 37–59, V.A. represents vertical asymptote, H.A. represents horizontal asymptote, and O.A. represents oblique asymptote.
37. V.A.: $x = 5$; H.A.: $y = 0$
39. V.A.: $x = -\frac{1}{2}$; H.A.: $y = -\frac{3}{2}$
41. V.A.: $x = -3$; O.A.: $y = x - 3$
43. V.A.: $x = -2$, $x = \frac{5}{2}$; H.A.: $y = \frac{1}{2}$
45. V.A.: none; H.A.: $y = 1$
47. **(a)** $f(x) = \frac{2x - 5}{x - 3}$ **(b)** $\frac{5}{2}$ **(c)** H.A.: $y = 2$; V.A: $x = 3$
49. **(a)** $y = x + 1$ **(b)** at $x = 0$ and $x = 1$ **(c)** above
51. A   **53.** V.A.: $x = 2$; H.A.: $y = 4$; $(-\infty, 2) \cup (2, \infty)$
55. V.A.: $x = \pm 2$; H.A.: $y = -4$; $(-\infty, -2) \cup (-2, 2)$ $\cup (2, \infty)$ **57.** V.A.: none; H.A.: $y = 0$; $(-\infty, \infty)$
59. V.A.: $x = -1$; O.A.: $y = x - 1$; $(-\infty, -1) \cup (-1, \infty)$

61.

$f(x) = \frac{x+1}{x-4}$

63.

$f(x) = \frac{x+2}{x-3}$

65.

$f(x) = \frac{4 - 2x}{8 - x}$

67.

$f(x) = \frac{3x}{x^2 - x - 2}$

69.

$f(x) = \frac{5x}{x^2 - 1}$

71.

$f(x) = \frac{(x + 6)(x - 2)}{(x + 3)(x - 4)}$

73.

$f(x) = \frac{3x^2 + 3x - 6}{x^2 - x - 12}$

75.

$f(x) = \frac{9x^2 - 1}{x^2 - 4}$

77.

$f(x) = \frac{(x - 3)(x + 1)}{(x - 1)^2}$

79.

$f(x) = \frac{x}{x^2 - 9}$

81.

$$f(x) = \frac{1}{x^2 + 1}$$

83.

$$f(x) = \frac{(x+4)^2}{(x-1)(x+5)}$$

85.

$$f(x) = \frac{20 + 6x - 2x^2}{8 + 6x - 2x^2}$$

87.

$$f(x) = \frac{x^2 + 1}{x + 3}$$

89.

$$f(x) = \frac{x^2 + 2x}{2x - 1}$$

91.

$$f(x) = \frac{x^2 - 9}{x + 3}$$

93.

$$f(x) = \frac{2x^2 - 5x - 2}{x - 2}$$

95.

$$f(x) = \frac{x^2 - 1}{x^2 - 4x + 3}$$

97.

$$f(x) = \frac{(x^2 - 9)(2 + x)}{(x^2 - 4)(3 + x)}$$

99.

$$f(x) = \frac{x^4 - 20x^2 + 64}{x^4 - 10x^2 + 9}$$

101. $f(x) = \dfrac{(x-3)(x+2)}{(x-2)(x+2)}$, or $f(x) = \dfrac{x^2 - x - 6}{x^2 - 4}$

103. $f(x) = \dfrac{x-2}{x(x-4)}$, or $f(x) = \dfrac{x-2}{x^2 - 4x}$

105. $f(x) = \dfrac{-x(x-2)}{(x-1)^2}$, or $f(x) = \dfrac{-x^2 + 2x}{x^2 - 2x + 1}$

107. Several answers are possible. One answer is
$f(x) = \dfrac{(x-3)(x+1)}{(x-1)^2}$.

109. $f(1.25) = -0.8\overline{1}$

111. $f(1.25) = 2.708\overline{3}$

$$f(x) = \frac{x+1}{x-4}$$

$$f(x) = \frac{x^2 + 2x}{2x - 1}$$

113. (a) 26 per min   **(b)** 5 park attendants

For $r = x$,
$$y = T(r) = \frac{2x - 25}{2x^2 - 50x} \qquad y = 0.5$$

115. (a) approximately 52.1 mph

x	$d(x)$	x	$d(x)$
20	**34**	50	**273**
25	**56**	55	**340**
30	**85**	60	**415**
35	**121**	65	**499**
40	**164**	70	**591**
45	**215**		

117. All answers are given in tens of millions.
(a) \$65.5   **(b)** \$64   **(c)** \$60   **(d)** \$40   **(e)** \$0
(f)

$$R(x) = \frac{80x - 8000}{x - 110}$$

119. $y = 1$ **120.** $(x+4)(x+1)(x-3)(x-5)$
121. (a) $(x-1)(x-2)(x+2)(x-5)$

(b) $f(x) = \dfrac{(x+4)(x+1)(x-3)(x-5)}{(x-1)(x-2)(x+2)(x-5)}$

122. (a) $x - 5$ **(b)** 5   **123.** $(-4, 0), (-1, 0), (3, 0)$
124. $(0, -3)$ **125.** $x = 1, \ x = 2, \ x = -2$
126. $\left(\dfrac{7 + \sqrt{241}}{6}, 1 \right), \left(\dfrac{7 - \sqrt{241}}{6}, 1 \right)$
127.

$$f(x) = \frac{x^4 - 3x^3 - 21x^2 + 43x + 60}{x^4 - 6x^3 + x^2 + 24x - 20}$$

128. (a) $(-4, -2) \cup (-1, 1) \cup (2, 3)$
(b) $(-\infty, -4) \cup (-2, -1) \cup (1, 2) \cup (3, 5) \cup (5, \infty)$

Chapter 3 Quiz *(page 359)*
[3.1] 1. (a) vertex: $(-3, -1)$;
axis: $x = -3$; domain: $(-\infty, \infty)$;
range: $(-\infty, -1]$; increasing: $(-\infty, -3)$;
decreasing: $(-3, \infty)$

$$f(x) = -2(x+3)^2 - 1$$

(b) vertex: $(2, -5)$; axis: $x = 2$;
domain: $(-\infty, \infty)$; range: $[-5, \infty)$;
increasing: $(2, \infty)$; decreasing: $(-\infty, 2)$

$f(x) = 2x^2 - 8x + 3$

2. (a) $s(t) = -16t^2 + 64t + 200$ **(b)** between approximately 0.78 sec and 3.22 sec

[3.2] 3. no; 38 **4.** yes

[3.3] 5. $f(x) = x^4 - 7x^3 + 10x^2 + 26x - 60$

[3.4] 6.

$f(x) = x(x - 2)^3(x + 2)^2$

7.

$f(x) = 2x^4 - 9x^3 - 5x^2 + 57x - 45$
$= (x - 3)^2(2x + 5)(x - 1)$

8.

$f(x) = -4x^5 + 16x^4 + 13x^3 - 76x^2 - 3x + 18$
$= -(x + 2)(2x + 1)(2x - 1)(x - 3)^2$

[3.5] 9.

$f(x) = \dfrac{3x + 1}{x^2 + 7x + 10}$

10.

$f(x) = \dfrac{x^2 + 2x + 1}{x - 1}$

Summary Exercises on Solving Equations and Inequalities *(pages 360–361)*

1. $\left\{\frac{4}{3}\right\}$ **2.** $\{20\}$ **3.** $\left\{\frac{9}{2}, \frac{14}{3}, \frac{16}{3}, \frac{11}{2}\right\}$ **4.** $\{25, 64\}$

5. $\{3, 7\}$ **6.** $\{2\}$ **7.** $\left\{\pm\frac{1}{8}\right\}$ **8.** $\{13\}$

9. (a) $(-1, 1) \cup (1, \infty)$ **(b)** $(-\infty, -1] \cup \{1\}$

10. (a) $(-\infty, -1) \cup (-1, 1) \cup (1, \infty)$ **(b)** $\varnothing$

11. (a) $\{0\}$ **(b)** $(-3, 0) \cup (3, \infty)$

12. (a) $\{1, 3\}$ **(b)** $[1, 2) \cup (2, 3]$

13. inequality; $\left(-\infty, \frac{2}{5}\right) \cup \left(\frac{2}{5}, \infty\right)$ **14.** inequality; $(-\infty, -1] \cup [4, \infty)$ **15.** equation; $\{-1, 2\}$

16. inequality; $\{-3\} \cup [1, \infty)$ **17.** equation; $\{-4, 0, 3\}$

18. inequality; $\left(-\sqrt{3}, -\frac{1}{2}\right) \cup \left(\frac{1}{2}, \sqrt{3}\right)$

19. inequality; $(-\infty, 0) \cup \left[\frac{3}{2}, 3\right) \cup [5, \infty)$

20. inequality; $(-2, -1) \cup (2, \infty)$ **21.** equation; $\{1\}$

22. inequality; $(-2, 0) \cup (2, \infty)$

3.6 Exercises *(pages 365–369)*

1. The circumference of a circle varies directly as (or is proportional to) its radius. **3.** The speed varies directly as (or is proportional to) the distance traveled and inversely as the time. **5.** The strength of a muscle varies directly as (or is proportional to) the cube of its length. **7.** C **9.** A

11. -30 **13.** 60 **15.** $\frac{3}{2}$ **17.** $\frac{12}{5}$ **19.** $\frac{18}{125}$

21. 69.08 in. **23.** 850 ohms **25.** 12 km **27.** 16 in.

29. 90 revolutions per minute **31.** 0.0444 ohm **33.** \$875

35. 800 lb **37.** $\frac{8}{9}$ metric ton **39.** $\frac{66\pi}{17}$ sec **41.** 21

43. 365.24 **45.** 92; undernourished **47.** increases; decreases **49.** y is half as large as before. **51.** y is one-third as large as before. **53.** p is $\frac{1}{32}$ as large as before.

Chapter 3 Review Exercises *(pages 375–380)*

1. vertex: $(-4, -5)$; axis: $x = -4$;
x-intercepts: $\left(\frac{-12 \pm \sqrt{15}}{3}, 0\right)$;
y-intercept: $(0, 43)$; domain: $(-\infty, \infty)$;
range: $[-5, \infty)$; increasing on $(-4, \infty)$; decreasing on $(-\infty, -4)$

$f(x) = 3(x + 4)^2 - 5$

$(-4, -5)$

3. vertex: $(-2, 11)$; axis: $x = -2$;
x-intercepts: $\left(\frac{-6 \pm \sqrt{33}}{3}, 0\right)$;
y-intercept: $(0, -1)$; domain: $(-\infty, \infty)$;
range: $(-\infty, 11]$; increasing on $(-\infty, -2)$; decreasing on $(-2, \infty)$

$(-2, 11)$

$f(x) = -3x^2 - 12x - 1$

5. (h, k) **7.** $k \le 0$; $\left(h \pm \sqrt{\frac{-k}{a}}, 0\right)$ **9.** 90 m by 45 m

11. (a) 120 **(b)** 40 **(c)** 22 **(d)** 84 **(e)** 146

(f) The minimum occurs in August when $x = 8$.

$(8, 22)$

August

13. Because the discriminant is 67.3033, a positive number, there are two x-intercepts. **15. (a)** the open interval $(-0.52, 2.59)$ **(b)** $(-\infty, -0.52) \cup (2.59, \infty)$

17. $x^2 + 4x + 1 + \frac{-7}{x - 3}$ **19.** $2x^2 - 8x + 31 + \frac{-118}{x + 4}$

21. $(x - 2)(5x^2 + 7x + 16) + 26$ **23.** -1 **25.** 28

27. yes **29.** $7 - 2i$

In Exercises 31–35, other answers are possible.

31. $f(x) = x^3 - 10x^2 + 17x + 28$

33. $f(x) = x^4 - 5x^3 + 3x^2 + 15x - 18$

35. $f(x) = x^4 - 6x^3 + 9x^2 - 6x + 8$ **37.** $\frac{1}{2}, -1, 5$

39. (a) $f(-1) = -10 < 0$; $f(0) = 2 > 0$
(b) $f(2) = -4 < 0$; $f(3) = 14 > 0$ **(c)** 2.414

43. yes **45.** $f(x) = -2x^3 + 6x^2 + 12x - 16$

47. $1, -\frac{1}{2}, \pm 2i$ **49.** $\frac{13}{2}$

51. Any polynomial that can be factored into $a(x - b)^3$ satisfies the conditions. One example is $f(x) = 2(x - 1)^3$.

$f(x) = 2(x - 1)^3$

53. (a) $(-\infty, \infty)$ **(b)** $(-\infty, \infty)$ **(c)** $f(x) \to \infty$ as $x \to \infty$, $f(x) \to -\infty$ as $x \to -\infty$: ↗ **(d)** at most seven **(e)** at most six

55.

$f(x) = (x - 2)^2(x + 3)$

57.

$f(x) = 2x^3 + x^2 - x$

59.

$f(x) = x^4 + x^3 - 3x^2 - 4x - 4$

61. C **63.** E **65.** B
67. 7.6533119, 1, -0.6533119

69. (a)

(b) $f(x) = -0.0109x^2 + 0.8693x + 11.85$
(c) $f(x) = -0.00087x^3 + 0.0456x^2 - 0.2191x + 17.83$
(d) $f(x) = -0.0109x^2 + 0.8693x + 11.85$

$f(x) = -0.00087x^3 + 0.0456x^2 - 0.2191x + 17.83$

(e) Both functions approximate the data well. The quadratic function is probably better for prediction, because it is unlikely that the percent of out-of-pocket spending would decrease after 2025 (as the cubic function shows) unless changes were made in Medicare law.

71. 12 in. × 4 in. × 15 in.

73.

$f(x) = \dfrac{4}{x - 1}$

75.

$f(x) = \dfrac{6x}{x^2 + x - 2}$

77.

$f(x) = \dfrac{x^2 + 4}{x + 2}$

79.

$f(x) = \dfrac{-2}{x^2 + 1}$

81. (a)

(b) One possibility is $f(x) = \dfrac{(x - 2)(x - 4)}{(x - 3)^2}$.

$f(x) = \dfrac{(x - 2)(x - 4)}{(x - 3)^2}$

83. $f(x) = \dfrac{-3x + 6}{x - 1}$

85. (a) $C(x) = \dfrac{6.7x}{100 - x}$ **(b)** approximately $127.3 thousand

87. 35 **89.** 0.75 **91.** 27 **93.** 7500 lb

Chapter 3 Test *(pages 380–382)*

[3.1] 1. x-intercepts: $\left(\dfrac{3 \pm \sqrt{3}}{2}, 0\right)$; y-intercept: $(0, -3)$; vertex: $\left(\dfrac{3}{2}, \dfrac{3}{2}\right)$; axis: $x = \dfrac{3}{2}$; domain: $(-\infty, \infty)$; range: $\left(-\infty, \dfrac{3}{2}\right]$; increasing on $\left(-\infty, \dfrac{3}{2}\right)$; decreasing on $\left(\dfrac{3}{2}, \infty\right)$

$f(x) = -2x^2 + 6x - 3$

2. (a) 2.75 sec **(b)** 169 ft **(c)** 0.7 sec and 4.8 sec **(d)** 6 sec **[3.2] 3.** $3x^2 - 2x - 5 + \dfrac{16}{x + 2}$
4. $2x^2 - x - 5$ **5.** 53 **[3.3] 6.** It is a factor. The other factor is $6x^3 + 7x^2 - 14x - 8$.
7. $-2, -3 - 2i, -3 + 2i$
8. $f(x) = 2x^4 - 2x^3 - 2x^2 - 2x - 4$
9. Because $f(x) > 0$ for all x, the graph never intersects or touches the x-axis. Therefore, $f(x)$ has no real zeros.
[3.3, 3.4] 10. (a) $f(1) = 5 > 0$; $f(2) = -1 < 0$
(b)

Positive	Negative	Nonreal Complex
2	1	0
0	1	2

(c) 4.0937635, 1.8370381, -0.9308016

[3.4] 11.

$f(x) = x^4$
$(-5, 3)$
$g(x) = -2(x + 5)^4 + 3$

To obtain the graph of g, translate the graph of f 5 units to the left, stretch by a factor of 2, reflect across the x-axis, and translate 3 units up.

12. C **13.**

$f(x) = x^3 - 5x^2 + 3x + 9$

14.

$f(x) = 2x^2(x - 2)^2$

15.

$f(x) = -x^3 - 4x^2 + 11x + 30$

16. $f(x) = 2(x - 2)^2(x + 3)$, or
$f(x) = 2x^3 - 2x^2 - 16x + 24$

17. (a) 270.08 **(b)** increasing from $t = 0$ to $t = 5.9$ and $t = 9.5$ to $t = 15$; decreasing from $t = 5.9$ to $t = 9.5$

[3.5] 18.

$f(x) = \frac{3x - 1}{x - 2}$

19.

$f(x) = \frac{x^2 - 1}{x^2 - 9}$

20. (a) $y = 2x + 3$ **(e)**
(b) $(-2, 0), \left(\frac{3}{2}, 0\right)$
(c) $(0, 6)$
(d) $x = 1$

$y = 2x + 3$
$x = 1$
$f(x) = \frac{2x^2 + x - 6}{x - 1}$

[3.6] 21. 60 **22.** $\frac{640}{9}$ kg

Chapter 4 Inverse, Exponential, and Logarithmic Functions

4.1 Exercises *(pages 393–397)*

1. Yes, it is one-to-one, because every number in the list of registered passenger cars is used only once. **3.** one-to-one
5. not one-to-one **7.** one-to-one **9.** one-to-one
11. not one-to-one **13.** one-to-one **15.** one-to-one
17. not one-to-one **19.** one-to-one **23.** one-to-one
25. range; domain **27.** false **29.** -3
31. untying your shoelaces **33.** leaving a room
35. unscrewing a light bulb **37.** inverses
39. not inverses **41.** inverses **43.** not inverses
45. inverses **47.** not inverses **49.** inverses
51. $\{(6, -3), (1, 2), (8, 5)\}$ **53.** not one-to-one
55. inverses **57.** not inverses

59. (a) $f^{-1}(x) = \frac{1}{3}x + \frac{4}{3}$ **(b)**

(c) Domains and ranges of both f and f^{-1} are $(-\infty, \infty)$.

61. (a) $f^{-1}(x) = -\frac{1}{4}x + \frac{3}{4}$ **(b)**

(c) Domains and ranges of both f and f^{-1} are $(-\infty, \infty)$.

63. (a) $f^{-1}(x) = \sqrt[3]{x - 1}$ **(b)**

(c) Domains and ranges of both f and f^{-1} are $(-\infty, \infty)$.

65. not one-to-one

67. (a) $f^{-1}(x) = \frac{1}{x}, \ x \neq 0$ **(b)**

$f = f^{-1}$

(c) Domains and ranges of both f and f^{-1} are $(-\infty, 0) \cup (0, \infty)$.

69. (a) $f^{-1}(x) = \frac{1 + 3x}{x}, \ x \neq 0$ **(b)**

(c) Domain of f = range of $f^{-1} = (-\infty, 3) \cup (3, \infty)$.
Domain of f^{-1} = range of $f = (-\infty, 0) \cup (0, \infty)$.

71. (a) $f^{-1}(x) = \frac{3x + 1}{x - 1}, \ x \neq 1$ **(b)**

(c) Domain of f = range of $f^{-1} = (-\infty, 3) \cup (3, \infty)$.
Domain of f^{-1} = range of $f = (-\infty, 1) \cup (1, \infty)$.

73. (a) $f^{-1}(x) = \frac{3x + 6}{x - 2}, \ x \neq 2$ **(b)**

(c) Domain of f = range of $f^{-1} = (-\infty, 3) \cup (3, \infty)$.
Domain of f^{-1} = range of $f = (-\infty, 2) \cup (2, \infty)$.

75. (a) $f^{-1}(x) = x^2 - 6, \; x \geq 0$ **(b)**

(c) Domain of f = range of $f^{-1} = [-6, \infty)$.
Domain of f^{-1} = range of $f = [0, \infty)$.

77.

79.

81.

83. 4 **85.** 2 **87.** -2
89. It represents the cost, in dollars, of building 1000 cars.
91. $\frac{1}{a}$ **93.** not one-to-one

95. one-to-one; $f^{-1}(x) = \frac{-5 - 3x}{x - 1}, \; x \neq 1$
97. $f^{-1}(x) = \frac{1}{3}x + \frac{2}{3}$; MIGUEL HAS ARRIVED
99. 6858 124 2743 63 511 124 1727 4095;
$f^{-1}(x) = \sqrt[3]{x + 1}$

4.2 Exercises *(pages 409–413)*

1. 9 **3.** $\frac{1}{9}$ **5.** $\frac{1}{16}$ **7.** 16 **9.** $3\sqrt{3}$ **11.** $\frac{1}{8}$
13. 13.076 **15.** 10.267

17.

19.

21.

23.

25.

27.

29.

31.

33.

35.

37.

39.

41.

43.

45.

47.

49.

51.

53. $\frac{1}{a}$ **55.** $f(x) = 2^{x+3} - 1$ **57.** $f(x) = -2^{x+2} + 3$
59. $f(x) = 3^{-x} + 1$ **61.** $\left\{\frac{1}{2}\right\}$ **63.** $\{-2\}$ **65.** $\{0\}$
67. $\left\{\frac{1}{2}\right\}$ **69.** $\left\{\frac{1}{5}\right\}$ **71.** $\{-7\}$ **73.** $\left\{-\frac{2}{3}\right\}$ **75.** $\left\{\frac{4}{3}\right\}$
77. $\{3\}$ **79.** $\{\pm 8\}$ **81.** $\{4\}$ **83.** $\{\pm 2\}$
85. $\{-27\}$ **87. (a)** \$13,891.16; \$4984.62
(b) \$13,968.24; \$5061.70 **89.** \$21,223.33 **91.** \$3528.81
93. 4.5% **95.** Bank A (even though it has the greatest stated rate)
97. (a)

(b) exponential

(c)

$P(x) = 1013e^{-0.0001341x}$

(d) $P(1500) \approx 828$ mb; $P(11,000) \approx 232$ mb

99. (a) about 63,000 **(b)** about 42,000 **(c)** about 21,000
101. $\{0.9\}$ **103.** $\{-0.5, 1.3\}$ **107.** $f(x) = 2^x$
109. $f(x) = \left(\frac{1}{4}\right)^x$ **111.** $f(t) = 27 \cdot 9^t$
113. $f(t) = \left(\frac{1}{3}\right)9^t$ **115.** 2.717 (A calculator gives 2.718.)
117. yes; an inverse function
118.

119. $x = a^y$ **120.** $x = 10^y$
121. $x = e^y$ **122.** (q, p)

4.3 Exercises (pages 422–425)

1. (a) C **(b)** A **(c)** E **(d)** B **(e)** F **(f)** D
3. $\log_3 81 = 4$ **5.** $\log_{2/3} \frac{27}{8} = -3$ **7.** $6^2 = 36$
9. $\left(\sqrt{3}\right)^8 = 81$ **13.** $\{-4\}$ **15.** $\left\{\frac{1}{2}\right\}$ **17.** $\left\{\frac{1}{4}\right\}$
19. $\{8\}$ **21.** $\{9\}$ **23.** $\left\{\frac{1}{5}\right\}$ **25.** $\{64\}$ **27.** $\left\{\frac{2}{3}\right\}$
29. $\{243\}$ **31.** $\{13\}$ **33.** $\{3\}$ **35.** $\{5\}$
39. $(0, \infty); (-\infty, \infty)$ **41.** $(-3, \infty); [0, \infty)$

43. $(2, \infty); (-\infty, \infty)$ **45.** E **47.** B **49.** F

51.

53.

55.

57. $f(x) = \log_2(x + 1) - 3$
59. $f(x) = \log_2(-x + 3) - 2$
61. $f(x) = -\log_3(x - 1)$
63. $\log_2 6 + \log_2 x - \log_2 y$
65. $1 + \frac{1}{2}\log_5 7 - \log_5 3$

67. This cannot be simplified.
69. $\frac{1}{2}(\log_m 5 + 3\log_m r - 5\log_m z)$
71. $\log_2 a + \log_2 b - \log_2 c - \log_2 d$
73. $\frac{1}{2}\log_3 x + \frac{1}{3}\log_3 y - 2\log_3 w - \frac{1}{2}\log_3 z$
75. $\log_a \frac{xy}{m}$ **77.** $\log_a \frac{m}{nt}$ **79.** $\log_b \left(x^{-1/6}y^{11/12}\right)$
81. $\log_a[(z+1)^2(3z+2)]$ **83.** $\log_5 \frac{5^{1/3}}{m^{1/3}}$, or $\log_5 \sqrt[3]{\frac{5}{m}}$
85. 0.7781 **87.** 0.1761 **89.** 0.3522 **91.** 0.7386

93. (a)

(b) logarithmic

95. (a) -4 **(b)** 6 **(c)** 4 **(d)** -1
99.

$f(x) = x \log_{10} x$

101. $\{0.01, 2.38\}$

Summary Exercises on Inverse, Exponential, and Logarithmic Functions (pages 426–427)

1. They are inverses. **2.** They are not inverses.
3. They are inverses. **4.** They are inverses.
5.

6.

7. It is not one-to-one.
8.

9. B **10.** D **11.** C **12.** A

13. The functions in **Exercises 9 and 12** are inverses of one another. The functions in **Exercises 10 and 11** are inverses of one another. **14.** $f^{-1}(x) = 5^x$
15. $f^{-1}(x) = \frac{1}{3}x + 2$; Domains and ranges of both f and f^{-1} are $(-\infty, \infty)$. **16.** $f^{-1}(x) = \sqrt[3]{\frac{x}{2}} - 1$; Domains and ranges of both f and f^{-1} are $(-\infty, \infty)$.
17. f is not one-to-one. **18.** $f^{-1}(x) = \frac{5x+1}{2+3x}$; Domain of f = range of $f^{-1} = \left(-\infty, \frac{5}{3}\right) \cup \left(\frac{5}{3}, \infty\right)$. Domain of f^{-1} = range of $f = \left(-\infty, -\frac{2}{3}\right) \cup \left(-\frac{2}{3}, \infty\right)$.
19. f is not one-to-one. **20.** $f^{-1}(x) = \sqrt{x^2 + 9}$, $x \geq 0$; Domain of f = range of $f^{-1} = [3, \infty)$. Domain of f^{-1} = range of $f = [0, \infty)$. **21.** $\log_{1/10} 1000 = -3$
22. $\log_a c = b$ **23.** $\log_{\sqrt{3}} 9 = 4$ **24.** $\log_4 \frac{1}{8} = -\frac{3}{2}$
25. $\log_2 32 = x$ **26.** $\log_{27} 81 = \frac{4}{3}$ **27.** $\{2\}$ **28.** $\{-3\}$
29. $\{-3\}$ **30.** $\{25\}$ **31.** $\{-2\}$ **32.** $\left\{\frac{1}{3}\right\}$
33. $(0, 1) \cup (1, \infty)$ **34.** $\left\{\frac{3}{2}\right\}$ **35.** $\{5\}$ **36.** $\{243\}$
37. $\{1\}$ **38.** $\{-2\}$ **39.** $\{1\}$ **40.** $\{2\}$ **41.** $\{2\}$
42. $\left\{\frac{1}{9}\right\}$ **43.** $\left\{-\frac{1}{3}\right\}$ **44.** $(-\infty, \infty)$

4.4 Exercises *(pages 434–439)*

1. increasing **3.** $f^{-1}(x) = \log_5 x$ **5.** natural; common
7. There is no power of 2 that yields a result of 0.
9. $\log 8 = 0.90308999$ **11.** 12 **13.** −1 **15.** 1.7993
17. −2.6576 **19.** 3.9494 **21.** 0.1803 **23.** 3.9494
25. 0.1803 **29.** 3.2 **31.** 8.4 **33.** 2.0×10^{-3}
35. 1.6×10^{-5} **37.** poor fen **39.** bog **41.** rich fen
43. (a) 2.60031933 **(b)** 1.60031933 **(c)** 0.6003193298
(d) The whole number parts will vary, but the decimal
parts will be the same. **45.** 1.6 **47.** −2 **49.** 3.3322
51. −8.9480 **53.** 10.1449 **55.** 2.0200 **57.** 10.1449
59. 2.0200 **63. (a)** 20 **(b)** 30 **(c)** 50 **(d)** 60
(e) about 3 decibels **65. (a)** 3 **(b)** 6 **(c)** 8
67. $630{,}957{,}345\,I_0$ **69.** 94.6 thousand; We must assume
that the model continues to be logarithmic.
71. (a) 2 **(b)** 2 **(c)** 2 **(d)** 1 **73.** 1
75. between 7°F and 11°F **77.** 1.126 billion yr
79. 2.3219 **81.** −0.2537 **83.** −1.5850 **85.** 0.8736
87. 1.9376 **89.** −1.4125 **91.** $4v + \frac{1}{2}u$ **93.** $\frac{3}{2}u - \frac{5}{2}v$
95. (a) 4 **(b)** 5^2, or 25 **(c)** $\frac{1}{e}$ **97. (a)** 6
(b) ln 3 **(c)** 2 ln 3, or ln 9 **99.** D
101. domain: $(-\infty, 0) \cup (0, \infty)$; range: $(-\infty, \infty)$;
symmetric with respect to the y-axis **103.** $f(x) = 2 + \ln x$,
so it is the graph of $g(x) = \ln x$ translated 2 units up.
105. $f(x) = \ln x - 2$, so it is the graph of $g(x) = \ln x$
translated 2 units down.

Chapter 4 Quiz *(page 439)*

[4.1] 1. $f^{-1}(x) = \frac{x^3 + 6}{3}$ **[4.2] 2.** $\{4\}$
3. **[4.3] 4.**

domain: $(-\infty, \infty)$; domain: $(-2, \infty)$;
range: $(-\infty, 0)$ range: $(-\infty, \infty)$

[4.2] 5. (a) $19{,}752.14 **(b)** $19{,}822.79 **(c)** $19{,}838.86
(d) $19{,}846.68 **[4.4] 6. (a)** 1.5386 **(b)** 3.5427
[4.3] 7. The expression $\log_6 25$ represents the exponent to
which 6 must be raised to obtain 25. **8. (a)** $\{4\}$ **(b)** $\{5\}$
(c) $\left\{\frac{1}{16}\right\}$ **[4.4] 9.** $\frac{1}{2}\log_3 x + \log_3 y - \log_3 p - 4\log_3 q$
10. 7.8137 **11.** 3.3578 **12.** 12

4.5 Exercises *(pages 446–450)*

1. $\log_7 19$; $\frac{\log 19}{\log 7}$; $\frac{\ln 19}{\ln 7}$ **3.** $\log_{1/2} 12$; $\frac{\log 12}{\log\left(\frac{1}{2}\right)}$; $\frac{\ln 12}{\ln\left(\frac{1}{2}\right)}$
5. $\{1.771\}$ **7.** $\{-2.322\}$ **9.** $\{-6.213\}$ **11.** $\{-1.710\}$

13. $\{3.240\}$ **15.** $\{\pm 2.146\}$ **17.** $\{9.386\}$ **19.** $\emptyset$
21. $\{32.950\}$ **23.** $\{7.044\}$ **25.** $\{25.677\}$
27. $\{2011.568\}$ **29.** $\{\ln 2, \ln 4\}$ **31.** $\left\{\ln\frac{3}{2}\right\}$
33. $\{\log_5 4\}$ **35.** $\{e^2\}$ **37.** $\left\{\frac{e^{1.5}}{4}\right\}$ **39.** $\left\{2 - \sqrt{10}\right\}$
41. $\{16\}$ **43.** $\{3\}$ **45.** $\{e\}$ **47.** $\{-6, 4\}$
49. $\{-8, 0\}$ **51.** $\{5\}$ **53.** $\{-5\}$ **55.** $\emptyset$ **57.** $\{-8\}$
59. $\{0\}$ **61.** $\{12\}$ **63.** $\{25\}$ **65.** $\emptyset$ **67.** $\left\{\frac{5}{2}\right\}$
69. $\{3\}$ **71.** $\left\{\frac{1 + \sqrt{41}}{4}\right\}$ **73.** $\{6\}$ **75.** $\{4\}$
77. $\{1, 100\}$ **79.** Proposed solutions that cause *any argu-
ment of a logarithm* to be negative or zero must be rejected.
The statement is not correct. For example, the solution set
of $\log(-x + 99) = 2$ is $\{-1\}$. **81.** $x = e^{kl(p-a)}$
83. $t = -\frac{1}{k}\log\left(\frac{T - T_0}{T_1 - T_0}\right)$ **85.** $t = -\frac{2}{R}\ln\left(1 - \frac{RI}{E}\right)$
87. $x = \dfrac{\ln\left(\frac{A + B - y}{B}\right)}{-C}$ **89.** $A = \frac{B}{x^C}$
91. $t = \dfrac{\log\left(\frac{A}{P}\right)}{n\log\left(1 + \frac{r}{n}\right)}$ **93.** $11{,}611.84 **95.** 2.6 yr
97. 5.02% **99. (a)** 10.9% **(b)** 35.8% **(c)** 84.1%
101. during 2013 **103. (a)** about 61% **(b)** 1989
105. (a) $P(T) = 1 - e^{-0.0034 - 0.0053T}$
(b) For $T = x$,
$P(x) = 1 - e^{-0.0034 - 0.0053x}$

(c) $P(60) \approx 0.275$, or 27.5%; The reduction in carbon
emissions from a tax of $60 per ton of carbon is 27.5%.
(d) $T = $130.14 **107.** $f^{-1}(x) = \ln(x + 4) - 1$;
domain: $(-4, \infty)$; range: $(-\infty, \infty)$ **109.** $\{1.52\}$
111. $\{0\}$ **113.** $\{2.45, 5.66\}$

4.6 Exercises *(pages 457–462)*

1. B **3.** C **5.** $\frac{1}{3}\ln\frac{1}{3}$ **7.** $\frac{1}{100}\ln\frac{1}{2}$ **9.** $\frac{1}{2}\ln\frac{1}{4}$
11. (a) 440 g **(b)** 387 g **(c)** 264 g **(d)** 21.66 yr
13. 1611.97 yr **15.** 3.57 g **17.** about 9000 yr
19. about 15,600 yr **21. (a)** possible answer:
$f(x) = 31{,}000(0.97)^{x - 1970}$ **(b)** 3% **23.** 6.25°C
25. (a) 5% compounded quarterly **(b)** $837.73
27. about 27.73 yr **29.** about 21.97 yr **31. (a)** 315
(b) 229 **(c)** 142 **33. (a)** $P = 1$; $a \approx 1.01355$
(b) 1.3 billion **(c)** 2030 **35. (a)** $2969 **(b)** 2006
37. (a) $12{,}879 **(b)** $15{,}590 **(c)** $17{,}483

39. (a) 15,000 **(b)** 9098 **(c)** 5249 **41. (a)** 611 million **(b)** 746 million **(c)** 1007 million **43.** about 13.2 hr **45.** 2016 **47.** 6.9 yr **49.** 11.6 yr **51. (a)** 0.065; 0.82; Among people age 25, 6.5% have some CHD, while among people age 65, 82% have some CHD. **(b)** about 48 yr

Summary Exercises on Functions: Domains and Defining Equations *(pages 463–464)*

1. $(-\infty, \infty)$ **2.** $\left[\frac{7}{2}, \infty\right)$ **3.** $(-\infty, \infty)$
4. $(-\infty, 6) \cup (6, \infty)$ **5.** $(-\infty, \infty)$
6. $(-\infty, -3] \cup [3, \infty)$
7. $(-\infty, -3) \cup (-3, 3) \cup (3, \infty)$
8. $(-\infty, \infty)$ **9.** $(-4, 4)$ **10.** $(-\infty, -7) \cup (3, \infty)$
11. $(-\infty, -1] \cup [8, \infty)$ **12.** $(-\infty, 0) \cup (0, \infty)$
13. $(-\infty, \infty)$ **14.** $(-\infty, -5) \cup (-5, \infty)$ **15.** $[1, \infty)$
16. $\left(-\infty, -\sqrt{5}\right) \cup \left(-\sqrt{5}, \sqrt{5}\right) \cup \left(\sqrt{5}, \infty\right)$
17. $(-\infty, \infty)$ **18.** $(-\infty, -1) \cup (-1, 1) \cup (1, \infty)$
19. $(-\infty, 1)$ **20.** $(-\infty, 2) \cup (2, \infty)$
21. $(-\infty, \infty)$ **22.** $[-2, 3] \cup [4, \infty)$
23. $(-\infty, -2) \cup (-2, 3) \cup (3, \infty)$
24. $[-3, \infty)$ **25.** $(-\infty, 0) \cup (0, \infty)$
26. $\left(-\infty, -\sqrt{7}\right) \cup \left(-\sqrt{7}, \sqrt{7}\right) \cup \left(\sqrt{7}, \infty\right)$
27. $(-\infty, \infty)$ **28.** $\emptyset$ **29.** $[-2, 2]$ **30.** $(-\infty, \infty)$
31. $(-\infty, -7] \cup (-4, 3) \cup [9, \infty)$ **32.** $(-\infty, \infty)$
33. $(-\infty, 5]$ **34.** $(-\infty, 3)$ **35.** $(-\infty, 4) \cup (4, \infty)$
36. $(-\infty, \infty)$ **37.** $(-\infty, -5] \cup [5, \infty)$ **38.** $(-\infty, \infty)$
39. $(-2, 6)$ **40.** $(0, 1) \cup (1, \infty)$ **41.** A **42.** B
43. C **44.** D **45.** A **46.** B **47.** D **48.** C
49. C **50.** B

Chapter 4 Review Exercises *(pages 468–471)*

1. not one-to-one **3.** one-to-one **5.** not one-to-one
7. $f^{-1}(x) = \sqrt[3]{x} + 3$ **9.** It represents the number of years after 2004 for the investment to reach \$50,000.
11. one-to-one **13.** B **15.** C **17.** $\log_2 32 = 5$
19. $\log_{3/4} \frac{4}{3} = -1$ **21.** $10^3 = 1000$ **23.** $e^{1/2} = \sqrt{e}$
25. 2 **27.** $\log_3 m + \log_3 n - \log_3 5 - \log_3 r$
29. -1.3862 **31.** 11.8776 **33.** 1.1592 **35.** $\left\{\frac{22}{5}\right\}$
37. $\{3.667\}$ **39.** $\{-13.257\}$ **41.** $\{-0.485\}$
43. $\{2.102\}$ **45.** $\{-2.487\}$ **47.** $\{3\}$ **49.** $\{\ln 3\}$
51. $\{6.959\}$ **53.** $\{e^{13/3}\}$ **55.** $\left\{\frac{\sqrt[4]{10} - 7}{2}\right\}$ **57.** $\{3\}$
59. $\left\{-\frac{4}{3}, 5\right\}$ **61.** $\left\{1, \frac{10}{3}\right\}$ **63.** $\{2\}$ **65.** $\{-3\}$
67. $I_0 = \frac{I}{10^{d/10}}$ **69.** $\{1.315\}$ **71. (a)** about $200{,}000{,}000 I_0$ **(b)** about $13{,}000{,}000 I_0$ **(c)** The 1906 earthquake had a magnitude almost 16 times greater than the 1989 earthquake. **73.** 5.1% **75.** \$24,970.64 **77.** 17.3 yr **79.** 2018 **81. (a)** \$15,207 **(b)** \$10,716 **(c)** \$4491 **(d)** They are the same.

Chapter 4 Test *(pages 472–473)*

[4.1] **1. (a)** $(-\infty, \infty)$; $(-\infty, \infty)$ **(b)** The graph is a stretched translation of $y = \sqrt[3]{x}$, which passes the horizontal line test and is thus a one-to-one function.
(c) $f^{-1}(x) = \frac{x^3 + 7}{2}$ **(d)** $(-\infty, \infty)$; $(-\infty, \infty)$
(e)

The graphs are reflections of each other across the line $y = x$.

[4.2, 4.3] **2. (a)** B **(b)** A **(c)** C **(d)** D [4.2] **3.** $\left\{\frac{1}{2}\right\}$
[4.3] **4. (a)** $\log_4 8 = \frac{3}{2}$ **(b)** $8^{2/3} = 4$
[4.1–4.3] **5.** They are inverses.

[4.3] **6.** $2 \log_7 x + \frac{1}{4} \log_7 y - 3 \log_7 z$ [4.4] **7.** 3.3780
8. 7.7782 **9.** 1.1674 [4.2] **10.** $\{\pm 125\}$ **11.** $\{0\}$
[4.5] **12.** $\{0.631\}$ [4.2] **13.** $\{4\}$ [4.5] **14.** $\{12.548\}$
15. $\{2.811\}$ **16.** $\left\{0, \ln \frac{3}{2}\right\}$; $\{0, 0.405\}$ [4.3] **17.** $\left\{\frac{3}{4}\right\}$
[4.5] **18.** $\{0, 6\}$ **19.** $\{2\}$ **20.** $\emptyset$ **21.** $\left\{\frac{7}{2}\right\}$
[4.6] **23.** 10 sec **24. (a)** 33.8 yr **(b)** 33.7 yr
25. 28.9 yr **26. (a)** 329.3 g **(b)** 13.9 days

Chapter 5 Systems and Matrices

5.1 Exercises *(pages 486–492)*

1. 2006–2009 **3.** approximately $(2005.2, 1.26)$ **5.** year; population (in millions) **7.** $\{(-4, 1)\}$ **9.** $\{(48, 8)\}$
11. $\{(1, 3)\}$ **13.** $\{(-1, 3)\}$ **15.** $\{(3, -4)\}$
17. $\{(0, 2)\}$ **19.** $\{(0, 4)\}$ **21.** $\{(1, 3)\}$
23. $\{(4, -2)\}$ **25.** $\{(2, -2)\}$ **27.** $\{(4, 6)\}$
29. $\{(5, 2)\}$ **31.** $\emptyset$; inconsistent system
33. $\left\{\left(\frac{y + 9}{4}, y\right)\right\}$; infinitely many solutions
35. $\emptyset$; inconsistent system **37.** $\left\{\left(\frac{-2y + 6}{7}, y\right)\right\}$; infinitely many solutions **39.** $\emptyset$; inconsistent system
41. $x - 3y = -3$ **43.** $\{(0.820, -2.508)\}$
$\quad\ 3x + 2y = 6$
45. $\{(0.892, 0.453)\}$ **47.** $\{(1, 2, -1)\}$ **49.** $\{(2, 0, 3)\}$
51. $\{(1, 2, 3)\}$ **53.** $\{(4, 1, 2)\}$ **55.** $\left\{\left(\frac{1}{2}, \frac{2}{3}, -1\right)\right\}$
57. $\{(-3, 1, 6)\}$ **59.** $\{(x, -4x + 3, -3x + 4)\}$
61. $\left\{\left(x, \frac{5x + 6}{5}, \frac{-5x + 69}{5}\right)\right\}$ **63.** $\left\{\left(x, \frac{-7x + 41}{9}, \frac{-x + 47}{9}\right)\right\}$
65. $\emptyset$; inconsistent system **67.** $\left\{\left(-\frac{z}{9}, \frac{z}{9}, z\right)\right\}$; infinitely many solutions **69.** $\{(2, 2)\}$ **71.** $\left\{\left(\frac{1}{5}, 1\right)\right\}$
73. $\{(4, 6, 1)\}$ **75.** $k \neq -6$; $k = -6$
77. $y = -3x - 5$ **79.** $y = \frac{3}{4}x^2 + \frac{1}{4}x - \frac{1}{2}$ **81.** $y = 3x - 1$
83. $y = -\frac{1}{2}x^2 + x + \frac{1}{4}$ **85.** $x^2 + y^2 - 4x + 2y - 20 = 0$

87. $x^2 + y^2 + x - 7y = 0$

89. $x^2 + y^2 - \frac{7}{5}x + \frac{27}{5}y - \frac{126}{5} = 0$

91. (a) $a = 0.009923$, $b = 0.9015$, $c = 317$; $C = 0.009923x^2 + 0.9015x + 317$ **(b)** near the end of 2098 **93.** 23, 24 **95.** baseball: \$194.98; football: \$420.54 **97.** 120 gal of \$9.00; 60 gal of \$3.00; 120 gal of \$4.50 **99.** 28 in.; 17 in.; 14 in. **101.** \$100,000 at 3%; \$40,000 at 2.5%; \$60,000 at 1.5% **103. (a)** \$16 **(b)** \$11 **(c)** \$6 **104. (a)** 8 **(b)** 4 **(c)** 0

105. See the answer to **Exercise 107.**

106. (a) 0 **(b)** $\frac{40}{3}$ **(c)** $\frac{80}{3}$

107.

108. price: \$6; demand: 8

109. $\{(40, 15, 30)\}$

111. 11.92 lb of Arabian Mocha Sanani; 14.23 lb of Organic Shade Grown Mexico; 23.85 lb of Guatemala Antigua (Answers are approximations.)

5.2 Exercises *(pages 499–504)*

1. $\begin{bmatrix} 1 & 4 \\ 0 & -9 \end{bmatrix}$ **3.** $\begin{bmatrix} 1 & 5 & 6 \\ 0 & 13 & 11 \\ 4 & 7 & 0 \end{bmatrix}$ **5.** $\begin{bmatrix} 2 & 3 & | & 11 \\ 1 & 2 & | & 8 \end{bmatrix}$; 2×3

7. $\begin{bmatrix} 2 & 1 & 1 & | & 3 \\ 3 & -4 & 2 & | & -7 \\ 1 & 1 & 1 & | & 2 \end{bmatrix}$; 3×4 **9.** $3x + 2y + z = 1$
$2y + 4z = 22$
$-x - 2y + 3z = 15$

11. $x = 2$ **13.** $x + y = 3$ **15.** $\{(2, 3)\}$
$y = 3$ $\quad\quad$ $2y + z = -4$
$z = -2$ $\quad\quad$ $x - z = 5$

17. $\{(-3, 0)\}$ **19.** $\left\{\left(1, \frac{2}{3}\right)\right\}$ **21.** $\emptyset$

23. $\left\{\left(\frac{4y + 7}{3}, y\right)\right\}$ **25.** $\{(-1, -2, 3)\}$

27. $\{(-1, 23, 16)\}$ **29.** $\{(2, 4, 5)\}$

31. $\left\{\left(\frac{1}{2}, 1, -\frac{1}{2}\right)\right\}$ **33.** $\{(2, 1, -1)\}$ **35.** $\emptyset$

37. $\left\{\left(\frac{-15z - 12}{23}, \frac{z - 13}{23}, z\right)\right\}$ **39.** $\{(1, 1, 2, 0)\}$

41. $\{(0, 2, -2, 1)\}$ **43.** $\{(0.571, 7.041, 11.442)\}$

45. none **47.** $A = \frac{1}{2}$, $B = -\frac{1}{2}$ **49.** $A = \frac{1}{2}$, $B = \frac{1}{2}$

51. day laborer: \$152; concrete finisher: \$160 **53.** 12, 6, 2

55. 9.6 cm³ of 7%; 30.4 cm³ of 2% **59.** 44.4 g of A; 133.3 g of B; 222.2 g of C **61. (a)** 65 or older: $y = 0.0018x + 0.13$; ages 25–34: $y = -0.000175x + 0.135$ **(b)** $\{(2.5316, 0.1346)\}$; 2012; 13.5%

63. (a) using the first equation, approximately 245 lb; using the second equation, approximately 253 lb **(b)** for the first, 7.46 lb; for the second, 7.93 lb **(c)** at approximately 66 in. and 118 lb

65. $a + 871b + 11.5c + 3d = 239$
$a + 847b + 12.2c + 2d = 234$
$a + 685b + 10.6c + 5d = 192$
$a + 969b + 14.2c + 1d = 343$

66. $\begin{bmatrix} 1 & 871 & 11.5 & 3 & | & 239 \\ 1 & 847 & 12.2 & 2 & | & 234 \\ 1 & 685 & 10.6 & 5 & | & 192 \\ 1 & 969 & 14.2 & 1 & | & 343 \end{bmatrix}$; The solution is $a \approx -715.457$, $b \approx 0.34756$, $c \approx 48.6585$, and $d \approx 30.71951$.

67. $F = -715.457 + 0.34756A + 48.6585P + 30.71951W$

68. approximately 323; This estimate is very close to the actual value of 320.

69.

n	T
3	28
6	191
10	805
29	17,487
100	681,550
200	5,393,100
400	42,906,200
1000	668,165,500
5000	8.3×10^{10}
10,000	6.7×10^{11}
100,000	6.7×10^{14}

71. no; It increases by almost a factor of 8.

5.3 Exercises *(pages 512–516)*

1. -31 **3.** 7 **5.** 0 **7.** -26 **9.** 0 **11.** 2, -6, 4 **13.** -6, 0, -6 **15.** 186 **17.** 17 **19.** 166 **21.** 0 **23.** 0 **25.** 1 **27.** 2 **29.** $-144 - 8\sqrt{10}$ **31.** -5.5 **34.** 102 **35.** 102; yes **36.** no **37.** $\left\{-\frac{4}{3}\right\}$ **39.** $\{-1, 4\}$ **41.** $\{-4\}$ **43.** $\{13\}$ **45.** $\left\{-\frac{1}{2}\right\}$ **47.** 1 unit² **49.** 9.5 units² **51.** approximately 19,328.3 ft² **53.** -3 **55.** 15 **57.** 3 **59.** 0 **61.** 0 **63.** 16 **65.** 17 **67.** 54 **69.** 0 **71.** 0 **73.** -88 **75.** 298 **77. (a)** D **(b)** A **(c)** C **(d)** B **79.** $\{(2, 2)\}$ **81.** $\{(2, -5)\}$ **83.** $\{(2, 0)\}$ **85.** Cramer's rule does not apply, since $D = 0$; $\emptyset$ **87.** Cramer's rule does not apply, since $D = 0$; $\left\{\left(\frac{-2y + 4}{3}, y\right)\right\}$ **89.** $\{(-4, 12)\}$ **91.** $\{(-3, 4, 2)\}$ **93.** $\{(0, 0, -1)\}$ **95.** Cramer's rule does not apply, since $D = 0$; $\emptyset$ **97.** Cramer's rule does not apply, since $D = 0$; $\left\{\left(\frac{19z - 32}{4}, \frac{-13z + 24}{4}, z\right)\right\}$ **99.** $\{(0, 4, 2)\}$ **101.** $\left\{\left(\frac{31}{5}, \frac{19}{10}, -\frac{29}{10}\right)\right\}$ **103.** $W_1 = W_2 = \frac{100\sqrt{3}}{3} \approx 58$ lb **105.** $\{(-a - b, a^2 + ab + b^2)\}$ **107.** $\{(1, 0)\}$ **109.** $\{(-1, 2)\}$

5.4 Exercises *(pages 521–522)*

1. $\frac{5}{3x} + \frac{-10}{3(2x + 1)}$ **3.** $\frac{6}{5(x + 2)} + \frac{8}{5(2x - 1)}$

5. $\frac{5}{6(x + 5)} + \frac{1}{6(x - 1)}$ **7.** $\frac{4}{x} + \frac{4}{1 - x}$

9. $\frac{15}{x} + \frac{-5}{x + 1} + \frac{-6}{x - 1}$ **11.** $\frac{2}{(x + 2)^2} + \frac{-3}{(x + 2)^3}$

13. $1 + \frac{-2}{x + 1} + \frac{1}{(x + 1)^2}$

15. $x^3 - x^2 + \frac{-1}{3(2x + 1)} + \frac{2}{3(x + 2)}$

17. $\frac{1}{9} + \frac{-1}{x} + \frac{25}{18(3x+2)} + \frac{29}{18(3x-2)}$

19. $\frac{-3}{5x^2} + \frac{3}{5(x^2+5)}$ **21.** $\frac{-2}{7(x+4)} + \frac{6x-3}{7(3x^2+1)}$

23. $\frac{1}{4x} + \frac{-8}{19(2x+1)} + \frac{-9x-24}{76(3x^2+4)}$

25. $\frac{-1}{x} + \frac{2x}{2x^2+1} + \frac{2x+3}{(2x^2+1)^2}$

27. $\frac{-1}{x+2} + \frac{3}{(x^2+4)^2}$ **29.** $5x^2 + \frac{3}{x} + \frac{-1}{x+3} + \frac{2}{x-1}$

31. $\frac{-1}{4(x+1)} + \frac{1}{4(x-1)} + \frac{1}{2(x^2+1)}$

33. $\frac{2}{x} + \frac{-1}{x-1} + \frac{3}{x+2}$

Chapter 5 Quiz *(page 522)*

[5.1] **1.** $\{(-2, 0)\}$ **2.** $\{(x, \frac{-x+2}{2})\}$, or $\{(-2y+2, y)\}$

3. ∅ **4.** $\{(3, -4)\}$ [5.2] **5.** $\{(-5, 2)\}$ [5.3] **6.** $\{(-3, 6)\}$

[5.1] **7.** $\{(-2, 1, 2)\}$ [5.2] **8.** $\{(2, 1, -1)\}$

[5.3] **9.** Cramer's rule does not apply, since $D = 0$;

$\{(\frac{2y+6}{9}, y, \frac{23y+6}{9})\}$ [5.1] **10.** at home: \$3762.50; away

from home: \$2687.50 **11.** \$1000 at 3%; \$1500 at 4%;

\$2500 at 6% [5.3] **12.** -3 **13.** 59

[5.4] **14.** $\frac{7}{x-5} + \frac{3}{x+4}$ **15.** $\frac{-1}{x+2} + \frac{6}{x+4} + \frac{-3}{x-1}$

5.5 Exercises *(pages 528–532)*

7. Consider the graphs; a line and a parabola cannot intersect in more than two points. **9.** $\{(1, 1), (-2, 4)\}$

11. $\{(2, 1), (\frac{1}{3}, \frac{4}{9})\}$ **13.** $\{(2, 12), (-4, 0)\}$

15. $\{(-\frac{3}{5}, \frac{7}{5}), (-1, 1)\}$ **17.** $\{(2, 2), (2, -2), (-2, 2), (-2, -2)\}$ **19.** $\{(0, 0)\}$ **21.** $\{(i, \sqrt{6}), (-i, \sqrt{6}), (i, -\sqrt{6}), (-i, -\sqrt{6})\}$ **23.** $\{(1, -1), (-1, 1), (1, 1), (-1, -1)\}$ **25.** ∅ **27.** $\{(\sqrt{6}, 0), (-\sqrt{6}, 0)\}$

29. $\{(-3, 5), (\frac{15}{4}, -4)\}$ **31.** $\{(4, -\frac{1}{8}), (-2, \frac{1}{4})\}$

33. $\{(3, 4), (-3, -4), (\frac{4\sqrt{3}}{3}i, -3i\sqrt{3}), (-\frac{4\sqrt{3}}{3}i, 3i\sqrt{3})\}$

35. $\{(\sqrt{5}, 0), (-\sqrt{5}, 0), (\sqrt{5}, \sqrt{5}), (-\sqrt{5}, -\sqrt{5})\}$

37. $\{(3, 5), (-3, -5), (5i, -3i), (-5i, 3i)\}$

39. $\{(5, 0), (-5, 0), (0, -5)\}$ **41.** $\{(3, -3), (3, 3)\}$

43. $\{(2, 2), (-2, -2), (2, -2), (-2, 2)\}$

45. $\{(-0.79, 0.62), (0.88, 0.77)\}$ **47.** $\{(0.06, 2.88)\}$

49. 14 and 3 **51.** 8 and 6, 8 and -6, -8 and 6, -8 and -6 **53.** 27 and 6, -27 and -6

55. 5 m and 12 m **57.** yes **59.** $y = 9$

61. *First answer:* length = width: 6 ft; height: 10 ft

Second answer: length = width: 12.780 ft; height: 2.204 ft

65. (a) 160 units **(b)** \$3 **67. (a)** The carbon dioxide emissions from consumption of fossil fuels are increasing in both countries, but they are increasing much more rapidly in China than in the United States. **(b)** 1996 to 2000 **(c)** The emissions were equal in 2006, when the levels were approximately 6000 million metric tons in each country. **(d)** year: 2007; emissions level: approximately 6107 million metric tons **68.** Translate the graph of $y = |x|$ one unit to

the right. **69.** Translate the graph of $y = x^2$ four units down.

70. $y = \begin{cases} x - 1 & \text{if } x \geq 1 \\ 1 - x & \text{if } x < 1 \end{cases}$

71. $x^2 - 4 = x - 1$ $(x \geq 1)$; $x^2 - 4 = 1 - x$ $(x < 1)$

72. $\frac{1 + \sqrt{13}}{2}; \frac{-1 - \sqrt{21}}{2}$

73. $\left\{\left(\frac{1 + \sqrt{13}}{2}, \frac{-1 + \sqrt{13}}{2}\right), \left(\frac{-1 - \sqrt{21}}{2}, \frac{3 + \sqrt{21}}{2}\right)\right\}$

Summary Exercises on Systems of Equations *(page 533)*

1. $\{(-3, 2)\}$ **2.** $\{(-5, -4)\}$

3. $\{(0, \sqrt{5}), (0, -\sqrt{5})\}$ **4.** $\{(-10, -6, 25)\}$

5. $\{(\frac{4}{3}, 3), (-\frac{1}{2}, -8)\}$ **6.** $\{(-1, 2, -4)\}$

7. $\{(5z - 13, -3z + 9, z)\}$ **8.** $\{(-2, -2), (\frac{14}{5}, -\frac{2}{5})\}$

9. ∅ **10.** $\{(-3, 3, -3)\}$ **11.** $\{(-3 + i\sqrt{7}, 3 + i\sqrt{7}), (-3 - i\sqrt{7}, 3 - i\sqrt{7})\}$ **12.** $\{(2, -5, 3)\}$

13. $\{(1, 2), (-5, 14)\}$ **14.** ∅ **15.** $\{(0, 1, 0)\}$

16. $\{(\frac{3}{8}, -1)\}$ **17.** $\{(-1, 0, 0)\}$

18. $\{(-8z - 56, z + 13, z)\}$

19. $\{(0, 3), (-\frac{36}{17}, -\frac{3}{17})\}$ **20.** $\{(2y + 3, y)\}$

21. $\{(1, 2, 3)\}$ **22.** $\{(2 + 2i\sqrt{2}, 5 - i\sqrt{2}), (2 - 2i\sqrt{2}, 5 + i\sqrt{2})\}$ **23.** $\{(-3, 5, -6)\}$ **24.** ∅

25. $\{(4, 2), (-1, -3)\}$ **26.** $\{(2, -3, 1)\}$

27. $\{(\sqrt{13}, i\sqrt{2}), (-\sqrt{13}, i\sqrt{2}), (\sqrt{13}, -i\sqrt{2}), (-\sqrt{13}, -i\sqrt{2})\}$ **28.** $\{(-z - 6, 5, z)\}$

29. $\{(1, -3, 4)\}$ **30.** $\{(-\frac{16}{3}, -\frac{11}{3}), (4, 1)\}$

31. $\{(-1, -2, 5)\}$ **32.** ∅ **33.** $\{(1, -3), (-3, 1)\}$

34. $\{(1, -6, 2)\}$ **35.** $\{(-6, 9), (-1, 4)\}$

36. $\{(0, -3), (\frac{12}{5}, \frac{9}{5})\}$ **37.** $\{(2, 2, 5)\}$

38. $\{(-2, -4, 0)\}$ **39.** $\{(2, 3)\}$

5.6 Exercises *(pages 540–544)*

1.

3.

5.

7.

9.

11.

13.

15.

17.

21. above **23.** B **25.** C **27.** A

29.

31.

33.

35.

37.

39.

41.

43. The solution set is ∅.

45.

47.

49.

51.

53.

55.

57.

59.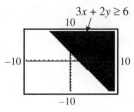

61. $x + y \ge 2, x + y \le 6$
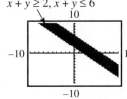

63. $x + 2y \le 4$
$3x - 4y \le 12$
65. $x^2 + y^2 \le 16$
$y \ge 2$

67. $x > 0$
$y > 0$
$x^2 + y^2 \le 4$
$y > \frac{3}{2}x - 1$

69. maximum of 65 at $(5, 10)$; minimum of 8 at $(1, 1)$
71. maximum of 66 at $(7, 9)$; minimum of 3 at $(1, 0)$

73. maximum of 100 at $(1, 10)$; minimum of 0 at $(1, 0)$
75. Let $x =$ the number of Brand X pills and $y =$ the number of Brand Y pills. Then $3000x + 1000y \ge 6000$, $45x + 50y \ge 195$, $75x + 200y \ge 600$, $x \ge 0$, $y \ge 0$.

77. **(a)** 300 cartons of food and 400 cartons of clothes **(b)** 6200 people
79. 8 of A and 3 of B, for a maximum storage capacity of 100 ft³
81. $3\frac{3}{4}$ servings of A and $1\frac{7}{8}$ servings of B, for a minimum cost of $1.69

5.7 Exercises *(pages 553–558)*
1. $a = 0, b = 4, c = -3, d = 5$ **3.** $x = -4, y = 14$, $z = 3, w = -2$ **5.** This cannot be true.
7. $w = 5, x = 6, y = -2, z = 8$
9. $z = 18, r = 3, s = 3, p = 3, a = \frac{3}{4}$ **11.** dimension; equal **13.** 2×2; square **15.** 3×4 **17.** 2×1; column

21. $\begin{bmatrix} -2 & -5 \\ 17 & 4 \end{bmatrix}$ **23.** $\begin{bmatrix} -2 & -7 & 7 \\ 10 & -2 & 7 \end{bmatrix}$

25. They cannot be added. **27.** $\begin{bmatrix} -6 & 8 \\ 4 & 2 \end{bmatrix}$ **29.** $\begin{bmatrix} 4 \\ -5 \\ 4 \end{bmatrix}$

31. They cannot be subtracted. **33.** $\begin{bmatrix} -\sqrt{3} & -13 \\ 4 & -2\sqrt{5} \\ -1 & -\sqrt{2} \end{bmatrix}$

35. $\begin{bmatrix} 5x + y & y \\ 6x + 2y & x + 3y \end{bmatrix}$ **37.** $\begin{bmatrix} -4 & 8 \\ 0 & 6 \end{bmatrix}$

39. $\begin{bmatrix} -9 & 3 \\ 6 & 0 \end{bmatrix}$ **41.** $\begin{bmatrix} 14 & 2 \\ -12 & 6 \end{bmatrix}$ **43.** $\begin{bmatrix} -1 & -3 \\ 2 & -3 \end{bmatrix}$

45. yes; 2×5 **47.** no **49.** yes; 3×2 **51.** $\begin{bmatrix} 13 \\ 25 \end{bmatrix}$

53. $\begin{bmatrix} -17 \\ -1 \end{bmatrix}$ **55.** $\begin{bmatrix} 17\sqrt{2} & -4\sqrt{2} \\ 35\sqrt{3} & 26\sqrt{3} \end{bmatrix}$

57. $\begin{bmatrix} 3+4\sqrt{3} & -3\sqrt{2} \\ 2\sqrt{15}+12\sqrt{6} & -2\sqrt{30} \end{bmatrix}$

59. They cannot be multiplied. **61.** $\begin{bmatrix} 2 & 7 & -4 \end{bmatrix}$

63. $\begin{bmatrix} -15 & -16 & 3 \\ -1 & 0 & 9 \\ 7 & 6 & 12 \end{bmatrix}$ **65.** $\begin{bmatrix} 23 & -9 \\ -6 & -2 \\ 33 & 1 \end{bmatrix}$

67. $\begin{bmatrix} -25 & 23 & 11 \\ 0 & -6 & -12 \\ -15 & 33 & 45 \end{bmatrix}$ **69.** They cannot be multiplied.

71. $\begin{bmatrix} 10 & -10 \\ 15 & -5 \end{bmatrix}$ **73.** $BA \neq AB,\ BC \neq CB,\ AC \neq CA$

75. (a) $\begin{bmatrix} 38 & -8 \\ -7 & -2 \end{bmatrix}$ **(b)** $\begin{bmatrix} 18 & 24 \\ 19 & 18 \end{bmatrix}$

77. (a) $\begin{bmatrix} 0 & 1 & -1 \\ 0 & 1 & 0 \\ 0 & 0 & 1 \end{bmatrix}$ **(b)** $\begin{bmatrix} 0 & 1 & -1 \\ 0 & 1 & 0 \\ 0 & 0 & 1 \end{bmatrix}$ **79.** 1

81. (a) $\begin{bmatrix} 50 & 100 & 30 \\ 10 & 90 & 50 \\ 60 & 120 & 40 \end{bmatrix}$ **(b)** $\begin{bmatrix} 12 \\ 10 \\ 15 \end{bmatrix}$ **(c)** $\begin{bmatrix} 2050 \\ 1770 \\ 2520 \end{bmatrix}$

(d) $6340 **83.** Answers will vary a little if intermediate steps are rounded. **(a)** 2940, 2909, 2861, 2814, 2767 **(b)** The northern spotted owl will become extinct. **(c)** 3023, 3051, 3079, 3107, 3135

5.8 Exercises *(pages 565–569)*

3. yes **5.** no **7.** no **9.** yes

11. $\begin{bmatrix} -\frac{1}{5} & -\frac{2}{5} \\ \frac{2}{5} & -\frac{1}{5} \end{bmatrix}$ **13.** $\begin{bmatrix} 2 & 1 \\ -\frac{3}{2} & -\frac{1}{2} \end{bmatrix}$ **15.** The inverse does

not exist. **17.** $\begin{bmatrix} -1 & 1 & 1 \\ 0 & -1 & 0 \\ 2 & -1 & -1 \end{bmatrix}$ **19.** $\begin{bmatrix} \frac{7}{8} & -\frac{3}{8} & -\frac{3}{8} \\ -\frac{1}{8} & \frac{5}{8} & -\frac{3}{8} \\ -\frac{1}{8} & -\frac{3}{8} & \frac{5}{8} \end{bmatrix}$

21. $\begin{bmatrix} -\frac{15}{4} & -\frac{1}{4} & -3 \\ \frac{5}{4} & \frac{1}{4} & 1 \\ -\frac{3}{2} & 0 & -1 \end{bmatrix}$ **23.** $\begin{bmatrix} \frac{1}{2} & 0 & \frac{1}{2} & -1 \\ \frac{1}{10} & -\frac{2}{5} & \frac{3}{10} & -\frac{1}{5} \\ -\frac{7}{10} & \frac{4}{5} & -\frac{11}{10} & \frac{12}{5} \\ \frac{1}{5} & \frac{1}{5} & -\frac{2}{5} & \frac{3}{5} \end{bmatrix}$

25. $\begin{bmatrix} 0 & -\frac{1}{3} & \frac{1}{3} & \frac{2}{3} \\ \frac{1}{3} & \frac{2}{3} & -\frac{1}{3} & -1 \\ \frac{2}{3} & 1 & -\frac{4}{3} & -\frac{4}{3} \\ -\frac{1}{3} & \frac{1}{3} & \frac{1}{3} & 0 \end{bmatrix}$ **27.** $\begin{bmatrix} 4 & 7 \\ -2 & 2 \end{bmatrix}$

29. $ad - bc$ is the determinant of A.

30. $\begin{bmatrix} \frac{d}{|A|} & \frac{-b}{|A|} \\ \frac{-c}{|A|} & \frac{a}{|A|} \end{bmatrix}$ **31.** $A^{-1} = \frac{1}{|A|}\begin{bmatrix} d & -b \\ -c & a \end{bmatrix}$

33. $A^{-1} = \begin{bmatrix} -\frac{3}{2} & 1 \\ \frac{7}{2} & -2 \end{bmatrix}$ **34.** zero **35.** $\{(2,3)\}$

37. $\{(-2,4)\}$ **39.** $\{(-2,\frac{3}{4})\}$ **41.** $\{(-\frac{1}{2},\frac{2}{3})\}$

43. $\{(4,-9)\}$ **45.** $\{(6,-\frac{5}{6})\}$

47. The inverse of $\begin{bmatrix} 7 & -2 \\ 14 & -4 \end{bmatrix}$ does not exist.

49. $\{(3,1,2)\}$ **51.** $\{(\frac{5}{4},\frac{1}{4},-\frac{3}{4})\}$ **53.** $\{(11,-1,2)\}$

55. $\{(1,0,2,1)\}$

57. (a) $602.7 = a + 5.543b + 37.14c$
$656.7 = a + 6.933b + 41.30c$
$778.5 = a + 7.638b + 45.62c$
(b) $a \approx -490.547,\ b = -89,\ c = 42.71875$
(c) $S = -490.547 - 89A + 42.71875B$
(d) $S \approx 843.5$ **(e)** $S \approx 1547.5$

59. Answers will vary.

61. $\begin{bmatrix} -0.1215875322 & 0.0491390161 \\ 1.544369078 & -0.046799063 \end{bmatrix}$

63. $\begin{bmatrix} 2 & -2 & 0 \\ -4 & 0 & 4 \\ 3 & 3 & -3 \end{bmatrix}$

65. $\{(1.68717058,\ -1.306990242)\}$

67. $\{(13.58736702,\ 3.929011993,\ -5.342780076)\}$

71. $A = \begin{bmatrix} 1 & 0 \\ 1 & 1 \end{bmatrix},\ B = \begin{bmatrix} 1 & 1 \\ 0 & 1 \end{bmatrix}$ (Other answers are

possible.) **73.** $\begin{bmatrix} \frac{1}{a} & 0 & 0 \\ 0 & \frac{1}{b} & 0 \\ 0 & 0 & \frac{1}{c} \end{bmatrix}$ **75.** $I_n,\ -A^{-1},\ \frac{1}{k}A^{-1}$

Chapter 5 Review Exercises *(pages 575–580)*

1. $\{(0,1)\}$ **3.** $\{(-5y+9, y)\}$; infinitely many solutions

5. $\emptyset$; inconsistent system **7.** $\{(\frac{1}{3},\frac{1}{2})\}$ **9.** $\{(3,2,1)\}$

11. $\{(5,-1,0)\}$ **13.** One possible answer is $\begin{array}{l} x + y = 2 \\ x + y = 3 \end{array}$.

15. $\frac{1}{3}$ cup of rice, $\frac{1}{5}$ cup of soybeans **17.** 5 blankets,
3 rugs, 8 skirts **19.** $x \approx 177.1,\ y \approx 174.9$; If an athlete's maximum heart rate is 180 beats per minute (bpm), then it will be about 177 bpm 5 sec after stopping and 175 bpm 10 sec after stopping.

21. $y = \frac{12}{5}x^2 - \frac{31}{5}x + \frac{3}{2}$, or $y = 2.4x^2 - 6.2x + 1.5$

23. $\{(x, -2x+1, -11x+6)\}$ **25.** $\{(-2,0)\}$

27. $\{(-3,2)\}$ **29.** $\{(0,3,3)\}$ **31.** 10 lb of $4.60 tea, 8 lb of $5.75 tea, 2 lb of $6.50 tea **33.** 1982; approximately 312 thousand **35.** -25 **37.** 0 **39.** -1 **41.** $\{\frac{8}{19}\}$

43. $\{(-4,2)\}$ **45.** $\emptyset$; inconsistent system

47. $\{(14,-15,35)\}$ **49.** $\frac{2}{x-1} - \frac{6}{3x-2}$

51. $\frac{1}{x-1} - \frac{x+3}{x^2+2}$ **53.** $\{(-3,4),(1,12)\}$

55. $\{(-4,1),(-4,-1),(4,-1),(4,1)\}$

57. $\{(5,-2),(-4,\frac{5}{2})\}$ **59.** $\{(-2,0),(1,1)\}$

61. $\{(\sqrt{3},2i),(\sqrt{3},-2i),(-\sqrt{3},2i),(-\sqrt{3},-2i)\}$

63. yes; $\left(\frac{8-8\sqrt{41}}{5}, \frac{16+4\sqrt{41}}{5}\right), \left(\frac{8+8\sqrt{41}}{5}, \frac{16-4\sqrt{41}}{5}\right)$

65.

$x + y \le 6$
$2x - y \ge 3$

67. maximum of 24 at $(0, 6)$

69. 3 units of food A and 4 units of food B; Minimum cost is $1.02 per serving. **71.** $a = 5$, $x = \frac{3}{2}$, $y = 0$, $z = 9$

73. $\begin{bmatrix} -4 \\ 6 \\ 1 \end{bmatrix}$ **75.** They cannot be subtracted.

77. $\begin{bmatrix} 3 & -4 \\ 4 & 48 \end{bmatrix}$ **79.** $\begin{bmatrix} -9 & 3 \\ 10 & 6 \end{bmatrix}$ **81.** $\begin{bmatrix} -2 & 5 & -3 \\ 3 & 4 & -4 \\ 6 & -1 & -2 \end{bmatrix}$

83. $\begin{bmatrix} 3 & -1 \\ -5 & 2 \end{bmatrix}$ **85.** $\begin{bmatrix} \frac{2}{3} & 0 & -\frac{1}{3} \\ \frac{1}{3} & 0 & -\frac{2}{3} \\ -\frac{2}{3} & 1 & \frac{1}{3} \end{bmatrix}$

87. $\left\{ \left(\frac{2}{3}, \frac{7}{12} \right) \right\}$ **89.** $\{(-3, 2, 0)\}$

Chapter 5 Test *(pages 580–582)*

[5.1] **1.** $\{(4, 3)\}$ **2.** $\left\{ \left(\frac{-3y - 7}{2}, y \right) \right\}$; infinitely many solutions **3.** $\{(1, 2)\}$ **4.** $\emptyset$; inconsistent system **5.** $\{(2, 0, -1)\}$ [5.2] **6.** $\{(5, 1)\}$ **7.** $\{(5, 3, 6)\}$ [5.1] **8.** $y = 2x^2 - 8x + 11$ **9.** 22 units from Toronto, 56 units from Montreal, 22 units from Ottawa

[5.3] **10.** -58 **11.** -844 **12.** $\{(-6, 7)\}$

13. $\{(1, -2, 3)\}$ [5.4] **14.** $\frac{2}{x} + \frac{-2}{x + 1} + \frac{-1}{(x + 1)^2}$

[5.5] **15.** $y = x^2 - 4x + 4$
$x - 3y = -2$

16. $\{(1, 2), (-1, 2), (1, -2), (-1, -2)\}$

17. $\{(3, 4), (4, 3)\}$ **18.** 5 and -6

[5.6] **19.** $x - 3y \ge 6$ $y^2 \le 16 - x^2$ **20.** maximum of 42 at $(12, 6)$ **21.** 0 VIP rings and 24 SST rings; Maximum profit is $960.

[5.7] **22.** $x = -1$; $y = 7$; $w = -3$ **23.** $\begin{bmatrix} 8 & 3 \\ 0 & -11 \\ 15 & 19 \end{bmatrix}$

24. The matrices cannot be added. **25.** $\begin{bmatrix} -5 & 16 \\ 19 & 2 \end{bmatrix}$

26. The matrices cannot be multiplied. **27.** A

[5.8] **28.** $\begin{bmatrix} -2 & -5 \\ -3 & -8 \end{bmatrix}$ **29.** The inverse does not exist.

30. $\begin{bmatrix} -9 & 1 & -4 \\ -2 & 1 & 0 \\ 4 & -1 & 1 \end{bmatrix}$ **31.** $\{(-7, 8)\}$ **32.** $\{(0, 5, -9)\}$

Index of Applications

Agriculture

Animal feed, 485
Cost of goats and sheep, 490
Cost of nutrients for animals, 579
Harvesting a cherry orchard, 124
Number of farms in the United States, 31
Profit from farm animals, 579

Astronomy

Distances and velocities of galaxies, 228–229
Magnitude of a star, 457
Planets' distances from the sun, 438
Planets' periods of revolution, 438
Temperature of a planet, 86, 155
Weight on and above Earth, 382
Weight on the moon, 366

Automotive

Accident rate, 298
Antifreeze, 89
Antique-car competition, 379
Automobile stopping distance, 299
Braking distance, 357
Conversion of methanol to gasoline, 155
Distance traveled, 362
Gallons of gasoline, 170, 188
Highway design, 336
Motor vehicle sales, 217
Pickup truck market share, 241
Registered passenger cars, 393
Skidding car, 380
Tire sales, 568
Traffic congestion, 502
Traffic intensity, 351, 356

Biology

Bacterial growth, 460
Deer population, 412
Diversity of species, 433, 437
Evolution of language, 460
Maximum number of mosquitos, 295
Northern spotted owl population, 556–557
Number of fawns, 503
Predator-prey relationship, 557–558
Tree growth, 462

Business

Apartments rented, 123–124
Break-even point, 139, 146, 165, 216
Buying and selling land, 95
Car sales quota, 501
Cost, revenue and profit analysis, 210, 216, 282
Day care center, 364
Filing cabinet storage capacity, 544
Financing an expansion, 501
Gasoline revenues, 544
Income from yogurt, 555
Internet publishing and broadcasting, 123
Lumber costs, 242
Manufacturing, 120, 363
Maximum profit, 537–538
Maximum revenue, 295
Minimum cost, 295, 539–540
Number of shopping centers, 295
Nut mixture, 167
Online retail sales, 162
Order quantities, 576
Ordering supplies, 581
Profit, 544, 581
Purchasing costs, 556
Revenue of fast-food restaurants in U.S., 174–175
Sales, 215, 448, 461, 567–568
Shipping requirements, 543

Chemistry

Acid mixture, 92, 94, 95, 501
Alcohol mixture, 89, 95, 156–157, 161
Carbon 14 dating, 454–455, 458
Dissolving a chemical, 458
Finding pH, 428–429, 435
Hydronium ion concentration, 428–429, 435
Newton's law of cooling, 445, 455–456, 458
Pressure exerted by a liquid, 366, 380
Radioactive decay, 453–454, 457, 473
Saline solution, 95
Volume of a gas, 367

Construction

Grade of a walkway, 212
Kitchen flooring, 163
Length of a walkway, 121
Model plans for a subdivision, 552
Painting a house, 134
Pitch of a roof, 212
Rain gutter, 334
Roof trusses, 515, 516
Sports arena construction, 368

Consumer Issues

Average monthly rate for basic cable TV, 209
Buying a refrigerator, 578
Cable TV subscribers, 123
Catering cost, 271
Cell phone charges, 282
Cellular telephone subscribers, 214
Comparing prices, 578
Cost of products, 462, 543
Express mail company charges, 237
Family income, 277
Gasoline usage, 241
Housing costs, 460
Long distance call charges, 258
Postage charges, 240
Sale price, 92, 263
Spending on food, 522
Spending on shoes and clothing, 298
Warehouse club membership, 95, 96

Economics

Adjusted poverty threshold, 78
Consumer Price Index, 461
Consumption expenditures, 229
Cost from inflation, 470
E-filing tax returns, 278
Equilibrium demand and price, 531
Poverty level income cutoffs, 179
Social Security, 165, 295, 333, 568
Supply and demand, 491–492, 576
Tax revenue, 358

Education

Ages of college undergraduates, 74
Associate's degrees earned, 266–267
Average annual public university costs, 449
Bachelor's degree attainment, 179
Bachelor's degrees in psychology, 436
Classroom ventilation, 96
College enrollments, 96–97
Cost of private college education, 228
Cost of public colleges, 123
Grade in mathematics, 170, 188
High school dropouts, 215
Honor roll, 166
Master's degrees, 577
Public college enrollment, 180
Teacher retirement, 166
Tuition and fees for in-state students, 224, 228
Two-year college enrollment, 216

Energy

Coal consumption in the U.S., 445–446
Electricity consumption, 462
Electricity usage, 202
Power of a windmill, 380
Solar panel frame, 121
Wind power extraction tests, 155

Engineering

Alternating current, 104
Circuit gain, 531
Current in a circuit, 367
Height of a dock, 120
Length of a walkway, 121
Number of computer calculations, 503–504
Oil pressure, 382
Resistance of a wire, 366
Speed of a pulley, 367
Transistors on computer chips, 471

Entertainment

Box office receipts, 148
Growth in science centers/museums, 295
Number of cable TV subscribers, 209
U.S. radio stations, 214

Environmental Issues

Atmospheric carbon dioxide, 298, 375–376, 408, 451–452, 490
Atmospheric pressure, 411
Carbon dioxide emissions, 155, 532
Carbon monoxide exposure, 122
Classifying wetlands, 429, 435
Cost-benefit model for a pollutant, 49
Emission of pollutants, 271
Environmental pollution, 380
Global temperature increase, 431, 437
Hazardous waste sites, 393
Indoor air pollution, 96
Methane gas emissions, 123
Number of named hurricanes, 213
Oil leak, 270
Ozone concentration, 165
Pollution in a river, 134
Prevention of indoor pollutants, 91
Radiative forcing, 431, 450
Recovery of solid waste, 148
Recycling aluminum cans, 124
Sulfur dioxide levels, 458
Toxic waste, 161
Water consumption for snowmaking, 167
Water pollution, 336

Finance

Amount saved, 213
Bank debit card transactions, 328–329
Comparing interest earned, 406, 407
Comparing loans, 411
Compound interest, 404, 407, 410, 411, 448

Continuous compounding, 406, 467
Cookbook royalties, 95
Doubling time, 452, 458, 459
Earning interest, 114
Expense categories, 170, 188
Financial planning, 471
Future value, 410, 411, 439
Growth of an account, 458, 459, 470, 473
Interest rate, 161, 411, 448, 470
Interest rates of treasury securities, 425
Investment, 90, 92, 95, 448, 458, 491, 522
Monetary value of coins, 92
Money denominations, 490
Present value, 405, 411
Real estate financing, 95
Retirement planning, 95
Simple interest, 83, 84, 85, 367
Tripling time, 473

General Interest

Blending coffee beans, 492
Broken bamboo, 122
Butane gas storage, 335
Counting marshmallows, 75
Distance to the horizon, 366
Filling a pool, 134, 135
Filling a sink, 135
Height of a kite, 121
Height of the Eiffel Tower, 449
Indian weavers, 576
Length of a ladder, 121
Long-distance phone calls, 368
Meal planning, 576
Mixing glue, 491
Mixing nuts, 501
Mixing teas, 577
Mixing water, 490
Mobile homes, 214–215
Photography, 369
Range of receivers, 121
Recreation expenditures, 460
Swimming pool levels, 198–199, 241
U.S. salmon catch, 295
Value of a machine, 213
Water in a tank, 240
Weight of an object, 367
World's largest ice cream cake, 93
World's smallest Quran, 93

Geology

Age of rocks, 430–431, 437, 458
Distance from the origin of the Nile River, 49
Earthquake intensity, 436, 469, 470
Epicenter of an earthquake, 184, 187
Filling a settling pond, 134

Geometry

Area of a rectangular region, 118, 119, 296, 335, 362, 375
Area of a square, 271
Area of a triangle, 364, 513

Area of an equilateral triangle, 270
Border around a rectangular region, 118, 163
Circumference of a circle, 366
Diagonal of a rectangle, 118
Dimensions of a box-shaped object, 93, 120, 379, 527–528, 531
Dimensions of a cube, 378
Dimensions of a cylinder, 280, 531
Dimensions of a rectangular object or region, 87, 92, 119, 120, 163, 167
Dimensions of a right triangle, 119, 158, 164, 530
Dimensions of a square, 87, 119, 120, 161
Dimensions of a triangular object or region, 93, 115–116, 120, 121, 491
Height of a cylinder, 93
Perimeter of a rectangle, 92, 280
Perimeter of a square, 270
Radius covered by a circular lawn sprinkler, 121
Radius of a can, 120
Relationship of measurement units, 270, 280
Sides of a right triangle, 118, 335
Surface area of a rectangular solid, 167
Volume of a box, 114–115, 120, 296, 297, 334, 335
Volume of a cylinder, 364, 367
Volume of a sphere, 280
Volume of the Great Pyramid, 31

Government

Aid to disaster victims, 544
Carbon dioxide emissions tax, 450
Encoding and decoding a message, 392–393, 397
Health care spending, 459
Legislative turnover, 459
Nuclear bomb detonation, 368
Science and space/technology spending, 267
Spending on health research and training, 336
U.S. government spending on medical care, 164

Health and Medicine

AIDS, 298–299
Blood alcohol concentration (BAC), 19–20
Blood donation, 166
Body mass index, 368
Diet requirements, 544
Drug levels in the bloodstream, 203, 470
Health care costs, 91
Heart disease, 462
Heart rate, 576
Hospital outpatient visits, 291–292
Lead intake, 161
Malnutrition measure, 369

Medication effectiveness, 460–461
Mixing drug solutions, 577
Newborns with AIDS, 295
Planning a diet, 501
Poiseuille's law, 368
Spread of disease, 460
Systolic blood pressure, 15, 18
Vitamin requirements, 539–540, 543
Weights of babies, 155

Labor

Chicago Cubs' payroll, 470–471
Daily wages, 500
Employee training, 412
Food bank volunteers, 375
Job market, 277
Medicare beneficiary spending, 378
Minimum wage, 114, 162
Percent of pay raise, 213
Salaries, 480–481
Software author royalties, 270
Women in the labor force, 449
Working alone to complete a job,
    127–128, 134
Working together to complete a job,
    127, 134

Physics

Current flow, 367
Decibel levels, 429, 435, 470
Floating ball, 335
Flow rates, 241
Height of a projected object, 290–291,
    295, 359
Height of a projectile, 116–117, 122,
    142, 148, 163, 165, 168, 202,
    375, 380
Hooke's law for a spring, 366
Illumination, 367
Period of a pendulum, 78, 337, 368

Stefan-Boltzmann law, 368
Velocity of an object, 148, 149
Vibrations of a guitar string, 364–365
Water emptied by a pipe, 366
Water pressure on a diver, 229

Social Issues

Births to unmarried women, 297
Strength of a habit, 445
Temporary assistance for needy
    families, 179
Waiting in line, 356–357

Sports

Athlete's weight and height, 502
Attendance at NCAA women's college
    basketball games, 199
Fan Cost Index, 490
Golf scores, 19
Height of a baseball, 148
Holding time of athletes, 57
Olympic times for 5000-meter run, 213,
    215
Passing rating for NFL quarterbacks,
    20, 77
Race speed, 449
Rowing speed, 69
Running times, 94
Shooting a free throw, 297
Skydiver fall speed, 473
Super Bowl viewers, 178
Total football yardage, 19
Track, 94

Statistics and Probability

Lottery, 95
Percent of high school students who
    smoke, 178

Transportation

Aircraft speed and altitude, 490
Airline carry-on baggage size, 161
Airline passengers, 124, 168
Airplane landing speed, 164
Airport parking charges, 240
Average rate, 88
Boat speed, 94, 161
Bus passengers, 124
Distance between cities, 93
Distance from home, 241, 277
Distance, 88, 92, 94, 161
Domestic leisure travel, 436–437
Mobility of Americans, 162
Speed of a plane, 94, 167
Trolley ridership, 117–118
Visits to U.S. National Parks, 199

U.S. and World Demographics

Age distribution in the United States, 501
Foreign-born Americans, 299
Nothing, AZ, 187
Population decline, 459
Population growth, 412, 453, 457, 459
Population size, 461
Population of metropolitan areas, 486
Populations of minorities in the United
    States, 576

Weather

Duration of a storm, 57
Force of wind, 367
Global temperature increase, 431, 437
Snow levels, 242, 448
Temperature, 86, 203, 229
Windchill, 69
Wind speed, 94

Index

A

Absolute value
  definition of, 14, 149
  evaluating, 14
  properties of, 15
Absolute value equations, 150
Absolute value function
  definition of, 234–235
  modeling with, 152
Absolute value inequalities, 151
Addition
  of complex numbers, 100
  of functions, 259
  identity property for, 12
  of matrices, 545
  of polynomials, 24
  of radicals, 63
  of rational expressions, 44
Addition property of equality, 80
Additive inverse
  of a matrix, 547
  of a real number, 12
Agreement on domain, 191
Algebraic expression, 23
Algorithm for division, 27
Analytic geometry, 171
Applied problems
  solving, 86
  steps to solve, 86
Approximating real zeros of
    polynomials, 328
Argument of a logarithm, 413
Array of signs for matrices, 507
Arrhenius, Svante, 431
Associative properties, 12, 13
Asymptotes
  determining, 344
  horizontal, 340, 343–344
  oblique, 344–345
  vertical, 340, 343–344, 349
Atanasoff, John, 504
Augmented matrix, 493
Average rate of change, 209
Axis
  of a coordinate system, 170
  of symmetry of a parabola,
    285
  symmetry with respect to, 245

B

Babbage, Charles, 421
Base of a logarithm, 413
Base of an exponent, 9
Binomials
  definition of, 23
  expansion of, 26
  factoring, 35

multiplication of, 25
  square of, 27
Blood alcohol concentration (BAC), 19
Body mass index (BMI), 368
Boundary of a half-plane, 534
Boundedness theorem, 327
Brackets, 10
Break-even point, 139, 216

C

Carbon dating, 454
Cardano, Girolamo, 317
Cartesian coordinate system, 170
Celsius-Fahrenheit relationship, 86
Center of a circle, 180
Center-radius form of the equation of a
    circle, 181
Chain rule, 265
Change in x, 206
Change in y, 206
Change-of-base theorem for logarithms,
    432
Circle(s)
  applications of, 184
  definition of, 180
  equations of, 180–181, 182
  general form of the equation of, 182
  graphing by calculator, 182
  graph of, 181
  radius of, 180
  tangent line to, 187
  unit, 181
Circular function, 181
Closed interval, 138
Closure properties, 12
Coefficient
  correlation, 328
  definition of, 23
  leading, 284
Cofactor of an element of a matrix,
    506
Collinear points, 173
Column matrix, 545
Column of a matrix, 493, 506
Combined variation, 363
Common factors of a rational
    expression, 41, 42
Common logarithms
  applications of, 428
  definition of, 427
  modeling with, 428
Commutative properties, 12, 13
Complement of a set, 4
Completing the square, 106, 286
Complex conjugates
  definition of, 101
  property of, 101, 311

Complex fractions
  definition of, 46
  simplifying, 46, 55
Complex numbers
  conjugates of, 101
  definition of, 97
  imaginary part of, 97
  operations on, 98, 101
  real part of, 97
  standard form of, 98
Composite function
  definition of, 263
  derivative of, 265
Composition of functions, 263
Compound amount, 405
Compounding continuously, 406
Compound interest
  definition of, 404
  general formula for, 404
Concours d'elegance, 379
Conditional equation, 82
Conditional statement
  converse of, 172
  definition of, 172
Conjugate(s)
  complex, 101, 311
  properties of, 311
  of a radical, 66
Conjugate zeros theorem, 311
Consistent systems, 476
Constant function, 197, 204
Constant of variation, 361
Constant velocity problems, 87
Constraints, 538
Continuity of a function, 231
Continuous compounding, 406
Continuous function, 231
Contradiction equations, 82
Converse of a conditional statement,
    172
Coordinate plane, 170
Coordinates of a point on a line,
    8, 171
Coordinate system, 8
Corner point of a region of feasible
    solutions, 538
Correlation coefficient, 328
Cost
  fixed, 209
  per item, 209
Cost-benefit model, 49
Cost function, 209
Counting numbers, 2
Cramer's rule for solving linear systems
  derivation of, 509
  general form of, 510
Cryptography, 392
Cube root function, 234

Cubes
    difference of, 35
    sum of, 35
Cubic equation, 109
Cubing function, 233

D

Decibel, 429
Decimals, 8
Decimal window, 341
Decomposition
    into partial fractions, 516, 521
    of rational expressions, 516
Decreasing function, 197
De Fermat, Pierre, 170
Definite integral, 289
Degree of a polynomial, 23
Delta x, 206
Delta y, 206
Denominator
    least common, 44
    rationalizing, 64
Dependent equations, 476
Dependent variable, 188
Derivative
    of a function, 207, 262, 265, 308
    power rule for, 60
Descartes, René, 170–171
Descartes' rule of signs, 313
Descending order, 24
Determinant(s)
    definition of, 504
    evaluating, 506
    expansion of, 506
    theorems for, 508
    of a 2×2 matrix, 504
    of a 3×3 matrix, 505
Diagonal form of a matrix, 494
Difference
    of cubes, 35
    of squares, 35
Difference quotient, 261
Differential equations, 451
Differentiation, 265
Dimensions of a matrix, 493
Directly proportional
    definition of, 361
    as nth power, 362
Direct variation
    definition of, 361
    as nth power, 362
Discontinuity of the graph of a function,
        232, 339, 349
Discriminant, 110
Disjoint interval, 138
Disjoint sets, 5
Distance, rate, and time problems, 87
Distance formula, 171–172
Distributive property, 12, 13
Division
    of complex numbers, 101
    of functions, 259
    of polynomials, 27, 300
    of rational expressions, 42
    synthetic, 301
Division algorithm
    for polynomials, 27, 300
    special case of, 303

Domain
    agreement on, 191
    of a function, 193, 259
    of a rational expression, 41
    of a relation, 190
Dominating term, 321
Doubling time, 451

E

e
    continuous compounding and, 406
    value of, 406
Element
    cofactor of, 506
    of a matrix, 493
    minor of, 505
    of a set, 2
Elimination method
    definition of, 477
    for solving linear systems, 477, 482
    for solving nonlinear systems, 524
Empty set
    definition of, 3, 82
    symbol for, 3, 82
End behavior of graphs of polynomial
        functions, 321–322
Equality
    addition property of, 80
    multiplication property of, 80
Equal matrices, 545
Equal sets, 4
Equation(s)
    absolute value, 150
    of an inverse function, 389
    of a circle, 182
    for compound interest, 404
    conditional, 82
    contradiction, 82
    cubic, 109
    definition of, 80
    dependent, 476
    differential, 451
    equivalent, 80
    exponential. *See* Exponential
        equations
    first-degree, 81, 476
    graphing of, 176
    graph of, 176
    of a horizontal line, 220
    identity, 82
    independent, 476
    of a line, 217
    linear. *See* Linear equations
    literal, 83
    logarithmic. *See* Logarithmic
        equations
    matrix, 564
    nonlinear systems of. *See* Nonlinear
        systems of equations
    power property of, 128
    quadratic in form, 131–132
    quadratic in one variable. *See*
        Quadratic equations in one
        variable
    radical, 128, 129
    rational, 124
    with rational exponents, 131
    roots of, 80, 304

second-degree, 104
solution set of, 80
solutions of, 80
steps to graph, 176
types of, 82
of a vertical line, 220
Equilibrium demand, 491
Equilibrium price, 491
Equilibrium supply, 491
Equivalent equations, 80
Equivalent systems of linear equations,
        477
Even-degree polynomial functions, 320
Even function, 248
Eves, Howard, 171
Expansion
    of binomials, 26
    of determinants, 506
    by a given row or column, 506
Exponent(s)
    base of, 9
    definition of, 9, 397
    integer, 21
    negative, 49, 53
    notation for, 9, 53
    properties of, 398
    rational, 52, 53, 60, 131
    rules for, 21, 50, 53
    zero, 22
Exponential decay function
    definition of, 451
    modeling with, 453–454
Exponential equations
    applications of, 445
    definition of, 402
    solving, 402, 439–441
    steps to solve, 444
Exponential expressions
    definition of, 10
    evaluating, 10, 399
Exponential function(s)
    characteristics of graphs of, 401
    graph of, 400
    modeling with, 407
    standard form of, 399
Exponential growth function
    definition of, 451
    modeling with, 451–453
Expression
    algebraic, 23
    exponential, 10, 399
    rational, 41–45, 516

F

Factored form of a polynomial, 32
Factoring
    binomials, 35
    definition of, 32
    difference of cubes, 35
    difference of squares, 35
    greatest common factor, 32
    by grouping, 33
    perfect square trinomial, 35
    polynomials, 32
    to solve quadratic equations, 105
    substitution method for, 37
    sum of cubes, 35
    trinomials, 34–35

Factors
    of a number, 9
    of a polynomial function, 325
Factor theorem, 306–307
Fahrenheit-Celsius relationship, 86
Fan Cost Index (FCI), 490
Finite set, 2
First-degree equations
    definition of, 81, 476
    in n unknowns, 476
Fixed cost, 209
FOIL, 25
Fraction bar, 10
Fractions
    complex, 46, 55
    decimal form of, 8
    finding least common denominator, 44
    fundamental principle of, 41
    partial, 516, 521
Frustum of a pyramid
    formula for volume of, 31
Function(s)
    absolute value, 234–235
    alternative definition of, 194
    circular, 181
    composite, 263, 265
    composition of, 263
    constant, 197, 204
    continuity of, 231
    continuous, 231
    cost, 209
    cube root, 234
    cubing, 233
    decreasing, 197
    definition of, 188
    derivative of, 207, 262, 265, 308
    discontinuity of, 232, 339, 349
    domain of, 193, 259
    even, 248
    exponential. *See* Exponential
        function(s)
    exponential decay, 451, 453–454
    exponential growth, 451–453
    greatest integer, 236
    identity, 232
    increasing, 197
    inverse. *See* Inverse function(s)
    inverse of, 389
    limit of, 15, 194
    linear. *See* Linear functions
    linear cost, 209
    logarithmic. *See* Logarithmic
        function(s)
    logistic, 461
    notation, 194, 388
    objective, 538
    odd, 248
    one-to-one, 384–386
    operations on, 259
    piecewise-defined, 235
    polynomial. *See* Polynomial
        function(s)
    profit, 210
    quadratic. *See* Quadratic function
    range of, 193
    rational. *See* Rational functions
    reciprocal, 340
    revenue, 210
    square root, 234

    squaring, 233
    step, 237
    tests for one-to-one, 386
    vertical line test for, 191
Fundamental principle of fractions, 41
Fundamental theorem of algebra, 309
Fundamental theorem of linear
    programming, 539
Future value, 83, 405
$f(x)$ notation, 194

G

Galilei, Galileo, 116
Gauss, Carl Friedrich, 309
Gauss-Jordan method for solving linear
    systems, 494
General form of the equation of a circle,
    182
Geometry problems, 87, 114
Grade, 206, 357
Graph(s)
    of a circle, 181
    of equations, 176
    of exponential functions, 400
    of a horizontal line, 204
    horizontal translation of, 250, 319
    of inequalities, 536
    intercepts of, 176
    of inverse functions, 391
    of linear inequalities in two
        variables, 535, 536
    of logarithmic functions, 415, 416
    of a parabola, 233
    of polynomial functions, 318
    of quadratic functions, 285
    of rational functions, 341, 342
    of the reciprocal function, 340
    reflecting across a line, 244
    reflecting across an axis, 244
    of a relation, 189
    shrinking of, 242, 243
    stretching of, 242, 243
    symmetry of, 245
    of systems of inequalities, 536
    translations of, 249, 319
    of a vertical line, 205
    vertical translation of, 249, 319
Graphing calculator(s)
    decimal window on, 341
    for fitting exponential curves to
        scatter diagrams, 408
    performing row operations, 497
    square viewing window on, 182
    standard viewing window on, 177
Graphing calculator method
    for circles, 182
    finding the determinant of a matrix,
        504
    for rational functions, 341, 350
    for the reciprocal function, 340
    for solving linear equations in one
        variable, 225
    for solving nonlinear systems,
        523
    for solving systems of equations
        using matrix inverses, 564
    for solving systems of linear equa-
        tions in two variables, 477, 479

Graphing techniques
    even and odd functions, 248
    reflecting, 244
    stretching and shrinking, 242–243
    summary of, 252
    symmetry, 245–247
    translations, 249–252
Greatest common factor, 32
Greatest integer function, 236
Grouping, factoring by, 33

H

Half-life, 453
Half-plane
    boundary of, 534
    definition of, 534
Height of a projected object, 116
Holding time, 57
Hooke's Law, 366
Horizontal asymptote
    definition of, 340, 343
    determining, 344, 345
Horizontal line
    equation of, 220, 223
    graph of, 204
    slope of, 207
Horizontal line test for one-to-one
    functions, 385
Horizontal shrinking of the graph of a
    function, 243
Horizontal stretching of the graph of a
    function, 243
Horizontal translation
    of a graph, 250
    of a graph of a polynomial function,
        319
Hubble constant, 229
Hypotenuse of a right triangle, 115

I

i
    definition of, 97
    powers of, 102
Identity equations, 82
Identity function, 232
Identity matrix, 558
Identity property
    for addition, 12
    for multiplication, 12
Imaginary numbers, 98
Imaginary part of a complex number, 97
Imaginary unit, 97
Inconsistent systems, 476, 497
Increasing function, 197
Independent equations, 476
Independent variable, 188
Index of a radical, 59
Inequalities
    absolute value, 151
    definition of, 14, 137
    graphing, 534, 536
    linear, 137, 139, 534, 535
    nonstrict, 142, 535
    properties of, 137
    quadratic, 140
    rational, 142–143
    strict, 142, 535
    three-part, 139

Infinite discontinuity, 349
Infinite set, 2
Infinity, limits at, 348
Integer exponents, 21
Integers
    definition of, 8
    relatively prime, 205
Integral, definite, 289
Intercepts of a graph, 176
Interest
    compound, 404
    simple, 83
Intermediate value theorem, 326
Intersection of sets, 4
Interval
    closed, 138
    of convergence, 149
    definition of, 138
    disjoint, 138
    notation, 138
    open, 138
Inverse
    additive, 12
    multiplicative, 559
Inverse function(s)
    definition of, 387
    equation of, 389
    facts about, 392
    graphs of, 391
    notation, 388
    theorem on, 422
Inverse properties, 12
Inverse variation
    definition of, 363
    as nth power, 363
Irrational numbers, 8
Is greater than, 14
Is less than, 14

J

Joint variation, 363
Jump discontinuity, 349

L

Laffer curve, 358
Leading coefficient, 284
Leading term, 284
Least common denominator
    in fractions, 44
    steps to find, 44
Legs of a right triangle, 115
Leibniz notation, 430
Like radicals, 63
Like terms, 23
Limit(s)
    definition of, 15, 149
    of a function, 15, 194
    at infinity, 348
    notation, 15, 194
    one-sided, 344
Line(s)
    equation of, 217
    horizontal, 204, 207, 220, 223
    parallel, 220–221
    perpendicular, 221
    secant, 262

slope of, 206
vertical, 205, 207, 220, 223
Linear cost function, 209
Linear equations
    definition of, 80, 476
    in n unknowns, 476
    standard form of, 205, 223
Linear equations in one variable
    definition of, 80
    graphing calculator method for
        solving, 225
    modeling with, 90
    solution of, 80
    solving by graphing, 225
Linear equations in two variables
    graphing calculator method for solv-
        ing, 477, 479
    modeling with, 223–225
    point-slope form of, 217
    slope-intercept form of, 218–219
    standard form of, 205
    summary of, 223
Linear functions
    definition of, 203
    graphing, 203
    point-slope form of, 217
    slope-intercept form of, 218–219
    x-intercept of graph of, 204
    y-intercept of graph of, 204
Linear inequalities in one variable
    applications of, 139
    solutions of, 137
    standard form of, 137
Linear inequalities in two variables
    graphs of, 535
    standard form of, 534
Linear models, 90, 209
Linear programming
    constraints for, 538
    definition of, 537
    fundamental theorem of, 539
    steps to solve problems with,
        539
Linear regression, 225
Linear systems, 476. See also Systems
    of linear equations
Literal equation, 83
Logarithmic differentiation, 420
Logarithmic equations
    definition of, 414
    solving, 414, 442
    steps to solve, 444
Logarithmic function(s)
    characteristics of graphs of, 417
    graphs of, 415, 416
    modeling with, 428, 445
    standard form of, 415
Logarithms
    argument of, 413
    base of, 413
    change-of-base theorem for,
        432
    common, 427
    definition of, 413
    natural, 430
    properties of, 419, 439
Logistic function, 461
Lowest terms of a rational expression,
    41

M

Mapping, 189
Marginal propensity to consume, 229
Mathematical model, 90, 209
Matrix (Matrices)
    addition of, 545
    additive inverse of, 547
    array of signs for, 507
    augmented, 493
    cofactor of an element of, 506
    column, 545
    column of, 493
    definition of, 493
    determinant of, 504–505
    diagonal form of, 494
    dimensions of, 493
    element of, 493
    equal, 545
    equation, 564
    expansion by a given row or column,
        506
    identity, 558
    minor of an element of, 505
    modeling with, 552
    multiplication of. See Matrix
        multiplication
    multiplicative inverse of, 559
    negative of, 547
    order of, 493
    properties of, 545–553
    reduced-row echelon form of, 494
    row, 545
    row of, 493
    row transformations of, 493
    scalar multiplication of, 548
    singular, 563
    size of, 493
    square, 545
    steps to find inverse of, 561
    subtraction of, 547
    zero, 546
Matrix inverses method for solving
    linear systems, 564
Matrix method for solving linear
    systems, 494
Matrix multiplication
    applications of, 552
    definition of, 549
    properties of, 548, 551
    by scalars, 548
Maturity value, 83
Maximum point of a parabola, 288
Members of a set, 2
Midpoint formula, 173
Minimum point of a parabola, 287
Minor of a matrix element, 505
Mixture problems, 88
Models
    absolute value, 152
    cost-benefit, 49
    exponential, 407
    linear, 90, 209
    linear systems, 483
    logarithmic, 428, 445
    matrix, 552
    polynomial, 328
    quadratic, 117, 290
    rational, 351

Monomial, 23
Motion problems, 87
Multiplication
    of binomials, 25
    of complex numbers, 100
    of functions, 259
    identity property for, 12
    of matrices. *See* Matrix multiplication
    of polynomials, 25
    of radicals, 64
    of rational expressions, 42
Multiplication property of equality, 80
Multiplication property of zero, 12
Multiplicative inverse
    definition of, 559
    of a matrix, 559
    of a real number, 12, 559
Multiplicity of zeros, 310

N

Napier, John, 421
Natural logarithms
    applications of, 430
    definition of, 430
    modeling with, 430
Natural numbers, 2, 8
Negative
    of a matrix, 547
    of a real number, 12
Negative exponents, 49, 53
Negative reciprocals, 221
Negative slope, 209
Newton's law of cooling, 455
Non-constant velocity, 87
Nonlinear systems of equations
    application of, 527
    with complex solutions, 527
    definition of, 523
    elimination method for solving, 524
    graphing calculator method for solving, 523
    with nonreal solutions, 527
Nonstrict inequalities, 142, 535
Notation
    exponential, 9, 53
    function, 194
    interval, 138
    inverse function, 387
    Leibniz, 430
    limit, 15, 194
    radical, 59
    set-builder, 2
    for sets of numbers, 8
Null set
    definition of, 3, 82
    symbol for, 3, 82
Number line, 8
    distance between points on, 16
    order on, 14
Number of zeros theorem, 310
Numbers
    complex, 97, 98
    conjugate of, 66
    counting, 2
    decimals, 8
    factors of, 9

imaginary, 97
integers, 8
irrational, 8
natural, 2, 8
negative reciprocals, 221
rational, 8
real, 8, 12, 559
sets of, 2, 8
whole, 8
Numerator, rationalizing, 66
Numerical coefficient, 23

O

Objective function, 538
Oblique asymptote
    definition of, 345
    determining, 344
Odd-degree polynomial functions, 320
Odd function, 248
One-sided limits, 344
One-to-one functions
    definition of, 384
    horizontal line test for, 385
    tests to determine, 386
Open interval, 138
Operations
    on complex numbers, 98
    on functions, 259
    order of, 10
    with radicals, 63
    on sets, 4, 6
Order
    descending, 24
    of a matrix, 493
    on a number line, 14
    of operations, 10
Ordered pair, 170, 175
Ordered triple, 172, 482
Order symbols, 14
Origin
    of rectangular coordinate system, 170
    symmetry with respect to, 246

P

Pair, ordered, 170, 175
Parabola(s)
    axis of, 285
    definition of, 233, 284
    graphing by completing the square, 286
    graph of, 233
    maximum point of, 288
    minimum point of, 287
    vertex formula for, 289
    vertex of, 233, 285, 289
Parallel lines
    definition of, 220
    slope of, 220–221
Partial fractions
    decomposition of, 516–517
    definition of, 516
Pascal, Blaise, 421
Pendulum, period of, 78
Perfect square trinomial, 35
Period of a pendulum, 78
Perpendicular lines
    definition of, 221
    slope of, 221

pH, 428
Phelps, Phlash, 187
Pi (π), 69
Piecewise-defined function, 235
Point-slope form of the equation of a line, 217, 223
Poiseuille's law, 368
Polynomial(s)
    addition of, 24
    definition of, 23
    degree of, 23
    descending order of, 24
    division algorithm for, 27, 300
    factored form of, 32
    factoring of, 32
    multiplication of, 25
    prime, 32
    special products of, 26
    subtraction of, 24
    synthetic division for, 301
    term of, 23
    in x, 23
    zero, 284
Polynomial function(s)
    approximating real zeros of, 328
    boundedness theorem for, 327
    comprehensive graph of, 323
    conjugate zeros theorem for, 311
    of degree n, 284
    derivative of, 308
    dominating term of, 321
    end behavior of graph of, 321
    of even degree, 320
    factors of, 325
    factor theorem for, 306–307
    graph of, 318
    intermediate value theorem for, 326
    leading coefficient of, 284
    modeling with, 328
    number of zeros theorem for, 310
    of odd degree, 320
    rational zeros theorem for, 308
    roots of, 304
    solutions of, 304, 325
    steps to graph, 323
    summary of end behavior of graphs of, 322
    turning points of graph of, 321
    zero, 284
    zeros of, 304
Polynomial interpolation, 568
Polynomial models, 328–329
Positive slope, 209
Power property of equations, 128
Power rule for derivatives, 60
Power rules for exponents, 21, 53
Power series, 149
Powers of i, 102
Present value, 405
Prime polynomial, 32
Principal root of a radical, 59
Product of the sum and difference of two terms, 26
Product rule for exponents, 21, 53
Profit function, 210
Proportional
    directly, 361
    as nth power, 362–363

Proposed solutions, 124
Pyramid
    frustum of, 31
    volume of a frustum of, 31
Pythagorean theorem, 115

Q

Quadrants of a coordinate system, 170
Quadratic equations in one variable
    applications of, 114
    completing the square to solve, 106
    discriminant of, 110
    factoring to solve, 105
    modeling with, 117
    quadratic formula for solving,
        107–108
    square root method for solving, 105
    standard form of, 104
    types of solutions for, 111
Quadratic formula
    discriminant of, 110
    standard form of, 108
Quadratic function
    definition of, 284
    graph of, 284–285
    maximum of, 288
    minimum of, 287
    steps to graph, 289
Quadratic inequalities
    standard form of, 140
    steps to solve, 140
Quadratic models, 117, 290
Quadratic regression, 292
Queuing theory, 356
Quotient rule for exponents, 50, 53

R

Radiative forcing, 431
Radical(s)
    addition of, 63
    conjugate of, 66
    index of, 59
    like, 63
    multiplication of, 64
    notation, 59
    operations with, 63
    principal root of, 59
    rational exponent form of, 59
    rationalizing the denominator of, 64
    rationalizing the numerator of, 66
    simplifying, 62
    steps to simplify, 62
    subtraction of, 63
    summary of rules for, 61
    symbol for, 59
Radical equations
    power property and, 128
    steps to solve, 129
Radical notation, 59
Radical symbol, 59
Radicand, 59
Radius of a circle, 180
Range
    of a function, 193
    of a relation, 190
Rate of change, average, 209
Rational equations, 124
Rational exponents

definition of, 52–53
    equations with, 131
    radical form of, 59
Rational expressions
    addition of, 44
    common factors of, 41
    decomposition of, 516–517
    definition of, 41
    division of, 42
    domain of, 41
    least common denominator of, 44
    lowest terms of, 41
    multiplication of, 42
    subtraction of, 44
Rational functions
    features of, 342
    graphing by calculator, 341, 350
    graph of, 341, 342
    modeling with, 351
    steps to graph, 346
Rational inequalities
    definition of, 142
    steps to solve, 143
Rationalizing the denominator, 64
Rationalizing the numerator, 66
Rational numbers, 8
Rational zeros theorem, 308
Real numbers
    additive inverse of, 12
    definition of, 8
    multiplicative inverse of, 12, 559
    negative of, 12
    properties of, 12
    reciprocal of, 12
    summary of properties of, 12
Real part of a complex number, 97
Reciprocal functions
    graphing by calculator, 340
    graph of, 340
    important features of, 340
Reciprocal of a real number, 12
Rectangular coordinate system, 170
Reduced-row echelon form of a matrix,
    494
Reflection of a graph across a line,
    244
Reflection of a graph across an axis,
    244
Region of feasible solutions
    corner point of, 538
    definition of, 538
    vertex of, 538
Regression
    linear, 225
    quadratic, 292
Relation
    definition of, 188
    domain of, 190
    graph of, 189
    range of, 190
Relatively prime integers, 205
Remainder theorem, 303
Removable discontinuity, 349
Revenue function, 210
Right triangle
    hypotenuse of, 115
    legs of, 115
    in Pythagorean theorem, 115
Rise, 206

Roots
    of an equation, 80, 304
    of polynomial functions,
        304
Row matrix, 545
Row of a matrix, 493, 506
Row transformations, 493
Rules for exponents, 21, 50, 53
Run, 206

S

Scalar, 547
Scalar multiplication
    definition of, 547
    properties of, 548
Scatter diagrams, 224, 408
Secant line, 262
Second-degree equation, 104
Set(s)
    complement of, 4
    definition of, 2
    disjoint, 5
    elements of, 2
    empty, 3, 82
    equal, 4
    finite, 2
    infinite, 2
    intersection of, 4
    members of, 2
    notation, 2
    null, 3, 82
    of numbers, 8
    operations on, 4, 6
    subset of, 3
    summary of operations on, 6
    union of, 5, 141
    universal, 3
Set braces, 2
Set-builder notation, 2
Shrinking graphs
    graphing techniques for,
        242
    horizontally, 243
    vertically, 243
Simple interest formula, 83
Simplifying complex fractions,
    46, 55
Simplifying radicals, 62
Singular matrix, 563
Size of a matrix, 493
Slope(s)
    of a curve at a point, 206
    definition of, 206
    formula for, 206
    of a horizontal line, 207
    of a line, 206
    negative, 209
    of parallel lines, 220–221
    of perpendicular lines, 221
    point-slope form, 217, 223
    positive, 209
    slope-intercept form, 218–219,
        223
    undefined, 207
    of a vertical line, 207
    zero, 207, 209
Slope-intercept form of the equation of a
    line, 218–219, 223

Solutions
    of an equation, 80
    of a polynomial function, 304, 325
Solution set, 80
Solving applied problems, 86
Solving for a specified variable, 83, 109
Special products of polynomials, 26
Speed, 87
Square(s)
    of a binomial, 26
    completing the, 106, 286
    difference of, 35
Square matrix, 545
Square root function, 234
Square root method for solving
        quadratic equations, 105
Square root property, 105
Square viewing window on calculators,
        182
Squaring function, 233
Standard form
    of a complex number, 98
    of the equation of a line, 218
    of a linear equation, 205, 223
    of a quadratic equation, 104
Standard viewing window on
        calculators, 177
Stefan-Boltzmann law, 368
Step function, 237
Stretching graphs
    graphing techniques for, 242
    horizontal, 243
    vertical, 243
Strict inequalities, 142, 535
Subset, 3
Substitution method
    definition of, 476
    for factoring, 37
    for solving linear systems, 477
Subtraction
    of complex numbers, 100
    of functions, 259
    of matrices, 547
    of polynomials, 24
    of radicals, 63
    of rational expressions, 44
Sum of cubes, 35
Supply and demand, 491–492
Symbols of order, 14
Symmetry
    example of, 169
    graphing techniques for, 245
    of graphs, 245
    with respect to origin, 246
    with respect to x-axis, 245–246
    with respect to y-axis, 245–246
    tests for, 245–247
Synthetic division for polynomials, 301
Systems of equations
    definition of, 476
    solutions of, 476
Systems of inequalities
    definition of, 536
    graphing, 536
Systems of linear equations in three
        variables
    applications of, 485
    Cramer's rule for solving, 510
    elimination method for solving, 482

Gauss-Jordan method for solving,
        496
    geometric interpretation of, 481
    matrix inverses method for solving,
        564
    modeling with, 483
    solving, 481
Systems of linear equations in two
        variables
    applications of, 480
    consistent, 476
    Cramer's rule for solving, 509
    elimination method for solving, 477
    equivalent, 477
    Gauss-Jordan method for solving, 494
    graphing calculator method for
        solving, 477, 479
    inconsistent, 476
    matrix method for solving, 494
    solution set for, 476
    substitution method for solving, 476
    transformations of, 477

T

Tangent line
    to a circle, 187
    to a curve, 218
Tartaglia, Niccolo, 317
Term(s)
    coefficient of, 23
    definition of, 23
    like, 23
    of a polynomial, 23
Tests for symmetry, 245–247
Three-part inequalities, 139
Threshold sound, 429
Tolerance, 152
Transformations of linear systems, 477
Translation(s)
    definition of, 249
    of graphs, 249, 319
    horizontal, 250
    summary of, 252
    vertical, 249
Triangular form, 502
Trinomials
    definition of, 23
    factoring, 33
    perfect square, 35
Triple, ordered, 172, 482
Turning points of polynomial function
        graphs, 321

U

Undefined slope, 207
Union of sets, 5, 141
Unit, imaginary, 97
Unit circle, 181
Universal set, 3

V

Value
    absolute, 14–15, 149
    future, 83, 405
    maturity, 83
    present, 405

Variable
    dependent, 188
    independent, 188
    solving for a specified, 83, 109
Variation
    combined, 363
    constant of, 361
    direct, 361
    directly as the nth power, 362–363
    inverse, 363
    inversely as the nth power, 363
    joint, 363
    steps to solve problems with, 362
Vector cross products, 505
Velocity
    constant, 87
    in motion problems, 87
    non-constant, 87
Venn diagram, 3
Vertex (Vertices)
    of a parabola, 233, 285, 289
    of a polygonal region, 538
Vertex formula for a parabola, 289
Vertical asymptote
    behavior of graphs near, 349
    definition of, 340, 343
    determining, 344
Vertical line
    equation of, 220, 223
    graph of, 205
    slope of, 207
Vertical line test for functions, 191
Vertical shrinking of the graph of a
        function, 243
Vertical stretching of the graph of a
        function, 243
Vertical translation of a graph, 249, 319

W

Waiting-line theory, 356
Whole numbers, 8
Windchill formula, 69
Work rate problems, 126–128

X

x-axis
    definition of, 170
    symmetry with respect to, 245–246
x-intercept, 176, 325
xy-plane, 170

Y

y-axis
    definition of, 170
    symmetry with respect to, 245–246
y-intercept, 176

Z

Zero(s)
    multiplicity of, 310
    of polynomial functions, 304
    as whole number, 8
Zero exponent, 22
Zero-factor property, 41, 105
Zero matrix, 546
Zero polynomial, 284
Zero slope, 207

Photo Credits

Front endpapers: (graduates) Shutterstock; (students at laptop) Sturti/E+/Getty Images **page xvii:** Sturti/E+/Getty Images
page 1: Diseñador/Fotolia **page 15:** Brian Maudsley/Shutterstock **page 19:** Don Emmert/AFP/Getty Images/Newscom
page 20: Photo from the collection of John Hornsby **page 31:** Gregory Johnston/Shutterstock **page 49:** Bestimagesever.com/
Shutterstock **page 57:** Jami Garrison/Shutterstock **page 69:** Danshutter/Shutterstock **page 74:** Corepics/Shutterstock
page 75: David Anderson/Shutterstock **page 79:** Gunter Marx/Dorling Kindersley, Ltd. **page 86:** Barry Blackburn/
Shutterstock **page 88:** AP Images **page 91:** Michaeljung/Shutterstock **page 93:** (cake) Tarzan Wong/EyePress/
Newscom; (bin) Scanrail/Fotolia **page 95:** Digital Vision/Thinkstock **page 96:** Armadillo Stock/Shutterstock
page 115: Photodisc/Digital Vision/Thinkstock **page 116:** Photos/Getty Images/Thinkstock **page 117:** Medioimages/
Photodisc/Thinkstock **page 118:** (parking lot) David Phillips/Shutterstock; (picture) Thomas Bethge/ Shutterstock
page 119: (carpet) Marco Mayer/Shutterstock; (floor) Kutlayev Dmitry/Shutterstock **page 123:** Thomas Barrat/Shutterstock
page 134: Columbia Pictures/Everett Collection **page 148:** (Avatar) Trademark™ and copyright © 20th Century Fox. All
Rights Reserved/Everett Collection; (refuse) Picsfive/Shutterstock **page 155:** Kapu/Shutterstock **page 161:** Discpicture/
Shutterstock **page 162:** (lawn mower) Suzanne Tucker/Shutterstock; (U-Haul truck) Beth Anderson/Pearson Education, Inc.
page 167: Sergei Butorin/Shutterstock **page 169:** Peter Radacsi/Shutterstock **page 171:** Photos/Getty Images/Thinkstock
page 174: Mipstudio/Shutterstock **page 184:** Evlakhov Valeriy/Shutterstock **page 187:** Photo from the collection of
John Hornsby **page 190:** Courtesy of John Hornsby **page 224:** Karen Roach/Shutterstock **page 228:** Wolfgang Kloehr/
Shutterstock **page 237:** Samuel Acosta/Shutterstock **page 263:** Imagemore/Getty Images **page 271:** Dudarev
Mikhail/Shutterstock **page 283:** Cloudrain/Fotolia **page 291:** Avava/Shutterstock **page 299:** Yuri Arcurs/Shutterstock
page 309: Bilwissedition Ltd. & Co. KG/Alamy **page 328:** EML/Shutterstock **page 351:** Michael Fritzen/Fotolia
page 366: Ocean Image Photography/Shutterstock **page 367:** Poznukhov Yuriy/Shutterstock **page 369:** Cristi Lucaci/
Shutterstock **page 379:** Stocker1970/Shutterstock **page 383:** Lee Prince/ Shutterstock **page 412:** Guy J.
Sagi/Shutterstock **page 421:** Dave King/The Science Museum, London, Dorling Kindersley, Ltd. **page 429:** Etienjones/
Shutterstock **page 433:** Kletr/Shutterstock **page 436:** Jiri Flogel/Shutterstock **page 448:** Lukich/Shutterstock
page 453: Dariush M/Shutterstock **page 455:** Gosphotodesign/Shutterstock **page 457:** Konstantin Mironov/Shutterstock
page 459: Steven Frame/Shutterstock **page 460:** Crok Photography/Shutterstock **page 462:** Courtesy of B. K. Tomlin
page 470: Dgareri/Dreamstime **page 475:** Haak78/Shutterstock **page 480:** Ampyang/Shutterstock **page 490:** Courtesy
of Callie Daniels **page 492:** James Thew/Shutterstock **page 500:** Christina Richards/Shutterstock **page 501:** Stockbyte/
Thinkstock **page 502:** Yuri Arcurs/Shutterstock **page 503:** Jeff Banke/Shutterstock **page 504:** U.S. Dept. of Energy's
Ames Laboratory **page 522:** Tyler Olson/Shutterstock **page 543:** Mangostock/Shutterstock **page 544:** Snyderdf/
Dreamstime **page 552:** Chad McDermott/Shutterstock **page 556:** Pix2go/Shutterstock **page 557:** Tom Reichner/
Shutterstock **page 576:** Andrey Yurlov/Shutterstock **page 578:** U.S. Environmental Protection Agency

Geometry Formulas

Square Perimeter: $P = 4s$ Area: $\mathcal{A} = s^2$ 	**Rectangle** Perimeter: $P = 2L + 2W$ Area: $\mathcal{A} = LW$ 	**Triangle** Perimeter: $P = a + b + c$ Area: $\mathcal{A} = \dfrac{1}{2}bh$
Parallelogram Perimeter: $P = 2a + 2b$ Area: $\mathcal{A} = bh$ 	**Trapezoid** Perimeter: $P = a + b + c + B$ Area: $\mathcal{A} = \dfrac{1}{2}h(B + b)$ 	**Circle** Diameter: $d = 2r$ Circumference: $C = 2\pi r = \pi d$ Area: $\mathcal{A} = \pi r^2$ 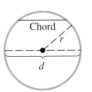
Cube Volume: $V = e^3$ Surface area: $S = 6e^2$ 	**Rectangular Solid** Volume: $V = LWH$ Surface area: $S = 2HW + 2LW + 2LH$ 	**Sphere** Volume: $V = \dfrac{4}{3}\pi r^3$ Surface area: $S = 4\pi r^2$
Cone Volume: $V = \dfrac{1}{3}\pi r^2 h$ Surface area: $S = \pi r\sqrt{r^2 + h^2}$ (excludes the base) 	**Right Circular Cylinder** Volume: $V = \pi r^2 h$ Surface area: $S = 2\pi rh + 2\pi r^2$ (includes top and bottom) 	**Right Pyramid** Volume: $V = \dfrac{1}{3}Bh$ B = area of the base